APPLIED MATHEMATICS
FOR ENGINEERS
AND PHYSICISTS

INTERNATIONAL SERIES IN PURE AND APPLIED MATHEMATICS

William Ted Martin and E. H. Spanier

CONSULTING EDITORS

APPLIED MATHEMATICS FOR ENGINEERS AND PHYSICISTS

Third Edition

LOUIS A. PIPES

Professor of Engineering
University of California

LAWRENCE R. HARVILL

Associate Professor of Engineering Science
University of Redlands

INTERNATIONAL STUDENT EDITION

McGRAW-HILL INTERNATIONAL BOOK COMPANY

Auckland Bogotá Guatemala Hamburg Johannesburg Lisbon
London Madrid Mexico New Delhi Panama Paris San Juan
São Paulo Singapore Sydney Tokyo

**APPLIED MATHEMATICS FOR
ENGINEERS AND PHYSICISTS**

INTERNATIONAL STUDENT EDITION

Copyright © 1971
Exclusive rights by McGraw-Hill Kogakusha, Ltd. for
manufacture and export. This book cannot be re-exported
from the country to which it is consigned by McGraw-Hill.

8th Printing 1982

Library of Congress Catalog Card Number 78-101383

When ordering this title use ISBN 0-07-085577-3

TOSHO PRINTING CO., LTD., TOKYO, JAPAN

PREFACE

The first edition of this text was published 23 years ago and the second edition 11 years ago. It was stated in the Preface to the second edition that the tremendous development of high-speed computing devices was a major factor guiding the changes and revisions presented in that edition. The developments which have come about in digital computers since the publication of the second edition have been all the more spectacular. Computation speeds and memory size have increased by two orders of magnitude during this period, while machine costs have dropped by an order of magnitude. A single modern center, such as the Western Data Processing Center at the University of California at Los Angeles, has as much computing capacity today as all the combined installations in the United States a decade ago. There is also no apparent reason why one should not expect these trends to continue over the next decade.

Another factor to be considered is the recent development of time-sharing computer systems which, through large numbers of remote terminals, provide engineers and scientists with direct and immediate access to a computer for problem solving. It may not be too long until the remote terminal has replaced the slide rule as a readily available and sophisticated computing tool. As a result the analyst will be spending more time on developing realistic mathematical models for a physical problem and less time on the computational details. Also, the magnitude of the computations are of lesser concern because of the great speed of digital machines. However, computers have not eliminated the problem of choosing between accuracy and speed; they have only shifted the break-even point to a higher level.

We would like to suggest two particular areas of concern which we feel should be reviewed from time to time by every analyst. The first is to maintain an awareness of the limitations of any mathematical model resulting from the various approximations imposed during the modeling. process. This is very important in order to avoid predicting the behavior of a system by a solution obtained from a model based on postulates which are invalid in the region of interest. Our second concern is that even though computers have given us the ability to study complex models, we should not stop seeking simpler representations, as it is an easy matter to overcomplicate a problem.

Perhaps the greatest factor creating the need for the extensive changes included in this third edition has been the widespread changes in engineering curricula which will continue for several years to come. Coupled with this are improvements in teaching mathematics at the high school level, which have enabled the shifting of more advanced material into the lower-division mathematics sequences in many colleges. These two situations have increased the mathematics requirements for engineering students and raised the level of mathematical rigor. The authors feel that both of these situations are desirable, but it was decided not to trade off the valuable physical applications for increased rigor in this edition. It is felt that most instructors teaching an advanced course based on this text can easily add any desired amount of rigor during the lectures but that it is usually more difficult to add a wide range of physical applications during the lecture period.

Those readers who are familiar with the previous two editions of this book will rapidly become aware of the extensive changes which have been made in this third edition. The arrangement of the chapters has been changed greatly, with those on series, special functions, vectors and tensors, transcendental equations, and partial differentiation being moved to appendixes. This change is an attempt to make the material more flexible for the variety of courses in which this book could be employed as a text. Since the material in the body of the text is designed for use in a one-year course, it is hoped that each instructor will feel free to add from, or exchange with, the material in the appendixes to meet the level and individual requirements of his course.

Other changes are the combining of the two previous tables of Laplace transforms into a single one in Appendix A, to which a few additional transform pairs have been added. The p-multiplied Laplace-transform notation has been dropped in favor of the more common s notation. Also additional problems have been added to each chapter, with answers and hints to selected problems supplied in Appendix G. Fourier, Hankel, and Mellin transforms have been added to the chapter on operational methods, and a new chapter on statistics and probability has been included.

As previously mentioned, this text is designed for use in a one-year course, and the chapters have been written to make them as independent as possible to give each instructor freedom in designing his own course. Chapters 1 to 8 are primarily concerned with the analysis of lumped parameter systems and could comprise the material for the first semester's course with or without some combinations of material contained in the appendixes. Chapters 9 to 13 deal with distributed parameter systems, while Chapters 14 to 16 cover various important areas of applied mathematics. As such, these eight chapters could comprise the material for the second-semester's course.

The authors would like to extend their appreciation to colleagues and students for their helpful assistance and insights in the preparation of this material. Thanks are also due to the editors of McGraw-Hill for their patience and assistance, and to the reviewers who contributed several helpful suggestions for improvement. Special thanks are due to Professor S. Takeda, Hosei University, Tokyo, for his detailed errata for the second edition which have been included in this present edition. Finally the greatest acknowledgment is due to our wives, Johanna and Doris, for their continued encouragement and understanding.

LOUIS A. PIPES
LAWRENCE R. HARVILL

CONTENTS

Chapter 2
LINEAR DIFFERENTIAL EQUATIONS   53

Chapter 3
LINEAR ALGEBRAIC EQUATIONS,
DETERMINANTS, AND MATRICES   82

Chapter 4
LAPLACE TRANSFORMS   143

Chapter 14
APPROXIMATE METHODS IN APPLIED
MATHEMATICS   561

Chapter 15
THE ANALYSIS OF NONLINEAR SYSTEMS 595

The Reversion Method for Solving Nonlinear
Differential Equations   653

Forced Oscillations of Nonlinear Circuits   665

Matrix Solution of Equations of the Mathieu-Hill Type   681

**Chapter 16
STATISTICS AND PROBABILITY**   733

APPLIED MATHEMATICS
FOR ENGINEERS
AND PHYSICISTS

1
The Theory of Complex Variables

1 INTRODUCTION

It is fairly safe to state that the subject of complex variables is fundamental to the subject of applied mathematics. A solid background in this area is essential for the application of many useful methods of analysis to obtain solutions to a wide range of practical problems.

This chapter is intended as a review and brief summary of many of the useful results of complex function theory. It has been assumed by the authors that the reader has some familiarity with complex numbers.† A more thorough coverage of this topic can be found in any of the references at the end of this chapter.

2 FUNCTIONS OF A COMPLEX VARIABLE

From the basic theory of complex numbers it may be shown how the elementary transcendental functions, for example, $\sin x$, $\cos x$, e^x, and so on, are defined

---

† Readers who would like a brief review of complex numbers are referred to Pipes, 1958, chap. 2 (see References at the end of this chapter).

for complex values of their arguments simply by allowing the variable in the power series expansions of these functions to take on complex values, for example, $\sin z$, $\cos z$, e^z, and so on. Any function of the complex variable

$$z = x + jy \qquad j = -1^{\frac{1}{2}} \tag{2.1}$$

is termed a *complex function* and is frequently expressed in the following form:

$$w(z) = u(x,y) + jv(x,y) \tag{2.2}$$

In this expression the functions u and v are real functions of the real variables x and y. Just as x and y are the real and imaginary parts of z, $u(x,y)$ and $v(x,y)$ are the real and imaginary parts of the complex function $w(z)$:

$$u(x,y) = \operatorname{Re} w(z) \qquad v(x,y) = \operatorname{Im} w(z) \tag{2.3}$$

As specific examples consider the complex functions $\sin z$ and e^z. The real and imaginary parts may be found as follows:

$$\sin z = \sin(x + jy) = \sin x \cos jy + \cos x \sin jy$$
$$= \sin x \cosh y + j \cos x \sinh y \tag{2.4}$$

where use has been made of the well-known formulas

$$\cos jx = \cosh x \qquad \sin jy = j \sinh y \tag{2.5}$$

From (2.4) we find that

$$u(x,y) = \sin x \cosh y$$
$$v(x,y) = \cos x \sinh y \tag{2.6}$$

For e^z we have

$$e^z = e^{x+jy} = e^x e^{jy}$$
$$= e^x(\cos y + j \sin y) \tag{2.7}$$

from which we find that

$$u(x,y) = e^x \cos y \qquad v(x,y) = e^x \sin y \tag{2.8}$$

In deriving (2.7) we have made use of the first of the following three Euler formulas:

$$e^{jx} = \cos x + j \sin x$$
$$\cos x = \frac{e^{jx} + e^{-jx}}{2} \tag{2.9}$$
$$\sin x = \frac{e^{jx} - e^{-jx}}{2j}$$

Let us now consider the expression

$$w(z) = u(x,y) + jv(x,y) \tag{2.10}$$

and determine what conditions it must satisfy in order that it may be a function of z. If we speak in the broadest sense of the word *function*, then w is always a function of z since if z is given then x and y are determined and it follows that u and v are determined in terms of x and y.

This definition is too broad, and in the theory of functions of a complex variable it is restricted by demanding that the function w shall have a *definite derivative* for a given value of z.

It must be realized that if $w(z)$ is one of the analytic elementary transcendental functions, such as z^n, $\sin z$, $\cosh z$, $\ln z$, etc., this condition is usually met since many of the operations used in the calculus of real variables to obtain the derivatives are still valid for the complex variable and hence the derivative is uniquely determined. We thus have

$$\frac{d}{dz}\cosh z = \sinh z \qquad \frac{d}{dz}\ln z = \frac{1}{z} \qquad \text{etc.} \tag{2.11}$$

However, it will now be shown that the uniqueness of the derivative requires the functions u and v to satisfy certain conditions.

3 THE DERIVATIVE AND THE CAUCHY-RIEMANN DIFFERENTIAL EQUATIONS

In the theory of functions of a real variable, if the difference quotient

$$\frac{F(x+h) - F(x)}{h} = F'(x) + \phi(x,h) \tag{3.1}$$

where $F(x)$ is a real function of the real variable x, can be resolved into two terms in such a way that the first is independent of h and the second one is such that

$$\lim_{h\to 0} \phi(x,h) = 0 \tag{3.2}$$

then by definition it is possible to differentiate $F(x)$, and $F'(x)$ is called the *derivative* of $F(x)$. Hence

$$F'(x) = \lim_{h\to 0} \frac{F(x+h) - F(x)}{h} \tag{3.3}$$

This definition is transferred to complex functions as follows:

$$\frac{w(z+q) - w(z)}{q} = w'(z) + \phi(z,q) \tag{3.4}$$

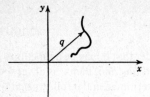

Fig. 3.1

where

$$\lim_{q \to 0} \phi(z,q) = 0 \tag{3.5}$$

In this case, however, q represents a *vector* in the xy plane of Fig. 3.1. The end point of the vector representing the complex number q can converge toward the origin as $q \to 0$ along any arbitrary curve. The derivative $w'(z)$ is defined by

$$w'(z) = \lim_{q \to 0} \frac{w(z + q) - w(z)}{q} \tag{3.6}$$

and must be independent of the manner in which q approaches zero in order to be unique.

In order to carry out the limiting process (3.6), let

$$q = \Delta z = \Delta x + j \Delta y \tag{3.7}$$

and

$$\Delta w = w(z + \Delta z) - w(z)$$
$$= \Delta u + j \Delta v \tag{3.8}$$

Now, since u and v are functions of x and y, then, if u and v have continuous partial derivatives of the first order, we have

$$\Delta u = \left(\frac{\partial u}{\partial x} + \epsilon_1\right) \Delta x + \left(\frac{\partial u}{\partial y} + \epsilon_2\right) \Delta y$$
$$\Delta v = \left(\frac{\partial v}{\partial x} + \epsilon_3\right) \Delta x + \left(\frac{\partial v}{\partial y} + \epsilon_4\right) \Delta y \tag{3.9}$$

where

$$\lim_{\substack{\Delta x \to 0 \\ \Delta y \to 0}} \epsilon_r = 0 \qquad r = 1, 2, 3, 4 \tag{3.10}$$

Substituting (3.7), (3.8), and (3.9) in (3.6), we have

$$\frac{dw}{dz} = \lim_{\Delta z \to 0} \frac{\Delta w}{\Delta z} = \frac{[(\partial u/\partial x) + j(\partial v/\partial x)] + [(\partial u/\partial y) + j(\partial v/\partial y)] \, dy/dx}{1 + j(dy/dx)},$$
$$= \frac{A + Bm}{1 + mj} \tag{3.11}$$

where

$$A = \frac{\partial u}{\partial x} + j \frac{\partial v}{\partial x}$$

$$B = \frac{\partial u}{\partial y} + j \frac{\partial v}{\partial y} \qquad (3.12)$$

$$m = \frac{dy}{dx}$$

Now, if dw/dz is to be unique, it must be independent of the manner in which Δz approaches zero and hence must be independent of $m = dy/dx$. If $dw/dz = w'$ is independent of m, we must have

$$\frac{\partial}{\partial m}(w') = \frac{\partial}{\partial m} \left(\frac{A + Bm}{1 + jm} \right) = 0$$

$$= \frac{(1 + jm)B - (A + Bm)j}{(1 + jm)^2} = 0 \qquad (3.13)$$

Hence

$$B = jA \qquad (3.14)$$

or

$$\frac{\partial u}{\partial y} + j \frac{\partial v}{\partial y} = j \left(\frac{\partial u}{\partial x} + j \frac{\partial v}{\partial x} \right) \qquad (3.15)$$

Equating the coefficients of the real and imaginary terms of (3.15), we obtain

$$\frac{\partial u}{\partial y} = - \frac{\partial v}{\partial x} \qquad (3.16)$$

and

$$\frac{\partial u}{\partial x} = \frac{\partial v}{\partial y} \qquad (3.17)$$

Equations (3.16) and (3.17) are the conditions that the real and imaginary parts of $w(z)$ must satisfy in order that $w(z)$ may have a unique derivative $w'(z)$. Such a function is said to be *analytic* at the point z. These equations are called the *Cauchy-Riemann* differential equations. We thus see that the real and imaginary parts of an analytic function satisfy Cauchy-Riemann differential equations. Conversely, it is easily verified that, if the real and imaginary parts have continuous first partial derivatives satisfying Cauchy-Riemann differential equations, then the function is analytic.

If we differentiate Eq. (3.16) with respect to y and Eq. (3.17) with respect to x and add the results, we obtain

$$\frac{\partial^2 u}{\partial x^2} + \frac{\partial^2 u}{\partial y^2} = 0 \qquad (3.18)$$

Differentiating (3.17) with respect to y and (3.16) with respect to x and subtracting the results, we obtain

$$\frac{\partial^2 v}{\partial x^2} + \frac{\partial^2 v}{\partial y^2} = 0 \tag{3.19}$$

Thus we see that the real and imaginary parts of an *analytic* function $w(z)$ of a complex variable are solutions of the two-dimensional Laplace equation. It is therefore true that $w(z)$ is itself a solution to Laplace's equation.

The functions u and v that satisfy the Cauchy-Riemann equations (3.16) and (3.17) are called *conjugate functions*. It is also true that if one of these functions is known, the other may be found to within an arbitrary constant by integrating the Cauchy-Riemann equations. This implies that $w(z)$ may be determined to within an arbitrary complex constant. As an example let us consider that u is given and it is desired to find the conjugate function v. Assuming that

$$u = x^2 - y^2 \tag{3.20}$$

and integrating (3.17) yields

$$\begin{aligned}
v(x,y) &= \int \frac{\partial u}{\partial x} dy + c(x) \\
&= \int 2x\, dy + c(x) \\
&= 2xy + c(x)
\end{aligned} \tag{3.21}$$

Now, if we integrate (3.16) we find

$$\begin{aligned}
v(x,y) &= -\int \frac{\partial u}{\partial y} dx + c'(y) \\
&= -\int 2y\, dx + c'(y) \\
&= 2xy + c'(y)
\end{aligned} \tag{3.22}$$

Comparing (3.21) and (3.22), it may be deduced that

$$v(x,y) = 2xy + c''$$

where $c'' = $ a constant.

If Eqs. (3.20) and (3.23) are substituted into (2.2) there results

$$\begin{aligned}
w(z) &= x^2 - y^2 + j2xy + jc'' \\
&= (x + jy)^2 + jc'' \\
&= z^2 + jc''
\end{aligned} \tag{3.24}$$

It should be mentioned at this point that once Eq. (3.21) has been obtained the result could have been substituted into Eq. (3.16) to yield a first-order differential equation for the undetermined function $c(x)$. This method is more direct but requires the solution of a first-order differential equation.

In a large number of applications only the derivatives of $w(z)$ are required and therefore the arbitrary constant may be dropped.

Substituting Eqs. (3.16) and (3.17) into (3.11) yields

$$\frac{dw}{dz} = w'(z) = \frac{\partial u}{\partial x} + j\frac{\partial v}{\partial x} = \frac{1}{j}\left(\frac{\partial u}{\partial y} + j\frac{\partial v}{\partial y}\right) \tag{3.25}$$

The use of these functions in the solution of two-dimensional potential problems will be discussed in Chap. 9.

4 LINE INTEGRALS OF COMPLEX FUNCTIONS

The concept of the line integral of a vector field is familiar to those who have had a brief introduction to vector algebra. In this chapter the concept of a line integral of a complex function will be developed.

Let $w(z)$ be a complex function of the complex variable z defined in the region B (Fig. 4.1), and let C be a smooth curve in the region B having end points z_0 and z. Let $z_1, z_2, z_3, \ldots, z_{n-1}$ be an arbitrary number of intermediate points on C and z_n chosen to be the point z.

Let

$$\Delta z_r = z_r - z_{r-1} \qquad r = 1, 2, \ldots, n \tag{4.1}$$

represent chord vectors of the curve C. Let m_r be a point on the curve C located between z_{r-1} and z_r (this point may also coincide with z_{r-1} or with z_r). Let the following summation be performed:

$$\sum_{r=1}^{r=n} w(m_r)\Delta z_r = L_n \tag{4.2}$$

If the curve C is divided into smaller and smaller parts so that $n \to \infty$, $|\Delta z_r| \to 0$, and if the summation tends to a limit that is independent of the choice of the intermediate points and of the manner in which the division is performed, then the above limit L_n is called the *definite integral* of the complex

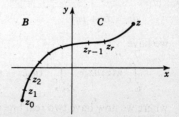

Fig. 4.1

function $w(z)$ taken along the path C and between the limits $z = z_0$ and $z = z$ and is denoted by

$$\lim_{n \to \infty} L_n = \int_{z_0}^{z} {}_C \, w(\zeta) \, d\zeta \tag{4.3}$$

As in the case of real quantities, it may be shown that the limit L_n exists if $w(z)$ is continuous along the path of integration.

The value of this integral *in general* depends on $w(z)$ and on the limits of the integral as well as on the form of the path C.

ESTIMATION OF THE INTEGRAL VALUE

Let the curve C be a curve of finite length (rectifiable curve); then, if

$$|w(z)| \leqslant M \tag{4.4}$$

where M represents a fixed real quantity throughout the curve C, the following estimation of the integral along the curve C exists,

$$\left| \int_{z_0}^{z} {}_C \, w(\zeta) \, d\zeta \right| \leqslant Ms \tag{4.5}$$

where s is the length of the curve C from z_0 to z.

To prove (4.5), we substitute (4.4) in Eq. (4.2) and obtain

$$\left| \int_{z_0}^{z} {}_C \, w(\zeta) \, d\zeta \right| \leqslant M \lim_{n \to \infty} \left| \sum_{r=1}^{r=n} \varDelta z_r \right| \tag{4.6}$$

However,

$$\lim_{u \to \infty} \left| \sum_{r=1}^{r=n} \varDelta z_r \right| = \left| \int_{z_0}^{z} {}_C \, dz \right| \leqslant \int_{z_0}^{z} {}_C \, \sqrt{dx^2 + dy^2} = s \tag{4.7}$$

This proves Eq. (4.5).

If we decompose $w(z)$ into its real and imaginary parts

$$w(z) = u + jv \tag{4.8}$$

and write

$$dz = dx + j \, dy \tag{4.9}$$

we have

$$\int_{z_0}^{z} {}_C \, w(z) \, dz = \int_{x_0, y_0}^{x, y} {}_C \, (u \, dx - v \, dy) + j \int_{x_0, y_0}^{x, y} {}_C \, (v \, dx + u \, dy) \tag{4.10}$$

where we now have two real line integrals.

5 CAUCHY'S INTEGRAL THEOREM

Let the function $w(z)$ be single-valued and continuous and possess a definite derivative throughout a region R; that is, let $w(z)$ be analytic in R.

Cauchy's integral theorem states that

$$\oint_c w(z)\,dz = 0 \tag{5.1}$$

where the above notation signifies that the line integral is taken along an arbitrary closed path lying inside the region R.

The proof of this theorem can be made to depend on Stokes's theorem of Appendix E, Sec. 10. Stokes's theorem states that, if we have a vector field $\mathbf{A}$ whose components possess the continuous partial derivatives involved in the calculation of $\nabla \times \mathbf{A}$, then

$$\oint_c \mathbf{A} \cdot d\mathbf{l} = \iint_s (\nabla \times \mathbf{A}) \cdot d\mathbf{s} \tag{5.2}$$

where the line integral is taken along the curves bounding the open surface s. If $\mathbf{A}$ is a two-dimensional vector field having the components A_x and A_y above, then (5.2) reduces to

$$\oint_c (A_x\,dx + A_y\,dy) = \iint_s \left(\frac{\partial A_y}{\partial x} - \frac{\partial A_x}{\partial y} \right) dx\,dy \tag{5.3}$$

where c is a closed curve lying in the xy plane and s is the surface bounded by this curve. As a consequence of (4.10) we may write

$$\oint_c w(z)\,dz = \oint_c (u\,dx - v\,dy) + j \oint_c (v\,dx + u\,dy) \tag{5.4}$$

By means of (5.3) we may transform both the integrals of the right-hand member of (5.4) into surface integrals. To transform the first one, let

$$
\begin{aligned}
A_z &= u \\
A_y &= -v
\end{aligned} \tag{5.5}
$$

We thus obtain

$$\oint_c (u\,dx - v\,dy) = -\iint_s \left(\frac{\partial v}{\partial x} + \frac{\partial u}{\partial y} \right) dx\,dy \tag{5.6}$$

However, as a consequence of the first Cauchy-Riemann equation (3.16), we have

$$\frac{\partial v}{\partial x} + \frac{\partial u}{\partial y} = 0 \tag{5.7}$$

Hence the first integral of the right-hand member of (5.4) vanishes. Using Eq. (5.3) to transform the second integral of the right-hand member of

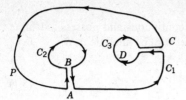

Fig. 5.1

(5.4) and the second Cauchy-Riemann equation (3.16), this integral may be shown to vanish also.   Hence Cauchy's integral theorem (5.1) is proved.

As a consequence of this theorem it follows that the path of the line integral, whether closed or between fixed limits, may be deformed without changing the value of the integral, provided that in the deformation no point is encountered at which $w(z)$ ceases to be analytic.

MULTIPLY CONNECTED REGIONS

Cauchy's integral theorem has been deduced under the assumption that the closed curve is the boundary of a *simply connected* region.   However, as we shall see, the theorem still holds if the enclosed region is *multiply connected*. Consider the region of Fig. 5.1.

This region requires three curves c_1, c_2, and c_3 to divide it into two separate parts and is therefore a triply connected region.   By introducing the crosscuts AB and CD the region may be transformed into a simply connected region. We thus have

$$\oint_s w(z)\,dz = 0 \tag{5.8}$$

where the curve s includes the outer curve c_1 traversed in the mathematically positive direction, the curves c_2 and c_3 traversed in the negative direction, and the crosscuts AB and CD.   We thus have

$$\oint_s w(z)\,dz = \oint_{c_1} w(z)\,dz + \oint_{c_2} w(z)\,dz + \oint_{c_3} w(z)\,dz + \int_A^B w(z)\,dz$$
$$+ \int_B^A w(z)\,dz + \int_C^D w(z)\,dz + \int_C^D w(z)\,dz \tag{5.9}$$

Since the function is analytic along the crosscuts and the integral from A to B is traversed in the opposite direction from the integral B to A, etc., the integrals along the crosscuts cancel out in pairs.   Using this fact and transposing, we obtain

$$\oint_{c_1} w(z)\,dz = \oint_{c_2} w(z)\,dz + \oint_{c_3} w(z)\,dz \tag{5.10}$$

where we have reversed the direction of integration along curves c_2 and c_3.

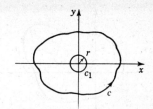

Fig. 5.2

As an example of the use of Cauchy's integral theorem, let it be required to compute the integral

$$I = \oint_c \frac{dz}{z} \tag{5.11}$$

where c is a simple closed curve.

The function $w(z) = 1/z$ is analytic for any value of z except for $z = 0$. If, therefore, the simple closed curve c encloses the origin, let us draw an arc c_1 of small radius r with center at the origin as shown in Fig. 5.2.

Since the function $1/z$ is analytic in the region between c_1 and c, we have by Eq. (5.10)

$$\oint_c \frac{dz}{z} = \oint_{c_1} \frac{dz}{z} \tag{5.12}$$

Now on the circle c_1 we have

$$z = r e^{j\theta} \qquad dz = jr e^{j\theta} d\theta \tag{5.13}$$

Hence

$$\oint_{c_1} \frac{dz}{z} = \int_{\theta=0}^{\theta=2\pi} j \, d\theta = 2\pi j \tag{5.14}$$

We thus have the result

$$\oint \frac{dz}{z} = \begin{cases} 0 & \text{if } c \text{ does not surround origin} \\ 2\pi j & \text{if } c \text{ surrounds origin} \end{cases} \tag{5.15}$$

The implication of Cauchy's integral theorem is that, provided no singularities are crossed, any contour integral may be deformed into any other. As seen in the example above, the contour c was deformed into the contour c_1.

6 CAUCHY'S INTEGRAL FORMULA

Let $w(z)$ be analytic in a region including a point $z = a$ and bounded by a curve c. Let us draw a small circle c_1 of radius r and center at a, as shown in Fig. 6.1.

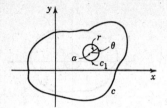

Fig. 6.1

Then, in the area bounded by the circle c_1 and the curve c, the function

$$\phi(z) = \frac{w(z)}{z-a} \tag{6.1}$$

is analytic.   Hence, by Cauchy's theorem, we have

$$\oint_c \frac{w(z)\,dz}{z-a} = \oint_{c_1} \frac{w(z)\,dz}{z-a} \tag{6.2}$$

Now on the circle c_1 we have

$$z = a + r\,e^{j\theta} \qquad dz = jr\,e^{j\theta}\,d\theta$$
$$w(z) = w(a) + [w(z) - w(a)] \tag{6.3}$$

By continuity of $w(z)$ we may make $|w(z) - w(a)| < \epsilon$, for any given $\epsilon > 0$, by making r sufficiently small.   Hence

$$\left| \oint_{c_1} \frac{w(z)\,dz}{z-a} - \int_0^{2\pi} w(a)\,j\,d\theta \right|$$
$$= \left| \int_0^{2\pi} j[w(z) - w(a)]\,d\theta \right| \leqslant 2\pi\epsilon \qquad \text{by (4.5)} \tag{6.4}$$

Since by (5.10) the integral on the right may be evaluated on any circle enclosing a and since $w(z)$ is continuous at a, it can be made arbitrarily small in magnitude.   In other words,

$$\left| \oint_{c_1} \frac{w(z)\,dz}{z-a} - 2\pi j w(a) \right| = 0 \tag{6.5}$$

Hence, from (6.4) and (6.2), we obtain

$$\oint_c \frac{w(z)\,dz}{z-a} \doteq 2\pi j w(a) \tag{6.6}$$

or

$$w(a) = \frac{1}{2\pi j} \oint_c \frac{w(z)}{z-a}\,dz \tag{6.7}$$

This is Cauchy's integral formula; it is remarkable in that it enables one to compute the value of a function $w(z)$ inside a region in which it is analytic from the values of the function on the boundary.

If we apply the formula to a circle with center at a of radius R, we have

$$z = Re^{j\theta} + a \qquad dz = jRe^{j\theta}\, d\theta \tag{6.8}$$

and we obtain

$$w(a) = \frac{1}{2\pi} \int_{\theta=0}^{\theta=2\pi} w(z)\, d\theta \tag{6.9}$$

This shows that the value of the function $w(z)$ at the center of a circle is equal to the average value of its boundary values. If M denotes the maximum of the absolute value of $w(z)$ on the boundary of the circle, then

$$\left| \int_{\theta=0}^{\theta=2\pi} w(z)\, dz \right| \leqslant 2\pi M \tag{6.10}$$

Hence

$$|w(a)| \leqslant M \tag{6.11}$$

Therefore the maximum of the absolute value of an analytic function cannot be situated inside a circular region.

It is of interest to note that since $w(z) = u + jv$, the above relations derived for $w(z)$, that is, Eqs. (6.9) and (6.10), also hold for u and v separately. Thus we may infer that solutions of Laplace's equation must satisfy the stated relations. This means that the values of the solutions to Laplace's equation may be computed by an integration over the boundary conditions via Eq. (6.7).

The feature that the value of a solution at the center of a circular region is the average of the boundary conditions has great significance for obtaining numerical solutions to Laplace's equation which should be familiar to those with some background in numerical analysis.

Another form of Cauchy's integral formula is obtained by replacing a by z and letting $z = t$ in (6.7). We then have

$$w(z) = \frac{1}{2\pi j} \oint_c \frac{w(t)}{t - z}\, dt \tag{6.12}$$

where now z is held fixed in the integration and t traverses the curve c.

DERIVATIVES OF AN ANALYTIC FUNCTION

It may be shown that the integral (6.12) may be differentiated under the integral sign and that the result thus obtained may be differentiated in the same way. This will be assumed. Then we have

$$\frac{dw}{dz} = w'(z) = \frac{1}{2\pi j} \oint_c \frac{w(t)}{(t - z)^2}\, dt \tag{6.13}$$

$$w''(z) = \frac{2!}{2\pi j} \oint \frac{w(t)}{(t-z)^3} dt \tag{6.14}$$

$$w^{(n)}(z) = \frac{n!}{2\pi j} \oint_c \frac{w(t)}{(t-z)^{n+1}} dt \qquad n = 0, 1, 2, 3, \ldots \tag{6.15}$$

From this it follows that if a function is analytic all its derivatives exist. This is not necessarily true of a function of a real variable.

7 TAYLOR'S SERIES

The Taylor's series expansion of a function of a real variable should be quite familiar to those who have had only a slight background in differential calculus. By means of Cauchy's integral formula we may develop the Taylor's series expansion of a function of a complex variable. Let us consider the function $w(z)$ and let z be replaced by $z + h$ in Eq. (6.12), which yields

$$w(z + h) = \frac{1}{2\pi j} \oint_c \frac{w(t) dt}{t - (z + h)} \tag{7.1}$$

where we assume that $w(z)$ is analytic in a circle R with the boundary c and that the points z and $z + h$ are inside of this circle. Let us now expand $1/[t - (z + h)]$ into a power series in h in the form

$$\frac{1}{t - (z + h)} = \frac{1}{(t - z)\left(1 - \dfrac{h}{t - z}\right)}$$

$$= \frac{1}{t - z}\left[1 + \frac{h}{t - z} + \frac{h^2}{(t - z)^2} + \cdots \right.$$

$$\left. + \frac{h^n}{(t - z)^n} + \frac{h^{n+1}}{(t - z)^n (t - z - h)}\right] \tag{7.2}$$

Hence

$$\frac{1}{2\pi j} \oint_c \frac{w(t) dt}{t - (z + h)} = \frac{1}{2\pi j}\left[\oint_c \frac{w(t) dt}{t - z} + h \oint_c \frac{w(t) dt}{(t - z)^2} + h^2 \oint_c \frac{w(t) dt}{(t - z)^3}\right.$$

$$\left. + \cdots + h^n \oint \frac{w(t) dt}{(t - z)^{n+1}} + h^{n+1} \oint \frac{w(t) dt}{(t - z)^{n+1}(t - z - h)}\right] \tag{7.3}$$

Using (6.15) and (6.1), we have

$$w(z + h) = w(z) + \frac{h w'(z)}{1!} + \frac{h^2 w''(z)}{2!} + \cdots + \frac{h^n}{n!} w^{(n)}(z) + R_{n+1} \tag{7.4}$$

where R_{n+1} is the remainder after $n + 1$ terms and is given by

$$R_{n+1} = \frac{h^{n+1}}{2\pi j} \oint_c \frac{w(t) dt}{(t - z)^{n+1}(t - z - h)} \tag{7.5}$$

Now, by the estimation formula (4.5), it may be shown that

$$\lim_{n \to \infty} R_{n+1} = 0 \tag{7.6}$$

The series (7.4) converges inside of a circle with z as its center, the radius of the circle being equal to the distance from z to the nearest point $z + h$ at which $w(z + h)$ is no longer analytic. Series (7.4) may also be represented as follows:

$$w(z + h) = \sum_{r=0}^{\infty} a_r h^r \tag{7.7}$$

where

$$a_r = \frac{1}{2\pi j} \oint \frac{w(t)\, dt}{(t - z)^{r+1}} \qquad r = 0, 1, 2, 3, \ldots \tag{7.8}$$

Thus the coefficients may be obtained by means of integration.

8 LAURENT'S SERIES

Let $w(z)$ be an analytic function in the annular region of Fig. 8.1 including the boundary of the region. The annular region is formed by two concentric circles c and c_1 whose center is z_0. Let $z_0 + h = z$ be located inside the annular region. If we apply Cauchy's integral theorem to this annular region, we have

$$w(z_0 + h) = \frac{1}{2\pi j} \oint \frac{w(t)\, dt}{t - (z_0 + h)} \tag{8.1}$$

The integration is carried out over the entire boundary of the annular region composed of the circles c and c_1. We thus have symbolically

$$w(z_0 + h) = \frac{1}{2\pi j} \left(\oint_c - \oint_{c_1} \right) \tag{8.2}$$

where the integration over both circles must be taken in the positive direction. Each integral is calculated individually. Now, as in Eq. (7.2), we have

$$\frac{1}{t - z_0 - h} = \frac{1}{t - z_0} + \frac{h}{(t - z_0)^2} + \frac{h^2}{(t - z_0)^3} + \cdots$$
$$+ \frac{h^n}{(t - z_0)^{n+1}} + \frac{h^{n+1}}{(t - z_0)^{n+1}(t - z_0 - h)} \tag{8.3}$$

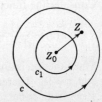

Fig. 8.1

If we interchange $t - z_0$ and h, we have

$$-\frac{1}{t - z_0 - h} = \frac{1}{h} + \frac{t - z_0}{h^2} + \frac{(t - z_0)^2}{h^3} + \cdots$$

$$+ \frac{(t - z_0)^n}{h^{n+1}} + \frac{(t - z_0)^{n+1}}{h^{n+1}(t - z_0 - h)} \qquad (8.4)$$

If we now substitute these expansions in Eq. (8.2), we have

$$w(z_0 + h) = \sum_{s=-(n+1)}^{s=n} a_s h^s + R_n' + R_n'' \qquad (8.5)$$

where

$$a_s = \frac{1}{2\pi j} \oint_c \frac{w(t)\,dt}{(t - z_0)^{s+1}} \qquad s = 0, +1, +2, +3, \ldots \qquad (8.6)$$

$$a_s = \frac{1}{2\pi j} \oint_{c_1} \frac{w(t)\,dt}{(t - z_0)^{s+1}} \qquad s = -1, -2, -3, -4, \ldots \qquad (8.7)$$

$$R_n' = \frac{1}{2\pi j} \oint_c \frac{h^{n+1} w(t)\,dt}{(t - z_0)^{n+1}(t - z_0 - h)}$$

$$R_n'' = \frac{1}{2\pi j} \oint_{c_1} \frac{(t - z_0)^{n+1} w(t)\,dt}{h^{n+1}(t - z_0 - h)} \qquad (8.8)$$

By Cauchy's estimation formula (4.5) it is easy to show that

$$\lim_{n \to \infty} R_n' = 0$$

$$\lim_{n \to \infty} R_n'' = 0 \qquad (8.9)$$

Hence we obtain the Laurent's series

$$w(z_0 + h) = \sum_{s=-\infty}^{s=+\infty} a_s h^s \qquad (8.10)$$

where the coefficients a_s are given by (8.6) and (8.7). It may be noted that if all the coefficients of negative index have the value zero, then (8.10) reduces to Taylor's series.

EXAMPLES OF LAURENT'S-SERIES EXPANSIONS
Consider the function

$$w(z) = \frac{1}{z(z - 1)} \qquad (8.11)$$

This function is analytic at all points of the z plane except at the points $z = 0$ and $z = 1$. Therefore it is possible to expand this function in a Laurent's series in an annular region about $z = 0$ or $z = 1$.

To expand about $z = 0$, we have

$$\frac{1}{z(z-1)} = -\frac{1}{z}(1-z)^{-1}$$

$$= -\frac{1}{z}(1 + z + z^2 + z^3 + \cdots)$$

$$= -\frac{1}{z} - 1 - z - z^2 - z^3 - \cdots \qquad 0 < |z| < 1 \qquad (8.12)$$

To expand about $z = 1$, write

$$\frac{1}{z(z-1)} = \frac{1}{z-1}\frac{1}{(z-1)+1}$$

$$= \frac{1}{z-1}[1 - (z-1) + (z-1)^2 - (z-1)^3 + \cdots]$$

$$= \frac{1}{z-1} - 1 + (z-1) - (z-1)^2 + \cdots$$

$$0 < |z-1| < 1 \quad (8.13)$$

As another example, consider the function

$$w(z) = e^{1/z} \qquad (8.14)$$

This function may be expanded in any annular region enclosing the origin in the form

$$e^{1/z} = 1 + \frac{1}{z} + \frac{1}{z^2 \times 2!} + \frac{1}{z^3 \times 3!} + \cdots \qquad z \neq 0 \qquad (8.15)$$

In this case all the terms of the Laurent's series have a negative index.

9 RESIDUES; CAUCHY'S RESIDUE THEOREM

RESIDUES

In the Laurent's series (8.10) let

$$z = z_0 + h \qquad \text{or} \qquad h = z - z_0 \qquad (9.1)$$

We then obtain

$$w(z) = \sum_{s=-\infty}^{s=+\infty} a_s(z - z_0)^s \qquad (9.2)$$

The coefficient a_{-1} is given by Eq. (8.7), and it is

$$a_{-1} = \frac{1}{2\pi j} \oint_{c_1} w(t)\, dt \qquad (9.3)$$

where c_1 is a curve surrounding the point z_0. a_{-1} is in general a complex number and is called the *residue* of $w(z)$ at the point $z = z_0$. It is usually denoted by

$$a_{-1} = \operatorname{Res} w(z)_{z=z_0} \tag{9.4}$$

It sometimes happens that the Laurent's series for $w(z)$ is known in the neighborhood of $z = z_0$, or the series is easier to compute than the integral. In this case from a knowledge of a_{-1} we may compute the integral (9.3).

As a very simple example, consider the function $w(z) = 1/z$. In this case the Laurent's series about the origin consists of only one term, and we have

$$a_{-1} = 1 \tag{9.5}$$

Hence

$$\oint \frac{dz}{z} = 2\pi j \times 1 = 2\pi j \tag{9.6}$$

where the integral is taken about a curve surrounding the origin. In the same manner we have

$$\oint e^{1/z} \, dz = 2\pi j \tag{9.7}$$

where the path of integration encloses the origin. This follows from the expansion of the function $e^{1/z}$ in the series (8.15). Hence we see that

$$a_{-1} = \operatorname{Res} e^{1/z} = 1 \qquad z = 0 \tag{9.8}$$

CAUCHY'S RESIDUE THEOREM

Let $w(z)$ be an analytic function inside a region R at all points except at the points $z_1, z_2, \ldots, z_n$. Let $w(z)$ be analytic at all points on the boundary c of the region R. Let us surround the points $z_1, z_2, \ldots, z_n$ by the closed curves $c_1, c_2, \ldots, c_n$. Then by Cauchy's integral theorem of Sec. 5, we have (see Fig. 9.1)

$$\oint_c w(z)\,dz = \oint_{c_1} w(z)\,dz + \oint_{c_2} w(z)\,dz + \cdots + \oint_{c_n} w(z)\,dz \tag{9.9}$$

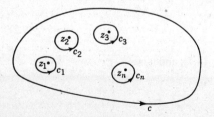

Fig. 9.1

However, by (9.3), we have

$$\oint_{c_r} w(z)\,dz = 2\pi j \operatorname{Res} w(z) \qquad z = z_r \tag{9.10}$$

Hence, substituting this in (9.9), we obtain

$$\oint_c w(z)\,dz = 2\pi j \sum_{r=1}^{r=n} \operatorname{Res} w(z) \qquad z = z_r \tag{9.11}$$

This is *Cauchy's residue theorem*. It is of extreme importance in evaluating definite integrals and in the theory of functions in general. Applications of this theorem to the evaluation of definite integrals will be considered in a later section.

10 SINGULAR POINTS OF AN ANALYTIC FUNCTION

All the points of the z plane at which an analytic function does not have a unique derivative are said to be *singular points*. They are the points at which the function ceases to be analytic. If we concern ourselves only with single-valued functions of the complex variable $w(z)$, then $w(z)$ may have two types of singularities:

1. *Poles*, or *nonessential singular points*
2. *Essential singular points*

The distinction between these two types of singularities will now be explained.

Let z_0 be a singular point of $w(z)$. Let us expand $w(z)$ in a Laurent's series in powers of $z - z_0$. This expansion will contain powers of $z - z_0$ with negative exponents, for otherwise $z = z_0$ would not be a singular point. Hence we have

$$w(z) = \frac{a_{-m}}{(z - z_0)^m} + \frac{a_{-m+1}}{(z - z_0)^{m-1}} + \cdots + \frac{a_{-1}}{z - z_0}$$
$$+ a_0 + a_1(z - z_0) + \cdots \tag{10.1}$$

There are two possibilities:

a. The expansion (10.1) has only a finite number of powers of $z - z_0$ with negative exponents.

In this case $w(z)$ is said to have a *pole* at $z = z_0$. If m is the largest of the negative exponents and if the function

$$\phi(z) = (z - z_0)^m w(z) \qquad m \text{ a positive integer} \tag{10.2}$$

behaves regularly (is analytic) and is not zero at the point z_0, m is called the *order of the pole z_0* and $w(z)$ is said to have a pole of the mth order at the point z_0.

The sum of the terms with negative exponents

$$\frac{a_{-1}}{z - z_0} + \frac{a_{-2}}{(z - z_0)^2} + \cdots + \frac{a_{-m}}{(z - z_0)^m} \tag{10.3}$$

is called the *principal part* of the function $w(z)$ at z_0.

b. The other possibility is that the Laurent's expansion of the function $w(z)$ about the point z_0 will have an infinite number of negative powers of $z - z_0$ and is of the form

$$w(z) = \sum_{r=-\infty}^{r=+\infty} a_r(z - z_0)^r \tag{10.4}$$

In this case the point $z = z_0$ is said to be an *essential singular point*, and $w(z)$ is said to have an *essential singularity* at $z = z_0$.

As examples of these possibilities, consider the function

$$w(z) = \frac{1}{(z - 2)^2(z - 5)^3(z - 1)} \tag{10.5}$$

This function has a pole of the second order at $z = 2$, a pole of the third order at $z = 5$, and a pole of the first order at $z = 1$.

$$w(z) = e^{1/z} \tag{10.6}$$

has the Laurent's series (8.15) in the neighborhood of the origin. Its Laurent's-series expansion contains an infinite number of negative powers of z and has, therefore, an essential singularity at $z = 0$.

MEROMORPHIC FUNCTIONS

If a function $w(z)$ has poles only in the finite part of the z plane, it is said to be a *meromorphic function*.

11   THE POINT AT INFINITY

In the theory of the complex variable it is convenient to regard infinity as a single point. The behavior of $w(z)$ "at infinity" is considered by making the substitution

$$z = \frac{1}{t} \tag{11.1}$$

and examining $w(1/t)$ at $t = 0$. We then say that $w(z)$ is analytic or has a pole or an essential singularity at infinity according as $w(1/t)$ has the corresponding property at $t = 0$.

It may thus be shown that $1/z^2$ is analytic at infinity, z^3 has a pole of the third order at infinity, and the function

$$\sin z = z - \frac{z^3}{3!} + \frac{z^5}{5!} - \frac{z^7}{7!} + \cdots \tag{11.2}$$

has an essential singularity at infinity.

THE RESIDUE AT INFINITY

The residue of $w(z)$ at infinity is defined as

$$\frac{1}{2\pi j} \oint_c w(z)\,dz \tag{11.3}$$

where c is a large circle that encloses all the singularities of $w(z)$ except at $z = \infty$. The integration is taken around c in the *negative* sense, that is, negative with respect to the origin, provided that this integral has a definite value.

If we apply the transformation

$$z = \frac{1}{t} \tag{11.4}$$

to the integral (11.3), it becomes

$$\frac{1}{2\pi j} \oint_s \left[-w\left(\frac{1}{t}\right) \right] \frac{dt}{t^2} \tag{11.5}$$

where the integration is performed in a positive sense about a small circle whose center is at the origin. It follows that, if

$$\lim_{t \to 0} \left[-w\left(\frac{1}{t}\right)\frac{1}{t} \right] = \lim_{z \to \infty} \left[-zw(z) \right] \tag{11.6}$$

has a definite value, then that value is the residue of $w(z)$ at infinity.

For example, the function

$$w(z) = \frac{z}{(z-a)(z-b)} \tag{11.7}$$

behaves like $1/z$ for large values of z and is therefore analytic at $z = \infty$. However,

$$\lim_{z \to \infty} \left[-zw(z) \right] = -1 \tag{11.8}$$

Hence the residue of $w(z)$ at infinity is -1. We thus see that a function may be analytic at infinity and still have a residue there.

12   EVALUATION OF RESIDUES

The calculation of the residues of a function $w(z)$ at its poles may be performed in several ways.   By the definition of the residue of the function $w(z)$ at a simple pole $z = z_0$ is meant the coefficient a_{-1} in the Laurent's expansion of $w(z)$ in the form

$$w(z) = \frac{a_{-1}}{z - z_0} + a_0 + a_1(z - z_0) + a_2(z - z_0)^2 + \cdots \tag{12.1}$$

where z_0 is a simple pole.   If we now multiply Eq. (12.1) by $z - z_0$ and take the limit $z \to z_0$, we have

$$\lim_{z \to z_0}(z - z_0)\,w(z) = a_{-1}$$

$$= \operatorname{Res} w(z) \qquad z = z_0 \tag{12.2}$$

For example, the function

$$w(z) = \frac{e^z}{z^2 + a^2} = \frac{e^z}{(z + ja)(z - ja)} \tag{12.3}$$

has two simple poles, one at $z = ja$ and another at $z = -ja$.   To evaluate the residue at $z = ja$, we form the limit

$$\lim_{z \to ja}(z - ja)\,w(z) = \lim_{z \to ja}\frac{e^z}{z + ja} = \frac{e^{ja}}{2ja} \tag{12.4}$$

Similarly, the limit at $z = -ja$ is $-e^{-ja}/2ja$.

RESIDUES AT SIMPLE POLES OF $w(z) = F(z)/G(z)$

Frequently it is required to evaluate residues of a function $w(z)$ that has the form

$$w(z) = \frac{F(z)}{G(z)} \tag{12.5}$$

where $G(z)$ has simple zeros and hence $w(z)$ has simple poles.   If $z = z_0$ is a simple pole of $w(z)$, then by (12.2) we have

$$\operatorname{Res} w(z)_{z=z_0} = \lim_{z \to z_0}[(z - z_0)\,w(z)]$$

$$= \lim_{z \to z_0}\left[(z - z_0)\frac{F(z)}{G(z)}\right] \tag{12.6}$$

Since $z = z_0$ is a simple pole of $w(z)$, we must have

$$G(z_0) = 0 \tag{12.7}$$

so that expression (12.6) becomes 0/0. To evaluate it, we use L'Hospital's rule and obtain

$$\text{Res } w(z)_{z=z_0} = \lim_{z \to z_0} \frac{F(z) + (z - z_0)\,F'(z)}{G'(z)} = \frac{F(z_0)}{G'(z_0)} \tag{12.8}$$

As an example of the use of this formula, let it be required to compute the residue of $w(z) = e^{jz}/(z^2 + a^2)$ at the simple pole $z = ja$. Using (12.8), we have

$$\text{Res}\left(\frac{e^{jz}}{z^2 + a^2}\right)_{z=ja} = \frac{e^{-a}}{2ja} \tag{12.9}$$

EVALUATION OF A RESIDUE AT A MULTIPLE POLE

If the function $w(z)$ has a multiple pole at $z = z_0$ of order m, then the Laurent's expansion of $w(z)$ is

$$w(z) = \frac{a_{-m}}{(z - z_0)^m} + \frac{a_{-m+1}}{(z - z_0)^{m-1}} + \cdots + \frac{a_{-1}}{z - z_0}$$
$$+ a_0 + a_1(z - z_0) + a_2(z - z_0)^2 + \cdots \tag{12.10}$$

The residue at $z = z_0$ is a_{-1}, and to obtain it we multiply (12.10) by $(z - z_0)^m$ and obtain

$$(z - z_0)^m\, w(z) = a_{-m} + a_{-m+1}(z - z_0) + \cdots + a_{-1}(z - z_0)^{m-1}$$
$$+ (z - z_0)^m\, a_0 + a_1(z - z_0)^{m+1} + a_2(z - z_0)^{m+2} + \cdots \tag{12.11}$$

If we differentiate both sides of (12.11), with respect to z, $m - 1$ times and place $z = z_0$, we obtain

$$\frac{d^{m-1}}{dz^{m-1}}\left[(z - z_0)^m\, w(z)\right]_{z=z_0} = a_{-1}(m - 1)! \tag{12.12}$$

Hence the residue a_{-1} at the multiple pole is

$$a_{-1} = \frac{1}{(m - 1)!}\frac{d^{m-1}}{dz^{m-1}}\left[(z - z_0)^m\, w(z)\right]_{z=z_0} \tag{12.13}$$

For example, let it be required to find the residue of

$$w(z) = \frac{z e^z}{(z - a)^3} \tag{12.14}$$

at the third-order pole $z = a$. Applying (12.13), we have

$$a_{-1} = \frac{1}{2!}\frac{d^2}{dz^2}(z e^z)_{z=a} = e^a\left(1 + \frac{a}{2}\right) \tag{12.15}$$

13 LIOUVILLE'S THEOREM

A very interesting and useful theorem may be established by the aid of the notion of residues. Let $w(z)$ be a function that is analytic at all points of the complex z plane and finite at infinity. Then, if a and b are any two distinct points, the only singularities of the function

$$\phi(z) = \frac{w(z)}{(z-a)(z-b)} \tag{13.1}$$

are a and b and possibly infinity. However, since $w(z)$ is by hypothesis finite at infinity, we have

$$\lim_{z \to \infty} z\phi(z) = 0 \tag{13.2}$$

Hence the residue of $\phi(z)$ is zero at infinity. However, in Sec. 11 we saw that the residue $\phi(z)$ at infinity is defined by

$$\frac{1}{2\pi j} \oint_c \phi(z)\, dz \tag{13.3}$$

where c is a large circle that encloses all the singularities of $\phi(z)$ except the one at $z = \infty$ and the integration is performed in the negative sense. However, by Cauchy's residue theorem, we have

$$\frac{1}{2\pi j} \oint_c \phi(z)\, dz = \sum \text{Res inside } c \tag{13.4}$$

Hence the sum of the residues of $\phi(z)$, including that at infinity, is zero. Now the residue of $\phi(z)$ at $z = a$ is

$$\lim_{z \to a} (z-a) \frac{w(z)}{(z-a)(z-b)} = \frac{w(a)}{a-b} \tag{13.5}$$

and, similarly, the residue at $z = b$ is $w(b)/(b-a)$. Since the sum of the residues vanishes, we have

$$\frac{w(a)}{a-b} + \frac{w(b)}{b-a} = 0 \tag{13.6}$$

Hence

$$w(a) = w(b) \tag{13.7}$$

and since a and b are arbitrary points, $w(z)$ is a constant. We have thus proved Liouville's theorem, which states:

A function that is analytic at all points of the z plane and finite at infinity must be a constant.

As a corollary of this theorem, it follows that every function that is not a constant must have at least one singularity.

It also follows that if $w(z)$ is a polynomial in z, the equation

$$w(z) = 0 \tag{13.8}$$

has a root, because if it had not, the function $1/w(z)$ would be finite and analytic for all values of z and would therefore be a constant; then $w(z)$ would be a constant. This contradicts the original hypothesis. *This is the fundamental theorem of algebra.*

THE INTEGRAL OF A LOGARITHMIC DERIVATIVE

A theorem due to Cauchy will now be discussed. This theorem is very useful in determining the number of zeros and poles of a function $W(z)$ within a closed contour by an inspection of the behavior of the function $W(z)$ itself as the point z traverses the given contour. The results of the theorem can be applied to the problem of determining the approximate location of the roots of algebraic equations.

Consider the function $W(z)$ that has both zeros and poles within some prescribed contour C of the z plane and no other singularities. Let

$$\ln W(z) = A + jB \tag{13.9}$$

If we differentiate (13.9) with respect to z, we have

$$\frac{1}{W}\frac{dW}{dz} = \frac{dA}{dz} + j\frac{dB}{dz} \tag{13.10}$$

Now let z_0 be either a zero of the nth order of $W(z)$ or a pole of the nth order of $1/W$. In such a case we can write

$$W(z) = (z - z_0)^n g(z) \tag{13.11}$$

where the function $g(z)$ is analytic and nonzero at $z = z_0$ and n is a positive integer if z_0 is a zero of $W(z)$ or n is a negative integer if z_0 is a pole of $W(z)$.

Let (13.11) be differentiated with respect to z. We then have

$$\frac{dW}{dz} = n(z - z_0)^{n-1} g(z) + (z - z_0)^n \frac{dg}{dz} \tag{13.12}$$

Hence

$$\frac{1}{W}\frac{dW}{dz} = \frac{n}{z - z_0} + \frac{1}{g}\frac{dg}{dz} = \frac{d}{dz}[\ln W(z)] \tag{13.13}$$

Equation (13.13) shows that the function $(1/W)(dW/dz)$ has a *simple* pole at z_0 and the *residue* at this pole is n. Now consider the closed contour C and the integral (13.13):

$$\oint_C \frac{1}{W}\frac{dW}{dz}\,dz = 2\pi j \sum \operatorname{Res} \frac{1}{W}\frac{dW}{dz} \text{ inside } C \tag{13.14}$$

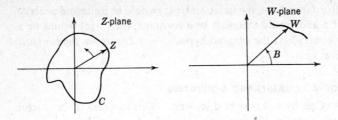

Fig. 13.1

It is apparent from (13.12) and (13.13) that

$$\oint_c \frac{1}{W} \frac{dW}{dz} dz = 2\pi j(N - P) \tag{13.15}$$

where N is the *total* number of zeros and P the *total* number of poles of $W(z)$ within the closed contour C, provided multiple zeros and poles are weighted according to their multiplicity.   Now

$$\oint_c \frac{1}{W} \frac{dW}{dz} dz = \oint_c \frac{d}{dz} [\ln W(z)] dz = \Delta_c \ln W(z) \tag{13.16}$$

where $\Delta_c \ln W(z)$ is the change in $\ln W(z)$ as the closed contour C is traversed. But

$$\ln W(z) = \ln |W(z)| + j \arg W(z) = A + jB \tag{13.17}$$

by (13.9), where $\arg W(z)$ is the argument of $W(z)$.   But $\ln |W(z)|$ is a single-valued function.   Hence

$$\Delta_c \ln |W(z)| = 0 \tag{13.18}$$

Combining the results of (13.15), (13.16), and (13.18), we get

$$\oint_c \frac{1}{W} \frac{dW}{dz} dz = j\Delta_c \arg W(z) = j\Delta_c B = 2\pi j(N - P) \tag{13.19}$$

or

$$\Delta_c B = 2\pi(N - P) \tag{13.20}$$

where $\Delta_c B$ indicates the change in B as the closed contour C is traversed. As the point z moves in the z plane and traces the contour C, the point W moves in the W plane, tracing another contour. B is the angle that the representative vector of the complex variable $W(z)$ makes with the real axis of the complex W plane.

Equation (13.20) states that the number of revolutions that the representative vector in the W plane makes about the origin of the W plane as the point z

traverses the contour C in the z plane is equal to $N - P$. This result may be summarized as follows:

If a function $W(z)$ is analytic, except for possible poles within a contour C, then the number of times that the representative point of $W(z)$ encircles the origin in the $W(z)$ plane in the positive direction while the point z traverses the contour C in a positive direction in the z plane equals $N - P$. N is the number of zeros and P the number of poles lying within the contour C. Each zero and pole must be counted in accordance with its multiplicity.

In recent years this theorem has been applied extensively to the study of the stability of electrical and mechanical dynamical systems. In order that these systems shall be stable, it is necessary that their characteristic equations have roots whose real parts are negative (see Chap. 6). The above theorem provides a method of determining whether the roots of characteristic equations have positive real parts without solving the equations. An adaptation of this theorem due to Nyquist in the study of the stability of feedback amplifiers and servomechanisms is called *Nyquist's criterion* in the literature.†

14 EVALUATION OF DEFINITE INTEGRALS

By the use of Cauchy's residue theorem many definite integrals may be evaluated. It should be observed that a definite integral that can be evaluated by the use of Cauchy's residue theorem may be evaluated by other methods, although not so easily. However, some simple integrals such as

$$\int_0^\infty e^{-u^2}\,du$$

cannot be evaluated by Cauchy's method.

a INTEGRATION AROUND THE UNIT CIRCLE

An integral of the type

$$\int_0^{2\pi} F(\cos\theta, \sin\theta)\,d\theta \tag{14.1}$$

where the integrand is a rational function of $\cos\theta$ and $\sin\theta$ that is finite in the range of integration, may be evaluated by the transformation

$$e^{j\theta} = z \tag{14.2}$$

Since

$$\cos\theta = \frac{1}{2}\left(z + \frac{1}{z}\right) \qquad \sin\theta = \frac{1}{2j}\left(z - \frac{1}{z}\right) \tag{14.3}$$

the integral takes the form

$$\oint_c S(z)\,dz = 2\pi j \sum \operatorname{Res} S(z) \text{ inside } c \tag{14.4}$$

† See H. Nyquist, Regeneration Theory, *Bell System Technical Journal*, p. 126, January, 1932.

where $S(z)$ is a rational function of z, finite on the path of integration and c is a circle of unit radius and center at the origin.

As an example of the general procedure, let it be required to prove that if $a > b > 0$

$$I = \int_0^{2\pi} \frac{\sin^2 \theta \, d\theta}{a + b \cos \theta} = \frac{2\pi}{b^2} (a - \sqrt{a^2 - b^2}) \tag{14.5}$$

If we let $e^{j\theta} = z$, this becomes

$$I = \frac{j}{2b} \oint_c \frac{(z^2 - 1) \, dz}{z^2 (z - p)(z - q)} \tag{14.6}$$

where

$$p = \frac{-a + \sqrt{a^2 - b^2}}{b} \qquad q = \frac{-a - \sqrt{a^2 - b^2}}{b} \tag{14.7}$$

These are the two roots of the quadratic

$$z^2 + \frac{2az}{b} + 1 = 0 \tag{14.8}$$

It is seen that p is the only simple pole of the integrand inside the unit circle c, and the origin is a pole of order 2. We must now compute the residues of

$$S(z) = \frac{(z^2 - 1)^2}{z^2 (z - p)(z - q)} \tag{14.9}$$

at the poles $z = p$ and $z = 0$.

We do this by the methods of Sec. 12. The residue at $z = p$ may be evaluated by formula (12.2). We thus obtain

$$\operatorname*{Res}_{z=p} S(z) = \lim_{z \to p} \frac{(z^2 - 1)^2}{z^2 (z - q)} = \frac{(p^2 - 1)^2}{p^2 (p - q)}$$

$$= \frac{(p - 1/p)^2}{p - q} = \frac{(p - q)^2}{p - q} = p - q = \frac{2\sqrt{a^2 - b^2}}{b} \tag{14.10}$$

The residue at the double pole $z = 0$ may be evaluated by Eq. (12.12); it is

$$\operatorname*{Res}_{z=0} S(z) = \frac{d}{dz} [z^2 S(z)]_{z=0} = -\frac{2a}{b} \tag{14.11}$$

Now, by Cauchy's residue theorem (9.11), we have

$$I = \frac{j}{2b} \oint_c S(z) \, dz$$

$$= \frac{j}{2b} 2\pi j \left(-\frac{2a}{b} + \frac{2\sqrt{a^2 - b^2}}{b} \right)$$

$$= \frac{2\pi}{b^2} (a - \sqrt{a^2 - b^2}) \tag{14.12}$$

This proves the result.

b EVALUATION OF CERTAIN INTEGRALS BETWEEN THE LIMITS $-\infty$ AND $+\infty$

We shall now consider the evaluation of integrals of the type

$$\int_{-\infty}^{+\infty} Q(x)\,dx = I \tag{14.13}$$

where $Q(z)$ is a function that satisfies the following restrictions:

1. It is analytic in the upper half plane except at a finite number of poles.
2. It has no poles on the real axis.
3. $zQ(z) \to 0$ uniformly as $|z| \to \infty$ for $0 \leqslant \arg z \leqslant \pi$.
4. When x is real, $xQ(x) \to 0$ as $x \to \pm\infty$ in such a way that

$$\int_{\infty}^{0} Q(x)\,dx$$

and

$$\int_{-\infty}^{0} Q(x)\,dx$$

both converge. Then

$$\int_{-\infty}^{\infty} Q(x)\,dx = 2\pi j \sum R^+ \tag{14.14}$$

where $\sum R^+$ denotes the sum of the residues of $Q(z)$ at its poles in the upper half plane.

To prove this, choose as a contour a semicircle c with center at the origin and radius R in the upper half plane, as shown in Fig. 14.1. Then, by Cauchy's residue theorem, we have

$$\int_{-R}^{R} Q(x)\,dx + \int_{c} Q(z)\,dz = 2\pi j \sum R^+ \tag{14.15}$$

Now by condition 3, if R is large enough, we have

$$|zQ(z)| < \delta \tag{14.16}$$

for all points on c, and so

$$\left| \int_{c} Q(z)\,dz \right| = \left| \int_{0}^{\pi} Q(Re^{j\theta}) jRe^{j\theta}\,d\theta \right| < \delta \int_{0}^{\pi} d\theta = \pi\delta \tag{14.17}$$

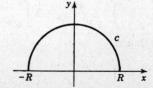

Fig. 14.1

Hence as $R \to \infty$, the integral around c tends to zero, and if (4) is satisfied, we have Eq. (14.14).

This theorem is particularly useful in the case when $Q(x)$ is a rational function. As an example of this theorem, let it be required to prove that if $a > 0$

$$\int_0^\infty \frac{dx}{x^4 + a^4} = \frac{\pi}{2\sqrt{2}\,a^3} \tag{14.18}$$

Consider

$$Q(z) = \frac{1}{z^4 + a^4} \tag{14.19}$$

This function has simple poles at $ae^{\pi j/4}$, $ae^{3\pi j/4}$, $ae^{5\pi j/4}$, $ae^{7\pi j/4}$. Only the first two of these poles are in the upper half plane. The function $Q(z)$ clearly satisfies the conditions of the theorem; therefore

$$\int_{-\infty}^{+\infty} \frac{dx}{x^4 + a^4} = 2\pi j \sum \text{Res at } ae^{j\pi/4} \text{ and } ae^{3j\pi/4} \tag{14.20}$$

By the methods of Sec. 12, we have

$$\operatorname*{Res}_{z=z_0} Q(z) = \lim_{z \to z_0} \frac{1}{4z^3} \tag{14.21}$$

Hence

$$\operatorname*{Res}_{z=ae^{j\pi/4}} Q(z) = \frac{1}{4a^3} e^{-j3\pi/4} \tag{14.22}$$

and

$$\operatorname*{Res}_{z=ae^{j3\pi/4}} Q(z) = \frac{1}{4a^3} e^{-j9\pi/4} \tag{14.23}$$

Therefore

$$\int_{-\infty}^{+\infty} \frac{dx}{x^4 + a^4} = 2\pi j \left(\frac{1}{4a^3} e^{-j3\pi/4} + \frac{1}{4a^3} e^{-j9\pi/4} \right)$$

$$= \frac{\pi}{\sqrt{2}\,a^3} \tag{14.24}$$

Since the function $Q(x)$ is an even function of x, we have

$$\int_{-\infty}^{+\infty} \frac{dx}{x^4 + a^4} = 2 \int_0^\infty \frac{dx}{x^4 + a^4} \tag{14.25}$$

Hence

$$\int_0^\infty \frac{dx}{x^4 + a^4} = \frac{\pi}{2\sqrt{2}\,a^3} \tag{14.26}$$

15 JORDAN'S LEMMA

A very useful and important theorem will now be proved.   It is usually known as *Jordan's lemma*.

Let $Q(z)$ be a function of the complex variable z that satisfies the following conditions:

1. It is analytic in the upper half plane except at a finite number of poles.
2. $Q(z) \to 0$ uniformly as $|z| \to \infty$ for $0 < \arg z < \pi$.
3. m is a positive number.

Then

$$\lim_{R \to \infty} \int_c e^{jmz} \, Q(z) \, dz = 0 \tag{15.1}$$

where c is a semicircle with its center at the origin and radius R.

Proof.   For all points on c we have

$$z = R e^{j\theta} = R(\cos \theta + j \sin \theta)$$
$$dz = jR e^{j\theta} \, d\theta \tag{15.2}$$

Now

$$|e^{jmz}| = |e^{jmR(\cos \theta + j \sin \theta)}| = |e^{-mR \sin \theta}| \tag{15.3}$$

By condition 2, if R is sufficiently large, we have for all points on c

$$|Q(z)| < \delta \tag{15.4}$$

Hence

$$\left| \int_c Q(z) \, e^{jmz} \, dz \right| = \left| \int_0^\pi Q(z) \, e^{jmz} \, R e^{j\theta} j \, d\theta \right| < \delta \int_0^\pi R e^{-mR \sin \theta} \, d\theta$$
$$= 2R\delta \int_0^{\pi/2} e^{-mR \sin \theta} \, d\theta \tag{15.5}$$

It can be proved that $\sin \theta / \theta$ decreases steadily from 1 to $2/\pi$ as θ increases from 0 to $\pi/2$.   Hence

$$\frac{\sin \theta}{\theta} \geqslant \frac{2}{\pi} \qquad \text{when } 0 \leqslant \theta \leqslant \frac{\pi}{2} \tag{15.6}$$

Therefore

$$\left| \int_c Q(z) \, e^{mjz} \, dz \right| \leqslant 2R\delta \int_0^{\pi/2} e^{-2m(R\theta/\pi)} \, d\theta = \frac{\pi\delta}{m}(1 - e^{-mR}) < \frac{\pi\delta}{m} \tag{15.7}$$

from which (15.1) follows.

By the use of Jordan's lemma the following type of integrals may be evaluated: Let

$$Q(z) = \frac{N(z)}{D(z)} \tag{15.8}$$

where $N(z)$ and $D(z)$ are polynomials and $D(z)$ has no real zeros. Then if (i) the degree of $D(z)$ exceeds that of $N(z)$ by at least 1, and (ii) $m > 0$, we have

$$\int_{-\infty}^{+\infty} Q(x) e^{jmz} \, dx = 2\pi j \sum R^+ \tag{15.9}$$

where $\sum R^+$ denotes the sum of the residues of $Q(z)e^{jmz}$ at its poles in the upper half plane. To prove this, integrate $Q(z)e^{jmz}$ around the closed contour of Fig. 15.1. We then have

$$\int_{-R}^{+R} Q(x) e^{jmx} \, dx + \int_C Q(z) e^{jmz} \, dz = 2\pi j \sum \text{Res inside the contour} \tag{15.10}$$

Since $Q(z)e^{jmz}$ satisfies the conditions of Jordan's lemma, we have on letting $R \to \infty$ the result (15.9) since the integral around the infinite semicircle vanishes. Taking the real and imaginary parts of (15.9), we can evaluate integrals of the type

$$\int_{-\infty}^{+\infty} Q(x) \cos mx \, dx \qquad \text{and} \qquad \int_{-\infty}^{+\infty} Q(x) \sin mx \, dx \tag{15.11}$$

As an example, let it be required to show that

$$\int_0^\infty \frac{\cos x \, dx}{x^2 + a^2} = \frac{\pi e^{-a}}{2a} \qquad \text{where } a > 0 \tag{15.12}$$

Here we consider the function $e^{jz}/(z^2 + a^2)$, and since it satisfies the above conditions, we have

$$\int_{-\infty}^{+\infty} \frac{e^{jx}}{x^2 + a^2} \, dx = 2\pi j \sum R^+ \tag{15.13}$$

The only pole of the integrand in the upper half plane is at ja; the residue there is $e^{-a}/2ja$. Hence

$$\int_{-\infty}^{+\infty} \frac{e^{jx} \, dx}{x^2 + a^2} = 2\pi j \cdot \frac{e^{-a}}{2ja} = \frac{\pi e^{-a}}{a} \tag{15.14}$$

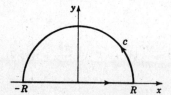

Fig. 15.1

Therefore, taking the real part of e^{jx}, we have

$$\int_{-\infty}^{+\infty} \frac{\cos dx}{x^2 + a^2} = \frac{\pi e^{-a}}{a} = 2 \int_0^{\infty} \frac{\cos dx}{x^2 + a^2} \tag{15.15}$$

Hence (15.12) follows.

INDENTING THE CONTOUR

By an extension of the above theorem it may be proved that

$$\int_0^{\infty} \frac{\sin mx}{x} dx = \frac{\pi}{2} \qquad \text{if } m > 0 \tag{15.16}$$

To prove this, let us consider the integral

$$\oint_c \frac{e^{jmz}}{z} dz \tag{15.17}$$

taken about the contour of Fig. 15.2, where c_1 is a semicircle of radius R and s is a semicircle of radius r.

We notice that the integrand

$$Q(z) = \frac{e^{jmz}}{z} \tag{15.18}$$

has a simple pole at $z = 0$ and none in the upper half plane. By Jordan's lemma, we have

$$\lim_{R \to \infty} \left| \int_{c_1} \frac{e^{jmz}}{z} dz \right| = 0 \tag{15.19}$$

where c_1 is a semicircle of radius R and center at the origin. Since the contour c does not enclose any singularities of the integrand, we have by Cauchy's residue theorem

$$\oint_c \frac{e^{jmz}}{z} dz = \int_{-R}^r Q(z) dz + \int_s Q(z) dz + \int_r^R Q(z) dz$$

$$+ \int_{c_1} Q(z) dz = 0 \tag{15.20}$$

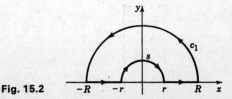

Fig. 15.2

Now, on the semicircle s, we have†

$$z = r\,e^{j\theta} \qquad \frac{dz}{z} = j\,d\theta \tag{15.21}$$

and

$$\lim_{r \to 0} \int_s Q(z)\,dz = \lim_{r \to 0} \int_\pi^0 e^{-mr\sin\theta}\,e^{jmr\cos\theta}\,j\,d\theta = -j\pi \tag{15.22}$$

Letting $R \to \infty$ and $r \to 0$ in (15.20), we have

$$\int_{-\infty}^{+\infty} \frac{e^{jmx}\,dx}{x} = j\pi \tag{15.23}$$

On equating real and imaginary parts we get

$$\int_{-\infty}^{+\infty} \frac{\cos mx\,dx}{x} = 0 \quad \text{and} \quad \int_{-\infty}^{+\infty} \frac{\sin mx}{x}\,dx = \pi \tag{15.24}$$

Hence, since the integrand of the second integral is an even function of x, we have

$$\int_0^\infty \frac{\sin mx}{x}\,dx = \frac{\pi}{2} \qquad m > 0 \tag{15.25}$$

16 BROMWICH CONTOUR INTEGRALS

A particular type of contour integral that occurs with great frequency in mathematical analysis is the Bromwich contour integral, which is illustrated in Fig. 16.1. This contour extends from $c - j\infty$ to $c + j\infty$ where $c \geqslant 0$. A typical integral involving the Bromwich integral is of the form

$$f(t) = \frac{1}{2\pi j} \int_{c-j\infty}^{c+j\infty} g(z)\,e^{zt}\,dz \tag{16.1}$$

† A very useful fact to remember is that an integral over a small semicircular contour centered at a *simple pole* is always equal to half the residue of the simple pole.

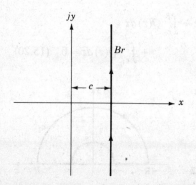

Fig. 16.1

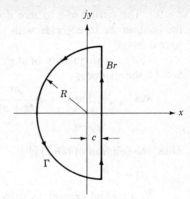

Fig. 16.2

which is known as the *inversion formula for Laplace transforms*.   The function
$f(t)$ is said to be the inverse Laplace transform of the function $g(z)$.†   The
constant c is adjusted such that all of the singularities of $g(z)$ lie to the left of the
Bromwich contour.   This requirement is necessary for the integral to exist.

The evaluation of any integral involving the contour indicated by Eq.
(16.1) is accomplished in the same manner as discussed in the previous sections.
In most cases the Bromwich contour is closed by adding a semicircular path Γ
on the left, as illustrated in Fig. 16.2.   Jordan's lemma can be applied in this
case to obtain conditions on $g(z)$ such that

$$\lim_{R\to\infty} \int_{\Gamma} g(z)\,e^{zt}\,dz \to 0 \tag{16.2}$$

and hence by Cauchy's residue theorem

$$\int_{c-j\infty}^{c+j\infty} g(z)\,e^{zt}\,dz = 2\pi j \sum \text{Res in } (Br + \Gamma) \tag{16.3}$$

Comparing this expression with (16.1) leads to the result that

$$f(t) = \sum \text{Res in } (Br + \Gamma) \tag{16.4}$$

To illustrate the procedure we shall find $f(t)$ corresponding to the complex function $g(z) = 1/(z + a)$, where a is a real positive constant.   From (16.1)
we have

$$f(t) = \frac{1}{2\pi j} \int_{c-j\infty}^{c+j\infty} \frac{e^{zt}}{z + a}\,dz \tag{16.5}$$

By (16.4) we may see that $f(t) = \sum \text{Res in } (Br + \Gamma)$ since it may be easily shown
that the contribution of $e^{zt}/(z + a)$ around Γ vanishes as $R \to \infty$ provided

† In standard transform literature the complex variable z is replaced by s.

$c = 0$. The setting of c to zero does not affect the problem since in moving the contour Br to coincide with the imaginary axis no singularities of $g(z)$ were crossed.

The only singularity of $g(z)$ occurs at $z = -a$; therefore, according to Sec. 12, the residue is

$$\text{Res}\,|_{z=-a} = \lim_{z \to -a} (z+a)\frac{e^{zt}}{(z+a)}$$

$$= e^{-at}$$

Thus, the solution of (16.5) is

$$f(t) = e^{-at} \tag{16.6}$$

As a second example consider the function

$$g(z) = \frac{1}{(z^2 + w^2)^2} \tag{16.7}$$

Again it may be easily verified that $g(z)$ satisfies the conditions of Jordan's lemma, and hence it may be stated that

$$\int_{c-j\infty}^{c+j\infty} g(z)e^{zt}\,dz = 2\pi j \sum \text{Res in } (Br + \Gamma) \tag{16.8}$$

or by (16.4)

$$f(t) = \sum \text{Res in } (Br + \Gamma) \tag{16.9}$$

By inspection of (16.7) it is easily seen that $g(z)$ has two double poles, one at $z = jw$ and the other at $z = -jw$. Since these poles lie on the imaginary axis, the positive constant c in the Bromwich contour integral must be chosen such that $c > 0$, or else the contour will cross the singularities.

All that remains now is to evaluate the residues of $e^{zt}/(z^2 + w^2)^2$ at the two double poles. To do this we shall employ the residue formula given by (12.13). For the double pole at $z = jw$ we have

$$\text{Res}\,|_{z=jw} = \frac{1}{1!}\frac{d}{dz}\left[(z-jw)^2\frac{e^{zt}}{(z^2+w^2)^2}\right]_{z=jw}$$

$$= \frac{d}{dz}\left[\frac{e^{zt}}{(z+jw)^2}\right]_{z=jw}$$

$$= \frac{e^{jwt}}{4jw^3} - \frac{te^{jwt}}{4w^2} \tag{16.10}$$

To evaluate the residue at $z = -jw$, simply take the complex conjugate of (16.10); that is, set $j = -j$. Performing this operation yields

$$\text{Res}\,|_{z=-jw} = -\frac{e^{-jwt}}{4jw^3} - \frac{te^{-jwt}}{4w^2} \tag{16.11}$$

Adding the results of (16.10) and (16.11) gives the sum of both residues and hence by (16.9)

$$f(t) = \frac{e^{jwt} - e^{-jwt}}{4jw^3} - \frac{t}{4w^3}(e^{jwt} + e^{-jwt}) \tag{16.12}$$

This expression is easily reduced by applying the Euler formulas (2.9). The final result is

$$f(t) = \frac{1}{2w^3} \sin wt - \frac{t}{2w^2} \cos wt \tag{16.13}$$

17 INTEGRALS INVOLVING MULTIPLE-VALUED FUNCTIONS (BRANCH POINTS)

Frequently integrals of the following type are encountered:

$$I = \int_0^\infty x^{a-1} Q(x)\, dx \tag{17.1}$$

where a is not an integer. These integrals may be evaluated by contour integration; however, since z^{a-1} is a multiple-valued function, it is necessary to introduce a barrier or cut in the z plane. The reason for this may be illustrated in the following manner.

As seen in Sec. 5, any contour integral over a closed contour containing a finite number of singularities may be reduced to a series of closed contour integrals, one around each singularity. Hence no loss in generality will occur if we consider a contour integral about the origin containing a single singular point at the origin, viz.,

$$I = \int_C \frac{dz}{z^b} \tag{17.2}$$

where C is a closed contour enclosing the origin.

Now let us examine the function $w(z) = z^{-b}$ which comprises the integrand of (17.2). If $z = re^{j\theta}$, then it is easy to see that

$$\arg w(z) = -b\theta \tag{17.3}$$

Hence, if a point on the contour C makes one complete circuit of the origin, then the change in the argument of $w(z)$ is

$$\Delta \arg w(z) = -2\pi b \tag{17.4}$$

Now if b is an integer, then $w(z)$ returns to the same value it had prior to circumnavigating the origin. However, if b is not an integer, then $w(z)$ does not return to its original value.

If $b = \frac{1}{2}$, then two complete circuits of the origin would be required in order to return $w(z)$ to its initial value. Because of this peculiarity, the function

$w(z)$ is said to have a branch point at the origin. The function $w(z)$ is said to have as many branches as the number of complete circuits around the branch point required to return the function to its original value. The number of branches may also be easily identified from b because if b is of the form $1/n$, where n is an integer, then the function $w(z)$ has n branches. In the example the function $w(z) = z^{-\frac{1}{2}}$ has two branches, one corresponding to $+z^{-\frac{1}{2}}$ and the other to $-z^{-\frac{1}{2}}$.

In order to avoid the problem of determining which branch the function is on when performing a contour integration about the origin, a *branch cut* is introduced into the complex z plane. This cut restricts the argument of $w(z)$ to a range of only 2π. This is accomplished by simply following the rule that once a branch cut has been introduced into the z plane, any contour of integration may not cross it. This artifice of an impassable cut in the z plane automatically restricts $w(z)$ to remain on only one of its branches.

Branch cuts always start at the branch point and terminate either at infinity or at another branch point.†

Returning to our original problem, Eq. (17.1), one method of evaluating integrals of this type is to use a contour consisting of a large circle C with center at the origin and radius R. The plane must be cut along the real axis from 0 to ∞ and the branch point at $z = 0$ enclosed by a small circle s of radius r, as shown in Fig. 17.1.

Now let $Q(x)$ be a rational function of x with no poles on the positive real axis. Let us write

$$w(z) = z^{a-1} Q(z) \tag{17.5}$$

† See McLachlan, 1955, p. 70 ff., for several examples of branch cuts (see References).

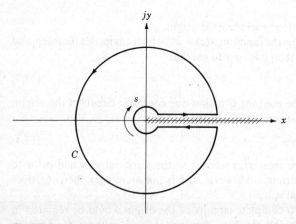

Fig. 17.1

and let us suppose that

$$\lim_{|z|\to\infty} zw(z) = 0 \tag{17.6}$$

and

$$\lim_{|z|\to 0} zw(z) = 0 \tag{17.7}$$

We then get the integral around C tending to zero as $R \to \infty$ and the integral around s tending to zero as $r \to 0$. Hence, on making $R \to \infty$ and $r \to 0$, we get

$$\int_0^\infty x^{a-1} Q(x)\,dx + \int_\infty^0 x^{a-1} e^{2\pi j(a-1)} Q(x)\,dx = 2\pi j \sum R \tag{17.8}$$

where $\sum R$ is the sum of the residues of $w(z)$ inside the contour. It must be noticed that the values of x^{a-1} at points on the upper and lower sides of the cut are not the same. This may be seen as follows.

If $z = re^{j\theta}$, we have

$$z^{a-1} = r^{a-1} e^{j\theta(a-1)} \tag{17.9}$$

and the values on the upper side of the cut correspond to $|z| = x$ and $\theta = 0$, and at the lower side they correspond to $|z| = x$ and $\theta = 2\pi$. Since

$$e^{2\pi j(a-1)} = e^{2\pi ja} \tag{17.10}$$

we get

$$\int_0^\infty x^{a-1} Q(x)\,dx = \frac{2\pi j \sum R}{1 - e^{2\pi ja}} \tag{17.11}$$

As an example of this method of integration, let us evaluate the integral

$$\int_0^\infty \frac{x^{a-1}\,dx}{1+x} = \Gamma(a)\,\Gamma(1-a) = \frac{\pi}{\sin a\pi} \qquad 0 < a < 1 \tag{17.12}$$

This integral is of importance in the theory of gamma functions discussed in Appendix B, Sec. 27.

In this case we have

$$w(z) = \frac{z^{a-1}}{1+z} \tag{17.13}$$

Hence, if $0 < a < 1$, conditions (17.6) and (17.7) are satisfied. $w(z)$ has a pole at $z = -1$. The residue at $z = -1$ is

$$\lim_{z\to -1} \left[(1+z)\frac{z^{a-1}}{1+z} \right] = (-1)^{a-1}$$

$$= (e^{j\pi})^{a-1} = e^{j\pi(a-1)}$$

$$= -e^{j\pi a} \tag{17.14}$$

Hence, by (17.11), we have

$$\int_0^\infty \frac{x^{a-1}\,dx}{1+x} = \frac{-2\pi j\,e^{j\pi a}}{1 - e^{2\pi ja}} = -2\pi j\,\frac{1}{e^{-j\pi a} - e^{j\pi a}} = \frac{\pi}{\sin a\pi} \tag{17.15}$$

18   FURTHER EXAMPLES OF CONTOUR INTEGRALS AROUND BRANCH POINTS

In this section we shall consider two more examples of Bromwich contour integrals of functions containing branch points.

The first example is a contour integral which is of importance in solving problems in heat conduction.

$$f(t) = \frac{1}{2\pi j}\int_{c-j\infty}^{c+j\infty} \frac{e^{-az^{\frac{1}{2}}}}{z}\,e^{zt}\,dz \tag{18.1}$$

The integrand of (18.1) has a branch point at the origin. Because there are no other singularities in the z plane, the Bromwich contour may be deformed into the one shown in Fig. 18.1, in which the arrows indicate the direction of integration.

It will be convenient to break the integration into three parts:

$I_1 =$ the contour below the branch cut on which $z = re^{-j\pi}$

$I_2 =$ the contour around the small circle at the origin on which $z = \epsilon e^{j\theta}$

$I_3 =$ the contour above the branch cut on which $z = re^{j\pi}$

Thus we may write

$$I_1 = \frac{1}{2\pi j}\int_\infty^\epsilon e^{jar^{\frac{1}{2}}}\,e^{-rt}\,\frac{dr}{r} \tag{18.2}$$

$$I_2 = \frac{1}{2\pi}\int_{-\pi}^\pi e^{-a\epsilon^{\frac{1}{2}}(\cos\theta/2 + j\sin\theta/2)}\,e^{\epsilon t(\cos\theta + j\sin\theta)}\,d\theta \tag{18.3}$$

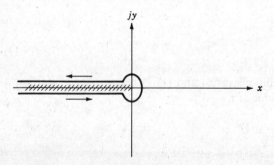

Fig. 18.1

$$I_3 = \frac{1}{2\pi j} \int_\epsilon^\infty e^{-jar^{\frac{1}{2}}} e^{-rt} \frac{dr}{r} \tag{18.4}$$

$$f(t) = \lim_{\epsilon \to 0} (I_1 + I_2 + I_3) \tag{18.5}$$

Examining I_2, the limit and integral may be interchanged. Performing this and the indicated operations yields the result $\lim_{\epsilon \to 0} I_2 = 1$. Since I_1 and I_3 involve ϵ only in the limits of integration, ϵ may be set directly to 0 and (18.2) and (18.4) combined to give

$$I_1 = I_2 = -\frac{1}{\pi} \int_0^\infty \sin ar^{\frac{1}{2}} e^{-rt} \frac{dr}{r} \tag{18.6}$$

Expanding the sine function in a Taylor's series

$$\sin ar^{\frac{1}{2}} = ar^{\frac{1}{2}} - \left(\frac{1}{3!}\right)(ar^{\frac{1}{2}})^3 + \left(\frac{1}{5!}\right)(ar^{\frac{1}{2}})^5 - \cdots$$

and integrating (18.6) term by term results in

$$I_1 + I_2 = -\frac{2}{\pi^{\frac{1}{2}}} \left[\frac{x}{2} - \frac{x^3}{1!}(2^3)(3) + \frac{x^5}{2!}(2^5)(5) - \cdots \right] \tag{18.7}$$

where $x = a/t^{\frac{1}{2}}$.

The series in (18.7) is an expansion of the function known as the error function, that is, erf $\frac{1}{2}x$, which permits (18.7) to be expressed in the following form:

$$I_1 + I_2 = -\mathrm{erf} \frac{a}{2t^{\frac{1}{2}}} \tag{18.8}$$

Combining all the components of the integral given in (18.1) and represented by (18.5) yields†

$$f(t) = 1 - \mathrm{erf} \frac{a}{2t^{\frac{1}{2}}} = \mathrm{erfc} \frac{a}{2t^{\frac{1}{2}}} \tag{18.9}$$

Our second example involves two branch points that are located at $z = \pm ja$. In this example we have

$$f(t) = \frac{1}{2\pi j} \int_{c-j\infty}^{c+j\infty} \frac{e^{zt} \, dz}{(z^2 + a^2)^{\frac{1}{2}}} \tag{18.10}$$

To evaluate this integral the Bromwich contour will be deformed into the dumbbell contour shown in Fig. 18.2. A branch cut is introduced between

---

† The notation erfc denotes the complementary error function which is defined in terms of erf as in (18.8) above. The function erf x has the following integral representation:

$$\mathrm{erf}\, x = \frac{2}{\pi^{\frac{1}{2}}} \int_0^x e^{-\xi^2} \, d\xi = \frac{1}{\pi^{\frac{1}{2}}} \int_0^{x^2} e^{-\xi} \xi^{-\frac{1}{2}} \, d\xi$$

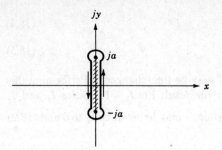

Fig. 18.2

$z = \pm ja$ since the integrand has two branch points, one at each of the designated points.

The evaluation of $f(t)$ is accomplished by breaking the integral into four parts:

$$I_1 = \frac{1}{2\pi} \int_{-a+\epsilon}^{a-\epsilon} \frac{e^{jyt}\,dy}{(a^2-y^2)^{\frac{1}{2}}}$$

$$I_2 = \frac{\epsilon}{2\pi} \int_{-\pi/2}^{\pi/2} \frac{e^{t(ja+\epsilon e^{j\theta})}\,e^{j\theta}\,d\theta}{[(ja+\epsilon e^{j\theta})^2+a^2]^{\frac{1}{2}}}$$

$$I_3 = \frac{-1}{2\pi} \int_{a-\epsilon}^{-a+\epsilon} \frac{e^{-jyt}\,dy}{(a^2-y^2)^{\frac{1}{2}}} \qquad (18.11)$$

$$I_4 = \frac{\epsilon}{2\pi} \int_{-3\pi/2}^{\pi/2} \frac{e^{t(-ja+\epsilon e^{j\theta})}\,e^{j\theta}\,d\theta}{[(-ja+\epsilon e^{j\theta})^2+a^2]^{\frac{1}{2}}}$$

If we now take the limit as $\epsilon \to 0$ we find that I_2 and I_4 both vanish. This leaves us with the result

$$f(t) = I_1 + I_3$$
$$= \frac{1}{2\pi} \int_{-a}^{a} \frac{e^{jyt}\,dy}{(a^2-y^2)^{\frac{1}{2}}} - \frac{1}{2\pi} \int_{a}^{-a} \frac{e^{-jyt}\,dy}{(a^2-y^2)^{\frac{1}{2}}} \qquad (18.12)$$

This latter expression may be reduced to

$$f(t) = \frac{1}{2\pi} \int_{-a}^{a} \frac{e^{jyt}+e^{-jyt}\,dy}{(a^2-y^2)^{\frac{1}{2}}}$$
$$= \frac{1}{\pi} \int_{-a}^{a} \frac{\cos ty\,dy}{(a^2-y^2)^{\frac{1}{2}}} \qquad (18.13)$$

To evaluate (18.13) we introduce the transformation $y = a\sin\theta$ and we find the following integral:

$$f(t) = \frac{1}{\pi} \int_{-\pi/2}^{\pi/2} \cos(at\sin\theta)\,d\theta \qquad (18.14)$$

This expression is a standard form for the representation of the Bessel function of the zeroth order, that is,

$$f(t) = J_0(at) \tag{18.15}$$

19 THE USE OF z AND $\bar{z}$ IN THE THEORY OF COMPLEX VARIABLES

At various times in mathematical analysis it becomes convenient to make use of many of the properties of complex conjugates; it is the purpose of this section to introduce a few of the interesting properties that result from the study of conjugates. It is well known that the complex variable z and its conjugate are defined by

$$z = x + jy \qquad \bar{z} = x - jy \tag{19.1}$$

Now if we consider a function ϕ which is a function of z and $\bar{z}$, that is, $\phi = \phi(z, \bar{z})$, its derivatives with respect to x and y are

$$\frac{\partial \phi}{\partial x} = \frac{\partial \phi}{\partial z}\frac{\partial z}{\partial x} + \frac{\partial \phi}{\partial \bar{z}}\frac{\partial \bar{z}}{\partial x} = \frac{\partial \phi}{\partial z} + \frac{\partial \phi}{\partial \bar{z}} \tag{19.2}$$

$$\frac{\partial \phi}{\partial y} = \frac{\partial \phi}{\partial z}\frac{\partial z}{\partial y} + \frac{\partial \phi}{\partial \bar{z}}\frac{\partial \bar{z}}{\partial y} = j\left(\frac{\partial \phi}{\partial z} - \frac{\partial \phi}{\partial \bar{z}}\right) \tag{19.3}$$

Now defining the operators

$$D_x = \frac{\partial}{\partial x} \qquad D_y = \frac{\partial}{\partial y} \qquad D_z = \frac{\partial}{\partial z} \qquad D_{\bar{z}} = \frac{\partial}{\partial \bar{z}} \tag{19.4}$$

we may rewrite Eqs. (19.2) and (19.3) as

$$D_x = D_z + D_{\bar{z}} \tag{19.5}$$

$$D_y = j(D_z - D_{\bar{z}}) \tag{19.6}$$

or

$$2D_{\bar{z}} = D_x + jD_y \qquad 2D_z = D_x - jD_y \tag{19.7}$$

From (19.7) we may readily obtain

$$(2D_{\bar{z}})(2D_z) = (D_x + jD_y)(D_x - jD_y) = D_x^2 + D_y^2 \tag{19.8}$$

Hence we see that the Laplace operator may be expressed in the form

$$\nabla^2 \phi = \frac{\partial^2 \phi}{\partial x^2} + \frac{\partial^2 \phi}{\partial y^2} = (D_x^2 + D_y^2)\phi \tag{19.9}$$

The expression of the Laplacian operator in the form of (19.9) allows us to obtain a class of solutions quite easily. Using the notation introduced, Laplace's equation becomes

$$(D_x^2 + D_y^2)\phi = 4D_z D_{\bar{z}} = 0 \tag{19.10}$$

which implies that Laplace's equation may be written as

$$D_z D_{\bar{z}} \phi = 0 \tag{19.11}$$

To obtain a solution to this equation we need only to integrate with respect to z first and then $\bar{z}$ as follows:

$$D_{\bar{z}} \phi = F_1'(\bar{z})$$

$$\phi = F_1(\bar{z}) + F_2(z)$$

or

$$\phi = F_1(x - jy) + F_2(x + jy) \tag{19.12}$$

This equation is known in classical literature as D'Alembert's solution.

As a second topic let us consider Green's theorem in two dimensions, which, from Appendix E, is

$$\oint_C (P\,dx + Q\,dy) = \iint_S \left(\frac{\partial Q}{\partial x} - \frac{\partial P}{\partial y} \right) dx\,dy \tag{19.13}$$

Using our notation, we have

$$2 D_{\bar{z}} F(x, y) = (D_x + j\,D_y) F(x, y) \tag{19.14}$$

If we now consider the double integral

$$2 \iint_S D_{\bar{z}} F\,dx\,dy = \iint_S (D_x + j\,D_y) F\,dx\,dy$$

$$= \iint_S \left(\frac{\partial F}{\partial x} + j \frac{\partial F}{\partial y} \right) dx\,dy \tag{19.15}$$

But from (19.13) we have

$$\iint_S \left(\frac{\partial Q}{\partial x} - \frac{\partial P}{\partial y} \right) dx\,dy = \oint_C (P\,dx + Q\,dy) \tag{19.16}$$

If we now let $Q = F$ and $-P = jF$, then (19.16) becomes

$$\iint_S \left(\frac{\partial F}{\partial x} + j \frac{\partial F}{\partial y} \right) dx\,dy = \oint_C (F\,dy - jF\,dx) \tag{19.17}$$

However,

$$\frac{\partial F}{\partial x} - j \frac{\partial F}{\partial y} = 2 \frac{\partial F}{\partial \bar{z}}$$

and

$$F\,dy - jF\,dx = F(dy - j\,dx) = -j\,dz$$

Therefore (19.17) may be written in the form

$$2 \iint_S \frac{\partial F}{\partial \bar{z}} dx\, dy = -j \oint_C F\, dz$$

or

$$\oint_C F\, dz = 2j \iint_S \frac{\partial F}{\partial \bar{z}} dx\, dy \tag{19.18}$$

which is known as the complex form of Green's theorem.

As examples of the application of the complex form of Green's theorem, we may consider the evaluation of area integrals. To accomplish this let

$$\frac{\partial F}{\partial \bar{z}} = 1$$

Then (19.18) becomes

$$\oint F\, dz = 2j \iint dx\, dy = 2jS \tag{19.19}$$

and since

$$\frac{\partial F}{\partial \bar{z}} = 1 \qquad F = \bar{z}$$

we find that

$$\text{Area} = S = \frac{1}{2j} \oint_C \bar{z}\, dz \tag{19.20}$$

As a specific example, consider a circle defined by $z = re^{j\theta}$. We thus have the relations

$$\bar{z} = re^{-j\theta} \qquad dz = jr\, e^{j\theta}\, d\theta$$

from which we quickly find that

$$S = \frac{1}{2j} \int_0^{2\pi} r^2 j\, d\theta = \pi r^2$$

Another application of the complex form of Green's theorem is the computation of moments of inertia of plane areas about the origin, or the polar moments of inertia. Consider an area S in the xy plane bounded by a simply connected contour C. By definition the polar moment of inertia is given by

$$I = \iint_S r^2\, dx\, dy = \iint_S (x^2 + y^2)\, dx\, dy$$

$$= \iint_S \bar{z}z\, dx\, dy \tag{19.21}$$

However, from (19.18) we have

$$\iint_S \frac{\partial F}{\partial \bar{z}} \, dx \, dy = \frac{1}{2j} \oint_C F \, dz \tag{19.22}$$

To apply (19.22) let

$$\frac{\partial F}{\partial \bar{z}} = z\bar{z} \qquad \text{therefore} \qquad F = \frac{z\bar{z}^2}{2} \tag{19.23}$$

Substituting into (19.22) yields the complex form for the polar moment of inertia.

$$I = \frac{1}{4j} \oint_C z\bar{z}^2 \, dz \tag{19.24}$$

Complex notation is also convenient for deriving Cauchy's integral theorem. Let us again consider Eq. (19.18):

$$\oint_C F \, dz = 2j \iint_S \frac{\partial F}{\partial \bar{z}} \, dx \, dy \tag{19.25}$$

If we let $F = w$, we have

$$\oint_C w \, dz = 2j \iint_S \frac{\partial w}{\partial \bar{z}} \, dx \, dy \tag{19.26}$$

Now if $\partial w/\partial \bar{z} = 0$ inside and on C we have

$$\oint_C w \, dz = 0$$

But

$$\frac{\partial w}{\partial \bar{z}} = D_{\bar{z}} w = \tfrac{1}{2}(D_x + j D_y)(u + jv) = 0 \tag{19.27}$$

However,

$$(D_x + j D_y)(u + jv) = (D_x + j D_y)u + (D_x + j D_y)jv$$
$$= (D_x u - D_y v) + j(D_y u + D_x v) = 0$$

and hence

$$D_x u = D_y v \qquad D_y u = -D_x v \tag{19.28}$$

which are the Cauchy-Riemann equations. This implies that the condition $\partial w/\partial \bar{z} = 0$ is the same as the Cauchy-Riemann equations and hence w is analytic whenever this condition is fulfilled. Thus if w is analytic then $\partial w/\partial \bar{z} = 0$ and by (19.26)

$$\oint_C w\,dz = 0 \qquad (19.29)$$

which is Cauchy's integral theorem.

Morera's theorem, which states that if (19.29) is true then $\partial w/\partial \bar{z} = 0$ and w is analytic also follows easily from the complex form of Green's theorem.

Other interesting items using the complex conjugate notation are the relations between $u(x,y)$, $v(x,y)$, and $w(z)$. To develop these relations let us consider

$$w(z) = w(x + jy) = u(x,y) + jv(x,y) \qquad (19.30)$$

If we let $y = 0$, then

$$w(x) = u(x,0) + jv(x,0) \qquad (19.31)$$

Placing $x = z$ in (19.31) yields

$$w(z) = u(z,0) + jv(z,0) \qquad (19.32)$$

Equation (19.32) states that the functional form of $w(z)$ may be formulated directly from $u(x,y)$ and $v(x,y)$. As an example, let us consider the functions

$$u(x,y) = -2xy \qquad v(x,y) = x^2 - y^2 \qquad (19.33)$$

Setting $y = 0$ gives

$$u(x,0) = 0 \qquad v(x,0) = x^2$$

and hence from (19.32) we have

$$w(z) = 0 + jz^2 = jz^2 \qquad (19.34)$$

The above result may easily be verified.

As a final item we will derive a method using the properties of complex conjugates for determining $w(z)$ when either $u(x,y)$ or $v(x,y)$ are given. First let us consider the case when $u(x,y)$ is known and it is desired to determine $w(z)$. We have

$$\frac{\partial w}{\partial x} = \frac{dw}{dz}\frac{\partial z}{\partial x} = \frac{dw}{dz}$$

hence

$$\frac{dw}{dz} = \frac{\partial w}{\partial x} = \frac{\partial}{\partial x}(u + jv)$$

$$= u_x + jv_x \qquad (19.35)$$

or

$$w'(z) = u_x(x,y) - ju_y(x,y) \tag{19.36}$$

by making use of the Cauchy-Riemann relation. Now if we place $y = 0$ in (19.36)

$$w'(x) = u_x(x,0) - ju_y(x,0)$$

Setting $x = z$ in this relation yields

$$w'(z) = u_x(z,0) - ju_y(z,0)$$

and therefore

$$w(z) = \int [u_x(z,0) - ju_y(z,0)]\, dz \tag{19.37}$$

In a similar fashion if $v(x,y)$ is known, then we may derive the relation

$$w(z) = \int [v_y(z,0) + jv_x(z,0)]\, dz \tag{19.38}$$

As an example let us consider the problem where

$$u(x,y) = y^3 - 3x^2 y$$

Now

$$u_x = -6xy \qquad u_y = 3y^2 - 3x^2$$

and hence

$$u_x(z,0) = 0 \qquad u_y(z,0) = -3z^2$$

From (19.37) we have

$$w(z) = \int (0 + 3jz^2)\, dz = jz^3 \tag{19.39}$$

As a further example consider

$$v(x,y) = 2xy$$

from which

$$v_x = 2y \qquad v_y = 2x$$

and from (19.38)

$$w(z) = \int (2z + 0)\, dz = \frac{z^2}{2} \tag{19.40}$$

PROBLEMS

1. Show that the function $w(z) = |z|^2$ has a unique derivative at the origin but nowhere else.

2. Show that the real and imaginary parts of $\sin z$ satisfy Laplace's equation in two dimensions.

3. Find the residues of $w(z) = e^z/(z^2 + a^2)$ at its poles.

4. Find the poles of the function $w(z) = e^z/\sin z$.

5. Find the sum of the residues of the function $w(z) = e^z/(z \cosh mz)$ at its poles.

6. Prove that

$$\int_0^\infty \frac{x \sin x \, dx}{x^2 + a^2} = \frac{\pi}{2} e^{-a} \qquad \text{where } a > 0$$

7. Prove that

$$\int_0^\infty \frac{\sinh ax}{\sinh \pi x} dx = \frac{1}{2} \tan \frac{a}{2} \qquad \text{where } -\pi < a < \pi$$

HINT: Integrate $e^{az}/\sinh \pi z$ around the rectangle of sides $y = 0, y = 1, x = \pm R$, indented at the origin and at j.

8. Show that, if m and n are positive integers and $m < n$,

$$\int_0^\infty \frac{x^{2m}}{x^{2n} + 1} dx = \frac{\pi}{2n \sin [(2m + 1)/2n] \pi}$$

9. Show that, if $m > 0, a > 0$,

$$\int_0^\infty \frac{x \sin mx}{x^4 + a^4} dx = \frac{\pi}{2a^2} e^{-ma/\sqrt{2}} \sin \frac{ma}{\sqrt{2}}$$

10. Show that

$$\int_0^\infty \frac{\cos mx}{x^4 + a^4} dx = \frac{\pi}{2a^3} e^{-ma/\sqrt{2}} \sin \left(\frac{ma}{\sqrt{2}} + \frac{\pi}{4} \right)$$

11. Prove that, if $a > 0$,

$$\int_{-\infty}^{+\infty} \frac{a \cos x + x \sin x}{x^2 + a^2} dx = 2\pi e^{-a}$$

HINT: Integrate $e^{jz}/(z - ja)$ over a suitable contour.

12. Prove that

$$\int_0^{2\pi} \cos^n \theta \, d\theta = \begin{cases} 0 & \text{if } n \text{ is odd} \\ \dfrac{1 \times 3 \times 5 \times \cdots \times (n-1)}{2 \times 4 \times 6 \times \cdots \times n} \times 2\pi & \text{if } n \text{ is even} \end{cases}$$

13. Show that, if $-\pi < a < \pi$,

$$\int_0^\infty \frac{\cosh ax}{\cosh \pi x} dx = \frac{1}{2} \sec \frac{a}{2}$$

HINT: Integrate $e^{az}/\cosh \pi z$ around the rectangle of sides $x = \pm R, y = 0, y = 1$.

14. Show that

$$\int_0^\infty \frac{x\,dx}{\sinh x} = \frac{\pi^2}{4}$$

15. By integrating e^{jz^2}/z around a suitable contour show that

$$\int_0^\infty \frac{\sin x^2\,dx}{x} = \frac{\pi}{4}$$

16. By integrating $e^{jz}/\sqrt{z}$ along a suitable path show that

$$\int_0^\infty \frac{\cos x\,dx}{\sqrt{x}} = \int_0^\infty \frac{\sin x\,dx}{\sqrt{x}} = \sqrt{\frac{\pi}{2}}$$

17. By taking as a contour a square whose corners are $\pm N$, $\pm N + 2Nj$, where N is an integer, and letting $N \to \infty$, prove that

$$\int_0^\infty \frac{dx}{(1 + x^2)\cosh(\pi x/2)} = \log 2$$

18. Show that

$$\int_0^\infty \frac{x^4\,dx}{x^6 + 1} = \frac{\pi}{3}$$

19. Show that

$$\int_0^\pi \frac{a\,d\theta}{a^2 + \sin^2\theta} = \frac{\pi}{\sqrt{1 + a^2}} \qquad a > 0$$

20. Prove that

$$\int_{-\infty}^{+\infty} \frac{\cos x\,dx}{(x^2 + a^2)(x^2 + b^2)} = \frac{\pi}{a^2 - b^2}\left(\frac{e^{-b}}{b} - \frac{e^{-a}}{a}\right) \qquad \text{where } a > b > 0$$

21. Show that

$$\int_{-\infty}^{+\infty} \frac{dx}{(x^2 + b^2)(x^2 + c^2)^2} = \frac{\pi(b + 2c)}{2bc^3(b + c)^2} \qquad b > 0, c > 0$$

22. Compute the first terms of ϵ_i, $i = 1, 2, 3, 4$, in (3.9).

23. Derive (3.11).

24. Find $u(x,y)$ and $v(x,y)$ for (a) $w = z^2$; (b) $w = e^z$; (c) $w = 1/z$; (d) $w = \cos z$; and (e) $w = \sinh z$.

25. Given $v(x,y) = -\cos x \sinh y$, find $u(x,y)$ and $w(z)$.

26. Given $u(x,y) = \ln(x^2 + y^2)^{\frac{1}{2}}$, find $v(x,y)$ and $w(z)$.

27. Rewrite (7.7) and (7.8) to represent a Taylor's series expansion about the origin.

28. Show that (7.6) is correct.

29. Show that the expansion of (8.15) is correct by applying (8.6) and (8.7) to the first few terms.

30. Derive (9.3) by integrating (9.2) around a unit circle centered at $z = z_0$.

HINT: Show that the only nonzero contribution of this integration is the result given by Eq. (9.3).

31. Rewrite the conditions in Jordan's lemma to apply to $g(z)$ in (16.2).

32. Derive the solution given by (16.14) by shrinking the contour $(Br + \Gamma)$ to unit circles around each of the singularities and performing each of the integrations.

33. Evaluate

$$f(t) = \frac{1}{2\pi j} \int_{c-j\infty}^{c+j\infty} \frac{e^{zt}\,dz}{z^{\frac{1}{2}}}$$

34. Evaluate each of the following integrals by enclosing *all* the singularities in the finite plane with a closed contour (the constants a and b may be real or complex):

(a) $\dfrac{1}{2\pi j} \displaystyle\int \dfrac{dz}{(z-a)(z-b)}$

(b) $\dfrac{1}{2\pi j} \displaystyle\int \dfrac{z^3\,dz}{(z+a)(z+b)^2}$

(c) $\dfrac{1}{2\pi j} \displaystyle\int \dfrac{dz}{z^2(z-a)}$

(d) $\dfrac{1}{2\pi j} \displaystyle\int \dfrac{z e^{zt}\,dz}{(z+a)(z-b)^2}$

(McLachlan, 1955, p. 320)

35. Show that

$$\int_0^\infty \frac{x \sin mx\,dx}{(x^2 + a^2)} = \frac{\pi}{2} e^{-ma}$$

for $a = \mathrm{Re} > 0$, $m = \mathrm{Re} > 0$.

36. Derive Eqs. (6.13) and (6.14) by the method stated.

37. Verify Eq. (8.9).

38. Equation (8.12) yields an expansion of $w(z) = \dfrac{1}{z(z-1)}$ about the origin which is valid in the interval $0 < z < 1$. Obtain an expansion for $w(z)$ about $z = 0$ that is valid in the interval $z > 1$.

39. Compute the polar moment of inertia of a circle of radius a about the origin.

40. Compute the polar moment of inertia of a square centered at the origin and side a.

41. Derive a formula for the moment of inertia of a plane area about the y axis in complex form:

$$I_y = \iint_S x^2\,dx\,dy$$

42. Derive (19.38).

43. Given $u(x,y) = e^x \cos y$, find $w(z)$ by the complex conjugate method.

44. Given

$$(x,y) = \sin x \cosh y + 2 \cos x \sinh y + x^2 - y^2 + 4xy$$

find $w(z)$ by the complex conjugate method.

45. Evaluate the contour integral

$$f(t) = \frac{1}{2\pi j} \int_{c-j\infty}^{c+j\infty} \frac{e^{zt}\,dz}{z(z+1)^{\frac{1}{2}}}$$

REFERENCES

1927. Whittaker, E. T., and G. N. Watson: "A Course in Modern Analysis," Cambridge University Press, New York.

1933. MacRobert, T. M.: "Functions of a Complex Variable," The Macmillan Company, New York.

1933. Rothe, R., F. Ollendorf, and K. Pohlhausen: "Theory of Functions as Applied to Engineering Problems," M.I.T. Press, Cambridge, Mass.

1955. McLachlan, N. W.: "Complex Variable and Operational Calculus," Cambridge University Press, New York.

1958. Pipes, L. A.: "Applied Mathematics for Engineers and Physicists," McGraw-Hill Book Company, New York.

2
Linear Differential Equations

1 INTRODUCTION

The analysis of any type of linear system generally leads to a mathematical model in the form of a differential equation. In applied mathematics the most important and frequently occurring differential equations are linear.

A linear differential equation of order n is one of the form

$$a_0(x)\frac{d^n y}{dx^n} + a_1(x)\frac{d^{n-1} y}{dx^{n-1}} + \cdots + a_{n-1}(x)\frac{dy}{dx} + a_n(x)y = f(x) \qquad (1.1)$$

where $a_0, a_1, \ldots, a_n$ and f are functions of the independent variable x and $a_0 \neq 0$.

If $n = 1$, we have the linear equation of the first order; this is written in the form

$$\frac{dy}{dx} + P(x)y = Q(x) \qquad (1.2)$$

53

If $Q(x) = 0$, we have

$$\frac{dy}{dx} + P(x)y = 0 \tag{1.3}$$

This equation is called a *homogeneous linear differential equation of the first order*.   It may be put in the form

$$\frac{dy}{y} = -P(x)\,dx \tag{1.4}$$

In this form the variables are said to be *separated*, and we may therefore integrate both members and obtain

$$\ln y = - \int P(x)\,dx + c \tag{1.5}$$

where c is an arbitrary constant of integration.   Therefore we have

$$y = e^{-\int P(x)\,dx + c} = e^{-\int P(x)\,dx}\,e^{c} \tag{1.6}$$

but since e^c is an arbitrary constant, we·may denote it by K.   Hence the solution of (1.3) is

$$y = Ke^{-\int P(x)\,dx} \tag{1.7}$$

To solve the more general differential equation (1.2), let us place

$$y = u(x)\,v(x) \tag{1.8}$$

where u and v are functions of x to be determined.   Placing this form for y in (1.2), we obtain

$$\frac{du}{dx}v + u\frac{dv}{dx} + P(x)uv = Q(x) \tag{1.9}$$

This may be written in the form

$$v\left[\frac{du}{dx} + P(x)u\right] + u\frac{dv}{dx} = Q(x) \tag{1.10}$$

Since u and v are at our disposal, let us place the term in brackets equal to zero.   We then obtain

$$\frac{du}{dx} + P(x)u = 0 \tag{1.11}$$

and

$$u\frac{dv}{dx} = Q(x) \tag{1.12}$$

However, (1.11) is of the same form as (1.3), and therefore its solution is

$$u = Ke^{-\int P(x)\,dx} \tag{1.13}$$

If we substitute this value of u into (1.12), we obtain

$$dv = \frac{1}{K} e^{+\int P(x)\,dx} Q(x)\,dx \tag{1.14}$$

Since the right-hand member is a function of x alone, we may integrate both sides and thus obtain

$$v = \frac{1}{K} \int e^{+\int P(x)\,dx}(Qx)\,dx + C_1 \tag{1.15}$$

where C_1 is an arbitrary constant.

Substituting these values of u and v into (1.8), we obtain

$$y = Ke^{-\int P(x)\,dx} \left[\frac{1}{K} \int e^{\int P(x)\,dx} Q(x)\,dx + C_1 \right] \tag{1.16}$$

This may be written in the form

$$y = Ce^{-\int P\,dx} + e^{-\int P\,dx} \int e^{\int P(x)\,dx} Q(x)\,dx \tag{1.17}$$

where C is an arbitrary constant.

We thus see that the solution of Eq. (1.2) consists of two parts. One part is the solution of the homogeneous equation with the right-hand member equal to zero. This is called the *complementary function*; it contains an arbitrary constant. The other part involves an integral of the right-hand member $Q(x)$. This is called the *particular integral*. The general solution is the sum of these two parts and is given by (1.17).

2   THE REDUCED EQUATION; THE COMPLEMENTARY FUNCTION

In the last section the solution of the general linear differential equation with variable coefficients of the first order was obtained. Equation (1.17) gives a formula by means of which the solution may be obtained, provided the indicated integrations may be performed.

If the linear differential equation with variable coefficients is of order higher than the first, it is not possible to obtain an explicit solution in closed form in the general case. In general a series solution must be resorted to in this case. Fortunately a great many of the problems of applied mathematics such as the study of small-amplitude mechanical oscillations and the analysis of electrical networks lead to the solution of linear differential equations with constant coefficients. Accordingly in this chapter we shall study methods of solution of this type of equation.

If the various coefficients $a_r(x)$, $r = 0, 1, 2, \ldots, n$ of (1.1) are constants, we may write this equation in the form

$$\frac{d^n y}{dx^n} + a_1 \frac{d^{n-1} y}{dx^{n-1}} + \cdots + a_{n-1} \frac{dy}{dx} + a_n y = F(x) \tag{2.1}$$

provided $a_0 = 1$.

It is convenient to introduce the symbol of operation

$$D^r = \frac{d^r}{dx^r} \qquad r = 1, 2, \ldots, n \tag{2.2}$$

We may then write (2.1) in the form

$$D^n y + a_1 D^{n-1} y + \cdots + a_n y = F(x) \tag{2.3}$$

This may also be written in the form

$$(D^n + a_1 D^{n-1} + \cdots + a_{n-1} D + a_n) y = F(x) \tag{2.4}$$

where the significance of the term in parentheses of the left-hand member is that it constitutes an operator that when operating on $y(x)$ leads to the left-hand member of (2.3).

To save writing, we may condense our notation further by letting

$$L_n(D) = D^n + a_1 D^{n-1} + a_2 D^{n-2} + \cdots + a_{n-1} D + a_n \tag{2.5}$$

We may then write (2.4) concisely in the form

$$L_n(D) y = F(x) \tag{2.6}$$

If $F(x)$ in (2.6) is placed equal to zero, we obtain the equation

$$L_n(D) y = 0 \tag{2.7}$$

This is called the reduced equation.

It will now be shown that the general solution of (2.6) consists of the sum of two parts y_c and y_P. y_c is the solution of the reduced equation and is called the complementary function. It then satisfies

$$L_n(D) y_c = 0 \tag{2.8}$$

The particular integral y_P satisfies the equation

$$L_n(D) y_P = F(x) \tag{2.9}$$

If we add (2.8) and (2.9), we obtain

$$L_n(D) y_c + L_n(D) y_P = F(x) \tag{2.10}$$

But this may be written in the form

$$L_n(D)(y_c + y_P) = F(x) \tag{2.11}$$

If we now let

$$y = y_c + y_P \tag{2.12}$$

we thus obtain

$$L_n(D) y = F(x) \tag{2.13}$$

This proves the proposition.

It thus follows that the general solution of a linear differential equation with constant coefficients is the sum of a particular integral y_P and the complementary function y_c, the latter being the solution of the equation obtained by substituting zero for the function $F(x)$.

3 PROPERTIES OF THE OPERATOR $L_n(D)$

GENERAL SOLUTION OF THE LINEAR DIFFERENTIAL EQUATION

We have seen that the general linear differential equation with constant coefficients may be written in the form

$$L_n(D)y = F(x) \tag{3.1}$$

The expression $L_n(D)$ is known as a *linear differential operator of order n*. It is not an algebraic expression multiplied by y but a symbol that expresses the fact that certain operations of differentiation are to be performed on the function y.

Consider the particular linear operator

$$L_2(D)y = 2\frac{d^2 y}{dx^2} + 5\frac{dy}{dx} + 2y$$
$$= (2D^2 + 5D + 2)y \tag{3.2}$$

We shall also write this in the factorized form

$$L_2(D)y = (2D + 1)(D + 2)y \tag{3.3}$$

factorizing the expression in D as if it were an ordinary algebraic quantity. Is this justifiable?

The operations of multiplication performed in ordinary algebra are based upon three laws:

1. The distributive law

$$m(a + b) = ma + mb$$

2. The commutative law

$$ab = ba$$

3. The index law

$$a^n a^m = a^{m+n}$$

Now D satisfies the first and third of these laws, for

$$D(u + v) = Du + Dv \tag{3.4}$$
$$D^m D^n u = D^{m+n} u \tag{3.5}$$

As for the second law,

$$D(cu) = cD(u) \tag{3.6}$$

is true if c is a constant, but not if c is a variable.   We also have

$$D^m(D^n u) = D^n(D^m u) \tag{3.7}$$

if m and n are positive integers.

Thus D satisfies the fundamental laws of algebra except in that it is not commutative with variables.   It follows that we are justified in performing any operations depending on the fundamental laws of algebra on the linear operator.

$$L_n(D) = D^n + a_1 D^{n-1} + \cdots + a_{n-1} D + a_n \tag{3.8}$$

In view of this, the solution of the general linear differential equation with constant coefficients may be written symbolically in the form

$$y = \frac{1}{L_n(D)} F(x) \tag{3.9}$$

We must now investigate the interpretation of the symbol $1/L_n(D)$ when operating on $F(x)$.   Let us consider the case $n = 1$.   That is,

$$y = \frac{1}{L_1(D)} F(x) = \frac{1}{D + a_1} F(x) \tag{3.10}$$

This is the solution of the equation

$$\frac{dy}{dx} + a_1 y = F(x) \tag{3.11}$$

This is a special case of the general linear equation of the first order (1.2) with $P(x) = a_1$, $Q(x) = F(x)$.   Accordingly the solution of this equation is given by (1.17) with the above values for $P(x)$ and $Q(x)$.   The solution is

$$y = Ce^{-a_1 x} + e^{-a_1 x} \int e^{a_1 x} F(x)\, dx \tag{3.12}$$

We see that the solution consists of two parts.   One part is the solution of Eq. (3.11) if $F(x) = 0$.   This is the complementary function, so that, using the notation of Sec. 2, we have

$$y_c = Ce^{-a_1 x} \tag{3.13}$$

This part contains the arbitrary constant C.   The second part, which involves $F(x)$, is the particular integral; so we have

$$y_P = e^{-a_1 x} \int e^{a_1 x} F(x)\, dx \tag{3.14}$$

$$\frac{1}{D - a} F(x) = Ce^{ax} + e^{ax} \int e^{-ax} F(x)\, dx \tag{3.15}$$

for the operator $1/(D - a)$ operating on $F(x)$.

DECOMPOSITION OF $L(D)$ INTO PARTIAL FRACTIONS [DISTINCT ROOTS OF $L_n(D) = 0$]

Let us return to the general problem of interpreting $1/L_n(D)F(x)$, where $L_n(D)$ is a linear operator of the nth order. Consider the equation

$$L_n(D)y = (D^n + a_1 D^{n-1} + \cdots + a_{n-1} D + a_n)y = 0 \tag{3.16}$$

regarding $L_n(D)$ as a polynomial in D. Now, if this equation has n *distinct* roots $(m_1, m_2, \ldots, m_n)$, it is known from the theory of partial fractions that we may decompose $1/L_n(D)$ into the simple factors

$$\frac{1}{L_n(D)} = \frac{A_1}{D - m_1} + \frac{A_2}{D - m_2} + \cdots + \frac{A_n}{D - m_n} \tag{3.17}$$

This is an algebraic identity, and the A_r $(r = 1, 2, \ldots, n)$ quantities are constant, given by

$$A_r = \frac{1}{(d/dD)L_n(D)}\bigg|_{D = m_r} = \frac{1}{L_n'(m_r)} \tag{3.18}$$

It should be noted that A_r is the residue of $1/L_n(z)$ at m_r.

In this case the solution of the equation becomes

$$y = \frac{1}{L_n(D)} F(x) = \sum_{r=1}^{r=n} \frac{A_r}{D - m_r} F(x) \tag{3.19}$$

But by (3.15) we have

$$\frac{1}{D - m_r} F(x) = C_r e^{m_r x} + e^{m_r x} \int e^{-m_r x} F(x)\, dx \tag{3.20}$$

Hence the general solution is

$$y = \frac{1}{L_n(D)} F(x) = \sum_{r=1}^{r=n} K_r e^{m_r x} + \sum_{r=1}^{r=n} A_r e^{m_r x} \int e^{-m_r x} F(x)\, dx \tag{3.21}$$

where the K_r quantities are *arbitrary* constants and the A_r quantities are given by (3.18).

THE CASE OF REPEATED ROOTS OF $L_n(D) = 0$

If the equation $L_n(D) = 0$ has repeated roots, then the above partial-fraction expansion of $1/L_n(D)$ is no longer possible. Let us first consider the case in which *all* the roots of $L_n(D)$ are repeated. Let the multiple root be equal to m. In this case the equation to be solved is

$$L_n(D)y = (D - m)^n y = F(x) \tag{3.22}$$

To solve this equation, let us assume a solution of the form

$$y = e^{mx} v(x) \tag{3.23}$$

where $v(x)$ is a function of x to be determined. Let us consider the effect of operating with the operator $D - m$ on $e^{mx} v(x)$. We have

$$(D - m) e^{mx} v(x) = m e^{mx} v(x) + e^{mx} Dv - m e^{mx} v \qquad (3.24)$$
$$= e^{mx} Dv$$

If we operate again with $D - m$, we obtain

$$(D - m)^2 e^{mx} v(x) = (D - m) e^{mx} Dv = e^{mx} D^2 v \qquad (3.25)$$

If we repeat this procedure n times, we obtain

$$(D - m)^n e^{mx} v = e^{mx} D^n v \qquad (3.26)$$

In view of this, we see that Eq. (3.22), because of the assumption (3.23), becomes

$$e^{mx} D^n v = F(x) \qquad (3.27)$$

In order to satisfy this, we must have

$$D^n v = e^{-mx} F(x) \qquad (3.28)$$

If we integrate Eq. (3.28) n times, we obtain

$$v = \underset{n}{\int\int \cdots \int\int} e^{-mx} F(x)\, dx \cdots dx$$
$$+ C_1 + C_2 x + \cdots + C_n x^{n-1} \qquad (3.29)$$

where the factor $e^{-mx} F(x)$ must be integrated n times and the quantities $C_r\ (r = 1, 2, \ldots, n)$ are *arbitrary* constants.

We thus see from (3.22) that the result of the operator $1/(D - m)^n$ operating on $F(x)$ may be written in the form

$$\frac{1}{(D - m)^n} F(x) = e^{mx} \underset{n}{\int\int \cdots \int\int} e^{-mx} F(x)\, dx \cdots dx$$
$$+ e^{mx}(C_1 + C_2 x + \cdots + C_n x^{n-1}) \qquad (3.30)$$

Here the term involving the integrals is the particular integral of Eq. (3.22), and the term involving the arbitrary constants is the complementary function.

Let us consider the case in which the operator $L_n(D)$ is such that $(D - m)^r$ is 'a factor of $L_n(D)$ and that $D - m_1$, $D - m_2$, etc., are simple factors of $L_n(D)$. To solve the equation $L_n(D) y = F(x)$, we must expand

$$\frac{1}{L_n(D)} = \frac{1}{(D - m)^r (D - m_1)(D - m_2) \cdots (D - m_s)} \qquad (3.31)$$

where $s = n - r$, into partial fractions. In this case the partial-fraction expansion is of the form

$$\frac{1}{L_n(D)} = \frac{A_1}{(D-m)^r} + \frac{A_2}{(D-m)^{r-1}} + \cdots + \frac{A_r}{D-m}$$

$$+ \frac{B_1}{D-m_1} + \frac{B_2}{D-m_2} + \cdots + \frac{B_s}{D-m_s} \qquad (3.32)$$

The coefficients A_P $(P = 1, 2, \ldots, r)$ are given by

$$A_P = \frac{\phi^{P-1}(m)}{(P-1)!} \qquad (3.33)$$

where

$$\phi(D) = \frac{(D-m)^r}{L_n(D)} \qquad (3.34)$$

and

$$\phi^{P-1}(m) = \frac{d^{P-1}}{dD^{P-1}} \phi(D) \Big|_{D=m} \qquad (3.35)$$

The coefficients B_r $(r = 1, 2, \ldots, s)$ are given by

$$B_r = \frac{1}{(d/dD)L_n(D)|_{D=m_r}} = \frac{1}{L_n'(m_r)} \qquad (3.36)$$

We thus see that the solution of the equation

$$y = \frac{1}{L_n(D)} F(x) \qquad (3.37)$$

when the equation

$$L_n(D) = 0 \qquad (3.38)$$

has multiple roots, contains terms of the form

$$\frac{B_r F(x)}{D-m_r} = C_r e^{m_r x} + B_r e^{m_r x} \int e^{-m_r x} F(x) \, dx \qquad (3.39)$$

as in the solution of (3.21). The term involving the repeated roots gives rise to terms of the form given by (3.30).

We thus have an explicit solution for the general linear differential equation of the nth order with constant coefficients. The difficulties that arise in using the general formulas are due to the difficulties in evaluating the integrals involved in various special cases.

As an example of the general theory, consider the equation

$$\frac{d^2 y}{dx^2} - 3\frac{dy}{dx} + 2y = x e^x$$

or

$$(D^2 - 3D + 2)y = x e^x$$

Here

$$L_2(D) = D^2 - 3D + 2$$
$$= (D - 2)(D - 1)$$

Accordingly the two roots are

$$\left.\begin{array}{l} m_1 = 2 \\ m_2 = 1 \end{array}\right\} \qquad \frac{d}{dD} L_2(D) = 2D - 3$$

By (3.18) we have

$$A_1 = \tfrac{1}{1} = 1$$

$$A_2 = \frac{1}{-1} = -1$$

$$y = \left(\frac{1}{D-2} - \frac{1}{D-1}\right) x e^x$$

By (3.21) we then have

$$y = K_1 e^{2x} + K_2 e^x + e^{2x} \int e^{-2x} x e^x \, dx - e^x \int e^{-x} x e^x \, dx$$

$$= K_1 e^{2x} + K_2 e^x - (1 + x) e^x - \frac{x^2}{2} e^x$$

$$= K_1 e^{2x} + K_2 e^x - \left(1 + x + \frac{x^2}{2}\right) e^x$$

As another example, consider

$$\frac{d^2 y}{dx^2} + 2 \frac{dy}{dx} + y = x$$

or

$$(D + 1)(D + 1)y = x$$

$$y = \frac{1}{(D+1)^2} x$$

This is a special case of (3.30), for

$$m = -1 \qquad \text{and} \qquad n = 2$$

We therefore have

$$y = e^{-x} \int\int e^x x (dx)^2 + e^{-x}(C_1 + C_2 x)$$

$$= x - 2 + e^{-x}(C_1 + C_2 x)$$

4   THE METHOD OF PARTIAL FRACTIONS

Before we continue with the discussion of differential equations it will be instructive to derive the partial-fraction relations stated in the last section. It should be noted that the results of this section apply whether the roots are real or complex.

First let us consider the case in which the linear operator contains only distinct roots; that is,

$$L_n(D) = (D - m_1)(D - m_2) \cdots (D - m_n) \tag{4.1}$$

As indicated it is desired to decompose $1/L_n(D)$ into partial fractions:

$$\frac{1}{L_n(D)} = \frac{A_1}{D - m_1} + \frac{A_2}{D - m_2} + \cdots + \frac{A_n}{D - m_n} \tag{4.2}$$

To determine the coefficients we shall derive the relation for the kth coefficient A_k. Equation (4.2) is multiplied by the term $D - m_k$ to obtain

$$\frac{D - m_k}{L_n(D)} = A_k + (D - m_k)\left(\frac{A_1}{D - m_1} + \cdots + \frac{A_{k-1}}{D - m_{k-1}} \right.$$
$$\left. + \frac{A_{k+1}}{D - m_{k+1}} + \cdots + \frac{A_n}{D - m_n} \right) \tag{4.3}$$

Now if we allow $D \to m_k$ the right-hand side of (4.3) $\to A_k$, we are left with the relation

$$A_k = \lim_{D \to m_k} \frac{D - m_k}{L_n(D)} \tag{4.4}$$

The ratio $(D - m_k)/L_n(D)$ is indeterminant and may be evaluated by means of L'Hospital's rule.   The result is

$$A_k = \frac{1}{L_n'(m_k)} \tag{4.5}$$

where

$$L_n'(m_k) = \frac{dL_n(D)}{dD} \bigg|_{D = m_k}$$

which was previously stated in (3.18).

As an example consider

$$L_3(D) = (D - 1)(D - 2)(D - 3)$$
$$= D^3 - 6D^2 + 11D - 6 \tag{4.6}$$

The roots are 1, 2, and 3.   To apply (4.5) we have

$$L_3'(D) = 3D^2 - 12D + 11 \tag{4.7}$$

Hence

$$L_3'(1) = 2 \qquad L_3'(2) = -1 \qquad L_3'(3) = 2 \tag{4.8}$$

and therefore

$$A_1 = \tfrac{1}{2} \qquad A_2 = -1 \qquad A_3 = \tfrac{1}{2}$$

From (4.2) we have

$$\frac{1}{L_3(D)} = \frac{1}{2(D-1)} - \frac{1}{D-2} + \frac{1}{2(D-3)} \tag{4.9}$$

Now let us consider the case of a multiple root; that is, we shall let

$$L_n(D) = (D-m)^r(D-m_1)(D-m_2) \cdots (D-m_s) \tag{4.10}$$

and attempt to determine the expansion

$$\frac{1}{L_n(D)} = \frac{A_1}{(D-m)^r} + \frac{A_2}{(D-m)^{r-1}} + \cdots + \frac{A_r}{D-m}$$
$$+ \frac{B_1}{D-m_1} + \cdots \frac{B_s}{D-m_s} \tag{4.11}$$

The coefficients B_k of the simple poles may be determined by the method described above.   To determine A_k multiply both sides of (4.11) by $(D-m)^r$:

$$\frac{(D-m)^r}{L_n(D)} = A_1 + A_2(D-m) + \cdots + A_k(D-m)^{k-1} + \cdots$$
$$+ A^r(D-m)^{r-1} + \frac{(D-m)^r B_1}{D-m_1} + \cdots + \frac{B_s}{D-m_s} \tag{4.12}$$

Differentiating both sides $k-1$ times, we have

$$\frac{d^{k-1}}{dD^{k-1}} \frac{(D-m)^r}{L_n(D)} = (k-1)! A_k + O(D-m) \tag{4.13}$$

where $O(D-m)$ denotes terms of order $(D-m)$ and higher.   Taking the limit of both sides as $D \to m$ and solving for A_k, we have

$$A_k = \frac{1}{(k-1)!} \frac{d^{k-1}}{dD^{k-1}} \frac{(D-m)^r}{L_n(D)} \bigg|_{D=m} \tag{4.14}$$

which agrees with (3.33), (3.34), and (3.35) in Sec. 3.

As an illustrative example let us consider the operator

$$L_4(D) = (D-1)^3(D-2) \tag{4.15}$$

From (4.11) we have

$$\frac{1}{L_4(D)} = \frac{A_1}{(D-1)^3} + \frac{A_2}{(D-1)^2} + \frac{A_3}{D-1} + \frac{B}{D-2} \tag{4.16}$$

To evaluate B we apply (4.5):

$$L_4'(D) = 3(D-1)^2(D-2) + (D-1)^3$$
$$L_4'(2) = (2-1)^3 = 1$$

Hence

$$B = \frac{1}{L_4'}(2) = 1$$

Now if we apply (4.14) to evaluate A_1, A_2, and A_3, we find

$$A_1 = \lim_{D \to 1} \frac{(D-1)^3}{L_4(D)}$$

$$= \lim_{D \to 1} \frac{1}{D-2} = -1$$

$$A_2 = \lim_{D \to 1} \frac{d}{dD} \frac{(D-1)^3}{L_4(D)}$$

$$= \lim_{D \to 1} \frac{d}{dD} \frac{1}{D-2} = -1$$

$$A_3 = \tfrac{1}{2} \lim_{D \to 1} \frac{d^2}{dD^2} \frac{1}{D-2} = -1$$

Thus we finally have

$$\frac{1}{L_4(D)} = -\frac{1}{(D-1)^3} - \frac{1}{(D-1)^2} - \frac{1}{D-1} + \frac{1}{D-2} \tag{4.17}$$

which the reader may verify as being the correct partial-fraction expansion of the inverse of (4.15).

As a concluding note it should be realized that this method is not restricted to operational polynomials but may be applied to polynomials of any type where it is desired to obtain partial-fraction expansions. To accomplish this the operator D in the above relations may be replaced by any variable that appears in the polynomial.

5 LINEAR DEPENDENCE: WRONSKIAN

At this point in our discourse it is convenient to discuss the concept of linear dependence of a set of functions, say $f_n(x)$. Linear dependence is formally defined as follows. If for a given set of functions $f_n(x)$ a set of constants c_n can be found that satisfy the relation

$$\sum_{i=1}^{n} c_i f_i(x) = 0 \qquad \text{not all } c_i = 0 \tag{5.1}$$

then the set of functions is said to be linearly dependent. If, on the other hand, the only set of constants that will satisfy (5.1) is

$$c_i = 0 \qquad \text{for all } i$$

then the set of functions are said to be linearly independent. In terms of only two functions $f(x)$ and $g(x)$ this implies that if

$$\frac{f(x)}{g(x)} = \text{constant} \tag{5.2}$$

then $f(x)$ and $g(x)$ are linearly dependent. Conversely if

$$\frac{f(x)}{g(x)} = \text{function of } x \tag{5.3}$$

then the two functions are linearly independent.

A convenient tool for testing for linear dependence of a set of functions is the Wronskian. The development of this test function proceeds as follows. Equation (5.1) is differentiated a total of $n - 1$ times, yielding the following system of n homogeneous equations for the n unknown coefficients c_i:

$$c_1 f_1(x) + c_2 f_2(x) + \cdots + c_n f_n(x) = 0$$
$$c_1 f_1'(x) + c_2 f_2'(x) + \cdots + c_n f_n'(x) = 0$$
$$\cdots \cdots \cdots \cdots \cdots \cdots \cdots \cdots \cdots \cdots \cdots \cdots$$
$$c_1 f_1^{n-1}(x) + c_2 f_2^{n-1}(x) + \cdots + c_n f_n^{n-1}(x) = 0 \tag{5.4}$$

Using matrix notation, this system may be written as follows:

$$[W(x)]\{c\} = \{0\} \tag{5.5}$$

where

$$[W(x)] = \begin{bmatrix} f_1 & f_2 & \cdots & f_n \\ f_1' & f_2' & \cdots & f_n' \\ \cdots & \cdots & \cdots & \cdots \\ f_1^{n-1} & f_2^{n-1} & \cdots & f_n^{n-1} \end{bmatrix}$$

$$\{c\} = \begin{Bmatrix} c_1 \\ c_2 \\ \vdots \\ c_n \end{Bmatrix}$$

From Sec. 18 of Chap. 3 it is seen that a nontrivial solution for $\{c\}$ exists only when $|W| = 0$. It may thus be stated that the set of functions $f_n(x)$ are linearly dependent whenever $|W| = 0$.†

The determinant $|W|$ is called the *Wronskian* of the system of functions $f_n(x)$.

---

† Note that for $|W|$ to exist, the functions $f_n(x)$ must be differentiable at least $n - 1$ times, which means that this definition is more restrictive than (5.1). However, since the Wronskian is used mainly with differential equations, this restriction does not affect the situation because the same order of differentiability is required for the existence of solutions.

6 THE METHOD OF UNDETERMINED COEFFICIENTS

The labor involved in performing the integrations in the general method to obtain the particular integral may sometimes be avoided by the use of a method known as the *method of undetermined coefficients.*

This may be illustrated by an example. Consider the differential equation

$$\frac{d^2 y}{dx^2} + 3\frac{dy}{dx} + 2y = x^3 + x \tag{6.1}$$

To obtain the particular integral, let us assume a general polynomial of the third degree of the form

$$y_p = ax^3 + bx^2 + cx + d \tag{6.2}$$

Substituting this into (6.1), we obtain

$$(6ax + 2b) + 3(3ax^2 + 2bx + c) + 2(ax^3 + bx^2 + cx + d) = x^3 + x \tag{6.3}$$

Equating coefficients of like powers of x, we obtain

$$\begin{aligned} 2a &= 1 \\ 9a + 2b &= 0 \\ 6a + 6b + 2c &= 1 \\ 2b + 3c + 2d &= 0 \end{aligned} \tag{6.4}$$

Solving these equations, we obtain

$$\begin{aligned} a &= \tfrac{1}{2} \\ b &= -\tfrac{9}{4} \\ c &= \tfrac{23}{4} \\ d &= -\tfrac{51}{8} \end{aligned} \tag{6.5}$$

Substituting these values into (6.2), we obtain the particular integral

$$y_p = \tfrac{1}{8}(4x^3 - 18x^2 + 46x - 51) \tag{6.6}$$

The substitution (6.2) was successful because it did not give in the first member of (6.1) any new types of terms. Both members are linear combinations of the functions x^3, x^2, x, and 1; hence we could equate coefficients of like powers of x.

The method of undetermined coefficients for obtaining the particular integral is particularly well adapted when the function $F(x)$ is a sum of terms such as sines, cosines, exponentials, powers of x, and their products whose derivatives are combinations of a finite number of functions. In this case we assume for y a linear combination of all terms entering with undetermined coefficients and then substitute it into the equation and equate coefficients of like terms.

As another example, consider the equation

$$\frac{d^3 y}{dx^3} + y = e^{2x} \cos 3x \tag{6.7}$$

In this case we assume

$$y_p = a e^{2x} \cos 3x + b e^{2x} \sin 3x \tag{6.8}$$

Substituting this into the equation, we obtain

$$(9b - 45a) e^{2x} \cos 3x - (9a + 45b) e^{2x} \sin 3x = e^{2x} \cos 3x \tag{6.9}$$

We equate coefficients of like terms and obtain

$$
\begin{aligned}
9b - 45a &= 1 \\
9a + 45b &= 0
\end{aligned} \tag{6.10}
$$

Solving for a and b and substituting them into (6.8), we obtain

$$y_p = \frac{e^{2x}}{234} (\sin 3x - 5 \cos 3x) \tag{6.11}$$

Whenever the driving function $f(x)$ contains terms that appear in the homogeneous solution, the method must be altered in the following manner. The term or family of terms that are included in the assumed particular integral which are also contained in the homogeneous solution must be multiplied by x^m, where m is of large enough order so that all of these particular terms are of the form xy_h.

As an example consider the following differential equation:

$$y'' + y' = 2x + 3 \cos x + e^{-x} \tag{6.12}$$

If the standard procedure were followed, the particular integral would be assumed to have the form

$$y_p = Ax + B + C \cos x + D \sin x + E e^{-x} \tag{6.13}$$

However, since the homogeneous solution is

$$y_h = c_1 + c_2 e^{-x}$$

the assumed particular integral must be modified according to the proceeding rule; that is,

$$y_p = Ax^2 + Bx + C \cos x + D \sin x + Ex e^{-x} \tag{6.14}$$

The reason for this modification is to render the assumed particular integral linearly independent of the homogeneous solution. Substituting (6.14) into (6.12) and equating like coefficients yields

$$A = 1 \qquad B = -2 \qquad C = -\tfrac{3}{2} \qquad D = \tfrac{3}{2} \qquad E = -1$$

The complete solution to (6.12) may now be written as

$$y = c_1 + c_2 e^{-x} + x^2 - 2x - \tfrac{3}{2}\cos x + \tfrac{3}{2}\sin x - x - x e^{-x} \qquad (6.15)$$

If (6.13) had been used instead of (6.14), it would have turned out that $A = 0$ and E and B would be undetermined. This would mean that the terms x^2 and $x e^{-x}$ would not have been found as part of y_p.

7 THE USE OF COMPLEX NUMBERS TO FIND THE PARTICULAR INTEGRAL

In the analysis of electrical networks or mechanical oscillations we are usually interested in finding the particular integral of an equation of the type

$$L_n(D)y = B_0 \sin \omega x \qquad \text{or} \qquad B_0 \cos \omega x \qquad (7.1)$$

We can obtain the particular solution in this case by replacing the right-hand member by a complex exponential. The success of the method depends on the following theorem:

Consider the equation

$$L_n(D)y = F_1(x) + jF_2(x) \qquad (7.2)$$

where $F_1(x)$ and $F_2(x)$ are real functions of x and $j = \sqrt{-1}$. Then the particular integral of (7.2) is of the form

$$y = y_1 + jy_2 \qquad (7.3)$$

where y_1 satisfies

$$L_n(D)y_1 = F_1(x) \qquad (7.4)$$

and y_2 satisfies

$$L_n(D)y_2 = F_2(x) \qquad (7.5)$$

To prove this, it is necessary only to substitute (7.3) into (7.2), and, on equating the real and imaginary coefficients, we obtain (7.4) and (7.5). To illustrate the method, let us solve (6.7) by making the substitution

$$e^{2x}\cos 3x = \operatorname{Re} e^{(2+j3)x} \qquad (7.6)$$

where Re denotes the "real part of." We thus replace the right-hand member of (6.7) by $e^{(2+j3)x}$ and take the real part of the solution; that is, we have

$$\frac{d^3 y}{dx^3} + y = e^{(2+j3)x} \qquad (7.7)$$

instead of (6.7).

To solve this, assume

$$y = A e^{(2+j3)x} \tag{7.8}$$

where A is a complex constant to be determined. Substituting this into (7.7) and dividing both members by the common factor $e^{(2+j3)x}$, we obtain

$$A[(2+j3)^3 + 1] = 1 \tag{7.9}$$

We therefore have

$$A = \frac{1}{-45+9j} = \frac{-45-9j}{2,106} = \frac{-5-j}{234} \tag{7.10}$$

Substituting this into (7.8), we have

$$y = \frac{-5-j}{234} e^{2x}(\cos 3x + j\sin 3x) \tag{7.11}$$

If we take the real part of this expression, we have

$$y_1 = \frac{e^{2x}}{234}(\sin 3x - 5\cos 3x) \tag{7.12}$$

This is the required particular integral.

To solve the equation

$$L_n(D)y = B_0 \sin \omega x \tag{7.13}$$

we replace $\sin \omega x$ by $e^{j\omega x}$ and consider

$$L_n(D)y = B_0 e^{j\omega x} \tag{7.14}$$

We now assume a solution of the form

$$y = A e^{j\omega x} \tag{7.15}$$

We note that, if we operate on $A e^{j\omega x}$ with D, we have the result

$$DA e^{j\omega x} = A(j\omega) e^{j\omega x} \tag{7.16}$$

and

$$D^s A e^{j\omega x} = A(j\omega)^s e^{j\omega x} \tag{7.17}$$

We therefore note that the result of these operations merely replaces D by $j\omega$; accordingly we have

$$L_n(D) A e^{j\omega x} = L_n(j\omega) A e^{j\omega x} \tag{7.18}$$

Hence, on substituting (7.15) into (7.14), we have

$$L_n(j\omega) A e^{j\omega x} = B_0 e^{j\omega x} \tag{7.19}$$

and thus

$$A = \frac{B_0}{L_n(j\omega)} \tag{7.20}$$

provided $L_n(j\omega) \neq 0$.

Now $L_n(j\omega)$ is in general a complex number and may be written in the form

$$L_n(j\omega) = R\,e^{j\phi} \tag{7.21}$$

where

$$R = |L_n(j\omega)| \tag{7.22}$$

$$\phi = \tan^{-1}\frac{\operatorname{Im} L_n(j\omega)}{\operatorname{Re} L_n(j\omega)} \tag{7.23}$$

and Im denotes the "imaginary part of." Hence A may be written in the form

$$A = \frac{B_0}{R}\,e^{-j\phi} \tag{7.24}$$

Substituting this into (7.15), we have

$$y = \frac{B_0}{R}\,e^{j(\omega x - \phi)} \tag{7.25}$$

The solution of (7.13) is obtained by taking the imaginary part of (7.25) to correspond to $B_0 \sin \omega x$. Hence

$$y_2 = \frac{B_0}{R}\sin(\omega x - \phi) \tag{7.26}$$

is the required particular integral.

If we had taken the real part, we would obtain the solution of the equation

$$L_n(D)y = B_0 \cos \omega x \tag{7.27}$$

This method is of extreme importance in the field of electrical engineering and mechanical oscillations and forms the basis of the use of complex numbers in the field of alternating currents. These matters will be discussed more fully in Chaps. 5 and 6.

8 LINEAR SECOND-ORDER DIFFERENTIAL EQUATIONS
WITH VARIABLE COEFFICIENTS

Linear second-order differential equations with variable coefficients are probably the most frequently encountered differential equations in applied mathematics next to those with constant coefficients.   The first type of equation we will consider is the homogeneous equation, which is typified by the form

$$y'' + p(x)y' + q(x)y = 0 \qquad (8.1)$$

This equation may be transformed to the so-called "normal" form by the transformation†

$$y(x) = w(x)\exp\left[\tfrac{1}{2}\int p(x)\,dx\right] \qquad (8.2)$$

Applying this transformation to (8.1) yields

$$w'' + r(x)w = 0 \qquad (8.3)$$

where

$$r(x) = q(x) - \tfrac{1}{2}p'(x) - \tfrac{1}{4}p^2(x) \qquad (8.4)$$

If $r(x)$ happens to be a constant, then the solution to (8.3) is easy to obtain. If $r(x)$ does not happen to be a constant, then (8.3) may be transformed into other possible forms.   One may also recognize (8.1) as one of the standard equations that are well known in applied fields and readily obtain the solution. Equations of this type are Bessel's equation and Legendre's equation, which, among others, are discussed in Appendix B.

A theorem which is of great use for solving equations of the form of (8.3) is the comparison theorem.   This theorem states that if a solution to (8.3) is oscillatory, then the solution to any similar equation, say

$$w'' + s(x)w = 0 \qquad (8.5)$$

will be oscillatory as long as

$$s(x) \geqslant r(x)$$

for all values of x within the interval of interest.   For example, consider the equation

$$w'' + a^2 w = 0 \qquad a^2 > 0 \qquad (8.6)$$

This equation has an oscillatory solution of the form

$$w(x) = c_1 \sin ax + c_2 \cos ax \qquad (8.7)$$

The comparison theorem states that as long as

$$s(x) \geqslant a^2 > 0$$

then (8.5) will have an oscillatory solution.

† Cf. Bellman, 1953, p. 108 ff., or Kamke, 1959, p. 119 (see References).

9 THE METHOD OF FROBENIUS†

The most common method of solving homogeneous linear second-order differential equations is the method of Frobenius. This method is a power-series method by which solutions to

$$y'' + p(x)y' + q(x)y = 0 \tag{9.1}$$

near the origin, that is, $x = 0$, are found.‡

To apply the method of Frobenius to (9.1), it is assumed that the co-efficients $p(x)$ and $q(x)$ are in the form of polynomials or may be expanded as such by means of a Taylor's series expansion. For the moment we will assume that

$$p(x) = \sum_{n=0}^{\infty} a_n x^n \qquad q(x) = \sum_{n=0}^{\infty} b_n x^n \tag{9.2}$$

where the constants a_n and b_n are known.

The method proceeds by assuming a solution of the form

$$y(x) = \sum_{n=0}^{\infty} c_n x^n \tag{9.3}$$

substituting it into (9.1), and then expanding and equating coefficients of like powers of x. This procedure will provide a method of determining the successive constants c_n in terms of the prior ones. Because Eq. (9.1) is a second-order equation, two of these coefficients will remain as arbitrary constants. As an example, consider the equation

$$y'' + 2xy' + (x^2 + 2)y = 0 \tag{9.4}$$

Substituting (9.3) and equating the coefficients of like powers of x yields

$$c_2 = -c_0 \qquad c_3 = -\tfrac{2}{3}c_1 \qquad c_4 = \tfrac{5}{12}c_0 \qquad c_5 = \tfrac{13}{60}c_1$$

hence

$$y(x) = c_0(1 - x^2 + \tfrac{5}{12}x^4 - \cdots) + c_1(x - \tfrac{2}{3}x^3 + \tfrac{13}{60}x^5 - \cdots) \tag{9.5}$$

SINGULAR POINTS

If the coefficients in (9.1) are regular at $x = 0$, then this point is called an *ordinary point* of the differential equation. The point is termed a *regular singular point* if

$$\lim_{x \to 0} p(x) \approx \frac{1}{x}$$

$$\lim_{x \to 0} q(x) \approx \frac{1}{x^2}$$

† For a good detailed discussion of this method, see either Hildebrand, 1949, p. 132 ff., or Kreyszig, 1962, p. 180 ff. (see References).

‡ This method may be used to determine solutions of (9.1) about any point x_0 by expanding in terms of $x - x_0$. Throughout the remainder of this section it will be assumed that $x_0 = 0$.

The point $x = 0$ is an *irregular singular point* whenever

$$\lim_{x \to 0} p(x) \approx \frac{1}{x^n} \qquad n > 1$$

$$\lim_{x \to 0} q(x) \approx \frac{1}{x^m} \qquad m > 2$$

With this classification of singular points for (9.1) the following general rules may be stated.

1. If $p(x)$ and $q(x)$ have no singular points, then (9.1) possesses two distinct solutions of the form given by (9.3).
2. If (9.1) has only regular singular points, then there will always be at least one solution of the form

$$y(x) = \sum_{n=0}^{\infty} c_n x^{n+\alpha} \tag{9.6}$$

3. If (9.1) has an irregular singular point, then regular solutions may or may not exist and no general method exists for determining these solutions.

SOLUTIONS FOR REGULAR SINGULAR POINTS

If rule 2 applies to (9.1), it may then be written in the following form:

$$y'' + \frac{f(x)}{x} y' + \frac{g(x)}{x^2} y = 0 \tag{9.7}$$

where $f(x)$ and $g(x)$ are both regular at $x = 0$.   Assuming that $f(x)$ and $g(x)$ have the following expansions

$$f(x) = \sum_{n=0}^{\infty} a_n x^n \qquad g(x) = \sum_{n=0}^{\infty} b_n x^n$$

Eq. (9.6) may be substituted into (9.1) and the coefficients of like powers of x equated.   This procedure yields the indicial equation as the coefficient of the lowest power of x:

$$\alpha^2 + (a_0 - 1)\alpha + b_0 = 0 \tag{9.8}$$

The remaining coefficients yield the standard recursion relations to evaluate the succeeding powers of x.

The indicial equation yields the possible values of α for which a solution to (9.1) of the form given by (9.6) will exist.   There are three distinct cases for the roots of (9.8) that must be examined further.

1. The roots are distinct and do not differ by a real integer. This includes the case of complex conjugate roots. In this case the solutions are

$$y_1(x) = \sum_{n=0}^{\infty} c_n x^{n+\alpha_1} \qquad y_2(x) = \sum_{n=0}^{\infty} d_n x^{n+\alpha_2} \tag{9.9}$$

2. Equation (9.8) has a double root. In this case the solutions to (9.1) are

$$y_1(x) = \sum_{n=0}^{\infty} c_n x^{n+\alpha}$$

$$y_2(x) = \sum_{n=0}^{\infty} d_n x^{n+\alpha} + y_1(x) \ln x \tag{9.10}$$

3. The roots differ by an integer. In this case one solution is given by

$$y_1(x) = \sum_{n=0}^{\infty} c_n x^{n+\alpha_1} \tag{9.11}$$

where α_1 is the root with the largest real part. The second solution is given by

$$y_2(x) = A y_1(x) \ln x + \sum_{n=0}^{\infty} d_n x^{n+\alpha_2} \tag{9.12}$$

where the constant A may or may not vanish depending upon the original form of (9.1).

Of the equations that are classified under rule 2, a few common types occur frequently in applied mathematics. These are Legendre's, Bessel's, and the hypergeometric differential equations (cf. Appendix B).

10 VARIATION OF PARAMETERS

In Sec. 6 the method of undetermined coefficients was introduced for obtaining particular integrals of linear differential equations. This method is most useful for equations with constant coefficients. The method of variation of parameters, to be presented in this section, is more general in that it may be applied to linear differential equations with variable coefficients. However, it should be remembered that the method of undetermined coefficients is easier to apply in the case of equations with constant coefficients.

The method of variation of parameters may be illustrated by considering the following linear differential equation:

$$y'' + p(x)y' + q(x)y = r(x) \tag{10.1}$$

This method assumes that the general homogeneous solution is known a priori; that is,

$$y_h(x) = c_1 y_1(x) + c_2 y_2(x) \tag{10.2}$$

where y_1 and y_2 satisfy

$$y_i'' + p(x)\,y_i' + q(x)\,y_i = 0 \qquad i = 1, 2 \tag{10.3}$$

A particular integral to (10.1) is now assumed to have the form of (10.2), except the arbitrary constants c_1 and c_2 are replaced by arbitrary functions $A(x)$ and $B(x)$:

$$y_p(x) = A(x)\,y_1(x) + B(x)\,y_2(x) \tag{10.4}$$

At this point we must evaluate the first and second derivatives of (10.4):

$$y_p' = Ay_1' + By_2' + A'\,y_1 + B'\,y_2 \tag{10.5}$$

The last two terms in this expression are arbitrarily set to zero:

$$A'\,y + B'\,y = 0 \tag{10.6}$$

Applying (10.6) to (10.5) and differentiating again yields

$$y_p'' = Ay_1'' + By_2'' + A'\,y_1' + B'\,y_2' \tag{10.7}$$

Substituting Eqs. (10.4), (10.5), (10.6), and (10.7) into (10.1) gives

$$A'\,y_1' + B'\,y_2' = r \tag{10.8}$$

Equations (10.6) and (10.8) constitute two simultaneous differential equations for A and B. Solving these two equations by Cramer's rule yields

$$A' = -\frac{ry_2}{|W|} \qquad B' = \frac{ry_1}{|W|} \tag{10.9}$$

where $|W|$ is the Wronskian of the homogeneous solution. Integrating (10.9) and substituting into (10.4) yields the complete general solution:

$$y(x) = y_1\left[-\int \frac{ry_2}{|W|}\,dx + c_1\right] + y_2\left[\int \frac{ry_1}{|W|}\,dx + c_2\right] \tag{10.10}$$

As a first example consider the equation

$$y'' + y = \sin x \tag{10.11}$$

For this equation

$$y_1 = \sin x \qquad y_2 = \cos x \qquad |W| = 1 \qquad r = \sin x \tag{10.12}$$

Substituting these values into (10.10) and performing the indicated integrations yields the general solution

$$y = c_1 \sin x + c_2 \cos x - \tfrac{1}{4}(2\sin^3 x - 2x\cos x + \sin 2x \cos x) \tag{10.13}$$

For the second example let us consider the problem of obtaining a solution to the equation

$$x^2 y'' - 3xy' - 5y = 6x^5 \tag{10.14}$$

For this equation

$$y_1 = \frac{1}{x} \qquad y_2 = x^5 \qquad |W| = 6x^3 \qquad r = 6x^5 \qquad (10.15)$$

Thus, substituting into (10.10) and again performing the indicated integrations results in the following general solution:

$$y = c_1 \frac{1}{x} + c_2 x^5 + \tfrac{3}{8} x^7 \qquad (10.16)$$

11 THE STURM-LIOUVILLE DIFFERENTIAL EQUATION

An equation that occurs with great frequency in applied mathematics is the Sturm-Liouville differential equation. This equation is generally written in the form

$$[p(x)y']' + [q(x) + \mu r(x)]y = 0 \qquad (11.1)$$

In this equation μ is a real constant. Instead of initial conditions, this equation is usually subjected to the boundary conditions

$$ay(x_1) + by'(x_1) = 0$$
$$cy(x_2) + dy'(x_2) = 0 \qquad (11.2)$$

where a, b, c, and d are constants.

In finding solutions to (11.1) subject to the general boundary conditions (11.2), it turns out that nontrivial solutions exist only for specific values of the constant μ. These values have been defined as the *characteristic* or *eigenvalues* of the equation (11.1). The nontrivial solutions to (11.1) corresponding to the eigenvalues are likewise termed *eigenfunctions* or *characteristic functions*. The common notation is that μ_i represents the ith eigenvalue and y_i the ith eigenfunction which is the solution to (11.1) when $\mu = \mu_i$.

An important property of the eigenfunctions may now be demonstrated by considering the special case in which $b = 0$, and $d = 0$ in (11.2); that is, the boundary conditions become

$$y(x_1) = y(x_2) = 0 \qquad (11.3)$$

We now consider two distinct eigenvalues and their associated eigenfunctions, say μ_i, μ_j, y_i, and y_j where $\mu_i \neq \mu_j$. Since these are solutions to (11.1) we may write

$$[py_i']' + [q + \mu_i r]y_i = 0$$
$$[py_j']' + [q + \mu_j r]y_j = 0 \qquad (11.4)$$

Multiplying the first equation by y_j and the second by $-y_i$ and adding the two equations yields

$$(\mu_i - \mu_j)ry_i y_j = y_i[py_j']' - y_j[py_i']' \qquad (11.5)$$

Integrating from x_1 to x_2 gives

$$(\mu_i - \mu_j) \int_{x_1}^{x_2} r y_i y_j \, dx = \int_{x_1}^{x_2} y_i [p y_j']' \, dx - \int_{x_1}^{x_2} y_j [p y_i']' \, dx \qquad (11.6)$$

If the two integrals on the right-hand side of (11.6) are integrated by parts, the result is

$$(\mu_i - \mu_j) \int_{x_1}^{x_2} r y_i y_j \, dx = p(y_i y_j' - y_j y_i')|_{x_1}^{x_2} \qquad (11.7)$$

Imposing the boundary conditions given by (11.3), we have

$$\int_{x_1}^{x_2} r(x) y_i(x) y_j(x) \, dx = 0 \qquad i \neq j \qquad (11.8)$$

since $\mu_i \neq \mu_j$.

It may thus be stated that the eigenfunctions y_i and y_j are orthogonal with respect to the weighting function $r(x)$ over the interval x_1 to x_2.

Frequently $r(x) = 1$ and (11.8) becomes

$$\int_{x_1}^{x_2} y_i(x) y_j(x) \, dx = 0 \qquad i \neq j \qquad (11.9)$$

This is a common definition of orthogonal functions. We have thus shown that the eigenfunctions of the Sturm-Liouville differential equation† are either orthogonal functions or weighted orthogonal functions.

The theory of orthogonal functions is quite important in applied mathematics, and the interested reader is referred to any of the references listed at the end of this chapter.

Problems characterized by (11.1) are commonly referred to as *eigenvalue problems* or *boundary value problems*. The usual method of solution is to apply the method of Frobenius unless the equation is recognized to be one of the more common forms to which solutions are readily found. Examples would be differential equations with constant coefficients or Legendre's or Bessel's equations.

PROBLEMS

Solve the equations:

1. $\dfrac{dy}{dx} + y = x$

2. $\dfrac{dy}{dx} - \dfrac{2y}{x+1} = (x+1)^{\frac{3}{2}}$

3. $x\dfrac{dy}{dx} - 2y = 2x$

4. $\dfrac{dy}{dx} + y = x^2 + 2$

5. Show that the equation $dy/dx + Py = Qy^n$ is reduced to a linear equation by the substitution $v = y^{1-n}$.

---

† This has been shown only for the special boundary conditions given by (11.3). The results are also true for the general case, and the interested reader may refer to Kreyszig, 1962, p. 513 ff., for the proof of the general case (see References).

6. Solve $x(dy/dx) - 2y = 4x^3\sqrt{y}$.

7. Solve the equation $d^4 y/dx^4 + ky = 0$ subject to the initial conditions that $y = y_0$ at $x = 0$ and that the first three derivatives of y are zero at $x = 0$.

8. Find the general solution of $(D - a)^n y = \sin bx$, where $D = d/dx$, n is a positive integer, and a and b are real and unequal.

Find the solution of the following equations which satisfy the given conditions:

9. $\dfrac{d^2 x}{dt^2} + 3\dfrac{dx}{dt} + 2x = 0$ $\qquad \left.\begin{array}{c} x = 0 \\[2mm] \dfrac{dx}{dt} = 1 \end{array}\right\} \quad t = 0$

10. $\dfrac{d^2 x}{dt^2} + n^2 x = 0$ $\qquad \left.\begin{array}{c} x = a \\[2mm] \dfrac{dx}{dt} = 0 \end{array}\right\} \quad t = 0$

11. $\dfrac{d^2 x}{dt^2} + 9x = t + \tfrac{1}{2}$ $\qquad \left.\begin{array}{c} x = \tfrac{1}{18} \\[2mm] \dfrac{dx}{dt} = \tfrac{1}{9} \end{array}\right\} \quad t = 0$

12. $\dfrac{d^2 x}{dt^2} + 9x = 5\cos 2t$ $\qquad \left.\begin{array}{c} x = 1 \\[2mm] \dfrac{dx}{dt} = 3 \end{array}\right\} \quad t = 0$

13. $\dfrac{d^2 x}{dt^2} + 4\dfrac{dx}{dt} + 4x = 4e^{2t}$ $\qquad \left.\begin{array}{c} x = 0 \\[2mm] \dfrac{dx}{dt} = 0 \end{array}\right\} \quad t = 0$

14. Solve $d^3 x/dt^3 + d^2 x/dt^2 = 6t^2 + 4$.

15. Find the general solution of the equation $(D^{2n+1} - 1)y = 0$, where n is a positive integer.

16. Solve $d^2 x/dt^2 + 4x = \sin 2x$ subject to the conditions

$\left.\begin{array}{c} x = x_0 \\[2mm] \dfrac{dx}{dt} = v_0 \end{array}\right\} \quad t = 0$

17. Solve $d^2 x/dt^2 + b^2 x = k\cos bt$, if

$\left.\begin{array}{c} x = 0 \\[2mm] \dfrac{dx}{dt} = 0 \end{array}\right\} \quad t = 0$

18. Solve the following first-order differential equations:

    (a) $y' - y = 0$ $\qquad y(0) = 1$

    (b) $xy' + 5y = 6x^2$ $\qquad y(1) = 3$

    (c) $x^3 y' + 2y = x^3 + 2x$ $\qquad y(1) = e + 1$

    (d) $y' + y = x$ $\qquad y(0) = 0$

(Hochstadt, 1964)

19. Solve the following homogeneous equations:

    (a) $y'' - 3y' - 10y = 0$ $y(0) = 3$ $y'(0) = 15$

    (b) $y'' + y' - 12y = 0$ $y(0) = 0$ $y'(0) = 21$

    (c) $y''' - 7y'' - 2y' + 40y = 0$ $y(0) = 3$ $y'(0) = 7$ $y''(0) = 27$

    (d) $y''' - 3y'' + y' + 3y = 0$ $y(0) = 0$ $y'(0) = 8$ $y''(0) = 24$

    (e) $y''' - y'' + 4y' - 4y = 0$ $y(0) = 1$ $y'(0) = 0$ $y''(0) = 0$

                                              (Hochstadt, 1964)

20. Solve the following differential equations:

    (a) $y' + y = e^{-x}$ $y(0) = 1$

    (b) $y'' + 2y' + 2y = \sin x$ $y(0) = 0$ $y'(0) = 1$

    (c) $y''' + 2y'' + y' + 2y = -2e^x$ $y(0) = 1$ $y'(0) = 1$ $y''(0) = 1$

                                              (Hochstadt, 1964)

21. Using (5.1) show that (5.2) does in fact guarantee that $f_1(x)$ and $f_2(x)$ are linearly independent.

22. Solve (9.4) by reducing it to normal form; that is, apply (8.2).

23. Show how (10.8) results from (10.1) by the indicated substitutions.

24. In the method of undetermined coefficients a special modification must be applied whenever one or more of the terms in the forcing function are contained in the homogeneous solution. Why is it that a similar modification does not have to be made in the method of variation of parameters?

25. In Sec. 10 the method of variation of parameters is presented as a general method for computing a particular integral for a differential equation. However, the solution, as given by (10.10), is the complete general solution. Why?

26. Solve the following differential equations:

    (a) $y'' - 4y' + 3y = e^{2x}$ $y(0) = 0$ $y'(0) = 0$

    (b) $y'' - 2y' - 3y = 4$ $y(0) = 1$ $y'(0) = -1$

    (c) $y'' + y' - 2y = 3\cos 3x - 11\sin 3x$ $y(0) = 0$ $y'(0) = 6$

    (d) $y''' - 3y'' - 4y' + 12y = 12e^{-x}$ $y(0) = 4$ $y'(0) = 2$ $y''(0) = 18$

    (e) $y'' - 2y' + 2y = 0$ $y(0) = 0$ $y'(0) = 1$

    (f) $y'' - 2y' + 2y = 2\cos 2x - 4\sin 2x$ $y(0) = 0$ $y'(0) = 0$

    (g) $y'' + 8y' + 16y = 0$ $y(0) = 3$ $y'(0) = -14$

    (h) $y'' + 2y' + y = e^{-2x}$ $y(0) = 0$ $y'(0) = 0$

    (i) $y''' + 6y'' + 12y' + 8y = 0$ $y(0) = 4$ $y'(0) = -12$ $y''(0) = 34$

                                              (Kreyszig, 1962)

27. In each case, show that the given set is orthogonal on the given interval I with respect to the weighting function $r(x) = 1$:

    (a) $\sin x, \sin 2x, \sin 3x, \ldots, I: -\pi \leqslant x \leqslant \pi$

    (b) $\sin x, \sin 2x, \sin 3x, \ldots, I: 0 \leqslant x \leqslant \pi$

    (c) $1, \cos x, \cos 2x, \cos 3x, \ldots, I: 0 \leqslant x \leqslant \pi$

    (d) $1, \cos x, \sin x, \cos 2x, \sin 2x, \ldots, I: 0 \leqslant x \leqslant 2\pi$

    (e) $1, \cos 2x, \cos 4x, \cos 6x, \ldots, I: 0 \leqslant x \leqslant \pi$

    (f) $1, x, \frac{1}{2}(3x^2 - 1), \frac{1}{2}(5x^3 - 3x), I: -1 \leqslant x \leqslant 1$

                                              (Kreyszig, 1962)

28. Find the eigenvalues and eigenfunctions to the following Sturm-Liouville differential equations:

(a) $y'' + y = 0$ $y''(0) = 0$ $y'(l) = 0$

(b) $y'' + y = 0$ $y(0) = 0$ $y''(l) = 0$

(Kreyszig, 1962)

REFERENCES

1927. Jeffreys, H.: "Operational Methods in Mathematical Physics," Cambridge University Press, New York.
1929. Forsythe, A. R.: "A Treatise on Differential Equations," The Macmillan Co., New York.
1929. Piaggio, H. T. H.: "An Elementary Treatise on Differential Equations and their Applications," G. Bell & Sons, Ltd., London.
1933. Cohen, A.: "An Elementary Treatise on Differential Equations," 2d ed., D. C. Heath and Company, Boston.
1933. Ford, L. R.: "Differential Equations," McGraw-Hill Book Company, New York.
1939. McLachlan, N. W.: "Complex Variable and Operational Calculus with Technica Applications," Cambridge University Press, New York.
1939. Pipes, L. A.: The Operational Calculus, *Journal of Applied Physics*, vol. 10, pp. 172–180.
1941. Carslaw, H. S., and J. C. Jaeger: "Operational Methods in Applied Mathematics," Oxford University Press, New York.
1949. Hildebrand, F. B.: "Advanced Calculus for Engineers," Prentice-Hall, Inc., Englewood Cliffs, N.J.
1953. Bellman, R.: "Stability Theory of Differential Equations," McGraw-Hill Book Company, New York.
1956. Ince, E. L.: "Ordinary Differential Equations," Dover Publications, Inc., New York.
1958. Kaplan, W.: "Ordinary Differential Equations," Addison-Wesley Publishing Company, Inc., Reading, Mass.
1959. Kamke, E.: "Differentialgleichungen: Lösungemethoden und Lösungen," 2d ed., Akademische Verlagsgesellschaft, Leipzig.
1962. Kreyszig, E.: "Advanced Engineering Mathematics," John Wiley & Sons, Inc., New York.
1964. Hochstadt, H.: "Differential Equations," Holt, Rinehart and Winston, Inc., New York.

3
Linear Algebraic Equations, Determinants, and Matrices

1 INTRODUCTION

This chapter will be devoted to the discussion of the solution of linear algebraic equations and the related topics, determinants and matrices. These subjects are of extreme importance in applied mathematics, since a great many physical phenomena are expressed in terms of linear differential equations. By appropriate transformations the solution of a set of linear differential equations with constant coefficients may be reduced to the solution of a set of algebraic equations. Such problems as the determination of the transient behavior of an electrical circuit or the determination of the amplitudes and modes of oscillation of a dynamical system lead to the solution of a set of algebraic equations.

It is therefore important that the algebraic processes useful in the solution and manipulation of these equations should be known and clearly understood by the student.

2 SIMPLE DETERMINANTS

Before considering the properties of determinants in general, let us consider the solution of the following two linear equations:

$$a_{11}x_1 + a_{12}x_2 = k_1$$
$$a_{21}x_1 + a_{22}x_2 = k_2 \tag{2.1}$$

If we multiply the first equation by a_{22} and the second by $-a_{12}$ and add, we obtain

$$(a_{11}a_{22} - a_{21}a_{12})x_1 = k_1 a_{22} - k_2 a_{12} \tag{2.2}$$

The expression $(a_{11}a_{22} - a_{21}a_{12})$ may be represented by the symbol

$$\begin{vmatrix} a_{11} & a_{12} \\ a_{21} & a_{22} \end{vmatrix} = a_{11}a_{22} - a_{21}a_{12} \tag{2.3}$$

This symbol is called a *determinant of the second order*. The solution of (2.2) is

$$x_1 = \frac{k_1 a_{22} - k_2 a_{12}}{a_{11}a_{22} - a_{21}a_{12}} = \frac{\begin{vmatrix} k_1 & a_{12} \\ k_2 & a_{22} \end{vmatrix}}{\begin{vmatrix} a_{11} & a_{12} \\ a_{21} & a_{22} \end{vmatrix}} \tag{2.4}$$

Similarly, if we multiply the first equation of (2.1) by $-a_{21}$ and the second one by a_{11} and add, we obtain

$$x_2 = \frac{\begin{vmatrix} a_{11} & k_1 \\ a_{21} & k_2 \end{vmatrix}}{\begin{vmatrix} a_{11} & a_{12} \\ a_{21} & a_{22} \end{vmatrix}} \tag{2.5}$$

The expression for the solution of Eqs. (2.1) in terms of the determinants is very convenient and may be generalized to a set of n linear equations in n unknowns. We must first consider the fundamental definitions and rules of operation of determinants and matrices.

3 FUNDAMENTAL DEFINITIONS

Consider the square array of n^2 quantities a_{ij}, where the subscripts i and j run from 1 to n, and written in the form

$$|a| = \begin{vmatrix} a_{11} & a_{12} & a_{13} & \cdots & a_{1n} \\ a_{21} & a_{22} & a_{23} & \cdots & a_{2n} \\ a_{31} & a_{32} & a_{33} & \cdots & a_{3n} \\ \cdots & \cdots & \cdots & \cdots & \cdots \\ a_{n1} & a_{n2} & a_{n3} & \cdots & a_{nn} \end{vmatrix} \tag{3.1}$$

This square array of quantities, bordered by vertical bars, is a symbolical representation of a certain homogeneous polynomial of the nth order in the quantities a_{ij} to be defined later and constructed from the rows and columns of $|a|$ in a certain manner. This symbolical representation is called a *determinant*. The n^2 quantities a_{ij} are called the *elements* of the determinant.

In this brief treatment we cannot go into a detailed exposition of the fundamental theorems concerning the homogeneous polynomial that the determinant represents; only the essential theorems that are important to the solution of sets of linear equations will be considered. Before giving explicit rules concerning the construction of the homogeneous polynomial from the symbolic array of rows and columns, it will be necessary to define some terms that are of paramount importance in the theory of determinants.

a   MINORS

If in the determinant $|a|$ of (3.1) we delete the ith row and the jth column and form a determinant from all the elements remaining, we shall have a new determinant of $n - 1$ rows and columns. This new determinant is defined to be the *minor* of the element a_{ij}. For example, if $|a|$ is a determinant of the fourth order,

$$|a| = \begin{vmatrix} a_{11} & a_{12} & a_{13} & a_{14} \\ a_{21} & a_{22} & a_{23} & a_{24} \\ a_{31} & a_{32} & a_{33} & a_{34} \\ a_{41} & a_{42} & a_{43} & a_{44} \end{vmatrix} \tag{3.2}$$

the minor of the element a_{32} is denoted by M_{32} and is given by

$$M_{32} = \begin{vmatrix} a_{11} & a_{13} & a_{14} \\ a_{21} & a_{23} & a_{24} \\ a_{41} & a_{43} & a_{44} \end{vmatrix} \tag{3.3}$$

b   COFACTORS

The cofactor of an element of a determinant a_{ij} is the minor of that element with a sign attached to it determined by the numbers i and j which fix the position of a_{ij} in the determinant $|a|$. The sign is chosen by the equation

$$A_{ij} = (-1)^{i+j} M_{ij} \tag{3.4}$$

where A_{ij} is the cofactor of the element a_{ij} and M_{ij} is the minor of the element a_{ij}.

4   THE LAPLACE EXPANSION

We come now to a consideration of what may be regarded as the definition of the homogeneous polynomial of the nth order that the symbolical array of elements of the determinant represents. Let us, for simplicity, first consider the second-order determinant

$$|a| = \begin{vmatrix} a_{11} & a_{12} \\ a_{21} & a_{22} \end{vmatrix} \tag{4.1}$$

By definition this symbolical array represents the second-order homogeneous polynomial

$$|a| = a_{11}a_{22} - a_{21}a_{12} \tag{4.2}$$

The third-order determinant

$$|a| = \begin{vmatrix} a_{11} & a_{12} & a_{13} \\ a_{21} & a_{22} & a_{23} \\ a_{31} & a_{32} & a_{33} \end{vmatrix} \tag{4.3}$$

represents the unique third-order homogeneous polynomial defined by any one of the six equivalent expressions

$$|a| = \sum_{j=1}^{3} a_{ij} A_{ij} \quad \text{or} \quad \sum_{i=1}^{3} a_{ij} A_{ij} \tag{4.4}$$

where the elements a_{ij} in (4.4) must be taken from a *single* row or a *single* column of a. The A_{ij}'s are the cofactors of the corresponding elements a_{ij} as defined in Sec. 3. As an example of this definition, we see that the third-order determinant may be expanded into the proper third-order homogeneous polynomial that it represents, in the following manner:

$$|a| = a_{11} \begin{vmatrix} a_{22} & a_{23} \\ a_{32} & a_{33} \end{vmatrix} - a_{12} \begin{vmatrix} a_{21} & a_{23} \\ a_{31} & a_{33} \end{vmatrix} + a_{13} \begin{vmatrix} a_{21} & a_{22} \\ a_{31} & a_{32} \end{vmatrix} \tag{4.5}$$

$$|a| = a_{11}(a_{22}a_{33} - a_{32}a_{23}) - a_{12}(a_{21}a_{33} - a_{31}a_{23}) + a_{13}(a_{21}a_{32} - a_{31}a_{22}) \tag{4.6}$$

This expansion was obtained by applying the fundamental rule (4.4) and expanding in terms of the first row. Since one has the alternative of using any row or any column, it may be seen that (4.3) could be expanded in six different ways by the fundamental rule (4.4). It is easy to show that all six ways lead to the same third-order homogeneous polynomial (4.6).

The definition (4.4) may be generalized to the nth-order determinant (3.1), and this symbol is defined to represent the nth-order homogeneous polynomial given by

$$|a| = \sum_{j=1}^{n} a_{ij} A_{ij} \quad \text{or} \quad \sum_{i=1}^{n} a_{ij} A_{ij} \tag{4.7}$$

where the a_{ij} quantities must be taken from a *single* row or a *single* column. In this case the cofactors A_{ij} are determinants of the $(n-1)$st order, but they may in turn be expanded by the rule (4.7), and so on, until the result is a homogeneous polynomial of the nth order.

It is easy to demonstrate that, in general,

$$\sum_{j=1}^{n} a_{ji} A_{jk} - \sum_{j=1}^{n} a_{ij} A_{kj} = \begin{cases} |a| & \text{if } i = k \\ 0 & \text{if } i \neq k \end{cases} \tag{4.8}$$

5   FUNDAMENTAL PROPERTIES OF DETERMINANTS

From the basic definition (4.7), the following properties of determinants may be deduced:

a. If *all* the elements in a row or in a column are zero, the determinant is equal to zero. This may be seen by expanding in terms of that row or column, in which case each term of the expansion contains a factor of zero.

b. If all elements but one in a row or column are zero, the determinant is equal to the product of that element and its cofactor.

c. The value of a determinant is not altered when the rows are changed to columns and the columns to rows. This may be proved by developing the determinant by (4.7).

d. The interchange of any two columns or two rows of a determinant changes the sign of the determinant.

e. If two columns or two rows of a determinant are identical, the determinant is equal to zero.

f. If all the elements in any row or column are multiplied by any factor, the determinant is multiplied by that factor.

g. If each element in any column or any row of a determinant is expressed as the sum of two quantities, the determinant can be expressed as the sum of two determinants of the same order.

h. Is it possible, without changing the value of a determinant, to multiply the elements of any row (or any column) by the same constant and add the products to any other row (or column). For example, consider the third-order determinant

$$|a| = \begin{vmatrix} a_{11} & a_{12} & a_{13} \\ a_{21} & a_{22} & a_{23} \\ a_{31} & a_{32} & a_{33} \end{vmatrix} \tag{5.1}$$

With the aid of property g we have

$$\Delta = \begin{vmatrix} a_{11} + ma_{13} & a_{12} & a_{13} \\ a_{21} + ma_{23} & a_{22} & a_{23} \\ a_{31} + ma_{33} & a_{32} & a_{33} \end{vmatrix} = \begin{vmatrix} a_{11} & a_{12} & a_{13} \\ a_{21} & a_{22} & a_{23} \\ a_{31} & a_{32} & a_{33} \end{vmatrix}$$

$$+ m \begin{vmatrix} a_{13} & a_{12} & a_{13} \\ a_{23} & a_{22} & a_{23} \\ a_{33} & a_{32} & a_{33} \end{vmatrix} \tag{5.2}$$

Since the second determinant is zero because the first and third columns are identical, we have

$$|a| = \Delta \tag{5.3}$$

6   THE EVALUATION OF NUMERICAL DETERMINANTS

The evaluation of determinants whose elements are numbers is a task of frequent occurrence in applied mathematics.   This evaluation may be carried out by a direct application of the fundamental Laplacian expansion (4.7). This process, however, is most laborious for high-order determinants, and the expansion may be more easily effected by the application of the fundamental properties outlined in Sec. 5 and by the use of two theorems that will be mentioned in this section.

As an example, let it be required to evaluate the numerical determinant

$$|a| = \begin{vmatrix} 2 & -1 & 5 & 1 \\ 1 & 4 & 6 & 3 \\ 4 & 2 & 7 & 4 \\ 3 & 1 & 2 & 5 \end{vmatrix} \tag{6.1}$$

This determinant may be transformed by the use of property b of Sec. 5. The procedure is to make all elements but one in some row or column equal to zero.   The presence of the factor -1 in the second column suggests that we add four times the first row to the second, then add two times the first row to the third row, and finally add the first row to the fourth row.   By Sec. 5h these operations do not change the value of the determinant.   Hence we have

$$|a| = \begin{vmatrix} 2 & -1 & 5 & 1 \\ 9 & 0 & 26 & 7 \\ 8 & 0 & 17 & 6 \\ 5 & 0 & 7 & 6 \end{vmatrix} = \begin{vmatrix} 9 & 26 & 7 \\ 8 & 17 & 6 \\ 5 & 7 & 6 \end{vmatrix} \tag{6.2}$$

If we now subtract the elements of the last column from the first column, we have

$$|a| = \begin{vmatrix} 2 & 26 & 7 \\ 2 & 17 & 6 \\ -1 & 7 & 6 \end{vmatrix} \tag{6.3}$$

We now add two times the third row to the first row and two times the third row to the second row and obtain

$$|a| = \begin{vmatrix} 0 & 40 & 19 \\ 0 & 31 & 18 \\ -1 & 7 & 6 \end{vmatrix} = - \begin{vmatrix} 40 & 19 \\ 31 & 18 \end{vmatrix} \tag{6.4}$$

Now we may subtract the second row from the first row and obtain

$$|a| = -1 \begin{vmatrix} 9 & 1 \\ 31 & 18 \end{vmatrix} = -(162 - 31) = -131 \tag{6.5}$$

This procedure is much shorter than a direct application of the Laplacian expansion rule.

A USEFUL THEOREM

We now turn to a consideration of a theorem of great power for evaluating numerical determinants. The method of evaluation, based on the theorem to be considered, was found most successful at the Mathematical Laboratory of the University of Edinburgh and was due originally to F. Chiò.

To deduce the theorem in question, consider the fourth-order determinant

$$|b| = \begin{vmatrix} b_{11} & b_{12} & b_{13} & b_{14} \\ b_{21} & b_{22} & b_{23} & b_{24} \\ b_{31} & b_{32} & b_{33} & b_{34} \\ b_{41} & b_{42} & b_{43} & b_{44} \end{vmatrix} \tag{6.6}$$

We now notice whether any element is equal to unity. If not, we prepare the determinant in such a manner that one of the elements is unity. This may be done by dividing some row or column by a suitable number, say m, that will make one of the elements equal to unity and then placing the number m as a factor outside the determinant. The same result may also be effected in many cases with the aid of Sec. $5h$. For simplicity we shall suppose that this has been done and that, in this case, the element b_{23} is equal to unity.

Now let us divide the various columns of the determinant b by b_{21}, b_{22}, b_{23}, and b_{24}, respectively; then in view of the fact that the element b_{23} has been made unity, we have

$$|b| = b_{21} b_{22} b_{23} b_{24} \begin{vmatrix} \dfrac{b_{11}}{b_{21}} & \dfrac{b_{12}}{b_{22}} & b_{13} & \dfrac{b_{14}}{b_{24}} \\ 1 & 1 & 1 & 1 \\ \dfrac{b_{31}}{b_{21}} & \dfrac{b_{32}}{b_{22}} & b_{33} & \dfrac{b_{34}}{b_{24}} \\ \dfrac{b_{41}}{b_{21}} & \dfrac{b_{42}}{b_{22}} & b_{43} & \dfrac{b_{44}}{b_{24}} \end{vmatrix} \tag{6.7}$$

Now, if we subtract the elements of the third column from those of the other columns, we obtain

$$|b| = b_{21} b_{22} b_{23} b_{24} |a| \tag{6.8}$$

where

$$|a| = \begin{vmatrix} \dfrac{b_{11}}{b_{21}} - b_{13} & \dfrac{b_{12}}{b_{22}} - b_{13} & b_{13} & \dfrac{b_{14}}{b_{24}} - b_{13} \\[2mm] 0 & 0 & 1 & 0 \\[2mm] \dfrac{b_{31}}{b_{21}} - b_{33} & \dfrac{b_{32}}{b_{22}} - b_{33} & b_{33} & \dfrac{b_{34}}{b_{24}} - b_{33} \\[2mm] \dfrac{b_{41}}{b_{21}} - b_{43} & \dfrac{b_{42}}{b_{22}} - b_{43} & b_{43} & \dfrac{b_{44}}{b_{24}} - b_{43} \end{vmatrix}$$

Expanding in terms of the second row, we obtain

$$|a| = (-1)^{2+3} \begin{vmatrix} \dfrac{b_{11}}{b_{21}} - b_{13} & \dfrac{b_{12}}{b_{22}} - b_{13} & \dfrac{b_{14}}{b_{24}} - b_{13} \\[2mm] \dfrac{b_{31}}{b_{21}} - b_{33} & \dfrac{b_{32}}{b_{22}} - b_{33} & \dfrac{b_{34}}{b_{24}} - b_{33} \\[2mm] \dfrac{b_{41}}{b_{21}} - b_{43} & \dfrac{b_{42}}{b_{22}} - b_{43} & \dfrac{b_{44}}{b_{24}} - b_{43} \end{vmatrix} \qquad (6.9)$$

Substituting this value of $|a|$ into (6.8) and multiplying the various columns by the factors outside, we finally obtain

$$|b| = (-1)^{2+3} \begin{vmatrix} b_{11} - b_{21}b_{13} & b_{12} - b_{22}b_{13} & b_{14} - b_{24}b_{13} \\ b_{31} - b_{21}b_{33} & b_{32} - b_{22}b_{33} & b_{34} - b_{24}b_{33} \\ b_{41} - b_{21}b_{43} & b_{42} - b_{22}b_{43} & b_{44} - b_{24}b_{43} \end{vmatrix} \qquad (6.10)$$

The theorem may be formulated by means of the following rule.   The element that has been made unity at the start of the process is called the *pivotal element*. In this case it is the element b_{23}.   The rule says:

The row and column intersecting in the pivotal element of the original determinant, say the rth row and the sth column, are deleted; then every element u is diminished by the product of the elements which stand where the eliminated row and column are met by perpendiculars from u, and the whole determinant is multiplied by $(-1)^{r+s}$.

By the application of this theorem we reduce the order of the determinant by one unit.   Repeated application of this theorem reduces the determinant to one of the second order, and then its value is immediately written down.

Before turning to a consideration of linear algebraic equations and to the application of the theory of determinants to their solution, a brief discussion of matrix algebra and the fundamental operations involving matrices is necessary.   In this chapter the fundamental definitions and the most important properties of matrices will be outlined.

7 DEFINITION OF A MATRIX

By a square matrix a of order n is meant a system of elements that may be real or complex numbers arranged in a square formation of n rows and columns. That is, the symbol $[a]$ stands for the array

$$[a] = \begin{bmatrix} a_{11} & a_{12} & \cdots & a_{1n} \\ a_{21} & a_{22} & \cdots & a_{2n} \\ \cdots\cdots\cdots\cdots\cdots \\ a_{n1} & a_{n2} & \cdots & a_{nn} \end{bmatrix} \tag{7.1}$$

where a_{ij} denotes the element standing in the ith row and jth column. The determinant having the same elements as the matrix $[a]$ is denoted by $|a|$ and is called the *determinant* of the matrix.

Besides square arrays like (7.1) we shall have occasion to use rectangular arrays or matrices of m rows and n columns. Such arrays will be called *matrices of order* $(m \times n)$. Where there is only one row so that $m = 1$, the matrix will be termed a *vector of the first kind*, or a *prime*. As an example, we have

$$[a] = [a_{11} \quad a_{12} \cdots a_{1n}] \tag{7.2}$$

On the other hand, a matrix of a single column of n elements will be termed a *vector of the second kind*, or a *point*. To save space, it will be printed horizontally and not vertically and will be denoted by parentheses thus:

$$\{b\} = (b_{11} \quad b_{21} \cdots b_{n1}) \tag{7.3}$$

TRANSPOSITION OF MATRICES

The accented matrix $[a]' = [a_{ji}]$ obtained by a complete interchange of rows and columns in $[a]$ is called the *transposed matrix of* $[a]$. The ith row of $[a]$ is identical with the ith column of $[a]'$. For vectors we have

$$[u]' = \{u\} \qquad \{v\}' = [v], \text{ a row} \tag{7.4}$$

For rectangular matrices the transpose is

$$\begin{bmatrix} a_{11} & a_{12} & a_{13} \\ a_{21} & a_{22} & a_{23} \end{bmatrix}' = \begin{bmatrix} a_{11} & a_{21} \\ a_{12} & a_{22} \\ a_{13} & a_{23} \end{bmatrix} \tag{7.5}$$

8 SPECIAL MATRICES

a. *Square matrix:* If the numbers of rows and of columns of a matrix are equal to n, then such a matrix is said to be a *square matrix of order n*, or simply a *matrix of order n*.

b. *Diagonal matrix:* If all the elements other than those in the principal diagonal are zero, then the matrix is called a *diagonal matrix*.

c. *The unit matrix:* The unit matrix of order n is defined to be the diagonal matrix of order n which has 1's for all its diagonal elements. It is denoted by U_n or simply by U when its order is apparent.

d. *Symmetrical matrices:* If $a_{ij} = a_{ji}$ for all i and j, the matrix $[a]$ is said to be symmetrical and it is identical to its transposed matrix. That is, if $[a]$ is symmetrical, then $[a]' = [a]$, and conversely.

e. *Skew-symmetric matrix:* If $a_{ij} = -a_{ji}$ for all unequal i and j, but the elements a_{ii} are not all zero, then the matrix is called a *skew* matrix.

If $a_{ij} = -a_{ji}$ for all i and j so that all $a_{ii} = 0$, the matrix is called a *skew-symmetric* matrix. It may be noted that both symmetrical and skew matrices are necessarily square.

f. *Null matrices:* If a matrix has all its elements equal to zero, it is called a *null* matrix and is represented by $[0]$.

9   EQUALITY OF MATRICES; ADDITION AND SUBTRACTION

It is apparent from what has been said concerning matrices that a matrix is entirely different from a determinant. A determinant is a symbolic representation of a certain homogeneous polynomial formed from the elements of the determinant as described in Sec. 4. A matrix, on the other hand, is merely a square or rectangular array of quantities. By defining certain rules of operation that prescribe the manner in which these arrays are to be manipulated, a certain algebra may be developed that has a formal similarity to ordinary algebra but involves certain operations that are performed on the elements of the matrices. It is to a consideration of these fundamental definitions and rules of operation that we now turn.

a   EQUALITY OF MATRICES

The concept of equality is fundamental in algebra and is likewise of fundamental importance in matrix algebra. In matrix algebra two matrices $[a]$ and $[b]$ of the same order are defined to be equal if and only if their corresponding elements are identical; that is, we have

$$[a] = [b] \tag{9.1}$$

provided that

$$a_{ij} = b_{ij} \quad \text{all } i \text{ and } j \tag{9.2}$$

b   ADDITION AND SUBTRACTION

If $[a]$ and $[b]$ are matrices of the same order, then the sum of $[a]$ and $[b]$ is defined to be a matrix $[c]$, the typical element of which is $c_{ij} = a_{ij} + b_{ij}$. In other words, by definition

$$[c] = [a] + [b] \tag{9.3}$$

provided

$$c_{ij} = a_{ij} + b_{ij} \tag{9.4}$$

In a similar manner we have

$$[d] = [a] - [b] \tag{9.5}$$

provided that

$$d_{ij} = a_{ij} - b_{ij} \tag{9.6}$$

10   MULTIPLICATION OF MATRICES

a   SCALAR MULTIPLICATION

By definition, multiplication of a matrix $[a]$ by an ordinary number or scalar k results in a new matrix b defined by

$$k[a] = [b] \tag{10.1}$$

where

$$b_{ij} = ka_{ij} \tag{10.2}$$

That is, by definition, the multiplication of a matrix by a scalar quantity yields a new matrix whose elements are obtained by multiplying the elements of the original matrix by the scalar multiplier.

b   MATRIX MULTIPLICATION

The definition of the operation of multiplication of matrices by matrices differs in important respects from ordinary scalar multiplication. The rule of multiplication is such that two matrices can be multiplied only when the number of columns of the first is equal to the number of rows of the second. Matrices that satisfy this condition are termed *conformable matrices*.

Definition   The product of a matrix $[a]$ by a matrix $[b]$ is defined by the equation

$$[a][b] = [c] \tag{10.3}$$

where

$$c_{ij} = \sum_{k=1}^{k=p} a_{ik} b_{kj} \tag{10.4}$$

and the orders of the matrices $[a]$, $[b]$, and $[c]$ are $(m \times p)$, $(p \times n)$, and $(m \times n)$, respectively.   As an example of this definition, let us consider the multiplication of the matrix

$$[a] = \begin{bmatrix} a_{11} & a_{12} & a_{13} \\ a_{21} & a_{22} & a_{23} \\ a_{31} & a_{32} & a_{33} \end{bmatrix} \tag{10.5}$$

by

$$[b] = \begin{bmatrix} b_{11} & b_{12} \\ b_{21} & b_{22} \\ b_{31} & b_{32} \end{bmatrix} \tag{10.6}$$

By applying the definition we obtain

$$[a][b] = \begin{bmatrix} a_{11}b_{11} + a_{12}b_{21} + a_{13}b_{31} & a_{11}b_{12} + a_{12}b_{22} + a_{13}b_{32} \\ a_{21}b_{11} + a_{22}b_{21} + a_{23}b_{31} & a_{21}b_{12} + a_{22}b_{22} + a_{23}b_{32} \\ a_{31}b_{11} + a_{32}b_{21} + a_{33}b_{31} & a_{31}b_{12} + a_{32}b_{22} + a_{33}b_{32} \end{bmatrix} \tag{10.7}$$

If the matrices are square and each of order n, then the corresponding relation $|a||b| = |c|$ is true for the determinants of the matrices $[a]$, $[b]$, and $[c]$. The proof of this frequently useful result will be omitted.

Since matrices are regarded as equal only when they are element for element identical, it follows that, since a row-by-column rule will, in general, give different elements than a column-by-row rule, the product $[b][a]$, when it exists, is usually different from $[a][b]$. Therefore it is necessary to distinguish between *premultiplication*, as when $[b]$ is premultiplied by $[a]$ to yield the product $[a][b]$, and *postmultiplication*, as when $[b]$ is postmultiplied by $[a]$ to yield the product $[b][a]$. If we have the equality

$$[a][b] = [b][a] \tag{10.8}$$

the matrices a and b are said to *commute* or to be *permutable*. The unit matrix U, it may be noted, commutes with any square matrix of the same order. That is, we have for all such $[a]$

$$[a]U = U[a] = [a] \tag{10.9}$$

c CONTINUED PRODUCTS OF MATRICES

Except for the noncommutative law of multiplication (and therefore of division, which is defined as the inverse operation), all the ordinary laws of algebra apply to matrices. Of particular importance is the associative law of continued products,

$$([a][b][c]) = [a]([b][c]) \tag{10.10}$$

which allows one to dispense with parentheses and to write $[a][b][c]$ without ambiguity, since the double summation

$$\sum_k \sum_l a_{ik} b_{kl} c_{lj}$$

can be carried out in either of the orders indicated.

It must be noted, however, that the product of a chain of matrices will have meaning only if the adjacent matrices of the chain are conformable.

d POSITIVE POWERS OF A SQUARE MATRIX

If a square matrix is multiplied by itself n times, the resultant matrix is defined as $[a]^n$. That is,

$$[a]^n = [a][a] \cdots [a] \text{ to } n \text{ factors} \tag{10.11}$$

11 MATRIX DIVISION; THE INVERSE MATRIX

If the determinant $|a|$ of a square matrix $[a]$ does not vanish, $[a]$ is said to be *nonsingular* and possesses a *reciprocal*, or inverse, matrix $[R]$ such that

$$[a][R] = U = [R][a] \tag{11.1}$$

where U is the unit matrix of the same order as $[a]$.

a THE ADJOINT MATRIX OF A MATRIX

Let A_{ij} denote the cofactor of the element a_{ij} in the determinant $|a|$ of the matrix $[a]$. Then the matrix $[A_{ji}]$ is called the *adjoint* of the matrix $[a]$. This matrix exists whether $[a]$ is singular or not. Now by (4.8) we have

$$[a][A_{ji}] = |a|\ U = [A_{ji}][a] \tag{11.2}$$

It is thus seen that the product of $[a]$ and its adjoint is a special type of diagonal matrix called a *scalar matrix*. Each diagonal element $i = j$ is equal to the determinant $|a|$, and the other elements are zero.

If $|a| \neq 0$, we may divide through by the scalar $|a|$ and hence obtain at once the required form of $[R]$, the inverse of $[a]$. From (11.2) we thus have

$$\frac{[a][A_{ji}]}{|a|} = U \tag{11.3}$$

Therefore, comparing this with (11.1) we see that

$$[R] = \frac{[A_{ji}]}{|a|} = [a]^{-1} \tag{11.4}$$

The notation $[a]^{-1}$ is introduced to denote the inverse of $[a]$. By actual multiplication it may be proved that

$$[a][a]^{-1} = [a]^{-1}[a] = U \tag{11.5}$$

so that the name *reciprocal* and the notation $[a]^{-1}$ are justified.

If a square matrix is nonsingular, it possesses a reciprocal, and multiplication by the reciprocal is in many ways analogous to division in ordinary algebra. As an illustration, let it be required to obtain the inverse of the matrix

$$[a] = \begin{bmatrix} a_{11} & a_{12} & a_{13} \\ a_{21} & a_{22} & a_{23} \\ a_{31} & a_{32} & a_{33} \end{bmatrix} \tag{11.6}$$

Let $|a|$ denote the determinant of $[a]$. The next step in the process is to form the transpose of a.

$$[a]' = \begin{bmatrix} a_{11} & a_{21} & a_{31} \\ a_{12} & a_{22} & a_{32} \\ a_{13} & a_{23} & a_{33} \end{bmatrix} \tag{11.7}$$

We now replace the elements of $[a]'$ by their corresponding cofactors and obtain

$$[A_{ji}] = \begin{bmatrix} a_{22}a_{33} - a_{23}a_{32} & a_{13}a_{32} - a_{12}a_{33} & a_{12}a_{23} - a_{13}a_{22} \\ a_{23}a_{31} - a_{21}a_{33} & a_{11}a_{33} - a_{13}a_{31} & a_{13}a_{21} - a_{11}a_{23} \\ a_{21}a_{32} - a_{22}a_{31} & a_{12}a_{31} - a_{11}a_{32} & a_{11}a_{22} - a_{12}a_{21} \end{bmatrix} \tag{11.8}$$

This inverse $[a]^{-1}$ is therefore

$$\frac{[A_{ji}]}{|a|} \tag{11.9}$$

As has been mentioned, multiplication by an inverse matrix plays the same role in matrix algebra that division plays in ordinary algebra. That is, if we have

$$[a][b] = [c][d] \tag{11.10}$$

where $[a]$ is a nonsingular matrix, then, on premultiplying by $[a]^{-1}$, the inverse of $[a]$, we obtain

$$[a]^{-1}[a][b] = [a]^{-1}[c][d] \tag{11.11}$$

or

$$[b] = [a]^{-1}[c][d] \tag{11.12}$$

b NEGATIVE POWERS OF A SQUARE MATRIX

If $[a]$ is a nonsingular matrix, then negative powers of $[a]$ are defined by raising the inverse matrix of $[a]$, $[a]^{-1}$, to positive powers. That is,

$$[a]^{-n} = ([a]^{-1})^n \tag{11.13}$$

12   THE REVERSAL LAW IN TRANSPOSED AND RECIPROCATED PRODUCTS

One of the fundamental consequences of the noncommutative law of matrix multiplication is the *reversal law* exemplified in transposing and reciprocating a continued product of matrices.

a TRANSPOSITION

Let $[a]$ be a $(p \times n)$ matrix, that is, one having p rows and n columns, and let $[b]$ be an $(m \times p)$ matrix. Then the product $[c] = [b][a]$ is an $(m \times n)$ matrix of which the typical element is

$$c_{ij} = \sum_{r=1}^{p} b_{ir} a_{rj} \tag{12.1}$$

When transposed, $[a]'$ and $[b]'$ become $(n \times p)$ and $(p \times m)$ matrices; they are now conformable when multiplied in the order $[a]'[b]'$. This product is an $(n \times m)$ matrix, which may be readily seen to be the transpose of $[c]$ since its typical element is

$$c_{ji} = \sum_{r=1}^{p} a_{ri} b_{jr} \tag{12.2}$$

It thus follows that when a matrix product is transposed the order of the matrices forming the product must be reversed. That is,

$$([a][b])' = [b]'[a]' \tag{12.3}$$

and similarly

$$([a][b][c])' = [c]'[b]'[a]', \text{ etc.} \tag{12.4}$$

b RECIPROCATION

Let us suppose that in the equation $[c] = [b][a]$ the matrices are square and nonsingular. If we premultiply both sides of the equation by $[a]^{-1}[b]^{-1}$ and postmultiply by $[c]^{-1}$, then we obtain $[a]^{-1}[b]^{-1} = [c]^{-1}$. We thus get the rule that

$$[a]^{-1}[b]^{-1} = ([b][a])^{-1} \tag{12.5}$$

Similarly

$$([a][b][c])^{-1} = [c]^{-1}[b]^{-1}[a]^{-1} \tag{12.6}$$

13 PROPERTIES OF DIAGONAL AND UNIT MATRICES

Suppose that $[a]$ is a square matrix of order n and $[b]$ is a diagonal matrix, that is, a matrix that has all its elements zero with the exception of the elements in the main diagonal, and is of the same order as $[a]$. Then, if $[c] = [b][a]$, we have

$$c_{ij} = \sum_{r=1}^{n} b_{ir} a_{rj} = b_{ii} a_{ij} \tag{13.1}$$

since $b_{ir} = 0$ unless $r = i$.

It is thus seen that premultiplication by a diagonal matrix has the effect of multiplying every element in any row of a matrix by a constant. It can be similarly shown that postmultiplication by a diagonal matrix results in the multiplication of every element in any column of a matrix by a constant. The unit matrix plays the same role in matrix algebra that the number 1 does in ordinary scalar algebra.

14   MATRICES PARTITIONED INTO SUBMATRICES

It is sometimes convenient to extend the use of the fundamental laws of combinations of matrices to the case where a matrix is regarded as constructed from elements that are submatrices or minor matrices of elements.   As an example, consider

$$[a] = \begin{bmatrix} 1 & 9 & \vdots & 3 \\ 2 & 8 & \vdots & 4 \\ \hdashline 6 & 2 & \vdots & 7 \end{bmatrix} \tag{14.1}$$

This can be written in the form

$$[a] = \begin{bmatrix} P & Q \\ R & S \end{bmatrix} \tag{14.2}$$

where

$$[P] = \begin{bmatrix} 1 & 9 \\ 2 & 8 \end{bmatrix} \quad \{Q\} = \begin{bmatrix} 3 \\ 4 \end{bmatrix} \quad [R] = [6 \quad 2] \quad [S] = [7]$$

In this case the diagonal submatrices $[P]$ and $[S]$ are square, and the partitioning is diagonally symmetrical.   Let $[b]$ be a square matrix of the third order that is similarly partitioned.

$$[b] = \begin{bmatrix} 2 & 9 & \vdots & 4 \\ 3 & 6 & \vdots & 8 \\ \hdashline 1 & 5 & \vdots & 7 \end{bmatrix} = \begin{bmatrix} P_1 & Q_1 \\ R_1 & S_1 \end{bmatrix} \tag{14.3}$$

Then, by addition and multiplication, we have

$$[a] + [b] = \begin{bmatrix} P + P_1 & Q + Q_1 \\ R + R_1 & S + S_1 \end{bmatrix} \tag{14.4}$$

$$[a][b] = \begin{bmatrix} PP_1 + QR_1 & PQ_1 + QS_1 \\ RP_1 + SR_1 & RQ_1 + SS_1 \end{bmatrix} \tag{14.5}$$

as may be easily verified.

In each case the resulting matrix is of the same order that is partitioned in the same way as the original matrix factors.   As has been stated in Sec. 10, a rectangular matrix $[b]$ may be premultiplied by another rectangular matrix $[a]$ provided the two matrices are *conformable*, that is, the number of rows of $[b]$ is equal to the number of columns of $[a]$.   Now, if $[a]$ and $[b]$ are both partitioned into submatrices such that the grouping of columns in $[a]$ agrees with the grouping of rows in $[b]$, it can be shown that the product $[a][b]$ may be obtained by treating the submatrices as elements and proceeding according to the multiplication rule.

15 MATRICES OF SPECIAL TYPES

a. *Conjugate matrices:* To the operations $[a]'$ and $[a]^{-1}$, defined by transposition and inversion, may be added another one. This operation is denoted by $[\bar{a}]$ and implies that if the elements of $[a]$ are complex numbers, the corresponding elements of $[\bar{a}]$ are their respective complex conjugates. The matrix $[\bar{a}]$ is called the conjugate of $[a]$.

b. *The associate of* $[a]$: The transposed conjugate of $[a]$, $[\bar{a}]'$, is called the *associate* of $[a]$.

c. *Symmetric matrix:* If $[a] = [a]'$, the matrix $[a]$ is symmetric.

d. *Involutory matrix:* If $[a] = [a]^{-1}$, the matrix $[a]$ is involutory.

e. *Real matrix:* If $[a] = [\bar{a}]$, $[a]$ is a real matrix.

f. *Orthogonal matrix:* If $[a] = ([a]')^{-1}$, $[a]$ is an orthogonal matrix.

g. *Hermitian matrix:* If $[a] = [\bar{a}]'$, $[a]$ is a Hermitian matrix.

h. *Unitary matrix:* If $[a] = ([\bar{a}]')^{-1}$, $[a]$ is unitary.

i. *Skew-symmetric matrix:* If $[a] = -[a]'$, $[a]$ is skew symmetric.

j. *Pure imaginary:* If $[a] = -[\bar{a}]$, $[a]$ is pure imaginary.

k. *Skew Hermitian:* If $[a] = -[\bar{a}]'$, $[a]$ is skew Hermitian.

16 THE SOLUTION OF LINEAR ALGEBRAIC EQUATIONS

Prior to discussing the solution of a system of linear algebraic equations, the *rank* of a matrix must be defined. The rank† of a matrix is equal to the order of the largest nonvanishing determinant contained within the matrix. Thus the ranks of

$$\begin{bmatrix} 1 & 0 \\ 0 & 1 \end{bmatrix} \quad \begin{bmatrix} 0 & 0 & 0 \\ 0 & 0 & 0 \\ 0 & 0 & 0 \end{bmatrix} \quad \begin{bmatrix} 1 & -1 & 0 \\ 2 & 1 & -1 \end{bmatrix} \quad \begin{bmatrix} 1 & 0 & 0 \\ 0 & 2 & 0 \\ 0 & 0 & 3 \end{bmatrix}$$

and

$$\begin{bmatrix} 1 & 0 & -2 \\ 2 & 3 & -4 \\ -1 & 2 & 2 \end{bmatrix}$$

are 2, 0, 2, 3, and 2, respectively.

The concept of rank is not restricted to square matrices, as can be seen

† Rank may also be defined as the order of the largest nonvanishing minor contained in the matrix.

in the third example above.   For the case of square matrices, it is easily seen that if the rank equals the order of the matrix, then it necessarily has an inverse.

Now to move to the subject of this section we shall consider a non-homogeneous system of m linear algebraic equations and n unknowns:

$$
\begin{aligned}
a_{11}x_1 + a_{12}x_2 + \cdots + a_{1n}x_n + b_1 &= 0 \\
a_{21}x_1 + a_{22}x_2 + \cdots + a_{2n}x_n + b_2 &= 0 \\
\cdots \cdots \cdots \cdots \cdots \cdots \cdots \cdots \cdots \\
a_{m1}x_1 + a_{m2}x_2 + \cdots + a_{mn}x_n + b_m &= 0
\end{aligned}
\qquad (16.1)
$$

Using matrix notation, the above system reduces to

$$[A]\{x\} + \{b\} = \{0\} \qquad (16.2)$$

where $[A] = mxn$ coefficient matrix

$\{x\} = nx1$ vector of unknowns

$\{b\} = mx1$ vector of known constants

Another useful matrix is the $mx(n + 1)$-augmented matrix defined by

$$[\text{aug}] = [A,b] \qquad (16.3)$$

This particular matrix is formed by attaching the vector of known constants as an additional column on the coefficient matrix.

The system of equations defined by (16.2) is said to be consistent if a solution vector $\{x\}$ exists which satisfies the system.   If no solution exists, then the system is said to be inconsistent.†   We are now at a point where we may present the conditions, without proof,‡ under which the system (16.2) will be consistent.   These conditions may be grouped into the following three categories:

1. *Rank of $[A]$ = rank $[A,b]$ = $r \leqslant m$:* This situation yields a solution for any r unknowns in terms of the $n - r$ remaining ones.
2. *Rank of $[A]$ = rank of $[A,b]$ = $r = n$:* This situation yields a unique solution for the n unknowns.
3. *Rank of $[A]$ = rank of $[A,b]$ = $r < n$:* A solution always exists in this case.

It should be noted that in each of the above conditions the ranks of the coefficient and augmented matrices were equal.   This in fact is the basic and *only* requirement for the existence of a solution to (16.2).   Reiterating this statement, it can be proven§ that a system of linear algebraic equations is

---

† For an excellent and thorough theoretical discussion of the subject of linear algebraic equations, see Hohn, 1958, chap. 5 (see References).

‡ *Ibid.* for a detailed proof.

§ See Hohn, 1958, chap. 5, for a detailed proof (see References).

consistent if and only if the rank of the coefficient matrix equals the rank of the augmented matrix.

17 THE SPECIAL CASE OF n EQUATIONS AND n UNKNOWNS

When $m = n$ in (16.2), we have the special case of a system of n linear algebraic equations and n unknowns.   As in any case a solution will only exist when the ranks of the coefficient and augmented matrices are equal.   For convenience we shall assume that the ranks of both of these matrices equal n.   This allows us to premultiply by the inverse of the coefficient matrix to obtain the solution.

Suppose we have the set of n linear algebraic equations in n unknowns,

$$
\begin{aligned}
e_1 &= Z_{11} i_1 + Z_{12} i_2 + \cdots + Z_{1n} i_n \\
e_2 &= Z_{21} i_1 + Z_{22} i_2 + \cdots + Z_{2n} i_n \\
&\cdot\cdot\cdot\cdot\cdot\cdot\cdot\cdot\cdot\cdot\cdot\cdot\cdot\cdot\cdot\cdot\cdot\cdot \\
e_n &= Z_{n1} i_1 + Z_{n2} i_2 + \cdots + Z_{nn} i_n
\end{aligned}
\tag{17.1}
$$

The quantities e_j, Z_{ij}, and i_j may be real or complex numbers.   If the relation (17.1) exists between the variables e_j and i_j, then the set of variables e_j is said to be derived from the set i_j by a *linear transformation*.   The whole set of Eqs. (17.1) may be represented by the single matrix equation

$$
\{e\} = [Z]\{i\}
\tag{17.2}
$$

where $\{e\}$ and $\{i\}$ are column matrices whose elements are the variables e_j and i_j. The square matrix $[Z]$ is called the *matrix of the transformation*.   When $[Z]$ is nonsingular, Eq. (17.2) may be premultiplied by $[Z]^{-1}$ and we obtain

$$
[Z]^{-1}\{e\} = [Z]^{-1}[Z]\{i\} = \{i\}
\tag{17.3}
$$

We thus have a very convenient explicit solution for the unknowns $(i_1, i_2, \ldots, i_n)$.   The elements of the matrix $[Z]^{-1}$ may be calculated by Eq. (11.4).   As a simple example, let us solve the system of Eqs. (2.1) by using matrix notation.   If we introduce the matrices

$$
[a] = \begin{bmatrix} a_{11} & a_{12} \\ a_{21} & a_{22} \end{bmatrix} \qquad \{x\} = \begin{Bmatrix} x_1 \\ x_2 \end{Bmatrix} \qquad \{k\} = \begin{Bmatrix} k_1 \\ k_2 \end{Bmatrix}
\tag{17.4}
$$

then Eqs. (2.1) may be written in the form

$$
[a]\{x\} = \{k\}
\tag{17.5}
$$

and if $[a]$ is nonsingular, the solution is

$$
\begin{Bmatrix} x_1 \\ x_2 \end{Bmatrix} = \begin{bmatrix} a_{11} & a_{12} \\ a_{21} & a_{22} \end{bmatrix}^{-1} \begin{Bmatrix} k_1 \\ k_2 \end{Bmatrix}
\tag{17.6}
$$

But by (11.4) we have

$$\begin{bmatrix} a_{11} & a_{12} \\ a_{21} & a_{22} \end{bmatrix}^{-1} = \frac{1}{a_{11}a_{22} - a_{21}a_{12}} \begin{bmatrix} a_{22} & -a_{12} \\ -a_{21} & a_{11} \end{bmatrix} \tag{17.7}$$

Hence, carrying out the matrix multiplication expressed in (17.6), we obtain

$$x_1 = \frac{k_1 a_{22} - k_2 a_{12}}{a_{11}a_{22} - a_{21}a_{12}}$$

$$x_2 = \frac{k_2 a_{11} - k_1 a_{21}}{a_{11}a_{22} - a_{21}a_{12}} \tag{17.8}$$

Although the solution of a set of linear equations is very elegantly and concisely expressed in terms of the inverse matrix of the coefficients, it frequently happens in practice that it is necessary to solve for only one or two of the unknowns.

In this case it is frequently more convenient to use Cramer's rule to obtain the solution. The proof of Cramer's rule follows from the properties of the determinant expressed by Eq. (4.8) and will not be given. Let us consider the set of n equations in the unknowns $(x_1, x_2, \ldots, x_n)$,

$$\begin{aligned} a_{11}x_1 + a_{12}x_2 + \cdots + a_{1n}x_n &= k_1 \\ a_{21}x_1 + a_{22}x_2 + \cdots + a_{2n}x_n &= k_2 \\ \cdot\ \cdot\ \cdot\ \cdot\ \cdot\ \cdot\ \cdot\ \cdot\ \cdot\ \cdot\ \cdot\ \cdot\ \cdot\ \cdot\ \cdot\ \cdot\ \cdot& \\ a_{n1}x_1 + a_{n2}x_2 + \cdots + a_{nn}x_n &= k_n \end{aligned} \tag{17.9}$$

Then if the determinant of the system

$$|a| = \begin{vmatrix} a_{11} & a_{12} & \cdots & a_{1n} \\ \cdot & \cdot & \cdots & \cdot \\ a_{n1} & a_{n2} & \cdots & a_{nn} \end{vmatrix} \neq 0 \tag{17.10}$$

the system of Eqs. (17.9) has a unique solution given by

$$x_1 = \frac{D_1}{|a|} \qquad x_2 = \frac{D_2}{|a|} \qquad \cdots \qquad x_n = \frac{D_n}{|a|} \tag{17.11}$$

where D_r is the determinant formed by replacing the elements

$$a_{1r}, a_{2r}, \ldots, a_{nr}$$

of the rth column of $|a|$ by $(k_1, k_2, \ldots, k_n)$, respectively.

For example, let us solve the system

$$\begin{aligned} x_1 - 2x_2 + 3x_3 &= 2 \\ 2x_1 \qquad - 3x_3 &= 3 \\ x_1 + x_2 + x_3 &= 6 \end{aligned} \tag{17.12}$$

We have

$$|a| = \begin{vmatrix} 1 & -2 & 3 \\ 2 & 0 & -3 \\ 1 & 1 & 1 \end{vmatrix} = 19 \tag{17.13}$$

Hence

$$x_1 = \frac{\begin{vmatrix} 2 & -2 & 3 \\ 3 & 0 & -3 \\ 6 & 1 & 1 \end{vmatrix}}{19} = 3$$

$$x_2 = \frac{\begin{vmatrix} 1 & 2 & 3 \\ 2 & 3 & -3 \\ 1 & 6 & 1 \end{vmatrix}}{19} = 2 \tag{17.14}$$

$$x_3 = \frac{\begin{vmatrix} 1 & -2 & 2 \\ 2 & 0 & 3 \\ 1 & 1 & 6 \end{vmatrix}}{19} = 1$$

At the risk of repeating what has already been stated in Sec. 16, we shall review some special cases that may arise when working with a system of n equations and n unknowns.

1. If the determinant $|a| \neq 0$, there exists a unique solution that is given by Cramer's rule, that is, (17.11).
2. If a is of rank $r < n$ and any of the determinants $D_1, D_2, \ldots, D_n$ of (17.11) are of rank greater than r, there is no solution.
3. If $|a|$ is of rank $r < n$ and the rank of none of $D_1, D_2, \ldots, D_n$ exceeds r, then there exist infinitely many sets of solutions.

18 SYSTEMS OF HOMOGENEOUS LINEAR EQUATIONS

If in Eq. (16.2) the vector $\{b\}$ is a null vector, that is $\{b\} = \{0\}$, then (16.2) reduces to the following system of homogeneous linear equations:

$$[A]\{x\} = \{0\} \tag{18.1}$$

Regardless of the order of $[A]$, this system is consistent; hence a solution *always* exists. By inspection it is easy to see that the solution vector $\{x\} = \{0\}$ satisfies (18.1) identically. This particular solution is called the *trivial solution*. In most cases this solution is of little interest to the analyst, and other solutions called *nontrivial solutions* are sought.

In the most general case the matrix $[A]$ is of order $m \times n$; however, it usually is $n \times n$. In either case it may be stated that if the rank of $[A]$ is less than n, then a nontrivial solution to (18.1) exists. In the case when the rank of the coefficient matrix is n, the only solution that exists is the trivial one.

A few examples of solutions to homogeneous systems are given below.

Example 1

$$\begin{bmatrix} 1 & 2 \\ 3 & 4 \\ 5 & 6 \end{bmatrix}\{x\} = \{0\} \tag{18.2}$$

The rank of $[A]$ in this case is 2 and $n = 2$, hence only the trivial solution exists.

Example 2

$$\begin{bmatrix} 1 & 2 & 3 \\ 4 & 5 & 6 \end{bmatrix}\{x\} = \{0\} \tag{18.3}$$

In this case the rank of the coefficient matrix is 2, but $n = 3$; thus a nontrivial solution exists. To obtain the solution we proceed as follows. Since the rank of $[A]$ is 2 and there are three unknowns, then any two unknowns may be solved for in terms of the third. This may be accomplished in a variety of ways, one of which is the following.

By partitioning, write (18.3) as

$$\begin{bmatrix} 1 & 2 & \vdots & 3 \\ 4 & 5 & \vdots & 6 \end{bmatrix}\begin{Bmatrix} x_1 \\ x_2 \\ -- \\ x_3 \end{Bmatrix} = 0$$

$$\begin{bmatrix} 1 & 2 \\ 4 & 5 \end{bmatrix}\begin{Bmatrix} x_1 \\ x_2 \end{Bmatrix} + \begin{Bmatrix} 3 \\ 6 \end{Bmatrix}x_3 = \{0\} \tag{18.4}$$

If we premultiply this equation through by

$$\begin{bmatrix} 1 & 2 \\ 4 & 5 \end{bmatrix}^{-1}$$

and solve for the vector

$$\begin{Bmatrix} x_1 \\ x_2 \end{Bmatrix}$$

we find the following solution:

$$\begin{Bmatrix} x_1 \\ x_2 \end{Bmatrix} = -\begin{bmatrix} 1 & 2 \\ 4 & 5 \end{bmatrix}^{-1}\begin{Bmatrix} 3 \\ 6 \end{Bmatrix}x_3$$

$$= \tfrac{1}{3}\begin{bmatrix} 5 & -2 \\ -4 & 1 \end{bmatrix}\begin{Bmatrix} 3 \\ 6 \end{Bmatrix}x_3$$

or

$$x_1 = x_3 \qquad x_2 = -2x_3 \tag{18.5}$$

The complete solution may be written in vector form as

$$\{x\} = \begin{Bmatrix} 1 \\ -2 \\ 1 \end{Bmatrix}x_3 = c\begin{Bmatrix} 1 \\ -2 \\ 1 \end{Bmatrix} \tag{18.6}$$

where c represents an arbitrary constant.

19 THE CHARACTERISTIC MATRIX AND THE CHARACTERISTIC EQUATION OF A MATRIX

If μ is a scalar parameter and $[M]$ is a square matrix of order n, the matrix

$$[K] = [\mu U - M] \tag{19.1}$$

$U =$ the nth-order unit matrix, is called the *characteristic matrix* of the matrix $[M]$. The determinant of the characteristic matrix, $\det[K]$, is called the *characteristic function* of $[M]$. That is,

$$\Delta(\mu) = \det[K] \tag{19.2}$$

is the characteristic function of the matrix $[M]$.
    The equation

$$\Delta(\mu) = \det[K] = 0 \tag{19.3}$$

is the *characteristic equation* of the matrix $[M]$. In general, if $[M]$ is a square matrix of the nth order, its characteristic equation is an equation of the nth degree in μ.

$$\Delta(\mu) = \mu^n + c_1\mu^{n-1} + \cdots + c_{n-1}\mu + c_n = 0 \tag{19.4}$$

The n roots of this equation are called the characteristic roots, or *eigenvalues*, of the matrix $[M]$.

20 EIGENVALUES AND THE REDUCTION OF A MATRIX TO DIAGONAL FORM

In matrix algebra, when a matrix contains only one column, it is called a *column vector*. If $\{x\}$ is a column vector of the nth order and $[M]$ is an nth-order square matrix, the product

$$\{y\} = [M]\{x\} \tag{20.1}$$

expresses the fact that the vector $\{y\}$ is derived from the vector $\{x\}$ by a linear transformation. The square matrix $[M]$ is called the *transformation matrix*. If it is required that

$$\{y\} = \mu\{x\} \tag{20.2}$$

then (20.1) reduces to

$$\mu\{x\} = [M]\{x\} \tag{20.3}$$

and the only effect of the transformation matrix $[M]$ on the vector $\{x\}$ is to multiply it by the scalar factor μ. Equation (20.3) may be written in the form

$$[\mu U - M]\{x\} = [K]\{x\} = 0 \tag{20.4}$$

This equation represents a set of homogeneous linear equations in the unknowns $x_1, x_2, \ldots, x_n$. Except in the trivial case where all the elements of $\{x\}$ are zero, the determinant of $[K]$ must vanish. Conversely, if $|K| = 0$, there exist nontrivial solutions. In order for this to be the case, the scalar μ can take only the values $\mu_1, \mu_2, \ldots, \mu_n$ which are the roots of the characteristic equation (19.3). The possible values of μ which satisfy (20.3) are thus the eigenvalues of $[M]$. The nontrivial solutions of (20.4) associated with these eigenvalues are called *eigenvectors* of $[M]$.

To each eigenvalue μ_i there corresponds an equation of the form (20.3) which can be written as follows:

$$\mu_i\{x\}_i = [M]\{x\}_i \qquad i = 1, 2, 3, \ldots, n \tag{20.5}$$

where $\{x\}_i$ is any eigenvector associated with μ_i.

If (20.5) is premultiplied by $[M]$, it becomes

$$\mu_i[M]\{x\}_i = \mu_i^2\{x\}_i = [M]^2\{x\}_i \tag{20.6}$$

By repeated multiplications by $[M]$ it can be shown that

$$[M]^r\{x\}_i = \mu_i^r\{x\}_i \tag{20.7}$$

Let the ith eigenvector be denoted by

$$\{x\}_i = \begin{bmatrix} x_{1i} \\ x_{2i} \\ x_{3i} \\ \vdots \\ x_{ni} \end{bmatrix} \tag{20.8}$$

and let the square matrix $[x]$ be the matrix formed by grouping the column eigenvectors into a square matrix so that

$$[x] = \begin{bmatrix} x_{11} & x_{12} & x_{13} & \cdots & x_{1n} \\ x_{21} & x_{22} & x_{23} & \cdots & x_{2n} \\ \cdots & \cdots & \cdots & \cdots & \cdots \\ x_{n1} & x_{n2} & x_{n3} & \cdots & x_{nn} \end{bmatrix} \tag{20.9}$$

If $[d]$ denotes the square diagonal matrix

$$[d] = \begin{bmatrix} \mu_1 & 0 & \cdots & 0 \\ 0 & \mu_2 & \cdots & 0 \\ \cdots & \cdots & \cdots & \cdots \\ 0 & 0 & \cdots & \mu_n \end{bmatrix} \tag{20.10}$$

it can be seen that the set of n equations (20.5) may be written in the form

$$[x][d] = [M][x] \tag{20.11}$$

If the n eigenvalues of $[M]$ are distinct, it can be shown that the matrix $[x]$ is nonsingular and possesses an inverse $[x]^{-1}$.† Hence (20.11) may be written in the form

$$[d] = [x]^{-1}[M][x] \tag{20.12}$$

or

$$[M] = [x][d][x]^{-1} \tag{20.13}$$

It is thus seen that a matrix $[x]$ which diagonalizes $[M]$ may be found by grouping the eigenvectors of $[M]$ into a square matrix. This may be done when the eigenvalues of $[M]$ are all different. When two or more of the eigenvalues of $[M]$ are equal to each other, reduction to the diagonal form is not always possible.

21   THE TRACE OF A MATRIX

The sum of all the diagonal elements of a square matrix is called the *trace* of the matrix and is denoted by the following expression:

$$\mathrm{Tr}\,[M] = \sum_{i=1}^{i=n} M_{ii} \tag{21.1}$$

There exist some interesting relations between the trace of a matrix and the characteristic equation of the matrix.‡ The characteristic equation of $[M]$ has the form

$$\Delta(\mu) = \det[\mu U - M] = 0$$
$$= \mu^n + c_1 \mu^{n-1} + c_2 \mu^{n-2} + \cdots + c_{n-1}\mu + c_n = 0 \tag{21.2}$$

It can be shown that

$$c_1 = -(M_{11} + M_{22} + M_{33} + \cdots + M_{nn}) = -\mathrm{Tr}\,[M] \tag{21.3}$$

If $\mu_1, \mu_2, \ldots, \mu_n$ are the *eigenvalues* of $[M]$, then as a result of (21.3) we have

$$\mu_1 + \mu_2 + \mu_3 + \cdots + \mu_n = \mathrm{Tr}\,[M] \tag{21.4}$$

It can be shown that the product of the eigenvalues is given by the equation

$$\mu_1 \times \mu_2 \times \mu_3 \times \cdots \times \mu_n = \det[M] \tag{21.5}$$

† This restriction may be expanded to include multiple eigenvalues as long as there are n linear independent eigenvectors associated with $[M]$.

‡ M. Bôcher, "Introduction to Higher Algebra," The Macmillan Company, New York, 1931.

Maxime Bôcher has shown that there is an intimate connection between the traces of powers of a matrix and the coefficients of its characteristic equation (21.2). These results are of importance in obtaining the characteristic equations of matrices and in the computation of the inverses of matrices. The results of Bôcher are deduced in his book on higher algebra and are given here without proof.

The characteristic equation of a square matrix $[M]$ of the nth order is, in general, an equation of the nth degree as given by (21.2). Bôcher's results relate the coefficients c_i of the characteristic equation of $[M]$ to the traces of powers of $[M]$. Let

$$S_1 = \text{Tr}\,[M] \qquad S_2 = \text{Tr}\,[M]^2 \qquad \cdots \qquad S_n = \text{Tr}\,[M]^n \qquad (21.6)$$

Then the coefficients $c_1, c_2, \ldots, c_n$ of the characteristic equations are

$$c_1 = -S_1$$
$$c_2 = -\tfrac{1}{2}(c_1 S_1 + S_2)$$
$$c_3 = -\tfrac{1}{3}(c_2 S_1 + c_1 S_2 + S_3)$$
$$\cdots \cdots \cdots \cdots \cdots \cdots$$
$$c_n = -\frac{1}{n}(c_{n-1} S_1 + c_{n-2} S_2 + \cdots + c_1 S_{n-1} + S_n) \qquad (21.7)$$

These equations enable one to obtain the characteristic equation of a matrix in a more efficient manner than by the direct expansion of $\det[\mu U - M]$. If the determinant of $[M]$ itself is desired, it can be calculated by the use of the following equation:

$$\det[M] = (-1)^n c_n \qquad (21.8)$$

where c_n is as given above.

22   THE CAYLEY-HAMILTON THEOREM

One of the most important theorems of the theory of matrices is the Cayley-Hamilton theorem. By the use of this theorem many properties of matrices may be deduced, and functions of matrices can be simplified. The Cayley-Hamilton theorem is frequently stated in the following terse manner: Every square matrix satisfies its own characteristic equation in a matrix sense. This statement means that if the characteristic equation of the nth-order matrix $[M]$ is given by

$$\Delta(\mu) = \mu^n + c_1 \mu^{n-1} + c_2 \mu^{n-2} + \cdots + c_{n-1} \mu + c_n = 0 \qquad (22.1)$$

the Cayley-Hamilton theorem states that

$$\Delta([M]) = [M]^n + c_1 [M]^{n-1} + c_2 [M]^{n-2} + \cdots$$
$$+ c_{n-1}[M] + c_n U = [0] \qquad (22.2)$$

where U is the nth-order unit matrix.

A heuristic proof of this theorem may be given by the use of the following result of Sec. 20:

$$[M]^r\{x\}_i = \mu_i^r\{x\}_i \qquad r = 1, 2, 3, \ldots \qquad i = 1, 2, 3, \ldots, n \qquad (22.3)$$

where μ_i is the ith eigenvalue of $[M]$ and $\{x\}_i$ is the corresponding eigenvector. *The characteristic function* of $[M]$ is given by

$$\Delta(\mu) = \mu^n + c_1\mu^{n-1} + c_2\mu^{n-2} + \cdots + c_{n-1}\mu + c_n \qquad (22.4)$$

If in (22.4) the scalar μ is replaced by the square matrix $[M]$ and the coefficient of c_n is multiplied by the nth-order unit matrix U, the result may be written in the following form:

$$\Delta([M]) = [M]^n + c_1[M]^{n-1} + c_2[M]^{n-2} + \cdots + c_n U \qquad (22.5)$$

If (22.5) is postmultiplied by the eigenvector $\{x\}_i$, the result is

$$\Delta(M)\{x\}_i = ([M]^n + c_1[M]^{n-1} + c_2[M]^{n-2} + \cdots + c_n)\{x\}_i \qquad (22.6)$$

As a consequence of (22.3), this may be written in the following form:

$$\Delta([M])\{x\}_i = (\mu_i^n + c_1\mu_i^{n-1} + c_2\mu_i^{n-2} + \cdots + c_n)\{x\}_i \qquad (22.7)$$

However, since μ_i is one of the eigenvalues of $[M]$, we have

$$\mu_i^n + c_1\mu_i^{n-1} + c_2\mu^{n-2} + \cdots + c_n = 0 \qquad (22.8)$$

Hence (22.7) reduces to

$$\Delta([M])\{x\}_i = \{0\} \qquad i = 1, 2, 3, \ldots, n \qquad (22.9)$$

The set of equations (22.9) may be written as the single equation

$$\Delta([M])[x] = [0] \qquad (22.10)$$

where $[x]$ is the square matrix (20.9) and $[0]$ is an nth-order square matrix of zeros. If $[M]$ has n distinct eigenvalues, the matrix $[x]$ is nonsingular and has an inverse. If (22.10) is postmultiplied by $[x]^{-1}$, the result is

$$\Delta([M]) = [0] \qquad (22.11)$$

This important result is the Cayley-Hamilton theorem. It may be shown that the Cayley-Hamilton theorem holds for singular matrices and for matrices that have repeated eigenvalues.†

If the matrix $[M]$ is nonsingular, its inverse may be computed by the use of the Cayley-Hamilton theorem in the following manner: Expand (22.11) in the form

$$\Delta([M]) = [M]^n + c_1[M]^{n-1} + c_2[M]^{n-2} + \cdots + c_n U = [0] \qquad (22.12)$$

† M. Bôcher, "Introduction to Higher Algebra," p. 296, The Macmillan Company, New York, 1931.

If this equation is premultiplied by $[M]^{-1}$, the result is

$$[M]^{n-1} + c_1[M]^{n-2} + \cdots + c_n[M]^{-1} = [0] \tag{22.13}$$

Hence, if this equation is solved for $[M]^{-1}$, the result is

$$[M]^{-1} = -\frac{1}{c_n}([M]^{n-1} + c_1[M]^{n-2} + \cdots + c_{n-1}U) \tag{22.14}$$

This result expresses the inverse of $[M]$ in terms of the $n-1$ powers of $[M]$ and has been suggested as a practical method for the computation of the inverses of large matrices.† Hotelling gives the following application of (22.14) to the computation of the inverse of a numerical matrix: Let the matrix to be inverted be

$$[M] = \begin{bmatrix} 15 & 11 & 6 & -9 & -15 \\ 1 & 3 & 9 & -3 & -8 \\ 7 & 6 & 6 & -3 & -11 \\ 7 & 7 & 5 & -3 & -11 \\ 17 & 12 & 5 & -10 & -16 \end{bmatrix} \tag{22.15}$$

Then

$$[M]^2 = \begin{bmatrix} -40 & -9 & 105 & -9 & -40 \\ -76 & -43 & 32 & 44 & 23 \\ -55 & -22 & 62 & 20 & -10 \\ -61 & -25 & 65 & 20 & -7 \\ -40 & -9 & 110 & -14 & -40 \end{bmatrix} \tag{22.16}$$

$$[M]^3 = \begin{bmatrix} -617 & -380 & 64 & 499 & 256 \\ -260 & -189 & -316 & 355 & 280 \\ -443 & -279 & -106 & 415 & 259 \\ -464 & -300 & -136 & 439 & 292 \\ -617 & -385 & 69 & 499 & 256 \end{bmatrix} \tag{22.17}$$

$$[M]^4 = \begin{bmatrix} -1,342 & -978 & -2,963 & 2,444 & 2,006 \\ 944 & 522 & -1,982 & -10 & 503 \\ -358 & -333 & -2,435 & 1,307 & 1,334 \\ -175 & -243 & -2,645 & 1,247 & 1,355 \\ -1,312 & -963 & -2,978 & 2,444 & 1,991 \end{bmatrix} \tag{22.18}$$

From the diagonals of the above matrices, the following traces may be calculated:

$$\begin{aligned} S_1 &= \text{Tr}\,[M] = 5 \\ S_2 &= \text{Tr}\,[M]^2 = -41 \\ S_3 &= \text{Tr}\,[M]^3 = -217 \\ S_4 &= \text{Tr}\,[M]^4 = -17 \end{aligned} \tag{22.19}$$

† H. Hotelling, New Methods in Matrix Calculation, *The Annals of Mathematical Statistics*, vol. 14, no. 1, pp. 1–34, March, 1943.

By means of (21.7) the following coefficients of the characteristic equation of $[M]$ can be computed:

$$c_1 = -5 \qquad c_2 = 33 \qquad c_3 = -51 \qquad c_4 = 135 \qquad c_5 = 225 \qquad (22.20)$$

The characteristic equation of $[M]$ is therefore

$$\mu^5 - 5\mu^4 + 33\mu^3 - 51\mu^2 + 135\mu + 225 = 0 \qquad (22.21)$$

The inverse of $[M]$ is now computed by the use of (22.14) and is given by

$$[M]^{-1} = \frac{-1}{225}
\begin{bmatrix}
-207 & 64 & -124 & 111 & 171 \\
-315 & 30 & 195 & -180 & 270 \\
-315 & 30 & -30 & 45 & 270 \\
-225 & 75 & -75 & 0 & 225 \\
-414 & 53 & 52 & -3 & 342
\end{bmatrix} \qquad (22.22)$$

If the determinant of $[M]$ is desired, it may be obtained from the equation

$$\det [M] = (-1)^n c_n = -225 \qquad (22.23)$$

This method for the computation of the inverse of a matrix is straightforward and easily checked and is ideally adapted to matrix multiplication by means of punched cards. As a by-product of the computation the characteristic equation and the determinant of the matrix are obtained.

23 THE INVERSION OF LARGE MATRICES

With the development of many powerful modern instruments of calculation such as punched-card machines, electronic digital-computing machines, analogue computers, etc., the field of application of mathematical analysis has vastly increased. Many diversified problems in various fields may be reduced to one mathematical operation: that of the solution of systems of linear algebraic equations. A few of the problems that may be reduced to this mathematical common denominator are the following:

1. The distribution of electrical currents in multimesh electrical circuits.
2. The distribution of the rate of flow of water in complex hydraulic systems.
3. The approximate solution of potential problems.
4. The solution of the normal equations in the application of the method of least squares to various physical investigations.
5. The applications of statistical studies to psychology, sociology, and economics.
6. The determination of the amplitudes of motion in the forced vibrations of mechanical systems.
7. The approximate solution of certain problems of electromagnetic radiation.

The above list is far from complete, but it gives an indication of the diversified classes of problems that may be reduced to the solution of a system of linear equations.

It may be thought that the problem of solving systems of linear equations is an old one and that it can be effected by means of well-known techniques such as Cramer's rule (Sec. 16).   However, in studying problems of the type enumerated above, it is frequently necessary to solve 150 or 200 simultaneous equations.   This task presents practically insurmountable difficulties if undertaken by well-known elementary methods. . In the past few years many powerful and apparently unrelated methods have been devised to perform this task.   A very complete survey of these methods, with a bibliography containing 131 references, will be found in the paper by Forsythe.†

The magnitude of the task of solving a large system of linear equations by conventional elementary methods and the spectacular reduction of effort that may be effected by the use of modern techniques are vividly presented by Forsythe.   He points out that the conventional Cramer's-rule procedure for solving n simultaneous equations requires the evaluation of $n + 1$ determinants. If these determinants are evaluated by the Laplace expansion (Sec. 4), this requires $n!$ multiplications per determinant.   It is thus apparent that the conventional determinantal method of solution requires $(n + 1)!$ multiplications.   Therefore, if it is required to solve 26 equations, it would, in general, require $(n + 1)! = 27! \approx 10^{28}$ multiplications.   A typical high-speed digital machine such as the IBM/360 model 75 automatic computer can perform 133,000 multiplications per second.   Hence, it would take this machine approximately 10^{14} years to effect the solution of this problem by the use of the conventional Cramer's-rule technique.   By more efficient techniques the computational procedure may be arranged so that only about $n^3/3$ multiplications are required to solve the system of n equations.   If this efficient method is used, 6,000 multiplications would be required, and the IBM machine can solve the problem in about 0.05 sec.

The detailed theory of the various methods that have been developed in recent years will be found in the above survey article by Forsythe and should be consulted by those interested in the solution of large numbers of simultaneous linear equations.   A simple method that has been found valuable and that makes use of partitioned matrices will now be described.

Let the matrix whose inverse is desired be the matrix $[M]$, of order n. Then, by the methods of Sec. 14, the matrix $[M]$ can be partitioned into four submatrices in the following manner:

$$[M] = \begin{bmatrix} M_{11}(r,r) & M_{12}(r,s) \\ M_{21}(s,r) & M_{22}(s,s) \end{bmatrix} \tag{23.1}$$

† G. E. Forsythe, Solving Linear Equations Can Be Interesting, *Bulletin of the American Mathematical Society*, vol. 59, no. 4, pp. 299–329, July, 1953.

The orders of the various matrices are indicated by the letters in parentheses. The first letter denotes the number of rows and the second letter the number of columns. Since the original matrix is of order n, we have the relation

$$r + s = n \tag{23.2}$$

Now let the reciprocal of $[M]$ be the matrix $[B]$, and let it be correspondingly partitioned so that we have

$$[B] = [M]^{-1} = \begin{bmatrix} B_{11}(r,r) & B_{12}(r,s) \\ B_{21}(s,r) & B_{22}(s,s) \end{bmatrix} \tag{23.3}$$

Since $[B]$ is the inverse of $[M]$, we have

$$[B][M] = U_n \tag{23.4}$$

where U_n is the unit matrix of order n. If the multiplication indicated by (23.4) is carried out, we have

$$
\begin{aligned}
B_{11} M_{11} + B_{12} M_{21} &= U_r \\
B_{11} M_{12} + B_{12} M_{22} &= [0] \\
B_{21} M_{11} + B_{22} M_{21} &= [0] \\
B_{21} M_{12} + B_{22} M_{22} &= U_s
\end{aligned}
\tag{23.5}
$$

These equations may, in general, be solved to give the submatrices of the required matrix $[B]$ explicitly. The results may be expressed as follows: Let

$$
\begin{aligned}
X &= M_{11}^{-1} M_{12} \qquad Y = M_{21} M_{11}^{-1} \\
\theta &= M_{22} - Y M_{12} = M_{22} - M_{21} X
\end{aligned}
\tag{23.6}
$$

We then have explicitly

$$
\begin{aligned}
B_{11} &= M_{11}^{-1} + X\theta^{-1} Y \qquad B_{21} = \theta^{-1} Y \\
B_{12} &= -X\theta^{-1} \qquad\qquad\quad\; B_{22} = \theta^{-1}
\end{aligned}
\tag{23.7}
$$

These equations determine $[B] = [M]^{-1}$ provided that the reciprocals M_{11}^{-1} and θ^{-1} exist. If M_{11}^{-1} is known, then the matrices X, Y, and θ can be calculated by matrix multiplication and $[B]$ can be determined by the use of (23.7). The choice of the manner in which the original matrix $[M]$ is partitioned depends on the particular matrix to be inverted. From the above description of the method it is obvious that it yields to various modifications. The method of Sec. 22 may advantageously be applied to find M_{11}^{-1} and θ^{-1}. By this procedure the inversion of a large matrix may be reduced to the inversion of several smaller matrices.

CHOLESKI'S METHOD

Another very effective method for the inversion of matrices is Choleski's method. In its simplest form it applies to the inversion of symmetric matrices and is simple in theory and straightforward in practice. This method is sometimes called the *square-root method*, and an excellent exposition of this method is given in the paper by Fox, Huskey, and Wilkinson.† Choleski's method consists simply in expressing the symmetric matrix $[M]$ whose inverse is desired in the form

$$[M] = [L][L]_t \tag{23.8}$$

where $[L]$ is a lower triangular matrix and $[L]_t$, its transpose, an upper triangular matrix. Then the desired inverse is given by

$$[M]^{-1} = [L]_t^{-1}[L]^{-1} \tag{23.9}$$

and the problem is solved. To fix the fundamental ideas involved, consider the case of a third-order symmetric matrix $[M]$. We then have to find the third-order matrices $[L]$ and $[L]_t$ by means of the equation

$$[L][L]_t = \begin{bmatrix} a & 0 & 0 \\ b & d & 0 \\ c & e & f \end{bmatrix} \begin{bmatrix} a & b & c \\ 0 & d & e \\ 0 & 0 & f \end{bmatrix} = \begin{bmatrix} 1 & 2 & 3 \\ 2 & 3 & 4 \\ 3 & 4 & 4 \end{bmatrix} = [M] \tag{23.10}$$

By matrix multiplication of $[L][L]_t$ the following result is obtained:

$$\begin{bmatrix} a^2 & ab & ac \\ ab & b^2+d^2 & bc+de \\ ac & bc+de & c^2+e^2+f^2 \end{bmatrix} = \begin{bmatrix} 1 & 2 & 3 \\ 2 & 3 & 4 \\ 3 & 4 & 4 \end{bmatrix} \tag{23.11}$$

By equating the elements of $[L][L]_t$ to those of $[M]$ and solving the resulting equations one finds $a = 1$, $b = 2$, $c = 3$, $d = j$, $e = 2j$, $f = j$. In order to determine $[M]^{-1}$ by (23.9), it is necessary to invert $[L]_t$. This is easily done because $[L]_t^{-1}$ is also an upper triangular matrix of the form

$$[L]_t^{-1} = \begin{bmatrix} P & Q & R \\ 0 & S & T \\ 0 & 0 & U \end{bmatrix} \tag{23.12}$$

The elements of $[L]_t^{-1}$ may be determined by the equation

$$[L]_t[L]_t^{-1} = \begin{bmatrix} 1 & 2 & 3 \\ 0 & j & 2j \\ 0 & 0 & j \end{bmatrix} \begin{bmatrix} P & Q & R \\ 0 & S & T \\ 0 & 0 & U \end{bmatrix} = \begin{bmatrix} 1 & 0 & 0 \\ 0 & 1 & 0 \\ 0 & 0 & 1 \end{bmatrix} = U \tag{23.13}$$

† L. Fox, H. D. Huskey, and J. H. Wilkinson, Notes on the Solution of Algebraic Linear Simultaneous Equations, *Quarterly Journal of Mechanics and Applied Mathematics*, vol. 1, pp. 149–173, 1948.

From (23.13) we find that $P = 1$, $Q = 2j$, $R = -j$, $S = -j$, $T = 2j$, $U = -j$. The required inverse of $[M]$ is given by

$$[M]^{-1} = [L]_t^{-1}[L]^{-1} = \begin{bmatrix} 1 & 2j & -j \\ 0 & -j & 2j \\ 0 & 0 & -j \end{bmatrix} \begin{bmatrix} 1 & 0 & 0 \\ 2j & -j & 0 \\ -j & 2j & -j \end{bmatrix} = \begin{bmatrix} -4 & 4 & -1 \\ 4 & -5 & 2 \\ -1 & 2 & -1 \end{bmatrix}$$

(23.14)

The work involved in the Choleski method appears formidable in the above example but in practice can be greatly reduced. The method can be extended to the computations of inverses of unsymmetric matrices.

24 SYLVESTER'S THEOREM

A very important and useful theorem of matrix algebra is known as *Sylvester's theorem*. This theorem states that, if the n eigenvalues μ_1, μ_2, . . . , μ_n of the square matrix $[M]$ are all distinct and $P([M])$ is a polynomial in $[M]$ of the following form:

$$P([M]) = c_0[M]^k + c_1[M]^{k-1} + \cdots + c_{k-1}[M] + c_k U \tag{24.1}$$

where U is the nth-order unit matrix and the c_k's are constants, then the polynomial $P([M])$ may be expressed in the following form:

$$P([M]) = \sum_{r=1}^{n} P(\mu_r) Z_0(\mu_r) \tag{24.2}$$

where

$$Z_0(\mu_r) = \frac{[A(\mu_r)]}{\Delta'(\mu_r)} \tag{24.3}$$

In (24.3) $\Delta(\mu)$ is the characteristic function of $[M]$ as defined by (19.2); $[A(\mu)]$ is the adjoint matrix of the characteristic matrix of $[M]$ defined by

$$[A(\mu)] = \Delta(\mu)[\mu U - M]^{-1} \tag{24.4}$$

and

$$\Delta'(\mu_r) = \left(\frac{d\Delta}{d\mu}\right)_{\mu = \mu_r} \tag{24.5}$$

A proof of Sylvester's theorem as well as the confluent form of the theorem in the case that the matrix $[M]$ does not have distinct eigenvalues will be found in the book by Frazer, Duncan, and Collar.[†]

[†] R. A. Frazer, W. J. Duncan, and A. R. Collar, "Elementary Matrices," pp. 78–79, Cambridge University Press, New York, 1938.

Sylvester's theorem is very useful for simplifying the calculations of powers and functions of matrices. For example, let it be required to find the value of $[M]^{256}$, where

$$[M] = \begin{bmatrix} 1 & 0 \\ 0 & 3 \end{bmatrix} \tag{24.6}$$

To do this, consider the special polynomial

$$P([M]) = [M]^{256} \tag{24.7}$$

and apply the results of Sylvester's theorem. In this special case we have

$$[\mu U - M] = \begin{bmatrix} \mu - 1 & 0 \\ 0 & \mu - 3 \end{bmatrix} \tag{24.8}$$

and

$$\Delta(\mu) = (\mu - 1)(\mu - 3) = \mu^2 - 4\mu + 3 \tag{24.9}$$

The eigenvalues of $[M]$ are $\mu_1 = 1$ and $\mu_2 = 3$. $[A(\mu)]$ is now given by

$$[A(\mu)] = \begin{bmatrix} \mu - 3 & 0 \\ 0 & \mu - 1 \end{bmatrix} \tag{24.10}$$

and

$$\Delta'(\mu) = 2\mu - 4 \tag{24.11}$$

Hence we have

$$Z_0(1) = \begin{bmatrix} 1 & 0 \\ 0 & 0 \end{bmatrix} \qquad Z_0(3) = \begin{bmatrix} 0 & 0 \\ 0 & 1 \end{bmatrix} \tag{24.12}$$

Therefore, by Sylvester's theorem, we have

$$\begin{bmatrix} 1 & 0 \\ 0 & 3 \end{bmatrix}^{256} = 1^{256} \begin{bmatrix} 1 & 0 \\ 0 & 0 \end{bmatrix} + 3^{256} \begin{bmatrix} 0 & 0 \\ 0 & 1 \end{bmatrix} = \begin{bmatrix} 1 & 0 \\ 0 & 3^{256} \end{bmatrix} \tag{24.13}$$

$$\approx 3^{256} \begin{bmatrix} 0 & 0 \\ 0 & 1 \end{bmatrix}$$

As the last step above suggests, Sylvester's theorem is very useful in obtaining the approximate value of a matrix raised to a high power. Thus, if μ_m is the *largest* eigenvalue of the square matrix $[M]$, then, if k is very large, the term $P(\mu_m)Z_0(\mu_m)$ will dominate the right-hand member of (24.2) and we have the following *approximate* value of $[M]^k$:

$$[M]^k = \frac{\mu_m^k [A(\mu_m)]}{\Delta'(\mu_m)} \qquad \text{for } k \text{ very large} \tag{24.14}$$

This expression is useful in certain applications of matrix algebra to physical problems.

25 POWER SERIES OF MATRICES; FUNCTIONS OF MATRICES

A series of the type

$$S([M]) = c_0 U + c_1[M] + c_2[M]^2 + c_3[M]^3 + \cdots = \sum_{i=0}^{\infty} c_i[M]^i \quad (25.1)$$

where $[M]$ is a square matrix of order n, the coefficients c_i are ordinary numbers, and $[M]^0 = U$, the nth-order unit matrix, is called a *power series of matrices*. The general question of the convergence of such series will not be discussed here; however, if $[M]$ is a square matrix of order n, it is clear that by the use of the Cayley-Hamilton theorem (Sec. 22) it is possible to reduce any matrix polynomial of $[M]$ to a polynomial of maximum degree $n - 1$ regardless of the degree of the original polynomial. Hence, if the series (25.1) converges,† it may be expressed in the following form:

$$S([M]) = \sum_{r=1}^{r=n} S(\mu_r) Z_0(\mu_r) \quad (25.2)$$

where

$$Z_0(\mu_r) = \frac{[A(\mu_r)]}{\Delta'(\mu_r)} \quad (25.3)$$

as a consequence of Sylvester's theorem (Sec. 24). For example, it can be shown that the series

$$\exp[M] = \sum_{k=0}^{\infty} \frac{[M]^k}{k!} \qquad [M]^0 = U \quad (25.4)$$

converges for *any* square matrix $[M]$. In particular, suppose that the matrix $[M]$ is the following second-order matrix:

$$[M] = \begin{bmatrix} 1 & 0 \\ 0 & 2 \end{bmatrix} \quad (25.5)$$

The characteristic equation of this matrix is

$$\Delta(\mu) = \begin{vmatrix} \mu - 1 & 0 \\ 0 & \mu - 2 \end{vmatrix} = (\mu - 1)(\mu - 2) = \mu^2 - 3\mu + 2 = 0 \quad (25.6)$$

so that the eigenvalues of $[M]$ are $\mu_1 = 1$, $\mu_2 = 2$. The matrix $[A(\mu)]$ is now

$$[A(\mu)] = \begin{bmatrix} \mu - 2 & 0 \\ 0 & \mu - 1 \end{bmatrix} \quad (25.7)$$

Since $\Delta'(1) = -1$ and $\Delta'(2) = +1$, we have as a consequence of (25.2)

$$\exp[M] = e^2 \begin{bmatrix} 0 & 0 \\ 0 & 1 \end{bmatrix} - e \begin{bmatrix} -1 & 0 \\ 0 & 0 \end{bmatrix} = \begin{bmatrix} e & 0 \\ 0 & e \end{bmatrix} \quad (25.8)$$

---

† Convergence is guaranteed provided $S(\mu_r)$ converges for each eigenvalue. Cf. Frazer, Duncan, and Collar, 1957, p. 81 (see References).

Other functions of matrices such as trigonometric, hyperbolic, and logarithmic functions are defined by the same power series that define these functions for scalar values of their arguments and by replacing these scalar arguments by square matrices. For example, we have

$$\sin [M] = [M] - \frac{[M]^3}{3!} + \frac{[M]^5}{5!} - \cdots \tag{25.9}$$

$$\cos [M] = U - \frac{[M]^2}{2!} + \frac{[M]^4}{4!} - \cdots \tag{25.10}$$

$$\sinh [M] = [M] + \frac{[M]^3}{3!} + \frac{[M]^5}{5!} + \cdots \tag{25.11}$$

$$\cosh [M] = U + \frac{[M]^2}{2!} + \frac{[M]}{4!} + \cdots \tag{25.12}$$

Although every scalar power series has its matrix analogue, the matrix power series have, in general, more complex properties. The properties of the matrix functions may be deduced from their series expansions. For example, if t is a scalar variable and $[u]$ is a square matrix of constants, we have

$$\exp [u] t = [u] + \frac{[u]}{1!} t + [u]^2 \frac{t^2}{2!} + [u]^3 \frac{t^3}{3!} + \cdots \tag{25.13}$$

The series obtained by the term-by-term differentiation of (25.13) is

$$\frac{d}{dt} \exp [u] t = [u] + [u]^2 t + [u]^3 \frac{t^2}{2!} + \cdots$$
$$= [u] \exp [u] t \tag{25.14}$$

In the same way, it may be shown that, if

$$\sin [u] t = [u] t - [u]^3 \frac{t^3}{3!} + [u]^5 \frac{t^5}{5!} - \cdots \tag{25.15}$$

then

$$\frac{d}{dt} \sin [u] t = [u] - [u]^3 \frac{t^2}{2!} + [u]^5 \frac{t^4}{4!} - \cdots$$
$$= [u] \cos [u] t \tag{25.16}$$

26   ALTERNATE METHOD OF EVALUATING FUNCTIONS OF MATRICES

In Sec. 25 we saw how Sylvester's theorem could be used to evaluate functions of matrices. This process, though useful, is sometimes tedious in its application. In this section an alternate method is presented which is generally more convenient to use.

In Sec. 22 it was seen how any matrix polynomial of order n could be reduced to a polynomial of order $n-1$ by applying the Cayley-Hamilton theorem. Since any function of a matrix of order n may be represented by an infinite series, the Cayley-Hamilton theorem allows us to assert that any matrix function of order n may be expressed in terms of a polynomial of order $n-1$; that is,

$$F([A]) = a_0[u] + a_1[A] + a_2[A]^2 + \cdots + a_{n-1}[A]^{n-1} \qquad (26.1)$$

where $[A]$ is of order n.

The coefficients a_i in (26.1) are unknown, and the problem is to determine them. Resorting again to the Cayley-Hamilton theorem, we assert that for each eigenvalue of $[A]$, μ_i, we must have

$$F(\mu_i) = a_1 + a_2 \mu_i + \cdots + a_{n-1} \mu_i^{n-1} \qquad i = 1, 2, \ldots, n \qquad (26.2)$$

Equation (26.2) yields n equations for the n unknowns a_i. Employing matrix notation, (26.2) may be expressed as

$$\begin{Bmatrix} F(\mu_1) \\ F(\mu_2) \\ \vdots \\ F(\mu_n) \end{Bmatrix} = \begin{bmatrix} 1 & \mu_1 & \cdots & \mu_1^{n-1} \\ 1 & \mu_2 & \cdots & \mu_2^{n-1} \\ \cdots & \cdots & \cdots & \cdots \\ 1 & \mu_n & \cdots & \mu_n^{u-1} \end{bmatrix} \begin{Bmatrix} a_0 \\ a_1 \\ \vdots \\ a_{n-1} \end{Bmatrix} \qquad (26.3)$$

or more simply

$$\{F_i\} = [V]\{a\} \qquad (26.4)$$

where the element F_i denotes $F(\mu_i)$ and $[V]$ is the Vandermonde matrix of eigenvalues.

As long as all the eigenvalues are distinct, the matrix $[V]$ has an inverse and hence we may solve for the unknowns a_i:

$$\{a\} = [V]^{-1}\{F_i\} \qquad (26.5)$$

As a first example let us evaluate

$$e^{[M]} \qquad [M] = \begin{bmatrix} 1 & 0 \\ 0 & 2 \end{bmatrix}$$

which was done in Sec. 25 by application of Sylvester's theorem. According to (26.1), we may write

$$e^{[M]} = a_0[U] + a_1[M] \qquad (26.6)$$

The eigenvalues of $[M]$ are 1 and 2; thus by (26.2) we have

$$\begin{aligned} e^1 &= a_0 + a_1 \\ e^2 &= a_0 + 2a_1 \end{aligned} \qquad (26.7)$$

or

$$\begin{Bmatrix} e \\ e^2 \end{Bmatrix} = \begin{bmatrix} 1 & 1 \\ 1 & 2 \end{bmatrix} \begin{Bmatrix} a_0 \\ a_1 \end{Bmatrix} \qquad (26.8)$$

Inverting and solving for a_0 and a_1 yields

$$\begin{Bmatrix} a_0 \\ a_1 \end{Bmatrix} = \begin{bmatrix} 1 & 1 \\ 1 & 2 \end{bmatrix}^{-1} \begin{Bmatrix} e \\ e^2 \end{Bmatrix} = \begin{bmatrix} 2 & -1 \\ -1 & 2 \end{bmatrix} \begin{Bmatrix} e \\ e^2 \end{Bmatrix} \qquad (26.9)$$

$$a_0 = 2e - e^2 \qquad a_1 = -e + e^2$$

Substituting into (26.6) gives

$$e^{[M]} = \begin{bmatrix} e & 0 \\ 0 & e^2 \end{bmatrix} \qquad (26.10)$$

As a second example consider the problem of evaluating

$$\sin [A] \qquad [A] = \begin{bmatrix} 1 & 20 & 0 \\ -1 & 7 & 1 \\ 3 & 0 & -2 \end{bmatrix} \qquad (26.11)$$

The reader may verify that the eigenvalues of A are 1, 2, and 3, respectively. By (26.1) we may write

$$\sin [A] = a_0[U] + a_1[A] + a_2[A]^2 \qquad (26.12)$$

Substituting into (26.2) yields

$$\sin 1 = a_0 + a_1 + a_2$$
$$\sin 2 = a_0 + 2a_1 + 4a_2$$
$$\sin 3 = a_0 + 3a_1 + 9a_2$$

or in the form of (26.3)

$$\begin{Bmatrix} \sin 1 \\ \sin 2 \\ \sin 3 \end{Bmatrix} = \begin{bmatrix} 1 & 1 & 1 \\ 1 & 2 & 4 \\ 1 & 3 & 9 \end{bmatrix} \begin{Bmatrix} a_0 \\ a_1 \\ a_2 \end{Bmatrix} \qquad (26.13)$$

Solving for the vector $\{a\}$ gives

$$\begin{Bmatrix} a_0 \\ a_1 \\ a_2 \end{Bmatrix} = \tfrac{1}{2} \begin{bmatrix} 6 & -6 & 2 \\ -5 & 8 & -3 \\ 1 & -2 & 1 \end{bmatrix} \begin{Bmatrix} \sin 1 \\ \sin 2 \\ \sin 3 \end{Bmatrix} \qquad (26.14)$$

from which a_0, a_1, and a_2 may be found.   Substituting these values into (26.12) yields the final result:

$$\sin [A] = \begin{bmatrix} -9 & 30 & 10 \\ 0 & 0 & 0 \\ -9 & 30 & 10 \end{bmatrix} \sin 1 + \begin{bmatrix} 20 & -80 & -20 \\ 1 & -4 & -1 \\ 15 & -60 & -15 \end{bmatrix} \sin 2$$

$$+ \begin{bmatrix} -10 & 50 & 10 \\ -1 & 5 & 1 \\ -6 & 30 & 6 \end{bmatrix} \sin 3 \qquad (26.15)$$

27 DIFFERENTIATION AND INTEGRATION OF MATRICES

Up to this point we have considered the elements of the matrices as constants, which is the most common situation. However, in the theory of matrix differential equations, cases frequently arise wherein the elements of the matrices are functions of the independent variable, e.g., time.

If, for example, we consider a matrix whose elements are functions of time, then we can define the derivative and integral of such a matrix as follows:

$$\frac{d}{dt}[A(t)] = \frac{d}{dt}[a_{ij}(t)] = \left[\frac{d}{dt}a_{ij}(t)\right] \qquad (27.1)$$

$$\int [B(t)]\,dt = \int [b_{ij}(t)]\,dt = \left[\int b_{ij}(t)\,dt\right] \qquad (27.2)$$

In other words, the derivative or integral of a matrix is the matrix of the derivatives or integrals of the individual elements. The requirement for the existence of the derivative or integral of a matrix is simply that the derivative or integral of every element must exist.

28 ASSOCIATION OF MATRICES WITH LINEAR DIFFERENTIAL EQUATIONS

A very important field of application of the theory of matrices is the solution of systems of linear differential equations. Matrix notation and operations with matrices lead to concise formulations of otherwise cumbersome procedures. This conciseness is very important in many practical problems that involve several dependent variables. Such problems arise in dynamics, the theory of vibrations of structures and machinery, the oscillations of electrical circuits, electric filters, and many other branches of physics and engineering.

In order to apply matrix notation to a set of differential equations, it is advisable to express the equations in forms that involve only the first-order differential coefficients. Any set of differential equations can be easily changed into another set that involves only first-order differential coefficients by writing the highest-order coefficient as the differential coefficient of a new dependent variable. All the other differential coefficients are regarded as separate dependent variables, and, for each new variable introduced, another equation must be added to the set. As a very simple example, consider the following equation:

$$\frac{d^2x}{dt^2} + x = 0 \qquad (28.1)$$

In order to express this as a set of first-order equations, let $x = x_1$ and

$$\frac{dx_1}{dt} = x_2 \qquad (28.2)$$

With this notation (28.1) becomes

$$\frac{dx_2}{dt} = -x_1 \tag{28.3}$$

Therefore the original second-order differential equation (28.1) may be written in the following matrix form:

$$\frac{d}{dt}\begin{bmatrix} x_1 \\ x_2 \end{bmatrix} = \begin{bmatrix} 0 & 1 \\ -1 & 0 \end{bmatrix}\begin{bmatrix} x_1 \\ x_2 \end{bmatrix} \tag{28.4}$$

This is a special case of the first-order matrix differential equation

$$\frac{d}{dt}\{x\} = [A]\{x\} \tag{28.5}$$

where $\{x\}$ is in general a column vector of the n elements $x_1, x_2, \ldots, x_n$ and $[A]$ is a square matrix of the nth order. A general set of homogeneous linear differential equations may be expressed in the form (28.5). In general the elements a_{ij} of the matrix $[A]$ are functions of the independent variable t; however, in many special cases of practical importance the elements of $[A]$ are constants.

It will be left as an exercise for the reader to show that the general linear nth-order differential equation

$$x^n + a_{n-1}x^{n-1} + \cdots + a_1 x' + a_0 x = 0 \tag{28.6}$$

where

$$x^k = \frac{d^k x}{dt^k}$$

and its associated set of initial conditions

$$x(0) = x_0 \qquad x'(0) = x_0' \qquad \cdots \qquad x^{n-1}(0) = x_0^{n-1} \tag{28.7}$$

may be transformed into the linear first-order matrix differential equation

$$\{x\}' = [A]\{x\} \qquad \{x\}_{t=0} = \{x_0\} \tag{28.8}$$

by the transformation

$$x = x_1 \qquad x_1' = x_2 \qquad x_2' = x_3 \qquad \cdots \tag{28.9}$$

In (28.8) we have

$$\{x\} = \begin{Bmatrix} x_1 \\ x_2 \\ \vdots \\ x_n \end{Bmatrix} \qquad [A] = \begin{bmatrix} 0 & 1 & 0 & \cdots & 0 \\ 0 & 0 & 1 & \cdots & 0 \\ \cdot & \cdot & \cdot & \cdots & \cdot \\ 0 & 0 & 0 & \cdots & 1 \\ -a_0 & -a_1 & -a_2 & \cdots & -a_{n-1} \end{bmatrix} \tag{28.10}$$

For those cases where $[A]$ is a constant matrix there are two convenient forms for the solution of (28.8). By differentiation one may verify that the solution may be expressed in the convenient matrix form

$$\{x\} = e^{[A]t}\{x_0\} \tag{28.11}$$

This solution is particularly convenient in cases where numerical solutions are desired.

The second form of the solution is commonly called an *eigenvector expansion* and may be derived in the following way. Consider the linear transformation from $\{x\}$ to a new dependent vector $\{y\}$ by means of the modal matrix of $[A]$, $[M]$; that is, let

$$\{x\} = [M]\{y\} \tag{28.12}$$

Substituting this relation into (28.8) and premultiplying by $[M]^{-1}$ yields

$$\{y\}' = [M]^{-1}[A][M]\{y\} \tag{28.13}$$

Since $[M]$ is the modal matrix of $[A]$, the term $[M]^{-1}[A][M]$ is a diagonal matrix with the eigenvalues of $[A]$ on the main diagonal:

$$\{y\}' = \begin{bmatrix} \ddots & & \\ & \lambda_i & \\ & & \ddots \end{bmatrix}\{y\} \tag{28.14}$$

The result is that the transformed system is uncoupled, that is, (28.14) may be represented by the single scalar differential equation

$$y'_k = \lambda_k y_k \qquad k = 1, 2, \ldots, n \tag{28.15}$$

whose general solution is

$$y_k = c_k e^{\lambda_k t} \tag{28.16}$$

where c_k is an arbitrary constant. Returning to (28.14), the general solution may be written in the form

$$\{y\} = \{c_k e^{\lambda_k t}\} \tag{28.17}$$

From (28.12) the general solution to (28.8) now becomes

$$\{x\} = [M]\{c_k e^{\lambda_k t}\} \tag{28.18}$$

From Sec. 20 the modal matrix is defined as

$$[M] = [\{x\}_1, \{x\}_2, \ldots, \{x\}_n] \tag{28.19}$$

where $\{x\}_k$ is the kth eigenvector of $[A]$. Substituting (28.19) into (28.18) and expanding gives the general solution to (28.8) in terms of an eigenvector expansion:

$$\{x\} = \sum_{k=1}^{n} c_k e^{\lambda_k t}\{x\}_k \tag{28.20}$$

This equation shows that the general solution may be expressed as a linear combination of the eigenvectors. By applying the initial conditions the n constants c_k may be evaluated from (28.20):

$$\{x_0\} = \sum_{k=1}^{n} c_k \{x\}_k = [M]\{c\} \tag{28.21}$$

where

$$\{c\} = \begin{Bmatrix} c_1 \\ c_2 \\ \vdots \\ c_n \end{Bmatrix}$$

and therefore

$$\{c\} = [M]^{-1}\{x_0\} \tag{28.22}$$

As an example of this method consider (28.1) subject to the initial conditions:

$$x(0) = 0 \qquad \frac{dx}{dt}(0) = 1$$

The matrix form of this equation is given in (28.5), and the initial condition vector is

$$\{x_0\} = \begin{Bmatrix} 0 \\ 1 \end{Bmatrix} \tag{28.23}$$

The eigenvalues and eigenvectors of $[A]$ are

$$\lambda_1 = j \qquad \{x\}_1 = \begin{Bmatrix} 1 \\ j \end{Bmatrix}$$

$$\lambda_2 = -j \qquad \{x\}_2 = \begin{Bmatrix} 1 \\ -j \end{Bmatrix}$$

Using (28.20), the general solution has the form

$$\{x\} = c_1 e^{Jt} \begin{Bmatrix} 1 \\ j \end{Bmatrix} + c_2 e^{-Jt} \begin{Bmatrix} 1 \\ -j \end{Bmatrix} \tag{28.24}$$

Applying (28.22) to determine the arbitrary constants yields

$$\begin{Bmatrix} c_1 \\ c_2 \end{Bmatrix} = \begin{bmatrix} 1 & 1 \\ j & -j \end{bmatrix}^{-1} \begin{Bmatrix} 0 \\ 1 \end{Bmatrix} = \frac{1}{2j} \begin{Bmatrix} 1 \\ -1 \end{Bmatrix}$$

or

$$c_1 = \frac{1}{2j} \qquad c_2 = -\frac{1}{2j}$$

and the particular solution is

$$\{x\} = \frac{e^{Jt}}{2j} \begin{Bmatrix} 1 \\ j \end{Bmatrix} - \frac{e^{-Jt}}{2j} \begin{Bmatrix} 1 \\ -j \end{Bmatrix} \tag{28.25}$$

Since $x_1 = x$, the solution to (28.1) is obtained from the first row of (28.25):

$$x = \frac{1}{2j}(e^{Jt} - e^{-Jt}) = \sin t \qquad (28.26)$$

The only restriction on (28.20) is that the coefficient matrix must have distinct eigenvalues. Whenever multiple eigenvalues occur, a modified approach must be employed since all of the eigenvectors associated with the multiple root cannot be determined by usual methods. When this situation occurs, the alternate method is based on the method of undetermined coefficients.

For convenience, consider that the first eigenvalue of an nth-order system has a multiplicity of 3: that is, consider

$$\{x\}' = [A]\{x\} \qquad \{x\}_{t=0} = \{x_0\} \qquad (28.27)$$

where $[A]$ is nth order and

$$\lambda_1 = \lambda_2 = \lambda_3 = \lambda \qquad \lambda_4, \lambda_5, \ldots, \lambda_n \qquad \text{distinct}$$

To solve (28.27) assume a solution in the form†

$$\{x\} = (\{x\}_1 + t\{x\}_2 + t^2\{x\}_3)e^{\lambda t} + \sum_{k=4}^{n} e^{\lambda_k t}\{x\}_k \qquad (28.28)$$

Substituting this assumed solution into (28.27) gives

$$\lambda(\{x\}_1 + t\{x\}_2 + t^2\{x\}_3)e^{\lambda t} + (\{x\}_2 + 2t\{x\}_3)e^{\lambda t} + \sum_{k=4}^{n} c_k \lambda_k e^{\lambda_k t}\{x\}_k$$

$$= [A](\{x\}_1 + t\{x\}_2 + t^2\{x\}_3)e^{\lambda t} + \sum_{k=4}^{n} c_k e^{\lambda_k t}[A]\{x\}_k \quad (28.29)$$

By employing the definition of eigenvectors and eigenvalues, the summations on each side of the equality may be canceled. The remaining expression becomes

$$\lambda(\{x\}_1 + t\{x\}_2 + t^2\{x\}_3) + \{x\}_2 + 2t\{x\}_3 = [A](\{x\}_1 + t\{x\}_2 + t^2\{x\}_3)$$
$$(28.30)$$

Equating like powers of t yields

$$\lambda\{x_3\} = [A]\{x\}_3$$
$$\lambda\{x\}_2 + 2\{x\}_3 = [A]\{x\}_2 \qquad (28.31)$$
$$\lambda\{x\}_1 + \{x\}_2 = [A]\{x\}_1$$

which may be solved sequentially for the three unknown eigenvectors.

---

† Arbitrary constants have been omitted from the terms involving the unknown eigenvectors since at most an eigenvector is known up to an arbitrary constant.

Once the unknown eigenvectors have been evaluated, the general solution becomes

$$\{x\} = (c_1\{x\}_1 + c_2 t\{x\}_2 + c_3 t^2\{x\}_3) e^{\lambda t} + \sum_{k=4}^{n} c_k e^{\lambda_k t}\{x\}_k \tag{28.32}$$

for which the n arbitrary constants may be evaluated in the usual manner by application of the initial conditions.

29 METHOD OF PEANO-BAKER

When the matrix $[A]$ is a function of the independent variable t, the set of differential equations (28.7) may be integrated by means of the Peano-Baker method.† The principle of this method of integration is quite simple. Let the initial conditions of the system be such that

$$\{x\} = \{x_0\} \qquad \text{at } t = 0 \tag{29.1}$$

where $\{x_0\}$ is a column vector of initial values of the elements of $\{x\}$. Direct integration of (28.8) gives the following integral equation:

$$\{x\} = \{x_0\} + \int_0^t [A(t')]\{x\}\, dt' \tag{29.2}$$

where t' is a subsidiary variable of integration. This integral equation may be solved by repeated substitutions of $\{x\}$ from the left-hand member of (29.2) into the integral. If the integral operator is introduced

$$Q = \int_0^t (\quad) dt' \tag{29.3}$$

then (29.2) may be written in the following form:

$$\{x\} = \{x_0\} + Q[A]\{x\} \tag{29.4}$$

By repeated substitution, the following equation is obtained:

$$\{x\} = ([U] + Q[A] + Q[A]\, Q[A] + Q[A]\, Q[A]\, Q[A] + \cdots)\{x_0\} \tag{29.5}$$

Let

$$G([A]) = [U] + Q[A] + Q[A]\, Q[A] + \cdots \tag{29.6}$$

This expression has been called the *matrizant of* $[A]$. In the series (29.6) $[U]$ is the unit matrix of order n and the second term is the integral of $[A]$ taken between the limits 0 and t. To obtain $Q[A]\,Q[A]$, $[A]$ and $Q[A]$ are multiplied in the order $[A]\,Q[A]$ and the product is integrated between 0 and t. The remaining terms are formed in the same manner. If the elements of the matrix $[A(t)]$ remain bounded in the range 0 to t, it may be shown that the

† See E. L. Ince, "Ordinary Differential Equations," pp. 408–415, Longmans, Green & Co., Ltd., London, 1927.

series (29.6) is absolutely and uniformly convergent and that it defines the square matrix $G[A]$ in a unique manner.

The solution of the set of differential equations (28.8) is given by

$$\{x\} = G[A]\{x_0\} \tag{29.7}$$

If the elements of the matrix $[A]$ are constants, we have, by direct integration of the series (29.6),

$$G[A] = [U] + [A]\frac{t}{1!} + [A]^2\frac{t^2}{2!} + \cdots + [A]^n\frac{t^n}{n!} + \cdots$$
$$= e^{[A]t} \tag{29.8}$$

Therefore, if $[A]$ is a matrix of constants, the solution of (28.8) is

$$\{x\} = e^{[A]t}\{x_0\} \tag{29.9}$$

The interpretation of $e^{[A]t}$, for purposes of calculation, may be made by the use of Sylvester's theorem (24.2). To illustrate the general procedure, consider the second-order case in which the matrix $[A]$ has the form

$$[A] = \begin{bmatrix} a_{11} & a_{12} \\ a_{21} & a_{22} \end{bmatrix} \tag{29.10}$$

In this case the characteristic equation of $[A]$ is given by

$$\lambda^2 - (a_{11} + a_{22})\lambda + (a_{11}a_{22} - a_{12}a_{21}) = 0 \tag{29.11}$$

Let the two roots of (29.11) or the eigenvalues of (29.10) be given by

$$\lambda_1 = a + b \qquad \lambda_2 = a - b \tag{29.12}$$

By the use of Sylvester's theorem (24.2), the function may be expressed in the following form:

$$e^{[A]t} = \begin{bmatrix} A_{11} & A_{12} \\ A_{21} & A_{22} \end{bmatrix} \tag{29.13}$$

where

$$A_{11} = \left[\frac{a_{11} - a}{b}\sinh(bt) + \cosh(bt)\right]e^{at}$$

$$A_{12} = \frac{a_{12}}{b}e^{at}\sinh bt$$

$$A_{21} = \frac{a_{21}}{b}e^{at}\sinh bt \tag{29.14}$$

$$A_{22} = \left[\frac{a_{22} - a}{b}\sinh(bt) + \cosh(bt)\right]e^{at}$$

To illustrate this method of solution by an example of technical importance, consider the differential equations

$$L\frac{di}{dt} + Ri + \frac{q}{C} = 0 \qquad \frac{dq}{dt} = i \tag{29.15}$$

These differential equations govern the oscillation of a series electric circuit having a resistance R, an inductance L, and a capacitance C, all in series (see Chap. 5). The variable q is the charge on the capacitance, and i is the current that flows in the circuit. The free oscillations of such a circuit are completely determined when the initial charge q_0 and the initial current i_0 of the circuit at $t = 0$ are specified. If we let $x_1 = q$ and $x_2 = i$, Eqs. (29.15) may be written in the standard matrix form:

$$\frac{d}{dt}\begin{Bmatrix} x_1 \\ x_2 \end{Bmatrix} = \begin{bmatrix} 0 & 1 \\ \dfrac{-1}{CL} & \dfrac{-R}{L} \end{bmatrix}\begin{Bmatrix} x_1 \\ x_2 \end{Bmatrix} = [A]\{x\} \tag{29.16}$$

The characteristic equation of $[A]$ is, in this case,

$$\lambda^2 + \frac{R}{L}\lambda + \frac{1}{CL} = 0 \tag{29.17}$$

In this case

$$a = \frac{-R}{2L} \qquad b = \left(a^2 - \frac{1}{LC}\right)^{\frac{1}{2}} \tag{29.18}$$

The oscillations of the circuit are given by the following matrix equation:

$$\begin{Bmatrix} q \\ i \end{Bmatrix} = \begin{bmatrix} A_{11} & A_{12} \\ A_{21} & A_{22} \end{bmatrix}\begin{Bmatrix} q_0 \\ i_0 \end{Bmatrix} \tag{29.19}$$

where the elements A_{ij} are given by (29.14). If b is imaginary so that $b = j\omega$, then we have

$$\frac{\sinh bt}{b} = \frac{\sinh j\omega t}{j\omega} = \frac{\sin \omega t}{\omega} \tag{29.20}$$
$$\cosh bt = \cosh j\omega t = \cos \omega t$$

so that the hyperbolic functions are replaced by trigonometric functions and the circuit oscillates. If $b = 0$, the circuit is critically damped and $\sinh(bt)/b$ becomes indeterminate. When this is evaluated by L'Hospital's rule for indeterminate forms, it is seen that it must be replaced by t. The result (29.19) is thus seen to be a convenient expression that gives the current and charge of the circuit in terms of the initial conditions in concise form.

DIFFERENTIAL EQUATIONS WITH VARIABLE COEFFICIENTS

The Peano-Baker method is a convenient method for the integration of homogeneous linear differential equations with variable coefficients. As an example of the general procedure, consider the differential equation

$$\frac{d^2 y}{dt^2} + ty = 0 \tag{29.21}$$

This equation is known as Stokes's equation in the mathematical literature of the diffraction and reflection of electromagnetic waves and in quantum mechanics.†

In order to reduce this equation to the form of (28.8), let

$$x_1 = -\frac{dy}{dt} \qquad x_2 = y \tag{29.22}$$

The system

$$\frac{d}{dt}\begin{Bmatrix} x_1 \\ x_2 \end{Bmatrix} = \begin{bmatrix} 0 & t \\ -1 & 0 \end{bmatrix}\begin{Bmatrix} x_1 \\ x_2 \end{Bmatrix} = [A]\{x\} \tag{29.23}$$

replaces the original equation (29.21). Let the initial condition vector at $t = 0$ be

$$\{x_0\} = \begin{Bmatrix} x_1 \\ x_2 \end{Bmatrix}_{t=0} \tag{29.24}$$

The matrizant $G[A]$ is now computed by means of the series (29.6):

$$Q[A] = Q\begin{bmatrix} 0 & t \\ -1 & 0 \end{bmatrix} = \begin{bmatrix} 0 & \dfrac{t^2}{2} \\ -t & 0 \end{bmatrix} \tag{29.25}$$

$$Q[A]Q[A] = \begin{bmatrix} \dfrac{-t^3}{3} & 0 \\ 0 & \dfrac{-t^3}{(2)(3)} \end{bmatrix} \tag{29.26}$$

$$Q[A]Q[A]Q[A] = \begin{bmatrix} 0 & \dfrac{-t^5}{(2)(3)(5)} \\ \dfrac{t^4}{(3)(4)} & 0 \end{bmatrix} \tag{29.27}$$

† See "Tables of the Modified Hankel Functions of Order One-third," Harvard University Press, Cambridge, Mass., 1945.

The first few terms of the matrizant are, by (29.6),

$$
G[A] = \begin{bmatrix} 1 - \dfrac{t^3}{3} + \dfrac{t^6}{(3)(4)(6)} - \cdots \\ -t + \dfrac{t^4}{(3)(4)} - \dfrac{t^7}{(3)(4)(6)(7)} + \cdots \end{bmatrix}
$$

$$
\begin{aligned}
&+ \dfrac{t^2}{2} - \dfrac{t^5}{(2)(3)(5)} + \dfrac{t^8}{(2)(3)(5)(6)(8)} - \cdots \\
&+ 1 - \dfrac{t^3}{(2)(3)} + \dfrac{t^6}{(2)(3)(5)(6)} - \cdots
\end{aligned} \Bigg] \tag{29.28}
$$

The elements of the matrizant $G[A]$ are series expansions of the Bessel functions of order $\frac{1}{3}$ and $-\frac{2}{3}$. The solution of Stokes's equation may be expressed in terms of the initial conditions in the form

$$
\begin{Bmatrix} x_1 \\ x_2 \end{Bmatrix} = \begin{bmatrix} G_{11} & G_{12} \\ G_{21} & G_{22} \end{bmatrix} \begin{Bmatrix} x_1 \\ x_2 \end{Bmatrix}_0 \tag{29.29}
$$

30 ADJOINT METHOD

Besides the homogeneous first-order matrix differential equation considered in the last section, systems of nonhomogeneous equations are also of considerable interest. A typical equation of this type may be written in the following form:

$$
\frac{d}{dt}\{x\} = \{\dot{x}\} = [A]\{x\} + \{f\} \qquad \{x\}_{t=0} = \{x_0\} \tag{30.1}
$$

where $[A]$ = coefficient matrix

$\{f\}$ = vector of forcing functions

$\{x\}$ = unknown vector

As previously considered, both $[A]$ and $\{f\}$ may be functions of the independent variable.

The method of solving the system (30.1) presented in this section is the *adjoint method*. This method yields a solution to (30.1) by seeking a solution to the following set of *adjoint equations:*

$$
[\dot{Y}] = -[A]'[Y] \qquad [Y]_{t=0} = [U] \tag{30.2}
$$

where $[Y]$ is an unknown matrix. The reasons for choosing the particular form for the adjoint system given in (30.2) will become clear as the process of solution unfolds.

The method proceeds as follows: premultiply (30.1) by $[Y]'$ and postmultiply the transpose of (30.2) by $\{x\}$. Adding the two resulting equations together yields

$$
[\dot{Y}]'\{x\} + [Y]'\{\dot{x}\} = [Y]'\{f\} \tag{30.3}
$$

However, the two terms on the left are the derivative of a product and therefore (30.3) may be rewritten as

$$\frac{d}{dt}([Y]'\{x\}) = [Y]'\{f\} \tag{30.4}$$

Integrating this equation between the limits of 0 and t and recalling that

$$[Y]_{t=0} = [U]$$
$$\{x\}_{t=0} = \{x_0\}$$

gives

$$[Y]'\{x\} - \{x_0\} = \int_0^t [Y]'\{f\}\, dt \tag{30.5}$$

Hence, the solution to (30.1) is

$$\{x\} = ([Y]')^{-1}\{x_0\} + ([Y]')^{-1} \int_0^t [Y]'\{f\}\, dt \tag{30.6}$$

where $[Y]$ is the solution to the adjoint equation (30.2), which is homogeneous with convenient initial conditions.

If the coefficient matrix $[A]$ is a constant matrix, the solution to (30.1) as given by (30.6) may be reduced to the following result:

$$\{x\} = e^{[A]t}\{x_0\} + e^{[A]t} \int_0^t e^{-[A]t'}\{f(t')\}\, dt' \tag{30.7}$$

As an example consider the equation

$$\{\dot{x}\} = \begin{bmatrix} 1 & 0 \\ 0 & 2 \end{bmatrix}\{x\} + \begin{Bmatrix} e^t \\ 0 \end{Bmatrix} \qquad \{x_0\} = \begin{Bmatrix} 1 \\ -1 \end{Bmatrix} \tag{30.8}$$

For this equation

$$[A] = \begin{bmatrix} 1 & 0 \\ 0 & 2 \end{bmatrix}$$

Applying the method of Sec. 26 yields

$$e^{[A]t} = \begin{bmatrix} e^t & 0 \\ 0 & e^{2t} \end{bmatrix} \qquad \{f\} = \begin{Bmatrix} e^t \\ 0 \end{Bmatrix} \tag{30.9}$$

Substituting this result into (30.11) gives

$$\{x\} = \begin{bmatrix} e^t & 0 \\ 0 & e^{2t} \end{bmatrix}\begin{Bmatrix} 1 \\ -1 \end{Bmatrix} + \begin{bmatrix} e^t & 0 \\ 0 & e^{2t} \end{bmatrix} \int_0^t \begin{bmatrix} e^{-t} & 0 \\ 0 & e^{-2t} \end{bmatrix}\begin{Bmatrix} e^t \\ 0 \end{Bmatrix}\, dt$$

$$= \begin{Bmatrix} (1+t)e^t \\ -e^{2t} \end{Bmatrix} \tag{30.10}$$

31   EXISTENCE AND UNIQUENESS OF SOLUTIONS TO MATRIX DIFFERENTIAL EQUATIONS

In this section we shall simply state the conditions required for the existence and uniqueness of solutions to matrix differential equations without showing the proof.

Before presenting these conditions further notation will be introduced in order to simplify the equations.   Instead of writing the matrix differential equations as we have thus far, the following notation will be used:

$$[L]\{x\} = \{f\} \tag{31.1}$$

where $[L]$ = an operator matrix

$\{f\}$ = a matrix of forcing functions

This notation may be used to represent (30.1) as follows:

$$[L] = [U]\frac{d}{dt} - [A]$$

Likewise (28.10) may be represented by (31.1) if

$$[L] = [U]\frac{d^2}{dt^2} + [A]\frac{d}{dt} + [B]$$

and

$$\{f\} = 0$$

Hence any set of simultaneous differential equations may be written in the compact form of (31.1).

Besides (31.1) the corresponding homogeneous equation must be considered; that is,

$$[L]\{x\} = 0 \tag{31.2}$$

The conditions for the existence and uniqueness of a solution to (31.1) may be stated as follows:

1. A solution to (31.1) is unique if and only if the corresponding homogeneous equation (31.2) possesses only a trivial solution.
2. A solution to (31.1) exists† if and only if the vector $\{f\}$ is orthogonal to all solutions of the homogeneous adjoint equation $[L]'\{x\} = \{0\}$.

† To be precise, the operator matrix must have a closed range.   Cf. Friedman, 1956 (see References).

32   LINEAR EQUATIONS WITH PERIODIC COEFFICIENTS

The mathematical analysis of a considerable variety of physical problems leads to a formulation involving a differential equation that may be reduced to the following form:

$$\frac{d^2 y}{dt^2} + F(t)y = 0 \tag{32.1}$$

where $F(t)$ is a single-valued periodic function of fundamental period T. An equation of this type is known as *Hill's equation* in the mathematical literature. If a change of variable of the form

$$t = nT + \tau \qquad 0 \leqslant \tau \leqslant T \qquad n = 0, 1, 2, 3, \ldots \tag{32.2}$$

is introduced into the Hill equation (32.1), the equation is transformed into

$$\frac{d^2 y}{d\tau^2} + F(\tau)y = 0 \tag{32.3}$$

as a consequence of the periodicity of $F(t)$. Hence Hill's equation is *invariant* under the change of variable (32.2), and it is thus apparent that it is necessary only to obtain a solution of (32.1) in a fundamental interval $0 \leqslant t \leqslant T$. The final values of y and $\dot{y}$ at the end of one interval of the variation of $F(t)$ are the *initial* values of y and $\dot{y}$ in the following interval.

Let $u_1(t)$ and $u_2(t)$ be two linearly independent solutions of Hill's equation (32.1) in the fundamental interval $0 \leqslant t \leqslant T$. Let $y = x_1$ and $\dot{y} = x_2$. Then we have

$$\left. \begin{aligned} x_1(t) &= K_1 u_1(t) + K_2 u_2(t) \\ x_2(t) &= K_1 \dot{u}_1(t) + K_2 \dot{u}_2(t) \end{aligned} \right\} \qquad \text{for } 0 \leqslant t \leqslant T \tag{32.4}$$

where K_1 and K_2 are arbitrary constants. Equations (32.4) may be written in the following convenient matrix form:

$$\begin{Bmatrix} x_1(t) \\ x_2(t) \end{Bmatrix} = \begin{bmatrix} u_1(t) & u_2(t) \\ \dot{u}_1(t) & \dot{u}_2(t) \end{bmatrix} \begin{Bmatrix} K_1 \\ K_2 \end{Bmatrix} \tag{32.5}$$

The following notation will be introduced in order to simplify the writing:

$$[x(t)] = \begin{Bmatrix} x_1(t) \\ x_2(t) \end{Bmatrix} \qquad [K] = \begin{Bmatrix} K_1 \\ K_2 \end{Bmatrix}$$

$$[u(t)] = \begin{bmatrix} u_1(t) & u_2(t) \\ \dot{u}_1(t) & \dot{u}_2(t) \end{bmatrix} = \begin{bmatrix} u_{11}(t) & u_{12}(t) \\ u_{21}(t) & u_{22}(t) \end{bmatrix} \tag{32.6}$$

In this notation (32.5) may be expressed in the following concise form:

$$\{x(t)\} = [u(t)]\{K\} \tag{32.7}$$

The determinant

$$\det [u(t)] = u_1 \dot{u}_2 - \dot{u}_1 u_2 = W_0 \tag{32.8}$$

is called the *Wronskian* of the two solutions $u_1(t)$ and $u_2(t)$. It is a constant in the fundamental interval $0 \leqslant t \leqslant T$, as may be shown by differentiation. Since the two solutions $u_1(t)$ and $u_2(t)$ are linearly independent, $W_0 \neq 0$. The matrix $[u(t)]$ is therefore nonsingular and has the following inverse:

$$[u(t)]^{-1} = \frac{1}{W_0}\begin{bmatrix} u_{22}(t) & -u_{12}(t) \\ -u_{21}(t) & u_{11}(t) \end{bmatrix} \tag{32.9}$$

At $t = 0$, (32.7) reduces to

$$\{x(0)\} = [u(0)]\{K\} \tag{32.10}$$

If this is premultiplied by $[u(0)]^{-1}$, the result may be written in the following form:

$$K = [u(0)]^{-1} x(0) \tag{32.11}$$

This equation determines the column of arbitrary constants in terms of the *initial* conditions. If (32.11) is now substituted into (32.7), the result is

$$\{x(t)\} = [u(t)][u(0)]^{-1}\{x(0)\} \tag{32.12}$$

At the end of the fundamental interval, $t = T$, and the solution is given in terms of the initial conditions by the equation

$$\{x(T)\} = [u(T)][u(0)]^{-1}\{x(0)\} \tag{32.13}$$

CASCADING THE SOLUTION

Since, during the second fundamental interval, Hill's equation has the same form as it has during the first interval, the relation expressing the initial and final values of $[x(t)]$ in this interval has the same form as that given in (32.13). Therefore, at the end of the second cycle of the variation of $F(t)$ we have

$$\{x(2T)\} = [u(T)][u(0)]^{-1}\{x(T)\} \tag{32.14}$$

Introduce the following notation:

$$[M] = [u(T)][u(0)]^{-1} = \begin{bmatrix} A & B \\ C & D \end{bmatrix} \tag{32.15}$$

It is easy to see that, at the end of n periods of the variation of $F(t)$, the solution of Hill's equation is given by

$$\{x(nT)\} = [M]^n\{x(0)\} \qquad n = 0, 1, 2, 3, \ldots \tag{32.16}$$

and the solution within the $(n + 1)$st interval is given by

$$\{x(nT + \tau)\} = [u(\tau)][u(0)]^{-1}\{x(nT)\} \qquad 0 \leqslant \tau \leqslant T \tag{32.17}$$

This gives the solution of Hill's equation at any time $t > 0$ in terms of the *initial* conditions and two linearly independent solutions in the fundamental interval $0 \leqslant t \leqslant T$. The solution involves the computation of positive integral powers of the matrix $[M]$.

COMPUTATION OF INTEGRAL POWERS OF $[M]$

As a consequence of the constancy of the Wronskian determinant W_0 given by (32.8), the determinant of the matrix $[M]$ is equal to unity. We thus have

$$\det [M] = 1 \tag{32.18}$$

The characteristic equation of $[M]$ is

$$u^2 - (A + D)\mu + 1 = 0 \tag{32.19}$$

If $A + D \neq \pm 2$, the eigenvalues of $[M]$ are given by

$$\mu_1 = e^a \qquad \mu_2 = e^{-a} \qquad \cosh a = \frac{A + D}{2} \tag{32.20}$$

and in this case $[M]$ has distinct eigenvalues. By Sylvester's theorem (24.2) the nth power of $[M]$ may easily be computed. The result is

$$[M]^n = \frac{1}{S_1} \begin{bmatrix} S_{n+1} - DS_n & BS_n \\ CS_n & S_{n+1} - AS_n \end{bmatrix} \qquad n = 0, 1, 2, 3, \dots \tag{32.21}$$

where

$$S_n = \sinh an \qquad n = 0, 1, 2, 3, \dots \tag{32.22}$$

Equation (32.21) gives the proper form for $[M]^n$ only when $A + D \neq \pm 2$. When $A + D = \pm 2$, the two eigenvalues of $[M]$ are not distinct and a confluent form of Sylvester's theorem must be used to compute $[M]^n$.†

33 MATRIX SOLUTION OF THE HILL-MEISSNER EQUATION

If the function $F(t)$ has the form of the rectangular ripple of Fig. 33.1, Eq. (32.1) is known as the *Hill-Meissner* equation. The Hill-Meissner equation is of

† See L. A. Pipes, Matrix Solutions of Equations of the Mathieu-Hill Type, *Journal of Applied Physics*, vol. 24, no. 7, pp. 902–910, July, 1953.

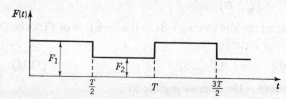

Fig. 33.1

importance in the study of the oscillations of the driving rods of electric locomotives, and its solution is easily effected by the use of matrix algebra. In this case Eq. (32.1) reduces to two equations of the form

$$\frac{d^2 y}{dt^2} + F_1 y = 0 \qquad 0 < t < \frac{T}{2}$$

$$\frac{d^2 y}{dt^2} + F_2 y = 0 \qquad \frac{T}{2} < t < T$$

$$(33.1)$$

The solution of Eqs. (33.1) at $t = T$ may be expressed in terms of the *initial* values of y and $\dot{y}$ at $t = 0$ by the equation

$$\begin{bmatrix} x_1(T) \\ x_2(T) \end{bmatrix} = \begin{bmatrix} \cos\theta_2 & \dfrac{\sin}{w_2}\theta_2 \\ -w_2\sin\theta_2 & \cos\theta_2 \end{bmatrix} \begin{bmatrix} \cos\theta_1 & \dfrac{\sin}{w_1}\theta_1 \\ -w_1\sin\theta_1 & \cos\theta_1 \end{bmatrix} \begin{bmatrix} x_1(0) \\ x_2(0) \end{bmatrix} \qquad (33.2)$$

where $x_1 = y$, $x_2 = \dot{y}$, and

$$\theta_1 = \frac{T}{2}\sqrt{F_1} \qquad w_1 = \sqrt{F_1}$$

$$\theta_2 = \frac{T}{2}\sqrt{F_2} \qquad w_2 = \sqrt{F_2}$$

$$(33.3)$$

If the indicated matrix multiplication in (33.2) is performed, the result may be written in the following form:

$$\begin{bmatrix} x_1(T) \\ x_2(T) \end{bmatrix} = \begin{bmatrix} A & B \\ C & D \end{bmatrix} \begin{bmatrix} x_1(0) \\ x_2(0) \end{bmatrix} = [M][x(0)] \qquad (33.4)$$

In this case the matrix $[M]$ has the form

$$[M] = \begin{bmatrix} A & B \\ C & D \end{bmatrix} \qquad (33.5)$$

where

$$A = \cos\theta_1\cos\theta_2 - \sin\theta_1\sin\theta_2\frac{w_1}{w_2}$$

$$B = \frac{\cos\theta_2\sin\theta_1}{w_1} - \frac{\cos\theta_1\sin\theta_2}{w_2}$$

$$C = -(w_2\cos\theta_1\sin\theta_2 + w_1\cos\theta_2\sin\theta_1)$$

$$D = \cos\theta_1\cos\theta_2 - \sin\theta_1\sin\theta_2\frac{w_2}{w_1}$$

$$(33.6)$$

The eigenvalues of $[M]$ are given by

$$\mu_1 = e^a \qquad \mu_2 = e^{-a} \qquad (33.7)$$

where

$$\cosh a = \frac{A+D}{2} = \cos \theta_1 \cos \theta_2 - \left(\frac{w_1}{w_2} + \frac{w_2}{w_1}\right) \sin \theta_1 \sin \theta_2 \qquad (33.8)$$

The values of y and $\dot{y}$ after the end of n complete periods of the square-wave variation are given by

$$[x(nT)] = [M]^n[x(0)] \qquad n = 0, 1, 2, 3, \ldots \qquad (33.9)$$

THE STABILITY OF THE SOLUTION

The nature of the solution of the Hill-Meissner equation depends on the form taken by the powers of the matrix $[M]$, as given by (32.22). There are two separate cases to be considered. If $A + D > 2$, a must have a positive real part and the elements of $[M]^n$ increase exponentially with the number of periods, n. This leads to an *unstable* solution in physical applications. If, on the other hand, $A + D < 2$, a must be a pure imaginary number and the elements of $[M]^n$ oscillate with n. This leads to *stable* solutions in physical applications. The dividing line leads to the case $A + D = 2$. In this case the eigenvalues of $[M]$ are repeated and both have the value $\mu_1 = \mu_2 = 1$. The form (32.22) is no longer applicable and must be replaced by one obtained from a confluent form of Sylvester's theorem. In general the condition $A + D = 2$ leads to *instability* in physical problems. The question of stability may be summarized by the following expressions:

$$
\begin{aligned}
A + D &> 2 \qquad \text{unstable} \\
A + D &< 2 \qquad \text{stable} \\
A + D &= 2 \qquad \text{unstable}
\end{aligned}
\qquad (33.10)
$$

In any special cases the conditions for stability or instability may be deduced from (33.8).

34 THE USE OF MATRICES TO DETERMINE THE ROOTS OF ALGEBRAIC EQUATIONS

The properties of matrices may be used to determine the roots of algebraic equations. As a simple example, consider the equation

$$z^3 + a_1 z^2 + a_2 z + a_3 = 0 \qquad (34.1)$$

Let the following third-order matrix be constructed:

$$[C] = \begin{bmatrix} 0 & 1 & 0 \\ 0 & 0 & 1 \\ -a_3 & -a_2 & -a_1 \end{bmatrix} \qquad (34.2)$$

This matrix is called the *companion matrix* of the cubic equation (34.1). It is easy to show that the characteristic equation of the companion matrix

$[C]$ is the cubic equation. Therefore the eigenvalues of $[C]$ are the roots of (34.1). The general algebraic equation of the nth degree

$$z^n + a_1 z^{n-1} + a_2 z^{n-2} + \cdots + a_n = 0 \tag{34.3}$$

has a companion matrix of the nth order

$$C = \begin{bmatrix} 0 & 1 & 0 & \cdots & 0 \\ 0 & 0 & 1 & \cdots & 0 \\ 0 & 0 & 0 & \cdots & 0 \\ \multicolumn{5}{c}{\dotfill} \\ 0 & 0 & 0 & \cdots & 1 \\ -a_n & -a_{n-1} & -a_{n-2} & \cdots & -a_1 \end{bmatrix} \tag{34.4}$$

The eigenvalues of $[C]$, μ_1, μ_2, μ_3, . . . , μ_n, are the roots of the equation (34.3). If the various roots of the nth-degree equation (34.3) are real and distinct and if μ_1 is the root having the largest absolute value, then it can be shown by Sylvester's theorem (24.2) that for a *sufficiently large m* we have

$$[C]^{m+1} \approx \mu_1 [C]^m \tag{34.5}$$

That is, for a sufficiently large-power m, a further multiplication of $[C]^m$ by $[C]$ yields a new matrix whose elements are μ_1 times the elements of $[C]^m$. This multiple, μ_1, is the dominant root of (34.3). A practical procedure for extracting the dominant root of an equation is to choose an arbitrary column vector of the nth order (x) and set up the following sequence:

$$[C](x) = (x)_1$$
$$[C]^m(x) = (x)_m \qquad [C](x)_1 = (x)_2 = [C]^2(x) \tag{34.6}$$

Then, for a sufficiently large m, we have

$$[C](x)_m = \mu_1 (x)_m \tag{34.7}$$

so that after a sufficient number of multiplications one more multiplication merely multiplies all the elements of the vector $(x)_m$ by a common factor. This common factor μ_1 is the largest root of the equation and the largest eigenvalue of the companion matrix. As a simple numerical example, consider the following cubic equation:

$$z^3 - 16z^2 + 68z - 80 = 0 \tag{34.8}$$

The companion matrix of this equation is

$$[C] = \begin{bmatrix} 0 & 1 & 0 \\ 0 & 0 & 1 \\ 80 & -68 & 16 \end{bmatrix} \tag{34.9}$$

We now choose the *arbitrary* column vector

$$(x) = \begin{bmatrix} 1 \\ 1 \\ 1 \end{bmatrix}$$

By direct multiplication we obtain

$$[C](x) = \begin{bmatrix} 1 \\ 1 \\ 28 \end{bmatrix} = (x)_1 \qquad [C](x)_1 = \begin{bmatrix} 1 \\ 28 \\ 460 \end{bmatrix} = (x)_2$$

$$[C](x)_2 = \begin{bmatrix} 28 \\ 460 \\ 5{,}540 \end{bmatrix} = 28 \begin{bmatrix} 1 \\ 16.4 \\ 198 \end{bmatrix} = (x)_3 \tag{34.10}$$

The factor 28 may be discarded since we are interested only in the ratios of similar elements after successive multiplication. Let $\doteqdot$ denote the fact that a factor has been discarded, and proceed.

$$[C](x)_3 \doteqdot \begin{bmatrix} 16.4 \\ 198 \\ 2{,}125 \end{bmatrix} \doteqdot \begin{bmatrix} 1 \\ 12.1 \\ 130 \end{bmatrix} \doteqdot (x)_4$$

$$[C](x)_4 \doteqdot \begin{bmatrix} 1 \\ 10.7 \\ 111 \end{bmatrix} \doteqdot (x)_5 \qquad [C](x)_5 \doteqdot \begin{bmatrix} 1 \\ 10.4 \\ 105 \end{bmatrix} = (x)_6 \tag{34.11}$$

$$[C](x)_6 \doteqdot \begin{bmatrix} 1 \\ 10.1 \\ 101 \end{bmatrix} = (x)_7 \qquad [C](x)_7 \doteqdot \begin{bmatrix} 1 \\ 10 \\ 100 \end{bmatrix}$$

Further multiplications only multiply the elements of the vectors by a factor of 10. Hence the *dominant* root is $\mu_1 = 10$. After having determined the dominant root, it may be divided out by dividing the polynomial $f(z) = z^3 - 16z^2 + 68z - 80$ by $z - 10$. This procedure may be performed by the synthetic division

$$
\begin{array}{r}
1 - 16 + 68 - 80 \quad \lfloor 10 \\
+\, 10 - 60 \quad\quad\quad \\
\hline
1 -6 +8 \quad\, \lfloor 0 \\
\end{array}
$$

Since the remainder is zero, the dominant root is exactly $\mu_1 = 10$. The polynomial $f(z)$ is seen to have the factors $f(z) = (z - 10)(z^2 - 6z + 8)$. The roots of the quadratic $z^2 - 6z + 8 = 0$ are $\mu_2 = 4$ and $\mu_3 = 2$. The eigenvalues of the companion matrix (34.9) are therefore $\mu_1 = 10$, $\mu_2 = 4$, $\mu_3 = 2$. If only the *smallest* root of the equation is desired, the substitution $z = 1/y$ may be made in Eq. (34.3) and the companion matrix of the resulting equation in y

may be constructed.   The iteration scheme (34.7) then yields the *largest* root in y and therefore the *smallest* root of the original equation.

The method of matrix iteration may be used to determine repeated and complex roots of algebraic equations.†

PROBLEMS

1. Solve the following system of equations by the use of determinants:

$$x - 2y + 3z = 2$$
$$2x - 3z = 3$$
$$x + y + z = 6$$

2. Evaluate the determinant

$$D = \begin{vmatrix} 3 & 1 & -1 & 2 & 1 \\ -2 & 3 & 1 & 4 & 3 \\ 1 & 4 & 2 & 3 & 1 \\ 5 & -2 & -3 & 5 & -1 \\ -1 & 1 & 2 & 3 & 2 \end{vmatrix}$$

3. Solve the following system of equations:

$$9x + 3y + 4z + 2w = 28$$
$$3x + 5z - 4w = 12$$
$$y + z + w = 5$$
$$6x - y + 3w = 19$$

4. Prove Cramer's rule for the general case of n equations.

5. Given

$$[a] = \begin{bmatrix} 1 & 2 \\ 3 & 4 \end{bmatrix}$$

compute $[a]^{10}$.

6. Given

$$[a] = \begin{bmatrix} 8 & 4 & 3 \\ 2 & 1 & 1 \\ 1 & 2 & 1 \end{bmatrix}$$

compute $[a]^{-2}$.

7. Find the characteristic equation and the eigenvalues of the following matrix:

$$[a] = \begin{bmatrix} 1 & 1 & 1 \\ 1 & 2 & 2 \\ 1 & 2 & 3 \end{bmatrix}$$

8. Show that the matrix

$$[A] = \begin{bmatrix} a & h \\ h & b \end{bmatrix}$$

is transformed to the diagonal form

$$B = T_\theta [A] T_\theta'$$

† See L. A. Pipes, Matrix Solution of Polynomial Equations, *Journal of the Franklin Institute*, vol. 225, no. 4, pp. 437–454, April, 1938.

where T_θ is the following matrix:

$$T_\theta = \begin{bmatrix} \cos\theta & \sin\theta \\ -\sin\theta & \cos\theta \end{bmatrix}$$

and $\tan 2\theta = 2h/(a-b)$ (Jacobi's transformation).

9. Show that

$$\begin{bmatrix} \cos\theta & -\sin\theta \\ \sin\theta & \cos\theta \end{bmatrix} = \begin{bmatrix} 1 & -\tan\dfrac{\theta}{2} \\ \tan\dfrac{\theta}{2} & 1 \end{bmatrix} \begin{bmatrix} 1 & -\tan\dfrac{\theta}{2} \\ -\tan\dfrac{\theta}{2} & 1 \end{bmatrix}^{-1}$$

10. Show that the matrix

$$[J] = \begin{bmatrix} 0 & -1 \\ 1 & 0 \end{bmatrix}$$

has properties similar to those of the imaginary number $j = (-1)^{\frac{1}{2}}$. Show that $[J]^2 = -U$, where U is the second-order unit matrix, and that $[J]^3 = -[J]$, $[J]^4 = U$, $[J]^5 = [J]$, etc.

11. Determine the eigenvalues and eigenvectors of the matrix

$$[M] = \begin{bmatrix} 2 & -2 & 3 \\ 1 & 1 & 1 \\ 1 & 3 & -1 \end{bmatrix}$$

12. Determine the roots of the equation $z^3 - 16z^2 + 65z - 50 = 0$ by the matrix iteration method.

13. Solve the differential equation $d^2 y/dt^2 + w^2 y = 0$, subject to the conditions $y(0) = y_0$ and $\dot{y}(0) = v_0$, by the Peano-Baker method.

14. Show that (18.1) is always consistent.

15. Using the matrix method illustrated in Sec. 18, solve (18.3) for x_1 and x_3 in terms of x_2.

16. Find the value of the constant c for which the system

$$\begin{bmatrix} 1 & 2 & 0 \\ 1 & -1 & 1 \\ -2 & 2 & c \end{bmatrix} \{x\} = \{0\}$$

is consistent, and find the solution.

17. Derive (28.7).

18. Given (30.2) and (30.7), derive (30.8).

19. If $[A]$ is a constant matrix, show that (30.10) reduces to (30.11).

20. If $[A]$ is a diagonal matrix, show that $f([A])$ must equal $[f(a_{ii})]$; for example,

$$[A] = \begin{bmatrix} 1 & 0 & 0 \\ 0 & 2 & 0 \\ 0 & 0 & 3 \end{bmatrix} \qquad f([A]) = \begin{bmatrix} f(1) & 0 & 0 \\ 0 & f(2) & 0 \\ 0 & 0 & f(3) \end{bmatrix}$$

21. Verify that (30.14) is the solution to (30.12).

22. Solve the following ([A] is the matrix in Prob. 7 above):

$$\{\dot{x}\} = [A]\{x\} + \{f\}$$

$$\{f\} = \begin{Bmatrix} t \\ e^{-t} \\ 0 \end{Bmatrix} \qquad \{x_0\} = \begin{Bmatrix} 0 \\ 0 \\ 1 \end{Bmatrix}$$

23. Solve the following systems:

(a) $\{\dot{x}\} = \begin{bmatrix} 2 & 1 \\ 3 & 4 \end{bmatrix} \{x\} \qquad \{x_0\} = \begin{Bmatrix} 2 \\ 2 \end{Bmatrix}$

(b) $\{\dot{x}\} = \begin{bmatrix} 5 & 1 \\ 4 & 2 \end{bmatrix} \{x\} \qquad \{x_0\} = \begin{Bmatrix} 5 \\ 0 \end{Bmatrix}$

(c) $\{\dot{x}\} = \begin{bmatrix} 4 & 2 & -2 \\ -5 & 3 & 2 \\ -2 & 4 & 1 \end{bmatrix} \{x\} \qquad \{x_0\} = \begin{Bmatrix} 3 \\ 0 \\ 4 \end{Bmatrix}$

(Hochstadt, 1964)

24. Solve the following differential equations by reducing them to first-order matrix equations:

(a) $y'' - 3y' - 10y = 0$
$\quad y(0) = 3 \qquad y'(0) = 15$

(b) $y''' - 7y'' - 2y' + 40y = 0$
$\quad y(0) = 3 \qquad y'(0) = 7 \qquad y''(0) = 27$

(Hochstadt, 1964)

25. Find the eigenvalues and eigenvectors of the following matrices:

(a) $\begin{bmatrix} 4 & -2 \\ 1 & 1 \end{bmatrix}$ (b) $\begin{bmatrix} 1 & 2 \\ -8 & 11 \end{bmatrix}$ (c) $\begin{bmatrix} 1 & 0 \\ 2 & -1 \end{bmatrix}$

(d) $\begin{bmatrix} 13 & -3 & 5 \\ 0 & 4 & 0 \\ -15 & 9 & -7 \end{bmatrix}$ (e) $\begin{bmatrix} 15 & -4 & -3 \\ -10 & 12 & -6 \\ -20 & 4 & -2 \end{bmatrix}$

(Kreyszig, 1962)

REFERENCES

1931. Bôcher, M.: "Introduction to Higher Algebra," The Macmillan Company, New York.
1932. Turnbull, H. W., and A. C. Aitken: "An Introduction to the Theory of Canonical Matrices," Blackie and Son, Ltd., Glasgow.
1937. Pipes, L. A.: Matrices in Engineering, *Electrical Engineering*, vol. 56, pp. 1177–1190.
1938. Frazer, R. A., W. J. Duncan, and A. R. Collar: "Elementary Matrices," Cambridge University Press, New York.
1946. MacDuffee, C. C.: "The Theory of Matrices," Chelsea Publishing Co., New York.
1949. Wedderburn, J. H. M.: "Lectures on Matrices," J. W. Edwards, Publisher, Incorporated, Ann Arbor, Mich.
1950. Zurmuhl, R.: "Matrizen," Springer-Verlag OHG, Berlin.
1951. Kaufmann, A., and M. Denis-Papin: "Cours de calcul matriciel," Editions Albin Michel, Paris.
1953. Paige, L. J., and Olga Taussky (eds.): "Simultaneous Linear Equations and the Determination of Eigenvalues," National Bureau of Standards, Applied Mathematics Series 29, August.
1956. Friedman, B.: "Principles and Techniques of Applied Mathematics," John Wiley & Sons, Inc., New York.
1958. Hohn, F. E.: "Elementary Matrix Algebra," The Macmillan Company, New York.
1960. Bellman, R.: "Introduction to Matrix Analysis," McGraw-Hill Book Company, New York.

1962. Kreyszig, E.: "Advanced Engineering Mathematics," John Wiley & Sons, Inc., New York.

1963. Pipes, L. A.: "Matrix Methods for Engineers," Prentice-Hall, Inc., Englewood Cliffs, N.J.

1964. Hochstadt, H.: "Differential Equations," Holt, Rinehart and Winston, Inc., New York.

4
Laplace Transforms

1 INTRODUCTION

In the theory of operational calculus a mathematical technique that has proved to be a powerful method for solving differential equations and general system analysis is the Laplace transform.   Before we proceed to a detailed exposition of the theory of Laplace transforms, it is of some interest to list some of the better-known uses of the Laplace-transform theory in applied mathematics.

1. *The solution of ordinary differential equations with constant coefficients:* The Laplace transform is of use in solving linear differential equations with constant coefficients.   By its use the differential equation is transformed into an algebraic equation which is readily solved.   The solution of the original differential equation is then reduced to the calculation of inverse transforms usually involving the ratio of two polynomials.   The inverse transforms required may be obtained by consulting a table of Laplace transforms or by evaluating the inverse integral by the calculus of residues.

The Laplace-transform method is particularly well adapted to the solution of differential equations whose boundary conditions are specified at one point. The solution of differential equations involving functions of an impulsive type may be solved by the use of the Laplace transform in a very efficient manner. Typical fields of application are the following:

a. Transient and steady-state analysis of electrical circuits [4, 7–9, 11, 15, 16].†
b. Applications to dynamical problems (impact, mechanical vibrations, acoustics, etc.) [5–8, 10, 17].
c. Applications to structural problems (deflection of beams, columns, determination of Green's functions and influence functions) [6–10, 13].

2. *The solution of linear differential equations with variable coefficients:* By the use of the Laplace-transform theory a linear differential equation whose coefficients are polynomials in the independent variable t is transformed into a differential equation whose coefficients are polynomials in the transform variables. In many cases the transformed equation is simpler than the original, and it can be readily solved. The inverse transform of the solution of this equation gives the solution of the original equation [17, 18]. Van der Pol [11, 18] has shown how this may be done in the case of Bessel functions and Legendre and Laguerre polynomials.

3. *The solution of linear partial differential equations with constant coefficients:* One of the most important uses of the Laplace-transform theory is its use in the solution of linear partial differential equations with constant coefficients that have two or more independent variables. The procedure employed is to take the Laplace transform with respect to one of the independent variables and thus reduce the partial differential equation to an ordinary one or one having a smaller number of independent variables. The transformed equation is then solved subject to the boundary conditions involved, and the inverse transform of this solution gives the required result. Problems involving impulsive forces and wave propagation are particularly well adapted for solution by the Laplace-transform method. In many cases it is possible to express the operational solution of the partial differential equation involved in a series of exponential terms and then interpret the solution as a succession of traveling waves. Typical physical problems that may be solved by the procedure outlined above are the following:

a. Transients and steady-state analysis of heat conduction in solids [5, 6, 13, 16, 17].
b Vibrations of continuous mechanical systems [6–8, 13, 17].
c. Hydrodynamics and fluid flow [6–8, 19].

---

† In this chapter we shall deviate from our usual method of indicating references and denote them by bracketed numbers which are keyed to the references at the end of this chapter.

d. Transient analysis of electrical transmission lines and also of cables [4, 6, 11, 13, 15, 16].

e. Transient analysis of electrodynamic fields [5, 7].

f. Transient analysis of acoustical systems [7, 8].

g. Analysis of static deflections of continuous systems (strings, beams, plates) [6, 9, 17].

4. *The solution of linear difference and difference-differential equations:* The Laplace-transform theory is very useful in effecting the solution of linear difference or mixed linear-difference–differential equations with constant coefficients.   Difference equations arise in systems where the variables change by finite amounts.   Typical fields of application of difference equations are the following:

a. Electrical applications of difference equations [6, 9, 11, 13] (artificial lines, wave filters, network representation of transformer and machine windings, multistage amplifiers, surge generators, suspension insulators, cyclic switching operations).

b. Mechanical applications of difference equations [6, 7, 9, 17].

c. Applications to economics and finance [9] (annuities, mortgages, interest, amortization, growth of populations).

d. Systems with hereditary characteristics [20] (retarded control systems, servomechanisms).

5. *The solution of integral equations of the convolution, or Faltung, type:* Equations in which the unknown function occurs inside an integral or integral equations are encountered quite frequently in applied mathematics.   If the integral equation happens to be of the convolution, or Faltung, type, the use of the Laplace-transformation theory transforms the equation into an algebraic equation in the transform variable and it is readily solved.   Typical problems leading to this type of equation are the following:

a. Mechanical problems such as the tautochrone [14, 17].

b. Economic problems (mortality of equipment, etc.) [17].

c. Electrical circuits involving variable parameters [4].

d. Problems involving heredity characteristics or aftereffect [21].

6. *Application of the Laplace transform to the theory of prime numbers:* In a remarkable paper [22] B. van der Pol applied the methods of the Laplace-transform theory to the theory of prime numbers and demonstrated the ease and simplicity by which results obtained with considerable difficulty by using conventional methods are simply obtained by the Laplace-transform theory. It is demonstrated in this paper that the properties of many discontinuous

functions belonging to the theory of numbers may be easily obtained by considering their Laplace transforms.

7. *Evaluation of definite integrals:* Certain types of definite integrals may be evaluated very easily by the use of the Laplace-transform theory. As a consequence of this fact, it is easy to establish integral relations involving Bessel functions, Legendre polynomials, and other higher transcendental functions in a simple manner by the use of the Laplace-transform theory [11, 13, 14, 18].

8. *Derivation of asymptotic series:* In some applications of the Laplace-transform theory to the solution of differential equations, it frequently happens that the solution is required in the form of an asymptotic series. If it is possible to expand the transform of the solution in fractional powers of s, it is easy to obtain the asymptotic-series expansion of the solution by obtaining the inverse transform of each term. Heaviside [23] used this procedure quite extensively. Carson [4] has clarified the mathematical validity of the procedure and demonstrated the utility of the method in the solution of electrical-circuit problems.

9. *Derivation of power series:* If the Laplace transform of a function is known, then the power-series expansion of the function may be obtained by expanding its transform in inverse powers of s. This procedure was used extensively by Heaviside [23] and Carson [4] to compute inverse Laplace transforms.

10. *Derivation of Fourier series:* If the Laplace transform of a periodic function is obtained, then the evaluation of the inverse integral (1.3) gives the Fourier expansion of the function [7, 8, 23]. In certain cases this procedure is simpler than the conventional method of obtaining Fourier-series expansions of functions.

11. *Summation of power series:* It is frequently possible, when a power series is given, to sum it and obtain the function that it represents by the use of the Laplace-transform theory. This is done by taking the Laplace transform of each term of the given series and thus obtaining the series of the transform of the function represented by the original series. If the new series is a geometric or other well-known series, it may be summed and thus the Laplace transform of the function required is obtained. A calculation of the inverse transform then gives the required function. This procedure was frequently used by Heaviside [23] in order to sum series.

12. *The summation of Fourier series:* The summation of certain Fourier series involving periodic discontinuous functions may be effected by the use of the Laplace-transform theory. The method consists in taking the Laplace transform of each term of the trigonometric series, thus obtaining a series of rational functions. In many cases the series of rational functions may be summed to obtain the transform of the function represented by the original Fourier series. A computation of the inverse transform of this function then gives the sum of the Fourier series in closed form. Pipes [24] has used this

method to advantage in obtaining a graphical representation of the function defined by a given Fourier series. The method is particularly well suited to cases in which the given Fourier series represents a function whose graph is composed of straight lines or a repetition of curves of simple form.

13. *The use of multiple Laplace transforms to solve partial differential equations:* By the use of multiple Laplace transforms a partial differential equation and its associated boundary conditions can be transformed into an algebraic equation in n independent complex variables. This algebraic equation can be solved for the multiple transform of the solution of the original partial differential equation. Multiple inversion of this transform then gives the desired solution. The theory and application of multiple Laplace transforms has been the subject of recent publications by van der Pol and Bremmer [11], Jaeger [25], Voelker and Doetsch [26] and Estrin and Higgins [27].

14. *The solution of nonlinear ordinary differential equations:* The Laplace-transform theory may be used to reduce the labor involved in solving certain nonlinear differential equations. The procedures used are of an iterative nature by which the solution of the original nonlinear differential equation is reduced to the solution of a set of simultaneous linear differential equations with constant coefficients. The Laplace-transform theory greatly reduces the labor involved in solving this set of simultaneous equations [28–32].

THE LAPLACE TRANSFORM (NOTATION)

It is unfortunate that not all the literature of the Laplace-transform theory is written in the same notation. This state of affairs sometimes leads to confusion, and it is well to mention the important differences of the various notations in use.

a **The p-multiplied notation** Several authors [4, 5, 7, 8, 11–13, 15] have attempted to retain the operational forms of the Heaviside method and use the following notation,

$$g(p) = p \int_0^\infty e^{-pt} h(t)\, dt \qquad \mathrm{Re}\, p > 0 \tag{1.1}$$

where p is a parameter whose real part is positive. Then, if $h(t)$ is such a function of the variable t for which the integral (1.1) exists, the function $g(p)$ is said to be the p-multiplied transform of $h(t)$. The relation (1.1) is expressed concisely in the form

$$g(p) = Lh(t) \tag{1.2}$$

If the function $h(t)$ is zero for negative values of t and has other restrictions which are usually satisfied in the application to most physical problems (see Sec. 2), then it may be shown that Eq. (1.2) has an inverse of the following form:

$$h(t) = \frac{1}{2\pi j} \int_{c-j\infty}^{c+j\infty} \frac{e^{pt} g(p) \, dp}{p} \tag{1.3}$$

where $j = (-1)^{\frac{1}{2}}$ and c is a positive constant.   In this case we write concisely

$$h(t) = L^{-1} g(p) \tag{1.4}$$

and $h(t)$ is said to be the *inverse p-multiplied Laplace transform* of $g(p)$.

The p-multiplied transform notation has the following advantages:

1. The forms of the "Heaviside calculus" are retained.   These forms are of long standing and are widespread in use.
2. The transform of a constant in this notation is the same constant.
3. The Laplace transform of t^n is $n!/p^n$ in this notation.   Hence, if t and p are considered to have dimensions of t and t^{-1}, respectively, and if $h(t)$ and its transform can be expanded in absolutely convergent series, the corresponding terms are dimensionally identical.   This is very useful for purposes of checking.

b   **The unmultiplied notation**   Most mathematicians and several engineers use the more concise notation [6, 9, 10, 14, 16, 17]

$$g(s) = \int_0^\infty e^{-st} h(t) \, dt \tag{1.5}$$

and

$$h(t) = \frac{1}{2\pi j} \int_{c-j\infty}^{c+j\infty} e^{st} g(s) \, ds \tag{1.6}$$

instead of (1.1) and (1.3).   This notation has certain advantages in some cases and has by far become the most commonly used notation [9, 10, 17].   The reader should note that the letter p has been replaced by the letter s and the definitions of the Laplace transform and its inverse as given by (1.5) and (1.6) will be the ones used throughout this text.

Because of the various differences in notation, when consulting various treatises and tables of Laplace transforms, the student is cautioned to assure himself as to which notation a particular writer is using.

2   THE FOURIER-MELLIN THEOREM

In this section the Fourier-Mellin theorem that is the foundation of the modern operational calculus will be derived.   The Fourier-Mellin theorem may be derived from Fourier's integral theorem.   The derivation of this theorem is heuristic in nature.   For a rigorous derivation the references at the end of this chapter should be consulted.

It was shown in Appendix C, Sec. 9, that, if $F(t)$ is a function that has a finite number of maxima and minima and ordinary discontinuities, it may be expressed by the integral

$$F(t) = \int_{-\infty}^{\infty} ds \int_{-\infty}^{\infty} F(u) e^{2\pi js(t-u)} du \qquad (2.1)$$

If we let

$$2\pi s = v \qquad (2.2)$$

then (2.1) becomes

$$F(t) = \frac{1}{2\pi} \int_{-\infty}^{+\infty} dv \int_{-\infty}^{+\infty} F(u) e^{jv(t-u)} du \qquad (2.3)$$

For the integrals involved in (2.1) to converge uniformly, it is also necessary for the integral

$$I = \int_{-\infty}^{\infty} |F(t)| dt \qquad (2.4)$$

to exist.

Let us now assume that the function $F(t)$ has the property that

$$F(t) = 0 \qquad \text{for } t < 0 \qquad (2.5)$$

In this case the Fourier-integral representation of the function is

$$F(t) = \frac{1}{2\pi} \int_{-\infty}^{\infty} e^{jvt} dv \int_{0}^{\infty} F(u) e^{-jvu} du \qquad (2.6)$$

where the lower limit of the second integral is zero as a consequence of the condition (2.5).

Let us now consider the function $\phi(t)$ defined by the equation

$$\phi(t) = e^{-ct} F(t) \qquad (2.7)$$

where c is a positive constant. Substituting this into (2.6), we obtain

$$\phi(t) = e^{-ct} F(t) = \frac{1}{2\pi} \int_{-\infty}^{+\infty} e^{jvt} dv \int_{0}^{\infty} F(u) e^{-uc} e^{-jvu} du \qquad (2.8)$$

Let us now make the substitution

$$s = c + jv \qquad (2.9)$$

Then (2.8) may be written in the form

$$e^{-ct} F(t) = \frac{1}{2\pi j} \int_{c-j\infty}^{c+j\infty} e^{(s-c)t} ds \int_{0}^{\infty} F(u) e^{-su} du \qquad (2.10)$$

On dividing both members of (2.10) by e^{-ct} we have

$$F(t) = \frac{1}{2\pi j} \int_{c-j\infty}^{c+j\infty} e^{st} ds \int_{0}^{\infty} F(u) e^{-su} du \qquad (2.11)$$

This modified form is more general than the Fourier integral (2.3) because if

$$I = \int_0^\infty |F(t)| \, dt \tag{2.12}$$

does not exist but

$$I' = \int_0^\infty e^{-c_0 t} |F(t)| \, dt \tag{2.13}$$

exists for some $c_0 > 0$, then (2.11) is valid for some $c > c_0$.

If we now let

$$g(s) = \int_0^\infty F(u) e^{-su} \, du \tag{2.14}$$

we have by (2.11)

$$F(t) = \frac{1}{2\pi j} \int_{c-j\infty}^{c+j\infty} g(s) e^{st} \, ds \tag{2.15}$$

Since the value of a definite integral is independent of the variable of integration, it is convenient to use the letter t instead of the letter u in the integral (2.14) and write

$$g(s) = \int_0^\infty e^{-st} F(t) \, dt \tag{2.16}$$

Equations (2.15) and (2.16) are the Fourier-Mellin equations. They are the foundations of the modern operational calculus. A more precise statement of the Fourier-Mellin theorem is the following one:

Let $F(t)$ be an arbitrary function of the real variable t that has only a finite number of maxima and minima and discontinuities and whose value is zero for negative values of t. If

$$g(s) = \int_0^\infty e^{-st} F(t) \, dt \qquad \mathrm{Re}\, s \geqslant c > 0 \tag{2.17}$$

then

$$F(t) = \frac{1}{2\pi j} \int_{c-j\infty}^{c+j\infty} e^{st} g(s) \, ds \tag{2.18}$$

provided

$$\int_0^\infty e^{-ct} F(t) \, dt \tag{2.19}$$

converges absolutely.†

---

† A rigorous discussion and derivation of this theorem will be found in E. C. Titchmarsh "Introduction to the Theory of Fourier Integrals," Oxford University Press, New York, 1937.

It is convenient to use the notation

$$g(s) = LF(t) \tag{2.20}$$

to denote the functional relation between $g(s)$ and $F(t)$ expressed in (2.17). We then say that $g(s)$ is the *direct Laplace transform of* $F(t)$. The relation (2.18) is expressed conveniently by the notation

$$F(t) = L^{-1} g(s) \tag{2.21}$$

We then say that $F(t)$ is the *inverse Laplace transform of* $g(s)$.

3   THE FUNDAMENTAL RULES

In this section some very powerful and useful general theorems concerning operations on transforms will be established. These theorems are of great utility in the solution of differential equations, evaluation of integrals, and other procedures of applied mathematics.

Theorem I   The Laplace transform of a constant is the same constant divided by s, that is,

$$L(k) = \frac{k}{s} \tag{3.1}$$

where k is a constant. To prove this, we have from the fundamental definition of the direct Laplacian transform

$$L(k) = \int_0^\infty e^{-st} k \, dt = k \left[-\frac{e^{-st}}{s} \right]_0^\infty = \frac{k}{s} \tag{3.2}$$

The integral vanishes at the upper limit since by hypothesis $\operatorname{Re} s > 0$.

Theorem II

$$Lk\phi(t) = kL\phi(t) \qquad \text{where } k \text{ is a constant} \tag{3.3}$$

This may be proved in the following manner:

$$Lk\phi(t) = \int_0^\infty e^{-st} k\phi(t) \, dt = k \int_0^\infty e^{-st} \phi(t) \, dt = kL\phi(t) \tag{3.4}$$

Theorem III   If F is a continuous differentiable function and if F and dF/dt are L-transformable,

$$L\frac{dF}{dt} = sLF(t) - F(0) \tag{3.5}$$

This theorem is very useful in solving differential equations with constant coefficients.   To prove it, we have

$$L\frac{dF}{dt} = \int_0^\infty e^{-st}\frac{dF}{dt}\,dt = Fe^{-st}\Big|_0^\infty + s\int_0^\infty e^{-st}F\,dt = sLF - F(0) \qquad (3.6)$$

where the integration has been performed by parts.

Theorem IV   If F is continuous and has derivatives of orders 1, 2, . . . , n which are L-transformable,

$$L\frac{d^n F}{dt^n} = s^n LF - \sum_{k=0}^{n-1} F^{(k)}(0)\,s^{n-k-1} \qquad (3.7)$$

where

$$F^{(k)}(0) = \frac{d^k F}{dt^k} \text{ evaluated at } t = 0$$

This theorem is an extension of Theorem III.   To prove this, we have

$$L\frac{d^n F}{dt^n} = sL\frac{d^{n-1} F}{dt^{n-1}} - \left(\frac{d^{n-1} F}{dt^{n-1}}\right)_{t=0} \qquad (3.8)$$

by Theorem III.   By repeated applications of Theorem III we finally obtain the result (3.7).   If we let

$$F_r = \frac{d^r F}{dt^r} \qquad \text{at } t = 0 \qquad (3.9)$$

evaluated at $t = 0$, we have for the transforms of the first four derivatives

$$L\frac{dF}{dt} = sLF - F_0 \qquad (3.10)$$

$$L\frac{d^2 F}{dt^2} = s^2 LF - sF_0 - F_1 \qquad (3.11)$$

$$L\frac{d^3 F}{dt^3} = s^3 LF - s^2 F_0 - sF_1 - F_2 \qquad (3.12)$$

$$L\frac{d^4 F}{dt^4} = s^4 LF - s^3 F_0 - s^2 F_1 - sF_2 - F_3 \quad \text{etc.} \qquad (3.13)$$

These expressions are of use in transforming differential equations.

Theorem V   The Faltung theorem   This theorem is known in the literature as the *Faltung* or *convolution theorem*.   In the older literature of the operational calculus it is sometimes referred to as the *superposition theorem*. Let

$$LF_1(t) = g_1(s) \qquad (3.14)$$

$$LF_2(t) = g_2(s) \tag{3.15}$$

Then the theorem states that

$$L \int_0^t F_1(y) F_2(t-y)\, dy = L \int_0^t F_2(y) F_1(t-y)\, dy = g_1 g_2 \tag{3.16}$$

To prove this theorem, let

$$LF_3(t) = g_1(s) g_2(s) \tag{3.17}$$

Then, by the Fourier-Mellin formula, we have

$$F_3(t) = \frac{1}{2\pi j} \int_{c-j\infty}^{c+j\infty} g_1(s) g_2(s) e^{st}\, ds \tag{3.18}$$

However, by hypothesis,

$$g_2(s) = \int_0^\infty e^{-sy} F_2(y)\, dy \tag{3.19}$$

Therefore

$$F_3(t) = \frac{1}{2\pi j} \int_0^\infty F_2(y)\, dy \int_{c-j\infty}^{c+j\infty} g_1(s) e^{s(t-y)}\, ds \tag{3.20}$$

if we do not question reversing the order of integration.

However, we have

$$\frac{1}{2\pi j} \int_{c-j\infty}^{c+j\infty} g_1(s) e^{s(t-y)}\, ds = F_1(t-y) \tag{3.21}$$

Hence

$$F_3(t) = \int_0^\infty F_2(y) F_1(t-y)\, dy \tag{3.22}$$

Now, by hypothesis, $F_1(t) = 0$ if t is less than 0.

$$F_1(t-y) = 0 \qquad \text{for } y > t \tag{3.23}$$

Consequently the infinite limit of integration may be replaced by the limit t. Therefore we may write (3.22) in the form

$$F_3(t) = \int_0^t F_2(y) F_1(t-y)\, dy \tag{3.24}$$

and, by symmetry,

$$F_3(t) = \int_0^t F_1(y) F_2(t-y)\, dy \tag{3.25}$$

Corollary to Theorem V   By applying Theorem III to Theorem V, we have

$$L \frac{d}{dt} \int_0^t F_2(y) F_1(t-y)\, dy = s g_1 g_2 \tag{3.26}$$

Theorem VI If

$$LF(t) = g(s) \tag{3.27}$$

then

$$Le^{-at} F(t) = g(s + a) \tag{3.28}$$

Proof Let

$$Le^{-at} F(t) = \phi(s) \tag{3.29}$$

Therefore

$$\phi(s) = \int_0^\infty e^{-st} e^{-at} F(t) \, dt$$

$$= \int_0^\infty e^{-t(s+a)} F(t) \, dt \tag{3.30}$$

But, as a consequence of (3.27), we have

$$g(s) = \int_0^\infty e^{-st} F(t) \, dt \tag{3.31}$$

Hence

$$g(s + a) = \int_0^\infty e^{-t(s+a)} F(t) \, dt \tag{3.32}$$

Comparing this with (3.30), we have

$$\phi(s) = g(s + a) = Le^{-at} F(t) \tag{3.33}$$

Theorem VII First shifting theorem If

$$LF(t) = g(s) \tag{3.34}$$

then

$$L^{-1} e^{-ks} g(s) = \begin{cases} 0 & t < k \\ F(t - k) & t > k \end{cases} \quad \text{provided } k > 0 \tag{3.35}$$

Proof Let

$$L\phi(t) = e^{-ks} g(s) \tag{3.36}$$

Hence

$$\phi(t) = \frac{1}{2\pi j} \int_{c-j\infty}^{c+j\infty} e^{st} e^{-ks} g(s) \, ds \tag{3.37}$$

while

$$F(t) = \frac{1}{2\pi j} \int_{c-j\infty}^{c+j\infty} e^{st} g(s) \, ds \tag{3.38}$$

Comparing (3.37) and (3.38), we have

$$\phi(t) = F(t - k) \tag{3.39}$$

However, $F(t) = 0$ for $t < 0$. Hence $F(t - k) = 0$ for $t < k$, since by hypothesis $k > 0$. This proves the theorem.

Theorem VIII   Second shifting theorem   If

$$LF(t) = g(s) \tag{3.40}$$

then

$$L^{-1} e^{ks} g(s) = F(t + k) \qquad k > 0 \tag{3.41}$$

provided that

$$F(t) = 0 \qquad \text{for } 0 < t < k \tag{3.42}$$

Proof   Let

$$LF(t + k) \, \phi(s) \tag{3.43}$$

Hence

$$\phi(s) = \int_0^\infty e^{-st} F(t + k) \, dt \tag{3.44}$$

Now let us make the change in variable

$$t + k = y \tag{3.45}$$

Hence

$$\begin{aligned}
\phi(s) &= e^{sk} \int_k^\infty e^{-sy} F(y) \, dy \\
&= e^{sk} \int_0^\infty e^{-sy} F(y) \, dy - e^{sk} \int_0^k e^{-sy} F(y) \, dy \\
&= e^{sk} g(s) - e^{sk} \int_0^k e^{-sy} F(y) \, dy
\end{aligned} \tag{3.46}$$

Now, if $F(y) = 0$ for $0 < y < k$, then the second integral vanishes and we have

$$LF(t + k) = e^{ks} g(s) \tag{3.47}$$

and the theorem is proved.

Theorem IX   (Final value theorem)   If

$$LF(t) = g(s) \tag{3.48}$$

and $sg(s)$ is analytic on the axis of imaginaries and in the right half plane, then

$$\lim_{t \to \infty} F(t) = \lim_{s \to 0} sg(s) \tag{3.49}$$

Theorem X   (Initial value theorem)   If

$$LF(t) = g(s) \tag{3.50}$$

then

$$\lim_{t \to 0} F(t) = \lim_{s \to \infty} sg(s) \tag{3.51}$$

For proofs of these last two theorems, the reader is referred to Gardner and Barnes [9].

Theorem XI   Real indefinite integration   If

$$LF(t) = g(s) \tag{3.52}$$

and

$$\int F(t)\,dt$$

is Laplace transformable, then

$$L \int F(t)\,dt = \frac{1}{s}g(s) + \frac{1}{s}\int^0 F(t)\,dt \tag{3.53}$$

Proof   By the definition of the Laplace transform, we have

$$L \int F(t)\,dt = \int_0^\infty e^{-st}\left[\int F(t)\,dt\right]dt \tag{3.54}$$

Now integrating by parts yields

$$L \int F(t)\,dt = -\frac{1}{s}e^{-st}\int F(t)\,dt\,\Big|_0^\infty + \frac{1}{s}\int_0^\infty e^{-st}F(t)\,dt \tag{3.55}$$

Evaluating the indicated limits and using (3.52) gives

$$L \int F(t)\,dt = \frac{1}{s}g(s) + \frac{1}{s}\int^0 F(t)\,dt \tag{3.56}$$

where the term

$$\int^0 F(t)\,dt$$

denotes the initial value of the integral.

Theorem XII   Real definite integration   If

$$LF(t) = g(s) \tag{3.57}$$

and

$$\int_0^t F(t')\,dt' \tag{3.58}$$

is Laplace transformable, then

$$L \int_0^t F(t')\,dt' = \frac{1}{s}g(s) \tag{3.59}$$

Proof The proof of this theorem follows the same steps as in the proof of Theorem XI, except after integrating by parts we obtain

$$L \int_0^t F(t')\,dt' = -\frac{1}{s}e^{-st}\int_0^t F(t')\,dt' \Big|_0^\infty + \frac{1}{s}g(s) \tag{3.60}$$

Now at the upper limit the first term vanishes because of the exponential function.   At the lower limit the first term also vanishes because of the definite integral.   Hence only the second term is left, and it is the result stated in (3.59).

Following the same steps it may be shown that

$$L \int_0^t \int_0^{t'} F(t'')\,dt''\,dt' = \frac{1}{s^2}g(s) \tag{3.61}$$

where

$$LF(t) = g(s)$$

Theorem XIII Complex differentiation If

$$LF(t) = g(s) \tag{3.62}$$

then

$$LtF(t) = -\frac{dg(s)}{ds} \tag{3.63}$$

Proof To prove this theorem consider the definition of $g(s)$ as indicated by (3.62):

$$g(s) = \int_0^t e^{-st}F(t)\,dt \tag{3.64}$$

Now we differentiate both sides of (3.64) with respect to s:

$$\frac{dg(s)}{ds} = \frac{d}{ds}\int_0^t e^{-st}F(t)\,dt \tag{3.65}$$

The order of integration and differentiation may be interchanged as the integration is over t only; thus performing this interchange of operations in (3.65) yields

$$\frac{dg(s)}{ds} = -\int_0^t e^{-st}tF(t)\,dt = -LtF(t) \tag{3.66}$$

and hence (3.63) is proved.

Following the same procedure one may easily prove that

$$Lt^n F(t) = -1^n \frac{d^n g(s)}{ds^n} \tag{3.67}$$

Theorem XIV   **Differentiation with respect to a second independent variable**   Consider the following function of two independent variables:

$$F = F(x,t)$$

If

$$LF(x,t) = g(x,s) \tag{3.68}$$

then

$$L \frac{\partial F(x,t)}{\partial x} = \frac{\partial g(x,s)}{\partial x} \tag{3.69}$$

Proof   The proof of this theorem follows directly from the definition of the Laplace transform:

$$L \frac{\partial F(x,t)}{\partial x} = \int_0^\infty e^{-st} \frac{\partial F(x,t)}{\partial x} dt \tag{3.70}$$

Since the variable x is not a variable of integration, then the order of differentiation and integration may be interchanged to give

$$L \frac{\partial F(x,t)}{\partial x} = \frac{\partial}{\partial x} \int_0^t e^{-st} F(x,t) dt$$

$$= \frac{\partial g(x,s)}{\partial x} \tag{3.71}$$

which completes the proof.

Theorem XV   **Integration with respect to a second independent variable**   Consider again a function of two independent variables

$$F = F(x,t)$$

If

$$LF(x,t) = g(x,s) \tag{3.72}$$

then

$$L \int_{x_1}^{x_2} F(x',t) dx' = \int_{x_1}^{x_2} g(x',s) dx' \tag{3.73}$$

Proof   The proof of this theorem will not be given as the procedure is identical to the proof of Theorem XIV.

4 CALCULATION OF DIRECT TRANSFORMS

The computation of transforms is based on the Fourier-Mellin integral theorem. If the function $F(t)$ is known and the integral

$$g(s) = \int_0^\infty e^{-st} F(t)\, dt \qquad\qquad (4.1)$$

can be computed, then the function $g(s)$ may be determined and the direct Laplace transform

$$g(s) = LF(t) \qquad\qquad (4.2)$$

obtained.

If, on the other hand, the function $g(s)$ is known, then to obtain the function $F(t)$ we must use the integral

$$F(t) = \frac{1}{2\pi j} \int_{c-j\infty}^{c+j\infty} e^{st} g(s)\, ds \qquad\qquad (4.3)$$

If this complex integral can be evaluated, then the inverse transform

$$F(t) = L^{-1} g(s) \qquad\qquad (4.4)$$

may be obtained.

The reader may verify that the computation of several direct transforms may be evaluated by performing the integration in (4.1).

FRACTIONAL POWERS OF s

In the applications of his operational calculus to problems of physical importance, Oliver Heaviside, in the course of his heuristic processes, encountered fractional powers of s. Since his use of the operational method was based on the interpretation of s as the derivative operator and $1/s$ as the integral operator, there was some trepidation by the followers of Heaviside in interpreting fractional powers of s.

In Appendix B a discussion of the gamma function based on the Euler integral

$$\Gamma(n) = \int_0^\infty x^{n-1} e^{-x}\, dx \qquad n > 0 \qquad\qquad (4.5)$$

was given. This integral converges for n positive and defines the function $\Gamma(n)$.

By direct evaluation we have

$$\Gamma(1) = 1 \qquad\qquad (4.6)$$

and, by an integration by parts, the recursion relation

$$\Gamma(n+1) = n\Gamma(n) \qquad\qquad (4.7)$$

The integral and the recursion formula taken together define the gamma function for all values of n.

If we now place

$$x = st \tag{4.8}$$

in (4.5), we have

$$\Gamma(n) = s^n \int_0^\infty e^{-st} t^{n-1} \, dt \tag{4.9}$$

or

$$\frac{\Gamma(n)}{s^n} = \int_0^\infty e^{-st} t^{n-1} \, dt \tag{4.10}$$

Comparing this with the basic Laplace-transform integral (4.1), we have

$$Lt^{n-1} = \frac{\Gamma(n)}{s^n} \qquad n > 0 \tag{4.11}$$

If we let

$$n - 1 = m \tag{4.12}$$

we have

$$L^{-1} \frac{1}{s^{m+1}} = \frac{t^m}{\Gamma(m+1)} \qquad m > -1 \tag{4.13}$$

It will be shown in a later section that this formula holds for *any m* not a positive integer. This equation is more symmetrically written in terms of Gauss's π function

$$\pi(m) = \Gamma(m+1) \tag{4.14}$$

in the form

$$L^{-1} s^{-(m+1)} = \frac{t^m}{\pi(m)} \tag{4.15}$$

When *m* is a *positive integer*, we have

$$\pi(m) = m! \tag{4.16}$$

and when *m* is a negative integer, $\pi(m)$ is not defined. We thus have

$$L^{-1} \frac{1}{s^{m+1}} = \frac{t^m}{m!} \qquad m \text{ a positive integer} \tag{4.17}$$

By the use of Eq. (4.13) and a table of gamma functions the transforms of fractional powers of *s* may be readily computed. As an example, since

$$\Gamma(\tfrac{1}{2}) = \sqrt{\pi} \tag{4.18}$$

we have from (4.13)

$$L^{-1} s^{-\frac{1}{2}} = \frac{t^{-\frac{1}{2}}}{\sqrt{\pi}} = \frac{1}{\sqrt{\pi t}} \tag{4.19}$$

5 CALCULATION OF INVERSE TRANSFORMS

The problem of computing the inverse transform of a function $g(s)$ by the use
of the equation

$$F(t) = \frac{1}{2\pi j} \int_{c-j\infty}^{c+j\infty} g(s) e^{st} \, ds \tag{5.1}$$

will now be considered.

The line integral for $F(t)$ is usually evaluated by transforming it into a
closed contour and applying the calculus of residues discussed in Chap 1.
The contour usually chosen is shown in Fig. 5.1.

Let us take as the closed contour Γ the straight line parallel to the axis
of imaginaries and at a distance c to the right of it and the large semicircle s_0
whose center is at $(c,0)$. We then have

$$\oint_\Gamma e^{st} g(s) \, ds = \int_{c+jR}^{c-jR} e^{st} g(s) \, ds + \int_{s_0} e^{st} g(s) \, ds \tag{5.2}$$

where c is chosen great enough so that all the singularities of the integral lie
to the left of the straight line along which the integral from $c - j\infty$ to $c + j\infty$ is
taken.

The evaluation of the contour integral along the contour Γ is greatly
facilitated by the use of Jordan's lemma (Chap. 1, Sec. 15), which in this case
may be stated in the following form:

Let $\phi(s)$ be an integrable function of the complex variable s such that

$$\lim_{|s|\to\infty} |\phi(s)| = 0 \tag{5.3}$$

Then

$$\lim_{R\to\infty} \left| \int_{s_0} e^{st} \phi(s) \, ds \right| = 0 \qquad t > 0 \qquad \operatorname{Re} s \leqslant 0 \tag{5.4}$$

It usually happens in practice that the function

$$\phi(s) = g(s) \tag{5.5}$$

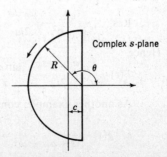

Fig. 5.1

has such properties that Jordan's lemma is applicable. In such a case the integral around the large semicircle in (5.2) vanishes as $R \to \infty$, and we have

$$F(t) = \frac{1}{2\pi j} \int_{c-j\infty}^{c+j\infty} g(s) e^{st}\, ds = \lim_{R\to\infty} \frac{1}{2\pi j} \oint_s e^{st} g(s)\, ds \tag{5.6}$$

Now, by Cauchy's residue theorem (Chap. 1, Sec. 9), we have

$$\oint_\Gamma e^{st} g(s)\, ds = 2\pi j \sum \operatorname{Res} e^{st} g(s) \text{ inside } \Gamma \tag{5.7}$$

Hence by (5.6) we have

$$F(t) = \sum \operatorname{Res} e^{st} g(s) \text{ inside } \Gamma \tag{5.8}$$

if R is large enough to include all singularities.

If the function $e^{st} g(s)$ is not single-valued within the contour Γ and possesses branch points within Γ, it may be made single-valued by introducing suitable cuts and then applying the residue theorem.

As an example of the computation of an inverse transform, let us consider the determination of the inverse transform of

$$g(s) = \frac{1}{s^2 + a^2} = LF(t) \tag{5.9}$$

This function clearly satisfies the condition imposed by Jordan's lemma. Hence $F(t)$ is given by (5.8) in the form

$$F(t) = \sum \operatorname{Res} \frac{e^{st}}{s^2 + a^2} \tag{5.10}$$

The poles of $e^{st}/(s^2 + a^2)$ are at

$$s = \pm ja \tag{5.11}$$

By Chap. 1, Sec. 12, we have

$$\operatorname{Res}_{s=ja} \frac{e^{st}}{s^2 + a^2} = \frac{e^{jat}}{2ja} \tag{5.12}$$

Similarly we have

$$\operatorname{Res}_{s=-ja} \frac{e^{st}}{s^2 + a^2} = \frac{-e^{-jat}}{2ja} \tag{5.13}$$

Hence

$$F(t) = \frac{1}{a} \frac{e^{jat} - e^{-jat}}{2j} = \frac{\sin at}{a} \tag{5.14}$$

As another example, consider

$$L^{-1} g(s) = \frac{\omega}{(s+a)^2 + \omega^2} = F(t) \tag{5.15}$$

This function also satisfies the condition of Jordan's lemma. We must compute the sum of the residues of

$$\phi(s) = \frac{e^{st}\,\omega}{(s+a)^2 + \omega^2} \qquad (5.16)$$

The poles of this function are at

$$s = -a \pm j\omega \qquad (5.17)$$

The sum of the residues at these poles is

$$\frac{e^{-at}\,e^{j\omega t}}{2j\omega} - \frac{e^{-at}\,e^{-j\omega t}}{2j\omega} = e^{-at}\sin\omega t = F(t) \qquad (5.18)$$

To conclude this section let us state that once the function $g(s)$ is known, its inverse may be readily evaluated, provided $g(s)$ satisfies the conditions of Jordan's lemma, by means of the residue theorem discussed in Sec. 12 of Chap. 1. Summarizing the results of the theory of residues as applied to (5.8) we may state that if $g(s)$ has a simple pole at $s = s_0$, then

$$\operatorname*{Res}_{s=s_0} g(s)\,e^{st} = \lim_{s \to s_0}(s - s_0)\,g(s)\,e^{st} \qquad (5.19)$$

or if $g(s)$ has a pole of order n at $s = s_0$, then

$$\operatorname*{Res}_{s=s_0} g(s)\,e^{st} = \lim_{s \to s_0}[1/(n-1)!](d^{n-1}/ds^{n-1})(s - s_0)^n g(s)\,e^{st} \qquad (5.20)$$

Thus these formulas may be used to evaluate the residues of $g(s)\,e^{st}$ at all of its poles and through the application of (5.8) the inverse is found.

6 THE MODIFIED INTEGRAL

Let us consider the fundamental integral of the inverse transform

$$F(t) = \frac{1}{2\pi j}\int_{c-j\infty}^{c+j\infty} g(s)\,e^{st}\,ds \qquad (6.1)$$

Now, if we perform a formal differentiation under the integral sign with respect to t, we have

$$\frac{dF}{dt} = \frac{1}{2\pi j}\int_{c-j\infty}^{c+j\infty} s g(s)\,e^{st}\,ds \qquad (6.2)$$

and we see that differentiation introduces a factor of s before $g(s)$. In many cases, however, the new integral does not converge uniformly.

In such cases the process of differentiation cannot be performed in this manner, since for the operation to be valid the new integral must converge uniformly. It is possible, however, to modify the original integral in such a way that the procedure is correct. This may be done provided that $g(s)$

satisfies certain conditions. The conditions, which are usually fulfilled in practice, are the following:

1. $g(s)$ is analytic in the region to the right of the straight lines L_1 and L_2. $s = c + re^{j\theta}$ and $s = c + re^{-j\theta}$, where $\pi/2 < \theta \leqslant \pi$.

2. $g(s)$ tends uniformly to zero as $|s| \to \infty$ in this region (see Fig. 6.1).

It is possible, when these conditions hold, to complete the line integral by the way of a large semicircle convex to the left and with center at $(c,0)$ as before. Then, since no singularities are traversed, the path of integration may be changed into two straight lines L_1 and L_2, as shown in Fig. 6.1. This integral along L_1 and L_2 is called the *modified integral*.

The modified integral is easily proved uniformly convergent for all positive values of t. Also, integrals of the type

$$I = \frac{1}{2\pi j} \int_{L_1, L_2} Q(s) g(s) e^{st} \, ds \qquad (6.3)$$

where $Q(s)$ is a function that in the region under consideration increases at most as a power of s when s tends to infinity, may be proved uniformly convergent.

In Sec. 4 the relation

$$L^{-1} \frac{1}{s^{m+1}} = \frac{t^m}{\Gamma(m+1)} \qquad m > -1 \qquad (6.4)$$

was derived.

We shall now remove the above restriction on m and obtain a general formula of the form

$$L^{-1} \frac{1}{s^{r+1}} = \frac{t^r}{\Gamma(r+1)} \qquad (6.5)$$

valid for all values of r.

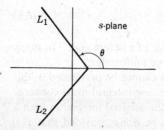

Fig. 6.1

If we begin with the restricted result (6.4), we have

$$\frac{t^m}{\Gamma(m+1)} = \frac{1}{2\pi j} \int_{c-j}^{c+j} \frac{e^{st}\,ds}{s^{m+1}} \qquad m > -1 \tag{6.6}$$

If $m > -1$, the function

$$g(s) = \frac{1}{s^{m+1}} \tag{6.7}$$

tends to zero uniformly as $|s| \to \infty$. Hence we are permitted to differentiate (6.6) k times with respect to t under the integral sign.

We thus obtain

$$\frac{m(m-1)\,\cdots\,(m-k+1)\,t^{m-k}}{\Gamma(m+1)} = \frac{1}{2\pi j} \int_{c-j\infty}^{c+j\infty} \frac{e^{st}\,s^k}{s^{m+1}}\,ds \tag{6.8}$$

Hence we have

$$L^{-1}\frac{s^k}{s^{m+1}} = \frac{m(m-1)\,\cdots\,(m-k+1)}{\Gamma(m+1)} t^{m-k} \tag{6.9}$$

where

$$m > -1 \tag{6.10}$$

Now by the fundamental recursion formula of the gamma functions, we have

$$m(m-1)\,\cdots\,(m-k+1) = \frac{\Gamma(m+1)}{\Gamma(m-k+1)} \tag{6.11}$$

Therefore we may write (6.9) in the form

$$L^{-1}\frac{1}{s^{m-k+1}} = \frac{\Gamma(m+1)t^{m-k}}{\Gamma(m+1)\,\Gamma(m-k+1)} \tag{6.12}$$

Now, if we let

$$m - k = r \tag{6.13}$$

we may write (6.12) in the form

$$L^{-1}\frac{1}{s^{r+1}} = \frac{t^r}{\Gamma(r+1)} \tag{6.14}$$

for all values of $r + 1$ except negative integers. If $r + 1$ is a negative integer, let us place

$$r + 1 = -q \tag{6.15}$$

We then have

$$L^{-1} s^q = \frac{1}{2\pi j} \int_{c-j\infty}^{c+j\infty} e^{st} s^q\,ds = 0 \qquad \begin{cases} \text{for} \quad t > 0 \\ \quad q = 1, 2, \ldots \end{cases} \tag{6.16}$$

This result follows from the fact that the integrand in (6.16) has no poles in the finite part of the s plane.

Although (6.16) holds for all values of r when $t > 0$, the inverse transforms of *positive powers* of s are not zero at $t = 0$ but are of the nature of impulsive functions.

7   IMPULSIVE FUNCTIONS

The careful treatment of impulsive functions presents considerable difficulties and is beyond the scope of this discussion.   However, many physical phenomena are of an impulsive nature.   In electrical-circuit theory, for example, we frequently encounter electromotive forces that are impulsive in character, and we desire to study the behavior of the electric currents produced by the applications of electromotive forces of this type to linear circuits.   In mechanics we frequently encounter problems in which a system is set in motion by the action of an impact or other impulsive force.   Because of the practical importance of forces of this type, a simple treatment of them will be given.

One of the most frequently encountered functions is the Dirac function $\delta(t)$.   This function is defined to be zero if $t \neq 0$ and to be infinite at $t = 0$ in such a way that

$$\int_{-\infty}^{+\infty} \delta(t)\, dt = 1 \tag{7.1}$$

This is a concise manner of expressing a function that is very large in a very small region and zero outside this region and has a unit integral.   Let us consider the function $\delta(t)$ that has the following values:

$$\delta(t) = \begin{cases} 0 & t \leqslant 0 \\ \dfrac{1}{\epsilon} & 0 < t < \epsilon \\ 0 & t > \epsilon \end{cases} \tag{7.2}$$

where ϵ may be made as small as we please.   This function possesses the property (7.1).

We now prove the following:

$$\int_{-\infty}^{+\infty} F(t)\, \delta(t - a)\, dt = F(a) \tag{7.3}$$

provided that $F(t)$ is continuous.   The proof of this follows from (7.2), so that we have

$$\int_{-\infty}^{+\infty} F(t)\, \delta(t - a)\, dt = \frac{1}{\epsilon} \int_{a}^{a+\epsilon} F(t)\, dt$$

$$= F(a + \theta\epsilon) \qquad \text{where } 0 < \theta < 1 \tag{7.4}$$

Now, since by hypothesis $F(t)$ is continuous, we have

$$\lim_{\epsilon \to 0} F(a + \theta\epsilon) = F(a) \tag{7.5}$$

Hence the result (7.3) follows. As an application of (7.3), let us compute the Laplace transform of $\delta(t)$. We have

$$L\delta(t) = \int_0^\infty e^{-st} \delta(t)\, dt = 1 \tag{7.6}$$

so that the Laplace transform of the Dirac function is 1.

As an example of the use of the $\delta(t)$ function, let us consider the effect produced when a particle of mass m situated at the origin is acted upon by an impulse F_0 applied to the mass at $t = 0$. If the impulse acts in the x direction, the equation of motion is

$$m \frac{d^2 x}{dt^2} = F_0 \delta(t) \tag{7.7}$$

Let

$$Lx = y \tag{7.8}$$

Now, if the mass is at $x = 0$ and $t = 0$ and has no initial velocity, we have

$$L \frac{d^2 x}{dt^2} = s^2 y \tag{7.9}$$

Hence Eq. (7.7) is transformed into

$$ms^2 y = F_0 \tag{7.10}$$

Therefore

$$y = \frac{F_0}{ms^2} \tag{7.11}$$

and hence

$$x = L^{-1} y = \frac{F_0}{m} t \tag{7.12}$$

As another example, consider a series circuit consisting of an inductance L_0 in series with a capacitance C. Initially the circuit has zero current, and the capacitor is discharged. At $t = 0$ an electromotive force of very large voltage is applied for a very short time so that it may be expressed in the form $E_0 \delta(t)$. It is required to determine the subsequent current in the system.

The differential equation governing the current in the circuit is

$$L_0 \frac{di}{dt} + \frac{q}{C} = E_0 \delta(t) \tag{7.13}$$

where

$$i = \frac{dq}{dt} \tag{7.14}$$

Now let

$$Li = I \qquad Lq = Q \tag{7.15}$$

Therefore (7.13) transforms into

$$\left(L_0 s + \frac{1}{Cs}\right) I = E_0 \tag{7.16}$$

Hence

$$I = \frac{E_0}{L_0} \frac{s}{s^2 + \omega_0^2} \tag{7.17}$$

where

$$\omega_0 = \frac{1}{\sqrt{L_0 C}} \tag{7.18}$$

From the table of transforms we have

$$L^{-1} \frac{s}{s^2 + \omega_0^2} = \cos \omega_0 t \tag{7.19}$$

Hence

$$L^{-1} I = i = \frac{E_0}{L_0} \cos \omega_0 t \tag{7.20}$$

8 HEAVISIDE'S RULES

In Heaviside's application of his operational method to the solution of physical problems, he made frequent and skillful use of certain procedures in order to interpret his "resistance operators." These procedures were known as *Heaviside's rules* in the early history of the subject.   At that time there appears to have been much controversy over the justification of these rules.   Viewed from the Laplacian-transform point of view, the proof of these powerful rules is seen to be most simple.

a HEAVISIDE'S EXPANSION THEOREM

Heaviside's first rule is commonly known as *Heaviside's expansion theorem*, or the *partial-fraction rule*.   In its usual form it may be stated as follows:

If $g(s)$ has the form

$$g(s) = \frac{N(s)}{\phi(s)} \tag{8.1}$$

where $N(s)$ and $\phi(s)$ are polynomials in s and the degree of $\phi(s)$ is at least one higher than the degree of $N(s)$ and, in addition, the polynomial $\phi(s)$ has n distinct zeros, $s_1, s_2, \ldots, s_n$, we then have

$$
L^{-1} \frac{N(s)}{\phi(s)} = \frac{1}{2\pi j} \int_{c-j\infty}^{c+j\infty} \frac{N(s) e^{st} \, ds}{\phi(s)}
$$

$$
= \sum \text{Res} \frac{N(s) e^{st}}{\phi(s)} \tag{8.2}
$$

by the Cauchy residue theorem since in this case the integrand satisfies the conditions of Jordan's lemma and the contribution over the large semicircle to the left of the path of integration vanishes as $R \to \infty$.

We now compute the residues by the methods of Chap. 1, Sec. 12. Now if $s_1, s_2, s_3, \ldots, s_n$ are n *distinct* zeros of $\phi(s)$, then the points $s = s_1$, $s = s_2, \ldots, s = s_n$ are simple poles of the function $N(s) e^{st}/\phi(s)$.

The residue at the simple pole $s = s_r$ is

$$
\lim_{s \to s_r} \frac{N(s) e^{st}}{\phi'(s)} = \frac{N(s_r) e^{s_r t}}{\phi'(s_r)} \tag{8.3}
$$

where

$$
\phi'(s_r) = \left(\frac{d\phi}{ds} \right)_{s=s_r} \tag{8.4}
$$

It follows from (8.2) that

$$
L^{-1} \frac{N(s)}{\phi(s)} = + \sum_{r=1}^{r=n} \frac{N(s_r) e^{s_r t}}{\phi'(s_r)} \tag{8.5}
$$

This result is called the *Heaviside expansion theorem* and is frequently used to obtain inverse transforms of the ratios of two polynomials. As a simple application of the expansion theorem, let it be required to determine the inverse transform of

$$
\frac{N(s)}{\phi(s)} = \frac{s}{s^2 + a^2} \tag{8.6}
$$

In this case the zeros of $\phi(s)$ are

$$
s = \pm ja \tag{8.7}
$$

We also have

$$
\phi'(s) = 2s \tag{8.8}
$$

Hence, substituting in (8.5), we obtain

$$L^{-1} \frac{s}{s^2 + a^2} = \frac{-a^2 e^{jat}}{-2a^2} + \frac{-a^2 e^{-jat}}{-2a^2}$$

$$= \frac{e^{jat} + e^{-jat}}{2} = \cos at \qquad (8.9)$$

AN EXTENSION OF HEAVISIDE'S EXPANSION THEOREM

Heaviside's expansion theorem as stated above is of use in evaluating the inverse transforms of ratios of polynomials in s. The expansion theorem is sometimes used to compute the inverse transforms of certain transcendental functions in s that have simple poles and that satisfy the conditions of Jordan's lemma so that the path of integration may be closed to the left by a large semicircle. In this case the result

$$L^{-1} g(s) = \sum \text{Res } g(s) e^{st} \qquad (8.10)$$

is applicable. If the poles of $g(s)$ are simple poles at $s = s_1$, $s = s_2$, . . . , $s = s_r$, . . . , then Eq. (8.10) is very similar to the Heaviside expansion theorem (8.5).

As an example of this, consider the function

$$g(s) = \frac{\tanh (Ts/2)}{s} = \frac{\sinh (Ts/2)}{s \cosh (Ts/2)} \qquad (8.11)$$

where T is real and positive. The poles of this function are at

$$s \cosh \frac{Ts}{2} = 0 \qquad (8.12)$$

or

$$s_k = \frac{(2k + 1) \pi j}{T} \qquad k = 0, \pm 1, \pm 2, . . . \qquad s_0 = 0 \qquad (8.13)$$

The fact that these poles are simple poles follows from the fact that the derivative of $s \cosh (Ts/2)$ with respect to s does not have zeros at these points. The function

$$\frac{e^{st} \tanh (Ts/2)}{s} \qquad (8.14)$$

satisfies the condition of Jordan's lemma since

$$\lim_{|s| \to \infty} \tanh \frac{Ts}{2} = 1 \qquad (8.15)$$

We thus have by (8.10)

$$L^{-1}g(s) = \sum_{k=-\infty}^{k=+\infty} \frac{e^{s_k t} \sinh(s_k T/2)}{s_k(T/2)\sinh(s_k T/2)} \qquad (8.16)$$

since the residue at $s = 0$ is 0.

This expression reduces to

$$L^{-1}g(s) = \frac{2}{T}\sum_{k=-\infty}^{k=+\infty} \frac{e^{s_k t}}{s_k} \qquad (8.17)$$

Substituting (8.13) into (8.17), we finally obtain

$$L^{-1}\frac{\tanh(Ts/2)}{s} = \frac{4}{\pi}\sum_{n=1}^{n=\infty}\frac{1}{n}\sin\frac{n\pi t}{T} \qquad n \text{ odd} \qquad (8.18)$$

It is thus seen that the result (8.10) is more general than the Heaviside expansion theorem, but when applied to functions whose poles are simple, then (8.10) has a close resemblance to the Heaviside expansion theorem.

b HEAVISIDE'S SECOND RULE

Heaviside made frequent use of the following result:

$$L^{-1}\sum_{n=1}^{\infty}\frac{a_n}{s^n} = \sum_{n=1}^{\infty} a_n \frac{t^{n-1}}{n-1!} \qquad (8.19)$$

where the series $\sum_{n=1}^{\infty} a_n/s^n$ is a convergent series of negative powers of s. This result is obtained by term-by-term substitution of

$$L^{-1}\frac{1}{s^n} = \frac{t^{n-1}}{n-1!} \qquad \text{for } n \text{ a positive integer} \qquad (8.20)$$

since the series converges uniformly.

c HEAVISIDE'S THIRD RULE

This concerns the transform of the series

$$g(s) = \sum_{n=0}^{\infty} a_n s^n + s^{-\frac{1}{2}}\sum_{n=1}^{\infty} b_n s^n \qquad (8.21)$$

By making use of the fact that

$$L^{-1}\frac{1}{s^{r+1}} = \frac{t^r}{\Gamma(r+1)} \qquad r \neq \text{negative integer} \qquad (8.22)$$

and the facts that

$$\Gamma(\tfrac{1}{2}) = \sqrt{\pi} \qquad (8.23)$$

and

$$L^{-1} s^k = 0 \qquad \text{for } t > 0 \qquad k = 0, 1, 2, \ldots \tag{8.24}$$

we have directly

$$L^{-1} g(s) = + \frac{1}{\sqrt{\pi t}} \left[b_0 - \frac{b_1}{2t} + \frac{1 \times 3 b_2}{(2t)^2} \right.$$

$$\left. - \frac{1 \times 3 \times 5 b_3}{(2t)^3} + \cdots \right] \qquad \text{for } t > 0 \tag{8.25}$$

By arranging the transform of a function in the form (8.21) Heaviside was able to obtain the asymptotic expansion of certain functions valid for large values of t.

9 THE TRANSFORMS OF PERIODIC FUNCTIONS

Let $F(t)$ be a periodic function with fundamental period T, that is,

$$F(t + T) = F(t) \qquad t > 0 \tag{9.1}$$

If $F(t)$ is sectionally continuous over a period $0 < t < T$, then its direct Laplace transform is given by

$$LF(t) = \int_0^\infty e^{-st} F(t)\, dt$$

$$= \sum_{n=0}^\infty \int_{nT}^{(n+1)T} e^{-st} F(t)\, dt \tag{9.2}$$

If we now let

$$u = t - nT \tag{9.3}$$

and realize that, as a consequence of the periodicity of the function $F(t)$, we have

$$F(u + nT) = F(u) \tag{9.4}$$

in view of this we may write (9.2) in the form

$$LF(t) = \sum_{n=0}^\infty e^{-nTs} \int_0^T e^{-su} F(u)\, du \tag{9.5}$$

We also have the well-known result

$$\sum_{n=0}^{n=\infty} e^{-nTs} = \frac{1}{1 - e^{-Ts}} \tag{9.6}$$

Hence we may write (9.5) in the form

$$LF(t) = \frac{1}{1 - e^{-Ts}} \int_0^T e^{-st} F(t)\, dt \tag{9.7}$$

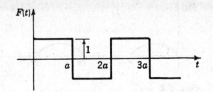

Fig. 9.1

Let us apply this formula to obtain the transform of the "meander function" of Fig. 9.1. In this case we have

$$\int_0^T F(t)e^{-st}\,dt = \int_0^a e^{-st}\,dt - \int_a^{2a} e^{-st}\,dt$$

$$= \frac{1 - e^{-sa} - e^{-sa} + e^{-2sa}}{s}$$

$$= \frac{1 - 2e^{-sa} + e^{-2sa}}{s} = \frac{(1 - e^{-sa})^2}{s} \tag{9.8}$$

Hence, substituting this into (9.7), we have

$$LF(t) = \frac{(1 - e^{-sa})^2}{s(1 - e^{-2as})} = \frac{1 - e^{-sa}}{s(1 + e^{-sa})}$$

$$= \frac{e^{sa/2} - e^{-sa/2}}{s(e^{sa/2} + e^{-sa/2})} = \frac{\sinh(sa/2)}{s\cosh(sa/2)} = \frac{\tanh(as/2)}{s} \tag{9.9}$$

Using the results of (8.18), we have

$$F(t) = L^{-1}\frac{\tanh(as/2)}{s} = \frac{4}{\pi}\sum_{n=1}^{n=\infty}\frac{1}{n}\sin\frac{n\pi t}{a} \qquad n \text{ odd} \tag{9.10}$$

This is the Fourier-series expansion of the meander function.

As another example, consider the function $F(t)$ defined by

$$F(t) = \begin{cases} \sin t & 0 < t < \pi \\ 0 & \pi < t < 2\pi \end{cases} \tag{9.11}$$

This function is given graphically by Fig. 9.2 and it is of importance in the theory of half-wave rectification. In this case the fundamental period is 2π. To obtain the transform of this function, we have

$$\int_0^\pi e^{-st}\sin t\,dt = \frac{1}{s^2 + 1}(1 + e^{-s\pi}) \tag{9.12}$$

Hence, by Eq. (9.7), we obtain

$$LF(t) = \frac{1}{s^2 + 1}\frac{1 + e^{-s\pi}}{1 - e^{-2s\pi}} = \frac{1}{(s^2 + 1)(1 - e^{-s\pi})} \tag{9.13}$$

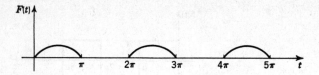

Fig. 9.2

10 THE SIMPLE DIRECT LAPLACE-TRANSFORM, OR OPERATIONAL, METHOD OF SOLVING LINEAR DIFFERENTIAL EQUATIONS WITH CONSTANT COEFFICIENTS

The method for solving differential equations mentioned in the heading of this section is essentially the same as that known under the name of *Heaviside's operational calculus*. The modern approach to this method is based on the Laplace transformation. This method provides a most convenient means for solving the differential equations of electrical networks and mechanical oscillations.

The chief advantage of this method is that it is very direct and does away with tedious evaluations of arbitrary constants. The procedure in a sense reduces the solution of a differential equation to a matter of looking up a particular transformation in a table of transforms. In a way this procedure is much like consulting a table of integrals in the process of performing integrations.

Consider the functional relation between a function $g(s)$ and another function $h(t)$ expressed in the form

$$g(s) = \int_0^\infty e^{-st} h(t)\,dt \qquad \mathrm{Re}\,s > 0 \tag{10.1}$$

where s is a complex number whose real part is greater than zero and $h(t)$ is a function such that the infinite integral of (10.1) converges and such that it satisfies the condition that

$$h(t) = 0 \qquad \text{for } t < 0 \tag{10.2}$$

In most of the modern literature on operational, or Laplace-transform, methods the functional relation expressed between $g(s)$ and $h(t)$ is written in the following form:

$$g(s) = Lh(t) \tag{10.3}$$

The L denotes the "Laplace transform of" and greatly shortens the writing. The relation between $h(t)$ and $g(s)$ is also written in the form

$$h(t) = L^{-1} g(s) \tag{10.4}$$

In this case we speak of $h(t)$ as being the inverse Laplace transform of $g(s)$.

THE TRANSFORMS OF DERIVATIVES

Let us suppose that we have the functional relation

$$y(s) = \int_0^\infty e^{-st} x(t)\, dt \tag{10.5}$$

or, symbolically,

$$y(s) = Lx(t) \tag{10.6}$$

Let us now determine $L(dx/dt)$ in terms of $y(s)$. To do this, we have

$$L\left(\frac{dx}{dt}\right) = \int_0^\infty e^{-st}\left(\frac{dx}{dt}\right) dt \tag{10.7}$$

By integrating by parts we obtain

$$\int_0^\infty e^{-st}\left(\frac{dx}{dt}\right) dt = e^{-st}\, x\Big|_0^\infty + s \int_0^\infty e^{-st} x\, dt \tag{10.8}$$

Now, if we assume that

$$\lim_{t \to \infty} (e^{-st} x) = 0 \tag{10.9}$$

and that

$$\int_0^\infty e^{-st} x\, dt$$

exists when s is greater than some fixed positive number, then (10.8) becomes

$$L\left(\frac{dx}{dt}\right) = -x_0 + sy \tag{10.10}$$

where

$$x_0 = x(0) \tag{10.11}$$

Equation (10.10) gives the value of the Laplace transform of dx/dt in terms of the transform of x and the value of x at $t = 0$.

In order to compute the Laplace transform of $d^2 x/dt^2$, let

$$u = \frac{dx}{dt} \tag{10.12}$$

Then in view of (10.10) we have

$$L\left(\frac{d^2 x}{dt^2}\right) = L\left(\frac{du}{dt}\right) = -u_0 + sL\left(\frac{dx}{dt}\right)$$

$$= -x_1 - s x_0 + s^2 y \tag{10.13}$$

where x_1 is the value of dx/dt at $t = 0$.

Repeating the process, we obtain

$$L\frac{d^3 x}{dt^3} = s^3 y - s^2 x_0 - sx_1 - x_2 \qquad (10.14)$$

$$L\frac{d^4 x}{dt^4} = s^4 y - s^3 x_0 - s^2 x_1 - sx_2 - x_3 \qquad (10.15)$$

and

$$L\frac{d^n x}{dt^n} = s^n y - (s^{n-1} x_0 + s^{n-2} x_1 + s^{n-3} x_2 + \cdots + x_{n-1}) \qquad (10.16)$$

where

$$x_n = \frac{d^n x}{dt^n} \qquad \text{evaluated at } x = 0 \qquad (10.17)$$

The above formulas for the transforms of derivatives are of great importance in the solution of linear differential equations with constant coefficients.

To illustrate the general theory, let it be required to solve the simple differential equation

$$\frac{dx}{dt} + ax = 0 \qquad (10.18)$$

subject to the initial condition that

$$x = x_0 \qquad \text{at } t = 0 \qquad (10.19)$$

To solve this by the Laplace-transform method, we let

$$y = Lx \qquad (10.20)$$

and use (10.10); then, in terms of y, the equation becomes

$$sy - x_0 + ay = 0 \qquad (10.21)$$

or

$$y = \frac{x_0}{s+a} = L(x) \qquad (10.22)$$

The solution of one equation could be written symbolically in the form

$$x = L^{-1}\frac{x_0}{s+a} \qquad (10.23)$$

Consulting the table of transforms in Appendix A, we find that transform 1.1 gives

$$L^{-1}\frac{1}{s+a} = e^{-at} \qquad (10.24)$$

Accordingly we have

$$x = x_0 e^{-at} \qquad (10.25)$$

as the solution of the differential equation (10.19) subject to the given initial conditions.

As another example, let us solve the equation

$$\frac{d^2 x}{dt^2} + \omega^2 x = \cos \omega t \qquad t > 0 \tag{10.26}$$

subject to the initial conditions that

$$\left. \begin{array}{l} x = x_0 \\ \dfrac{dx}{dt} = x_1 \end{array} \right\} \qquad \text{at } t = 0 \tag{10.27}$$

As before, we let $y = Lx$ and replace every member of (10.20) by its transform. Consulting the table of transforms, we find from No. 2.5 that

$$L(\cos \omega t) = \frac{s}{s^2 + \omega^2} \tag{10.28}$$

Using (10.13), Eq. (10.20) is transformed to

$$(s^2 y - s x_0 - x_1) + \omega^2 y = \frac{s}{s^2 + \omega^2} \tag{10.29}$$

or

$$y = \frac{s x_0}{s^2 + \omega^2} + \frac{x_1}{s^2 + \omega^2} + \frac{s}{(s^2 + \omega^2)^2} \tag{10.30}$$

Consulting the table of transforms in Appendix A, we find that Nos. 2.4, 2.5, and 4.10 give the required information, and we have

$$x = x_0 \cos \omega t + x_1 \frac{\sin \omega t}{\omega} + \frac{t}{2\omega} \sin \omega t \tag{10.31}$$

as the required solution.

THE GENERAL CASE

To solve the general equation with constant coefficients

$$\frac{d^n x}{dt^n} + a_1 \frac{d^{n-1} x}{dt^{n-1}} + \cdots + a_{n-1} \frac{dx}{dt} + a_n x = F(t) \tag{10.32}$$

we introduce $y = Lx$ and $\phi_1(s) = LF(t)$ and replace the various derivatives of x by their transforms given by (10.16).   We then obtain

$$(s^n + a_1 s^{n-1} + \cdots + a_n) y = \phi_1(s) + \phi_2(s) \tag{10.33}$$

where

$$\phi_2(s) = (x_{n-1} + sx_{n-2} + \cdots + s^{n-1}x_0)$$
$$+ a_1(x_{n-2} + sx_{n-3} + \cdots + s^{n-2}x_0)$$
$$+ a_2(x_{n-3} + sx_{n-4} + \cdots + s^{n-3}x_0)$$
$$+ \cdots \cdots \cdots \cdots \cdots \cdots \cdots$$
$$+ a_{n-1}(x_0) \tag{10.34}$$

If we write

$$L_n(s) = s^n + a_1 s^{n-1} + \cdots + a_n \tag{10.35}$$

Eq. (10.33) may be written concisely in the form

$$y(s) = \frac{\phi_1(s)}{L_n(s)} + \frac{\phi_2(s)}{L_n(s)} = Lx(t) \tag{10.36}$$

To obtain the solution of the differential equation (10.32), we must obtain in some manner the inverse transform of $y(s)$, and we would then have

$$x(t) = L^{-1} y(s) \tag{10.37}$$

If $F(t)$ is a constant, e^{at}, $\cos \omega t$, $\sin \omega t$, t^n where n is a positive integer, $e^{at} \sin \omega t$, $e^{at} \cos \omega t$, $t^n e^{at}$, $t^n \cos \omega t$, or $t^n \sin \omega t$, then $\phi_1(s)$ is a polynomial in s. The procedure in such a case is to decompose the expressions $\phi_1(s)/L_n(s)$ and $\phi_2(s)/L_n(s)$ into partial fractions, examine the table of transforms, and obtain the appropriate inverse transforms.

11   SYSTEMS OF LINEAR DIFFERENTIAL EQUATIONS WITH CONSTANT COEFFICIENTS

In the study of dynamical systems, whether electrical or mechanical, the analysis usually leads to the solution of a system of linear differential equations with constant coefficients. The Laplace-transform method is well adapted for solving systems of equations of this type. The method will be made clear by an example. Consider the system of linear differential equations with constant coefficients given by

$$3\frac{dx_1}{dt} + 2x_1 + \frac{dx_2}{dt} = 1 \qquad t \geqq 0$$
$$\frac{dx_1}{dt} + 4\frac{dx_2}{dt} + 3x_2 = 0 \tag{11.1}$$

Let us solve these equations subject to the initial conditions

$$\left.\begin{array}{l} x_1 = 0 \\ x_2 = 0 \end{array}\right\} \qquad \text{at } t = 0 \tag{11.2}$$

To solve these equations by the Laplace-transform method, let

$$Lx_1 = y_1$$
$$Lx_2 = y_2 \tag{11.3}$$

Then, in view of (10.10) and the fact that the Laplace transform of unity is 1, the equations transform to

$$(3s + 2)y_1 + sy_2 = \frac{1}{s}$$
$$sy_1 + (4s + 3)y_2 = 0 \tag{11.4}$$

Solving these two simultaneous equations for y_1, we obtain

$$y_1 = \frac{4s + 3}{(s + 1)(11s + 6)} = Lx_1 \tag{11.5}$$

Using the transform pair No. 3.5 of the table (Appendix A), we obtain

$$x_1 = \tfrac{1}{2} - \tfrac{1}{5}e^{-t} - \tfrac{3}{10}e^{-6t/11} \tag{11.6}$$

Solving for y_2, we have

$$y_2 = \frac{-1}{(11s + 6)(s + 1)} = Lx_2 \tag{11.7}$$

By the use of the transform pair No. 2.2, we obtain

$$x_2 = \tfrac{1}{5}(e^{-t} - e^{-6t/11}) \tag{11.8}$$

This example illustrates the general procedure.   In Chaps. 5 and 6 the systems of differential equations arising in the study of electrical networks and mechanical oscillations are considered in detail.

Further examples   To further illustrate the use of the table of Laplace transforms in the solution of differential equations, the following examples are appended.

1. Suppose it is required to solve the equation

$$\frac{d^2 y}{dt^2} + a^2 y = ce^{-bt}$$

subject to the initial conditions

$$\left. \begin{array}{l} y = 0 \\ \dfrac{dy}{dt} = y_1 \end{array} \right\} \quad \text{at } t = 0$$

To solve the equation, we let

$$Ly = Y(s)$$

By the basic theorem (III) we have

$$L\frac{d^2 y}{dt^2} = s^2 Y - y_1$$

and by No. 1.1 (Appendix A)

$$L(ce^{-bt}) = \frac{c}{s+b}$$

The equation to be solved is thus transformed to

$$s^2 Y - y_1 + a^2 Y = \frac{c}{s+b}$$

Therefore

$$Y(s) = \frac{c}{(s^2+a^2)(s+b)} + \frac{y_1}{s^2+a^2}$$

Using Nos. 3.16 and 2.4, we have the inverse transform of $Y(s)$:

$$y(t) = \frac{c}{a^2+b^2}\left(e^{-bt} - \cos at + \frac{b}{a}\sin at\right) + \frac{y_1}{a}\sin at$$

This is the required solution.

2. A constant electromotive force E is applied at $t = 0$ to an electrical circuit consisting of an inductance L, resistance R, and capacitance C in series. The initial values of the current i and the charge on the capacitor q are zero. It is required to find the current.

The current is given by the equation

$$L\frac{di}{dt} + Ri + \frac{q}{C} = E$$

where $dq/dt = i$. Let $\mathscr{L}i(t) = I(s)$, $\mathscr{L}q(t) = Q(s)$.

Note: The script $\mathscr{L}$ is used to denote the Laplace transform in order not to confuse it with the inductance parameter L.

By (III) the equations are transformed to

$$LsI + RI + \frac{Q}{C} = E$$

$$sQ = I$$

Therefore we have

$$LsI + RI + \frac{I}{sC} = E$$

Therefore

$$I = \frac{E}{L(s^2 + R/Ls + 1/LC)} = \frac{E}{L[(s+a)^2 + \omega^2]}$$

where

$$a = \frac{R}{2L} \qquad \omega^2 = \frac{1}{LC} - \frac{R^2}{4L^2}$$

By No. 2.12 we have

$$i = \frac{E}{\omega L}e^{-at}\sin \omega t \qquad \text{if } \omega^2 > 0$$

or by No. 18

$$i = \frac{E}{L}te^{-at} \qquad \text{if } \omega^2 = 0$$

$$i = \frac{E}{kL}e^{-at}\sinh kt \qquad \text{if } \omega^2 < 0 \qquad k^2 = -\omega^2$$

3. *Resonance of a pendulum*   A simple pendulum, originally hanging in equilibrium, is disturbed by a force varying harmonically.   It is required to determine the motion.   The differential equation is

$$\ddot{x} + \omega_0^2 x = F_0 \sin nt$$

Let the pendulum start from rest at its equilibrium position.   In such a case we have the initial conditions

$$\left.\begin{array}{c} x = 0 \\ \dot{x} = 0 \end{array}\right\} \quad \text{at } t = 0$$

Let $Lx = y$; then by (III) $L\ddot{x} = s^2 y$.   Using No. 2.4, the differential equation of motion is transformed to

$$(s^2 + \omega_0^2)\, y = \frac{F_0 n}{s^2 + n^2}$$

or

$$y = \frac{F_0 n}{(s^2 + n^2)(s^2 + \omega_0^2)} = \frac{F_0 n}{\omega_0^2 - n^2} \left(\frac{1}{s^2 + n^2} - \frac{1}{s^2 + \omega_0^2} \right)$$

By No. 2.4 we obtain

$$x = \frac{F_0 n}{\omega_0^2 - n^2} \left(\frac{\sin nt}{n} - \frac{\sin \omega_0 t}{\omega_0} \right) \qquad \text{if } \omega_0^2 \neq n^2$$

If $\omega_0^2 = n^2$,

$$y = \frac{F_0 n}{(s^2 + n^2)^2}$$

and by No. 4.9

$$x = \frac{F_0}{2n^2} (\sin nt - nt \cos nt) \qquad \text{the case of resonance}$$

4. *Differential equation with variable coefficients*   As the final example let us consider the differential equation

$$x \frac{d}{dx} xy' + x^2 y = 0$$

which is a standard form of Bessel's equation (cf. Appendix B).   If we let

$$Ly(x) = Y(s)$$

then

$$Lxy' = \frac{d}{ds} s Y(s)$$

and

$$L \frac{d}{dx} xy' = s \frac{d}{ds} s Y(s)$$

and hence

$$Lx \frac{d}{dx} xy' = \frac{d}{ds} s \frac{d}{ds} s Y(s)$$

Taking the Laplace transform of the original differential equation yields

$$\frac{d}{ds}s\frac{d}{ds}sY(s) + \frac{d^2\,Y(s)}{ds^2} = 0$$

Integrating this equation once with respect to s and setting the constant of integration to zero† gives

$$s\frac{d}{ds}sY(s) + \frac{dY(s)}{ds} = 0$$

This equation is separable, and performing this operation gives

$$\frac{dY(s)}{Y(s)} + \frac{s\,ds}{1+s^2} = 0$$

Integrating this equation yields

$$\ln Y(s) + \tfrac{1}{2}\ln(1+s^2) = \ln c$$

or

$$Y(s) = \frac{c}{(1+s^2)^{\frac{1}{2}}}$$

Using the table of transforms in Appendix A, we may find the inverse of the above expression from entry R-B.1 which gives

$$y(x) = cJ_0(x)$$

which is a Bessel function of the first kind of order zero.

The reader may also recall that the inverse of the above result for $Y(s)$ was computed by means of evaluating the inversion integral via integration in the complex plane as the second example in Chap. 1, Sec. 18.

PROBLEMS

Solve the following equations:

1. $y' + y = x$

2. $y' - \dfrac{2y}{x+1} = (x+1)^{\frac{3}{2}}$

3. $xy' - 2y = 2x$

4. $y' + y = x^2 + 2$

5. Find the solution of the following equations which satisfy the given initial conditions:

(a) $y'' + 3y' + 2y = 0$ $y(0) = 0$ $y'(0) = 1$

(b) $y'' + n^2 y = 0$ $y(0) = a$ $y'(0) = 0$

(c) $y'' + 9y = x + \tfrac{1}{2}$ $y(0) = \tfrac{1}{18}$ $y'(0) = \tfrac{1}{9}$

(d) $y'' + 9y = 5\cos 2x$ $y(0) = 1$ $y'(0) = 3$

(e) $y'' + 4y' + 4y = 4e^{2x}$ $y(0) = 0$ $y'(0) = 0$

(f) $y''' + y'' = 6x^2 + 4$ $y(0) = 0$ $y'(0) = 0$ $y''(0) = 0$

(g) $y'' + 4y = \sin 2x$ $y(0) = y_0$ $y'(0) = v_0$

(h) $y'' + b^2 y = k\cos bx$ $y(0) = 0$ $y'(0) = 0$

---

† The setting of this arbitrary constant to zero simply eliminates the second solution to the original equation.

6. Using the complex line integral of the Fourier-Mellin transformation, show that the following inverse transforms have the indicated values:

(a) $L^{-1} \dfrac{1}{s+a} = e^{-at}$

(b) $L^{-1} \dfrac{s}{s^2 + a^2} = \cos(at)$

(c) $L^{-1} \dfrac{a}{s^2 + a^2} = \sin(at)$

(d) $L^{-1} \dfrac{1}{(s+a)^n} = \dfrac{e^{-at} t^{n-1}}{(n-1)!}$ n a positive integer

(e) $L^{-1} \dfrac{1}{(s^2 + a^2)^2} = \dfrac{\sin at}{2a^3} - \dfrac{t \cos at}{2a^2}$

7. Solve the transforms in Prob. 6 by use of the residue theorem.

8. Show that

$$L^{-1} \frac{1}{(s^2 + a^2)^{\frac{1}{2}}} = J_0(at)$$

by expanding the left side into a series of inverse powers of s and inverting term by term to obtain the series expansion for $J_0(at)$.

9. Show that

$$L^{-1} \frac{\tanh as}{s} \frac{as}{2}$$

is the meander function given in Fig. 9.1.

10. Evaluate the following integrals operationally:

(a) $\displaystyle\int_0^\infty \frac{\sin tx \, dx}{x^{\frac{1}{2}}} = \left(\frac{\pi}{2t}\right)^{\frac{1}{2}}$ $t > 0$

(b) $\displaystyle\int_{-\infty}^\infty \frac{x \sin tx \, dx}{a^2 + x^2} = \pi e^{-at}$ $a > 0$

(c) $\displaystyle\int_0^\infty \frac{e^{-tx^2}}{(1 + x^2) \, dx} = \frac{\pi}{2} e^t \operatorname{erf} t^{\frac{1}{2}}$

11. Evaluate the Laplace transforms of the following functions by direct integration of the transform integral:

(a) t (b) t^2 (c) e^{-at}
(d) $\sin at$ (e) $\cos at + \phi$ (f) $e^{-at} \sin bt$

12. Solve the following homogeneous equations:

(a) $y'' - 3y' - 10y = 0$ $y(0) = 3$ $y'(0) = 15$
(b) $y'' + y' - 12y = 0$ $y(0) = 0$ $y'(0) = 21$
(c) $y''' - 7y'' - 2y' + 40y = 0$ $y(0) = 3$ $y'(0) = 7$ $y''(0) = 27$
(d) $y''' - 3y'' + y' - 3y = 0$ $y(0) = 0$ $y'(0) = 8$ $y''(0) = 24$
(e) $y''' - y'' + 4y' - 4y = 0$ $y(0) = 1$ $y'(0) = 0$ $y''(0) = 0$

(Hochstadt, 1964, see reference in chap. 2)

13. Solve the following differential equations:

(a) $y'' - 4y' + 3y = e^{2x}$ $y(0) = 0$ $y'(0) = 0$

(b) $y'' - 2y' - 3y = 4$ $y(0) = 1$ $y'(0) = -1$

(c) $y'' + y' - 2y = 3\cos 3x - 11\sin 3x$ $y(0) = 0$ $y'(0) = 6$

(d) $y''' - 3y'' - 4y' + 12y = 12e^{-x}$ $y(0) = 4$ $y'(0) = 2$ $y''(0) = 18$

(e) $y'' - 2y' + 2y = 0$ $y(0) = 0$ $y'(0) = 1$

(f) $y'' - 2y' + 2y = 2\cos 2x - 4\sin 2x$ $y(0) = 0$ $y'(0) = 0$

(g) $y' + 8y' + 16y = 0$ $y(0) = 3$ $y'(0) = -14$

(h) $y'' + 2y' + y = e^{-2x}$ $y(0) = 0$ $y'(0) = 0$

(i) $y''' + 6y'' + 12y' + 8y = 0$ $y(0) = 4$ $y'(0) = -12$ $y''(0) = 34$

(Kreyszig, 1962, see reference in chap. 2)

14. Following Theorem XIII, derive a relation for $Lt\ddot{y}(t)$ when $y(0)$ and $\dot{y}(0)$ are not zero.

REFERENCES

1. Davis, H. T.: "The Theory of Linear Operators," The Principia Press, Bloomington, Ind., 1936.
2. Whittaker, E. T.: Oliver Heaviside, *Bulletin of the Calcutta Mathematical Society*, vol. 20, p. 199, 1928–1929.
3. Laplace, P. S.: Sur les suites, oeuvres complètes, *Mémoires de l'académie des sci.*, vol. 10, pp. 1–89, 1779.
4. Carson, J. R.: "Electric Circuit Theory and the Operational Calculus," McGraw-Hill Book Company, New York, 1926.
5. Jeffreys, H.: "Operational Methods in Mathematical Physics," Cambridge University Press, New York, 1931.
6. Carslaw, H. S., and J. C. Jaeger: "Operational Methods in Applied Mathematics," Oxford University Press, New York, 1941.
7. McLachlan, N. W.: "Modern Operational Calculus," The Macmillan Company, New York, 1948.
8. McLachlan, N. W.: "Complex Variable Theory and Transform Calculus," Cambridge University Press, New York, 1953.
9. Gardner, M. S., and J. L. Barnes: "Transients in Linear Systems," John Wiley & Sons, Inc., New York, 1942.
10. Thomson, W. T.: "Laplace Transformation Theory and Engineering Applications," Prentice-Hall, Inc., New York, 1950.
11. van der Pol, B., and H. Bremmer: "Operational Calculus," Cambridge University Press, New York, 1950.
12. Pipes, L. A.: The Operational Calculus, *Journal of Applied Physics*, vol. 10, nos. 3–5, 1939.
13. Denis-Papin, M., and A. Kaufmann: "Cours de calcul opérationnel," Editions Albin Michel, Paris, 1950.
14. Doetsch, G.: "Theorie und Anwendung der Laplace-Transformation," Springer-Verlag OHG, Berlin, 1947.
15. Wagner, K. W.: "Operatorenrechnung," J. W. Edwards, Publisher, Inc., Ann Arbor, Mich., 1944.
16. Jaeger, J. C.: "An Introduction to the Laplace Transformation," John Wiley & Sons, Inc., New York, 1949.
17. Churchill, R. V.: "Modern Operational Mathematics in Engineering," McGraw-Hill Book Company, Inc., New York, 1944.

18. van der Pol, B., and K. F. Niessen: Symbolic Calculus, *Philosophical Magazine*, vol. 13, pp. 537–577, March, 1932.

19. Sears, W. R.: Operational Methods in the Theory of Airfoils in Non-uniform Motion, *Journal of the Franklin Institute*, vol. 230, pp. 95–111, 1940.

20. Pipes, L. A.: The Analysis of Retarded Control Systems, *Journal of Applied Physics*, vol. 19, no. 7, pp. 617–623, 1948.

21. Gross, B.: On Creep and Relaxation, *Journal of Applied Physics*, vol. 18, no. 2, pp. 212–221, 257–264, 1947.

22. van der Pol, B.: Application of the Operational or Symbolic Calculus to the Theory of Prime Numbers, *Philosophical Magazine*, ser. 7, vol. 26, pp. 912–940, 1938.

23. Heaviside, Oliver: "Electromagnetic Theory," Dover Publications, Inc., New York, 1950.

24. Pipes, L. A.: The Summation of Fourier Series by Operational Methods, *Journal of Applied Physics*, vol. 21, no. 4, pp. 298–301, 1950.

25. Jaeger, J. C.: The Solution of Boundary Value Problems by a Double Laplace Transformation, *Bulletin of the American Mathematical Society*, vol. 46, p. 687, 1940.

26. Voelker, D., and G. Doetsch: Die zweidimensionale Laplace-Transformation," Verlag Virkhauser, Basel, 1950. [Contains an extensive table of two-dimensional Laplace transforms.]

27. Estrin, T. A., and T. J. Higgins: The Solution of Boundary Value Problems by Multiple Laplace Transformations, *Journal of the Franklin Institute*, vol. 252, no. 2, pp. 153–167, 1951.

28. Pipes, L. A.: An Operational Treatment of Nonlinear Dynamical Systems, *Journal of the Acoustical Society of America*, vol. 10, pp. 29–31, 1938.

29. Pipes, L. A.: Operational Analysis of Nonlinear Dynamical Systems, *Journal of Applied Physics*, vol. 13, no. 2, February, 1942.

30. Pipes, L. A.: The Reversion Method for Solving Nonlinear Differential Equations, *Journal of Applied Physics*, vol. 23, no. 2, pp. 202–207, 1952.

31. Pipes, L. A.: Operational Methods in Nonlinear Mechanics, Report 51-10, University of California, Los Angeles, 1951.

32. Pipes, L. A.: Applications of Integral Equations to the Solution of Nonlinear Electric Circuit Problems, *Communication and Electronics*, no. 8, pp. 445–450, September, 1953.

33. McLachlan, N. W., and P. Humbert: Formulaire pour le calcul symbolique, *Memorial des sciences mathématiques*, fasc. 100, 1941.

34. Erdelyi, A. (ed.): "Tables of Integral Transforms," vol. I, McGraw-Hill Book Company, New York, 1954.

35. Fodor, G.: "Laplace Transforms in Engineering," Akademiai Kiado, publishing house of the Hungarian Academy of Sciences, Budapest, 1965.

36. Roberts, G. E., and H. Kaufman: "Table of Laplace Transforms," W. B. Saunders Company, Philadelphia, 1966.

37. Ryshik, I. M., and I. S. Gradstein: "Tables of Series, Products, and Integrals," 2d ed., VEB Deutscher verlag der Wissenschaften, Berlin, 1963.

38. McCollum, P. A., and B. F. Brown: "Laplace Transform Tables and Theorems," Holt, Rinehart and Winston, Inc., New York, 1965.

5
Oscillations of Linear Lumped Electrical Circuits

1 INTRODUCTION

A large part of the analysis in engineering and physics is concerned with the study of vibrating systems.   The electrical circuit is the most common example of a vibrating system.   By analogy the electrical circuit serves as a model for the study of mechanical and acoustical vibrating systems.   Historically the equations of motion of mechanical systems were developed a long time before any attention was given to the equations for electrical circuits.   It was because of this that in the early days of electrical-circuit theory it was natural to explain the action in terms of mechanical phenomena.   At the present time electrical-circuit theory has been developed to a much higher state than the theory of corresponding mechanical systems.   Mathematically the elements in an electrical network are the coefficients in the differential equations describing the network.   In the same way the coefficients in the differential equations of a mechanical or acoustical system may be looked upon as mechanical or acoustical elements.   Kirchhoff's electromotive-force law plays the same role in setting up the electrical equations as D'Alembert's principle does in setting up the mechanical and acoustical equations.   Therefore a mechanical or acoustical system may be represented by an analogous electrical network, and the problem may be solved by electrical-circuit theory.

2   ELECTRICAL-CIRCUIT PRINCIPLES

The differential equations for electrical circuits with lumped parameters are of the same form as the equations for mechanical systems.   Kirchhoff's first law is the application of the conservation-of-electricity principle to the circuit and may be stated in the following form:

1. The algebraic sum of all the currents into the junction point of a network is zero.

    Kirchhoff's second law is a statement concerning the conservation of energy in the circuit and is usually stated in the following form:

2. The algebraic sum of the electromotive forces around a closed circuit is zero.

    Let us apply the above Kirchhoff laws to the simple series circuit of Fig. 2.1.   The parameters R, L, and S of the circuit are expressed symbolically in the diagram below and are called the *resistance*, *inductance*, and *elastance* coefficients, respectively.   From basic principles, we have

$$E_R = iR = \text{electromotive-force drop due to resistance}$$

$$E_L = L\frac{di}{dt} = \text{electromotive-force drop due to inductance} \qquad (2.1)$$

$$E_S = Sq = \text{electromotive-force drop due to elastance}$$

Where q is the charge on the capacitor whose elastance is S, it is related to the current i by the equation

$$i = \frac{dq}{dt} \qquad (2.2)$$

The capacitance of the capacitor C is related to S by the equation

$$C = \frac{1}{S} \qquad (2.3)$$

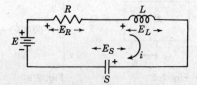

Fig. 2.1

Applying Kirchhoff's second law to the circuit, we have

$$L\frac{di}{dt} + Ri + Sq = e(t) \tag{2.4}$$

or, in terms of q, it is written in the form

$$L\ddot{q} + R\dot{q} + Sq = e(t) \tag{2.5}$$

where $\ddot{q} = d^2q/dt^2$ and $\dot{q} = dq/dt$.

If $R = 0$ and $S = 0$, the above equation reduces to

$$L\ddot{q} = e(t) \tag{2.6}$$

This equation has the same form as that governing the displacement x of a mass M when it is acted upon by a force F, as shown in Fig. 2.2. In the mechanical case, we have by Newton's law $(F = Ma)$ the equation

$$M\ddot{x} = F \tag{2.7}$$

If the mass is attached to a linear spring as shown in Fig. 2.3, we have by Newton's law

$$M\ddot{x} + kx = F \tag{2.8}$$

where k is the spring constant of the spring. This is analogous to Eq. (2.5) if we place $R = 0$. If in addition to the spring the mass of Fig. 2.3 were retarded by a force proportional to its velocity of the form $B\dot{x}$, then its equation of motion would be given by

$$M\ddot{x} + B\dot{x} + kx = F \tag{2.9}$$

This equation has exactly the same form as Eq. (2.5); we therefore see that we have the following equivalence between electrical and mechanical quantities:

$$
\begin{aligned}
e &\to F \\
L &\to M \\
R &\to B \\
S &\to k \\
q &\to x \\
i &\to \dot{x} = v
\end{aligned}
\tag{2.10}
$$

This correspondence between electrical and mechanical quantities is the basis of the electrical and mechanical analogies.

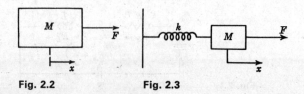

Fig. 2.2   Fig. 2.3

3 ENERGY CONSIDERATIONS

The energy of an oscillating electrical or mechanical system is of importance in studying the behavior of the system.   The potential energy stored in the spring of Fig. 2.3 when the mass has been displaced by a distance x is given by

$$V_M = \int_0^x F\,dx = \int_0^x kx\,dx = \frac{kx^2}{2} \tag{3.1}$$

In the electrical case the energy involved in charging the capacitor plays the same role as the elastic energy in the mechanical case.   This is given by

$$V_E = \int_0^t ei\,dt = \int_0^t Sqi\,dt = S \int_0^q q\,dq = \frac{Sq^2}{2} \tag{3.2}$$

The kinetic energy of motion of the mass is given by

$$T_M = \int_0^x F\,dx = \int_0^x M\ddot{x}\,dx = \int_0^v Mv\,dv = \frac{Mv^2}{2} \tag{3.3}$$

where $v = \dot{x}$ is the velocity of the mass.

In the electrical case we have

$$T_E = \tfrac{1}{2}Li^2 \tag{3.4}$$

The power dissipated in friction in the mechanical case is given by

$$P_M = v^2\,B \tag{3.5}$$

and the power dissipated in the electrical circuit is

$$P_E = i^2\,R \tag{3.6}$$

4 ANALYSIS OF GENERAL SERIES CIRCUIT

We have seen in Sec. 2 that the equation governing the current of the general series circuit of Fig. 2.1 is given by

$$L\frac{di}{dt} + Ri + Sq = e(t) \tag{4.1}$$

where

$$i = \dot{q} \tag{4.2}$$

This is a linear differential equation with constant coefficients, and we may solve it by the methods discussed in Chap. 2 or 4.   The Laplace-transform method is particularly well suited for solving this equation.   To do this, we let

$$\mathscr{L}i = I \qquad \mathscr{L}q = Q \tag{4.3}$$

Note: In order to avoid confusion with the inductance parameter, the Laplace transforms will be denoted by a script $\mathscr{L}$.

We therefore have from Chap. 4

$$\mathscr{L}\left(\frac{di}{dt}\right) = sI - i_0 \tag{4.4}$$

From (4.2) we have

$$\mathscr{L}i = \mathscr{L}\dot{q} = I = sQ - q_0 \tag{4.5}$$

where i_0 is the initial current flowing in the circuit and q_0 is the initial charge on the capacitor.   From (4.5) we obtain

$$Q = \frac{I}{s} + \frac{q_0}{s} \tag{4.6}$$

Equation (4.1) is transformed to

$$L(sI - i_0) + RI + SQ = \mathscr{L}\,e(t) \tag{4.7}$$

Eliminating Q, we obtain

$$\left(Ls + R + \frac{S}{s}\right)I + \frac{Sq_0}{s} - Li_0 = \mathscr{L}\,e(t) \tag{4.7a}$$

Solving for I, we obtain

$$I = \frac{1}{s^2 + (R/L)s + S/L}\,\frac{s\mathscr{L}e(t) + Lsi_0 - Sq_0}{L} \tag{4.8}$$

THE CASE OF FREE OSCILLATIONS

Let us first consider the case when there is no external electromotive force applied to the system.   In this case $e(t)$ is equal to zero.   To simplify the equation, let

$$a = \frac{R}{2L} \tag{4.9}$$

$$\omega_0 = \sqrt{\frac{S}{L}} = \sqrt{\frac{1}{LC}} \tag{4.10}$$

In this case we have

$$I = \frac{1}{s^2 + 2as + \omega_0^2}\,\frac{Lsi_0 - Sq_0}{L} \tag{4.11}$$

To obtain the inverse transform of this expression, we use Nos. 2.22 and 2.23 of our table of transforms (Appendix A).   There are three cases to consider:

(a) $\omega_0^2 > a^2$
(b) $\omega_0^2 = a^2$
(c) $\omega_0^2 < a^2$

Case a is called the *oscillatory case*. For this case we obtain from the table

$$i(t) = -\frac{i_0 \omega_0}{\omega_s} e^{-at} \sin(\omega_s t - \phi) - \frac{q_0}{LC} \frac{e^{-at} \sin \omega_s t}{\omega_s} \tag{4.12}$$

where

$$\omega_s = \sqrt{\omega_0^2 - a^2} \qquad \tan \phi = \frac{\omega_s}{a} \tag{4.13}$$

We thus see that the current is given as a damped oscillation, where i_0 and q_0 are the initial current and charge of the system.

If $\omega_0 = a$, we have case b. This is called the *critically damped case*. We then obtain

$$i(t) = i_0 e^{-at}(1 - at) - \frac{q_0}{LC} t e^{-at} \tag{4.14}$$

In this case there are no oscillations, and the current dies down in an exponential manner.

Case c is the *overdamped case*, and the solution for this case may be easily obtained from Nos. 2.22 and 2.23 of the table of transforms.

If there is no resistance present in the circuit, then $a = 0$ and our transform becomes

$$I = \frac{1}{s^2 + \omega_0^2} \frac{Lsi_0 - Sq_0}{L} \tag{4.15}$$

We may compute the inverse transform of this equation by the use of transform No. 2.6 of the table of transforms. We then have

$$i(t) = i_0 \cos w_0 t - \omega_0 q_0 \sin \omega_0 t \tag{4.16}$$

where now

$$\omega_0 = \frac{1}{\sqrt{LC}} \tag{4.17}$$

The foregoing equations give the velocity of the mass of the analogous mechanical system. In this case q_0 corresponds to the initial displacement x_0, and i_0 corresponds to the initial velocity of the mass of the mechanical system. The quantity $\omega_0 = 1/\sqrt{LC}$ in the case where we have no resistance is called the *natural angular frequency* of the electrical system. This corresponds to

$$\omega_0 = \sqrt{\frac{k}{M}} \tag{4.18}$$

for the mechanical system. Here k is the spring constant, and M is the mass of the system.

FORCED OSCILLATIONS

If $e(t)$ is not equal to zero, we must add to the preceding expressions the transform of

$$I = \frac{s\mathscr{L} e(t)}{L(s^2 + 2as + \omega_0^2)} \qquad (4.19)$$

The inverse transform of (4.19) gives the current in a general series circuit that has no initial current and no initial charge but has impressed on it an electromotive force $e(t)$ at $t = 0$. Let us compute this for the case in which $e(t)$ is an alternating potential of the form

$$e(t) = E_0 \sin \omega t \qquad (4.20)$$

By No. 2.4 of the tables we obtain

$$\mathscr{L} e(t) = \frac{E_0 \omega}{s^2 + \omega^2} \qquad (4.21)$$

Substituting this into (4.19), we obtain

$$I = \frac{E_0 \omega}{L} \frac{s}{(s^2 + \omega^2)(s^2 + 2as + \omega_0^2)} \qquad (4.22)$$

The inverse transform of this expression may be most simply computed by the use of No. T.17 of the table of transforms. For this case, we have

$$N(s) = s$$
$$D(s) = (s^2 + \omega^2)(s^2 + 2as + \omega_0^2) \qquad (4.23)$$

The roots of $D(s)$ are

$$
\begin{aligned}
s_1 &= -a + j\omega_s \\
s_2 &= -a - j\omega_s \qquad \text{where } \omega_s = \sqrt{\omega_0^2 - a^2} \\
s_3 &= +j\omega \\
s_4 &= -j\omega
\end{aligned} \qquad (4.24)
$$

We also have

$$D'(s) = 2s(s^2 + 2as + \omega_0^2) + (s^2 + \omega^2)(2s + a2) \qquad (4.25)$$

Hence, substituting into No. T.17 of the table of transforms, we obtain

$$i = \frac{E_0 \omega}{2L} \left[\frac{s_1 e^{s_1 t}}{(s_1 + a)(s^2 + \omega^2)} + \frac{s_2 e^{s_2 t}}{(s_2 + a)(s_2^2 + \omega^2)} \right]$$
$$+ \frac{E_0 \omega}{2L} \left(\frac{e^{j\omega t}}{\omega_0^2 + 2aj\omega - \omega^2} + \frac{e^{-j\omega t}}{\omega_0^2 - 2aj\omega - \omega^2} \right) \qquad (4.26)$$

Since the roots s_1 and s_2 have a negative real part, the expression within the brackets vanishes ultimately as time increases. The second expression may be denoted by i_s. It may be easily transformed to the form

$$i_s = \frac{E_0 \sin(\omega t - \theta)}{\sqrt{R^2 + (\omega L - 1/\omega C)^2}} \qquad \theta = \tan^{-1} \frac{\omega L - 1/\omega C}{R} \tag{4.27}$$

This term persists with the passage of time and is called the *steady-state term*. If R is zero so that the circuit is devoid of resistance, this becomes

$$i_s = \frac{E_0 \sin(\omega t \pm \pi/2)}{\omega L - 1/\omega C} \tag{4.28}$$

If $\omega = 1/\sqrt{LC} = \omega_0$, the denominator of (4.28) vanishes and the steady-state current amplitude becomes indefinitely great. This is the phenomenon of resonance and occurs when the impressed electromotive force has a frequency equal to that of the natural frequency of the circuit. If the circuit has resistance, then the denominator of (4.27) does not vanish and we do not have true resonance.

It may be noted that the steady-state current may be obtained more simply directly from the differential equation of the circuit

$$L\frac{di}{dt} + Ri + \frac{q}{C} = E_0 \sin \omega t \tag{4.29}$$

by the method of undetermined coefficients explained in Chap. 2, Sec. 6. The steady-state current is the particular integral of this equation. We replace the right-hand member of (4.29) by $\text{Im } E_0 e^{j\omega t}$, where "Im" means the "imaginary part of."

We let

$$i = \text{Im } A e^{j\omega t} \tag{4.30}$$

where A is a complex number to be determined by (4.22). Suppressing the Im symbol and realizing that

$$q = \int i \, dt = \int A e^{j\omega t} \, dt = \frac{A e^{j\omega t}}{j\omega} \tag{4.31}$$

substitution into (4.29) gives

$$\left(j\omega L + R + \frac{1}{j\omega C} \right) A = E_0 \tag{4.32}$$

where we have divided both sides by the common factor $e^{j\omega t}$. Hence we have

$$A = \frac{E_0}{R + j(\omega L - 1/\omega C)} \tag{4.33}$$

It is convenient to introduce the notation

$$Z = R + j\left(\omega L - \frac{1}{\omega C}\right) \tag{4.34}$$

This complex number is called the *complex impedance* of the circuit. It may be written in the polar form

$$Z = |Z|\, e^{i\theta} \tag{4.35}$$

where

$$|Z| = \sqrt{R^2 + (\omega L - 1/\omega C)^2} \quad\text{and}\quad \tan\theta = \frac{\omega L = 1/\omega C}{R} \tag{4.36}$$

We thus have

$$A = \frac{E_0}{|Z|}\, e^{-j\theta} \tag{4.37}$$

The steady-state current is now given by (4.30) in the form

$$i = \operatorname{Im} E_0 \frac{e^{-\theta j}\, e^{j\omega t}}{|Z|} = \frac{E_0}{|Z|} \sin(\omega t - \theta) \tag{4.38}$$

This is the steady-state current given in (4.27) obtained more directly. The solution (4.26), however, contains both the steady-state solution and the transient response of the system produced by the sudden application of the potential $E_0 \sin \omega t$ on the system at $t = 0$.

5   DISCHARGE AND CHARGE OF A CAPACITOR

An interesting application of the differential equations governing the distribution of charges and currents in electrical networks is the following one.

Consider the electrical circuit of Fig. 5.1. Let a charge q_0 be placed on the capacitor, and let the switch S be closed at $t = 0$. Let it be required to determine the charge on the capacitor at any instant later.

When the switch is closed, we have, by Kirchhoff's law, the equation

$$L\frac{d^2q}{dt^2} + R\frac{dq}{dt} + \frac{q}{C} = 0 \tag{5.1}$$

To solve this, let us introduce the transform

$$\mathscr{L}q = Q \tag{5.2}$$

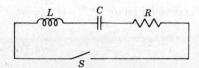

Fig. 5.1

The initial conditions of the problem are

$$\left. \begin{array}{l} q = q_0 \\ i = \dfrac{dq}{dt} = 0 \end{array} \right\} \quad \text{at } t = 0 \tag{5.3}$$

Hence we have

$$\mathscr{L}\frac{dq}{dt} = sQ - q_0$$

$$\mathscr{L}\frac{d^2q}{dt^2} = s^2 Q - sq_0 \tag{5.4}$$

Hence Eq. (5.1) transforms to

$$L(s^2 Q - sq_0) + R(sQ - q_0) + \frac{Q}{C} = 0 \tag{5.5}$$

or

$$\left(s^2 L + sR + \frac{1}{C} \right) Q = Lsq_0 + Rq_0 \tag{5.6}$$

As before, let

$$a = \frac{R}{2L} \qquad \omega_0 = \sqrt{\frac{1}{LC}} \tag{5.7}$$

We, therefore, have

$$Q = \frac{sq_0}{s^2 + 2as + \omega_0^2} + \frac{2aq_0}{s^2 + 2as + \omega_0^2} \tag{5.8}$$

By the use of transforms Nos. 2.22 and 2.23 of the table of transforms, we obtain

$$q = q_0 e^{-at}\left(\cosh \beta t + \frac{a}{\beta} \sinh \beta t \right) \qquad \text{if } a > \omega_0$$

$$q = q_0 e^{-at}(1 + at) \qquad \text{if } a = \omega_0 \tag{5.9}$$

$$q = q_0 e^{-at}\left(\cos \omega_s t + \frac{a}{\omega_s} \sin \omega_s t \right) \qquad \text{if } a < \omega_0$$

where $\omega_s = \sqrt{\omega_0^2 - a^2}$ and $\beta = \sqrt{a^2 - \omega_0^2}$.

THE CHARGING OF A CAPACITOR

Let us consider the circuit of Fig. 5.2. In this case, at $t = 0$, the switch is closed and the potential E of the battery is impressed on the circuit. It is required to determine the manner in which the charge on the capacitor behaves.

The equation satisfied by the charge is now

$$\ddot{q} + 2a\dot{q} + \omega_0^2 q = \frac{E}{L} \tag{5.10}$$

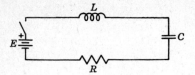

Fig. 5.2

To solve this equation, we again let

$$\mathscr{L}q = Q \tag{5.11}$$

and since E is a constant, we have

$$\mathscr{L}\frac{E}{L} = \frac{E}{Ls} \tag{5.12}$$

The initial conditions are now

$$\left.\begin{array}{l} q = 0 \\ \dot{q} = 0 \end{array}\right\} \quad \text{at } t = 0 \tag{5.13}$$

Hence we have

$$\begin{aligned} \mathscr{L}\ddot{q} &= s^2 Q \\ \mathscr{L}\dot{q} &= sQ \end{aligned} \tag{5.14}$$

Equation (5.10) transforms to

$$(s^2 + 2as + \omega_0^2) Q = \frac{E}{Ls} \tag{5.15}$$

and hence

$$Q = \frac{E}{L(s^2 + 2as + \omega_0^2)} \tag{5.16}$$

To obtain the inverse transform of (5.16), we must use transform No. 3.23 of the table of transforms and thus obtain

$$q = CE\left[1 - e^{-at}\left(\cosh \beta t + \frac{a}{\beta}\sinh \beta t\right)\right] \qquad \text{if } a > \omega_0$$

$$q = CE[1 - e^{-at}(1 + at)] \qquad \text{if } a = \omega_0 \tag{5.17}$$

$$q = CE\left[1 - e^{-at}\left(\cos \omega_s t + \frac{a}{\omega_s}\sin \omega_s t\right)\right] \qquad \text{if } a < \omega_0$$

where β and ω_s are as defined in (5.9).

In each case, the charging current is given by $i = \dot{q}$. The analogous mechanical problem is that of determining the motion of a mass when it has been given an initial displacement and is acted upon by a spring and retarded by viscous friction or if the mass has a sudden force applied to it.

6 CIRCUIT WITH MUTUAL INDUCTANCE

Let us consider the circuits of Fig. 6.1.   In this case we have two circuits coupled magnetically.   The coefficient L_{12} is termed the *mutual inductance coefficient*.   It is positive if the magnetic fields of i_1 and i_2 add.   If they are opposed, then the coefficient L_{12} is negative.   In any case the equations governing the currents in the two circuits are given by applying Kirchhoff's laws to the two loops and are

$$L_{11}\frac{di_1}{dt} + L_{12}\frac{di_2}{dt} + R_{11}i_1 = E$$

$$L_{22}\frac{di_2}{dt} + L_{12}\frac{di_1}{dt} + R_{22}i_2 = 0 \tag{6.1}$$

We wish to determine the currents i_1 and i_2 on the supposition that at $t = 0$ the switch s is closed and the initial currents are zero.   Let us introduce the transforms

$$\mathscr{L}i_1 = I_1$$

$$\mathscr{L}i_2 = I_2 \tag{6.2}$$

Now since we have

$$\left.\begin{array}{l} i_1 = 0 \\ i_2 = 0 \end{array}\right\} \quad \text{at } t = 0 \tag{6.3}$$

and also E is a constant, Eq. (6.1) transforms to

$$sL_{11}I_1 + sL_{12}I_2 + R_{11}I_1 = \frac{E}{s} \tag{6.4}$$

$$sL_{22}I_2 + sL_{12}I_1 + R_{22}I_2 = 0$$

We now solve these two algebraic equations by using Cramer's rule and obtain

$$I_1 = \frac{\begin{vmatrix} \dfrac{E}{s} & sL_{12} \\ 0 & sL_{22} + R_{22} \end{vmatrix}}{\begin{vmatrix} sL_{11} + R_{11} & sL_{12} \\ sL_{12} & sL_{22} + R_{22} \end{vmatrix}}$$

$$I_2 = \frac{\begin{vmatrix} sL_{11} + R_{11} & \dfrac{E}{s} \\ sL_{12} & 0 \end{vmatrix}}{\begin{vmatrix} sL_{11} + R_{11} & sL_{12} \\ sL_{12} & sL_{22} + R_{22} \end{vmatrix}} \tag{6.5}$$

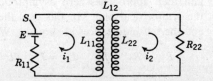

Fig. 6.1

Hence we have

$$I_1 = \frac{E(sL_{22} + R_{22})}{[(L_{11}L_{22} - L_{12}^2)s^2 + (R_{11}L_{22} + R_{22}L_{11})s + R_{11}R_{22}]s}$$

$$I_2 = \frac{-EL_{12}}{(L_{11}L_{22} - L_{12}^2)s^2 + (R_{11}L_{22} + R_{22}L_{11})s + R_{11}R_{22}}$$

(6.6)

If we let

$$a = \frac{R_{11}L_{22} + R_{22}L_{11}}{2(L_{11}L_{22} - L_{12}^2)}$$

(6.7)

and

$$\omega_0^2 = \frac{R_{11}R_{22}}{L_{11}L_{22} - L_{12}^2}$$

(6.8)

we then have

$$I_1 = \frac{E}{L_{11}L_{22} - L_{12}^2} \frac{sL_{22} + R_{22}}{(s^2 + 2as + \omega_0^2)s}$$

$$I_2 = \frac{-E}{L_{11}L_{22} - L_{12}^2} \frac{sL_{12}}{s^2 + 2as + \omega_0^2}$$

(6.9)

In this case

$$a^2 > \omega_0^2 \qquad \sqrt{a^2 - \omega_0^2} = \beta$$

(6.10)

Using the transforms Nos. 2.23 and 3.23 of the table of transforms, we obtain after some algebraic reductions

$$i_1 = \frac{E}{R_{11}}\left[1 - e^{-at}\cosh\beta t + \frac{(a^2 - \beta^2)L_{22} - aR_{22}}{\beta R_{22}}e^{-at}\sinh\beta t\right]$$

$$i_2 = \frac{(\beta^2 - a^2)L_{12}E}{\beta R_{11}R_{22}}e^{-at}\sinh\beta t$$

(6.11)

for the transforms of I_1 and I_2. We see that, as time elapses, i_1 approaches its final value E/R_{11}. If we set $di_2/dt = 0$ and solve for t, we find that i_2 rises to a maximum value when

$$t = \frac{1}{\beta}\tanh^{-1}\frac{\beta}{a}$$

(6.12)

and then approaches zero asymptotically. An interesting special case is the symmetrical one. In this case the resistances of each mesh are equal, and the self-inductances are equal. We then have

$$R_{11} = R_{22} = R \qquad L_{11} = L_{22} = L$$

$$L_{12} = M$$

(6.13)

Equations (6.4) then become

$$sLI_1 + sMI_2 + RI_1 = \frac{E}{s}$$

$$sLI_2 + sMI_1 + RI_2 = 0$$

(6.14)

If we add the two equations, we obtain

$$sL(I_1 + I_2) + sM(I_1 + I_2) + R(I_1 + I_2) = \frac{E}{s}$$

(6.15)

If we subtract the second equation from the first one, we have

$$sL(I_1 - I_2) - sM(I_2 - I_1) + R(I_1 - I_2) = \frac{E}{s}$$

(6.16)

If we now let

$$x_1 = I_1 + I_2 \qquad x_2 = I_1 - I_2$$

(6.17)

we have

$$s(L + M)x_1 + Rx_1 = \frac{E}{s}$$

$$s(L - M)x_2 + Rx_2 = \frac{E}{s}$$

(6.18)

Hence

$$x_1 = \frac{E}{[s(L + M) + R]s} \qquad x_2 = \frac{E}{[s(L - M) + R]s}$$

(6.19)

If we let

$$a_1 = \frac{R}{L + M} \qquad a_2 = \frac{R}{L - M}$$

(6.20)

we obtain

$$x_1 = \frac{E}{L + M}\frac{1}{(s + a_1)s} \qquad x_2 = \frac{E}{L - M}\frac{1}{(s + a_2)s}$$

(6.12)

Using transform No. 2.1 of the table of transforms, we have

$$\mathscr{L}^{-1}x_1 = \frac{E}{R}(1 - e^{-a_1 t}) \qquad \mathscr{L}^{-1}x_2 = \frac{E}{R}(1 - e^{-a_2 t})$$

(6.22)

Hence

$$i_1 + i_2 = \frac{E}{R}(1 - e^{-a_1 t})$$

$$i_1 - i_2 = \frac{E}{R}(1 - e^{-a_2 t})$$

(6.23)

and adding the two equations we obtain

$$i_1 = \frac{E}{R}\frac{2 - e^{-a_1 t} - e^{-a_2 t}}{2} \tag{6.24}$$

Subtracting the second equation from the first equation, we obtain

$$i_2 = \frac{E}{2R}(e^{-a_2 t} - e^{-a_1 t}) \tag{6.25}$$

These are the currents in the symmetrical case.

7 CIRCUITS COUPLED BY A CAPACITOR

Let us consider the circuit of Fig. 7.1. In this case, we have two coupled circuits. The coupling element is now a capacitor. Let the switch S be closed at $t = 0$, and let it be required to determine the current in the system. We write Kirchhoff's law for both meshes. We then obtain

$$L_1 \frac{di_1}{dt} + R_1 i_1 + \frac{q_1}{C_1} + \frac{q_1 - q_2}{C_{12}} = E$$
$$L_2 \frac{di_2}{dt} + R_2 i_2 + \frac{q_2}{C_2} + \frac{q_2 - q_1}{C_{12}} = 0 \tag{7.1}$$

where

$$i_1 = \frac{dq_1}{dt} \qquad i_2 = \frac{dq_2}{dt} \tag{7.2}$$

Let us introduce the transforms

$$\mathscr{L}i_1 = I_1 \qquad \mathscr{L}q_1 = Q_1$$
$$\mathscr{L}i_2 = I_2 \qquad \mathscr{L}q_2 = Q_2 \tag{7.3}$$

If we assume

$$\left.\begin{array}{ll} i_1 = 0 & q_1 = 0 \\ i_2 = 0 & q_2 = 0 \end{array}\right\} \quad \text{at } t = 0 \tag{7.4}$$

Eqs. (7.1) transform into

$$sL_1 I_1 + R_1 I_1 + \frac{I_1}{sC_1} + \frac{I_1 - I_2}{sC_{12}} = \frac{E}{s}$$
$$sL_2 I_2 + R_2 I_2 + \frac{I_2}{sC_2} + \frac{I_2 - I_1}{sC_{12}} = 0 \tag{7.5}$$

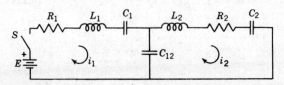

Fig. 7.1

We now solve these equations algebraically for the transforms I_1 and I_2. We thus obtain

$$
I_1 = \frac{\begin{vmatrix} \dfrac{E}{s} & -\dfrac{1}{sC_{12}} \\[3mm] 0 & sL_2 + R_2 + \dfrac{1}{sC_2} + \dfrac{1}{sC_{12}} \end{vmatrix}}{\begin{vmatrix} sL_1 + R_1 + \dfrac{1}{sC_1} + \dfrac{1}{sC_{12}} & -\dfrac{1}{sC_{12}} \\[3mm] -\dfrac{1}{sC_{12}} & sL_2 + R_2 + \dfrac{1}{sC_2} + \dfrac{1}{sC_{12}} \end{vmatrix}}
$$

$$
I_2 = \frac{\begin{vmatrix} sL_1 + R_1 + \dfrac{1}{sC_1} + \dfrac{1}{sC_{12}} & \dfrac{E}{s} \\[3mm] -\dfrac{1}{sC_{12}} & 0 \end{vmatrix}}{\begin{vmatrix} sL_1 + R_1 + \dfrac{1}{sC_1} + \dfrac{1}{sC_{12}} & -\dfrac{1}{sC_{12}} \\[3mm] -\dfrac{1}{sC_{12}} & sL_2 + R_2 + \dfrac{1}{sC_2} + \dfrac{1}{sC_{12}} \end{vmatrix}}
$$

$$\tag{7.6}$$

Expanding the determinants, we obtain the transforms I_1 and I_2 as the ratios of polynomials in s. The inverse transforms of I_1 and I_2 give the currents in the system. In this case, the determinant in the denominator of (7.6) of the system is a polynomial of the fourth degree in s. The inverse transforms may now be calculated by the use of No. T.17 in the table of transforms. This entails the solution of a quartic equation in s. If numerical values are given, this may be done by the Graeffe root-squaring method described in Appendix D. The trend of the general solution may be determined by solving the symmetrical case in which we have

$$
R_1 = R_2 = R \qquad C_1 = C_2 = C \qquad L_1 = L_2 = L \tag{7.7}
$$

In this case Eqs. (7.5) reduce to

$$
sLI_1 + RI_1 + \frac{I_1}{sC} + \frac{I_1}{sC_{12}} - \frac{I_2}{sC_{12}} = \frac{E}{s}
$$

$$
sLI_2 + RI_2 + \frac{I_2}{sC} + \frac{I_2}{sC_{12}} - \frac{I_1}{sC_{12}} = 0
$$

$$\tag{7.8}$$

If we add the two equations, we obtain

$$
sL(I_1 + I_2) + R(I_1 + I_2) + \frac{1}{sC}(I_1 + I_2) = \frac{E}{s} \tag{7.9}
$$

If we subtract the second equation from the first, we have

$$sL(I_1 - I_2) + R(I_1 + I_2) + \frac{1}{sC}(I_1 - I_2)$$

$$+ \frac{1}{sC_{12}}(I_1 - I_2) + \frac{1}{sC_{12}}(I_1 - I_2) = \frac{E}{s} \qquad (7.10)$$

If we let

$$x_1 = I_1 + I_2$$
$$x_2 = I_1 - I_2 \qquad\qquad\qquad (7.11)$$

we obtain

$$sLx_1 + Rx_1 + \frac{1}{sC}x_1 = \frac{E}{s}$$

$$sLx_2 + Rx_2 + \frac{1}{s}\left(\frac{1}{C} + \frac{2}{C_{12}}\right)x_2 = \frac{E}{s} \qquad (7.12)$$

If we let

$$\frac{R}{2L} = a \qquad \omega_1^2 = \frac{1}{LC} \qquad \omega_2^2 = \frac{1}{L}\left(\frac{1}{C} + \frac{2}{C_{12}}\right) \qquad (7.13)$$

the two equations become

$$(s^2 + 2as + \omega_1^2)x_1 = \frac{E}{L}$$

$$(s^2 + 2as + \omega_2^2)x_2 = \frac{E}{L} \qquad\qquad (7.14)$$

The inverse transforms of x_1 and x_2 may now be calculated by No. 2.22 in the table of transforms. In the case

$$\omega_1^2 > a^2 \qquad \omega_2^2 > a^2 \qquad\qquad\qquad (7.15)$$

we have

$$\mathscr{L}^{-1}x_1 = \frac{E}{L\omega_a}(e^{-at}\sin \omega_a t) \qquad \omega_a = \sqrt{\omega_1^2 - a^2}$$

$$\mathscr{L}^{-1}x_2 = \frac{E}{L\omega_b}(e^{-at}\sin \omega_b t) \qquad \omega_b = \sqrt{\omega_2^2 - a^2} \qquad (7.16)$$

Hence, adding the two equations (7.16), we have

$$i_1 = \frac{E}{2L}e^{-at}\left(\frac{\sin \omega_a t}{\omega_a} + \frac{\sin \omega_b t}{\omega_b}\right)$$

$$i_2 = \frac{E}{2L}e^{-at}\left(\frac{\sin \omega_a t}{\omega_a} - \frac{\sin \omega_b t}{\omega_b}\right) \qquad (7.17)$$

If there is no resistance in the circuit, then $a = 0$ and the currents oscillate without loss of amplitude with the angular frequencies ω_1 and ω_2.

8 THE EFFECT OF FINITE POTENTIAL PULSES

It frequently happens that the response of an electrical circuit is desired when a potential pulse is applied to it. The general procedure may be illustrated by the following example.

Consider the circuit of Fig. 8.1. Let $e(t)$ be a pulse of the form given by Fig. 8.2. It is required to find the current in the circuit. For generality, let us assume that the capacitor has an initial charge q_0 at $t = 0$. The current satisfies the equation

$$Ri + \frac{q}{C} = e(t) \qquad i = \frac{dq}{dt} \tag{8.1}$$

We introduce the transforms

$$\begin{aligned} \mathscr{L}i &= I \\ \mathscr{L}q &= Q \\ \mathscr{L}e &= E \end{aligned} \tag{8.2}$$

We now have

$$\mathscr{L}\frac{dq}{di} = sQ - q_0 = I \tag{8.3}$$

Hence

$$Q = \frac{I}{s} + \frac{q_0}{s} \tag{8.4}$$

The transform of the potential $e(t)$ is given by

$$E = \int_{t_1}^{t_2} E_0 e^{-st}\, dt = \frac{E_0(e^{st_1} - e^{-st_2})}{s} \tag{8.5}$$

Accordingly, Eq. (8.1) is transformed:

$$RI + \frac{1}{C}\left(\frac{I}{s} + \frac{q_0}{s}\right) = \frac{E_0(e^{-st_1} - e^{-st_2})}{s} \tag{8.6}$$

If we let

$$a = \frac{1}{RC} \tag{8.7}$$

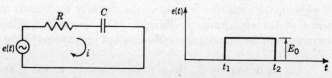

Fig. 8.1   **Fig. 8.2**

we obtain

$$I = \frac{E_0}{R} \frac{e^{-st_1} - e^{-st_2}}{s+a} - aq_0 \frac{1}{s+a} \tag{8.8}$$

To obtain the inverse transform of I, we use Nos. 1.1 and T.6 of the table of transforms. We thus obtain

$$i = -aq_0 e^{-at} \qquad\qquad\qquad\qquad 0 < t < t_1$$

$$i = -aq_0 e^{-at} + \frac{E_0}{R} e^{-a(t-t_1)} \qquad\qquad t_1 < t < t_2 \tag{8.9}$$

$$i = -aq_0 e^{-at} + \frac{E_0}{R} e^{-a(t-t_1)} - \frac{E_0}{R} e^{-a(t-t_2)} \qquad t > t_2$$

This example illustrates the general procedure.

9 ANALYSIS OF THE GENERAL NETWORK

In this section the analysis of a general n-mesh network will be considered. Given a network, we can draw n independent circulating currents so that they permit a different current in each branch of the network. We shall use the following notation:

    a. The resistance coefficients: R_{rs} is the resistance common to the i_r and i_s circuits. R_{rr} is the total resistance in the i_r circuit.

    b. The inductance coefficients: The inductance notation is complicated by the possibility of mutual inductance. If L_{rs}^1 is the total self-inductance common to i_r and i_s, and M_{rs} is the mutual inductance between the i_r and i_s circuits, then we define L_{rs} to be

$$L_{rs} = L_{rr}^1 \pm M_{rs} \tag{9.1}$$

The negative sign is used if M_{rs} opposes L_{rs}^1. We define L_{rr} as the *total self-inductance* in the i_r circuit.

    c. The elastance coefficients: In the circuits that we are considering, every capacitor will be traversed by one or more circulating currents. For a given current, capacitors appear in series but not in parallel. It is convenient if we write the equations in terms of the elastance coefficients rather than in terms of the capacitance coefficients. The elastance coefficients are the reciprocals of the capacitance coefficients and are denoted by the symbol S.

    Let us denote by S_{rs} the elastance common to the i_r and i_s circuits. We shall denote by S_{rr} the sum of the elastances in the i_r circuit.

THE RECIPROCITY RELATIONS

With the notations defined above, we have

$$R_{rs} = R_{sr} \qquad L_{rs} = L_{sr} \qquad S_{rs} = S_{sr} \tag{9.2}$$

These coefficients are called the *mutual parameters* of the network and S_{rr}, L_{rr}, and R_{rr} are called the *mesh parameters*.

THE GENERAL EQUATIONS

The set of differential equations describing the behavior of the general n-mesh networks may be written conveniently if we introduce the operators

$$Z_{rs}(D) = L_{rs}\frac{d}{dt} + R_{rs} + S_{rs} \int (\quad) \, dt$$

$$= L_{rs} D + R_{rs} + \frac{S_{rs}}{D} \tag{9.3}$$

If now $(e_1, e_2, \ldots, e_n)$ denote the potentials impressed on the contours of the meshes $1, 2, \ldots, n$, and $(i_1, i_2, \ldots, i_n)$ are the mesh currents of the corresponding n meshes, we have

$$
\begin{aligned}
Z_{11}(D)i_1 + Z_{12}(D)i_2 + \cdots + Z_{1n}(D)i_n &= e_1(t) \\
Z_{21}(D)i_1 + Z_{22}(D)i_2 + \cdots + Z_{2n}(D)i_n &= e_2(t) \\
\cdots \cdots \cdots \cdots \cdots \cdots \cdots \cdots \cdots \cdots \cdots \cdots & \\
Z_{n1}(D)i_1 + Z_{n2}(D)i_2 + \cdots + Z_{nn}(D)i_n &= e_n(t)
\end{aligned}
\tag{9.4}
$$

A concise form of these equations may be obtained by the use of matrix notation, as explained in Chap. 4. To do this, we introduce the following square matrices:

$$
[L] = \begin{bmatrix} L_{11} & L_{12} & \cdots & L_{1n} \\ L_{21} & L_{22} & \cdots & L_{2n} \\ \cdots & \cdots & \cdots & \cdots \\ L_{n1} & L_{n2} & \cdots & L_{nn} \end{bmatrix}
\qquad
[R] = \begin{bmatrix} R_{11} & R_{12} & \cdots & R_{1n} \\ R_{21} & R_{22} & \cdots & R_{2n} \\ \cdots & \cdots & \cdots & \cdots \\ R_{n1} & R_{n2} & \cdots & R_{nn} \end{bmatrix}
$$

$$
[S] = \begin{bmatrix} S_{11} & S_{12} & \cdots & S_{1n} \\ S_{21} & S_{22} & \cdots & S_{2n} \\ \cdots & \cdots & \cdots & \cdots \\ S_{n1} & S_{n2} & \cdots & S_{nn} \end{bmatrix}
\tag{9.5}
$$

In terms of these matrices we construct the square matrix $[Z(D)]$ by

$$[Z(D)] = [L] D + [R] + \frac{[S]}{D} \tag{9.6}$$

We also introduce the column matrices

$$\{e\} = \begin{Bmatrix} e_1 \\ e_2 \\ \vdots \\ e_n \end{Bmatrix} \qquad \{i\} = \begin{Bmatrix} i_1 \\ i_2 \\ \vdots \\ i_n \end{Bmatrix} \tag{9.7}$$

The set of differential equations (9.4) may then be written concisely in the form

$$[Z(D)]\{i\} = \{e\} \tag{9.8}$$

In the usual network problem we are given the various mesh charges and mesh currents of the system at $t = 0$, and the various mesh potentials $(e_1 e_2 \cdots e_n)$ are assumed impressed on the system at $t = 0$. We desire to find the subsequent distribution of currents in the system.

Let us introduce a column matrix $\{E(s)\}$ whose elements are the transforms of the elements of the matrix $\{e\}$. That is,

$$\mathcal{L}\{e\} = \{E(s)\} \tag{9.9}$$

In the same way introduce the column matrix $\{I(s)\}$.

$$\{I(s)\} = \mathcal{L}\{i\} \tag{9.10}$$

If we now introduce the column matrices

$$\{i^0\} = \begin{Bmatrix} i_1^0 \\ i_2^0 \\ i_3^0 \\ \vdots \\ i_n^0 \end{Bmatrix} \qquad \{q^0\} = \begin{Bmatrix} q_1^0 \\ q_2^0 \\ \vdots \\ q_n^0 \end{Bmatrix} \tag{9.11}$$

where $(i_1^0, i_2^0, \ldots, i_n^0)$ are the initial currents of the corresponding meshes of the system at $t = 0$ and $(q_1^0, q_2^0, \ldots, q_n^0)$ are the corresponding initial mesh charges of the system at $t = 0$, then the set of differential equations (9.8) transforms into

$$[Z(s)]\{I\} = \{E(s)\} + [L]\{i^0\} - \frac{1}{s}[S]\{q^0\} \tag{9.12}$$

If we now premultiply both sides of this equation by the inverse matrix $[Z(s)]^{-1}$, we obtain

$$I = [Z(s)]^{-1}\{E(s)\} + [Z(s)]^{-1}[L]\{i^0\} - \frac{1}{s}[Z(s)]^{-1}[S]\{q^0\} \tag{9.13}$$

We now obtain the various mesh currents from the equation

$$\{i\} = \mathcal{L}^{-1}\{I\} \tag{9.14}$$

This formally completes the solution of the problem.   In general the above algebraic procedure will give the elements of the matrix I as the ratios of polynomials in s.   The inverse transforms of these elements must be evaluated by using transform No. T.17 of the table of transforms.   The procedure involves the computation of the roots of the polynomial equation

$$|Z(s)| = 0 \tag{9.15}$$

In the general case this polynomial cannot be factored, and it is of the $2n$th degree.   The roots may be determined by the Graeffe root-squaring method of Appendix D.

If there are no initial mesh charges and mesh currents in the system, we have

$$\{i^0\} = \{q^0\} = \{0\} \tag{9.16}$$

and we have

$$\{I\} = [Z(s)]^{-1}\{E(s)\} \tag{9.17}$$

The solution in this case is that of a system initially at rest that has the potentials $\{e\}$ impressed on it at $t = 0$.

10   THE STEADY-STATE SOLUTION
ALTERNATING CURRENTS

The equations of the last section give the complete solution of a system having initial charges and potentials and also impressed electromotive forces of arbitrary form at $t = 0$.

There is one case of extreme practical importance in electrical engineering. This is the case where it is desired to find the so-called "steady-state current distribution" when the various mesh electromotive forces have the form

$$\{e\} = \begin{Bmatrix} E_1 \sin(\omega t + \phi_1) \\ E_2 \sin(\omega t + \phi_2) \\ E_3 \sin(\omega t + \phi_3) \\ \cdots \cdots \cdots \\ E_n \sin(\omega t + \phi_n) \end{Bmatrix} = \text{Im} \begin{Bmatrix} E_1\, e^{j\phi_1} \\ E_2\, e^{j\phi_2} \\ E_3\, e^{j\phi_3} \\ \cdots \\ E_n\, e^{j\phi_n} \end{Bmatrix} e^{j\omega t} \tag{10.1}$$

This is the case that occurs in alternating-current theory.   For generality, we shall assume that the various mesh potentials have the same frequency but differ in phase.   To determine the steady-state mesh currents, it is necessary to determine the particular integral of the set of equations

$$[Z(D)]\{i\} = \{e\} \tag{10.1a}$$

It is convenient to write

$$\{e\} = \text{Im}\{e\}_a\, e^{j\omega t} \tag{10.2}$$

where Im denotes the "imaginary part of" and

$$\{e\}_a = \begin{pmatrix} E_1\, e^{j\phi_1} \\ E_2\, e^{j\phi_2} \\ \cdots \cdots \\ E_n\, e^{j\phi_n} \end{pmatrix} \tag{10.3}$$

$$\{i\} = \mathrm{Im}\,\{I\}_a\, e^{j\omega t} \tag{10.4}$$

where $\{I\}_a$ is a column matrix whose elements are the unknown complex amplitudes of the steady-state currents to be determined by Eq. (10.1). Substituting (10.4) into (10.1) and suppressing the Im symbol, we obtain

$$[Z(j\omega)]\{I\}_a = \{e\}_a \tag{10.5}$$

Hence we have

$$\{I\}_a = [Z(j\omega)]^{-1}\{e\}_a \tag{10.6}$$

The steady-state currents are now given by (10.4). The elements of the column matrix $\{I\}_a$ are called the *complex currents* of the system. The above procedure is a generalization of the method of Sec. 4 for the single-mesh case. We have

$$[Z(j\omega)] = j\omega[L] + [R] + \frac{[S]}{j\omega} \tag{10.7}$$

This is called the *impedance matrix*. The terms

$$Z_{rr} = j\omega L_{rr} + R_{rr} + \frac{S_{rr}}{j\omega} \tag{10.8}$$

are called the *mesh impedances*, and the terms

$$Z_{rs} = j\omega L_{rs} + R_{rs} + \frac{S_{rs}}{j\omega} \qquad r \neq s$$

$$= Z_{sr} \tag{10.9}$$

are called the *mutual impedances*.

11   FOUR-TERMINAL NETWORKS IN THE ALTERNATING-CURRENT STEADY STATE

A special form of the general network discussed in Secs. 9 and 10 is one that has a pair of input and a pair of output terminals. It is customary to refer to such a network as a *four-terminal network* or a *two-terminal-pair network*. A four-terminal network is illustrated schematically by Fig. 11.1 as a box with two pairs of terminals.

The reference directions for voltage and current will be taken as those shown in Fig. 11.1. The four-terminal network will be supposed to consist

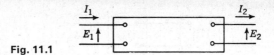

Fig. 11.1

of an n-mesh network having two pairs of accessible terminals as shown. The potentials E_1 and E_2 are complex potentials produced by sources external to the network, and the currents I_1 and I_2 are the complex currents in the corresponding meshes. The internal structure of the four-terminal network will be assumed to be quite general, and the n meshes will be supposed to have impedances Z_{kk} and complex mutual impedances Z_{ks} between the meshes. The restrictions on the component elements of the network are that they be linear and constant. It is further assumed that the network is passive; that is, the only sources of electromotive force are E_1 and E_2 produced by external agencies. It can be shown that, as a consequence of the general equations (10.5), we have the following linear relations:

$$E_1 = AE_2 + BI_2 \qquad I_1 = CE_2 + DI_2 \tag{11.1}$$

The coefficients A, B, C, D of Eqs. (11.1) are called *general four-terminal-network parameters*. It can be seen that as a consequence of the form of (11.1) A and D are in general complex numbers and that B has the dimension of an impedance and C the dimension of an admittance, or reciprocal impedance. As a result of Eq. (10.9) it can be shown that the following relation between the network parameters exists:

$$AD - BC = 1 \tag{11.2}$$

For many purposes it is convenient to express the pair of equations (11.1) as the following matrix equation:

$$\begin{bmatrix} E_1 \\ I_1 \end{bmatrix} = \begin{bmatrix} A & B \\ C & D \end{bmatrix} \begin{bmatrix} E_2 \\ I_2 \end{bmatrix} = [T] \begin{bmatrix} E_2 \\ I_2 \end{bmatrix} \tag{11.3}$$

where $[T]$ is the square matrix

$$[T] = \begin{bmatrix} A & B \\ C & D \end{bmatrix} \tag{11.4}$$

The square matrix $[T]$ is usually called the *transmission matrix* of the network. The determinant of $[T]$ is equal to unity as a consequence of the relation (11.2). The transmission matrix $[T]$ has the following inverse:

$$[T]^{-1} = \begin{bmatrix} D & -B \\ -C & A \end{bmatrix} \tag{11.5}$$

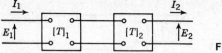

Fig. 11.2

If two four-terminal networks of transmission matrices T_1, T_2 are con-nected in cascade as shown in Fig. 11.2, then the over-all relations between the input and output quantities are given by

$$\begin{bmatrix} E_1 \\ I_1 \end{bmatrix} = [T]_1[T]_2 \begin{bmatrix} E_2 \\ I_2 \end{bmatrix} = [T]\begin{bmatrix} E_2 \\ I_2 \end{bmatrix} \tag{11.6}$$

where $[T]$ is the transmission matrix of the over-all network. It is given by

$$[T] = [T]_1[T]_2 = \begin{bmatrix} A_1 & B_1 \\ C_1 & D_1 \end{bmatrix}\begin{bmatrix} A_2 & B_2 \\ C_2 & D_2 \end{bmatrix}$$
$$= \begin{bmatrix} A_1 A_2 + B_1 C_2 & A_1 B_2 + B_1 D_2 \\ C_1 A_2 + D_1 C_2 & C_1 B_2 + D_1 D_2 \end{bmatrix} \tag{11.7}$$

It is thus seen that, in the cascade connection of several four-terminal networks, the over-all transmission matrix of the network is the matrix product of the transmission matrices of the individual networks taken in the order of connection. If a four-terminal network is *symmetrical* so that its input and output terminals may be interchanged without altering the current and potential distribution of the network, it may be shown that its transmission matrix has the property that $A = D$.

If the transmission matrices of several fundamental types of electrical circuits are known, then by matrix multiplication it is easy to obtain many useful properties of more complex structures formed by a cascade connection of fundamental circuits. The transmission matrices $[T]$ of several basic electrical circuits are listed in Table 1.

WAVE PROPAGATION ALONG A CASCADE OF SYMMETRICAL STRUCTURES

Many important problems of electrical-circuit theory such as those involving electric filters, delay lines, and transducers involve the determination of the nature of the current and potential distribution along a chain of identical symmetric four-terminal networks. Consider the cascade of n identical four-terminal networks as shown in Fig. 11.3. Let each of the four-terminal networks be a symmetrical one with the following transmission matrix:

$$[T] = \begin{bmatrix} A & B \\ C & A \end{bmatrix} \tag{11.8}$$

Table 1 Transmission matrices of fundamental four-terminal structures

No.	Network	Transmission matrix $\begin{bmatrix} A & B \\ C & D \end{bmatrix} = [T]$
1	Z Series impedance	$\begin{bmatrix} 1 & Z \\ 0 & 1 \end{bmatrix}$
2	Y Shunt admittance	$\begin{bmatrix} 1 & 0 \\ Y & 1 \end{bmatrix}$
3	L_1 M L_2 Coupled circuits	$\begin{bmatrix} \dfrac{L_1}{M} & \dfrac{jw(L_1 L_2 - M^2)}{M} \\ \dfrac{-j}{wM} & \dfrac{L_2}{M} \end{bmatrix}$
4	N_1 N_2 Ideal transformer $a = \dfrac{N_1}{N_2}$	$\begin{bmatrix} \dfrac{1}{a} & 0 \\ 0 & a \end{bmatrix}$
5	Y Z	$\begin{bmatrix} 1 & Z \\ Y & 1 + YZ \end{bmatrix}$
6	Z Y	$\begin{bmatrix} 1 + YZ & Z \\ Y & 1 \end{bmatrix}$

Table 1—*continued*

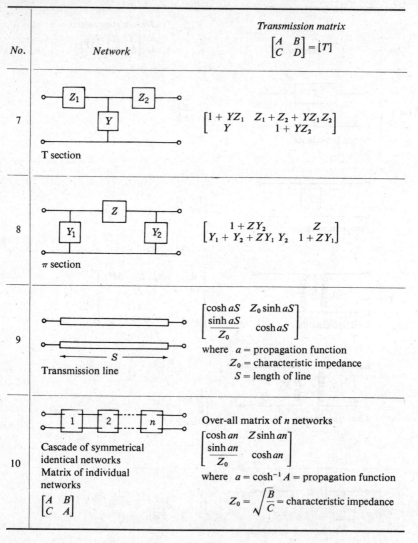

No.	Network	Transmission matrix $\begin{bmatrix} A & B \\ C & D \end{bmatrix} = [T]$
7	T section	$\begin{bmatrix} 1 + YZ_1 & Z_1 + Z_2 + YZ_1Z_2 \\ Y & 1 + YZ_2 \end{bmatrix}$
8	π section	$\begin{bmatrix} 1 + ZY_2 & Z \\ Y_1 + Y_2 + ZY_1Y_2 & 1 + ZY_1 \end{bmatrix}$
9	Transmission line	$\begin{bmatrix} \cosh aS & Z_0 \sinh aS \\ \dfrac{\sinh aS}{Z_0} & \cosh aS \end{bmatrix}$ where a = propagation function Z_0 = characteristic impedance S = length of line
10	Cascade of symmetrical identical networks Matrix of individual networks $\begin{bmatrix} A & B \\ C & A \end{bmatrix}$	Over-all matrix of n networks $\begin{bmatrix} \cosh an & Z \sinh an \\ \dfrac{\sinh an}{Z_0} & \cosh an \end{bmatrix}$ where $a = \cosh^{-1} A$ = propagation function $Z_0 = \sqrt{\dfrac{B}{C}}$ = characteristic impedance

Since each of the structures of the chain has the same matrix, the output potential and current E_n and I_n of the nth structure are related to the input potential and current E_0 and I_0 of the first structure by the following equation:

$$\begin{bmatrix} E_0 \\ I_0 \end{bmatrix} = \begin{bmatrix} A & B \\ C & A \end{bmatrix}^n \begin{bmatrix} E_n \\ I_n \end{bmatrix} \tag{11.9}$$

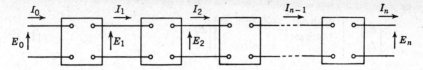

Fig. 11.3

In order to obtain a form for the transmission matrix $[T]$ that is convenient for computing powers of $[T]$, introduce the new variables a and Z_0 by the equations

$$a = \cosh^{-1} A \tag{11.10}$$

and

$$Z_0 = \left(\frac{B}{C}\right)^{\frac{1}{2}} \tag{11.11}$$

With this notation the transmission matrix $[T]$ takes the following form:

$$[T] = \begin{bmatrix} A & B \\ C & A \end{bmatrix} = \begin{bmatrix} \cosh a & Z_0 \sinh a \\ \dfrac{\sinh a}{Z_0} & \cosh a \end{bmatrix} \tag{11.12}$$

If the matrix $[T]$ is multiplied by itself, the following result is obtained:

$$[T][T] = [T]^2 = \begin{bmatrix} \sinh^2 a + \cosh^2 a & Z_0(2 \sinh a \cosh a) \\ \dfrac{2 \sinh a \cosh a}{Z_0} & \sinh^2 a + \cosh^2 a \end{bmatrix}$$

$$= \begin{bmatrix} \cosh 2a & Z_0 \sinh 2a \\ \dfrac{\sinh 2a}{Z_0} & \cosh 2a \end{bmatrix} \tag{11.13}$$

Similarly, by direct multiplication and by the use of the identities of hyperbolic trigonometry, it can be shown that

$$[T]^r = \begin{bmatrix} \cosh ra & Z_0 \sinh ra \\ \dfrac{\sinh ra}{Z_0} & \cosh ra \end{bmatrix} \qquad r = 0, \pm 1, \pm 2, \pm 3, \ldots \tag{11.14}$$

The result (11.14) is very useful in the study of the behavior of four-terminal networks and associated structures. By means of (11.14) Eq. (11.9) may be written in the form

$$\begin{bmatrix} E_0 \\ I_0 \end{bmatrix} = \begin{bmatrix} \cosh an & Z_0 \sinh an \\ \dfrac{\sinh an}{Z_0} & \cosh an \end{bmatrix} \begin{bmatrix} E_n \\ I_n \end{bmatrix} \tag{11.15}$$

or

$$\begin{bmatrix} E_n \\ I_n \end{bmatrix} = \begin{bmatrix} \cosh an & -Z_0 \sinh an \\ \dfrac{-\sinh an}{Z_0} & \cosh an \end{bmatrix} \begin{bmatrix} E_0 \\ I_0 \end{bmatrix} \tag{11.16}$$

The potential E_k and the current I_k along the chain of four-terminal structures of Fig. 11.3 are given by the equation

$$\begin{bmatrix} E_k \\ I_k \end{bmatrix} = \begin{bmatrix} \cosh ak & -Z_0 \sinh ak \\ \dfrac{-\sinh ak}{Z_0} & \cosh ak \end{bmatrix} \begin{bmatrix} E_0 \\ I_0 \end{bmatrix} \tag{11.17}$$

If the chain of four-terminal networks is terminated by an impedance equal to Z_0, then

$$I_n = \frac{E_n}{Z_0} \tag{11.18}$$

Then the impedance Z looking into the chain of structures may be obtained from (11.15) in the form

$$Z_i = \frac{E_0}{I_0} = \frac{E_n \cosh an + Z_0 I_n \sinh an}{(E_n/Z_0)\sinh an + I_n \cosh an} = Z_0 \tag{11.19}$$

The quantity Z_0 is called the *characteristic impedance* of the chain of four-terminal networks.   Hence, if the cascade of networks is terminated by the characteristic impedance Z_0, the impedance Z_i looking into the cascade is also equal to the characteristic impedance.   In this case the input current of the chain is

$$I_0 = \frac{E_0}{Z_0} \tag{11.20}$$

The potential and currents along the cascade may be obtained by the use of (11.17) if I_0 is eliminated by (11.20).   The results are

$$E_k = E_0 e^{-ak} \qquad I_k = \frac{E_0}{Z_0} e^{-ak} \qquad k = 0, 1, 2, 3, \ldots, n \tag{11.21}$$

The quantity a is the *propagation function* of the chain of four-terminal structures of Fig. 11.3.

ATTENUATION AND PASSBANDS

In general the propagation function a is a complex function of the angular frequency ω.   It has the form

$$a = \alpha + j\beta \tag{11.22}$$

The quantity α is called the *attenuation function*, and β is the *phase function* of the cascade. β gives the change of phase per section as one progresses along the cascade.   If the cascade of four-terminal networks is to pass a certain band of frequencies without attenuation, the real part of the propagation function must be zero and hence the propagation function a must be a pure imaginary quantity.   Therefore for a *passband* it is necessary that

$$\alpha = 0 \qquad a = j\beta \tag{11.23}$$

The relation (11.10) is then

$$\cosh a = \cosh j\beta = \cos \beta = A \tag{11.24}$$

It follows that the network constant A of the fundamental four-terminal network of the cascade must satisfy the two following conditions for a *passband*,

$$A = \cos \beta = \text{real} \tag{11.25}$$

and since $\cos \beta$ can vary only between -1 and $+1$, the following inequality must be satisfied:

$$|A| \leq 1 \tag{11.26}$$

A very important class of circuits are those which are devoid of resistance. Circuits of this class are called *dissipationless* circuits.   It can be shown that the parameter A is a real number if the fundamental circuit in the box is dissipationless so that (11.25) is always satisfied for circuits of this class. If the inequality (11.26) is satisfied, the cascade passes currents and potentials without attenuation.   If the inequality is not satisfied, the currents and potentials are attenuated along the cascade.   In general the parameter A is a function $A(\omega)$ of the angular frequency of the applied alternating potential. The ranges in frequencies that are attenuated or passed without attenuation may be determined by the use of the inequality (11.26).   This relation is fundamental in the theory of electrical filter circuits.

12   THE TRANSMISSION LINE AS A FOUR-TERMINAL NETWORK

A very important type of circuit is the transmission line shown schematically in Fig. 12.1.

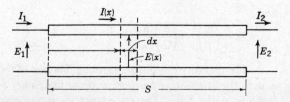

Fig. 12.1

This figure depicts a two-wire transmission line. In the alternating steady state this transmission line may be regarded as a continuous distribution of series impedance Z and shunt admittance Y per unit length, defined by the equations

$$Z = R + j\omega L \qquad Y = G + j\omega C \tag{12.1}$$

where R is the resistance of the line per unit length, L the inductance per unit length, C the capacitance per unit length, and G the leakage conductance per unit length.

Consider a length dx of the transmission line at a distance x from the sending end as shown in Fig. 12.1. This length of line contains a shunt admittance $Y\,dx$ and a series impedance $Z\,dx$. Let dI be the complex current that flows from one conductor to the other through the admittance $Y\,dx$, and let dE be the change in the complex potential difference between the conductors caused by the current $I(x)$ flowing in the impedance $Z\,dx$. If higher-order infinitesimals are neglected, it is seen that, by an application of Kirchhoff's laws to a section of the line of length, dx, we obtain the following equations:

$$dI = -EY\,dx \qquad dE = -IZ\,dx \tag{12.2}$$

The complex current $I(x)$ and the complex potential $E(x)$ therefore satisfy the following two differential equations:

$$\frac{dI}{dx} = -YE \qquad \frac{dE}{dx} = -ZI \tag{12.3}$$

If these two simultaneous first-order differential equations are solved subject to the boundary conditions that $E = E_1$ and $I = I_1$ at $x = 0$, the result is

$$E(x) = E_1 \cosh ax - I_1 Z_0 \sinh ax$$

$$I(x) = \frac{-E_1}{Z_0} \sinh ax + I_1 \cosh ax \tag{12.4}$$

where

$$a = (ZY)^{\frac{1}{2}} = [(R + j\omega L)(G + j\omega C)]^{\frac{1}{2}} = \alpha + j\beta \tag{12.5}$$

and

$$Z_0 = \left(\frac{Z}{Y}\right)^{\frac{1}{2}} = \left(\frac{R + j\omega L}{G + j\omega C}\right)^{\frac{1}{2}} \tag{12.6}$$

The quantity a is the *propagation function* of the line. The real part of a, α, is the *attenuation function*, and the imaginary part of a, β, is the *phase function* of the line. The impedance Z_0 is the *characteristic impedance* of the line.

If S is the length of the line and I_2 and E_2 the output complex current and potential of the line, Eqs. (12.4) yield the following values for E_2 and I_2:

$$E(S) = E_2 = E_1 \cosh aS - I_1 Z_0 \sinh aS$$

$$I(S) = I_2 = -\frac{E_1}{Z_0} \sinh aS + I_1 \cosh aS \tag{12.7}$$

The two equations (12.7) may be written in the convenient matrix form

$$\begin{bmatrix} E_2 \\ I_2 \end{bmatrix} = \begin{bmatrix} \cosh \theta & -Z_0 \sinh \theta \\ -\dfrac{\sinh \theta}{Z_0} & \cosh \theta \end{bmatrix} \begin{bmatrix} E_1 \\ I_2 \end{bmatrix} \qquad \theta = aS \tag{12.8}$$

Equation (12.8) may also be written in the alternative form

$$\begin{bmatrix} E_1 \\ I_1 \end{bmatrix} = \begin{bmatrix} \cosh \theta & Z_0 \sinh \theta \\ \dfrac{\sinh \theta}{Z_0} & \cosh \theta \end{bmatrix} \begin{bmatrix} E_2 \\ I_2 \end{bmatrix} \tag{12.9}$$

It is thus evident from (12.9) that the transmission matrix of the two-wire line of Fig. 12.1 is given by

$$[T] = \begin{bmatrix} \cosh \theta & Z_0 \sinh \theta \\ \dfrac{\sinh \theta}{Z_0} & \cosh \theta \end{bmatrix} \tag{12.10}$$

This matrix has the following inverse:

$$[T]^{-1} = \begin{bmatrix} \cosh \theta & -Z_0 \sinh \theta \\ -\dfrac{\sinh \theta}{Z_0} & \cosh \theta \end{bmatrix} \tag{12.11}$$

Calculations involving transmission lines may be greatly simplified by the use of the matrices (12.10) and (12.11). The quantity $\theta = aS$ is called the *angle of the line* and is, in general, a complex number. As an example of the use of the transmission matrix $[T]$, consider a transmission line short-circuited at the end $x = S$. In this case $E_2 = 0$, and from Eq. (12.9) we have the relation

$$E_1 = I_2 Z_0 \sinh \theta \tag{12.12}$$

Hence the current at the short-circuited end is given by the equation

$$I_2 = \frac{E_1}{Z_0 \sinh \theta} \tag{12.13}$$

If, on the other hand, the end $x = S$ is open-circuited, $I_2 = 0$ and (12.9) give $E_2 = E_1/\cosh \theta$ for the open-circuit potential.

PROBLEMS

1.   What is the analogous electrical circuit for a mechanical pendulum that is subjected to very small displacements?

2.   A circuit consisting of an inductance L in series with a capacitance C has impressed on it at $t = 0$ a potential $e(t) = E_0 t/T_0$ $(0 < t < T_0)$ and $e(t) = 0$, $t > T_0$.   Find the current in the system.

3.   Two circuits are coupled with a mutual inductance M.   One contains an impressed potential E_0, R, L_1, and a switch in series; the other circuit has L_2 and C in series.   For what value of the circuit constants are oscillations possible?

4.   Each side of an equilateral triangular circuit contains a capacitance C, and each vertex is connected to a common central point by an inductance L.   Show that the possible oscillations of this circuit have the period $T = 2\pi\sqrt{3LC}$.

5.   Two points are connected by three branches, two of which contain both a capacitance C and an inductance L, and the third only a capacitance C.   Show that the angular frequencies of oscillation of the network are $1/\sqrt{LC}$ and $\sqrt{3/LC}$.

6.   Two circuits L_1, R_1 and L_2, R_2 are coupled by a mutual inductance M, where $M^2 = L_1 L_2$. A constant electromotive force E is applied at $t = 0$ in the primary circuit.   The initial currents are zero.   Determine the currents in the circuits.

7.   An electromotive force $E\cos(\omega t + \theta)$ is applied at $t = 0$ to a circuit consisting of capacitance C and inductance L in series.   The initial charge and current are zero.   Find the current at time t.

8.   An electromotive force $E\sin\omega t$, where $\omega = 1/\sqrt{LC}$, is applied at $t = 0$ to a circuit consisting of capacitance C and inductance L in series.   The initial current and charge are zero.   Find the current.   (This is the case of resonance.)

9.   Show that a combination of capacitance C shunted by resistance R in series with a combination of inductance L shunted by resistance R behaves as a pure resistance for all forms of applied electromotive force if $L = CR^2$.

10.   Two resistanceless circuits L_1, C_1 and L_2, C_2 are coupled by mutual inductance M. If at $t = 0$, when the currents and charges are zero, a battery of electromotive force E_0 is applied in the primary, find the current in the secondary.

11.   A circuit consists of a resistance, an inductance, and a capacitance in series.   An electromotive force $e = E_1\cos\omega_1 t + E_2\cos\omega_2 t$ is impressed on the circuit.   Find the steady-state current.

12.   A series circuit consisting of an inductance capacitance and resistance has an electromotive force $e(t)$ of the "meander" type, as shown in the figure, impressed on it.   Find the steady-state current.

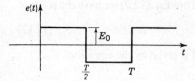

Fig. P 12

13.   Two transmission lines of series impedances Z_1 and Z_2 per length and shunt admittances Y_1 and Y_2 are connected in cascade.   The transmission lines have lengths S_1 and S_2, respectively.   Find the over-all transmission matrix of the combination.   The receiving end of the second transmission line is short-circuited.   Find the current at this end in terms of the complex potential applied to the sending end of the first transmission line.

14.   A circuit is composed of a cascade connection of n four-terminal networks.   All the four-terminal networks of the cascade are identical and contain the structure shown in Fig. P 14.   Obtain the over-all transmission matrix of the cascade, and determine the range in ω that gives the passband and the stopband.

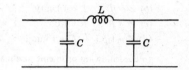

Fig. P 14

15.   The same as Prob. 14 for the case in which all four-terminal networks contain the structure shown in Fig. P 15.

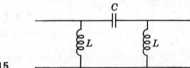

Fig. P 15

16.   The cascade of four-terminal networks of Prob. 14 is short-circuited at the receiving end.   A potential $E_0 \sin \omega t$ is impressed at the sending end of the cascade.   Determine the instantaneous value of the receiving current when ω lies in a stopband and when ω lies in a passband.

17.   Using (11.1) and the standard short- and open-circuit test conditions, derive

    (*a*) Transmission matrix No. 1, Table 1

    (*b*) Transmission matrix No. 2, Table 2

    (*c*) Transmission matrix No. 5, Table 2

    (*d*) Transmission matrix No. 6, Table 2

    (*e*) Transmission matrix No. 7, Table 2

    (*f*) Transmission matrix No. 8, Table 2

18.   Using transmission matrices Nos. 1 and 2 in Table 1 and matrix multiplication, derive

    (*a*) Transmission matrix No. 5

    (*b*) Transmission matrix No. 6

    (*c*) Transmission matrix No. 7

    (*d*) Transmission matrix No. 8

    Bode plots are plots of amplitude ratio (in decibels) and the phase angle (in degrees) versus the logarithm (base 10) of the frequency.   In terms of voltages, the amplitude ratio is given by $20 \log$ *magnitude of output/input voltage* (decibels), and the phase angle is the angle by which the output voltage leads (+angle) or lags (−angle) the input voltage.   To solve the following problems, assume that the input voltage E_1 is $E_0 \sin \omega t$, and that the output is an open circuit, that is, $I_2 = 0$.

19. Draw Bode plots for the transmission matrix No. 1, Table 2, when

$\left.\begin{array}{lll} (a)\ Z = j\omega L & L = 1 \text{ henry} \\ (b)\ Z = R & R = 1 \text{ ohm} \\ (c)\ Z = 1/j\omega C & C = 1 \text{ farad} \end{array}\right\}\quad 0.01 \leqslant \omega \leqslant 100$

20. Draw Bode plots for the transmission matrix No. 6, Table 2, for $0.01 \leqslant \omega \leqslant 100$ when

$(a)\ Z = R \qquad R = 1 \text{ ohm} \qquad Y = jC \qquad C = 1 \text{ farad}$

$(b)\ Z = jL \qquad L = 1 \text{ henry} \qquad Y = jC \qquad C = 1 \text{ farad}$

$(c)\ Z = R \qquad R = 1 \text{ ohm} \qquad Y = 1/jL \qquad L = 1 \text{ henry}$

$(d)\ Z = 1/jC \qquad C = 1 \text{ farad} \qquad Y = 1/R \qquad R = 1 \text{ ohm}$

In the following problems, use matrix methods to solve for the indicated circulating steady-state currents.

21.

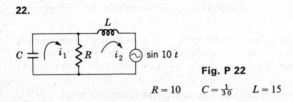

Fig. P 21

$$R_1 = 5 \qquad R_2 = 2 \qquad R_3 = 3$$

22.

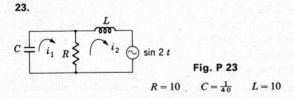

Fig. P 22

$$R = 10 \qquad C = \tfrac{1}{30} \qquad L = 15$$

23.

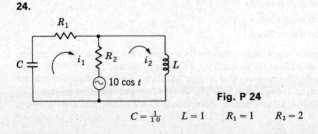

Fig. P 23

$$R = 10 \qquad C = \tfrac{1}{40} \qquad L = 10$$

24.

R_1

C — i_1 R_2 i_2 L

$10 \cos t$

Fig. P 24

$$C = \tfrac{1}{10} \qquad L = 1 \qquad R_1 = 1 \qquad R_2 = 2$$

25.

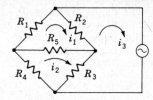

Fig. P 25

$$R_1 = 1 \qquad R_2 = 3 \qquad R_3 = 1 \qquad R_4 = 2 \qquad R_5 = 4$$

26.

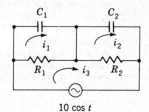

Fig. P 26   10 cos t

$$R_1 = 2 \qquad R_2 = 1 \qquad C_1 = \tfrac{1}{4} \qquad C_2 = \tfrac{1}{2}$$

In the following problems assume that all the initial conditions are zero and solve for the transient and steady-state currents when the switch S is closed at $t = 0$.

27.

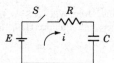

Fig. P 27

28.

Fig. P 28

29.

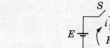

Fig. P 29

30.

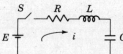
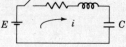

Fig. P 30

31.

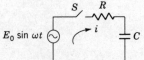

Fig. P 31

32.

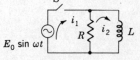

Fig. P 32 $E_0 \sin \omega t$

33.

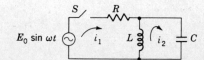

Fig. P 33

REFERENCES

1920. Carson, J. R.: "Electric Circuit Theory and the Operational Calculus," McGraw-Hill Book Company, New York.
1920. Pierce, G. W.: "Electric Oscillations and Electric Waves," McGraw-Hill Book Company, New York.
1929. Berg, E. J.: "Heaviside's Operational Calculus," McGraw-Hill Book Company, New York.
1929. Bush, V.: "Operational Circuit Analysis," John Wiley & Sons, Inc., New York.
1931. Guillemin, E. A.: "Communication Networks," John Wiley & Sons, Inc., New York.
1939. Pipes, L. A.: The Operational Calculus, *Journal of Applied Physics*, vol. 10, Nos. 3–5.
1940. Pipes, L. A.: The Matrix Theory of Four-terminal Networks, *Philosophical Magazine*, ser. 7, vol. 30, pp. 370–395, November.
1943. Jackson, L. S.: "Wave Filters," John Wiley & Sons, Inc., New York.
1943. Kerchner, R. M., and F. Corcoran: "Alternating Current Circuits," John Wiley & Sons, Inc., New York.
1945. Bode, H. W.: "Network Analysis and Feedback Amplifier Design," D. Van Nostrand Company, Inc., Princeton, N.J.
1946. Josephs, H. J.: "Heaviside's Electric Circuit Theory," John Wiley & Sons, Inc., New York.
1952. LePage, W., and S. Seely: "General Network Analysis," McGraw-Hill Book Company, New York.
1955. *IRE Transactions on Circuit Theory*, vol. CT-2, no. 2, June.
1963. Paskusz, G. F., and B. Bussel: "Linear Circuit Analysis," Prentice-Hall, Inc., Englewood Cliffs, N.J.
1963. Pipes, L. A.: "Matrix Methods for Engineering," Prentice-Hall, Inc., Englewood Cliffs, N.J.
1965. Skilling, H. H.: "Electrical Engineering Circuits," John Wiley & Sons, Inc., New York.
1965. Toro, V. D.: "Principles of Electrical Engineering," Prentice-Hall, Inc., Englewood Cliffs, N.J.

6
Oscillations of Linear Mechanical Systems

1 INTRODUCTION

One of the most important and interesting subjects of applied mathematics is the theory of small oscillations of mechanical systems in the neighborhood of an equilibrium position or a state of uniform motion. In the last chapter we considered the oscillations of a very important vibrating system, the electrical circuit. By the analogy between electrical and mechanical systems all the methods discussed in the last chapter may be used in the analysis of mechanical systems. In the mechanical system we are usually concerned with the determination of the natural frequencies and modes of oscillation rather than the complete solution for the amplitudes subject to the initial conditions of the system. In this chapter we shall use the classical method of solution rather than the Laplace-transform method. By comparing the analysis of the equivalent electrical circuits of the last chapter which were analyzed by the

Laplacian-transformation method and the classical analysis of the mechanical systems in this chapter, a proper perspective of the utility of the two methods will be apparent.

2 OSCILLATING SYSTEMS WITH ONE DEGREE OF FREEDOM

Let us consider the vibrating systems of Fig. 2.1. System a represents a mass that is constrained to move in a linear path. It is attached to a spring of spring constant k and is acted upon by a dashpot mechanism that introduces a frictional constraint proportional to the velocity of the mass. The mass has exerted upon it an external force $P_0 \sin \omega t$. By Newton's law we have

$$M\ddot{x} = -Kx - R\dot{x} + P_0 \sin \omega t \qquad \begin{cases} \dot{x} = \dfrac{dx}{dt} \\[2mm] \ddot{x} = \dfrac{d^2 x}{dt^2} \end{cases} \qquad (2.1)$$

where K is the spring constant and R is the friction coefficient of the dashpot.

System b represents a system undergoing torsional oscillations. It consists of a massive disk of moment of inertia J attached to a shaft of torsional stiffness K. The disk undergoes torsional damping proportional to its angular velocity $\dot{\theta}$. The disk has exerted upon it an oscillatory torque $T_0 \sin \omega t$. By Newton's law we have

$$J\ddot{\theta} = -K\theta - R\dot{\theta} + T_0 \sin \omega t \qquad (2.2)$$

System c is a series electrical circuit having inductance, resistance, and elastance. By Kirchhoff's law the equation satisfied by the mesh charge q is

$$L\ddot{q} + R\dot{q} + Sq = E_0 \sin \omega t \qquad (2.3)$$

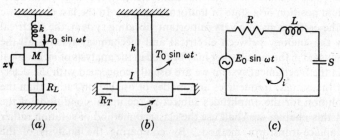

(a) $\qquad\qquad$ (b) $\qquad\qquad$ (c)

Fig. 2.1

By comparing these three equations, we obtain the following table of analogues:

Linear		Torsional		Electrical	
Mass	M	Moment of inertia	J	Inductance	L
Stiffness	K	Torsional stiffness	K	Elastance	$S = 1/C$
Damping	R	Torsional damping	R	Resistance	R
Impressed force	$F_0 \sin \omega t$	Impressed torque	$T_0 \sin \omega t$	Impressed potential	$E_0 \sin \omega t$
Displacement	x	Angular displacement	θ	Capacitor charge	q
Velocity	$\dot{x} = v$	Angular velocity	$\dot{\theta} = \omega$	Current	$i = \dot{q}$

We see from this table of analogues that it is necessary only for us to analyze one system and then by means of the table we may obtain the corresponding solution for the others.

FREE VIBRATIONS

Let us consider the system a when no impressed force is present. In this case, Eq. (2.1) reduces to

$$M\ddot{x} + R\dot{x} + Kx = 0 \tag{2.4}$$

This equation describes the free vibrations of the mass of Fig. 2.1. To find the general solution of (2.4), we find two nontrivial solutions. An arbitrary linear combination of the two particular solutions will then be the desired solution.

Since Eq. (2.4) is linear and homogeneous, we know from the general theory of Chap. 2 that we can obtain a particular solution of the form e^{st}, where s is a constant to be determined. If we substitute e^{st} for x in (2.4) and divide out the factor e^{st}, we obtain the quadratic equation

$$Ms^2 + Rs + K = 0 \tag{2.5}$$

This equation determines the quantity s. Let us denote the two roots of the quadratic equation (2.5) by s_1 and s_2. We then have

$$x = c_1 e^{s_1 t} + c_2 e^{s_2 t} \tag{2.6}$$

for the general solution of (2.5), where c_1 and c_2 are arbitrary constants. The nature of the solution depends on the nature of the roots. There are three cases to consider.

a. The case $R^2 - 4MK > 0$: In this case s_1 and s_2 are real and unequal. The general solution is

$$x = c_1 e^{s_1 t} + c_2 e^{s_2 t} \tag{2.7}$$

Both roots are negative, and (2.7) represents a disturbance that vanishes as t approaches infinity.

b. The case $R^2 - 4MK = 0$: In this case the quadratic equation (2.5) has a double root

$$s_1 = s_2 = \frac{-R}{2M} \tag{2.8}$$

Thus $c_1 e^{(-R/2M)t}$ is the only solution obtainable from (2.5). However, the general theory of repeated roots as discussed in Chap. 2 shows that there exists a second solution of the form $x = t e^{(-R/2M)t}$. The general solution is therefore

$$x = (c_1 + c_2 t) e^{(-R/2M)t} \tag{2.9}$$

In both these cases we have the so-called "aperiodic motion." As time increases, the displacement of the mass approaches zero asymptotically without oscillating about $x = 0$. This means that the effect of damping is so great that it prevents the elastic force from setting up oscillatory motions.

c. The case $R^2 - 4MK < 0$: In this case, the roots s_1 and s_2 are complex conjugates. If we let

$$\omega_s^2 = \frac{K}{M} - \frac{R^2}{4M^2} \tag{2.10}$$

we may write the general solution in the form

$$x = c_1 e^{s_1 t} + c_2 e^{s_2 t} = e^{(-R/2M)t}(c_1 e^{j\omega_s t} + c_2 e^{-j\omega_s t}) \tag{2.11}$$

By using the Euler relation, this becomes

$$x = e^{(-R/2M)t}(c_1' \cos \omega_s t + c_2' \sin \omega_s t) \tag{2.12}$$

where c_1' and c_2' are two new arbitrary constants. If we let

$$c_1' = A \cos \omega_s \delta \quad \text{and} \quad c_2' = A \sin \omega_s \delta \tag{2.13}$$

where A and δ are two new arbitrary constants, the (2.12) becomes

$$x = A e^{(-R/2M)t} \cos \omega_s(t - \delta) \tag{2.14}$$

The constant A is called the *amplitude of the motion*, and $\omega_s \delta$ is called the *phase*.

$$\omega_s = \sqrt{\frac{K}{M} - \frac{R^2}{4M^2}} \tag{2.15}$$

is called the *angular frequency of the motion*. The motion represented by (2.14) is quite different from the aperiodic motion discussed in cases *a* and *b*. In this case the damping is small compared with the elastic force, and the motion of the mass is oscillatory. The damping is not so small as to be considered negligible. The motion is one of damped harmonic oscillations. The oscillations behave like cosine waves with an angular frequency of ω_s except that the maximum value of the displacement attained with each oscillation is not constant. The "amplitude" is given by the expression $A e^{(-R/2M)t}$ and decreases exponentially as t increases.

The quantity $R/2M$ is called the *logarithmic decrement*. It indicates that the logarithm of the maximum displacement decreases at the rate $R/2M$. If there is no damping present $(R = 0)$, we obtain harmonic oscillations with natural angular frequency

$$\omega_0 = \sqrt{\frac{K}{M}} \tag{2.16}$$

The amplitude A and the phase $\omega_s \delta$ must be determined from the initial state of the system. If, for example,

$$\left.\begin{array}{l} x = 0 \\ \dot{x} = \dot{x}_0 \end{array}\right\} \quad \text{at } t = 0 \tag{2.17}$$

then

$$x = \frac{\dot{x}_0}{\omega_s} e^{-Rt/2M} \sin \omega_s t \tag{2.18}$$

is the particular solution.

FORCED OSCILLATIONS

If an external force $F(t)$ is impressed upon the physical system a above, the equation governing the displacement x of the mass is given by

$$M\ddot{x} + R\dot{x} + Kx = F(t) \tag{2.19}$$

From the general theory of Chap. 2 we know that the general solution of (2.19) is the superposition of the general solution of the homogeneous equation (2.4) and the particular solution of (2.19). Let the force (Ft) be a periodic force of the form

$$F_0 e^{j\omega t} = F_0 \cos \omega t + j F_0 \sin \omega t$$
$$= F(t) \tag{2.20}$$

Accordingly we have

$$M\ddot{x} + R\dot{x} + Kx = F_0 e^{j\omega t} \tag{2.21}$$

Since this equation is linear, we attempt to find a solution of the form

$$x = A e^{j\omega t} \tag{2.22}$$

We realize that the real part of the solution corresponds to the force function $F_0 \cos \omega t$ and the imaginary part of the force function $F_0 \sin \omega t$. Substituting the assumed form of the solution in Eq. (2.21), we obtain

$$-M\omega^2 A + jR\omega A + KA = F_0 \tag{2.23}$$

or

$$A = \frac{F_0}{K - M\omega^2 + jR\omega} \tag{2.24}$$

The complex number

$$Z = K - M\omega^2 + jR\omega \tag{2.25}$$

may be written in the polar form

$$Z = |Z| e^{j\theta} \tag{2.26}$$

where

$$|Z| = \sqrt{(K - M\omega^2)^2 + R^2 \omega^2}$$
$$\theta = \tan^{-1} \frac{R\omega}{K - M\omega^2} \tag{2.27}$$

We then have

$$A = \frac{F_0}{|Z|} e^{-j\theta} \tag{2.28}$$

Substituting this into Eq. (2.22), we have

$$x = \frac{F_0}{|Z|} e^{j(\omega t - \theta)} \tag{2.29}$$

The physical meaning of this solution is that if a force of the form $F_0 \cos \omega t$ is impressed on the vibrating system, the resulting steady-state motion is given by

$$x = \frac{F_0}{|Z|} \cos (\omega t - \theta) \tag{2.30}$$

If a force $F_0 \sin \omega t$ is impressed on the system, then the steady-state response is

$$x = \frac{F_0}{|Z|} \sin (\omega t - \theta) \tag{2.31}$$

The complex number Z is called the *mechanical impedance* of the system. It is a generalization of the spring constant to the case of oscillatory motion. It is convenient to introduce the quantity μ, defined by

$$\mu = \frac{1}{|Z|} \tag{2.32}$$

μ is called the *distortion factor*. The angle θ is the phase displacement. The solution (2.29) may be written in the form

$$x = F_0\,\mu\,e^{j(\omega t - \theta)} \tag{2.33}$$

We see that the resulting steady-state motion is given by a function of the same type as that of the impressed force. However, it differs from it in amplitude by the factor μ and in phase by the angle θ. The solution (2.33) represents the steady-state asymptotic motion after the superimposed free vibrations which are damped have disappeared.

The general solution is of the form

$$x = A\,e^{-Rt/2M}\cos\omega_s(t - \delta) + F_0\,\mu\,e^{j(\omega t - \theta)} \tag{2.34}$$

If the initial conditions at $t = 0$ are given, the arbitrary constants A and δ may be determined.

THE CASE OF GENERAL PERIODIC EXTERNAL FORCES

A very important case in practice is the one in which the impressed external force is periodic of fundamental period T. That is,

$$F(t + T) = F(t) \tag{2.35}$$

In this case we may express $F(t)$ in the complex Fourier series

$$F(t) = \sum_{n=-\infty}^{n=+\infty} c_n e^{jn\omega t} \qquad \omega = \frac{2\pi}{T} \tag{2.36}$$

as explained in Appendix C.

The coefficients c_n are given by

$$c_n = \frac{1}{T}\int_0^T F(t)e^{-jn\omega t}\,dt \tag{2.37}$$

To determine the response in this case, we assume a solution of the form

$$x = \sum_{n=-\infty}^{n=+\infty} a_n e^{jn\omega t} \tag{2.38}$$

On substituting this assumed form of solution and equating the coefficients of the term $e^{jn\omega t}$ we obtain

$$a_n = \frac{c_n}{Z_n} \tag{2.39}$$

where

$$Z_n = K - Mn^2\omega^2 + jRn\omega \tag{2.40}$$

This may be written in the polar form

$$Z_n = |Z_n|\,e^{j\theta_n} \tag{2.41}$$

where

$$|Z_n| = \sqrt{(K - Mn^2\omega^2)^2 + R^2 n^2 \omega^2} \qquad \theta_n = \tan^{-1}\frac{Rn\omega}{K - Mn^2\omega^2} \qquad (2.42)$$

If we write

$$\mu_n = \frac{1}{|Z_n|} \qquad (2.43)$$

then we have

$$a_n = c_n \mu_n e^{-j\theta_n} \qquad (2.44)$$

Substituting this into (2.38), we obtain

$$x = \sum_{n=-\infty}^{n=+\infty} c_n \mu_n e^{j(n\omega t - \theta_n)} \qquad (2.45)$$

This represents the steady-state response to the general external periodic force $F(t)$. In this case the general harmonic n is magnified by the distortion factor μ_n. Equation (2.45) is of extreme importance in the theory of recording instruments. If the parameters M, K, and R are adjusted so that the μ_n's are as large as possible, then the apparatus is as sensitive as possible. If for the various frequencies $n\omega$ the μ_n's have approximately the same value, then there is a minimum amount of distortion. The phase displacements θ_n are of secondary importance in acoustical instruments since they are imperceptible to the human ear.

RESONANCE PHENOMENA

If we study the distortion factor μ of Eq. (2.32) as a function of the frequency ω, that is, if we write

$$\mu = \frac{1}{|Z|} = \mu(\omega) \qquad (2.46)$$

we notice that

$$\lim_{\omega \to \infty} \mu(\omega) = 0 \qquad (2.47)$$

$$\lim_{\omega \to 0} \mu(\omega) = \frac{1}{K} \qquad (2.48)$$

That is, if we impress on the system a force of very high frequency, the amplitude of vibration tends to zero. If we apply a steady force, the motion is of constant magnitude F_0/K.

Between $\omega = 0$ and $\omega = \infty$ there is a value where $\mu(\omega)$ has a maximum value. If we place

$$\frac{d}{d\omega}\mu(\omega) = 0 \qquad (2.49)$$

we obtain

$$2M^2\omega^2 = 2MK - R^2 \tag{2.50}$$

If we let ω_r be the value of ω that satisfies this equation, we have

$$\omega_r = \sqrt{\frac{K}{M} - \frac{R^2}{2M^2}} \tag{2.51}$$

This value of ω_r makes $\mu(\omega)$ a maximum. If there is no friction, $R = 0$ and the above analysis fails. In this case the equation of motion is

$$M\ddot{x} + Kx = F_0 e^{j\omega t} \tag{2.52}$$

If

$$\omega = \sqrt{\frac{K}{M}} = \omega_0 \tag{2.53}$$

this equation has the particular solution

$$x = \frac{F_0 t e^{j\omega t}}{2jM\omega} = \frac{F_0 t e^{j\omega t}}{2j\sqrt{KM}} \tag{2.54}$$

We then see that in this case if the impressed force coincides with the natural frequency of the system, ω_0, then the amplitude increases without limit as t increases. If friction is present, $\mu(\omega)$ is always finite and has its maximum value when $\omega = \omega_r$. In the literature the frequency ω_r is called the *resonance frequency* of the system.

3 TWO DEGREES OF FREEDOM

The system considered in the above section consisted of a mass restrained to move in a linear manner, and its position at any instant was specified by the parameter x measured from the position of equilibrium.

Let us now consider the motion of the system of Fig. 3.1. This system is analogous to the electrical circuit of Fig. 3.2. The state of both of these systems is determined by two quantities. These are the linear displacements of the two masses of the mechanical system (x_1, x_2) or the mesh charges of the electrical system (q_1, q_2). We speak of a system whose motion and position are characterized by two independent quantities as a system having *two*

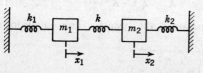

Fig. 3.1

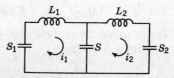

Fig. 3.2

degrees of freedom. If the position and motion of a system are characterized by n independent quantities, then the system is said to be one of *n degrees of freedom.*

The equations of motion of the systems of Figs. 3.1 and 3.2 may be obtained by D'Alembert's principle in the mechanical case or Kirchhoff's laws for the electrical case. We may pass from one system to the other one by means of the table of analogues of Sec. 2. It is usually simpler to write Kirchhoff's laws for the system and then translate the electrical quantities to mechanical ones.

Writing Kirchhoff's law for the fall of potential in the two meshes of the circuit, we have

$$L_1 \ddot{q}_1 + S_1 q_1 + S(q_1 - q_2) = 0$$
$$L_2 \ddot{q}_2 + S_2 q_2 + S(q_2 - q_1) = 0 \tag{3.1}$$

Translated to mechanical quantities, we have

$$M_1 \ddot{x}_1 + K_1 x_1 + K(x_1 - x_2) = 0$$
$$M_2 \ddot{x}_2 + K_2 x_2 + K(x_2 - x_1) = 0 \tag{3.2}$$

These equations are linear and homogeneous of the second order of the type discussed in Chap. 2. Their solutions are of the exponential type. To solve them, let us place

$$x_1 = A_1 e^{\alpha t}$$
$$x_2 = A_2 e^{\alpha t} \tag{3.3}$$

where A_1, A_2, and α may be real or complex. Substituting this into (3.2) and dividing out the factor $e^{\alpha t}$, we obtain

$$A_1(M_1 \alpha^2 + K_1 + K) - A_2 K = 0$$
$$-A_1 K + A_2(M_2 \alpha^2 + K_2 + K) = 0 \tag{3.4}$$

These are two homogeneous linear equations of the type discussed in Chap. 4. This system of equations has a solution other than the trivial one $A_1 = A_2 = 0$ if

$$\Delta(\alpha) = \begin{vmatrix} M_1 \alpha^2 + K_1 + K & -K \\ -K & M_2 \alpha^2 + K_2 + K \end{vmatrix} = 0 \tag{3.5}$$

Expanding the determinant, we have

$$\Delta(\alpha) = (M_1 \alpha^2 + K_1 + K)(M_2 \alpha^2 + K_2 + K) - K^2 = 0 \tag{3.6}$$

This is called the *characteristic equation* of the system. This equation may be written in the form

$$\left(\alpha^2 + \frac{K_1 + K}{M_1}\right)\left(\alpha^2 + \frac{K_2 + K}{M_2}\right) - \frac{K^2}{M_1 M_2} = 0 \tag{3.7}$$

It is convenient to introduce the notation

$$\omega_{11}^2 = \frac{K_1 + K}{M_1} \qquad \omega_{22}^2 = \frac{K_2 + K}{M_2} \qquad \omega_{12}^2 = \frac{K}{\sqrt{M_1 M_2}} \tag{3.8}$$

We may then write the characteristic equation in the form

$$(\alpha^2 + \omega_{11}^2)(\alpha^2 + \omega_{22}^2) - \omega_{12}^4 = 0 \tag{3.9}$$

or

$$\alpha^4 + \alpha^2(\omega_{11}^2 + \omega_{22}^2) + \omega_{11}^2 \omega_{22}^2 - \omega_{12}^4 = 0 \tag{3.10}$$

We therefore have

$$\alpha^2 = -\tfrac{1}{2}(\omega_{11}^2 + \omega_{22}^2) \pm \tfrac{1}{2}\sqrt{(\omega_{11}^2 - \omega_{22}^2)^2 + 4\omega_{12}^2} \tag{3.11}$$

The expression under the radical is positive, and the absolute value of the second term is less than that of the first term. Hence α^2 is real and negative. Accordingly let us write

$$\alpha^2 = -\omega^2 \qquad \text{or} \qquad \alpha = \pm j\omega \tag{3.12}$$

Hence

$$\omega^2 = \frac{\omega_{11}^2 + \omega_{22}^2}{2} \pm \frac{1}{2}\sqrt{(\omega_{11}^2 - \omega_{22}^2)^2 + 4\omega_{12}^4} \tag{3.13}$$

Equation (3.13) gives two values for ω^2; let us call them ω_1 and ω_2. From (3.12) we obtain *four* values for α: $j\omega_1$, $-j\omega_1$, $j\omega_2$, and $-j\omega_2$. We find from our original assumption (3.3) that the solutions for x_1 and x_2 are of an exponential form, and we obtained *four* values of α. The general solution may be written in the form

$$\begin{aligned} x_1 &= A_{11} e^{j\omega_1 t} + \bar{A}_{11} e^{-j\omega_1 t} + A_{12} e^{j\omega_2 t} + \bar{A}_{12} e^{-j\omega_2 t} \\ x_2 &= A_{21} e^{j\omega_1 t} + \bar{A}_{21} e^{-j\omega_1 t} + A_{22} e^{j\omega_2 t} + \bar{A}_{22} e^{-j\omega_2 t} \end{aligned} \tag{3.14}$$

Since x_1 and x_2 are real, $\bar{A}_{11}$ must be the conjugate of A_{11} and similarly for A_{12} and $\bar{A}_{12}$, etc. If we write

$$A_{11} = \frac{c_{11} e^{j\theta_1}}{2} \qquad \bar{A}_{11} = \frac{c_{11} e^{-j\theta_1}}{2}$$

$$A_{12} = \frac{c_{12} e^{j\theta_2}}{2} \qquad \bar{A}_{12} = \frac{c_{12} e^{-j\theta_2}}{2}$$

$$A_{21} = \frac{c_{21} e^{j\theta_1}}{2} \qquad \bar{A}_{21} = \frac{c_{21} e^{-j\theta_1}}{2}$$

$$A_{22} = \frac{c_{22} e^{j\theta_2}}{2} \qquad \bar{A}_{22} = \frac{c_{22} e^{-j\theta_2}}{2}$$

$$\tag{3.15}$$

where c_{11}, c_{12}, c_{21}, c_{22}, θ_1, and θ_2 are new arbitrary constants, we may then write the solution in the form

$$
\begin{aligned}
x_1 &= c_{11} \cos(\omega_1 t + \theta_1) + c_{12} \cos(\omega_2 t + \theta_2) \\
x_2 &= c_{21} \cos(\omega_1 t + \theta_1) + c_{22} \cos(\omega_2 t + \theta_2)
\end{aligned}
\tag{3.16}
$$

The ratios c_{11}/c_{21} and c_{12}/c_{22} are determined by Eqs. (3.4), since for each value of α the ratios of the quantities A_{11}/A_{21} and A_{12}/A_{22} are determined by these equations. It appears that in Eqs. (3.16) there are *four independent* arbitrary constants. We may take them to be c_{11}, c_{12}, θ_1, and θ_2. Equations (3.16) show that the most general solution of the system is made up of the superposition of two pure harmonic oscillations. In each of these oscillations the two masses oscillate with the same frequency and in the same phase. The amplitudes of oscillation are in a definite ratio given by Eqs. (3.4).

The pure harmonic oscillations are called the *principal oscillations* or the *principal modes of oscillation* of the system.

SPECIAL CASE $M_1 = M_2 = M$, $K_1 = K_2 = K_0$

A very interesting and illuminating special case of the above general theory occurs when the two masses of the system are equal, that is,

$$
M_1 = M_2 = M
\tag{3.17}
$$

and

$$
K_1 = K_2 = K_0
\tag{3.18}
$$

This is a very symmetric case. Rather than apply the above general theory, it is more convenient to begin with the differential equations of the system. The general equations (3.2) reduce in this case to

$$
\begin{aligned}
M\ddot{x}_1 + (K_0 + K)x_2 - Kx_2 &= 0 \\
M\ddot{x}_2 + (K_0 + K)x_2 - Kx_1 &= 0
\end{aligned}
\tag{3.19}
$$

If we add the two equations, we have

$$
M(\ddot{x}_1 + \ddot{x}_2) + K_0(x_1 + x_2) = 0
\tag{3.20}
$$

Let us introduce the new coordinate y_1 defined by

$$
y_1 = x_1 + x_2
\tag{3.21}
$$

We then have

$$
M\ddot{y}_1 + K_0 y_1 = 0
\tag{3.22}
$$

Hence the coordinate y_1 performs simple harmonic oscillations of the form

$$
y_1 = A_1 \sin \omega_1 t + B_1 \cos \omega_1 t
\tag{3.23}
$$

where

$$\omega_1 = \sqrt{\frac{K_0}{M}} \tag{3.24}$$

This represents a motion where the two masses swing to the left and right with equal amplitudes in such a manner that the coupling spring is not stressed. If we now subtract the second equation from the first one and let

$$y_2 = x_1 - x_2 \tag{3.25}$$

we obtain

$$M\ddot{y}_2 + (K_0 + 2K)y_2 = 0 \tag{3.26}$$

This equation has a solution

$$y_2 = A_2 \sin \omega_2 t + B_2 \cos \omega_2 t \tag{3.27}$$

where

$$\omega_2 = \sqrt{\frac{K_0 + 2K}{M}} \tag{3.28}$$

This case represents the one in which the two masses move in opposite directions with the same amplitude. We may write the transformation from the x coordinates to the y coordinates in the matrix form

$$\begin{Bmatrix} y_1 \\ y_2 \end{Bmatrix} = \begin{bmatrix} 1 & 1 \\ 1 & -1 \end{bmatrix} \begin{Bmatrix} x_1 \\ x_2 \end{Bmatrix} \tag{3.29}$$

and also

$$\begin{Bmatrix} x_1 \\ x_2 \end{Bmatrix} = \begin{bmatrix} 1 & 1 \\ 1 & -1 \end{bmatrix}^{-1} \begin{Bmatrix} y_1 \\ y_2 \end{Bmatrix} = \frac{1}{2} \begin{bmatrix} 1 & 1 \\ 1 & -1 \end{bmatrix} \begin{Bmatrix} y_1 \\ y_2 \end{Bmatrix} \tag{3.30}$$

In this very simple case we see that by a linear transformation of the coordinates (x_1, x_2) to the coordinates (y_1, y_2) we have effected a separation of the variables so that the motions of the y_1 and y_2 coordinates are uncoupled. These new coordinates y_1 and y_2 are called *normal coordinates*. They will be considered in greater detail in a later section.

The general solution of the system may be written in the form

$$\begin{aligned} x_1 &= \tfrac{1}{2}(A_1 \sin \omega_1 t + B_1 \cos \omega_1 t + A_2 \sin \omega_2 t + B_2 \cos \omega_2 t) \\ x_2 &= \tfrac{1}{2}(A_1 \sin \omega_1 t + B_1 \cos \omega_1 t - A_2 \sin \omega_2 t - B_2 \cos \omega_2 t) \end{aligned} \tag{3.31}$$

Let us suppose that the motion begins in such a manner that

$$\left. \begin{aligned} x_1 &= x_0 \\ x_2 &= 0 \\ \dot{x}_1 &= 0 \\ \dot{x}_2 &= 0 \end{aligned} \right\} \quad \text{at } t = 0 \tag{3.32}$$

That is, the system is initially at rest, but at $t = 0$ the mass 1 is displaced a distance x_0. In this case the general solution (3.31) reduces to

$$x_1 = \frac{x_0}{2}(\cos \omega_1 t + \cos \omega_2 t)$$

$$x_2 = \frac{x_0}{2}(\cos \omega_1 t - \cos \omega_2 t)$$

(3.33)

This solution may also be written in the form

$$x_1 = x_0 \cos \frac{\omega_1 + \omega_2}{2} t \cos \frac{\omega_1 - \omega_2}{2} t$$

$$x_2 = x_0 \sin \frac{\omega_1 + \omega_2}{2} t \sin \frac{\omega_1 - \omega_2}{2} t$$

(3.34)

If the coupling spring k is weak, then k is small and we may write

$$\omega_2 = \omega_1 + 2\delta$$

(3.35)

where δ is small compared with ω_1, and we have approximately

$$\omega_1 + \omega_2 = 2\omega_1$$

$$\omega_2 - \omega_1 = 2\delta$$

(3.36)

We may then write (3.34) in the form

$$x_1 = x_0 \cos \omega_1 t \cos \delta t$$

$$x_2 = x_0 \sin \omega_1 t \sin \delta t$$

(3.37)

The motions of the two masses in this case produce a phenomenon called *beats*. Each mass executes a rapid vibration of angular frequency ω_1 with an amplitude that changes slowly with an angular frequency δ. The two masses move in opposite phases so that the amplitude of one reaches its maximum when the other is at rest.

4 LAGRANGE'S EQUATIONS

In this section a simple exposition of Lagrange's equations for conservative systems will be given. For a more complete and detailed treatment the reader is referred to the standard treatises on mechanics.

Let us consider the simple mass-and-spring system of Fig. 4.1. x is a linear coordinate measured from the position of equilibrium of the mass. If the mass is moving with a velocity $v = \dot{x}$, its kinetic energy T is given by

$$T = \tfrac{1}{2}Mv^2 = \tfrac{1}{2}M(\dot{x})^2$$

(4.1)

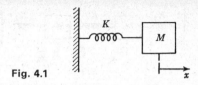

Fig. 4.1

When the spring is stretched a distance x from its position of equilibrium, then the elastic, or potential, energy stored in the spring is

$$V = \int_0^x F\,dx = \int_0^x Kx\,dx = \frac{Kx^2}{2} \tag{4.2}$$

By Newton's second law of motion the differential equation governing the motion of the mass is

$$Ma = -F \tag{4.3}$$

where $a = \dot{v}$ is the acceleration of the mass and F is the restoring force of the spring. The minus sign is introduced because the restoring force F acts in the opposite direction from x, since x is measured from the position of equilibrium. We may also write Eq. (4.3) in the more fundamental form

$$\frac{d}{dt}(Mv) = -F \tag{4.4}$$

If we differentiate the kinetic energy T of (4.1) with respect to $v = \dot{x}$, we obtain

$$\frac{\partial}{\partial \dot{x}} T = \frac{\partial}{\partial \dot{x}} \frac{1}{2} M(\dot{x})^2 = M\dot{x} = Mv \tag{4.5}$$

If we differentiate the potential energy V of (4.2) with respect to x, we have

$$\frac{\partial V}{\partial x} = \frac{\partial}{\partial x}\left(\frac{Kx^2}{2}\right) = Kx = F \tag{4.6}$$

Hence, in terms of the energy functions, the equation of motion may be written in the form

$$\frac{d}{dt}\left(\frac{\partial T}{\partial \dot{x}}\right) + \frac{\partial V}{\partial x} = 0 \tag{4.7}$$

This form of the equation of motion is called *Lagrange's equation of motion* and is simply a convenient way of writing Newton's second law of motion.

Let us now consider the system of Fig. 4.2. This is the system analyzed in Sec. 3. In this case the kinetic energy is given by

$$T = \tfrac{1}{2}M_1 \dot{x}_1^2 + \tfrac{1}{2}M_2 \dot{x}_2^2 \tag{4.8}$$

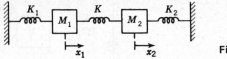

Fig. 4.2

The potential energy stored in the springs V is

$$V = \tfrac{1}{2}K_1 x_1^2 + \tfrac{1}{2}K(x_1 - x_2)^2 + \tfrac{1}{2}K_2 x_2^2 \qquad (4.9)$$

In this case the kinetic and potential energies are quadratic functions of the linear displacements x_1 and x_2. In this case, if we form the partial derivative

$$\frac{\partial T}{\partial \dot{x}_1} = M_1 \dot{x}_1 = M_1 v_1 \qquad (4.10)$$

we obtain the momentum of the first mass.

The derivative

$$\frac{\partial V}{\partial x_1} \; K_1 x_1 + K(x_1 - x_2) \qquad (4.11)$$

is exactly the restoring force acting on the mass M_1. We thus see that the first of Eqs. (3.2) may be written in the form

$$\frac{d}{dt}\left(\frac{\partial T}{\partial \dot{x}_1}\right) + \frac{\partial V}{\partial x_1} = 0 \qquad (4.12)$$

In the same manner the second equation of motion may be written in the form

$$\frac{d}{dt}\left(\frac{\partial T}{\partial \dot{x}_2}\right) + \frac{\partial V}{\partial x_2} = 0 \qquad (4.13)$$

In this case the motion of the system is given by the two Lagrange equations (4.12) and (4.13).

The importance of Lagrange's equations is that they hold with respect to any sort of coordinates, not merely the linear coordinates which we have used above. For example, in the symmetric case of Sec. 3 the kinetic and potential energies are given by

$$T = \tfrac{1}{2}M(\dot{x}_1^2 + \dot{x}_2^2)$$
$$V = \tfrac{1}{2}K_0(x_1^2 + x_2^2) + \tfrac{1}{2}K(x_1 - x_2)^2 \qquad (4.14)$$

We saw that the analysis was vastly simplified by introducing the new coordinates y_1 and y_2 by the linear transformation

$$y_1 = x_1 + x_2$$
$$y_2 = x_1 - x_2 \qquad (4.15)$$

or

$$x_1 = \tfrac{1}{2}(y_1 + y_2)$$
$$x_2 = \tfrac{1}{2}(y_1 - y_2)$$

(4.16)

In terms of these new coordinates the kinetic and potential energies become

$$T = \tfrac{1}{4}M(\dot{y}_1^2 + \dot{y}_2^2)$$
$$V = \tfrac{1}{4}K_0(y_1^2 + y_2^2) + \tfrac{1}{2}Ky_2^2$$

(4.17)

It was *asserted* that Lagrange's equations would hold for the y_1 and y_2 coordinates. We therefore have

$$\frac{d}{dt}\left(\frac{\partial T}{\partial \dot{y}_1}\right) + \frac{\partial V}{\partial y_1} = 0$$

(4.18)

or

$$\frac{M}{2}\ddot{y}_1 + \frac{K_0}{2}y_1 = 0$$

(4.19)

and

$$\frac{d}{dt}\left(\frac{\partial T}{\partial \dot{y}_2}\right) + \frac{\partial V}{\partial y_2} = 0$$

(4.20)

or

$$\frac{M}{2}\ddot{y}_2 + \tfrac{1}{2}K_0 y_2 + Ky_2 = 0$$

(4.21)

Hence the equations of motion are

$$M\ddot{y}_1 + K_0 y_1 = 0$$

(4.22)

and

$$M\ddot{y}_2 + (K_0 + 2K)y_2 = 0$$

(4.23)

These were the equations of motion of the symmetric system obtained directly.

In this simple example we see that the introduction of the two new coordinates y_1 and y_2 reduces the potential- and kinetic-energy expressions to sums of squares. This is a general property of *normal coordinates*. Each normal coordinate executes simple harmonic oscillations independently of the others.

As another example, let us consider the motion of three pendulums of equal masses M and length l connected by springs at a distance h from the suspension points A, B, and C as shown in Fig. 4.3. The masses of the springs and the bars of the pendulums are assumed so small that they can be neglected. The motion of the pendulums may be expressed in terms of the angles θ_1, θ_2, and θ_3 of the pendulums measured from the vertical in a clockwise direction. The kinetic energy of the system is

$$T = \tfrac{1}{2}Ml^2(\dot{\theta}_1^2 + \dot{\theta}_2^2 + \dot{\theta}_3^2)$$

(4.24)

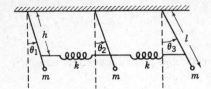

Fig. 4.3

The potential energy of the system consists of two parts: the energy due to the gravitational force and the strain energy of the springs. If we limit ourselves to a consideration of small oscillations, then θ_1, θ_2, and θ_3 are small quantities. The energy due to gravity is

$$V_g = Mgl(1 - \cos\theta_1) + Mgl(1 - \cos\theta_2) + Mgl(1 - \cos\theta_2)$$
$$\doteq \tfrac{1}{2}Mgl(\theta_1^2 + \theta_2^2 + \theta_3^2) \tag{4.25}$$

For small oscillations the springs may be assumed to remain always horizontal. The elongations of the springs are then given by

$$h(\sin\theta_2 - \sin\theta_1) \doteq h(\theta_2 - \theta_1) \tag{4.26}$$

and

$$h(\sin\theta_3 - \sin\theta_2) \doteq h(\theta_3 - \theta_2)$$

respectively. The strain energy of the springs is accordingly

$$V_s = \frac{K}{2}h^2[(\theta_2 - \theta_1)^2 + (\theta_3 - \theta_2)^2] \tag{4.27}$$

The total potential energy of the system is

$$V = \tfrac{1}{2}Mgl(\theta_1^2 + \theta_2^2 + \theta_3^2) + \frac{Kh^2}{2}[(\theta_2 - \theta_1)^2 + (\theta_3 - \theta_2)^2] \tag{4.28}$$

This mechanical system is analogous to the electrical circuit of Fig. 4.4. The magnetic energy of the electrical circuit is

$$T_M = \tfrac{1}{2}L(\dot{q}_1^2 + \dot{q}_2^2 + \dot{q}_3^2) \tag{4.29}$$

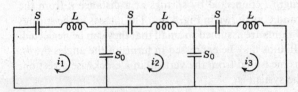

Fig. 4.4

where $\dot{q}_r = i_r$, the mesh currents of the system. The electric energy of the system is

$$V_E = \tfrac{1}{2}S(q_1^2 + q_2^2 + q_3^2) + \tfrac{1}{2}S_0(q_2 - q_1)^2 + \tfrac{1}{2}S_0(q_3 - q_2)^2 \tag{4.30}$$

These expressions are completely analogous to the kinetic and potential energies of the mechanical system. In this case we have

$$\begin{aligned} Mgl &\to S & Kh^2 &\to S_0 \\ l^2 M &\to L & \theta_r &\to q_r \end{aligned} \tag{4.31}$$

for the analogous electrical and mechanical quantities. The currents are analogous to the angular velocities of the pendulums. In this case there are three Lagrangian equations of motion of the system:

$$\frac{d}{dt}\left(\frac{\partial T}{\partial \dot{\theta}_r}\right) + \frac{\partial V}{\partial \theta_r} = 0 \qquad r = 1, 2, 3 \tag{4.32}$$

Performing the differentiation, we obtain

$$\begin{aligned} l^2 M\ddot{\theta}_1 + Mgl\theta_1 + Kh^2(\theta_1 - \theta_2) &= 0 \\ l^2 M\ddot{\theta}_2 + Mgl\theta_2 + Kh^2(\theta_2 - \theta_1) + Kh^2(\theta_2 - \theta_3) &= 0 \\ l^2 M\ddot{\theta}_3 + Mgl\theta_3 + Kh^2(\theta_3 - \theta_2) &= 0 \end{aligned} \tag{4.33}$$

We may find the normal coordinates in this case rather simply. If we add the three equations, we obtain

$$l^2 M(\ddot{\theta}_1 + \ddot{\theta}_2 + \ddot{\theta}_3) + Mgl(\theta_1 + \theta_2 + \theta_3) = 0 \tag{4.34}$$

If we let

$$y_1 = \theta_1 + \theta_2 + \theta_3 \tag{4.35}$$

Eq. (4.34) becomes

$$\ddot{y}_1 + \frac{g}{l}y_1 = 0 \tag{4.36}$$

The solution of this equation is

$$y_1 = A_1 \sin\sqrt{\frac{g}{l}}\,t + B_1 \cos\sqrt{\frac{g}{l}}\,t \tag{4.37}$$

and represents an oscillation of all three pendulums in synchronism with an angular frequency of

$$\omega_1 = \sqrt{\frac{g}{l}} \tag{4.38}$$

If we subtract the last equation (4.33) from the first one and let

$$y_2 = \theta_1 - \theta_3 \tag{4.39}$$

we obtain

$$Ml^2 \ddot{y}_2 + (Mgl + Kh^2) y_2 = 0 \tag{4.40}$$

The solution of this equation is

$$y_2 = A_2 \sin \omega_2 t + B_2 \cos \omega_2 t \tag{4.41}$$

where

$$\omega_2 = \sqrt{\frac{g}{l} + \frac{Kh^2}{Ml^2}} \tag{4.42}$$

If we now add the first and last equations (4.33) and subtract twice the second equation and let

$$y_3 = \theta_1 - 2\theta_2 + \theta_3 \tag{4.43}$$

we obtain

$$Ml^2 \ddot{y}_3 + (Mgl + 3Kh^2) y_3 = 0 \tag{4.44}$$

This equation has the solution

$$y_3 = A_3 \sin \omega_3 t + B_3 \cos \omega_3 t \tag{4.45}$$

where

$$\omega_3 = \sqrt{\frac{Mgl + 3Kh^2}{Ml^2}} \tag{4.46}$$

We have thus succeeded in obtaining the three normal coordinates $y_1, y_2,$ and y_3. The three angular frequencies of the system are $\omega_1, \omega_2,$ and ω_3. If we know the initial displacement and angular velocities of the pendulums, we may determine the six arbitrary constants $A_r, B_r, r = 1, 2, 3$. Then the θ_r coordinates are given by

$$\theta_1 = \frac{y_1}{3} + \frac{y_2}{2} + \frac{y_3}{6}$$

$$\theta_2 = \frac{y_1}{3} - \frac{y_3}{3} \tag{4.47}$$

$$\theta_3 = \frac{y_1}{3} - \frac{y_2}{2} + \frac{y_3}{6}$$

These equations are obtained by solving the θ_r coordinates in terms of the y_r coordinates. It may be shown that the kinetic and potential energies (4.25) and (4.28) are reduced to sums of squares in the y_r coordinates by this transformation.

5 PROOF OF LAGRANGE'S EQUATIONS

In the last section we considered a method of writing the equation of motion of oscillating systems in terms of the expressions for the kinetic and potential energy of the system.   We saw that the state of motion of the system could be expressed in terms of different parameters or coordinates.

In this section a proof of Lagrange's equations will be given for a particle; the proof may be easily generalized to a system of particles and to rigid bodies.

Let us consider a particle of mass M.   According to Newton's second law the equations of free motion of this particle referred to a set of rectangular coordinates are given by

$$M\ddot{x} = F_x \qquad M\ddot{y} = F_y \qquad M\ddot{z} = F_z \tag{5.1}$$

where F_x, F_y, F_z represent the components of the effective force acting on the particle in the x, y, and z directions.   Suppose we desire to express these equations of motion in terms of another set of coordinates (q_1, q_2, q_3) related functionally to the rectangular coordinates (x, y, z).   We may then express the coordinates (x, y, z) in terms of the coordinates (q_1, q_2, q_3) by

$$x = F_1(q_1, q_2, q_3) \qquad y = F_2(q_1, q_2, q_3) \qquad z = F_3(q_1, q_2, q_3) \tag{5.2}$$

With this notation, on differentiation with respect to time, the expression for the component velocities of the particle $(\dot{x}, \dot{y}, \dot{z})$ may be put in the form

$$
\begin{aligned}
\dot{x} &= \frac{\partial x}{\partial q_1}\dot{q}_1 + \frac{\partial x}{\partial q_2}\dot{q}_2 + \frac{\partial x}{\partial q_3}\dot{q}_3 \\[1mm]
\dot{y} &= \frac{\partial y}{\partial q_1}\dot{q}_1 + \frac{\partial y}{\partial q_2}\dot{q}_2 + \frac{\partial y}{\partial q_3}\dot{q}_3 \\[1mm]
\dot{z} &= \frac{\partial z}{\partial q_1}\dot{q}_1 + \frac{\partial z}{\partial q_2}\dot{q}_2 + \frac{\partial z}{\partial q_3}\dot{q}_3
\end{aligned}
\tag{5.3}
$$

where in general $\dot{x}$, $\dot{y}$, $\dot{z}$ are explicit functions of $(q_1, q_2, q_3, \dot{q}_1, \dot{q}_2, \dot{q}_3)$.

Now since

$$x = F_1(q_1, q_2, q_3) \tag{5.4}$$

then

$$\frac{\partial x}{\partial q_1} = \phi_1(q_1, q_2, q_3) \tag{5.5}$$

Hence

$$
\begin{aligned}
\frac{d}{dt}\left(\frac{\partial x}{\partial q_1}\right) &= \frac{\partial \phi_1}{\partial q_1}\dot{q}_1 + \frac{\partial \phi_1}{\partial q_2}\dot{q}_2 + \frac{\partial \phi_1}{\partial q_3}\dot{q}_3 \\[1mm]
&= \frac{\partial^2 x}{\partial q_1^2}\dot{q}_1 + \frac{\partial^2 x}{\partial q_2\,\partial q_1}\dot{q}_2 + \frac{\partial^2 x}{\partial q_3\,\partial q_1}\dot{q}_3
\end{aligned}
\tag{5.6}
$$

Differentiating the first equation (5.3) with respect to q_1, we obtain

$$\frac{\partial \dot{x}}{\partial q_1} = \frac{\partial^2 x}{\partial q_1^2}\dot{q}_1 + \frac{\partial^2 x}{\partial q_1\,\partial q_2}\dot{q}_2 + \frac{\partial^2 x}{\partial q_1\,\partial q_3}\dot{q}_3 \tag{5.7}$$

If we now compare (5.6) and (5.7), we obtain

$$\frac{d}{dt}\left(\frac{\partial x}{\partial q_1}\right) = \frac{\partial \dot{x}}{\partial q_1} \tag{5.8}$$

with similar relations for y and z.

It is also evident by differentiating the first equation (5.3) with respect to $\dot{q}_1$ that

$$\frac{\partial \dot{x}}{\partial \dot{q}_1} = \frac{\partial x}{\partial q_1} \tag{5.9}$$

with similar relations for y and z.

Let us now assume that we hold the coordinates q_2 and q_3 fixed and give the coordinate q_1 an infinitesimal increment ∂q_1. If δx, δy, and δz are the increments that this produced in x, y, and z, then, if we let δW_1 be the work done by the effective forces when the particle undergoes this infinitesimal displacement, we have

$$\begin{aligned}
\delta W_1 &= F_x\,\delta x + F_y\,\delta y + F_z\,\delta z \\
&= M(\ddot{x}\,\delta x + \ddot{y}\,\delta y + \ddot{z}\,\delta z)
\end{aligned} \tag{5.10}$$

as a consequence of Eqs. (5.1). We also have

$$\delta x = \frac{\partial x}{\partial q_1}\delta q_1 \qquad \delta y = \frac{\partial y}{\partial q_1}\delta q_1 \qquad \delta z = \frac{\partial z}{\partial q_1}\delta q_1 \tag{5.11}$$

Hence we may write (5.10) in the form

$$\delta W_1 = M\left(\ddot{x}\frac{\partial x}{\partial q_1} + \ddot{y}\frac{\partial y}{\partial q_1} + \ddot{z}\frac{\partial z}{\partial q_1}\right)\delta q_1 \tag{5.12}$$

The rule for the differentiation of a product yields

$$\frac{d}{dt}\left(\dot{x}\frac{\partial x}{\partial q_1}\right) = \ddot{x}\frac{\partial x}{\partial q_1} + \dot{x}\frac{d}{dt}\left(\frac{\partial x}{\partial q_1}\right) \tag{5.13}$$

or

$$\ddot{x}\frac{\partial x}{\partial q_1} = \frac{d}{dt}\left(\dot{x}\frac{\partial x}{\partial q_1}\right) - \dot{x}\frac{d}{dt}\left(\frac{\partial x}{\partial q_1}\right) \tag{5.14}$$

If we substitute the values of $\partial x/\partial q_1$ and $d/dt(\partial x/\partial q_1)$ given by (5.9) and (5.8), we obtain

$$\begin{aligned}
\ddot{x}\frac{\partial x}{\partial q_1} &= \frac{d}{dt}\left(\dot{x}\frac{\partial \dot{x}}{\partial \dot{q}_1}\right) - \dot{x}\frac{\partial \dot{x}}{\partial q_1} \\
&= \frac{d}{dt}\frac{\partial}{\partial \dot{q}_1}\left(\frac{\dot{x}^2}{2}\right) - \frac{\partial}{\partial q_1}\left(\frac{\dot{x}^2}{2}\right)
\end{aligned} \tag{5.15}$$

and similar relations for y and z. However, the kinetic energy of the particle is given by

$$T = \frac{M}{2}(\dot{x}^2 + \dot{y}^2 + \dot{z}^2) \tag{5.16}$$

Hence as a consequence of (5.15) and (5.16) the expression (5.12) may be written in the form

$$\delta W_1 = \left(\frac{d}{dt}\frac{\partial T}{\partial \dot{q}_1} - \frac{\partial T}{\partial q_1}\right)\delta q_1 \tag{5.17}$$

Now, if $Q_1\,\delta q_1$ is the work done in the specified displacement of the particle, then it is convenient to regard Q_1 as a sort of generalized force. We may write

$$\delta W_1 = Q_1\,\delta q_1 \tag{5.18}$$

and we have

$$\frac{d}{dt}\left(\frac{\partial T}{\partial \dot{q}_1}\right) - \frac{\partial T}{\partial q_1} = Q_1 \tag{5.19}$$

By the same reasoning, if q_1 and q_3 are held constant and q_2 is given an increment δq_2, we obtain the equation

$$\frac{d}{dt}\left(\frac{\partial T}{\partial \dot{q}_2}\right) - \frac{\partial T}{\partial q_2} = Q_2 \tag{5.20}$$

and in the same manner, holding q_1 and q_2 constant and giving q_3 an increment δq_3, we obtain

$$\frac{d}{dt}\left(\frac{\partial T}{\partial \dot{q}_3}\right) - \frac{\partial T}{\partial q_3} = Q_3 \tag{5.21}$$

We see that there are as many equations as there are degrees of freedom. The Q_r quantities are called *generalized forces.* The q_r quantities are the *generalized coordinates.*

CONSERVATIVE SYSTEMS

If there is no loss of energy in the dynamical system under consideration, the generalized forces may be derived from the potential energy of the system V in the form

$$Q_1 = -\frac{\partial V}{\partial q_1} \qquad Q_2 = -\frac{\partial V}{\partial q_2} \qquad Q_3 = -\frac{\partial V}{\partial q_3} \tag{5.22}$$

In this case the free motion of the particle is given by the three Lagrangian equations

$$\frac{d}{dt}\left(\frac{\partial T}{\partial \dot{q}_r}\right) - \frac{\partial T}{\partial q_r} + \frac{\partial V}{\partial q_r} = 0 \qquad r = 1, 2, 3 \tag{5.23}$$

By an extension of the above argument it may be shown that in the case of a conservative system having n degrees of freedom its Lagrangian equations are of the form (5.23). In this case there are n generalized coordinates $(q_1, q_2, \ldots, q_n)$, and there are n equations.

We notice that in the simple examples of Lagrange's equations discussed in Sec. 4 we had

$$\frac{\partial T}{\partial q_r} = 0 \qquad r = 1, 2, \ldots, n \tag{5.24}$$

The reason for this was that the kinetic energy in the systems discussed in that section was not a function of the coordinates, but only of the velocities of the system. As an example of a system where the kinetic energy is a function of the generalized coordinates q_r as well as the generalized velocities $\dot{q}_r$, consider the two-dimensional motion of a particle in the xy plane as shown in Fig. 5.1.

In this case we have

$$T = \tfrac{1}{2} M (\dot{x}^2 + \dot{y}^2) \tag{5.25}$$

Let us assume that the particle is attracted to the origin by a force proportional to the distance so that the particle has a radial force F_r directed toward the origin given by

$$F_r = Kr \tag{5.26}$$

In this case the potential energy is given by

$$V = \frac{Kr^2}{2} \tag{5.27}$$

To describe the motion of the particle it is convenient to use the polar coordinates r and θ as the generalized coordinates. These coordinates are related to the cartesian coordinates x and y by the equations

$$x = r \cos \theta \qquad y = r \sin \theta \tag{5.28}$$

In terms of these coordinates the kinetic energy becomes

$$T = \frac{M}{2} (\dot{r}^2 + r^2 \dot{\theta}^2) \tag{5.29}$$

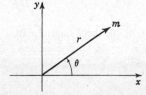

Fig. 5.1

The Lagrangian equations of motion are

$$\frac{d}{dt}\left(\frac{\partial T}{\partial \dot{r}}\right) - \frac{\partial T}{\partial r} + \frac{\partial V}{\partial r} = 0 \tag{5.30}$$

or

$$M\ddot{r} - Mr\dot{\theta}^2 + Kr = 0 \tag{5.31}$$

and

$$\frac{d}{dt}\left(\frac{\partial T}{\partial \dot{\theta}}\right) - \frac{\partial T}{\partial \theta} + \frac{\partial V}{\partial \theta} = 0 \tag{5.32}$$

or

$$\frac{d}{dt}(Mr^2\dot{\theta}) = 0 \tag{5.33}$$

Equations (5.31) and (5.33) are the equations of motion of the system. Equation (5.33) expresses the conservation of angular momentum of the system. The term $\partial T/\partial r$ is seen to give rise to the centrifugal-force term. Terms of the form $\partial T/\partial q_r$, which are absent in rectangular coordinates, may be regarded as a sort of fictitious force introduced by using generalized coordinates.

6 SMALL OSCILLATIONS OF CONSERVATIVE SYSTEMS

A very important class of problems in the theory of vibrations is that in which a dynamical system is performing free or forced oscillations without friction. The determination of the natural frequencies and modes of oscillations of the torsional vibrations of an engine, the oscillations of spring-mounted machinery, the oscillations of electrical networks whose resistance is so small as to be neglected, etc., are problems that fall in this category.

The Lagrangian equations of motion of a general conservative system of n degrees of freedom oscillating in the neighborhood of a position of equilibrium may be written in the form

$$
\begin{aligned}
M_{11}\ddot{q}_1 + \cdots + M_{1n}\ddot{q}_n + K_{11}q_1 + \cdots + K_{1n}q_n &= 0 \\
M_{21}\ddot{q}_1 + \cdots + M_{2n}\ddot{q}_n + K_{21}q_1 + \cdots + K_{2n}q_n &= 0 \\
\cdots\cdots\cdots\cdots\cdots\cdots\cdots\cdots\cdots\cdots\cdots\cdots \\
M_{n1}\ddot{q}_1 + \cdots + M_{nn}\ddot{q}_n + K_{n1}q_1 + \cdots + K_{nn}q_n &= 0
\end{aligned}
\tag{6.1}
$$

where $(q_1, \ldots, q_n)$ are generalized coordinates, M_{rs} are inertia coefficients, and K_{rs} are stiffness coefficients. These are the equations governing the motion of the most general conservative system performing free oscillations in the neighborhood of a position of equilibrium. The examples discussed in

Secs. 3 and 4 are special cases of this general system. The analysis of these equations is greatly facilitated by writing them in matrix notation. If we let

$$[M] = \begin{bmatrix} M_{11} & \cdots & M_{1n} \\ \cdots & \cdots & \cdots \\ M_{n1} & \cdots & M_{nn} \end{bmatrix} = \text{inertia matrix} \tag{6.2}$$

$$[K] = \begin{bmatrix} K_{11} & \cdots & K_{1n} \\ \cdots & \cdots & \cdots \\ K_{n1} & \cdots & K_{nn} \end{bmatrix} = \text{stiffness matrix} \tag{6.3}$$

$$\{q\} = \begin{Bmatrix} q_1 \\ \vdots \\ q_n \end{Bmatrix} = \text{coordinate matrix} \tag{6.4}$$

the set of Eqs. (6.1) may then be written in the convenient form

$$[M]\{\ddot{q}\} + [K]\{q\} = \{0\} \tag{6.5}$$

The coefficients M_{rs} and K_{rs} have the important property that

$$K_{rs} = K_{sr} \qquad M_{rs} = M_{sr} \tag{6.6}$$

Hence the matrices $[M]$ and $[K]$ are symmetric. The kinetic and potential energies of the system may be written in the convenient matrix form

$$T = \tfrac{1}{2}\{\dot{q}\}'[M]\{\dot{q}\} \qquad V = \tfrac{1}{2}\{q\}'[K]\{q\} \tag{6.7}$$

where $\{\dot{q}\}'$ and $\{q\}'$ are the transposed matrices of the column matrices $\{\dot{q}\}$ and $\{q\}$ and are hence row matrices. This is the notation used in Chap. 4.

Let us consider solutions of Eq. (6.5) corresponding to pure harmonic motion of the form

$$\{q\} = \{A\}\sin(\omega t + \theta) \tag{6.8}$$

where $\{A\}$ is a column matrix of amplitude constants, ω is the angular frequency of the oscillation, and θ is an arbitrary phase angle. Substituting this into Eq. (6.5), we obtain

$$([K] - \omega^2[M])\{A\} = \{0\} \tag{6.9}$$

This represents a set of homogeneous algebraic equations in the arbitrary amplitude constants $\{A\}$.

It is convenient to premultiply both sides of Eq. (6.9) by $[K]^{-1}$, the inverse of K. We then obtain

$$(I - \omega^2[K]^{-1}[M])\{A\} = \{0\} \tag{6.10}$$

where I is the unit matrix of the nth order. The matrix $[K]^{-1}[M]$ is usually called the *dynamical matrix*. That is, we have

$$[U] = [K]^{-1}[M] = \text{dynamical matrix} \tag{6.11}$$

In terms of the dynamical matrix the set of Eq. (6.10) may be written in the form

$$(I - \omega^2[U])\{A\} = \{0\} \tag{6.12}$$

If we now write

$$Z = \frac{1}{\omega^2} \tag{6.13}$$

then (6.12) may be written in the form

$$(ZI - [U])\{A\} = \{0\} \tag{6.14}$$

This set of equations will have solutions other than the trivial one of zero if the determinant of the system vanishes. That is,

$$|ZI - U| = \begin{vmatrix} Z - U_{11} & -U_{12} & -U_{13} - & \cdots & -U_{1n} \\ -U_{21} & Z - U_{22} & -U_{23} - & \cdots & -U_{2n} \\ -U_{n1} & -U_{n2} & - & \cdots & \cdots & Z - U_{nn} \end{vmatrix} = 0 \tag{6.15}$$

This is an equation of the nth degree in Z. It may be proved that all the roots are real and positive. Ordinarily this equation will have n distinct roots $Z_1, Z_2, Z_3, \ldots, Z_n$. To each root Z_r there corresponds a value ω_r of ω, given by

$$\omega_r = \sqrt{\frac{1}{Z_r}} \qquad r = 1, 2, \ldots, n \tag{6.16}$$

These are the natural angular frequencies of the system. We thus see that our original assumed solution (6.8) has led us to n values of ω. Each value ω_r of ω gives a solution of the form

$$\{A^{(r)}\} \sin(\omega_r t + \theta_r) \tag{6.17}$$

Since the original set of equations is linear, we may write the general solution by summing solutions of the form (6.17). We then have the general solution

$$\{q\} = \sum_{r=1}^{r=n} \{A^{(r)}\} \sin(\omega_r t + \theta_r) \tag{6.18}$$

The column matrices $\{A^{(r)}\}$ are called the *modal columns*. Every oscillation represented by each modal column is called a *principal mode of oscillation* of the system. The number of principal oscillations is equal to the number of degrees of freedom of the system.

Every principal oscillation is a pure harmonic motion. The most general form of oscillation of a system of n degrees of freedom consists of the superposition of n pure harmonic motions. The frequencies of the principal oscillations are called the *natural frequencies* of the system. The lowest frequency is called the *fundamental frequency*.

ORTHOGONALITY OF THE PRINCIPAL OSCILLATIONS

The modal columns satisfy Eq. (6.9) with the proper value of ω. For example, the rth modal column satisfies the equation

$$\omega_r^2[M]\{A^{(r)}\} = [K]\{A^{(r)}\} \tag{6.19}$$

This equation fixes the *ratios* of the numbers of the rth modal column. If, for example, the first number $A_1^{(r)}$ is chosen arbitrarily, then by Eq. (6.19) the numbers $A_2^{(r)}, A_3^{(r)}, \ldots, A_n^{(r)}$ may be expressed in terms of $A_1^{(r)}$. We thus see that the general solution (6.18) contains only $2n$ arbitrary constants, since any number in any modal column may be specified arbitrarily and the phase angles θ_r are arbitrary.

The modal columns possess a very interesting and important property that will now be derived. Let us write the equation satisfied by the mode $\{A^{(s)}\}$. This equation is

$$\omega_s^2[M]\{A^{(s)}\} = [K]\{A^{(s)}\} \tag{6.20}$$

Let us premultiply Eq. (6.19) by $\{A^{(s)}\}'$ and Eq. (6.20) by $\{A^{(r)}\}'$. We then obtain

$$\omega_r^2\{A^{(s)}\}'[M]\{A^{(r)}\} = \{A^{(s)}\}'[K]\{A^{(r)}\} \tag{6.21}$$

$$\omega_s^2\{A^{(r)}\}'[M]\{A^{(s)}\} = \{A^{(r)}\}'[K]\{A^{(s)}\} \tag{6.22}$$

Now, by a fundamental theorem of matrix algebra, if we have the product of three conformable matrices $[a][b][c]$, then the transpose of the product is given by

$$([a][b][c])' = [c]'[b]'[a]' \tag{6.23}$$

(see Chap. 3). If we then take the transpose of Eq. (6.21) and use the reversal law of transposed products (6.23), we obtain

$$\omega_r^2\{A^{(r)}\}'[M]\{A^{(s)}\} = \{A^{(r)}\}'[K]\{A^{(s)}\} \tag{6.24}$$

in view of the fact that the matrices $[M]$ and $[K]$ are symmetric, and hence $[M]' = [M]$ and $[K]' = [K]$.

If we now subtract Eq. (6.24) from Eq. (6.22), we obtain

$$(\omega_s^2 - \omega_r^2)\{A^{(r)}\}'[M]\{A^{(s)}\} = 0 \tag{6.25}$$

Now, by hypothesis, ω_s and ω_r are two *different* natural frequencies of the system; hence $\omega_s \neq \omega_r$. It follows that

$$\{A^{(r)}\}'[M]\{A^{(s)}\} = 0 \tag{6.26}$$

This relation is known as the *orthogonality* relation for the principal modes of oscillation.

7 SOLUTION OF THE FREQUENCY EQUATION AND CALCULATION OF THE NORMAL MODES BY THE USE OF MATRICES

A very useful and important method for the determination of the roots of the frequency equation (6.15) and the determination of the normal modes of a conservative system has been presented by W. J. Duncan and A. R. Collar.† This method is most convenient in that it avoids the expansion of the determinantal equation (6.15) and the solution of the resulting high-degree equation. The modal columns $\{A^{(r)}\}$ are also most simply obtained. Because of its great utility and increasing usefulness a brief treatment of the method will be given in this section.

We see from Eq. (6.14) that the rth modal column satisfies the equation

$$[U]\{A^{(r)}\} = Z_r\{A^{(r)}\} \tag{7.1}$$

where Z_r is the rth root of the frequency or determinantal equation (6.15). $[U]$ is the dynamical matrix of the system. If we premultiply Eq. (7.1) by U, we obtain

$$[U]^2\{A^{(r)}\} = Z_r[U]\{A^{(r)}\} = Z_r^2\{A^{(r)}\} \tag{7.2}$$

If we premultiply (7.1) by $[U]$ s times, we obtain

$$[U]^s\{A^{(r)}\} = Z_r^s\{A^{(r)}\} \qquad r = 1, 2, \ldots, n \tag{7.3}$$

We have n equations of this type, one for each modal column $\{A^{(r)}\}$ and its associated root Z_r. Let us now construct a square matrix $[A]$ from the various modal columns $\{A^{(r)}\}$ in the following manner:

$$[A] = [A^{(1)} A^{(2)} \cdots A^{(n)}] = A_{rs}] \tag{7.4}$$

The set of Eqs. (7.3) may be conveniently written in the form

$$[U]^s[A] = [A]\begin{bmatrix} Z_1^s & 0 & 0 & \cdots & & 0 \\ 0 & Z_2^s & 0 & \cdots & & 0 \\ 0 & 0 & Z_3^s & 0 & \cdots & 0 \\ \cdots\cdots\cdots\cdots\cdots\cdots\cdots \\ 0 & 0 & \cdots\cdots\cdots\cdots & Z_n^s \end{bmatrix} \tag{7.5}$$

Postmultiplying (7.5) by $[A]^{-1}$, we obtain

$$[U]^s = [A]\begin{bmatrix} Z_1^s & 0 & 0 & \cdots & 0 \\ 0 & Z_2^s & 0 & \cdots & 0 \\ 0 & 0 & Z_3^s & \cdots & 0 \\ \cdots\cdots\cdots\cdots\cdots\cdots \\ 0 & 0 & 0 & \cdots & Z_n^s \end{bmatrix}[A]^{-1} \tag{7.6}$$

† W. J. Duncan and A. R. Collar, A Method for the Solution of Oscillation Problems by Matrices, *Philosophical Magazine*, ser. 7, vol. 17, p. 865, 1934.

For convenience let

$$[B] = [A]^{-1} \tag{7.7}$$

Now let the roots Z_r of the determinantal equation be arranged in descending order of magnitude; that is, let Z_1 be the largest root of the determinantal equation, Z_2 the next largest, etc., or

$$Z_1 > Z_2 > Z_3 \cdots > Z_n \tag{7.8}$$

assuming that the roots are all distinct.   Now let us assume that s in Eq. (7.6) is so great that

$$Z_1^s \gg Z_2^s \quad \text{etc.} \tag{7.9}$$

If this is true, then only the terms corresponding to the dominant root Z_1 need be retained.   By direct multiplication we have for s sufficiently large

$$[U]^s \cong Z_1^s \begin{bmatrix} A_{11} B_{11} & A_{11} B_{12} & \cdots & A_{11} B_{1n} \\ A_{21} B_{11} & A_{21} B_{12} & \cdots & A_{21} B_{1n} \\ \cdots\cdots\cdots\cdots\cdots\cdots\cdots \\ A_{n1} B_{11} & A_{n1} B_{12} & \cdots & A_{n1} B_{1n} \end{bmatrix} \tag{7.10}$$

As a consequence of this property of the dynamical matrix $[U]$ we may perform the following procedure, which will yield the dominant root Z_1 of the system as well as the modal column associated with this root.

Let us select an *arbitrary* column matrix $\{x\}_0$ and form the following sequence:

$$\begin{aligned} [U]\{x\}_0 &= \{x\}_1 \\ [U]\{x\}_1 &= [U]^2\{x\}_0 = \{x\}_2 \\ [U]\{x\}_2 &= [U]^3\{x\}_0 = \{x\}_3 \\ &\cdots\cdots\cdots\cdots\cdots \\ [U]\{x\}_{s-1} &= [U]^s\{x\}_0 = \{x\}_s \end{aligned} \tag{7.11}$$

In view of Eq. (7.10) we have *for a sufficiently large s*

$$\{x\}_s = [U]^s\{x\}_0 = [U]^s \begin{Bmatrix} x_{10} \\ x_{20} \\ \vdots \\ x_{n0} \end{Bmatrix}$$

$$\cong Z_1^s \begin{Bmatrix} A_{11} R_1 \\ A_{21} R_1 \\ \cdots\cdots \\ A_{n1} R_1 \end{Bmatrix} \tag{7.12}$$

where

$$R_1 = B_{11} x_{10} + B_{12} x_{20} + \cdots + B_{1n} x_{n0} \tag{7.13}$$

That is, by repeated multiplication of the column matrix $\{x\}_0$ by the dynamical matrix $[U]$, we eventually reach a stage where further multiplication by $[U]$ merely multiplies every element of the column matrix $\{x\}_{s-1}$ by a common factor. This common factor is Z_1, the dominant root of the determinantal equation. By (6.16) the fundamental angular frequency ω_1 is given by

$$\omega_1 = \sqrt{\frac{1}{Z_1}} \tag{7.14}$$

We also see that the elements of the column matrix $\{x\}_s$ are proportional to those of the modal matrix $\{A^{(1)}\}$.

We now turn to a procedure by which we may obtain the next largest root Z_2 and its appropriate mode. To do this, let us place $s = 1$ in Eq. (7.6) and premultiply both sides by $[A]^{-1}$. We then obtain, since $[A]^{-1} = [B]$,

$$[B][U] = \begin{bmatrix} Z_1 & 0 & 0 & \cdots & 0 \\ 0 & Z_2 & 0 & \cdots & 0 \\ 0 & 0 & Z_3 & \cdots & 0 \\ \cdots\cdots\cdots\cdots\cdots\cdots \\ 0 & 0 & \cdots\cdots\cdots & Z_n \end{bmatrix} [B] \tag{7.15}$$

The matrix B is given by

$$[B] = \begin{bmatrix} B_{11} & B_{12} & B_{13} & \cdots & B_{1n} \\ B_{21} & B_{22} & B_{23} & \cdots & B_{2n} \\ \cdots\cdots\cdots\cdots\cdots\cdots \\ B_{n1} & B_{n2} & B_{n3} & \cdots & B_{nn} \end{bmatrix} \tag{7.16}$$

We thus see that if we take a typical *row* $[B_r]$ of the square matrix $[B]$ in the form

$$[B_r] = [B_{r1} \quad B_{r2} \quad B_{r3} \quad \cdots \quad B_{rn}] \tag{7.17}$$

then as a consequence of Eq. (7.15) we have

$$[B_r][U] = Z_r[B_r] \tag{7.18}$$

We also have, by (7.1), the relation

$$[U]\{A^{(r)}\} = Z_r\{A^{(r)}\} \tag{7.19}$$

a relation satisfied by the modal column $\{A^{(r)}\}$. Now since the dynamical matrix $[U]$ is equal to $[K]^{-1}[M]$, we may write (7.19) in the form

$$[K]^{-1}[M]\{A^{(r)}\} = Z_r\{A^{(r)}\} \tag{7.20}$$

If we premultiply both sides of this equation by M, take the transpose of both sides, and use the reversal law of transposed products, we obtain

$$\{A^{(r)}\}'[M][U] = Z_r\{A^{(r)}\}'[M] \tag{7.21}$$

Comparing Eqs. (7.21) and (7.18), we see that they are identical in form; hence we have

$$[B_r] = a_r\{A^{(r)}\}'[M]$$ (7.22)

where a_r is a proportionality factor. Equation (7.22) is of great importance in determining the higher roots of the determinantal equation as well as the modal columns corresponding to these roots.

We notice that since $[B] = [A]^{-1}$ we have

$$[B][A] = I$$ (7.23)

where I is the unit matrix of the nth order. Hence the row matrices $[B_r]$ and the modal column matrices $\{A^{(r)}\}$ have the property that

$$[B_r]\{A^{(s)}\} = \begin{cases} 0 & \text{if } r \neq s \\ 1 & \text{if } r = s \end{cases}$$ (7.24)

The general solution of the problem under consideration is given in terms of the natural angular frequencies ω_r and the modal columns $\{A^{(r)}\}$ in the form

$$\{q\} = \sum_{r=1}^{r=n} C_r\{A^{(r)}\} \sin(\omega_r t + \theta_r)$$ (7.25)

where the θ_r are the arbitrary phase angles determined from the initial conditions of the system at $t = 0$. The C_r are arbitrary constants also determined from the initial conditions of the system.

This equation may also be written in the convenient matrix form

$$\{q\} = [A] \begin{bmatrix} \sin(\omega_1 t + \theta_1) & 0 & \cdots & 0 \\ 0 & \sin(\omega_2 t + \theta_2) & 0 & \cdots & 0 \\ \cdots & \cdots & \cdots & \cdots \\ 0 & \cdots & & \sin(\omega_n t + \theta_n) \end{bmatrix} \{C\}$$ (7.26)

NORMAL COORDINATES

If we premultiply both sides of (7.26) by $[B] = [A]^{-1}$, we obtain

$$\{y\} = [B]\{q\} = \begin{bmatrix} \sin(\omega_1 t + \theta_1) & 0 & \cdots & 0 \\ 0 & \sin(\omega_2 t + \theta_2) & \cdots & 0 \\ \cdots & \cdots & \cdots & \cdots \\ 0 & \cdots & & \sin(\omega_n t + \theta_n) \end{bmatrix} \{C\}$$ (7.27)

Hence we have

$$y_1 = C_1 \sin(\omega_1 t + \theta_1)$$
$$y_2 = C_2 \sin(\omega_2 t + \theta_2)$$
$$\cdots \cdots \cdots \cdots \cdots$$
$$y_n = C_n \sin(\omega_n t + \theta_n)$$ (7.28)

The y_r quantities are the normal coordinates of the system.

CONTINUING THE SOLUTION

By Eq. (7.12) we have seen how the fundamental frequency ω_1 and the fundamental mode $\{A^{(1)}\}$ may be obtained. We shall now develop a procedure by which the higher angular frequencies and the corresponding modal columns may be obtained.

If we premultiply (7.25) by the row matrix $[B_1]$, we have in view of (7.24)

$$[B_1]\{q\} = C_1 \sin(\omega_1 t + \theta_1) \tag{7.29}$$

If now the fundamental mode is absent, we have

$$[B_1]\{q\} = 0 \tag{7.30}$$

Hence, expanding this equation, we obtain

$$B_{11}q_1 + B_{12}q_2 + \cdots + B_{1n}q_n = 0 \tag{7.31}$$

or

$$q_1 = -\frac{B_{12}}{B_{11}}q_2 - \frac{B_{13}}{B_{11}}q_3 - \cdots - \frac{B_{1n}}{B_{11}}q_n$$

$$q_2 = q_2$$

$$q_3 = q_3 \tag{7.32}$$

$$\cdots$$

$$q_n = q_n$$

This set of equations may be written in the matrix form

$$\begin{Bmatrix} q_1 \\ q_2 \\ \vdots \\ q_n \end{Bmatrix} = \begin{bmatrix} 0 - \dfrac{B_{12}}{B_{11}} - \dfrac{B_{13}}{B_{11}} & \cdots & -\dfrac{B_{1n}}{B_{11}} \\ 0 & 1 & 0 & \cdots & 0 \\ 0 & \cdots\cdots\cdots\cdots & 1 \end{bmatrix} \begin{Bmatrix} q_1 \\ \vdots \\ q_n \end{Bmatrix} \tag{7.33}$$

Or if we call the square matrix above $[S]$, we have

$$\{q\} = [S]\{q\} \tag{7.34}$$

This constrains the coordinates in such a manner that the fundamental mode is absent. The original differential equations of the system (6.5) may be written in terms of the *dynamical matrix* in the form

$$[U]\{\ddot{q}\} + I\{q\} = \{0\} \tag{7.35}$$

where I is the unit matrix of the nth order. If we now differentiate Eq. (7.34) twice with respect to time, we have

$$\{\ddot{q}\} = [S]\{\ddot{q}\} \tag{7.36}$$

Substituting this into (7.35), we obtain

$$[U][S]\{\ddot{q}\} + I\{q\} = 0 \qquad (7.37)$$

If we let

$$[U][S] = [U]_1 \qquad (7.38)$$

Eq. (7.37) becomes

$$[U]_1\{\ddot{q}\} + I(q) = \{0\} \qquad (7.39)$$

This set of equations has the same *form* as the original set (7.35). It represents a system whose dynamical matrix is $[U]_1$ and whose natural frequencies and modes are the same as those of the original system, but since we have used the constraint (7.34), the fundamental frequency and mode are now absent.

By carrying out the same procedure with the matrix $[U]_1$ that was performed with the matrix $[U]$, we obtain the root Z_2 and hence the next higher natural frequency ω_2 together with the corresponding mode. We then obtain a new row matrix $[B_2]$ by Eq. (7.22) and repeat the procedure until all the angular frequencies and all the modal columns of the system have been found. An example of the general procedure will now be given.

8 NUMERICAL EXAMPLE: THE TRIPLE PENDULUM

As a simple numerical example of the above theory, let us consider the oscillations of a triple pendulum under gravity in a vertical plane. This example is given by Duncan and Collar in their fundamental paper referred to in Sec. 7. The dynamical system under consideration is given by Fig. 8.1.

We shall consider the case of small oscillations, and for the coordinates of the system we shall take the small horizontal displacements of the masses M_1, M_2, M_3, respectively, from the equilibrium position. The first step in the procedure is to compute the dynamical matrix $[U]$.

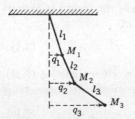

Fig. 8.1

THE FLEXIBILITY MATRIX

If we apply a set of static forces F_1, F_2, F_3 in the direction of the coordinates q_1, q_2, and q_3, we may write the relation between the displacements q_1, q_2, and q_3 and the forces F_1, F_2, F_3 in the form

$$\begin{Bmatrix} F_1 \\ F_2 \\ F_3 \end{Bmatrix} = \begin{bmatrix} K_{11} & K_{12} & K_{13} \\ K_{21} & K_{22} & K_{23} \\ K_{31} & K_{32} & K_{33} \end{bmatrix} \begin{Bmatrix} q_1 \\ q_2 \\ q_3 \end{Bmatrix} \tag{8.1}$$

where the square matrix $[K]$ is the stiffness matrix. We may premultiply by $[K]^{-1}$ and obtain the displacements in terms of the forces in the form

$$\begin{Bmatrix} q_1 \\ q_2 \\ q_3 \end{Bmatrix} = \begin{bmatrix} \phi_{11} & \phi_{12} & \phi_{13} \\ \phi_{21} & \phi_{22} & \phi_{23} \\ \phi_{31} & \phi_{32} & \phi_{33} \end{bmatrix} \begin{Bmatrix} F_1 \\ F_2 \\ F_3 \end{Bmatrix} \tag{8.2}$$

where the matrix $[\phi] = [K]^{-1}$ is the flexibility matrix. Since $K_{rs} = K_{sr}$, we have

$$\phi_{rs} = \phi_{sr} \tag{8.3}$$

That is, the flexibility matrix is a symmetric matrix.

In terms of the flexibility matrix the dynamical matrix is given by

$$[U] = [K]^{-1}[M] = [\phi][M] \tag{8.4}$$

To determine the elements of the flexibility matrix, it is necessary only to impose a unit force F_1 on the system and compute or measure the corresponding deflections $(q_1 q_2 q_3)$. This gives the elements of the first column of $[\phi]$. Applying a unit force F_2 and obtaining the corresponding deflections yields the second column of $[\phi]$, etc.

If a static unit force is applied horizontally to mass M_1, then the three masses will each be displaced a distance a given by

$$a = \frac{l_1}{g(M_1 + M_2 + M_3)} \tag{8.5}$$

Hence the first column of the flexibility matrix is given by

$$a = \phi_{11} = \phi_{21} = \phi_{31} \tag{8.6}$$

When a unit force is applied horizontally to M_2, M_1 will again be displaced a distance a, but M_2 and M_3 will each be displaced a distance $a + b$, where

$$b = \frac{l_2}{g(M_2 + M_3)} \tag{8.7}$$

Hence the second column of the matrix $[\phi]$ is given by

$$\phi_{12} = a \qquad \phi_{22} = \phi_{32} = a + b \tag{8.8}$$

Applying a unit horizontal force to M_3 displaces M_1 a distance a; M_2 a distance $a + b$; and M_3 a distance $a + b + c$, where

$$c = \frac{l_3}{gM_3} \tag{8.9}$$

The third column of the flexibility matrix is

$$\phi_{13} = a \qquad \phi_{23} = a + b \qquad \phi_{33} = a + b + c \tag{8.10}$$

Hence the flexibility matrix is given by

$$[\phi] = \begin{bmatrix} a & a & a \\ a & a+b & a+b \\ a & a+b & a+b+c \end{bmatrix} \tag{8.11}$$

The mass matrix, in this case, has the diagonal form

$$[M] = \begin{bmatrix} M_1 & 0 & 0 \\ 0 & M_2 & 0 \\ 0 & 0 & M_3 \end{bmatrix} \tag{8.12}$$

Hence the dynamical matrix is given by

$$\begin{aligned} [U] &= [\phi][M] \\ &= \begin{bmatrix} M_1 a & M_2 a & M_3 a \\ M_1 a & M_2(a+b) & M_3(a+b) \\ M_1 a & M_2(a+b) & M_3(a+b+c) \end{bmatrix} \end{aligned} \tag{8.13}$$

As a numerical example, let us take the case where all the masses are equal and the lengths of the pendulums are equal. In this case we have

$$M_1 = M_2 = M_3 = M \qquad l_1 = l_2 = l_3 = l$$

$$a = \frac{l}{3Mg} \qquad b = \frac{l}{2Mg} \qquad c = \frac{l}{Mg} \tag{8.14}$$

Hence the dynamical matrix in this case becomes

$$U = \frac{l}{6g} \begin{bmatrix} 2 & 2 & 2 \\ 2 & 5 & 5 \\ 2 & 5 & 11 \end{bmatrix} = \frac{l}{6g}[u] \tag{8.15}$$

Equation (6.12) reduces in this case to

$$\left(I - \frac{\omega^2 l}{6g}[u]\right)\{A\} = \{0\} \tag{8.16}$$

If we let

$$\frac{6g}{l}\frac{1}{\omega^2} = Z \tag{8.17}$$

then Eq. (6.14) becomes

$$(ZI - [u])\{A\} = \{0\} \tag{8.18}$$

where $[u]$ is the numerical part of the dynamical matrix $[U]$, and the factor $6g/l$ has been absorbed into Z.

The angular frequencies are given by

$$\omega_r = \sqrt{\frac{6g}{Z_r l}} \tag{8.19}$$

To find the roots Z_r, we begin the iterative procedure of Eq. (7.11). If we choose for our *arbitrary* column $\{x\}_0$ the column

$$\{x\}_0 = \begin{Bmatrix} 1 \\ 1 \\ 1 \end{Bmatrix} \tag{8.20}$$

we begin the sequence

$$[u]\{x\}_0 = \begin{bmatrix} 2 & 2 & 2 \\ 2 & 5 & 5 \\ 2 & 5 & 11 \end{bmatrix} \begin{Bmatrix} 1 \\ 1 \\ 1 \end{Bmatrix}$$

$$= \begin{Bmatrix} 6 \\ 12 \\ 18 \end{Bmatrix} = 18 \begin{Bmatrix} \frac{1}{3} \\ \frac{2}{3} \\ 1 \end{Bmatrix} \tag{8.21}$$

It is unnecessary to carry the common factor 18 in the further operations since it is the *ratios* of the successive elements in the multiplications that are important. Dropping the factor 18 and continuing, we have

$$\begin{bmatrix} 2 & 2 & 2 \\ 2 & 5 & 5 \\ 2 & 5 & 11 \end{bmatrix} \begin{Bmatrix} \frac{1}{3} \\ \frac{2}{3} \\ 1 \end{Bmatrix} = \begin{Bmatrix} 4 \\ 9 \\ 15 \end{Bmatrix} = 15 \begin{Bmatrix} 0.26 \\ 0.6 \\ 1 \end{Bmatrix}$$

$$\begin{bmatrix} 2 & 2 & 2 \\ 2 & 5 & 5 \\ 2 & 5 & 11 \end{bmatrix} \begin{Bmatrix} 0.26 \\ 0.6 \\ 1 \end{Bmatrix} = 14.53 \begin{Bmatrix} 0.25688 \\ 0.58716 \\ 1 \end{Bmatrix} \tag{8.22}$$

After nine multiplications we have

$$\begin{bmatrix} 2 & 2 & 2 \\ 2 & 5 & 5 \\ 2 & 5 & 11 \end{bmatrix} \begin{Bmatrix} 0.254885 \\ 0.584225 \\ 1 \end{Bmatrix} = 14.4309 \begin{Bmatrix} 0.254885 \\ 0.584225 \\ 1 \end{Bmatrix} \tag{8.23}$$

Repeating the process merely multiplies the column matrix by the factor 14.4309; we therefore have

$$Z_1 = 14.4309 \tag{8.24}$$

The fundamental frequency of the oscillation is given by

$$f_1 = \frac{\omega_1}{2\pi} = \frac{1}{2\pi}\sqrt{\frac{6g}{Z_1 l}} = 0.102624\sqrt{\frac{g}{l}} \tag{8.25}$$

The modal column $\{A^{(1)}\}$ is proportional to the column

$$\begin{pmatrix} 0.254885 \\ 0.584225 \\ 1 \end{pmatrix}$$

To obtain the higher harmonics, we first make use of Eq. (7.22) to obtain the row B_1. In this case we have

$$[B_1] = a_1[0.254885,\ 0.584225,\ 1]\begin{bmatrix} M & 0 & 0 \\ 0 & M & 0 \\ 0 & 0 & M \end{bmatrix} \tag{8.26}$$

Since a_1 is an arbitrary factor and we are interested only in a row proportional to $[B_1]$, we may take B_1 equal to $[0.254885,\ 0.584225,\ 1]$. The matrix $[S]$ of (7.34) is now given by

$$[S] = \begin{bmatrix} 0 & -2.29211 & -3.92334 \\ 0 & 1 & 0 \\ 0 & 0 & 1 \end{bmatrix} \tag{8.27}$$

We then obtain

$$[u]_1 = [u][S] = \begin{bmatrix} 0 & -2.58422 & -5.84668 \\ 0 & 0.41578 & -2.84668 \\ 0 & 0.41578 & 3.15332 \end{bmatrix} \tag{8.28}$$

This is the dynamical matrix that has the fundamental mode absent. We now repeat the iterative procedure by again choosing an arbitrary column matrix $\{x\}_0$; we thus find

$$[u]_1 \begin{pmatrix} 1 \\ 1 \\ 1 \end{pmatrix} = \begin{pmatrix} \cdots \\ -2.4309 \\ 3.5691 \end{pmatrix} = 3.5691\begin{pmatrix} \cdots \\ -0.68110 \\ 1 \end{pmatrix}$$

$$[u]_1 \begin{pmatrix} \cdots \\ -0.68110 \\ 1 \end{pmatrix} = \begin{pmatrix} \cdots \\ -3.1299 \\ 2.8701 \end{pmatrix} = 2.8701\begin{pmatrix} \cdots \\ -1.09049 \\ 1 \end{pmatrix} \tag{8.29}$$

After 15 multiplications the column repeats itself, and the multiple factor is

$$Z_2 = 2.6152 \tag{8.30}$$

The modal column $\{A^{(2)}\}$ for this approximation may be taken to be

$$\{A^{(2)}\} = \begin{Bmatrix} -0.95670 \\ -1.29429 \\ 1 \end{Bmatrix} \tag{8.31}$$

Hence the first overtone has this mode and a frequency given by

$$f_2 = \frac{\omega_2}{2\pi} = \frac{1}{2\pi}\sqrt{\frac{6g}{Z_2 l}} = 0.24107\sqrt{\frac{g}{l}} \tag{8.32}$$

Again by Eq. (7.22) we may take the row $\{B_2\}$ equal to $\{A^{(2)}\}'$.

The condition for the absence of the second overtone is

$$[B_2]\{q\} = 0 = [-0.95670, -1.29429, 1]\{q\} \tag{8.33}$$

Solving this for q_1, we have

$$q_1 = -1.35287q_2 + 1.04526q_3 \tag{8.34}$$

We may eliminate q_1 between this equation and the equation

$$[B_1]\{q\} = 0.254885q_1 + 0.584225q_2 + q_3 = 0 \tag{8.35}$$

We thus obtain

$$q_2 = -5.2900q_3 \tag{8.36}$$

This ensures that both the fundamental and first-overtone modes are absent from the oscillation. Equation (8.36) may be written in the convenient matrix form

$$\begin{Bmatrix} q_1 \\ q_2 \\ q_3 \end{Bmatrix} = \begin{bmatrix} 1 & 0 & 0 \\ 0 & 0 & -5.2900 \\ 0 & 0 & 1 \end{bmatrix} \begin{Bmatrix} q_1 \\ q_2 \\ q_3 \end{Bmatrix}$$

$$= [S]_1\{q\} \tag{8.37}$$

We now construct a new dynamical matrix $[u]_2$ that has the fundamental and overtone modes absent; this new matrix is given by

$$[u]_2 = [u]_1[S]_1 = \begin{bmatrix} 0 & 0 & 7.8238 \\ 0 & 0 & -5.0461 \\ 0 & 0 & 0.9539 \end{bmatrix} \tag{8.38}$$

We again repeat the iterative process and obtain

$$[u]_2\{x\}_0 = \begin{bmatrix} 0 & 0 & 7.8238 \\ 0 & 0 & -5.0461 \\ 0 & 0 & 0.9539 \end{bmatrix} \begin{Bmatrix} 1 \\ 1 \\ 1 \end{Bmatrix}$$

$$= \begin{Bmatrix} 7.8238 \\ -5.0461 \\ 0.9539 \end{Bmatrix} = 0.9539 \begin{Bmatrix} 8.2019 \\ -5.2900 \\ 1 \end{Bmatrix} \tag{8.39}$$

Repeating the process merely repeats the factor 0.9539; so it is unnecessary to go further. We thus have

$$Z_3 = 0.9539 \tag{8.40}$$

The highest frequency is given by

$$f_3 = \frac{1}{2\pi} \sqrt{\frac{6g}{Z_3 l}} = 0.39916 \sqrt{\frac{g}{l}} \tag{8.41}$$

The mode $\{A^{(3)}\}$ corresponding to this frequency may be taken to be

$$\{A^{(3)}\} = \begin{pmatrix} 8.2019 \\ -5.2900 \\ 1 \end{pmatrix} \tag{8.42}$$

to a factor of proportionality. The modal matrix may be taken to be

$$[A] = \begin{bmatrix} 0.254885 & -0.95670 & 8.2019 \\ 0.584225 & -1.29429 & -5.2900 \\ 1 & 1 & 1 \end{bmatrix} \tag{8.43}$$

In this case, because of the simplicity of the mass matrix $[M]$, we may take

$$[B] = [A]' \tag{8.44}$$

the transpose of matrix $[A]$.

The normal coordinates are then given by

$$\{y\} = [B]\{q\} \tag{8.45}$$

9 NONCONSERVATIVE SYSTEMS: VIBRATIONS WITH VISCOUS DAMPING

In the last few sections the general theory of conservative systems has been considered. These systems are characterized completely by their kinetic- and potential-energy functions and are of great practical importance. In many problems that arise in practice the frictional forces are relatively small and may be disregarded and the system treated as a conservative one.

The mathematical analyses of vibrating systems that contain frictional forces of a general nature usually lead to a formulation involving *nonlinear* differential equations (see Chap. 15). However, in a great many practical problems, the frictional forces involved are proportional to the velocities of the moving parts of the system. Frictional forces of this type are present in systems that contain viscous friction. The mathematical solution of vibration problems of systems involving viscous friction leads usually to the solution of sets of *linear* equations and is relatively simple.

The Lagrangian equations of Sec. 4 can be generalized to take into account viscous friction. In order to do this, it is necessary to compute the rate at which energy is dissipated by the viscous frictional forces. This may be done by considering a single particle moving along the x axis undergoing damping so that the resisting frictional force impeding its motion is

$$f_r = c\dot{x} \tag{9.1}$$

The work done by the frictional force during a small displacement δx is $-(c\dot{x}\,\delta x)$, and the amount of energy δW dissipated is

$$\delta W = f_r\,\delta x = c\dot{x}\,\delta x = (c\dot{x}^2)\,\delta t \qquad \delta x = \dot{x}\,\delta t \tag{9.2}$$

The time rate at which energy is dissipated in this case is $\delta W/\delta t = c\dot{x}^2$. Let the function F be defined by the equation

$$F = \frac{c\dot{x}^2}{2} \tag{9.3}$$

This function is called the *dissipation* function and represents half the rate at which the energy of the particle is dissipated by the viscous frictional force. The frictional force f_r may therefore be obtained by the following equation:

$$f_r = -\frac{dF}{d\dot{x}} = -c\dot{x} \tag{9.4}$$

By an extension of the above analysis to the case of a system having n degrees of freedom undergoing viscous friction, it can be shown that the dissipation function of the entire system in general has the form

$$F = \tfrac{1}{2}(C_{11}\dot{q}_1^2 + 2C_{12}\dot{q}_1\dot{q}_2 + C_{22}\dot{q}_2^2 + \cdots) \tag{9.5}$$

where $(\dot{q}_1, \dot{q}_2, \dot{q}_3, \ldots)$ are the generalized velocities of the system and the coefficients C_{11}, C_{12}, C_{22}, $\ldots$ in general depend on the configuration of the system, though for the case of small vibrations in the neighborhood of a configuration of stable equilibrium these coefficients can be treated with good accuracy as being constant. In such cases it is convenient to introduce the *damping matrix*

$$[C] = \begin{bmatrix} C_{11} & C_{12} & C_{13} & \cdots & C_{1n} \\ C_{21} & C_{22} & C_{23} & \cdots & C_{2n} \\ \cdot\cdot\cdot\cdot\cdot\cdot\cdot\cdot\cdot\cdot\cdot\cdot\cdot\cdot\cdot \\ C_{n1} & C_{n2} & C_{n3} & \cdots & C_{nn} \end{bmatrix} \qquad C_{ij} = C_{ji} \tag{9.6}$$

The elements of the damping matrix $[C]$ are the viscous-friction damping coefficients C_{ij}. The matrix $[C]$ is a symmetric matrix. In terms of this matrix the dissipation function F may be written in the following concise form,

$$F = \tfrac{1}{2}(\dot{q})'[C](\dot{q}) \tag{9.7}$$

where $(\dot{q})'$ is the transpose of the generalized velocity matrix $(\dot{q})$.

The dissipation function F represents half the rate at which energy is being dissipated from the entire system as it moves under the influence of viscous friction.   The generalized forces f_i that arise from the viscous friction of the system are

$$f_i = -\frac{\partial F}{\partial \dot{q}_i} \qquad i = 1, 2, 3, \ldots, n \tag{9.8}$$

The Lagrangian equations of the system then take the form

$$\frac{d}{dt}\left(\frac{\partial T}{\partial \dot{q}_i}\right) - \frac{\partial T}{\partial q_i} = -\frac{\partial V}{\partial q_i} - \frac{\partial F}{\partial \dot{q}_i} + Q_i \qquad i = 1, 2, 3, \ldots \tag{9.9}$$

The left-hand member of (9.9) represents the effect of the inertia terms of the system; the terms of the right-hand member are the various generalized forces of the system.   (1) The term $-(\partial V/\partial q_i)$ takes into account forces that may be derived from a potential function V such as gravitational forces, forces due to the presence of elastic springs, etc.   (2) The term $-(\partial F/\partial \dot{q}_i)$ takes into account the effect of retarding forces due to viscous friction.   (3) The term Q_i includes all other forces acting on the system, such as time-dependent disturbing forces.

In terms of the inertia matrix $[M]$, the stiffness matrix $[K]$, and the damping matrix $[C]$ of the system the Lagrangian equations (9.9) may be written in the following concise form:

$$[M]\{\ddot{q}\} + [C]\{\dot{q}\} + [K]\{q\} = \{Q\} \tag{9.10}$$

The elements of the column matrix (Q) are the generalized forces Q_i of the system.

The electrical networks of Chap. 5 may be regarded as special dynamical systems that are characterized by the following functions:

$$T = \tfrac{1}{2}\{\dot{q}\}'[L]\{\dot{q}\} \qquad V = \tfrac{1}{2}\{q\}'[S]\{q\} \qquad F = \tfrac{1}{2}\{\dot{q}\}'[R]\{\dot{q}\} \tag{9.11}$$

In these expressions the $\{\dot{q}\}$ are the various mesh currents of the circuit, the elements of $\{q\}$ are the various mesh charges of the circuit, and the square matrices $[L]$, $[S]$, and $[R]$ are the inductance, elastance, and resistance matrices of the circuit.   In this case T represents the magnetic energy of the system, V the electric energy of the system, and F is one-half the power dissipated by the resistances of the circuit.   The Lagrangian equations (9.9) now take the following form:

$$[L]\{\ddot{q}\} + [R]\{\dot{q}\} + [S]\{q\} = \{E\} \tag{9.12}$$

The elements of the column matrix $\{E\}$ are the various mesh potentials of the electrical circuit.

CLASSICAL SOLUTION, FREE OSCILLATIONS

If the generalized forces Q_i driving the nonconservative system are all equal to zero, then the set of Eqs. (9.10) reduce to

$$[M]\{\ddot{q}\} + [C]\{\dot{q}\} + [K]\{q\} = \{0\} \tag{9.13}$$

This set of differential equations determines the nature of the free oscillations of the system. In order to solve this set of equations, assume a solution of the exponential form

$$\{q\} = \{A\}\,e^{mt} \tag{9.14}$$

where m is a number to be determined and $\{A\}$ is a column of constants. If this assumed form of solution is substituted into (9.13), the following result is obtained:

$$([M]\,m^2 + [C]\,m + [K])\{A\}\,e^{mt} = \{0\} \tag{9.15}$$

If (9.15) is divided by the scalar factor e^{mt}, a set of linear homogeneous equations in the column of constants $\{A\}$ is obtained. For this set of equations to have a nontrivial solution, it is necessary that the determinant of the coefficients must vanish; hence we must have

$$\Delta(m) = |[M]\,m^2 + [C]\,m + [K]| = 0 \tag{9.16}$$

This equation is called *Lagrange's determinantal equation for m*. In general it is of degree $2n$. The following properties concerning the roots of $\Delta(m)$ may be proved:†

1. None of the roots is real and positive.
2. If $[C] = [0]$, there is no dissipation of energy and the system is a conservative one. In this case the roots are all pure imaginaries.
3. If $[M] = [0]$ or $[K] = [0]$ so that the system is devoid of inertia or of stiffness and $[C] \neq [0]$, the roots are real and negative so that the motion dies away exponentially.
4. If the elements of the damping matrix $[C]$ are not too large, all the roots are conjugate complex numbers with a negative real part. This is the most frequent case in practice.

Let it be supposed that Eq. (9.16) has $2n$ distinct roots $(m_1, m_2, m_3, \ldots, m_{2n})$. Then as a consequence of (9.15) there are $2n$ equations of the following type:

$$([M]\,m_r^2 + [C]\,m_r + [K])\{A_r\} = \{0\} \qquad r = 1, 2, 3, \ldots, 2n \tag{9.17}$$

---

† See A. G. Webster, "The Dynamics of Particles and of Rigid Bodies," Teubner Verlagsgesellschaft, Leipzig, 1925.

where $\{A_r\}$ represents a column of constants associated with the root m_r. Each column $\{A_r\}$ has the form

$$\{A_r\} = \begin{bmatrix} A_{1r} \\ A_{2r} \\ \vdots \\ A_{nr} \end{bmatrix} \tag{9.18}$$

Equation (9.5) fixes the ratios

$$A_{1r}:A_{2r}:A_{3r}: \cdots :A_{nr} \tag{9.19}$$

The theory of linear differential equations indicates that to obtain a general solution of the set of Eqs. (9.13) we must take the sum of the particular solutions $\{A_r\}e^{m_r t}$ for all the roots m_r. We thus obtain the following general solution:

$$\{q\} = \sum_{r=1}^{2n} \{A_r\} e^{m_r t} \tag{9.20}$$

It must be noted that the ratios of the A's in any one column are determined by the linear equations (9.17); therefore there is still a factor that is arbitrary for each column and hence $2n$ in all. Thus the general solution (9.20) contains $2n$ arbitrary constants, as it should.

If the roots of (9.16) are complex, they occur in conjugate complex pairs of the form

$$m_r = -a_r + j\omega_r \qquad \bar{m}_r = -a_r = j\omega_r \qquad r = 1, 2, 3, \ldots , n \tag{9.21}$$

In this case the general solution (9.20) contains terms of the following type:

$$\{A_r\} e^{m_r t} + \{\tilde{A}_r\} e^{\bar{m}_r t}$$
$$= e^{-a_r t} \{B_r\} \cos (\omega_r t + \theta_r) \qquad r = 1, 2, 3, \ldots , n \tag{9.22}$$

where $\{B_r\}$ represents a column of constants and the θ_r are phase angles. In this case the general solution of (9.13) takes the form

$$\{q\} = \sum_{r=1}^{n} \{B_r\} e^{-a_r t} \cos (\omega_r t + \theta_r) \tag{9.23}$$

It may easily be shown that the $\{B_r\}$ columns satisfy (9.17) in the same manner as do the $\{A_r\}$ columns, and hence the ratios of the B's in each column are fixed. Each column then contains an arbitrary constant in the phase angle θ_r belonging to the column. The following results may be stated:

If the roots of the determinantal equation (9.16) are distinct and conjugate complex quantities, then the motion of a dynamical system having n degrees of freedom slightly displaced from a position of stable equilibrium may be described as follows:

Each coordinate performs the resultant of n damped harmonic oscillations of different periods. The phase and damping factors of any simple oscillation of a particular period are the same for all the coordinates. The absolute value of the amplitude of any particular coordinate is arbitrary, but the ratios of the amplitudes for a particular period for the different coordinates are determined solely by the nature of the system. The $2n$ arbitrary constants determine the n amplitudes and phases of the system. These arbitrary constants are found from the values of the n coordinates $\{q\}$ and the n velocities $\{\dot{q}\}$ for a particular instant of time.

The classical method of solution for the general nonconservative system consists in obtaining the determinantal equation (9.16), expanding the determinantal equation, then solving it for the $2n$ roots $(m_1, m_2, m_3, \ldots, m_{2n})$ by the Graeffe method of Appendix D. Once the roots m_r of the determinantal equation have been found, the ratios of the columns (B_r) can be calculated by the use of (9.17). Finally the arbitrary constants are determined from a knowledge of the initial conditions of the system.

10 A MATRIX ITERATIVE METHOD FOR THE ANALYSIS OF NONCONSERVATIVE SYSTEMS

In a fundamental paper by Duncan and Collar† an iterative method similar to that discussed in Sec. 7 for the determination of the modes, decrement factors a_r, and frequencies ω_r of nonconservative systems is given. The basic principle that underlies the method is to write the system of differential equations (9.13) as a first-order system by the introduction of the following partitioned matrix:

$$\{y\} = \begin{bmatrix} (q) \\ (\dot{q}) \end{bmatrix} \tag{10.1}$$

The column matrix $\{y\}$ thus contains $2n$ elements, the n coordinates (q), and the n velocities $\{\dot{q}\}$. If the following square partitioned matrix of the $2n$th order is introduced,

$$[v] = \begin{bmatrix} [0] & [I] \\ -[M]^{-1}[K] & -[M]^{-1}[C] \end{bmatrix} \tag{10.2}$$

where $[0]$ is the nth-order zero matrix and $[I]$ is the nth-order unit matrix, then the system of differential equations (9.13) may be written in the following form:

$$\{\dot{y}\} = [v]\{y\} \tag{10.3}$$

---

† W. J. Duncan and A. R. Collar, Matrices Applied to the Motions of Damped Systems, *Philosophical Magazine*, ser. 7, vol. 19, p. 197, 1935.

In order to solve the first-order system of differential equation (10.3), assume a solution of the form

$$\{y\} = e^{mt}\{A\} \tag{10.4}$$

If the assumed solution (10.4) is substituted into the differential equation (10.3), the following result is obtained:

$$m e^{mt}\{A\} = [v] e^{mt}\{A\} \tag{10.5}$$

If the scalar factor e^{mt} is divided out of Eq. (10.5), the result may be written in the form

$$(mI - [v])\{A\} = \{0\} \tag{10.6}$$

where I is the unit matrix of order $2n$. The matrix

$$[c(m)] = mI - [v] \tag{10.7}$$

is the characteristic matrix of the matrix $[v]$. The characteristic equation of $[v]$ is

$$\det [c(m)] = |mI - [v]| = 0 \tag{10.8}$$

The roots $m_1, m_2, m_3, \ldots, m_{2n}$ are the *eigenvalues* of the matrix $[v]$, and the columns $\{A_1\}, \{A_2\}, \{A_3\}, \ldots, \{A_{2n}\}$ satisfy equations of the form

$$(m_r I - [v])\{A_r\} = \{0\} \qquad r = 1, 2, 3, 4, \ldots, 2n \tag{10.9}$$

The columns $\{A_r\}$ are therefore the *eigenvectors* of the matrix $[v]$.

The root m_r, having the largest modulus, may be determined by setting up a sequence similar to that of (7.11) of the form

$$\{x\}_s = [v]^s\{x\}_0 \qquad s = 1, 2, 3, 4, \ldots \tag{10.10}$$

where $\{x\}_0$ is an *arbitrary* column matrix of the $2n$th order. It will be found that for a sufficiently large s continued multiplications give rise merely to multiplicative factors. These factors enable the eigenvalue having the largest modulus to be determined, and also the associated eigenvector can be found. The details of the practical numerical procedure are somewhat lengthy and will be found in the above paper and in the book by Frazer, Duncan, and Collar (1938).

11   FORCED OSCILLATIONS OF A NONCONSERVATIVE SYSTEM

Let us suppose that on each coordinate of the general nonconservative system there is impressed a harmonically varying force

$$F_r \cos \omega t \qquad r = 1, 2, \ldots, n \tag{11.1}$$

In this case the differential equations (9.13) become

$$[M]\{\ddot{q}\} + [R]\{\dot{q}\} + [K]\{q\} = \{F\} \cos \omega t \tag{11.2}$$

where $\{F\}$ is a column matrix whose elements are the amplitudes of the impressed forces F_r. We replace $\cos \omega t$ by $\mathrm{Re}\, e^{j\omega t}$ and solve the equation

$$[M]\{\ddot{q}\} + [R]\{\dot{q}\} + [K]\{q\} = \{F\} e^{j\omega t} \tag{11.3}$$

retaining only the real part of the particular solution. To do this, let us assume

$$\{q\} = \{Q\} e^{j\omega t} \tag{11.4}$$

where $\{Q\}$ is a column of amplitude constants to be determined. Substituting this assumed solution into (11.4) and dividing by the common factor $e^{j\omega t}$, we have

$$([K] - \omega^2[M] + j\omega[R])\{Q\} = \{F\} \tag{11.5}$$

If we let

$$[Z] = [K] - \omega^2[M] + j\omega[R] = [Y]^{-1} \tag{11.6}$$

then on premultiplying both sides of (11.6) by $[Z]^{-1} = [Y]$, we have

$$\{Q\} = [Y]\{F\} \tag{11.7}$$

The steady-state solution of the forced oscillation is now given by

$$\{q\} = \mathrm{Re}\,([Y]\{F\} e^{j\omega t}) \tag{11.8}$$

where Re signifies "the real part of." The matrix $[Z]$ is sometimes termed in the literature the *mechanical impedance matrix* and $[Y]$ the *mechanical admittance matrix*. Equation (11.8) represents the forced oscillations after the damped free oscillations have vanished.

It must be noted that, if $[R] = [0]$ and the system is conservative, then, if ω happens to coincide with one of the natural frequencies of the system, $[Z]$ as given by Eq. (11.6) vanishes and we have the case of resonance.

PROBLEMS

1. Obtain the flexibility matrix of the three-pendulum system of Fig. 4.3. Write the dynamical matrix of the system.

2. An electric train is made up of three units: a locomotive and two passenger cars. Each unit has a mass M, and the spring constants of the coupling connecting them are equal to K. Write the Lagrangian equation of motion of the system. Draw the equivalent electrical circuit, and determine the natural frequencies of the system.

3. If in the system of Prob. 2 identical shock absorbers that act by viscous friction are placed between the three units, determine the smallest value of the damping factor R so that the relative motion of the locomotive and cars is not oscillatory.

4. A long train of n identical units of mass M coupled by springs of spring constants all equal to K is oscillating. Draw the equivalent electrical circuit, and write the frequency equation of the system.

5. A uniform shaft free to rotate in bearings carries five equidistant disks. The moments of inertia of four disks are equal to J, while the moment of inertia of one of the end disks is equal to $2J$. Set up the dynamical matrix, and obtain the lowest natural frequency by the iterative matrix method of Sec. 7. Draw the equivalent electrical circuit.

6. A particle of mass M is attracted to a center by a force proportional to the distance, or $F_z = -ax$, $F_y = -ay$. Write the equations of motion of the particle. Show that x and y execute independent simple harmonic vibrations of the same frequency.

7. Solve Prob. 6 by using polar coordinates.

8. Two balls, each of mass M, and three weightless springs, one of length $2d$ and the others of length d, are connected together in the arrangement spring d—ball—spring $2d$—ball—spring d, and the whole device is stretched in a straight line between two points, with a given tension in the spring. Gravity is neglected.

    Investigate the small vibrations of the balls at right angles to the straight line, assuming motion only in one plane. Set up the equations of motion. Determine the natural frequencies and the normal modes. What are the normal coordinates?

9. One simple pendulum is hung from another; that is, the string of the lower pendulum is tied to the bob of the upper one. Discuss the small oscillations of the resulting system, assuming arbitrary lengths and masses. Determine the natural frequencies and obtain the normal coordinates in the case of equal masses and equal lengths of strings.

10. Consider the case of the three coupled pendulums of Fig. 4.3. In this case assume that each pendulum is retarded by a viscous force proportional to its angular velocity. Consider an equal retarding force on each pendulum. Draw the equivalent electrical circuit, and obtain the general solution of the equations of motion. Assume that the damping is so small that oscillations are possible.

11. A particle subject to a linear restoring force and a viscous damping is acted on by a periodic force whose frequency differs from the natural frequency of the system by a small quantity.

    The particle starts from rest at $t = 0$ and builds up the motion. Discuss the whole problem, including initial conditions. Consider what happens when the frequency gets nearer and nearer the natural frequency and the damping gets smaller and smaller.

12. Show that for a particle subject to a linear restoring force and viscous damping the maximum amplitude occurs when the applied frequency is less than the natural frequency. Find this resonance frequency. Show that maximum energy is attained when the applied frequency is equal to the natural frequency.

13. Two small metal spheres A and B of equal mass m are suspended from two points A' and B' at a distance a apart on the same horizontal line by insulating threads of equal length s. They are then charged with equal charges $+q$. If A is pulled aside a *short* distance s_1 and B a *short* distance s_2 (both small in comparison with a) in the plane formed by the equilibrium position of both threads, determine the resulting motion in this plane. Determine the frequencies of oscillation of the system.

14. Two particles of mass m connected by a weightless spring of unstressed length s and spring constant k are set moving in an arbitrary manner on a smooth horizontal plane. Set up and solve Lagrange's equations for this case.

15. Discuss the oscillations of the coupled triple pendulum of Sec. 4 by the matrix-iteration method of Sec. 7. Determine the lowest and highest frequencies of oscillating by matrix iteration assuming that $Mgl/kh^2 = 1$.

REFERENCES

1905. Routh, E. J.: "Advanced Rigid Dynamics," The Macmillan Company, New York.

1926. Maxfield, J. P., and H. C. Harrison: Methods of High Quality Recording and Reproducing of Music and Speech Based on Telephone Research, *Bell System Technical Journal*, vol. 5, pp. 493–523, July.

1937. Timoshenko, S.: "Vibration Problems in Engineering," D. Van Nostrand Company, Inc., Princeton, N.J.

1938. Frazer, R. A., W. J. Duncan, and A. R. Collar: "Elementary Matrices and Some Applications to Dynamics and Differential Equations," Cambridge University Press, New York.

1940. Den Hartog, J. P.: "Mechanical Vibrations," 3d ed., McGraw-Hill Book Company, New York.

1940. Kármán, T. V., and M. A. Biot: "Mathematical Methods in Engineering," McGraw-Hill Book Company, New York.

1943. Olson, H. F.: "Dynamical Analogies," D. Van Nostrand Company, Inc., Princeton, N.J.

1948. Brown, G. S., and D. P. Campbell: "Principles of Servomechanisms," John Wiley & Sons, Inc., New York.

1948. Mason, W. P.: "Electromechanical Transducers and Wave Filters," D. Van Nostrand Company, Inc., Princeton, N.J.

1948. Timoshenko, S. P., and D. H. Young: "Advanced Dynamics," McGraw-Hill Book Company, New York.

1951. Goldstein, H.: "Classical Mechanics," Addison-Wesley Publishing Company, Inc., Reading, Mass.

1951. Scanlan, R. H., and R. Rosenbaum: "Aircraft Vibration and Flutter," The Macmillan Company, New York.

1957. Morrill, B.: "Mechanical Vibrations," The Ronald Press Company, New York.

1963. Paskusz, G. F., and B. Bussel: "Linear Circuit Analysis," Prentice-Hall, Inc., Englewood Cliffs, N.J.

1963. Pipes, L. A.: "Matrix Methods for Engineering," Prentice-Hall, Inc., Englewood Cliffs, N.J.

1965. Thomson, W. T.: "Vibration Theory and Applications," Prentice-Hall, Inc., Englewood Cliffs, N.J.

7
The Calculus of Finite Differences and Linear Difference Equations with Constant Coefficients

1 INTRODUCTION

A great many electrical and mechanical systems encountered in practice consist of many identical component parts. Such problems as the determination of the potential and current distribution along an electrical network of the ladder type or the determination of the natural frequencies of the torsional oscillations of systems consisting of identical disks attached to each other by identical lengths of shafting may be most simply solved by the use of difference equations.

The calculus of finite differences is of extreme importance in the theory of interpolation and numerical integration and differentiation. In this chapter some of the elementary procedures of the calculus of finite differences will be considered, and methods for the solution of linear difference equations will be developed.

2 THE FUNDAMENTAL OPERATORS OF THE CALCULUS OF FINITE DIFFERENCES

The calculus of finite differences is greatly facilitated by the use of certain *operators*. An operator may be defined as a symbol placed before a function to indicate the application of some process to the function to produce a new function. The symbol $D = d/dx$ is an example; we have

$$Df(x) = \frac{df}{dx} = f'(x) \tag{2.1}$$

In the calculus of finite differences it is convenient to use the operators E, Δ, D, and k, any constant. These operators indicate the following processes:

$$
\begin{aligned}
EF(x) &= F(x + h) \\
\Delta F(x) &= F(x + h) - F(x) \\
DF(x) &= F'(x) \\
kF(x) &= kF(x)
\end{aligned} \tag{2.2}
$$

The operator E when applied to a function means that the function is to be replaced by its value h units to the right, D indicates differentiation, and the constant operator k merely multiplies the function by a given constant.

If now an operator is applied to a function and a second operator is applied to the resulting function, etc., the several operators are written as a product. Each new operator is written to the left of those preceding it. It may be shown that the order in which the operators are applied is immaterial. For example,

$$\Delta DF(x) = \Delta F'(x) = F'(x + h) - F'(x) = D\Delta F(x) \tag{2.3}$$

If an operator is repeated n times, this is indicated by an exponent. For example,

$$E \times E \times E \times F(x) = E^3 F(x) \tag{2.4}$$

In this manner all positive and integral powers of operators may be defined. An operator with power zero produces no change in the function; for example,

$$D^0 F(x) = F(x) \tag{2.5}$$

Products of powers of operators combine according to the law of exponents; that is,

$$D^2 D^3 F(x) = D^5 F(x) \tag{2.6}$$

At present we shall restrict the powers of D and $\varDelta$ to be integral and nonnegative numbers. However, all real powers of E may be admitted. The general power of E is defined by the equation

$$E^n F(x) = F(x + nh) \tag{2.7}$$

These powers combine according to the law of exponents:

$$E^m E^n F(x) = F(x + nh + mh) = E^{m+n} F(x) \tag{2.8}$$

The sum or difference of two operators applied to a function is defined to be the sum or difference of the functions resulting from the application of each operator; that is,

$$(E + D) F(x) = EF(x) + DF(x) = F(x + h) + F'(x) \tag{2.9}$$

3 THE ALGEBRA OF OPERATORS

In the above section the definitions of the operators E, $\varDelta$, D, and k were given. The meaning of all operators found from E, $\varDelta$, D, and k was given. The meaning of all operators found from E, $\varDelta$, D, and k by addition, subtraction, and multiplication was given. The question of separating these operators from the functions to which they apply and working with them as if they were algebraic quantities will now be considered.

Two operators are said to be *equal* if, when applied to an arbitrary function, they produce the same results. For example,

$$\varDelta = E - 1 \tag{3.1}$$

These operators, which we may call A, B, C, etc., may be combined as if they were algebraic quantities provided they conform to the following five laws of algebra:

(a) $A + B = B + A$

(b) $A + (B + C) = (A + B) + C$

(c) $AB = BA$

(d) $A(BC) = (AB)C$

(e) $A(B + C) = AB + AC$

It is easy to show that the operators satisfy these fundamental laws of algebra. For example, to prove that

$$E^n D = DE^n \tag{3.2}$$

we have

$$E^n \, DF(x) = E^n \, F'(x) = F'(x + nh)$$
$$= DF(x + nh) = DE^n \, F(x) \tag{3.3}$$

Since the operators satisfy the laws of algebra, operators can be combined according to the usual algebraic rules.   For example,

$$(E^{\frac{1}{2}}\Delta - D)(E^{\frac{1}{2}}\Delta + D) = E\Delta^2 - D^2$$
$$(D - \Delta)(D - \Delta) = D^2 - 2D\Delta + \Delta^2 \tag{3.4}$$

etc.

4   FUNDAMENTAL EQUATIONS SATISFIED BY THE OPERATORS

As a consequence of the definition of the operators E and Δ, we have

$$E = 1 + \Delta \tag{4.1}$$

The connection between the operator E and the derivative operator D may be obtained by means of the symbolic form of Taylor's series given in Appendix C, Sec. 16.   We see there that Taylor's expansion could be written in the form

$$EF(x) = F(x + h)$$
$$= \left(1 + hD + \frac{h^2 \, D^2}{2!} + \frac{h^3 \, D^3}{3!} + \cdots \right) F(x)$$
$$= e^{hD} \, F(x) \tag{4.2}$$

Comparing both members of Eq. (4.2), we see that we may write symbolically

$$E = e^{hD} \tag{4.3}$$

We also have from (4.1)

$$1 + \Delta = e^{hD} \tag{4.4}$$

or

$$\Delta = e^{hD} - 1 \tag{4.5}$$

5   DIFFERENCE TABLES

If a function is known for equally spaced values of the argument, the differences of the various entries may be obtained by subtraction.   It is very convenient to write these differences in tabular form.   For example, let us consider the function $F(x) = x^3$, and let the tabular difference be $h = 1$; we then construct the following table of differences:

$F(x) = x^3$

x	$F(x)$	$\Delta F(x)$	$\Delta^2 F(x)$	$\Delta^3 F(x)$	$\Delta^4 F(x)$
1	1	7	12	6	0
2	8	19	18	6	0
3	27	37	24	6	
4	64	61	30		
5	125	91			
6	216				

We see that the third differences of this function are constant and the fourth differences are zero. This is a special case of the following theorem:

The nth differences of a polynomial of nth degree are constant, and all higher differences are zero. The proof of this theorem is not difficult and is left as an exercise for the reader.

Difference tables are of extreme importance in the theory of interpolation. Interpolation in its most elementary aspects is sometimes described as "the science of reading between the lines of a mathematical table." However, by the use of the theory of interpolation, it is possible to find the derivative and the integral of a function specified by a table taken between any limits. The utility of a difference table depends on the fact that in the case of practically all tabular functions the differences of a certain order are all zero.

6 THE GREGORY-NEWTON INTERPOLATION FORMULA

Let us consider a function $F(x)$ whose values at $x = a$, $x = a + h$, $x = a + 2h$, $x = a + 3h$, etc., are given. Suppose that from these given values of $F(x)$ we construct a difference table and that the differences of order m are constant. We desire to compute the value of the function at some intermediate value of the argument $a + nh$. By the use of the operator E of Sec. 2 we have

$$E^n F(a) = F(a + nh) \tag{6.1}$$

By Eq. (4.1) we have

$$E^n = (1 + \Delta)^n \tag{6.2}$$

But, by the binomial theorem, we obtain

$$E^n = (1 + \Delta)^n = 1 + n\Delta + \frac{n(n-1)}{2}\Delta^2 + \cdots \tag{6.3}$$

Hence we have from Eq. (6.1)

$$F(a + nh) = F(a) + n\Delta F(a) + \frac{n(n-1)}{2}\Delta^2 F(a) + \cdots \tag{6.4}$$

Now, in order to compute the value of $F(x)$ corresponding to any intermediate value of the argument such as $a = \frac{1}{2}h$, we simply substitute the value of $n = \frac{1}{2}$ in Eq. (6.4). Equation (6.4) is often referred to as *Newton's formula of interpolation*. It was discovered by James Gregory in 1670.

7 THE DERIVATIVE OF A TABULATED FUNCTION

From Eq. (4.5) giving the relation between the derivative operator D and the difference operator Δ, we have

$$e^{hD} = 1 + \Delta \tag{7.1}$$

Hence we have

$$D = \frac{1}{h}\ln(1 + \Delta) \tag{7.2}$$

If we expand $\ln(1 + \Delta)$ into a formal Maclaurin series in powers of Δ, we have

$$D = \frac{1}{h}\left(\Delta - \frac{\Delta^2}{2} + \frac{\Delta^3}{3} + \frac{\Delta^4}{4} + \cdots\right) \tag{7.3}$$

We thus have

$$DF(a) \doteq F'(a) = \frac{1}{h}\Delta F(a) - \frac{1}{2h}\Delta^2 F(a) + \frac{1}{3h}\Delta^3 F(a) - \cdots \tag{7.4}$$

for the derivative of the function $F(x)$ at $x = a$.

To obtain higher derivatives, we have, from (7.2),

$$D^r = \frac{1}{h^r}[\ln(1 + \Delta)]^r \qquad r = 1, 2, \ldots \tag{7.5}$$

The second member is expanded and applied to $F(a)$.

8 THE INTEGRAL OF A TABULATED FUNCTION

It is sometimes required to obtain the integral of a tabulated function over a certain range of the variable. To do this, it is convenient to introduce the operator D^{-1} or $1/D$. This operator is defined as the operator which when followed by D leaves the function unchanged; that is, we have

$$D^{-1}F(x) = \int F(x)\,dx + c \tag{8.1}$$

since if we operate on (8.1) with D, we obtain

$$D \times D^{-1} F(x) = \frac{d}{dx} \int F(x)\, dx + \frac{d}{dx} c = F(x) \tag{8.2}$$

We also have the definite integral

$$\int_a^{a+h} F(x)\, dx = \frac{1}{D} F(x) \Big|_a^{a+h} = \frac{1}{D}[F(a+h) - F(a)]$$

$$= \frac{1}{D} \Delta F(a) \tag{8.3}$$

If we use Eq. (7.2) for D, we may write (8.3) in the form

$$\int_a^{a+h} F(x)\, dx = \frac{h\Delta}{\ln(1+\Delta)} F(a)$$

$$= \frac{h\Delta}{\Delta - \Delta^2/2 + \Delta^3/3 - \Delta^4/4 + \cdots} F(a) \tag{8.4}$$

By division the right-hand member of (8.4) may be written in the form

$$\int_a^{a+h} F(x)\, dx = h\left(1 + \frac{\Delta}{2} - \frac{\Delta^2}{12} + \frac{\Delta^3}{24} - \frac{19}{720}\Delta^4 + \cdots\right) F(a) \tag{8.5}$$

By similar reasoning we have

$$\int_a^{a+nh} F(x)\, dx = \frac{1}{D}[F(a+nh) - F(a)] = \frac{1}{D}(E^n - 1) F(a)$$

$$= \frac{h[(1+\Delta)^n - 1] F(a)}{\ln(1+\Delta)}$$

$$= nh\left[1 + \frac{n}{2}\Delta + \frac{n(2n-3)}{12}\Delta^2 + \frac{n(n-2)^2}{24}\Delta^3\right.$$

$$\left. + \frac{n(6n^3 - 45n^2 + 110n - 90)}{720}\Delta^4 + \cdots\right] F(a) \tag{8.6}$$

If $n = 2$, we have

$$\int_a^{a+2h} F(x)\, dx = h\left(2 + 2\Delta + \frac{\Delta^2}{3} - \frac{\Delta^4}{90} + \cdots\right) F(a) \tag{8.7}$$

If $F(x)$ is such that its fourth and higher differences may be neglected, we have

$$\int_a^{a+2h} F(x)\, dx = h\left(2 + 2\Delta + \frac{\Delta^2}{3}\right) F(a)$$

$$= h[2 + 2(E-1) + \tfrac{1}{3}(E-1)^2] F(a)$$

$$= \frac{h}{3}(1 + 4E + E^2) F(a) \tag{8.8}$$

This is known as *Simpson's rule* for approximate integration.

If $F(x)$ is a polynomial of degree less than 4, formula (8.8) is *exact*. If we place $n = 3$ in (8.6) and neglect fourth and higher differences, we obtain

$$\int_{a}^{a+3h} F(x)\, dx = \frac{3h}{8}(1 + 3E + 3E^2 + E^3) F(a)$$

$$= \frac{3h}{8}[F(a) + 3F(a + h) + 3F(a + 2h) + F(a + 3h)] \qquad (8.9)$$

This is known as the *three-eighths rule of Cotes*.

9 A SUMMATION FORMULA

A formula of great utility for the summation of polynomials may be easily obtained by the use of the finite-difference operators.   Consider the sum of n terms:

$$S_n = F(a) + F(a + h) + \cdots + F[a + (n - 1)h]$$

$$= (1 + E + E^2 + \cdots + E^{n-1}) F(a) \qquad (9.1)$$

The geometric progression in E may be summed, and we have

$$S_n = \frac{E^n - 1}{E - 1} F(a)$$

$$= \frac{(1 + \Delta)^n - 1}{\Delta} F(a) \qquad (9.2)$$

Expanding $(1 + \Delta)^n$ by the binomial theorem and dividing by Δ, we obtain

$$S_n = \left[n + \frac{n(n - 1)}{2} \Delta + \frac{n(n - 1)(n - 2)}{6} \Delta^2 + \cdots \right] F(a) \qquad (9.3)$$

As an application of this equation, let it be required to find the sum of the first n cubes.   To do this, we let $F(x) = x^3$, $h = 1$, and we use the difference table of Sec. 5.   We then have

$$1^3 + 2^3 + 3^3 + \cdots + n^3 = n + \frac{n(n - 1)}{2} \times 7$$

$$+ \frac{n(n - 1)(n - 2)}{6} \times 12 + \frac{n(n - 1)(n - 2)(n - 3)}{24} \times 6 = \frac{n^2}{4}(n + 1)^2 \qquad (9.4)$$

The summation formula is exact if $F(x)$ is a polynomial.

0 DIFFERENCE EQUATION WITH CONSTANT COEFFICIENTS

An equation relating an unknown function $u(x)$ and its first n differences of the form

$$a_0 \Delta^n u(x) + a_1 \Delta^{n-1} u(x) + \cdots + a_{n-1} \Delta u(x) + a_n u(x) = \phi(x) \qquad (10.1)$$

where the a_r's are constants, is called a linear *difference* equation of order n with constant coefficients.

This type of equation is of frequent occurrence in technical applications. It has many striking analogies with the linear differential equations discussed in Chap. 6.   If $\phi(x) = 0$, the equation is said to be *homogeneous*.   By the use of the relation (3.1), Eq. (10.1) may be written in terms of the operator E in the form

$$(b_0 E^n + b_1 E^{n-1} + b_2 E^{n-2} + \cdots + b_n) u(x) = \phi(x) \tag{10.2}$$

where the b's are constants.   This is the form in which difference equations occur in practice.

THE COMPLEMENTARY FUNCTION

As in the case of linear *differential* equations, the solution of the difference equation (10.2) consists of the sum of the particular integral and the complementary function.   The complementary function is the solution of the homogeneous equation

$$(b_0 E^n + b_1 E^{n-1} + \cdots + b_n) u(x) = 0 \tag{10.3}$$

In the usual applications of difference equations $h = 1$; that is,

$$Eu(x) = u(x + 1) \tag{10.4}$$

etc.

To solve the homogeneous difference equation (10.3), we assume a solution of the exponential form

$$u(x) = c\,e^{mx} \tag{10.5}$$

where c is an arbitrary constant and m is a number to be determined.   If we operate on Eq. (10.5) with E, we obtain

$$Eu(x) = Ec\,e^{mx} = c\,e^{m(x+1)} = c\,e^{mx}\,e^m \tag{10.6}$$

In the same manner, we have

$$E^2 u(x) = Ec\,e^{mx}\,e^m = c\,e^{mx}\,e^{2m} \tag{10.7}$$

and in general

$$E^s u(x) = c\,e^{mx}\,e^{sm} \tag{10.8}$$

We therefore see that if we substitute the assumed solution (10.5) into the homogeneous difference equation (10.3), we obtain

$$c\,e^{mx}(b_0 e^{nm} + b_1 e^{(n-1)m} + \cdots + b_n) = 0 \tag{10.9}$$

If we let

$$q = e^m \tag{10.10}$$

Eq. (10.9) may be written in the form

$$cq^x(b_0 q^n + b_1 q^{n-1} + \cdots + b_n) = 0 \tag{10.11}$$

If we exclude the trivial solution $u(x) = 0$, then

$$cq^x \neq 0 \tag{10.12}$$

Hence the term in parentheses in (10.11) must vanish, and we have

$$b_0 q^n + b_1 q^{n-1} + \cdots + b_n = 0 \tag{10.13}$$

This is an algebraic equation that determines the possible values of q.   There are three cases to be considered.

  a. The case of distinct real roots: If the algebraic equation (10.13) has n distinct roots $(q_1, q_2, \ldots, q_n)$, then the general solution of the homogeneous difference equation (10.3) is

$$u(x) = c_1 q_1^x + c_2 q_2^x + \cdots + c_n q_n^x \tag{10.14}$$

where the c's are arbitrary constants.
  For example, let it be required to solve the equation

$$(2E^2 + 5E + 2)u(x) = 0 \tag{10.15}$$

In this case the algebraic equation determining the possible values of q is

$$2q^2 + 5q + 2 = 0 \tag{10.16}$$

The two roots of this equation are

$$q_1 = -\tfrac{1}{2} \qquad q_2 = -2 \tag{10.17}$$

Hence the solution of (10.15) is

$$u(x) = c_1(-2)^x + c_2(-\tfrac{1}{2})^x \tag{10.18}$$

where c_1 and c_2 are arbitrary constants.
  b. The case of complex roots: Let us suppose that the algebraic equation (10.13) has pairs of conjugate complex roots.   Let

$$q_1 = Re^{j\phi} \qquad q_2 = Re^{-j\phi} \tag{10.19}$$

be a pair of complex roots. The solutions of the difference equation corresponding to these terms are of the form

$$c_1(q_1)^x + c_2(q_2)^x = c_1 R^x e^{j\phi x} + c_2 R^x e^{-j\phi x}$$
$$= R^x(A \cos \phi x + B \sin \phi x) \tag{10.20}$$

where A and B are two new arbitrary constants. As an example, consider the equation

$$(E^2 - 2E + 4)u(x) = 0 \tag{10.21}$$

The roots in this case are

$$q_1 = 2\,e^{j\pi/3} \quad \text{and} \quad q_2 = 2\,e^{-j\pi/3} \tag{10.22}$$

Hence the solution of (10.21) is

$$u = 2^x \left(A\cos\frac{\pi x}{3} + B\sin\frac{\pi x}{3} \right) \tag{10.23}$$

where A and B are arbitrary constants.

    c. The case of repeated roots: Suppose we have the difference equation

$$(E - a)^2 u(x) = 0 \tag{10.24}$$

In this case the algebraic equation has repeated roots equal to a. To solve this equation, assume the solution

$$u(x) = a^x v(x) \tag{10.25}$$

We now have

$$\begin{aligned}
(E - a)u(x) &= (E - a)\,a^x v(x) \\
&= a^{x+1} v(x + 1) - a^{x+1} v(x) \\
&= a^{x+1}(E - 1)v(x)
\end{aligned} \tag{10.26}$$

Similarly

$$(E - a)^2 u(x) = a^{x+2}(E - 1)^2 v(x) \tag{10.27}$$

Hence the function $v(x)$ satisfies the difference equation

$$(E^2 - 2E + 1)v(x) = 0 \tag{10.28}$$

This equation is obviously satisfied by

$$v(x) = A + Bx \tag{10.29}$$

where A and B are arbitrary constants. Hence the solution of (10.24) is given by

$$u(x) = a^x(A + Bx) \tag{10.30}$$

    In the same manner it may be demonstrated that if the algebraic equation (10.13) has a root a that is repeated r times then the part of the solution due to this root has the form

$$u(x) = (c_1 + c_2 x + \cdots + c_r x^{r-1})\,a^x \tag{10.31}$$

where the c's are arbitrary constants.

THE PARTICULAR INTEGRAL

The difference equations most frequently encountered in practice are of the linear homogeneous type. The methods used for obtaining the particular integrals of linear differential equations with constant coefficients have their counterpart in the theory of difference equations. To illustrate the general procedure of determining the particular integral, let us write the linear inhomogeneous difference equation (10.2) in the form

$$\mathscr{L}(E)\,u(x) = \phi(x) \tag{10.32}$$

where $\mathscr{L}(E)$ denotes the linear operator involving the various powers of E expressed in Eq. (10.2).

a. $\phi(x)$ *is of the exponential form* e^{mx}*:* In this case, to obtain the particular integral, assume the solution

$$u(x) = A\,e^{mx} \tag{10.33}$$

where A is to be determined. On substituting this into (10.32) we have

$$A\,e^{mx}\,\mathscr{L}(e^m) = e^{mx} \tag{10.34}$$

Hence, provided that

$$\mathscr{L}(e^m) \neq 0 \tag{10.35}$$

we have

$$A = \frac{1}{\mathscr{L}(e^m)} \tag{10.36}$$

This case enables one to determine the solution when $\phi(x)$ is $\sin mx$ or $\cos mx$ by the use of Euler's relation.

b. $\phi(x)$ *is of the form* a^x*:* In this case we assume a solution of the form

$$u = A a^x \tag{10.37}$$

On substituting this into (10.32) we obtain

$$A a^x \mathscr{L}(a) = a^x \tag{10.38}$$

Hence

$$A = \frac{1}{\mathscr{L}(a)} \tag{10.39}$$

provided that $\mathscr{L}(a) \neq 0$.

c. Decomposition of $1/\mathscr{L}(E)$ *into partial fractions:* It is sometimes convenient to decompose the operator $1/\mathscr{L}(E)$ into partial fractions. For example, consider the difference equation

$$(E^2 - 5E + 6)\,u(x) = 5^x \tag{10.40}$$

In this case the operator $\mathscr{L}(E)$ has the factored form

$$\mathscr{L}(E) = (E - 3)(E - 2) \tag{10.41}$$

We may write

$$u(x) = \frac{5^x}{\mathscr{L}(E)} = \left(\frac{1}{E-3} - \frac{1}{E-2} \right) 5^x$$

$$= \frac{5^x}{2} + c_1 3^x - \frac{5^x}{3} + c_2 2^x$$

$$= \frac{5^x}{6} + c_1 3^x + c_2 2^x \tag{10.42}$$

The decomposition of the operator $1/\mathscr{L}(E)$ into partial fractions frequently facilitates the determination of the particular integral. Methods for the determination of the particular integral when $\phi(x)$ contains functions of special form will be found in the references at the end of the chapter.

11 OSCILLATIONS OF A CHAIN OF PARTICLES CONNECTED BY STRINGS

The calculus of finite differences may be applied to the solution of dynamical problems, whether electrical or mechanical. This calculus has great power when the system under consideration has a great many component parts arranged in some order. It may be that there are so many component parts that to write down all their equations of motion would be impossible. If there exists a certain amount of similarity between the successive component parts of the system, it may be possible to write down a few difference equations and in this manner include all the equations of motion. This may be illustrated by the following problem:

Consider a string of length $(n+1)a$, whose mass may be neglected, which is stretched between two fixed points with a force τ and is loaded at intervals a with n equal masses M not under the influence of gravity, and which is slightly disturbed. Let it be required to determine the natural frequencies of the system and the modes of oscillation.

Let (A,B) (Fig. 11.1) be the fixed points and $(y_1, y_2, \ldots, y_n)$ be the ordinates at time t of the n particles. Consider small displacements only and hence the tensions of all the strings as equal to τ. The force on the kth mass is given by

$$F_k = \frac{\tau}{a} [(y_{k-1} - y_k) + (y_{k+1} - y_k)] \tag{11.1}$$

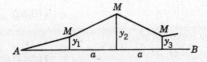

Fig. 11.1

This is a restoring force acting in the negative direction. By Newton's second law the equation of motion of the kth particle is given by

$$\frac{M\,d^2 y_k}{dt^2} + \frac{\tau}{a}(-y_{k-1} + 2y_k - y_{k+1}) = 0 \tag{11.2}$$

Since each particle is vibrating, let us place

$$y_k = A_k \cos(\omega t + \phi) \tag{11.3}$$

Substituting this into (11.2) and dividing out the common cosine term, we obtain

$$-\omega^2 M A_k + \frac{\tau}{a}(-A_{k-1} + 2A_k - A_{k+1}) = 0 \tag{11.4}$$

If we let

$$c = 2 - \frac{\omega^2 M a}{\tau} \tag{11.5}$$

we may write (11.4) in the convenient form

$$-A_{k+1} + cA_k - A_{k-1} = 0 \tag{11.6}$$

This is a homogeneous difference equation of the second order with constant coefficients in the unknown amplitude A_k. To solve it, we use the method of Sec. 10 and assume a solution of the form

$$A_k = B e^{\theta k} \tag{11.7}$$

where B is an arbitrary constant and θ is to be determined. If we substitute this assumed form of the solution into (11.6), we obtain

$$B e^{\theta k}(-e^{\theta} + c - e^{-\theta}) = 0 \tag{11.8}$$

Therefore, for a nontrivial solution, we must have

$$\frac{c}{2} = \frac{e^{\theta} + e^{-\theta}}{2} = \cosh \theta \tag{11.9}$$

This equation determines two values of θ since $\cosh \theta$ is an even function, and we also have

$$\cosh(-\theta) = \frac{c}{2} = \cosh \theta \tag{11.10}$$

Hence, $B_1 e^{\theta k}$ and $B_2 e^{-\theta k}$ are solutions of the difference equation (11.6), and the general solution is the sum of these two solutions. It is convenient to write the general solution in terms of hyperbolic functions rather than in terms of exponential functions. We thus write

$$A_k = P \sinh \theta k + Q \cosh \theta k \tag{11.11}$$

where P and Q are arbitrary constants. Equation (11.2) represents the amplitude of the motion of every particle except the first and last. In order that it may represent these also, it is necessary to suppose that y_0 and y_{n+1} are both zero, although there are no particles corresponding to the values of k equal to 0 and $n+1$. With this understanding the solution (11.11) represents the amplitude of the oscillation of every particle from $k=1$ to $k=n$.

Now, since $A_0 = 0$, we have from (11.11)

$$Q = 0 \tag{11.12}$$

The fact that $A_{n+1} = 0$ gives

$$\sinh \theta(n+1) = 0 \tag{11.13}$$

This equation fixes the possible values of θ; they are

$$\theta = \frac{r\pi j}{n+1} \qquad r = 1, 2, 3, \ldots, n \tag{11.14}$$

Having determined the possible values of θ, we now turn to Eq. (11.5) to determine the possible values of ω. We thus obtain

$$c = 2 - \frac{\omega^2 Ma}{\tau} = 2 \cosh \frac{r\pi j}{n+1}$$

$$= 2 \cos \frac{r\pi}{n+1} \tag{11.15}$$

Hence

$$\omega^2 = \frac{2\tau}{Ma}\left(1 - \cos\frac{r\pi}{n+1}\right)$$

$$= \frac{4\tau}{Ma}\sin^2\frac{r\pi}{2(n+1)} \tag{11.16}$$

It is convenient to write ω_r to denote the value of ω that corresponds to the number r. We thus write

$$\omega_r = 2\sqrt{\frac{\tau}{Ma}}\sin\frac{r\pi}{2(n+1)} \qquad r = 1, 2, 3, \ldots, n \tag{11.17}$$

The amplitude of the motion of each particle is given by Eq. (11.11) in the form

$$A_k = P \sinh \theta k \tag{11.18}$$

The amplitude of the motion of the first particle is given by

$$A_1 = P \sinh \theta \tag{11.19}$$

If, now, the value $r = 0$ were admitted, this would preclude the motion of the first particle. Similarly the value $r = n+1$ is not admitted. The values of ω

given by Eq. (11.17) are the n natural angular frequencies of the system. Giving r other values not multiples of $n + 1$, we merely repeat these frequencies. To each value θ_r of θ there corresponds a term of the amplitude of the kth particle. This may be written in the form

$$A_k = P_r \sinh \theta_r k = P_r' \sin \frac{r\pi k}{n+1} \tag{11.20}$$

where P_r' is an arbitrary constant.

By (11.3) the coordinate of the kth particle corresponding to the value θ_r of θ is given by

$$y_k = P_r' \sin \frac{r\pi k}{n+1} \cos (\omega_r t + \phi_r) \tag{11.21}$$

The general solution is then of the form

$$y_k = \sum_{r=1}^{r=n} P_r' \sin \frac{r\pi k}{n+1} \cos (\omega_r t + \phi_r) \tag{11.22}$$

The $2n$ arbitrary constants P_r' and ϕ_r are determined by the initial displacements and velocities of the particles.

The problem of the loaded string is of historical interest. It is discussed by Lagrange in his "Mécanique analytique." He deduced the solution from his own equations of motion. By means of an extension of the above analysis Pupin treated the problem of the vibration of a *heavy* string loaded with beads, both for free and for forced vibrations, and by an electrical application solved a very important telephonic problem.[†]

12   AN ELECTRICAL LINE WITH DISCONTINUOUS LEAKS

The following interesting electrical problem may be solved by the use of difference equations.

Let us suppose that the current for a load of resistance R is carried from the generator to the load by a single wire with an earth return. The wire is supported by n equally spaced identical insulators, as shown in Fig. 12.1.

† M. Pupin, Wave Propagation over Non-uniform Electrical Conductors, *Transactions of the American Mathematical Society*, vol. 1, p. 259, 1900.

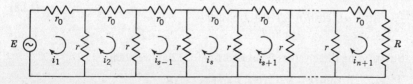

Fig. 12.1

The resistance of the sections of wire between insulators is r_0. The resistance of the earth return is negligible. Let us suppose that in dry weather when the insulators are perfect the current supplied by the generator is I_d but in wet weather it is necessary to supply a current I_ω in order to receive a current I_d at the load R. Assuming that the leakage of all the insulators is the same, let it be required to find the resistance r of each.

If we write Kirchhoff's second law for the sth loop, we obtain the difference equation

$$i_s r - i_{s-1} r + i_s r_0 + i_s r - i_{s+1} r = 0 \tag{12.1}$$

This may be written in the form

$$i_{s+1} - \left(2 + \frac{r_0}{r}\right) i_s + i_{s-1} = 0 \tag{12.2}$$

If we assume a solution of the form

$$i_s = c\,e^{\theta s} \tag{12.3}$$

where c is an arbitrary constant, and substitute this into the difference equation, we find that

$$\cosh \theta = 1 + \frac{r_0}{2r} \tag{12.4}$$

The general solution of (12.2) may be written in the form

$$i_s = A \cosh s\theta + B \sinh s\theta \tag{12.5}$$

To determine the arbitrary constants A and B, we have for the current in the first loop

$$I_\omega = i_1 = A \cosh \theta + B \sinh \theta \tag{12.6}$$

Also in the $(n + 1)$st loop we have

$$I_d = A \cosh (n + 1)\,\theta + B \sinh (n + 1)\,\theta \tag{12.7}$$

If we solve these two simultaneous equations for A and B, we obtain

$$A = \frac{I_\omega \sinh (n + 1)\,\theta - I_d \sinh \theta}{\sinh n\theta}$$
$$B = \frac{-I_\omega \cosh (n + 1)\,\theta + I_d \cosh \theta}{\sinh n\theta} \tag{12.8}$$

Substituting these values for A and B into (12.5), we obtain

$$i_s = \frac{I_\omega \sinh (n - s + 1)\,\theta + I_d \sinh (s - 1)\,\theta}{\sinh n\theta} \tag{12.9}$$

If we write Kirchhoff's second law around the last loop, we obtain

$$I_d\left(1 + \frac{r_0 + R}{r}\right) = i_n \tag{12.10}$$

If we solve Eq. (12.4) for r, we obtain

$$r = \frac{r_0}{2(\cosh\theta - 1)} = \frac{r_0}{4\sinh^2(\theta/2)} \tag{12.11}$$

Substituting this value of r, and i_n from (12.9), into (12.10), we obtain after some reductions

$$2I_d R \sinh n\theta \sinh\frac{\theta}{2} + r_0 I_d \cosh\frac{(2n+1)\theta}{2} - r_0 I_\omega \cosh\frac{\theta}{2} = 0 \tag{12.12}$$

This equation may be solved graphically for θ by the methods of Appendix D. Substituting the value of θ determined by this equation into Eq. (12.11) gives the desired value of r. If, as is usually the case, the line resistance r_0 is small in comparison with the insulator resistance, that is, if

$$\frac{r_0}{r} \ll 1 \tag{12.13}$$

then we have

$$4\sinh^2\frac{\theta}{2} \ll 1 \tag{12.14}$$

and we may make the approximation

$$4\sinh^2\frac{\theta}{2} \doteq \theta^2 \tag{12.15}$$

realizing that if x is small we have

$$\sinh x \doteq x \qquad \cosh x \doteq 1 + \frac{x^2}{2} \tag{12.16}$$

Using these approximations in Eq. (12.12) and solving for θ, we obtain

$$\theta^2 = \frac{8r_0(I_\omega - I_d)}{8nRI_d + r_0[I_d(2n+1)^2 - I_\omega]} = \frac{r_0}{r} \tag{12.17}$$

Hence

$$r = \frac{8nRI_d + r_0[I_d(2n+1)^2 - I_\omega]}{8(I_\omega - I_d)} \tag{12.18}$$

for the required value of r.

13 FILTER CIRCUITS

In Chap. 7 the mathematical technique for the determination of the steady-state behavior of alternating-current networks was explained. A very

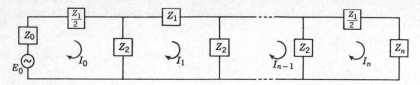

Fig. 13.1

important type of network is one in which numbers of similar impedance elements are assembled to form a recurrent structure. Networks of this type are called *filters* because they pass certain frequencies freely and stop others. Various forms of structure may be employed. A very common structure is the so-called "ladder" structure. This is shown in Fig. 13.1.

Except for the end meshes this structure consists of a number of identical series elements of complex impedance Z_1 and a number of shunt elements of complex impedance Z_2. The input and output impedances are Z_0 and Z_n, respectively, and there is a complex applied electromotive force

$$e(t) = E_0 e^{j\omega t} \qquad j = \sqrt{-1} \tag{13.1}$$

applied to the first mesh. The real and imaginary parts of this complex electromotive force correspond to actual electromotive forces of the type $E_0 \cos \omega t$ or $E_0 \sin \omega t$. We assume that the instantaneous currents in the various meshes have the form

$$i_s(t) = I_s e^{j\omega t} \tag{13.2}$$

where the I_s are the ordinary complex currents of steady-state alternating-current theory. The real or imaginary parts of (13.2) correspond to an applied potential of the form $E_0 \cos \omega t$ or $E_0 \sin \omega t$, respectively, with proper phase.

The circuit of Fig. 13.1 is composed of nT sections of the type shown in Fig. 13.2. As shown, the filter ends with half-series elements and is said to have *mid-series terminations*. It is sometimes more convenient to arrange the filter circuit as shown in Fig. 13.3. This circuit is said to have *mid-shunt terminations*. In this case we regard the filter as made up of $n - 1$ so-called "sections," as shown in Fig. 13.4.

The type of termination affects the values of the currents in the different sections of the filter for given input and output impedances. However, it

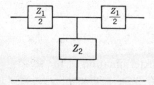

Fig. 13.2

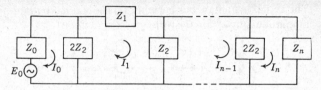

Fig. 13.3

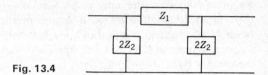

Fig. 13.4

does not affect the frequency characteristics of the filter. It is therefore necessary to analyze only one case, and we shall choose the case of the mid-series termination. If we apply Kirchhoff's second law to the various meshes of Fig. 13.1, we obtain the following set of equations:

$$Z_0 I_0 + \frac{Z_1 I_0}{2} + Z_2(I_0 - I_1) = E_0$$

$$-Z_2 I_0 + (Z_1 + 2Z_2) I_1 - Z_2 I_2 = 0$$

$$\cdots\cdots\cdots\cdots\cdots\cdots\cdots\cdots\cdots$$

$$-Z_2 I_{s-1} + (Z_1 + 2Z_2) I_s - Z_2 I_{s+1} = 0 \qquad (13.3)$$

$$\cdots\cdots\cdots\cdots\cdots\cdots\cdots\cdots\cdots$$

$$-Z_2 I_{n-1} + \left(Z_2 + Z_n + \frac{Z_1}{2}\right) I_n = 0$$

The equation for the sth mesh may be written in the form

$$-I_{s-1} + \frac{Z_1 + 2Z_2}{Z_2} I_s - I_{s+1} = 0 \qquad (13.4)$$

This difference equation has a general solution of the form

$$I_s = A e^{as} + B e^{-as} \qquad (13.5)$$

where A and B are arbitrary constants. The number a is given by

$$\cosh a = \frac{Z_1 + 2Z_2}{2Z_2} \qquad (13.6)$$

The quantity a is, in general, complex and is called the *propagation constant* of the filter. Since a is complex, let us write

$$a = -\alpha - j\phi \qquad (13.7)$$

The actual instantaneous current $i_s(t)$ is obtained by multiplying the complex amplitude I_s by $e^{j\omega t}$ in the form

$$i_s(t) = I_s e^{j\omega t} = (A e^{as} + B e^{-as}) e^{j\omega t}$$
$$= A e^{-\alpha s} e^{j(\omega t - \phi s)} + B e^{+\alpha s} e^{j(\omega t + \phi s)} \tag{13.8}$$

Assuming that $\alpha > 0$, we thus see that there is an attenuation factor $e^{-\alpha}$ introduced into the amplitude of the first term by each section of the filter as we move away from the input end. Similarly, for the second term of (13.8), there is an attenuation for each section of the same amount as we move *toward* the input end. It is thus apparent that if α differs from zero for any frequency and the filter has an appreciable number of sections the current transmitted by the filter is effectively zero. On the other hand a current whose frequency is such that $\alpha = 0$ is freely transmitted. The quantity α is called the *attenuation constant.*

In the same manner ϕ is called the *phase constant* since it gives the change of phase per section (measured as a lag) in the currents as we move along the filter.

The physical significance of Eq. (13.8) is now clear. The first term represents a *space wave* of current traveling away from the input end, attenuated from section to section much as the time waves of the free oscillations of a circuit are attenuated from instant to instant. The second term represents a similar wave traveling back toward the input end, arising from reflection at the output end.

From Eq. (13.8) we see that there are two separate points of interest to be considered:

1. The dependence of the frequency characteristics upon the filter construction, that is, on the nature of Z_1 and Z_2.
2. The dependence of the transmission characteristics upon the terminating impedances Z_0 and Z_n.

The first matter involves only Eqs. (13.6) and (13.7), while the second matter involves the determination of the arbitrary constants A and B.

FREQUENCY CHARACTERISTICS

If the resistances of the elements Z_1 and Z_2 of the filter are so small that they may be neglected, then Z_1 and Z_2 are pure imaginary quantities. It follows, therefore, that

$$\cosh a = 1 + \frac{Z_1}{2Z_2} \tag{13.9}$$

is *real.*

If we now express $\cosh a$ in terms of α and ϕ, we have

$$\cosh a = \cosh(-a) = \cosh(\alpha + j\phi)$$
$$= \cosh \alpha \cos \phi + j \sinh \alpha \sin \phi \tag{13.10}$$

Now since $\cosh a$ is *real*, either α must be zero or ϕ must be an integral multiple of π. Hence, since $\cosh \alpha$ is never less than unity and $\cos \phi$ is never greater than unity, we have three cases to consider.

(a) $-1 \leqslant \cosh a \leqslant 1$ $\alpha = 0$ $a = -j\phi$

(b) $\cosh a > 1$ $\phi = 0$ $a = -\alpha$

(c) $\cosh a < -1$ $\phi = \pm\pi$ $a = -\alpha \pm j\pi$

In the frequency range corresponding to case a, currents are transmitted freely without attenuation. The range of frequencies for which this is the case is called a *passband*. Frequency ranges corresponding to the other two cases are called *stopbands*.

In terms of Z_1 and Z_2, the passbands are given by

$$-1 \leqslant \frac{Z_1 + 2Z_2}{2Z_2} \leqslant 1 \tag{13.11}$$

or

$$0 \leqslant -\frac{Z_1}{Z_2} \leqslant 4 \tag{13.12}$$

Stopbands are given by all other ranges of values.

The simplest filters of the types we are considering are given by Figs. 13.5 and 13.6. For the arrangement of Fig. 13.5, we have

$$Z_1 = j\omega L_1 \qquad Z_2 = \frac{-j}{\omega C_2} \tag{13.13}$$

The passband is given by

$$0 \leqslant L_1 C_2 \omega^2 \leqslant 4 \tag{13.14}$$

or

$$0 \leqslant \omega \leqslant \omega_c \qquad \omega_c = \frac{2}{\sqrt{L_1 C_2}} \tag{13.15}$$

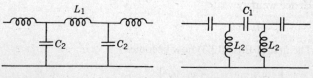

Fig. 13.5   **Fig. 13.6**

Since all frequencies from 0 to a critical, or *cutoff*, frequency given by

$$\omega_c = \frac{2}{\sqrt{L_1 C_2}}$$

are passed without attenuation, we have a *low-pass* filter.

In the arrangement of Fig. 13.6 we have

$$Z_1 = \frac{-j}{C_1 \omega} \qquad Z_2 = j\omega L_2 \tag{13.16}$$

The passband is now determined by

$$0 \leqslant \frac{1}{L_2 C_1 \omega^2} \leqslant 4 \tag{13.17}$$

or

$$\omega_c \leqslant \omega \leqslant \infty \qquad \omega_c = \frac{1}{2\sqrt{L_2 C_1}} \tag{13.18}$$

Here all frequencies above a critical frequency are passed without attenuation. This arrangement is called a *high-pass* filter. More complicated arrangements may be treated in a similar manner.

TRANSMISSION CHARACTERISTICS

The values of the arbitrary constants A and B of the solution of the difference equation (13.5) are most simply expressed in terms of the *characteristic impedance* Z_k of the given filter.

By definition the characteristic impedance of a filter of the type we are considering is the input impedance of the filter when it has an *infinite* number of sections. It is evident that, in this case, the wave due to reflection at the output end, which is represented by the second term of (13.5), vanishes. The current in each section is now independent of the terminal impedance Z_n, and the general solution (13.5) reduces to

$$I_s = A e^{as} \tag{13.19}$$

If $s = 0$, we have

$$I_0 = A \tag{13.20}$$

Hence we may write

$$I_s = I_0 e^{as} \tag{13.21}$$

The first of Eqs. (13.3) now becomes

$$\left[-Z_2(e^a - 1) + \frac{Z_1}{2} + Z_0 \right] I_0 = E_0 \tag{13.22}$$

But by definition, since Z_k is the impedance at the sending end of the entire filter that has an infinite number of sections, we have

$$(Z_k + Z_0) I_0 = E_0 \tag{13.23}$$

Comparing Eqs. (13.22) and (13.23) and using (13.9), we have

$$Z_k = -Z_2(e^a - 1) + \frac{Z_1}{2} = -Z_2 \frac{e^a - e^{-a}}{2} \tag{13.24}$$

If we square this equation and use Eq. (13.6), we obtain

$$\left(\frac{Z_k}{Z_2}\right)^2 = \left(\frac{Z_1 + 2Z_2}{2Z_2}\right)^2 - 1 = \frac{Z_1}{Z_2} + \frac{Z_1^2}{4Z_2^2} \tag{13.25}$$

Hence, finally,

$$Z_k = \sqrt{Z_1 Z_2 + \frac{Z_1^2}{4}} \tag{13.26}$$

We can show that when we neglect the resistance parts of Z_1 and Z_2 then Z_k is *real* in the passbands and *imaginary* in the stopbands.

To determine the constants A and B for the actual filter with a finite number of sections, we substitute (13.5) in the first and last of the equations of (13.3). We then obtain

$$\left[-Z_2(e^a - 1) + \frac{Z_1}{2} + Z_0\right] A + \left[-Z_2(e^{-a} - 1) + \frac{Z_1}{2} + Z_0\right] B = E_0$$

$$e^{an}\left[-Z_2(e^{-a} - 1) + \frac{Z_1}{2} + Z_n\right] A \tag{13.27}$$

$$+ e^{-an}\left[-Z_2(e^a - 1) + \frac{Z_1}{2} + Z_n\right] B = 0$$

However, from Eq. (13.24) we have

$$Z_k = \frac{Z_1}{2} - Z_2(e^a - 1) \tag{13.28}$$

and, similarly,

$$-Z_k = \frac{Z_1}{2} - Z_2(e^{-a} - 1) \tag{13.29}$$

Hence Eqs. (13.27) may be written in the convenient form

$$(Z_k + Z_0) A - (Z_k - Z_0) B = E_0$$

$$-e^{an}(Z_k - Z_n) A + e^{-an}(Z_k + Z_n) B = 0 \tag{13.30}$$

Solving these simultaneous equations for A and B and substituting the result into (13.5), we obtain

$$I_s = \frac{E_0(e^{(n-s)a} - r_R e^{-(n-s)a})}{(Z_0 + Z_k)(e^{na} - r_s r_R e^{-na})} \tag{13.31}$$

where

$$r_s = \frac{Z_0 - Z_k}{Z_0 + Z_k} \qquad r_R = \frac{Z_n - Z_k}{Z_n + Z_k} \tag{13.32}$$

The quantities r_s and r_R are called the sending-end and receiving-end *reflection coefficients*, respectively.

If the output impedance matches the characteristic impedance of the line, then

$$Z_n = Z_k \tag{13.33}$$

Under this condition the filter with its terminating impedance behaves like an infinite filter. The reflection coefficient $r_R = 0$, and reflection is absent. In this case all the energy delivered to the filter at the input end is transmitted to Z_n, and the filter operates at maximum efficiency. In this case Eq. (13.31) reduces to

$$I_s = \frac{E_0 e^{-sa}}{Z_0 + Z_k} \tag{13.34}$$

In practice, one is more interested in the input and output currents than in the intermediate currents. The completeness of filtering is measured by the ratio

$$\frac{I_n}{I_0} = e^{an} \tag{13.35}$$

14 FOUR-TERMINAL-NETWORK CONNECTION WITH MATRIX ALGEBRA

In this section it will be shown that there exists an intimate connection between difference equations and matrix multiplication. To fix the ideas, we shall consider an application to electrical-circuit theory, and we shall see that in certain cases matrix multiplication has some advantages over the method of difference equations in that we need not solve for the arbitrary constants that appear in the solutions of the difference equations.

Consider the electrical circuit of Fig. 14.1. This circuit consists of a series impedance and a perfect conductor return path. We suppose that an electromotive force

$$e_1(t) = E_1 e^{j\omega t} \tag{14.1}$$

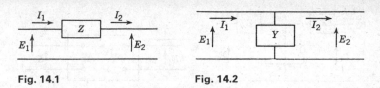

Fig. 14.1   Fig. 14.2

is impressed on one end of the circuit and that an electromotive force

$$e_2(t) = E_2 \, e^{j\omega t} \tag{14.2}$$

is impressed on the other end of the circuit.   This gives rise to a current

$$i_1(t) = I_1 \, e^{j\omega t} = I_2 \, e^{j\omega t} \tag{14.3}$$

Since we are interested in steady-state values, we may suppress the factor $e^{j\omega t}$ as is customary in electrical-circuit theory.   We then concern ourselves only with the complex amplitudes $E_1, E_2, I_1,$ and I_2.   By Kirchhoff's laws, we have the relations

$$\begin{aligned} E_1 &= E_2 + ZI_2 \\ I_1 &= I_2 \end{aligned} \tag{14.4}$$

This may be written in the convenient matrix form

$$\begin{Bmatrix} E_1 \\ I_1 \end{Bmatrix} = \begin{bmatrix} 1 & Z \\ 0 & 1 \end{bmatrix} \begin{Bmatrix} E_2 \\ I_2 \end{Bmatrix} \tag{14.5}$$

This matrix equation expresses a relation between the input quantities E_1 and I_1 and the output quantities E_2 and I_2.   We notice that the determinant of the square matrix of (14.5) is equal to unity.

Let us now consider the circuit of Fig. 14.2.   By Kirchhoff's laws, we now have the relations

$$\begin{aligned} E_1 &= E_2 \\ I_1 &= YE_2 + I_2 \end{aligned} \tag{14.6}$$

where Y is the admittance of the circuit and is hence the reciprocal of the impedance.   This relation may be written in the matrix form

$$\begin{Bmatrix} E_1 \\ I_1 \end{Bmatrix} = \begin{bmatrix} 1 & 0 \\ Y & 1 \end{bmatrix} \begin{Bmatrix} E_2 \\ I_2 \end{Bmatrix} \tag{14.7}$$

We notice that, in this case also, the determinant of the square matrix is equal to unity.

Let us now consider the network of Fig. 14.3.   We may regard this circuit as being formed by a circuit of the type of Fig. 14.1 in series with a circuit of the type of Fig. 14.2 and another circuit of the type of Fig. 14.1.

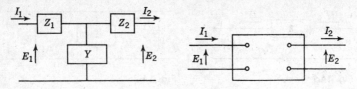

Fig. 14.3   **Fig. 14.4**

Since the output currents and voltage of one circuit are the input currents and voltages of the next network, we obtain

$$\begin{Bmatrix} E_1 \\ I_1 \end{Bmatrix} = \begin{bmatrix} 1 & Z_1 \\ 0 & 1 \end{bmatrix} \begin{bmatrix} 1 & 0 \\ Y & 1 \end{bmatrix} \begin{bmatrix} 1 & Z_2 \\ 0 & 1 \end{bmatrix} \begin{Bmatrix} E_2 \\ I_2 \end{Bmatrix}$$

$$= \begin{bmatrix} 1 + Z_1 Y & Z_1 + Z_1 Z_2 Y + Z_2 \\ Y & 1 + Z_2 Y \end{bmatrix} \begin{Bmatrix} E_2 \\ I_2 \end{Bmatrix} \qquad (14.8)$$

We thus may obtain the input and output quantities directly. The elements of the square matrix are functions of the impedances and admittances of the component parts of the network. Since the square matrix of (14.8) is the product of three matrices whose determinants are each equal to unity, its determinant also is equal to unity. The square matrix of Eq. (14.8) is called the *associated matrix* of the network.

Let us consider a box with four accessible terminals as shown in Fig. 14.4. Let us assume that within the box there exist various impedance and admittance elements joined together in a general manner. Let us also assume that there are no potential sources within the box. It may then be shown by a repeated multiplication of the associated matrices of the individual elements within the box that the input and output potentials and currents are related by the equation

$$\begin{Bmatrix} E_1 \\ I_1 \end{Bmatrix} = \begin{bmatrix} A & B \\ C & D \end{bmatrix} \begin{Bmatrix} E_2 \\ I_2 \end{Bmatrix} \qquad (14.9)$$

where in general A, B, C, D are complex numbers and the determinant of the square matrix of (14.9) satisfies the relation

$$AD - BC = 1 \qquad (14.10)$$

If we premultiply both sides of Eq. (14.9) by

$$\begin{bmatrix} A & B \\ C & D \end{bmatrix}^{-1}$$

we obtain

$$\begin{Bmatrix} E_2 \\ I_2 \end{Bmatrix} = \begin{bmatrix} A & B \\ C & D \end{bmatrix}^{-1} \begin{Bmatrix} E_1 \\ I_1 \end{Bmatrix} = \begin{bmatrix} D & -B \\ -C & A \end{bmatrix} \begin{Bmatrix} E_1 \\ I_1 \end{Bmatrix} \qquad (14.11)$$

If the network within the box is symmetrical so that it appears the same when viewed from the right as when viewed from the left, we have

$$A = D \tag{14.12}$$

CASCADE CONNECTION OF SYMMETRICAL NETWORKS

Let us suppose that we have a chain of n identical symmetrical networks connected as shown in Fig. 14.5. Since the networks are identical and symmetrical, we have the relation

$$\begin{Bmatrix} E_s \\ I_s \end{Bmatrix} = \begin{bmatrix} A & B \\ C & A \end{bmatrix}^n \begin{Bmatrix} E_r \\ I_r \end{Bmatrix} \tag{14.13}$$

There exists a very convenient manner of writing the matrix

$$\begin{bmatrix} A & B \\ C & A \end{bmatrix}$$

to enable one to raise it to a positive or negative integral power. To do this, we let

$$A = \cosh a \tag{14.14}$$

$$C = \frac{\sinh a}{Z_0} \tag{14.15}$$

Now since the determinant of

$$\begin{bmatrix} A & B \\ C & A \end{bmatrix}$$

equals unity, we have

$$A^2 - BC = 1 \tag{14.16}$$

or

$$B = \frac{A^2 - 1}{C} = \frac{(\cosh^2 a - 1)Z_0}{\sinh a} = Z_0 \sinh a \tag{14.17}$$

Hence with these substitutions we write

$$\begin{bmatrix} A & B \\ C & A \end{bmatrix} = \begin{bmatrix} \cosh a & Z_0 \sinh a \\ \dfrac{\sinh a}{Z_0} & \cosh a \end{bmatrix} \tag{14.18}$$

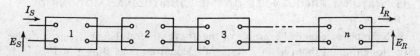

Fig. 14.5

We now have

$$\begin{bmatrix} A & B \\ C & A \end{bmatrix}^2 = \begin{bmatrix} \cosh a & Z_0 \sinh a \\ \dfrac{\sinh a}{Z_0} & \cosh a \end{bmatrix} \begin{bmatrix} \cosh a & Z_0 \sinh a \\ \dfrac{\sinh a}{Z_0} & \cosh a \end{bmatrix}$$

$$= \begin{bmatrix} \cosh^2 a + \sinh^2 a & 2Z_0(\cosh a \sinh a) \\ \dfrac{2}{Z_0} \cosh a \sinh a & \cosh^2 a + \sinh^2 a \end{bmatrix}$$

$$= \begin{bmatrix} \cosh 2a & Z_0 \sinh 2a \\ \dfrac{\sinh 2a}{Z_0} & \cosh 2a \end{bmatrix} \tag{14.19}$$

By mathematical induction it may be shown that

$$\begin{bmatrix} A & B \\ C & A \end{bmatrix}^n = \begin{bmatrix} \cosh an & Z_0 \sinh an \\ \dfrac{\sinh an}{Z_0} & \cosh an \end{bmatrix} \tag{14.20}$$

The result (14.20) holds for n a positive or negative integer.

We therefore have

$$\begin{Bmatrix} E_s \\ I_s \end{Bmatrix} = \begin{bmatrix} \cosh an & Z_0 \sinh an \\ \dfrac{\sinh an}{Z_0} & \cosh an \end{bmatrix} \begin{Bmatrix} E_r \\ I_r \end{Bmatrix} \tag{14.21}$$

This gives the relation between the receiving-end quantities and the sending-end quantities. If we premultiply Eq. (14.13) by

$$\begin{bmatrix} A & B \\ C & A \end{bmatrix}^{-n}$$

we obtain

$$\begin{Bmatrix} E_r \\ I_r \end{Bmatrix} = \begin{bmatrix} \cosh an & -Z_0 \sinh an \\ -\dfrac{\sinh an}{Z_0} & \cosh an \end{bmatrix} \begin{Bmatrix} E_s \\ I_s \end{Bmatrix} \tag{14.22}$$

Expressions (14.21) and (14.22) are extremely useful in the field of electrical-circuit theory, and, by the electrical and mechanical analogues, they are of use in the field of mechanical oscillations.

15 NATURAL FREQUENCIES OF THE LONGITUDINAL MOTIONS OF TRAINS

As a simple mechanical example of the above general theory, let us consider the longitudinal motions of a train of n equal units as shown in Fig. 15.1. For simplicity we shall assume that the mass of each unit is m and that the

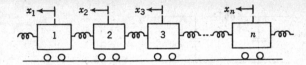

Fig. 15.1

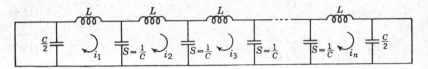

Fig. 15.2

units are coupled together by coupling whose spring constant equals k. Friction will be neglected.

By the principles explained in Chap. 6, the mechanical system of Fig. 15.1 is analogous to the electrical system of Fig. 15.2. This is a ladder network having n meshes. By the electrical-mechanical analogy principle, we have the following analogous quantities:

$$m \rightarrow L \tag{15.1}$$

$$k \rightarrow S = \frac{1}{C} \tag{15.2}$$

$$i_r \rightarrow \dot{x}_r = v_r \tag{15.3}$$

$$q_r \rightarrow x_r \tag{15.4}$$

where L is the inductance parameter of each mesh of the ladder network, S is the elastance of each capacitor of the network, i_r is the mesh current of the rth mesh, q_r is the mesh charge, x_r is the coordinate of the rth unit of the train, and v_r is the velocity of the rth unit.

The electrical circuit may be regarded as being composed of n units of the π type as shown in Fig. 15.3. The associated matrix of the circuit of Fig. 15.3 is

$$\begin{bmatrix} A & B \\ C & A \end{bmatrix} = \begin{bmatrix} 1 & 0 \\ \frac{j\omega C}{2} & 1 \end{bmatrix} \begin{bmatrix} 1 & j\omega L \\ 0 & 1 \end{bmatrix} \begin{bmatrix} 1 & 0 \\ \frac{j\omega C}{2} & 1 \end{bmatrix}$$

$$= \begin{bmatrix} 1 - \dfrac{\omega^2 LC}{2} & j\omega L \\ j\omega C - \dfrac{j\omega^3 LC^2}{4} & 1 - \dfrac{\omega^2 LC}{2} \end{bmatrix} \tag{15.5}$$

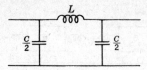

Fig. 15.3

We now let

$$\cosh a = 1 - \frac{\omega^2 LC}{2} \tag{15.6}$$

$$Z_0 = \frac{\sinh a}{j\omega C(1 - \omega^2 LC/4)} \tag{15.7}$$

in accordance with Eqs. (14.14) and (14.15).

We may then write the associated matrix of the circuit (15.5) in the form

$$\begin{bmatrix} A & B \\ C & A \end{bmatrix} = \begin{bmatrix} \cosh a & Z_0 \sinh a \\ \dfrac{\sinh a}{Z_0} & \cosh a \end{bmatrix} \tag{15.8}$$

To obtain the natural frequencies of the dynamical system of Fig. 15.1, we realize that when the train is oscillating freely there are no external forces exerted upon it at either end. Hence, in the equivalent electrical circuit, the sending-end and receiving-end potentials must be equal to zero. We thus have

$$\begin{Bmatrix} E_R \\ I_R \end{Bmatrix} = \begin{bmatrix} \cosh an & Z_0 \sinh an \\ \dfrac{\sinh an}{Z_0} & \cosh an \end{bmatrix} \begin{Bmatrix} E_s \\ I_s \end{Bmatrix} \tag{15.9}$$

Since $E_s = 0$, we have

$$E_R = I_s Z_0 \sinh an = 0 \tag{15.10}$$

where I_s is arbitrary. The frequency equation of the system is therefore given by

$$Z_0 \sinh an = 0 \tag{15.11}$$

If, in Eq. (15.6), we let

$$a = jb \qquad j = \sqrt{-1} \tag{15.12}$$

and solve for ω, we find

$$\omega = \frac{2}{\sqrt{LC}} \sin \frac{b}{2} \tag{15.13}$$

If we now substitute this into (15.7), we have

$$Z_0 = \frac{\sin b}{\omega C \cos^2 b/2} \tag{15.14}$$

The frequency equation (15.11) then becomes

$$\frac{j \sin b \sin bn}{\omega C \cos^2 b/2} = 0 \tag{15.15}$$

This equation is satisfied if

$$\sin bn = 0 \tag{15.16}$$

or

$$b = \frac{r\pi}{n} \quad . \quad r = 0, 1, 2, \ldots, n-1 \tag{15.17}$$

Substituting this into (15.13), we obtain the n natural angular frequencies ω_r:

$$\omega_r = \frac{2}{\sqrt{LC}} \sin \frac{r\pi}{2n} \quad r = 0, 1, 2, \ldots, n-1 \tag{15.18}$$

Translating this result into mechanical language, we obtain

$$\omega_r = 2 \sqrt{\frac{k}{m}} \sin \frac{r\pi}{2n} \quad r = 0, 1, 2, \ldots, n-1 \tag{15.19}$$

for the natural angular frequencies of oscillation of the train of n units.

PROBLEMS

1. Find the successive differences of $F(x) = 1/x$, the interval h being unity.
2. Evaluate

$$\pi/4 = \int_0^1 dx/(1 + x^2)$$

by integrating numerically.

3. A curve expressed by $F(x)$ has for $x = 0, 1, 2, 3, 4, 5, 6$ the ordinates 0, 1.17, 2.13, 2.68, 2.62, 1.77, −0.07, respectively. Find the slope of the curve at each of the seven points, and find the area under the curve from $x = 0$ to $x = 6$.

4. Show that $1^4 + 2^4 + 3^4 + \cdots + n^4 = (n/30)(6n^4 + 15n^3 + 10n^2 - 1)$.

5. Derive the formula

$$\int_a^{a+4h} F(x)\, dx = \frac{2h}{45} (7 + 32E + 12E^2 + 32E^3 + 7E^4) F(a)$$

6. Sum, to m terms, the series

$$(3^2 + 8) + (5^2 + 11) + (7^2 + 14) + (9^2 + 17) + \cdots$$

Solve the following difference equations:

7. $u(x + 2) - 3u(x + 1) - 4u(x) = m^x$.

8. $u(x + 2) + 4u(x + 1) + 4 = x$.

9. $\Delta u(x) + \Delta^2 u(x) = \sin x$.

10. $u(x + 2) + n^2 u(x) = \cos mx$.

11. A seed is planted. When it is one year old, it produces tenfold, and when two years old and upward, it produces eighteenfold. Every seed is planted as soon as produced. Find the number of grains at the end of the xth year.

12. A low-pass filter with mid-series termination is constructed of elements $L_1 = 1/\pi$ henry, $C_2 = 1/\pi$ μf. Find the cutoff frequency.

13. Draw the equivalent electrical circuit for the loaded string. Use the matrix method to compute the natural frequencies of the system.

14. Consider an infinitely extended string under tension F. The string carries equal equidistant masses m and is immersed in a viscous fluid. Draw the equivalent electrical circuit, and discuss the nature of wave propagation when a harmonic transverse force is impressed on the first mass.

15. A shaft of constant cross section carries n identical disks spaced at equal intervals of length a. The moment of inertia of each disk is denoted by J. The torsional stiffness of each section of shaft between two disks is determined by the constant c such that if the relative angular displacement of two neighboring disks is equal to θ, the torque transmitted by the section is equal to $c\theta$. Determine the natural frequencies of the torsional oscillations of the system. If an oscillatory torque is applied to the first disk, determine the motion of the last disk.

16. A transmission-line conductor carries an alternating current of angular frequency ω. It is supported by a string insulator of n identical units attached to a metallic transmission-line tower that is at zero potential. Assume that the metallic conductors between two insulators form an electrical capacitor of capacitance c_1; also, each metallic conductor and the tower form a capacitor of capacitance c_2. Determine the potential distribution along the chain of insulators.

17. A light elastic string of length ns and coefficient of elasticity E is loaded with n particles each of mass m ranged at intervals s along it, beginning at one extremity. If it is suspended by the other end, prove that the periods of its vertical oscillations are given by the formula

$$\pi \sqrt{\frac{sm}{E}} \csc \left(\frac{2r + 1}{2n + 1} \frac{\pi}{2} \right) \qquad \text{when } r = 0, 1, 2, \ldots, n - 1$$

18. A railway engine is drawing a train of equal carriages connected by spring couplings of strength μ, and the driving power is adjusted so that the velocity is $A + B \sin qt$. Show that, if $q^2[(M + 4m)b^2 + 4mk^2]$ is nearly equal to $2\mu b^2$, the couplings will probably break. M is the mass of a carriage that is supported on four equal wheels of mass m, radius b, and radius of gyration k. Are there any other values of q for which the couplings will probably break?

19. A regular polygon $A_1, A_2, \ldots, A_n$ is formed of n pieces of uniform wire, each of resistance r, and the center 0 is joined to each angular point by a straight piece of the same wire. Show that, if the point 0 is maintained at zero potential and the point A_1 at potential V, the current that flows in the conductor A_s, A_{s+1} is

$$I = \frac{2V \sinh \theta \sinh (n - 2s + 1) \theta}{r \cosh n\theta}$$

where θ is given by the equation $\cosh 2\theta = 1 + \sin (\pi/n)$.

REFERENCES

1880. Boole, George: "A Treatise on the Calculus of Finite Differences," The Macmillan Company, New York.
1905. Routh, E. J.: "Advanced Rigid Dynamics," The Macmillan Company, New York.
1920. Funk, P.: "Die Linearen Differenzengleichungen und ihre Anwendungen in der Theorie der Baukonstruktionen," Springer-Verlag OHG, Berlin.
1940. Kármán, T. V., and M. A. Biot: "Mathematical Methods in Engineering," McGraw-Hill Book Company, New York.

8
Transfer Functions and Impulse Responses

1 INTRODUCTION

In the field of circuit analysis and linear control systems the Laplace transform has come into extensive use. In particular, systems are commonly described by their transfer functions and analyzed by their response to impulsive input functions. It is the goal of this chapter to present these concepts and show their usefulness and interrelations.

The knowledgeable reader will recognize that the transfer function of a system and its impulse response are closely related to each other and to what are commonly termed *influence functions* and *Green's functions*. A brief section will be devoted to showing how these functions are interrelated.

2 TRANSFER FUNCTIONS OF LINEAR SYSTEMS

The concept of a *transfer function* is of fundamental importance in the analysis of control systems and feedback problems in general. As a means of introducing the idea of a transfer function, consider the simple *RC* electrical circuit, shown in Fig. 2.1. In this circuit e_i is the input voltage and e_o is the output voltage. If we assume that there is no current being drawn from the output, then applying Kirchhoff's current law to the only node gives

$$\frac{e_i - e_o}{R} + C\frac{de_o}{dt} = 0 \tag{2.1}$$

or

$$e_i = e_o + RC\frac{de_o}{dt} \tag{2.2}$$

Applying Laplace transforms to (2.2) and assuming homogeneous initial conditions yields

$$\bar{e}_i = \bar{e}_o(1 + RCs) \tag{2.3}$$

where the bar denotes the transforms of e_i and e_o.

From (2.3) the ratio of the output to the input can be found and is defined as the *transfer function* of the system; that is,

$$\frac{\bar{e}_o}{\bar{e}_i} = G(s) = \frac{1}{1 + RCs} \tag{2.4}$$

The term $G(s)$ is a common notation used to denote the transfer function of the system. Formally the transfer function may be defined as the ratio of the Laplace transform of the output of a system to its input.

Now consider a somewhat more complex circuit such as the one in Fig. 2.2. In this circuit Z_1 and Z_2 simply denote general impedances, as discussed in Chap. 5. Using the same assumptions as in the previous example and applying the same method leads us to the equation

$$\bar{e}_i = \bar{e}_o\left[1 + \frac{Z_1(s)}{Z_2(s)}\right] \tag{2.5}$$

from which the transfer function is easily identified to be

$$G(s) = \left[1 + \frac{Z_1(s)}{Z_2(s)}\right]^{-1}$$

$$= \frac{Z_2(s)}{Z_1(s) + Z_2(s)} \tag{2.6}$$

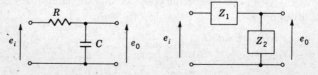

Fig. 2.1   **Fig. 2.2**

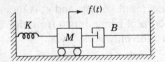

Fig. 2.3

The reader familiar with electrical circuits will recognize that (2.6) may be written down by inspection from Fig. 2.2, realizing that the circuit is a simple voltage divider.

As another example consider the mechanical system shown in Fig. 2.3. Applying Newton's laws, the equation of motion of the system is found to be

$$M\ddot{x} + B\dot{x} + Kx = f(t) \tag{2.7}$$

If $f(t)$ is considered to be the input (forcing function) to the system and $x(t)$ the resulting output (displacement of the mass), then the transfer function for this mechanical system is

$$G(s) = \frac{\bar{x}}{\bar{f}} = \frac{1}{M[s^2 + (B/M)s + K/M]} \tag{2.8}$$

Again the case of homogeneous initial conditions has been assumed.

3 SOLUTION TO PROBLEMS USING TRANSFER FUNCTIONS

In the previous section we saw that a system may be represented by its transfer function. The advantage of using transfer functions is that a complex system may be simply represented as a "blackbox" where the Laplace transform of the input and output variables, say $\bar{v}_i$ and $\bar{v}_o$, are related by the transfer function (cf. Fig. 3.1).

The usefulness of transfer functions lies in the fact that once $G(s)$ is known for a particular system, then the response of the system to any known input is readily found without solving the differential equation that describes the system. Thus for an arbitrary input variable v_i, the Laplace transform of the output is given by

$$\bar{v}_o = G(s)\,\bar{v}_i \tag{3.1}$$

or by applying the inverse transform

$$v_o(t) = L^{-1} G(s)\,\bar{v}_i \tag{3.2}$$

As an example let us consider that the circuit of Fig. 2.1 has an input given by

$$e_i(t) = E \sin wt \tag{3.3}$$

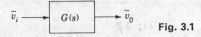

Fig. 3.1

for which it is desired to find the output voltage $e_o(t)$. Applying Laplace transforms to (3.3),

$$e_i = \frac{Ew}{s^2 + w^2} \tag{3.4}$$

Substituting (3.4) and (2.4) into (3.1) gives

$$e_o = \frac{Ew/RC}{(s^2 + w^2)(s + 1/RC)} \tag{3.5}$$

The inverse transform of this equation is given by No. 3.21 of the table of transforms (Appendix A).

$$e(t) = \left\{ \frac{E}{[1 + (RCw)^2]^{\frac{1}{2}}} \right\} [e^{-t/RC} \sin \beta + \sin (wt - \beta)] \tag{3.6}$$

where

$$\tan \beta = RCw \tag{3.7}$$

As another example consider the mechanical system shown in Fig. 2.3. The transfer function for this system is given by (2.8).   Assume that the system is at rest and a constant force of magnitude F is applied to the mass at $t = 0$. The input may therefore be represented by using the Heaviside unit step function:

$$f_i(t) = Fu(t) \tag{3.8}$$

Transforming this function and substituting the result into (3.1), we find

$$\bar{x} = \frac{F/M}{s[s^2 + (B/M)s + K/M]} \tag{3.9}$$

To find $x(t)$ we make use of transform pair No. 3.23 of the table of transforms in Appendix A.   It has been assumed that $K > B^2/4M$ in this particular case:

$$x(t) = \frac{F}{Mw^2} \left[1 - \frac{w_0}{w} e^{-Bt/2M} \sin (wt + \phi) \right] \tag{3.10}$$

where

$$w^2 = w_0^2 - \frac{B^2}{4M^2} \qquad \tan \phi = \frac{2Mw}{B}$$

$$w_0^2 = \frac{K}{M}$$

4   COMBINING TRANSFER FUNCTIONS OF SEVERAL SYSTEMS

Having defined the transfer function of a system and discussed how it may be derived for simple systems, we shall now consider the problem of obtaining the

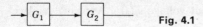

Fig. 4.1

transfer function for a complex system that is composed of two or more simple systems.   This problem is of particular importance in control systems where several subsystems with known transfer functions are combined to form the total system.

To illustrate this situation, consider two systems with transfer functions $G_1(s)$ and $G_2(s)$ which are to be connected as shown in Fig. 4.1.   If we assume that the systems do not interact, that is, $G_2(s)$ does not appreciably load $G_1(s)$, then it is easy to show that the over-all transfer function is $G(s) = G_1(s) G_2(s)$. Speaking in electrical terms, this implies that the input impedance $G_2(s)$ is much greater than the output impedance $G_1(s)$.[†]

As an example consider the circuit shown in Fig. 4.3.   From Sec. 2 we note that the circuit is similar to Fig. 4.1 where

$$G_1(s) = \frac{1}{\tau_1 s + 1} \qquad G_2(s) = \frac{1}{\tau_2 s + 1}$$
$$\tau_1 = R_1 C_1 \qquad \tau_2 = R_2 C_2 \tag{4.1}$$

At first one might be inclined to say that the over-all transfer function is given by

$$G = G_1 G_2 = \frac{1}{(\tau_1 s + 1)(\tau_2 s + 1)}$$
$$= \frac{1}{\tau_1 \tau_2 s^2 + (\tau_1 + \tau_2) s + 1} \tag{4.2}$$

This result is correct in two situations: The first is when the two circuits are not directly coupled together but separated by a buffer amplifier[‡]

[†] Savant, 1964, p. 71 (see References).

[‡] A buffer amplifier is one that has unity voltage gain but isolates the input and output circuits; that is, its transfer function is unity, and it does not load either the input or output circuit.

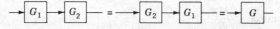

Fig. 4.2

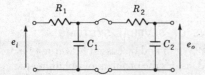

Fig. 4.3

(Fig. 4.4). The second situation may be seen if the correct over-all transfer function is derived from the circuit equations (Fig. 4.5). Applying Kirchhoff's voltage law to the three loops gives

$$R_1 i_1 + \frac{1}{C_1} \int (i_1 - i_2)\, dt = e_i$$

$$R_2 i_3 + \frac{1}{C_2} \int i_2\, dt + \frac{1}{C_1} \int (i_2 - i_1)\, dt = 0 \qquad (4.3)$$

$$\frac{1}{C_2} \int i_2\, dt = e_o$$

It has been assumed that the output does not load the circuit, that is, there is no current drawn from the circuit.

These equations are now Laplace transformed, assuming homogeneous initial conditions:

$$R_1 \bar{i}_1 + \frac{1}{C_1 s}(\bar{i}_1 - \bar{i}_2) = \bar{e}_i$$

$$R_2 \bar{i}_2 + \frac{1}{C_2 s}\bar{i}_2 + \frac{1}{C_1 s}(\bar{i}_2 - \bar{i}_1) = 0 \qquad (4.4)$$

$$\frac{1}{C_2 s}\bar{i}_2 = \bar{e}_o$$

These equations are now solved for the over-all transfer function which is

$$G(s) = \frac{\bar{e}_o}{\bar{e}_i} = \frac{1}{\tau_1 \tau_2 s^2 + [\tau_1 + \tau_2(1 + b)]s + 1} \qquad (4.5)$$

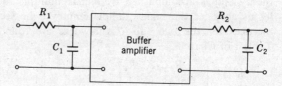

Fig. 4.4

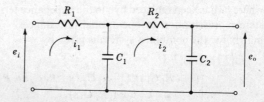

Fig. 4.5

where

$$b = \frac{R_1}{R_2} \qquad \tau_1 = R_1 C_1 \qquad \tau_2 = R_2 C_2$$

Only in the case when $R_2 \gg R_1$, that is, $b \ll 1$, does this equation reduce to (4.2).

Equation (4.5) is the correct over-all transfer function for the system of Fig. 4.5, and (4.2) is a valid representation only when either of the two special cases described above applies. In most control-system problems it is usually assumed that loading effects do not occur and hence the transfer functions may be multiplied to obtain the over-all system transfer function. However, in designing a control system one must always be certain to check to see that the loading effects are negligible.

5 MATRIX METHOD FOR EVALUATING OVER-ALL TRANSFER FUNCTIONS WHEN LOADING OCCURS

As mentioned in 'Sec. 4, over-all transfer functions for coupled systems are obtained by multiplying the transfer functions of the subsystems whenever loading effects are negligible. However, as the previous example illustrates, this direct combination cannot be applied when the circuits load each other. When loading effects must be accounted for, the over-all transfer function may be derived by applying the basic circuit equations to obtain the differential equations that describe the system. These equations are then Laplace transformed and solved for the over-all transfer function.

This procedure is usually laborious and time-consuming if the circuits are very complex. There is an alternative procedure that is frequently very convenient for finding the over-all transfer function in those cases where loading is to be considered. This alternate method arises from transmission matrix theory and is accomplished by matrix multiplication. As an illustration let us consider the same example employed in Sec. 4. The transmission matrices for the two sub-RC networks are found from entry No. 6 in Table 1 of Chap. 5.

$$[T_1] = \begin{bmatrix} 1 + R_1 C_1 s & R_1 \\ sC_1 & 1 \end{bmatrix}$$

$$[T_2] = \begin{bmatrix} 1 + R_2 C_2 s & R_2 \\ sC_2 & 1 \end{bmatrix} \tag{5.1}$$

where jw has been replaced by s in the impedance terms to convert from steady-state representation to operational impedances. The over-all transmission matrix for the combined system is given by

$$[T] = [T_1][T_2]$$

$$= \begin{bmatrix} (1 + R_1 C_1 s)(1 + R_2 C_2 s) + R_1 C_2 s & R_2(1 + R_1 C_1 s) + R_1 \\ C_1 s(1 + R_2 C_2 s) & 1 + R_2 C_1 s \end{bmatrix}$$

$$\tag{5.2}$$

Assuming that the output current of the over-all system is negligible, then the transfer function is simply the inverse of the T_{11} element:

$$G(s) = \frac{\bar{e}_o}{\bar{e}_i} = \frac{1}{T_{11}}$$

$$= \frac{1}{(1 + R_1 C_1 s)(1 + R_2 C_2 s) + R_1 C_2 s} \tag{5.3}$$

which reduces to

$$G(s) = \frac{1}{\tau_1 \tau_2 s^2 + [\tau_1 + (1 + b)\tau_2] + 1} \tag{5.4}$$

where

$$\tau_1 = R_1 C_1 \qquad \tau_2 = R_2 C_2 \qquad b = \frac{R_1}{R_2}$$

Equation (5.4) is identical to (4.4) and therefore represents the transfer function for the combined circuits with loading.

The reason that the output current has been neglected is that it has been assumed that the output of this system is connected to the high input impedance of some following circuit or amplifier.

Since a complex system can be reduced to a combination of series and parallel† simple systems with known transmission matrices, it is an easy matter of matrix multiplication to obtain the over-all transfer function.

6   DETERMINATION OF TRANSFER FUNCTIONS BY THE PERTURBATION METHOD

In addition to the above two methods of determining transfer functions, there is a third method which is particularly useful in studying small oscillations of systems that are nonlinear. To illustrate this method, let us return to the simple RC circuit of Fig. 2.1. The differential equation describing this system is given by (2.2):

$$e_i = e_o + RC \frac{de_o}{dt} \tag{6.1}$$

In accordance with perturbation theory, we assume the following form for the dependent variables.

$$e_i = E_i + E_i^* e^{st}$$
$$e_o = E_o + E_o^* e^{st} \tag{6.2}$$

---

† For a discussion of the connection of four-terminal networks in series and parallel combinations, see Pipes, 1963, Chap. 12 (see References).

where

E_i, E_o = steady-state direct-current potentials

E_i^*, E_o^* = perturbation potentials such that $E_i^* \ll E_i$ $E_o^* \ll E_o$

s = complex number

Substituting (6.2) into (6.1) and equating coefficients of e^{st} yield

$$E_i^* = E_o^* (1 + RCs)$$

To obtain the transfer function the variables should be normalized in the following manner. After (6.2) is substituted into (6.1) and the constant terms are equated, there results

$$E_i = E_o$$

which represents the steady-state direct-current condition. Dividing the previous relation by the steady-state equation yields

$$\frac{E_i^*}{E_i} = (1 + RCs)\frac{E_o^*}{E_o} \tag{6.3}$$

The ratio of the normalized output to the normalized input yields the desired transfer function:

$$G(s) = \frac{E_o^*/E_o}{E_i^*/E_i} = \frac{1}{1 + RCs} \tag{6.4}$$

which agrees with (2.4).

The reader may question at this point the usefulness of this approach when the Laplace-transform method is easier and more direct. The answer is that the Laplace-transform method is easier for linear systems; however, when nonlinear systems are encountered, the perturbation method will yield a transfer function whereas the Laplace-transform method will not. It should be noted that in the cases in which this method is applied to nonlinear systems, the resulting transfer function is valid *only* for small oscillations about the steady-state values. The choice of s as the complex perturbation variable is only for convenience in that the resulting transfer functions are identical with the result using Laplace transforms. The letter s in the perturbation method may be considered the Laplace variable.

As an example of this method as applied to a nonlinear system, consider the circuit of Fig. 2.1, with the exception that the capacitor will be a nonlinear capacitor with a charge versus voltage curve, as shown in Fig. 6.1. It will be assumed that the curve of Fig. 6.1 may be represented by the expression

$$q = C(e_o - be_o^3) \tag{6.5}$$

Fig. 6.1

where b is the nonlinearity parameter. Using the circulating charge instead of the current as a dependent variable, the differential equation representing the RC circuit is

$$e_i = R\frac{dq}{dt} + \frac{q}{C} \tag{6.6}$$

Substituting (6.5) into (6.6) gives

$$e_i = RC(1 - 3be_o^3)\frac{de_o}{dt} + e_o - be_o^3 \tag{6.7}$$

Assuming perturbation solutions as in (6.2)

$$e_i = E_i + E_i^* e^{st}$$
$$e_o = E_o + E_o^* e^{st} \tag{6.8}$$

substituting into (6.7), and neglecting terms of order E_i^{*2}, E_o^{*2}, and higher yields

$$E_i + E_i^* e^{st} = RCs(1 - 3bE_o^2) E_o^* e^{st} + E_o + E_o^* e^{st} - bE_o^3 - 3bE_o^2 E_o^* e^{st} \tag{6.9}$$

Equating constant terms and coefficients of e^{st} gives the following two equations:

$$E_i = E_o - bE_o^3 \tag{6.10}$$
$$E_i^* = (1 + RCs)(1 - 3bE_o^3) E_o^* \tag{6.11}$$

Equation (6.10) may be solved for the steady-state direct-current output E_o in terms of E_i, whereas (6.11), after dividing by (6.10) to obtain the proper normalization, yields the desired transfer function

$$G(s) = \frac{E_o^*/E_o}{E_i^*/E_i}$$

$$= \frac{1 - bE_o^2}{(1 + RCs)(1 - 3bE_o^3)} \tag{6.12}$$

When $b = 0$ the capacitor becomes linear and (6.12) reduces to the standard transfer function.

This example has been presented only to illustrate the method whereby the transfer function of a nonlinear system may be determined. The reader must be aware of the fact that the nonlinearity of this problem has been removed as evidenced by the transfer function. Our approach was to linearize the problem and obtain a transfer function that represents only small oscillations about some steady-state direct-current level. This result will not be valid for large amplitude oscillations. However, the method does find wide application for studying nonlinear systems which operate under steady-state direct-current conditions. The usefulness of the resulting transfer function for small oscillations is that the stability of the system may be studied with the result that the conditions under which a small perturbation will grow with time may be found. Thus, whenever the small oscillations are unstable, the system will be unstable or at least exhibit steady large amplitude oscillations. This method may not be employed to yield information as to the unstable response since the amplitude of such oscillations is usually large.

As a final example of this method let us consider the burning of a solid propellant in a combustion chamber† with a variable throat area nozzle. Figure 6.2 shows the combustion chamber. Applying conservation of mass within the chamber yields the following equation:

$$\rho_p a_b r = \dot{m} a_t + \frac{d}{dt}(\rho_g v_c) \tag{6.13}$$

where ρ_p = propellant density

ρ_g = gas density

a_b = burning surface area

a_t = throat area

r = propellant burning rate

$\dot{m}$ = mass flow rate at the throat

v_c = volume of the combustion chamber

In Eq. (6.13) the term on the left-hand side of the equality denotes the rate of consumption of propellant. The first term on the right-hand side represents the mass flow out the nozzle, and the last term gives the rate of change of mass

† R. Sehgal and L. Strand, A Theory of Low-frequency Combustion Instability in Solid Rocket Motors, *AIAA Journal*, vol. 2, no. 4, pp. 696–702, 1964.

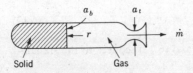

a_b a_t

r $\dot{m}$

Solid   Gas   **Fig. 6.2**

contained in the combustion chamber. Expanding the derivative in (6.13) and using the relation $m = pC_d/g$ yields

$$\rho_p a_b r = \frac{pC_d a_t}{g} + \rho_g \frac{dv_c}{dt} + v_c \frac{d\rho_g}{dt} \tag{6.14}$$

where p = pressure at throat which may be assumed to be equal to the combustion pressure

C_d = discharge coefficient of the nozzle

g = gravitational constant

Using the perfect-gas law to obtain

$$\frac{d\rho_g}{dt} = \frac{M}{RT_f}\frac{dp}{dt} \tag{6.15}$$

where M = molecular weight of gas

R = gas constant

T_f = combustion flame temperature

and noting that

$$\frac{dv_c}{dt} = ra_b \tag{6.16}$$

allows (6.14) to be rewritten as

$$\frac{ga_b r(\rho_p - \rho_g)}{C} = pa_t + \frac{gv_c M}{C_d RT_f}\frac{dp}{dt} \tag{6.17}$$

Since the gas density is much smaller than the propellant density, (6.17) may be approximated by

$$\frac{\rho_p a_b gr}{C_d} = pa_t + \frac{v_c Mg}{C_d RT_f}\frac{dp}{dt} \tag{6.18}$$

The throat area a_t, burning rate r, and combustion pressure p are the dependent variables in this problem.† Following the perturbation method, each of these variables will be assumed to have the following forms:

$$r = R + R^* e^{st}$$
$$a_t = A + A^* e^{st} \tag{6.19}$$
$$p = P + P^* e^{st}$$

$$|R^*| \ll |R| \qquad |A^*| \ll |A| \qquad |P^*| \ll |P|$$

† The flame temperature may also be considered as a system variable, but it will be assumed constant for this example.

Substituting these expressions into (6.18), neglecting products of small quantities, and collecting direct-current terms as well as coefficients of e^{st} gives

$$\frac{P}{R} = \frac{\rho_p a_b g}{C_d A} \tag{6.20}$$

$$\frac{R^*}{R} = \frac{P^*}{P}(1 + \tau s) + \frac{A^*}{A} \tag{6.21}$$

where

$$\tau = \frac{v_c M g}{C_d R T_f A} = \text{combustion time constant}$$

Let us assume that a similar procedure has been employed to derive a transfer function for the combustion process of the solid propellant; that is,

$$\frac{R^*/R}{P^*/P} = G_p(s) \tag{6.22}$$

where $G_p(s)$ represents the transfer function of the solid propellant. Substituting (6.22) into (6.21) yields the desired transfer function that relates perturbations in combustion pressure to perturbations in throat area:

$$G_m(s) = \frac{P^*/P}{A^*/A}$$

$$= \frac{G_p - 1}{1 + \tau s} \tag{6.23}$$

7 IMPULSE RESPONSES AND TRANSFER FUNCTIONS

The concepts of transfer functions and impulse responses are usually encountered in somewhat different areas of study. As we have already discussed, transfer functions are used widely in the subject of control systems whereas impulse responses are frequently alluded to in the field of circuit analysis. It is the purpose of this section to point out the interrelatedness of these concepts on a mathematical level.

An *impulse response* simply refers to the output of a system when it is excited by an impulsive input. The connection between impulse responses and transfer functions may be readily seen from the following example. Consider again the electrical circuit of Fig. 2.1 for which the system equation is

$$RC\frac{de_o}{dt} + e_o = e_i \tag{7.1}$$

Let us now assume that the input is an impulse function which is represented by the Dirac delta function:

$$e_i = \delta(t) \tag{7.2}$$

Substituting (7.2) into (7.1) gives

$$RC\frac{de_o}{dt} + e_o = \delta(t) \tag{7.3}$$

Taking Laplace transforms of this equation yields

$$(RCs + 1)\bar{e}_\delta = 1 \tag{7.4}$$

where $\bar{e}_\delta = Le_o(t)$. Solving (7.4) for $\bar{e}_\delta$,

$$\bar{e}_\delta = \frac{1}{1 + RCs} \tag{7.5}$$

Comparing this result with (2.4) shows that the Laplace transform of the response of a system excited by an impulse is identical to the transfer function.

To further illustrate the method of solving problems by the impulse-response method we shall continue with the same example. If the input voltage to the circuit is an arbitrary function rather than an impulse, then the transform solution is given by (2.4). Also making use of (7.5) we have

$$\bar{e}_o(s) = G(s)\bar{e}_i(s) = \bar{e}_\delta(s)\bar{e}_i(s) \tag{7.6}$$

where $G(s) = \bar{e}_\delta(s) = 1/(1 + RCs)$. Since the input function is arbitrary, this result may not be inverted directly. We may, however, apply the convolution theorem of Chap. 4 (Theorem V) to obtain the following general solution of (7.6):

$$e_o(t) = \int_0^t e_i(t')e_\delta(t - t')\,dt' \tag{7.7}$$

where $e_i(t)$ is the arbitrary input function and $e_\delta(t)$ is the so-called "impulse response," that is, the solution of (7.3). Equation (7.7) is a common form for expressing the solutions to problems in terms of impulse responses.

The impulse-response function also goes by the name of *weighting* or *influence function*. From (7.5) it is easily seen that the impulse response is the inverse Laplace transform to the systems transfer function.

We have seen the close mathematical relation that the impulse response and transfer function satisfy. This relation actually gives us two closely related methods of solving a given problem. One method is to multiply the transfer function by the Laplace transform of the input function and invert the resulting product as a single entity. The alternate method is to obtain the inverse Laplace transform of the transfer function, that is, the impulse response, and convolve (integrate) it with the input function according to (7.7). The former method is the most convenient when a problem is to be solved only once, whereas the second method is preferable if solutions are to be found for several various input functions.

As an illustration let us consider, for the last time, the simple RC circuit of Fig. 2.1. We have already found the delta-function response which is given by (7.5):

$$e_\delta = G(s) = \frac{1/RC}{s + 1/RC} \tag{7.8}$$

The inverse transform of (7.8) is obtained from No. 1.1 of the transform table (Appendix A) and is

$$e_\delta(t) = \frac{1}{RC} e^{-t/RC} \tag{7.9}$$

This is the impulse response for the circuit of Fig. 2.1.

Now let us assume that we wish to determine the response of our RC circuit to a sinusoidal input voltage of the form $e_i(t) = E\sin(wt)$. According to (7.7), the output voltage is

$$e_o(t) = \frac{E}{RC} \int_0^t \sin wt' \, e^{-(t-t')/RC} \, dt' \tag{7.10}$$

which may be written in the following form:

$$e_o(t) = \frac{E}{RC} e^{-t/RC} \int_0^t \sin wt' \, e^{t'/RC} \, dt' \tag{7.11}$$

Performing the indicated integration gives

$$e_o(t) = \frac{EwRC}{1 + (RCw)^2} e^{-t/RC} + \frac{E}{1 + (RCw)^2} \sin(wt) - RCw\cos(wt) \tag{7.12}$$

Introducing the phase angle ϕ as follows

$$\cos\phi = \frac{1}{[1 + (RCw)^2]^{\frac{1}{2}}}$$
$$\sin\phi = \frac{RCw}{[1 + (RCw)^2]^{\frac{1}{2}}} \tag{7.13}$$

gives

$$e_o(t) = \frac{E}{[1 + (RCw)^2]^{\frac{1}{2}}} [\sin(\phi) e^{-t/RC} + \sin(wt + \phi)] \tag{7.14}$$

This result is identical to (3.6) which was obtained by the alternate method.

8 FEEDBACK CONTROL IN LINEAR SYSTEMS

As mentioned earlier, the concept of transfer functions plays an important role in the field of feedback-control systems. In this section we will attempt to show a few basic principles of control systems; however we will not delve deeply into the theory and analysis because of the limited space.

The term *feedback* is used to describe situations in which a portion of the output of a system is fed back to the input. Feedback is employed for many purposes, a few of which are the following: to increase bandwidth, improve operating efficiency, obtain system stability, improve accuracy, and compensate for ignorance. Perhaps the largest application of feedback is found in the field of control systems.

A basic type of control system is one that controls the position of the output. As a simple example of a position control system consider the following system. It is desired to have a remote indicator, that is, slave, follow as closely as possible the position indicated by a master indicator. One way of accomplishing this operation is shown in Fig. 8.1.

The master or reference indicator is connected to a potentiometer that produces a voltage signal proportional to the reference position θ_r. A motor is used to drive the remote or output indicator which is also connected to a potentiometer that produces an output voltage proportional to its position θ_o. This output voltage is fed back and subtracted by means of a difference amplifier from the reference input signal to form the error signal $\epsilon = \theta_r - \theta_o$. The error signal is then amplified by the amplifier A which in turn drives the motor. Thus it may be seen that the motor will drive the slave indicator to the position for which the error signal is zero, that is, $\theta_o = \theta_r$.

Instead of working with the actual circuit and its governing differential equations, it is more customary to introduce Laplace transforms and redraw the circuit in terms of transfer functions. For example let

$$\bar{\theta}_r = L\theta_r(t)$$
$$\bar{\theta}_o = L\theta_o(t)$$
$$\bar{\epsilon} = L\epsilon(t)$$

Figure 8.1 may now be redrawn as shown in Fig. 8.2

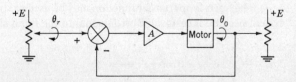

Fig. 8.1

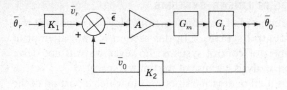

Fig. 8.2

where $\bar{v}_r$ = transformed reference voltage

$\bar{v}_o$ = transformed output voltage

$G_m(s)$ = motor transfer function

$G_1(s)$ = load transfer function

K_1, K_2 = potentiometer transfer functions

In order to introduce a few of the basic terms of control systems, let us now consider the general circuit of Fig. 8.3. In this figure $R(s)$ represents the transformed input or reference signal, $C(s)$ the output signal, and $\bar{\epsilon}(s)$ the error or actuating signal. The quantity $G(s)$ represents the *forward transfer function* which is the product of all the transfer functions in the forward loop. For example, the forward transfer function of the system of Fig. 8.2 is

$$G(s) = AG_m(s)G_1(s) \tag{8.1}$$

The quantity $H(s)$ is called the *feedback transfer function* and represents the product of all the transfer functions in the feedback loop.

In reference to the basic system of Fig. 8.3, the following standard relations may be derived:

$$\frac{\epsilon}{R} = \frac{1}{1 + GH} \tag{8.2}$$

$$\frac{C}{R} = \frac{G}{1 + GH} \tag{8.3}$$

These two terms relate, respectively, the error signal to the input signal and the output signal to the input signal. In addition to the above definitions, the term GH is called the *open-loop transfer function*, whereas the term $G/(1 + GH)$ is the *closed-loop transfer function*.

The open-loop transfer function may be used to study the behavior of the system because it relates the error signal to the input signal as given by (8.2).

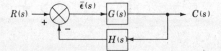

Fig. 8.3

In any control system a desirable performance is that in which the error signal goes to zero for all types of inputs. This implies that the output will precisely follow any input. In reality this cannot be accomplished because the response is not instantaneous due to many causes, such as inertia, etc. In fact, many systems will not follow certain types of inputs regardless of how long the input is applied. It is the function of the control-system designer to optimize by some preselected criterion a given system such that the response fulfills prescribed performance levels with maximum reliability and minimum cost.

There are three more quantities that are useful in analyzing control systems. These are the position-, velocity-, and acceleration-error constants. Respectively, they represent the response of a system to step position, velocity, and acceleration inputs.

The position-error constant is found from (8.2) when $R(t) = u(t)$, a Heaviside unit step function; that is,

$$\bar{\epsilon} = \frac{1}{1 + GH} \frac{1}{s} \tag{8.4}$$

Applying the final value theorem (Chap. 4, Sec. 3, Theorem IX) yields

$$\lim_{t \to \infty} \epsilon(t) = \lim_{s \to 0} s \frac{1}{s(1 + GH)}$$

$$= \lim_{s \to 0} \frac{1}{1 + GH}$$

$$= \frac{1}{1 + \lim_{s \to 0} GH} \tag{8.5}$$

The position-error constant is defined as

$$K_p = \lim_{s \to 0} GH \tag{8.6}$$

Thus the steady-state error signal becomes

$$\epsilon_p = \frac{1}{1 + K_p} \tag{8.7}$$

Equation (8.7) clearly shows that K_p must approach ∞ in order for the system to follow a step position change with zero steady-state error. Figure 8.4 illustrates a possible system response to a step position input.

To obtain the velocity- and acceleration-error constants one must proceed as before, except $R(t)$ is given by (8.8) for the velocity input and by (8.9) for the acceleration input:

$$R(t) = tu(t) \tag{8.8}$$

$$R(t) = \frac{t^2 u(t)}{2} \tag{8.9}$$

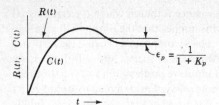

Fig. 8.4

Following the identical procedure, one obtains

$$\epsilon_v = \frac{1}{K_v} \qquad K_v = \lim_{s \to 0} sGH \tag{8.10}$$

for the velocity input and

$$\epsilon_a = \frac{1}{K_a} \qquad K_a = \lim_{s \to 0} s^2 GH \tag{8.11}$$

Figures 8.5 and 8.6 show typical system responses with finite error constants. (The reader is referred to Probs. 8 and 9 at the end of this chapter to discover the relation between these error constants.)

As a specific example, let us assume that the transfer functions for a specific system are

$$G = \frac{A}{s^2 + as + b} \qquad H = \frac{B}{s(s + c)} \tag{8.12}$$

Evaluating the error constants as defined in (8.6), (8.10), and (8.11) yields

$$K_p = \infty \qquad K_v = \frac{AB}{bc} \qquad K_a = 0 \tag{8.13}$$

These values imply that the system will follow any positional changes with zero steady-state error† and velocity changes with a finite steady-state error; acceleration changes cannot be followed at all.

† Even though the steady-state errors are zero, the system may never reach steady state due to further variations in the input during operation. One function of a control-system engineer is to attempt to optimize the system such that the steady state is attained as rapidly as possible. Frequently this is one item of performance that must be traded off against another in order to achieve a system that fulfills all of the desired performance characteristics.

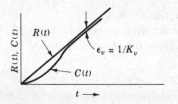

Fig. 8.5   **Fig. 8.6**

9 STABILITY OF LINEAR SYSTEMS

Consider a given dynamical system that is moving under the action of applied forces described by a set of differential equations. If any small disturbing influences are applied to the system, it may deviate only slightly from the previous condition of motion or it may depart from it further and further. If the deviation is slight, the system is said to be *dynamically stable;* otherwise, the system is *dynamically unstable.* The system is said to be dynamically stable if it is stable for all kinds of disturbances. In this section, criteria for the stability of linear dynamical systems will be given. These stability criteria are of great importance in determining the stability of many practical dynamical systems such as the stability of an aircraft in flight, the stability of certain electrical circuits used in communication and power engineering, and the stability of linear control systems.

Consider a linear dynamical system of n degrees of freedom whose differential equations of motion may be written in the following matrix form:

$$[a]\{\ddot{x}\} + [b]\{\dot{x}\} + [c]\{x\} = \{0\} \tag{9.1}$$

where $\{x\}$ is a column matrix whose elements are the n coordinates of the system and the $[a]$, $[b]$, and $[c]$ matrices are square matrices whose elements are constants. The dots indicate differentiation with respect to time in the usual manner. Let a solution of the form

$$\{x\} = \{A\}e^{st} \tag{9.2}$$

be assumed, where $\{A\}$ is a column of constants, e is the base of the natural logarithms, and s is a quantity to be determined. If this form of solution is substituted into (9.1), the result is

$$([a]s^2 + [b]s + [c])\{A\}e^{st} = \{0\} \tag{9.3}$$

If (9.3) is divided by the scalar factor e^{st}, the result is

$$([a]s^2 + [b]s + [c])\{A\} = \{0\} \tag{9.4}$$

This is a set of linear equations in the elements of the column matrix $\{A\}$. For a nontrivial solution the determinant of the coefficients $D(s)$ must be equal to zero so that

$$D(s) = |[a]s^2 + [b]s + [c]| = 0 \tag{9.5}$$

If the following notation is introduced,

$$[g(s)] = [a]s^2 + [b]s + [c] \tag{9.6}$$

the determinant $D(s)$ can be expressed in the form

$$D(s) = \det [g(s)] = \begin{vmatrix} g_{11}(s) & g_{12}(s) & \cdots & g_{1k}(s) \\ g_{21}(s) & g_{22}(s) & \cdots & g_{2k}(s) \\ \cdots & \cdots & \cdots & \cdots \\ g_{k1}(s) & g_{k2}(s) & \cdots & g_{kk}(s) \end{vmatrix} \tag{9.7}$$

If the determinant $D(s)$ is expanded, a polynomial in s of degree $n = 2k$ is obtained. This polynomial has the following form:

$$D(s) = a_0 s^n + a_1 s^{n-1} + a_2 s^{n-2} + \cdots + a_{n-1} s + a_n = 0 \qquad (9.8)$$

where the constants a_i are functions of the constants of the dynamical system.

The stability or instability of the dynamical system under consideration depends on the location of the roots of the polynomial $D(s)$ in the complex s plane. In general the roots of the polynomial equation $D(s) = 0$ will be complex numbers of the form $s_i = b_i + j\omega_i$, where $j = \sqrt{-1}$. If the real parts of all the roots b_i are negative numbers, there will be decrement factors $e^{b_i t}$ in the terms of the solution (9.2). This indicates that the solution is a decaying time function, and the solution will be *stable*. If, however, one or more roots s_i have a positive real part, then the solution (9.2) will contain one or more terms with the exponentially *increasing* factor $e^{b_i t}$, and the system will be *unstable*. An imaginary root of the form $s = j\omega$ will give rise to an oscillatory term of the form $e^{j\omega t}$ and thus represents a borderline case between stability and instability. If s_i is a multiple root, the above conclusions still hold unless $s_i = j\omega$ is a *multiple imaginary* root. In this case the solution contains the functions $e^{j\omega t}$, $te^{j\omega t}$, $t^2 e^{j\omega t}$, etc. The factor $te^{j\omega t}$ increases with time, and therefore a multiple root on the imaginary axis denotes instability. It is therefore apparent that, for the linear system characterized by the set of differential equations (9.1) to be stable, it is necessary and sufficient that the roots of the equation $D(s) = 0$ shall have nonpositive real parts and that, if any purely imaginary roots exist, they must not be multiple roots.

It is thus evident that the stability or instability of the dynamical system under consideration can be determined by obtaining the roots of $D(s) = 0$ by some procedure such as Graeffe's root-squaring method given in Appendix D. This is a laborious procedure, and information concerning the stability of the system may be obtained with much less labor by applying a criterion known in the mathematical literature as the *Routh-Hurwitz stability criterion*.[†] The Routh-Hurwitz criterion will be stated here without proof. The reader interested in the method of proof of the criterion is referred to the paper by Hurwitz.

Let the characteristic determinant $D(s)$ of the system of Eqs. (9.1) be written in the following form:

$$D(s) = a_0 s^n + a_1 s^{n-1} + a_2 s^{n-2} + \cdots + a_{n-1} s + a_n = 0 \qquad (9.9)$$

[†] See E. J. Routh, "Advanced Rigid Dynamics," pp. 223–231, The Macmillan Company, New York, 1905; and A. Hurwitz, Ueber die Bedingungen unter welchen eine Gleichung nur Wurzeln mit negativen reelen Teilen besitzt, *Mathematische Annalen*, vol. 46, pp. 273–284, 1895.

Then the conditions that *all* the roots of $D(s)$ have negative real parts are as follows:

1. A necessary but not sufficient condition is that all the a's in Eq. (9.9) have the same sign.
2. A necessary and sufficient condition for stability is that the following test functions T_i are all *positive* when the equation $D(s) = 0$ is put in such form that a_0 is positive:

$$T_1 = a_1 \qquad T_2 = \begin{vmatrix} a_1 & a_0 \\ a_3 & a_2 \end{vmatrix} \qquad T_3 = \begin{vmatrix} a_1 & a_0 & 0 \\ a_3 & a_2 & a_1 \\ a_5 & a_4 & a_3 \end{vmatrix}$$

$$T_4 = \begin{vmatrix} a_1 & a_0 & 0 & 0 \\ a_3 & a_2 & a_1 & a_0 \\ a_5 & a_4 & a_3 & a_2 \\ a_7 & a_6 & a_5 & a_4 \end{vmatrix} \qquad T_5 = \begin{vmatrix} a_1 & a_0 & 0 & 0 & 0 \\ a_3 & a_2 & a_1 & a_0 & 0 \\ a_5 & a_4 & a_3 & a_2 & a_1 \\ a_7 & a_6 & a_5 & a_4 & a_3 \\ a_9 & a_8 & a_7 & a_6 & a_5 \end{vmatrix} \qquad (9.10)$$

$$T_i = \begin{vmatrix} a_1 & a_0 & 0 & 0 & \cdots \\ a_3 & a_2 & a_1 & a_0 & \cdots \\ a_5 & a_4 & a_3 & a_2 & \cdots \\ \cdots & \cdots & \cdots & \cdots & \cdots \\ a_{2i-1} & a_{2i-2} & \cdots & \cdots & a_i \end{vmatrix}$$

In constructing these test functions all the coefficients a_r with $r > n$ or $r < 0$ are replaced by zeros. It is evident that, in the nth determinant, the bottom-row terms are all zero except the last term; hence

$$T_n = a_n T_{n-1} \qquad (9.11)$$

Consequently, if $a_n > 0$, then T_n has the same sign as T_{n-1} and it is therefore necessary to test the determinants only from T_1 to T_{n-1}. If any of these determinants have negative values, the system is *unstable*.

STABILITY OF A THIRD-ORDER SYSTEM

As an example of the use of the Routh-Hurwitz criterion, consider a system whose determinantal equation is the cubic equation

$$D(s) = s^3 + a_1 s^2 + a_2 s + a_3 = 0 \qquad (9.12)$$

where a_1, a_2, and a_3 are *all positive*. The Routh-Hurwitz criterion then assures stability if

$$T_2 = \begin{vmatrix} a_1 & 1 \\ a_3 & a_2 \end{vmatrix} = a_1 a_2 - a_3 > 0 \qquad (9.13)$$

Hence, if $a_1 a_2 > a_3$, the stability of the third-order system is assured.

STABILITY OF A FOURTH-ORDER SYSTEM

If a dynamical system under consideration has a determinantal equation of the fourth degree of the form

$$s^4 + a_1 s^3 + a_2 s^2 + a_3 s + a_4 = 0 \qquad (9.14)$$

then the Routh-Hurwitz criterion assures the stability of the system provided that $a_1, a_2, a_3,$ and a_4 are *all positive* and

$$T_2 = \begin{vmatrix} a_1 & 1 \\ a_3 & a_2 \end{vmatrix} > 0 \qquad T_3 = \begin{vmatrix} a_1 & 1 & 0 \\ a_3 & a_2 & a_1 \\ 0 & a_4 & a_3 \end{vmatrix} > 0 \qquad (9.15)$$

Since $T_3 = a_3 T_2 - a_4 a_1^2 > 0$, it is necessary that $T_2 > 0$ if a_3 is positive. Hence, if $T_3 > 0$, T_2 must be positive and hence the *complete* criterion for the stability of the fourth-order system is that all the a's must be positive and that

$$a_1 a_2 a_3 > a_1^2 a_4 + a_3^2 \qquad (9.16)$$

Extensive use of the Routh-Hurwitz criterion in the study of the stability of control systems or governors has been made since the time of Routh and Maxwell.[†] The Routh-Hurwitz criterion has been used extensively in recent years in studies involving the stability of control systems and servo-mechanisms of various sorts. Another method of determining the stability of dynamical systems based on the theory of the complex variable was given in Chap. 1.

STABILITY OF CONTROL SYSTEMS

It has already been mentioned that stability is an important consideration in control systems. In the remainder of this section we shall only attempt to introduce the reader to a few basic ideas concerning this topic.

Stability of a control system is investigated by considering (8.2), which relates the error signal to the input signal. The characteristic equation for a control system is given by

$$1 + GH = 0 \qquad (9.17)$$

The system will operate stably if the characteristic equation has no roots with real parts or multiple complex conjugate roots on the imaginary axis. If these conditions are satisfied, then the error signal will be a decreasing function with time or at most will be oscillatory with constant amplitude. As previously mentioned, this last condition is also undesirable and is generally referred to as *marginal stability*. Thus the only desirable case for stability is that one in which all the roots of the characteristic equation have roots with negative real parts. Therefore since G and H are known, then the Routh-Hurwitz criterion may be applied to control systems to examine stability.

[†] See the paper, On Governors, "Scientific Papers of J. C. Maxwell," vol. II, pp. 105–120, reprinted, Dover Publications, Inc., New York, 1952.

As an example consider the following system:

$$G = \frac{AK_1}{s(s+a)} \qquad H = \frac{K_2}{s+b} \tag{9.18}$$

The terms a, b, K_1, and K_2 are positive constants which represent various system parameters. The constant A represents the gain of an amplifier in the forward loop (see Fig. 9.1). Substituting (9.18) into (9.17) gives, after some manipulation,

$$s^3 + (a+b)s^2 + abs + AK_1 K_2 = 0 \tag{9.19}$$

The first condition of the Routh-Hurwitz criterion is satisfied since all the constants are positive.

From (9.10)

$$T_1 = a + b$$

$$T_2 = \begin{vmatrix} a+b & 1 \\ AK_1 K_2 & ab \end{vmatrix} = ab(a+b) - AK_1 K_2 \tag{9.20}$$

Since a and b are positive, then $T_1 > 0$. The second test function T_2 will be positive as long as the amplifier gain A satisfies the condition

$$A < \frac{ab(a+b)}{K_1 K_2} \tag{9.21}$$

This indicates that if the amplifier gain is increased beyond the limit prescribed by (9.21), the system will be unstable.

A classic example of this type of instability has most likely been witnessed by most of us at one time or another. If sound amplification is being used in an auditorium, the audio system will go unstable (as evidenced by intense high-frequency sound) when either the amplifier gain is increased or the microphone is brought too close to the speaker. Instability in the latter case is due to increased gain in the feedback loop. The ringing sound that occurs when someone speaks into a microphone is due to the fact that the roots of the system lie to the left of, but very close to, the imaginary axis. This is the case of positive damping of very small magnitude.

As a result of the intense interest in control systems, a number of other methods of analyzing system stability have been developed. Some of the

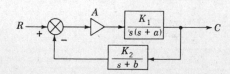

Fig. 9.1

major methods are the root locus plot, Nyquist plot, Bode plots, and Nichols charts.† These methods are based on examining (9.17) in the form

$$GH = -1 \tag{9.22}$$

The open-loop transfer function GH is usually written in the form

$$GH = K\frac{(s + z_1)(s + z_2) \cdots (s + z_n)}{(s + p_1)(s + p_2) \cdots (s + p_m)} \qquad m > n \tag{9.23}$$

where the p_i are the poles of GH, z_i are the zeros, and K is the total open-loop gain. Substituting (9.23) into (9.22) gives

$$K\frac{(s + z_1)(s + z_2) \cdots (s + z_n)}{(s + p_1)(s + p_2) \cdots (s + z_m)} = -1 \tag{9.24}$$

By examining (9.24) the reader may verify that for a fixed value of K there will be only m points in the complex s plane that will satisfy (9.24). As the open-loop gain varies, these points move and trace out the loci of points that satisfy (9.24). For $K = 0$, the loci must terminate on the poles p_i, and for infinite K the loci must terminate on the zeros z_i. Thus as the open-loop gain varies from 0 to ∞, the roots of (9.24) trace out paths from the poles of GH to its zeros. These curves are called *root locus plots* and are useful tools for examining the stability and performance of a control system.

† For detailed discussions of these methods, the reader is urged to consult any of the references at the end of this chapter.

PROBLEMS

1. Determine the transfer function relating the input voltage and output voltage of the following circuits.

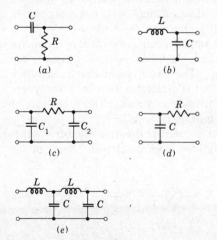

(a)

(b)

(c)

(d)

(e)

Fig. P 1

2. Verify that (8.6) reduces to (8.4) when $Z_1 = R$, $Z_2 = 1/sC$.

3. Derive (4.4) using (4.3).

4. Derive (4.4) by Kirchhoff's current law.

5. Determine the impulse responses of the circuits in Probs. 1a, b, c, and d, and write a general expression for the output voltages for an arbitrary input voltage.

6. In (6.12) discuss the significance of the fact that $G(s) \to \infty$ as $b \to \frac{1}{2}E_0^2$.

7. Formally derive K_v and K_a.

8. Show that if K_p is finite that both K_v and K_a must be zero.

9. Show that if K_a is finite then both K_p and K_v must be infinite.

10. An eighth-order system has the following characteristic equation:

$$D(s) = s^8 + 2s^7 + 4s^6 + 4s^5 + 6s^4 + 6s^3 + 7s^2 + 4s + 2 = 0$$

Determine whether the system is stable or unstable.

11. A sixth-order system has the following characteristic equation:

$$D(s) = s^6 + s^5 + 6s^4 + 5s^3 + 11s^2 + 6s + 6 = 0$$

Discuss the stability of the system.

12. Investigate the stability of the following system:

$$D(s) = s^5 + 9s^4 + 24s^3 + 12s^2 - 60s - 60 = 0$$

(Horowitz, 1963)

13. Determine the error constants for the following systems:

(a) $GH = \dfrac{K}{s(s + 3)(s + 4)}$

(b) $GH = \dfrac{K(s + 2)}{s(s + 1)(s + 3)(s + 4)}$

(c) $GH = \dfrac{K(s + 0.1)}{(s^2 + 10)(s^2 + 0.5s + 50)}$

14. Determine the values of K for which the following systems will be marginally stable:

(a) $GH = \dfrac{K}{s(s + 3)(s + 4)}$

(b) $GH = \dfrac{K(s + 2)}{s(s + 1)(s + 3)(s + 4)}$

(c) $GH = \dfrac{K(s + 0.1)}{(s^2 + 10)(s^2 + 0.5s + 50)}$

(d) $GH = \dfrac{K}{(s + 10)(s + 100)}$

(Savant, 1964)

(e) $GH = \dfrac{K}{s(s + 10)(s + 100)}$

(Savant, 1964)

(f) $GH = \dfrac{K}{s^2(s + 1)(s + 10)}$

(Savant, 1964)

(g) $GH = \dfrac{K}{s(s + 0.1)(s + 0.02)}$

(D'Azzo and Houpis, 1966)

(h) $GH = \dfrac{K}{s(s + 2)(s^2 + s + 10)}$

(D'Azzo and Houpis, 1966)

15. A certain control system is described by the following transfer functions:

$$G = \dfrac{A}{s(s + 3)} \qquad H = \dfrac{1}{s + 4}$$

Determine the time responses for both the error signal and output for amplifier gains of 10, 84, and 100 when the input signal is

(a) $R(t) = u(t)$

(b) $R(t) = tu(t)$

(c) $R(t) = A_0 \sin wt$

REFERENCES

1957. Pearson, E. B.: "Technology of Instrumentation," D. Van Nostrand Company, Inc., Princeton, N.J.

1963. Horowitz, I. M.: "Synthesis of Feedback Systems," Academic Press, Inc., New York.

1964. Savant, C. J., Jr.: "Control System Design," 2d ed., McGraw-Hill Book Company, New York.

1964. Shinners, S. M.: "Servomechanism Practice," John Wiley & Sons, Inc., New York.

1965. Dorf, R. C.: "Time-Domain Analysis and Design of Control Systems," Addison-Wesley Publishing Company, Reading, Mass.

1965. Langill, A. W., Jr.: "Automatic Control Systems Engineering," vols. I and II, Prentice-Hall, Inc., Englewood Cliffs, N.J.

1966. D'Azzo, J. J., and C. H. Houpis: "Feedback Control Systems Analysis and Synthesis," 2d ed., McGraw-Hill Book Company, New York.

9
Laplace's Equation

1 INTRODUCTION

Perhaps the most important partial differential equation of applied mathematics is the equation of Laplace,

$$\nabla^2 v = \frac{\partial^2 v}{\partial x^2} + \frac{\partial^2 v}{\partial y^2} + \frac{\partial^2 v}{\partial z^2} = 0 \tag{1.1}$$

As shown in Appendix E, if (x,y,z) are the rectangular coordinates of any point in space, this equation is satisfied by the following functions, which occur in various branches of applied mathematics:

1. The gravitational potential in regions not occupied by attracting matter.
2. The electrostatic potential in a uniform dielectric, in the theory of electrostatics.
3. The magnetic potential in free space, in the theory of magnetostatics.
4. The electric potential, in the theory of the steady flow of electric currents in solid conductors.

5. The temperature, in the theory of thermal equilibrium of solids.
6. The velocity potential at points of a homogeneous liquid moving irrotationally, in hydrodynamical problems.

In spite of the physical differences of the above subjects the mathematical investigations are much the same for all of them. For example, the problem of determining the temperature in a solid when its surface is maintained at a given temperature is mathematically identical with the problem of determining the electric intensity in a region when the points of its boundary are maintained at given potentials. In this chapter we shall discuss the solution of Laplace's equation by the method of separation of variables. The method of conjugate functions will be discussed in Chap. 10.

2 LAPLACE'S EQUATION IN CARTESIAN, CYLINDRICAL, AND SPHERICAL COORDINATE SYSTEMS

In Appendix E, Sec. 12, the Laplace operator ∇^2 is expressed in general orthogonal curvilinear coordinates. From the results of that expression we have the following forms of Laplace's equation:

a. Cartesian coordinates:

$$\nabla^2 v = \frac{\partial^2 v}{\partial x^2} + \frac{\partial^2 v}{\partial y^2} + \frac{\partial^2 v}{\partial z^2} = 0 \tag{2.1}$$

b. Cylindrical coordinates: Expressed in terms of the cylindrical coordinates of Fig. 2.1, Laplace's equation is

$$\nabla^2 v = \frac{1}{r}\frac{\partial}{\partial r}\left(r\frac{\partial v}{\partial r}\right) + \frac{1}{r^2}\frac{\partial^2 v}{\partial \theta^2} + \frac{\partial^2 v}{\partial z^2} = 0 \tag{2.2}$$

c. Spherical coordinates: Laplace's equation expressed in the spherical coordinates of Fig. 2.2 is

$$\nabla^2 v = \frac{1}{r^2}\frac{\partial}{\partial r}\left(r^2\frac{\partial v}{\partial r}\right) + \frac{1}{r^2\sin\theta}\frac{\partial}{\partial \theta}\left(\sin\theta\frac{\partial v}{\partial \theta}\right) + \frac{1}{r^2\sin^2\theta}\frac{\partial^2 v}{\partial \phi^2} = 0 \tag{2.3}$$

These are the most common coordinates usually encountered in practice.

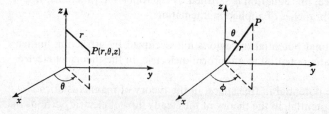

Fig. 2.1   **Fig. 2.2**

3 TWO-DIMENSIONAL STEADY FLOW OF HEAT

To illustrate the solution of Laplace's equation in a simple two-dimensional case, let us consider the following problem.

Suppose we have a thin plate that is bounded by the lines $x = 0$, $x = s$, $y = 0$, and $y = \infty$ (Fig. 3.1). Let the temperature of the edge $y = 0$ be constant with time and be given by $F(x)$. Let the temperature on the other edges be always zero. We shall suppose that heat cannot escape from either surface of the plate and that the effect of initial conditions has passed away. We assume that the temperature is everywhere independent of time. We wish to determine the temperature within the plate.

The problem is one of steady two-dimensional heat flow. In Appendix E, Sec. 14, we find that the temperature distribution v inside a homogeneous solid satisfies the equation

$$\frac{\partial v}{\partial t} = h^2 \nabla^2 v \tag{3.1}$$

where h^2 is the diffusivity of the substance and is a constant. In the case under consideration we have

$$\frac{\partial v}{\partial t} = 0 \tag{3.2}$$

since, by hypothesis, the temperature v does not depend on the time t. Equation (3.1) therefore reduces to

$$\nabla^2 v = \frac{\partial^2 v}{\partial x^2} + \frac{\partial^2 v}{\partial y^2} = 0 \tag{3.3}$$

subject to the boundary conditions

$$\begin{matrix} v=0 & v=0 & v=0 & v=F(x) \\ x=0 & x=s & y=\infty & y=0 \end{matrix} \tag{3.4}$$

In order to solve Eq. (3.3), we try a solution of the form

$$v = F_1(x) F_2(y) \tag{3.5}$$

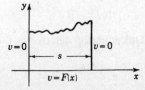

Fig. 3.1

where $F_1(x)$ is a function of x only and $F_2(y)$ is a function of y only. On substituting this assumed solution into (3.3) and rearranging, we obtain

$$\frac{1}{F_1(x)} \frac{d^2 F_1}{dx^2} = -\frac{1}{F_2} \frac{d^2 F_2}{dy^2} \tag{3.6}$$

Now a change in x will not change the right-hand member, and a change in y will not change the left-hand member. Hence each of the expressions of (3.6) must be independent of x and y and must, therefore, be equal to a constant. Let us call this constant $-k^2$. Equation (3.6) then breaks up into two equations:

$$\frac{d^2 F_1}{dx^2} = -k^2 F_1 \quad \text{and} \quad \frac{d^2 F_2}{dy^2} = k^2 F_2 \tag{3.7}$$

The variables are now said to be *separated*. This method of solving the partial differential equation (3.3) is called the *method of separation of variables*. Equations (3.7) are linear differential equations with constant coefficients, and their solutions are

$$\begin{aligned} F_1 &= c_1 \cos kx + c_2 \sin kx \\ F_2 &= c_3 e^{ky} + c_4 e^{-ky} \end{aligned} \tag{3.8}$$

where the c's are arbitrary constants. Substituting these into (3.5), we have

$$\begin{aligned} v &= (c_1 \cos kx + c_2 \sin kx)(c_3 e^{ky} + c_4 e^{-ky}) \\ &= e^{-ky}(A \cos kx + B \sin kx) + e^{ky}(M \cos kx + N \sin kx) \end{aligned} \tag{3.9}$$

where A, B, M, and N are *arbitrary constants*. We must now adjust this solution in order to satisfy the boundary conditions (3.4).

In the first place the temperature is zero when y is infinite. Hence we have

$$M = N = 0 \tag{3.10}$$

Now, since $v = 0$ for $x = 0$, we cannot have a cosine term in the solution; so $A = 0$. At $x = s$, $v = 0$ for all positive values of y. Hence

$$B \sin ks = 0 \tag{3.11}$$

For a nontrivial solution, $B \neq 0$; hence

$$\sin ks = 0 \tag{3.12}$$

This equation determines the possible values of k. They are

$$k = \frac{m\pi}{s} \quad m = 0, 1, 2, 3, \ldots \tag{3.13}$$

Hence to each value of m there is a solution

$$v_m = B_m e^{-m\pi y/s} \sin \frac{m\pi x}{s} \tag{3.14}$$

where B_m is an arbitrary constant.   If we take the sum of all possible solutions of the type (3.14), we construct the solution

$$v = \sum_{m=1}^{\infty} B_m e^{-m\pi y/s} \sin \frac{m\pi x}{s} \tag{3.15}$$

We now only have to satisfy the condition $v = F(x)$ at $y = 0$.   Placing $y = 0$ in (3.15), we have

$$F(x) = \sum_{m=1}^{m=\infty} B_m \sin \frac{m\pi x}{s} \tag{3.16}$$

This is the half-range sine series for v discussed in Appendix C.   The coefficients B_m are given by

$$B_m = \frac{2}{s} \int_0^s F(x) \sin \frac{m\pi x}{s} dx \tag{3.17}$$

Placing these values of the coefficients B_m in (3.15), we have the solution to the problem.

TEMPERATURE DISTRIBUTION OF A FINITE PLATE

As a simple extension of the above problem, let us consider the distribution of temperature inside the plate of Fig. 3.2.   The boundary conditions are

$$
\begin{array}{cccc}
v = 0 & v = 0 & v = 0 & v = F(x) \\
x = 0 & y = 0 & x = s & y = h
\end{array} \tag{3.18}
$$

Starting with the solution (3.9), we see that we cannot have any cosine terms present, and again the possible values of k are given by (3.13).   We thus have

$$v_m = e^{-m\pi y/s} B_m \sin \frac{m\pi x}{s} + e^{m\pi y/s} N_m \sin \frac{m\pi x}{s}$$

$$= (e^{-m\pi y/s} B_m + e^{m\pi y/s} N_m) \sin \frac{m\pi x}{s} \tag{3.19}$$

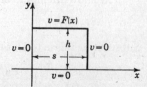

Fig. 3.2

Now at $y = 0$, $v = 0$, for $0 \leqslant x \leqslant s$; therefore

$$B_m + N_m = 0 \qquad \text{or} \qquad B_m = -N_m \tag{3.20}$$

Hence we can write the solution in the form

$$v_m = C_m \sinh \frac{m\pi y}{s} \sin \frac{m\pi x}{s} \tag{3.21}$$

where C_m is an arbitrary constant.

Summing over all possible values of m, we have

$$v = \sum_{m=1}^{m=\infty} C_m \sinh \frac{m\pi y}{s} \sin \frac{m\pi x}{s} \tag{3.22}$$

Now, at $y = h$, we have

$$F(x) = \sum_{m=1}^{m=\infty} C_m \sinh \frac{m\pi h}{s} \sin \frac{m\pi x}{s} \tag{3.23}$$

This is a half-range sine expansion for $F(x)$.   Hence

$$C_m \sinh \frac{m\pi h}{s} = \frac{2}{s} \int_0^s F(x) \sin \frac{m\pi x}{s} dx \tag{3.24}$$

or

$$C_m = \frac{2}{s \sinh (m\pi h/s)} \int_0^s F(x) \sin \frac{m\pi x}{s} dx \tag{3.25}$$

If we substitute this value of C_m into (3.22), we have the solution to the problem.

4   CIRCULAR HARMONICS

In its most general sense the term *harmonic* applies to any solution of Laplace's equation.   However, the term harmonic is ordinarily used in a more restricted sense to mean a solution of Laplace's equation in a specified coordinate system. If we write Laplace's equation in cylindrical coordinates and assume that v is independent of the coordinate z, we have from Eq. (2.2)

$$\frac{1}{r} \frac{\partial}{\partial r} \left(r \frac{\partial v}{\partial r} \right) + \frac{1}{r^2} \frac{\partial^2 v}{\partial \theta^2} = 0 \tag{4.1}$$

We now attempt to find a solution of this equation of the form

$$v = F_1(\theta) F_2(r) \tag{4.2}$$

Substituting this in (4.1), we have

$$\frac{F_1(\theta)}{r} \frac{d}{dr} \left(r \frac{dF_2}{dr} \right) + \frac{F_2(r)}{r^2} \frac{d^2 F_1(\theta)}{d\theta^2} = 0 \tag{4.3}$$

Multiplying by r^2 and dividing by $F_1 F_2$, we have

$$\frac{1}{F_2}\left(r^2\frac{d^2 F_2}{dr^2} + r\frac{dF_2}{dr}\right) = -\frac{1}{F_1}\frac{d^2 F_1}{d\theta^2} = n^2 \tag{4.4}$$

where, since the left-hand side of the equation is a function of r alone and the right-hand side is a function of θ alone, we conclude that both sides of the equation are equal to the same constant, which we have called n^2. We thus have the two equations

$$\frac{d^2 F_1}{d\theta^2} + n^2 F_1 = 0 \tag{4.5}$$

and

$$r^2\frac{d^2 F_2}{dr^2} + r\frac{dF_2}{dr} - n^2 F_2 = 0 \tag{4.6}$$

The variables are now separated. Equation (4.5) is the well-known equation of simple harmonic motion. Its solution is

$$F_1 = A\cos n\theta + B\sin n\theta \qquad \text{if } n \neq 0 \tag{4.7}$$

It is easily verified that the solution of (4.6) is

$$F_2 = Cr^n + Dr^{-n} \qquad \text{if } n \neq 0 \tag{4.8}$$

If $n = 0$, we have the solutions

$$F_1 = A_0\,\theta + B_0$$
$$F_2 = C_0\ln r + D_0 \tag{4.9}$$

where the quantities A, B, C, D are arbitrary constants.

The number n is called the *degree of harmonic*. The solutions of Laplace's equation in cylindrical coordinates when v is independent of the coordinate z are called *circular harmonics*. Circular harmonics are of extreme importance in the solution of two-dimensional problems having cylindrical symmetry. The circular harmonics are then

$$v_0 = (A_0\,\theta + B_0)(C_0\ln r + D_0) \qquad \text{degree zero}$$
$$v_n = (A_n\cos n\theta + B_n\sin n\theta)(C_n r^n + D_n r^{-n}) \qquad \text{degree } n \tag{4.10}$$

In most applications of circular harmonics to physical problems the function v is usually a *single-valued* function of θ. Since if we increase θ by 2π we reach the same point in the xy plane, in order for v_n to be single-valued, we must have

$$v_n(r, \theta + 2\pi) = v_n(r, \theta) \tag{4.11}$$

It is thus necessary for n to take only integral values. A general single-valued solution of Laplace's equation is obtained by summing the solutions (4.10) over all possible values of n in the form

$$v = a_0 \ln r + \sum_{n=1}^{\infty} r^n (a_n \cos n\theta + b_n \sin n\theta)$$

$$+ \sum_{n=1}^{\infty} \frac{1}{r^n} (q_n \cos n\theta + f_n \sin n\theta) + c_0 \quad (4.12)$$

where the quantities $a, b, q, f,$ and c_0 are arbitrary constants.

As a simple illustration of the use of circular harmonics, let us consider the following problem: Suppose a very long circular cylinder is composed of two halves as shown in Fig. 4.1. The two halves of the cylinder are thermally insulated from each other, and the upper half of the cylinder is kept at temperature v_1, while the lower half is kept at temperature v_2. It is required to find the steady-state temperature in the region inside the cylinder. It is assumed that the cylinder is so long in the z direction that the temperature is independent of z.

To solve this problem, we must solve the equation

$$\nabla^2 v = 0 \tag{4.13}$$

in the region inside the cylinder and satisfy the boundary conditions

$$v = v_1 \quad \text{at } r = R \quad 0 < \theta < \pi$$
$$v = v_2 \quad \text{at } r = R \quad \pi < \theta < 2\pi \tag{4.14}$$

We do this by taking the general solution (4.12) and specializing it to the boundary conditions (4.14). In the first place, we notice that the temperature must be finite at the origin $r = 0$. It is necessary, therefore, for the constants a_0, q_n, and f_n to be equal to zero. There remains the solution

$$v = \sum_{n=1}^{\infty} r^n (a_n \cos n\theta + b_n \sin n\theta) + c_0 \tag{4.15}$$

Before solving the problem under consideration, let us solve the more general problem in which the temperature is specified on the circumference of the cylinder $r = a$ as an arbitrary function of θ, so that

$$v = F(\theta) \quad \text{at } r = R \tag{4.16}$$

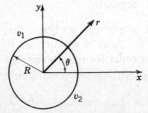

Fig. 4.1

We then have on placing $r = R$ in (4.15)

$$F(\theta) = \sum_{n=1}^{\infty} R^n(a_n \cos n\theta + b_n \sin n\theta) + c_0 \qquad (4.17)$$

The problem of determining the constants a_n, b_n, and c_0 reduces to expanding the function $F(\theta)$ in a Fourier series.   We therefore have

$$a_n = \frac{1}{R^n \pi} \int_0^{2\pi} F(\theta) \cos n\theta \, d\theta$$

$$b_n = \frac{1}{R^n \pi} \int_0^{2\pi} F(\theta) \sin n\theta \, d\theta \qquad (4.18)$$

$$c_0 = \frac{1}{2\pi} \int_0^{2\pi} F(\theta) \, d\theta$$

An interesting special case arises when the temperature of the upper half of the cylinder is kept at v_0 and the lower half is kept at a temperature equal to zero.   The function $F(\theta)$ is then given graphically by Fig. 4.2.   We then have

$$a_n = \frac{v_0}{R^n \pi} \int_0^{\pi} \cos n\theta \, d\theta = 0$$

$$b_n = \frac{v_0}{R^n \pi} \int_0^{\pi} \sin n\theta \, d\theta = \frac{2v_0}{R^n \pi n} \qquad n \text{ odd} \qquad (4.19)$$

$$c_0 = \frac{1}{2\pi} \int_0^{\pi} v_0 \, d\theta = \frac{v_0}{2}$$

Substituting this into (4.15), we obtain

$$v(r,\theta) = \frac{2v_0}{\pi} \sum_{n=1}^{\infty} \left(\frac{r}{R}\right)^n \frac{\sin n\theta}{n} + \frac{v_0}{2} \qquad n \text{ odd} \qquad (4.20)$$

Other two-dimensional problems having circular symmetry may be solved in a similar manner.

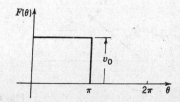

Fig. 4.2

5 CONDUCTING CYLINDER IN A UNIFORM FIELD

As another example of the application of circular harmonics, let us consider the following problem: An infinitely long uncharged conducting cylinder of circular cross section is placed in a uniform electric field E_0, with its axis at right angles to the lines of force.   Denote the radius of the cylinder by a, and take the x axis in the direction of the field as in Fig. 5.1 and the z axis along the axis of the cylinder.   We wish to determine the field induced by the presence of the cylinder.

It is clear from symmetry that the potential v is not a function of z. As explained in Appendix E, Sec. 16, the electrostatic field E is derived from the potential v by the equation

$$\mathbf{E} = -\operatorname{grad} v = -\nabla v \tag{5.1}$$

and v satisfies Laplace's equation.   The original potential outside the cylinder is of the form

$$v_0 = -E_0 x = -E_0 r \cos\theta \tag{5.2}$$

This field will induce charges on the perfectly conducting cylinder, and these induced charges will produce a potential v_i.   The total potential due to the external field and the induced separation of charges of the cylinder is given by

$$v = v_0 + v_i = -E_0 r \cos\theta + v_i \tag{5.3}$$

Since the cylinder is perfectly conducting, it is an equipotential surface and we may take its potential as zero.   We therefore have

$$v = 0 \qquad \text{at } r = a \tag{5.4}$$

Now, at infinity, the potential produced by the induced charges on the cylinder must vanish, since, by hypothesis, the cylinder is initially uncharged and the external field produces only a separation of charge.   Also, by symmetry, the induced potential must be symmetrical about the x axis; hence it must be an even function of θ.

It follows, therefore, that in the general circular harmonic solution of Laplace's equation we must have a_0, a_n, b_n equal to zero so that the potential

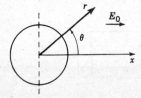

Fig. 5.1

will remain finite at infinity, and f_n must be equal to zero for a potential that is an even function of θ. The constant term may be taken equal to zero without loss of generality. We therefore take v_i in the form

$$v_i = \sum_{n=1}^{\infty} \frac{q_n}{r^n} \cos n\theta \tag{5.5}$$

The total potential is given by (5.3) in the form

$$v = -E_0 r \cos\theta + \sum_{n=1}^{\infty} \frac{q_n \cos n\theta}{r^n} \tag{5.6}$$

Now, at $r = a$, we have

$$0 = -E_0 a \cos\theta + \sum_{n=1}^{\infty} \frac{q_n \cos n\theta}{a^n} \tag{5.7}$$

or

$$\sum_{n=1}^{\infty} \frac{q_n \cos n\theta}{a^n} = E_0 a \cos\theta \tag{5.8}$$

To obtain the unknown coefficients q_n, we equate coefficients of like harmonic terms in θ and obtain

$$\frac{q_1}{a} = E_0 a$$
$$q_n = 0 \qquad \text{if } n > 1 \tag{5.9}$$

Hence

$$q_1 = E_0 a^2 \tag{5.10}$$

The complete potential (5.6) then becomes

$$v = -\left(1 - \frac{a^2}{r^2}\right) E_0 r \cos\theta \tag{5.11}$$

Since $\mathbf{E} = -\nabla v$, the radial and tangential components of the electric field are obtained by taking the gradient in cylindrical coordinates and we obtain

$$E_r = -\frac{\partial v}{\partial r} = \left(1 + \frac{a^2}{r^2}\right) E_0 \cos\theta$$

$$E_\theta = -\frac{\partial v}{r\,\partial\theta} = -\left(1 - \frac{a^2}{r^2}\right) E_0 \sin\theta \tag{5.12}$$

6 GENERAL CYLINDRICAL HARMONICS

We have considered solutions of Laplace's equation in cylindrical coordinates that were independent of the coordinate z. In certain three-dimensional problems it is necessary to obtain solutions of Laplace's equation when v is a function of the three cylindrical coordinates r, θ, and z:

$$\frac{1}{r}\frac{\partial}{\partial r}\left(r\frac{\partial v}{\partial r}\right) + \frac{1}{r^2}\frac{\partial^2 v}{\partial \theta^2} + \frac{\partial^2 v}{\partial z^2} = 0 \tag{6.1}$$

We begin by assuming a solution of the form

$$v = F_1(\theta)\,F_2(r)\,F_3(z) \tag{6.2}$$

where F_1 is a function of θ alone, etc. Substituting this assumed form of solution in (6.1), and after some reductions, we obtain

$$\frac{1}{F_3}\frac{d^2 F_3}{dz^2} = -\frac{1}{F_2}\frac{d^2 F_2}{dr^2} - \frac{1}{rF_2}\frac{dF_2}{dr} - \frac{1}{r^2 F_1}\frac{d^2 F_1}{d\theta^2} \tag{6.3}$$

Now by hypothesis F_3 is a function of z only, and by (6.3) a change in z does not change the right-hand side since it is not a function of z. Hence the left-hand side of (6.3) reduces to a constant, and we write

$$\frac{1}{F_3}\frac{d^2 F_3}{dz^2} = k^2 \tag{6.4}$$

Hence

$$F_3 = C_1 e^{ks} + C_2 e^{-ks} \tag{6.5}$$

From (6.3) and (6.4) we have

$$\frac{r^2}{F_2}\frac{d^2 F_2}{dr^2} + \frac{r}{F_2}\frac{dF_2}{dr} + k^2 r^2 = -\frac{1}{F_1}\frac{d^2 F_1}{d\theta^2} \tag{6.6}$$

By the same reasoning as above it follows that each member of (6.6) is a constant. Let us call this constant m^2. We therefore have

$$\frac{d^2 F_1}{d\theta^2} = -m^2 F_1 \tag{6.7}$$

Therefore

$$F_1 = C_3 \cos m\theta + C_4 \sin m\theta \tag{6.8}$$

and

$$r^2\frac{d^2 F_2}{dr^2} + r\frac{dF_2}{dr} + (k^2 r^2 - m^2) F_2 = 0 \tag{6.9}$$

If we place

$$kr = x \tag{6.10}$$

in (6.9), we obtain

$$x^2 \frac{d^2 F_2}{dx^2} + x \frac{dF_2}{dx} + (x^2 - m^2) F_2 = 0 \tag{6.11}$$

This is Bessel's differential equation, discussed in Appendix B. Its solution is

$$F_2 = C_5 J_m(x) + C_6 J_{-m}(x)$$
$$= C_5 J_m(kr) + C_6 J_{-m}(kr) \tag{6.12}$$

if m is fractional or

$$F_2 = C_5 J_m(kr) + C_6 Y_m(kr) \tag{6.13}$$

if m is integral.

Any of the above values of F_1, F_2, and F_3 substituted in (6.2) gives a solution of Laplace's equation. If we let k be a fixed constant and if we require v to be a single-valued function of θ, then m must take only integral values and we have the solution

$$v = \sum_{m=0}^{m=\infty} [e^{kz}(A_m \cos m\theta + B_m \sin m\theta)$$
$$+ e^{-kz}(C_m \cos m\theta + D_m \sin m\theta)] J_m(kr) \tag{6.14}$$

This solution remains finite at $r = 0$ and is useful in certain electrical problems and problems of steady heat conduction.

7 SPHERICAL HARMONICS

If the boundary conditions of a problem involving the solution of Laplace's equation are simply expressed in spherical polar coordinates, it is useful to have a general solution of Laplace's equation in this system of coordinates.

In this case we must find solutions of Eq. (2.3). This equation may be written in the form

$$\frac{\partial}{\partial r}\left(r^2 \frac{\partial v}{\partial r}\right) + \frac{1}{\sin \theta} \frac{\partial}{\partial \theta}\left(\sin \theta \frac{\partial v}{\partial \theta}\right) + \frac{1}{\sin^2 \theta} \frac{\partial^2 v}{\partial \phi^2} = 0 \tag{7.1}$$

We wish to find a solution of the form

$$v = R\Theta\Phi = RS \tag{7.2}$$

where R is a function of r only, Θ is a function of θ only, and Φ is a function of ϕ only.

The function

$$S(\theta, \phi) = \Theta\Phi \tag{7.3}$$

is called a *surface harmonic*. The function Θ, when ϕ is a constant, is called a *zonal* surface harmonic.

If we substitute (7.2) in (7.1) and divide through by RS, we have

$$\frac{1}{R}\frac{d}{dr}\left(r^2\frac{dR}{dr}\right) + \frac{1}{S\sin\theta}\frac{\partial}{\partial\theta}\sin\theta\frac{\partial S}{\partial\theta} + \frac{1}{S\sin^2\theta}\frac{\partial^2 S}{\partial\phi^2} = 0 \tag{7.4}$$

The first term is a function of r only, and the other ones involve only the angles. For all values of the coordinates, therefore, the equation can be satisfied only if

$$\frac{1}{R}\frac{d}{dr}\left(r^2\frac{dR}{dr}\right) = K \tag{7.5}$$

and

$$\frac{1}{S\sin\theta}\frac{\partial}{\partial\theta}\left(\sin\theta\frac{\partial S}{\partial\theta}\right) + \frac{1}{S\sin^2\theta}\frac{\partial^2 S}{\partial\phi^2} = -K \tag{7.6}$$

If we place

$$K = n(n+1) \tag{7.7}$$

the solution of Eq. (7.5) is easily seen to be

$$R = Ar^n + Br^{-n-1} \tag{7.8}$$

If we multiply Eq. (7.6) by S, we have

$$\frac{1}{\sin\theta}\frac{\partial}{\partial\theta}\left(\sin\theta\frac{\partial S}{\partial\theta}\right) + \frac{1}{\sin^2\theta}\frac{\partial^2 S}{\partial\phi^2} + n(n+1)S = 0 \tag{7.9}$$

Equation (7.2) therefore takes the form

$$v = (Ar^n + Br^{-n-1})S_n \tag{7.10}$$

The subscript on S_n signifies that the same value of n must be used in both terms of (7.10). Any sum of solutions of the type of (7.10) is also a solution.

A GENERAL PROPERTY OF SURFACE HARMONICS

By the use of Green's theorem (Appendix E, Sec. 9), which may be written in the form

$$\iiint_v (U\nabla^2 W - W\nabla^2 U)\,dv = \iint_s (U\nabla W - W\nabla U)\cdot d\mathbf{s} \tag{7.11}$$

we may derive an important property of the function S_n. To do this, let

$$U = r^m S_m \qquad W = r^n S_n \tag{7.12}$$

so that

$$\nabla^2 U = \nabla^2 W = 0 \tag{7.13}$$

and hence the volume integral vanishes. If we take for the surface, in Green's theorem, a unit sphere, we have

$$(\nabla U)_s = \frac{\partial}{\partial r}(r^m S_m) = mr^{m-1}S_m = mS_m \qquad \text{at } r = 1 \tag{7.14}$$

and similarly

$$(\nabla W)_s = nS_n \tag{7.15}$$

Since both $(\nabla U)_s$ and $(\nabla W)_s$ have the normal direction to the sphere, the dot product in (7.11) is absorbed and we have

$$\iint_s (nS_n S_m - mS_n S_m)\, ds = (n - m) \iint_s S_n S_m\, ds = 0 \tag{7.16}$$

and if $n \neq m$, we obtain the result

$$\iint_s S_n S_m\, ds = 0 \tag{7.17}$$

where the surface integration is taken over a unit sphere.

SURFACE ZONAL HARMONICS

A very important special case is the one in which v is independent of ϕ so that Φ is a constant and S_n is a function of θ only. In this case, we have

$$\frac{\partial^2 S}{\partial \phi^2} = 0 \tag{7.18}$$

and Eq. (7.9) reduces to

$$\frac{1}{\sin \theta} \frac{d}{d\theta} \left(\sin \theta \frac{dS_n}{d\theta} \right) + n(n + 1) S_n = 0 \tag{7.19}$$

If we write

$$\mu = \cos \theta \tag{7.20}$$

and transform Eq. (7.19) from the independent variable θ to the independent variable μ, we obtain

$$\frac{d}{d\mu} \left[(1 - \mu^2) \frac{dS_n}{d\mu} \right] + n(n + 1) S_n = 0 \tag{7.21}$$

This we recognize as the Legendre equation of Appendix B. If n is a positive integer, a solution of (7.21) is given by the Legendre polynomial

$$S_n = P_n(\mu) = P_n(\cos \theta) \tag{7.22}$$

By combining the solutions thus obtained, we have from (7.10) the solution

$$v = \sum_{n=0}^{n=\infty} \left(A_n r^n + \frac{B_n}{r^{n+1}} \right) P_n(\cos \theta) \tag{7.23}$$

where A_n and B_n are arbitrary constants.

This solution has many applications to problems in electrostatics, magnetostatics, and potential theory that have spherical symmetry such that the function v is symmetrical about the z axis so that it is independent of the angle ϕ.

8 THE POTENTIAL OF A RING

As shown in Appendix E, Sec. 15, if a particle of matter of mass m is at a point (a,b,c) of a cartesian reference frame, then the gravitational potential at (x,y,z) due to the mass is

$$v_m = \frac{m}{\sqrt{(x-a)^2 + (y-b)^2 + (z-c)^2}} \tag{8.1}$$

It is also shown in Appendix E, Sec. 15, that the gravitational potential v satisfies Laplace's equation

$$\nabla^2 v = 0 \tag{8.2}$$

in the region not occupied by matter.

Let us determine the potential at any point due to a uniform circular ring of small cross section lying in the plane of x, y with its center at O as shown in Fig. 8.1. Evidently the potential v is symmetric about the z axis and is independent of the angle ϕ. We know, therefore, that the potential v is of the form

$$v = \sum_{n=0}^{n=\infty} \left(A_n r^n + \frac{B_n}{r^{n+1}} \right) P_n(\cos \theta) \tag{8.3}$$

The problem is to determine the unknown coefficients A_n and B_n. We may determine these coefficients by realizing that any point Q on the z axis is at the same distance

$$\sqrt{a^2 + r^2}$$

from all points of the ring, where $\overline{OQ} = r$ and a is the radius of the ring. The cross section of the ring is assumed negligible. Hence the potential at Q is

$$\frac{M}{\sqrt{a^2 + r^2}} \tag{8.4}$$

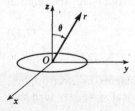

Fig. 8.1

where M is the total mass of the ring.   By the binomial theorem we have

$$\frac{M}{\sqrt{a^2 + r^2}} = \frac{M}{a}\left(1 - \frac{r^2}{2a^2} + \frac{1 \times 3r^4}{2 \times 4a^4} - \cdots\right) \tag{8.5}$$

when $r < a$, and

$$\frac{M}{\sqrt{a^2 + r^2}} = \frac{M}{a}\left(\frac{a}{r} - \frac{1a^3}{2r^3} + \frac{1 \times 3a^5}{2 \times 4r^5} - \cdots\right) \tag{8.6}$$

when $r > a$.

The general solution (8.3) must reduce to either (8.5) or (8.6) for a point on z where $\theta = 0$.   Now we have from Appendix B

$$P_n(\cos 0) = P_n(1) = 1 \tag{8.7}$$

Hence for points on the z axis (8.3) becomes

$$v = \sum_{n=0}^{\infty}\left(A_n r^n + \frac{B_n}{r^{n+1}}\right) \tag{8.8}$$

Comparing this with (8.5) and (8.6), we have $B_n = 0$ if $r < a$ and A_n are the coefficients of (8.5), while $A_n = 0$ if $r > a$ and B_n are the coefficients of (8.6).   Hence we have the solution

$$v = \frac{M}{a}\left[P_0(\cos\theta) - \frac{1}{2}\frac{r^2}{a^2}P_2(\cos\theta) + \frac{1 \times 3}{2 \times 4}\frac{r^4}{a^4}P_4(\cos\theta) - \cdots\right] \tag{8.9}$$

where $r < a$, and

$$v = \frac{M}{a}\left[\frac{a}{r}P_0(\cos\theta) - \frac{1}{2}\frac{a^3}{r^3}P_2(\cos\theta) + \frac{1 \times 3}{2 \times 4}\frac{a^5}{r^5}P_4(\cos\theta) - \cdots\right] \tag{8.10}$$

when $r > a$.

9 THE POTENTIAL ABOUT A SPHERICAL SURFACE

As another example of the use of the general solution (7.23), let us consider the following problem: Let a spherical surface be kept at a fixed distribution of electric potential of the form

$$v = F(\theta) \tag{9.1}$$

on the surface of the sphere of Fig. 9.1.   The space inside and outside the spherical surface is assumed to be free of charges, and it is desired to determine the potential inside and outside the spherical surface.

In this case we have the boundary conditions

$$v = F(\theta) \qquad \text{when } r = a \tag{9.2}$$

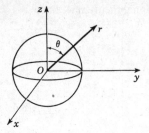

Fig. 9.1

where a is the radius of the sphere, and

$$\lim_{r \to \infty} v = 0 \tag{9.3}$$

that is, the potential vanishes at infinity.

a   THE REGION OUTSIDE THE SPHERICAL SURFACE

In the region outside the spherical surface because of the boundary condition (9.3) we cannot have any positive powers of r.   In this case the general solution (7.23) has $A_n = 0$, and we have

$$v = \sum_{n=0}^{\infty} \frac{B_n}{r^{n+1}} P_n(\cos \theta) \qquad r > a \tag{9.4}$$

To determine the unknown constants B_n, we use the boundary condition (9.2).   Placing $r = a$ in (9.4), we have

$$F(\theta) = f(\cos \theta) = \sum_{n=0}^{\infty} \frac{B_n}{a^{n+1}} P_n(\cos \theta) \tag{9.5}$$

If we let

$$\cos \theta = u \tag{9.6}$$

we have to expand $f(u)$ in a series of Legendre polynomials of the form

$$f(u) = \sum_{n=0}^{\infty} \frac{B_n}{a^{n+1}} P_n(u) \tag{9.7}$$

By the results of Appendix B we have

$$\frac{B_n}{a^{n+1}} = \frac{2n+1}{2} \int_{-1}^{+1} f(u) P_n(u) \, du$$

$$= \frac{2n+1}{2} \int_{0}^{\pi} F(\theta) P_n(\cos \theta) \sin \theta \, d\theta \tag{9.8}$$

Hence

$$B_n = a^{n+1} \frac{2n+1}{2} \int_0^\pi F(\theta) P_n(\cos \theta) \sin \theta \, d\theta \tag{9.9}$$

This determines the coefficients in the solution (9.4).

b   THE REGION INSIDE THE SPHERICAL SURFACE

In the region inside the spherical surface the potential cannot become infinite so that there cannot be any negative powers of r in the general solution (7.23). Hence the coefficients $B_n = 0$, and we have

$$v = \sum_{n=0}^\infty A_n r^n P_n(\cos \theta) \qquad r < a \tag{9.10}$$

To determine the unknown coefficients A_n, we place $r = a$, and we have

$$F(\theta) = \sum_{n=0}^\infty A_n a^n P_n(\cos \theta) \tag{9.11}$$

Using the formula for the determination of the coefficients of the expansion of an arbitrary function into a series of Legendre polynomials, we obtain

$$A_n = \frac{2n+1}{2a^n} \int_0^\pi F(\theta) P_n(\cos \theta) \sin \theta \, d\theta \tag{9.12}$$

These values of A_n when substituted into (9.10) give the solution to the problem.

A discussion of the general case when v is a function of all three spherical coordinates r, θ, and ϕ is beyond the scope of this book. For a detailed account of the theory of spherical harmonics the references at the end of this chapter should be consulted.

10   GENERAL PROPERTIES OF HARMONIC FUNCTIONS

In this section we shall discuss certain general properties of harmonic functions, that is, functions that satisfy Laplace's differential equation.

Let us suppose that we have a vector field $\mathbf{A}$ such that

$$\mathbf{A} = \nabla v \tag{10.1}$$

where v is a scalar point function that satisfies Laplace's equation

$$\nabla^2 v = 0 \tag{10.2}$$

Now by Gauss's theorem (Appendix E, Sec. 9), we have

$$\iint_S \mathbf{A} \cdot d\mathbf{s} = \iiint_V (\nabla \cdot \mathbf{A}) \, dv \tag{10.3}$$

Now, if v satisfies Laplace's equation at every point within the region bounded by the surface S, we have

$$\nabla \cdot \mathbf{A} = \nabla \cdot (\nabla v) = \nabla^2 v = 0 \tag{10.4}$$

Hence (10.3) becomes

$$\iint_S (\nabla v) \cdot d\mathbf{s} = 0 \tag{10.5}$$

If we take the curl of both sides of (10.1), we have

$$\nabla \times \mathbf{A} = \nabla \times (\nabla v) = 0 \tag{10.6}$$

Let us now apply Stokes's theorem (Appendix E, Sec. 10) to the vector field $\mathbf{A}$ and obtain

$$\oint_C \mathbf{A} \cdot d\mathbf{l} = \iint_S (\nabla \times \mathbf{A}) \cdot d\mathbf{s} = 0 \tag{10.7}$$

over the curve C bounding the open surface S. Substituting (10.6) in (10.7), we have

$$\oint_C (\nabla v) \cdot d\mathbf{l} = 0 \tag{10.8}$$

From Eqs. (10.5) and (10.8) certain important properties of harmonic functions may be deduced that are similar in the plane and in space.

By an application of Green's theorem

$$\iiint_V (U\nabla^2 W - W\nabla^2 U)\,dv = \iint_S (U\nabla W - W\nabla U) \cdot d\mathbf{s} \tag{10.9}$$

it may be shown that if $\nabla^2 v = 0$ in the region bounded by a sphere of radius r, the value of v at the center of the sphere v_0 is given by

$$v_0 = \frac{1}{4\pi r^2} \iint_S v\,ds \tag{10.10}$$

where the integral is taken over the surface of the sphere. The result may be stated as follows:

I The average value of a harmonic function on the surface of a sphere in which it has no singularities is equal to its value at the center of the sphere.

As a consequence of (10.5) we may deduce the following theorem:

II A nonconstant harmonic function without singularities in a given region cannot have a maximum value or a minimum value in the region.

To prove this statement, let us assume that v has a maximum value at the point p. Now, if we draw a small sphere s with p as its center, it is evident that the expression $(\nabla v)\,ds$, which is the normal derivative of v, is everywhere *negative* on *s* since, by hypothesis, *v* has a *greater* value at *p* than at any point in the neighborhood of *p*. Therefore the integral

$$\iint\limits_{s} (\nabla v) \cdot d\mathbf{s} \tag{10.11}$$

is negative. But by (10.5) this integral must vanish. This contradicts the initial assumption. In the same manner it may be shown that *v* cannot have a minimum at *p*, and the proposition is proved.

From the above proposition we deduce the following theorem:

III A harmonic function with no singularities within a region and constant everywhere on the bounding surface *S* of the region has the same constant value everywhere inside the region.

To prove this, assume that the function is *not* constant inside the region; it must, therefore, have maximum and minimum values. However, by proposition II, it cannot have any maximum or minimum values in the region, so that these maximum and minimum values must occur on the boundary of the region. However, by hypothesis, the function is *constant* on the boundary, and therefore its maximum and minimum values coincide, and the function is constant. This proves the proposition.

Another very important theorem is the following one:

IV Two harmonic functions that have identical values upon a closed contour and have no singularities within the contour are identical throughout the region bounded by the contour.

To prove this, let us suppose that we have two harmonic functions v_1 and v_2 which have the same values on the boundary of a closed region. Then $v_1 - v_2$ is a harmonic function which is zero on the boundary and hence, by Theorem III, is zero within the region under consideration. It follows, therefore, that $v_1 = v_2$, and the proposition is proved.

A practical result of this theorem is that if a solution of Laplace's equation has been found so as to take assigned values on a closed boundary no other solution is possible; that is, the solution is unique. Another important result may be obtained by using the first form of Green's theorem (Appendix E, Sec. 9) in

$$\iint\limits_{s} (U \nabla W) \cdot d\mathbf{s} = \iiint\limits_{v} [U \nabla^2 W + (\nabla U) \cdot (\nabla W)] \, dv \tag{10.12}$$

If we let

$$U = W \tag{10.13}$$

and
$$\nabla^2 U = 0 \tag{10.14}$$

in the region inside S, Eq. (10.12) reduces to

$$\iint_S (U\nabla U)\cdot d\mathbf{s} = \iiint_v (\nabla U)^2 \, dv \tag{10.15}$$

If the normal derivative of U vanishes on the surface S, we have $(\nabla U)\,ds = 0$ for every point of the surface, and hence the left-hand side of (10.15) vanishes. We then have

$$\iiint_v (\nabla U)^2 \, dv = 0 \tag{10.16}$$

Now since the volume v is arbitrary and the integrand $(\nabla U)^2$ is always positive, we must have

$$\nabla U = 0 \tag{10.17}$$

or hence

$$U = \text{const} \tag{10.18}$$

Hence we have the following theorem:

V If the normal derivative of a harmonic function is zero on a closed surface within which the function has no singularities, the function is a constant.

From this theorem comes the following one:

VI If two harmonic functions have the same normal derivatives on a closed surface within which they have no singularities, they differ at most by an additive constant.

These theorems are of great importance in the theory of electrostatics, magnetostatics, heat flow, etc. Applied to heat flow, Theorems IV and VI are physically evident. Theorem IV states that the temperature within a closed region is fully determined by the temperature on the boundary, and Theorem VI states that except for an additive constant the temperature inside the region is determined by the rate of flow across the boundary.

PROBLEMS

1. A homogeneous spherical shell is made of material of diffusivity h. The inner radius of the shell is a, and the outer radius is b. If the inner shell is kept at a constant temperature v_a and the outer shell is kept at a constant temperature v_b, show that the temperature of the shell is given by

$$v = (v_a - v_b)\frac{ab}{b-a}\frac{1}{r} + \frac{v_b b - v_a a}{b-a}$$

2. Given the temperatures at the boundaries of a rectangular plate as

$$v = 0 \qquad v = 0 \qquad v = \phi(x) \qquad v = F(x)$$
$$x = 0 \quad ' \quad x = s \qquad y = 0 \qquad y = h$$

show that the temperature of the region inside the plate is given by $v = U + W$, where

$$U = \sum_{n=1}^{\infty} a_n \frac{\sinh{(n\pi y/s)}}{\sinh{(n\pi h/s)}} \sin \frac{n\pi x}{s}$$

where

$$a_n = \frac{2}{s} \int_0^s F(x) \sin \frac{n\pi x}{s} dx$$

and

$$W = \sum_{n=1}^{\infty} b_n \frac{[\sinh \pi n(h-y)]/s}{\sinh{(n\pi h/s)}} \sin \frac{n\pi x}{s}$$

where

$$b_n = \frac{2}{s} \int_0^s \phi(x) \sin \frac{n\pi x}{s} dx$$

3. An infinitely long plane and uniform plate is bounded by two parallel edges and an end at right angles to these. The breadth is π, and the end is maintained at temperature v_0 at all points and the edge at temperature zero. Show that the steady-state temperature is given by

$$v = \frac{4v_0}{\pi} (e^{-y} \sin x + \tfrac{1}{3} e^{-3y} \sin 3x + \cdots)$$

Show by any method that this is equivalent to

$$v = \frac{2v_0}{\pi} \tan^{-1} \frac{\sin x}{\sinh y}$$

4. Find the temperature for a steady flow of heat in a semicircular plate of radius r. The circumference is kept at a temperature v_0 and the diameter at a temperature zero.

5. A long rectangular plate of width 1 cm with insulated surfaces has its temperature ϕ equal to zero on both the long sides and one of the short sides so that

$$v(0,y) = 0 \qquad v(a,y) = 0 \qquad v(x, \infty) = 0$$
$$v(x,0) = kx$$

Show that the steady-state temperature within the plate is given by

$$v(x,y) = \frac{2ak}{\pi} \sum_{n=1}^{\pi} \frac{(-1)^n}{n} e^{-n\pi y/a} \sin \frac{n\pi x}{a}$$

6. Show that, when the values of a potential function on the boundary of a circle of radius R are given by $v(R,\theta) = F(\theta)$, the potential at any interior point is given by

$$v(r,\theta) = \frac{1}{\pi} \int_{-\pi}^{\pi} \left[\frac{1}{2} + \sum_{n=1}^{\infty} \left(\frac{r}{R} \right)^n \cos n(\theta - u) \right] F(u) du$$

7. If the potential is a constant v_0 on the spherical surface of radius R, show that $v = v_0$ at all interior points and $v = v_0 R/r$ at each exterior point.

8. Find the steady temperatures inside a solid sphere of unit radius if one hemisphere of its surface is kept at temperature zero and the other at temperature unity.

9. Find the steady temperature inside a solid sphere of unit radius if the temperature of its surface is given by $U_0 \cos \theta$.

10. Find the gravitational potential due to a uniform circular disk of mass M and unit radius.

11. A square plate has its faces and its edge $y = 0$ insulated. Its edges $x = 0$ and $x = \pi$ are kept at temperature zero, and its edge $y = \pi$ at temperature $F(x)$. Show that its steady temperature is given by

$$v(x,y) = \frac{2}{\pi} \sum_{n=1}^{\infty} \frac{\cosh ny}{\cosh n\pi} \sin nx \int_0^{\pi} F(u) \sin nu \, du$$

12. Show that the steady temperature $v(r,z)$ in a solid cylinder bounded by the surfaces $r = 1$, $z = 0$, and $z = L$, where the first surface is insulated, the second kept at temperature zero, and the last at a temperature $F(r)$, is

$$v(r,z) = \frac{2z}{L} \int_0^1 uF(u) \, du + 2 \sum_{j=2}^{\infty} \frac{J_0(m_j r) \sinh m_j z}{J_0(m_j)^2 \sinh m_j L} \int_0^1 uJ_0(m_j u) F(u) \, du$$

where $m_2, m_3, \ldots$ are the positive roots of $J_1(m) = 0$.

13. Show that the surface $v = F(x,y,z) = C$ can be equipotential if $\nabla^2 C/(\nabla C)^2$ is a function of C only.

REFERENCES

1893. Byerly, W. E.: "Fourier Series and Spherical Harmonics," Ginn and Company, Boston, Mass.

1921. Carslaw, H. S.: "Mathematical Theory of Heat Conduction in Solids," The Macmillan Company, New York.

1925. Jeans, J. H.: "The Mathematical Theory of Electricity and Magnetism," Cambridge University Press, New York.

1927. MacRobert, T. M.: "Spherical Harmonics," E. P. Dutton & Co., Inc., New York.

1929. Kellog, O. D.: "Foundations of Potential Theory," Springer-Verlag OHG, Berlin.

1930. MacMillan, W. D.: "The Theory of the Potential," McGraw-Hill Book Company, New York.

1934. McLachlan, N. W.: "Bessel Functions for Engineers," Oxford University Press, New York.

1939. Smythe, W. R.: "Static and Dynamic Electricity," McGraw-Hill Book Company, New York.

1941. Churchill, R. V.: "Fourier Series and Boundary Value Problems," McGraw-Hill Book Company, New York.

10
The Solution of Two-dimensional Potential Problems by the Method of Conjugate Functions

1 INTRODUCTION

Many problems of applied mathematics involve the determination of two-dimensional vector fields that are *irrotational* and *solenoidal*. If we let **A** be such a vector field, then we see in Appendix E, Sec. 10, that if

$$\nabla \times \mathbf{A} = 0 \tag{1.1}$$

then

$$\mathbf{A} = \nabla V \tag{1.2}$$

That is, an irrotational field may be obtained by taking the gradient of a certain scalar function V.

If the vector field is also solenoidal, then

$$\nabla \cdot \mathbf{A} = \nabla \cdot (\nabla V) = \nabla^2 V = 0 \tag{1.3}$$

If the vector field **A** is two-dimensional, then

$$\mathbf{A} = iA_x + jA_y \tag{1.4}$$

357

where i is the unit vector in the x direction, j is the unit vector in the y direction, and A_x, A_y are the components of $\mathbf{A}$ in the x and y directions. In this case Eq. (1.3) reduces to

$$\nabla^2 V = \frac{\partial^2 V}{\partial x^2} + \frac{\partial^2 V}{\partial y^2} = 0 \qquad (1.5)$$

which is Laplace's equation in two dimensions.

In Chap. 9 we have solved Eq. (1.5) by the method of separation of variables. In this chapter the solution of (1.5) by complex-function theory will be discussed.

This method is particularly useful in determining the distribution of the following typical vector fields:

1. An electrostatic field between parallel cylindrical conductors of various shapes.
2. An electric current in a uniform conducting sheet.
3. The streamlines of fluid flowing in two dimensions through an unobstructed channel when the flow is irrotational.
4. The flow of heat in two dimensions through a homogeneous material that has arrived at a steady state.
5. The magnetic field around a straight conductor carrying current in the neighborhood of parallel ferromagnetic masses.
6. The streamlines of fluid flowing in two dimensions through a homogeneous porous obstructing medium when the flow has reached a steady state.

2 CONJUGATE FUNCTIONS

We have seen in Chap. 1 that if a function

$$w(z) = u + jv \qquad (2.1)$$

is an analytic function of the complex variable

$$z = x + jy \qquad (2.2)$$

then the real part u and the imaginary part v of $w(z)$ satisfy the Cauchy-Riemann equations

$$\frac{\partial u}{\partial x} = \frac{\partial v}{\partial y} \qquad \frac{\partial u}{\partial y} = -\frac{\partial v}{\partial x} \qquad (2.3)$$

and that as a consequence of these equations we have

$$\frac{\partial^2 u}{\partial x^2} + \frac{\partial^2 u}{\partial y^2} = 0 \qquad (2.4)$$

$$\frac{\partial^2 v}{\partial x^2} + \frac{\partial^2 v}{\partial y^2} = 0 \qquad (2.5)$$

so that the real and imaginary parts of $w(z)$ satisfy Laplace's differential equation in two dimensions. The functions u and v are called *conjugate functions*.

ORTHOGONALITY CONDITIONS

Let us construct the two families of curves

$$u(x,y) = c_1 \tag{2.6}$$

and

$$v(x,y) = c_2 \tag{2.7}$$

If x_0, y_0 is a point of intersection of these two curves, the slopes of the tangent lines at x_0, y_0 are, respectively,

$$\left(\frac{dy}{dx}\right)_{u=\text{const}} = -\frac{\partial u/\partial x}{\partial u/\partial y} \tag{2.8}$$

$$\left(\frac{dy}{dx}\right)_{v=\text{const}} = -\frac{\partial v/\partial x}{\partial v/\partial y} \tag{2.9}$$

But as a consequence of the Cauchy-Riemann equations (2.3), we have

$$\left(\frac{dy}{dx}\right)_{u=\text{const}} = \frac{-1}{(dy/dx)_{v=\text{const}}} \tag{2.10}$$

Hence the two slopes are negative reciprocals, and therefore the two curves intersect at right angles; that is, every curve of one family intersects every curve of the other family at right angles. This is expressed by saying that the families of curves corresponding to two conjugate functions form an orthogonal system.

As an example of this, consider the function

$$w(z) = \ln z = u + jv \tag{2.11}$$

If we write z in the polar form

$$z = r\,e^{j\theta} \tag{2.12}$$

we have

$$\ln r\,e^{j\theta} = \ln r + j\theta = u + jv \tag{2.13}$$

Hence

$$u = \ln \sqrt{x^2 + y^2} \tag{2.14}$$

and

$$v = \theta = \tan^{-1}\frac{y}{x} \tag{2.15}$$

are conjugate functions.　The curves

$$u = \ln \sqrt{x^2 + y^2} = \text{const} \tag{2.16}$$

are a family of circles, and the curves

$$v = \tan^{-1} \frac{y}{x} = \text{const} \tag{2.17}$$

are a family of straight lines passing through the origin and intersecting the family of circles at right angles.

3　CONFORMAL REPRESENTATION

Let
$$w(z) = u + jv \qquad z = x + jy \tag{3.1}$$

We can represent values of z in one Argand diagram called the z plane and values of w in another called the w plane, as shown in Fig. 3.1.

Any point P in the z plane corresponds to a definite value of z and, hence, by Eq. (3.1) may give one or more values of w depending on whether $w(z)$ is or is not a single-valued function of z.

If Q is a point in the w plane that represents one of these values of w, the points P and Q are said to *correspond*.

As P describes some curve S in the z plane, the point Q in the w plane that corresponds to P will describe some curve S_0 in the w plane and the curve S_0 is said to correspond to the curve S.

In particular, corresponding to any infinitesimal linear path PP' in the z plane, there will correspond a small linear element QQ' in the w plane. If we let the element PP' represent dz, then QQ' will represent dw, or

$$dw = \left(\frac{dw}{dz} \right) dz \tag{3.2}$$

if dw/dz exists and is nonzero.

In general dw/dz is complex and may be written in the polar form

$$\frac{dw}{dz} = a\, e^{j\phi} \tag{3.3}$$

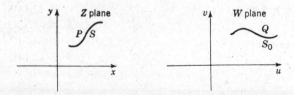

Fig. 3.1

where

$$a = \left| \frac{dw}{dz} \right| \tag{3.4}$$

and

$$\phi = \arg \frac{dw}{dz} \tag{3.5}$$

We can then write

$$dw = a\,e^{j\phi}\,dz \tag{3.6}$$

We thus find that the element dw may be obtained from the corresponding element dz by multiplying its length by a and turning it through an angle ϕ. It thus follows that any element of area in the z plane is represented in the w plane by an element of area that has the same form as the original element but whose linear dimensions are a times as great and whose orientation is obtained by turning the original element through an angle ϕ.

Because of the fact that the forms of two corresponding elements are the same, the process of passing from one plane to the other is known as *conformal representation*.

As an example of the above general principles, let us consider the function

$$w = \ln z = u + jv \tag{3.7}$$

Here we have

$$dw = \frac{dz}{z} \tag{3.8}$$

so that

$$a = \left| \frac{dw}{dz} \right| = \left| \frac{1}{z} \right| = \frac{1}{\sqrt{x^2 + y^2}} \tag{3.9}$$

$$\phi = -\tan^{-1} \frac{y}{x} \tag{3.10}$$

$$u = \ln \sqrt{x^2 + y^2} = \ln r \tag{3.11}$$

$$v = \tan^{-1} \frac{y}{x} \tag{3.12}$$

We thus see that a circular region in the z plane transforms conformally into a rectangular region in the w plane, as shown in Fig. 3.2. This may be seen from the relations

$$r = e^u \tag{3.13}$$

and

$$\tan v = \frac{y}{x} \tag{3.14}$$

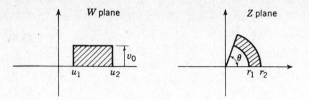

Fig. 3.2

We thus have the following values for the two radii:

$$r_1 = e^{u_1} \quad \text{and} \quad r_2 = e^{u_2} \tag{3.15}$$

and the angle θ is given by

$$\theta = \tan v_0 \tag{3.16}$$

The quantity a, which, as we have seen, measures the linear magnification produced in a small area on passing from the z plane to the w plane, is called the *modulus of the transformation.*

Now we have

$$\frac{\partial w}{\partial x} = \frac{dw}{dz}\frac{\partial z}{\partial x} \tag{3.17}$$

But

$$\frac{\partial z}{\partial x} = \frac{\partial}{\partial x}(x + jy) = 1 \tag{3.18}$$

Hence

$$\frac{dw}{dz} = \frac{\partial w}{\partial x} = \frac{\partial}{\partial x}(u + jv) \tag{3.19}$$

Therefore

$$\frac{dw}{dz} = \frac{\partial u}{\partial x} + j\frac{\partial v}{\partial x} \tag{3.20}$$

By the use of the Cauchy-Riemann equations this may be written in the form

$$\frac{dw}{dz} = \frac{\partial v}{\partial y} + j\frac{\partial v}{\partial x} \tag{3.21}$$

Hence the modulus of the transformation, a, may be written in the form

$$a = \left|\frac{dw}{dz}\right| = \sqrt{\left(\frac{\partial v}{\partial x}\right)^2 + \left(\frac{\partial v}{\partial y}\right)^2} \tag{3.22}$$

It is to be noted that the conformal property fails at points where a is zero or infinity.

4   BASIC PRINCIPLES OF ELECTROSTATICS

To show the utility of the method of conjugate functions in the solution of potential problems, the method will first be used to solve certain two-dimensional electrostatic problems.

As explained in Appendix E, Sec. 16, the basic equations of electrostatics are

$$\nabla \times \mathbf{E} = 0 \tag{4.1}$$

$$\nabla \cdot \mathbf{D} = \rho \tag{4.2}$$

$$\mathbf{D} = K\mathbf{E} \tag{4.3}$$

where $\mathbf{E}$ = electric intensity vector, volts per m

$\mathbf{D}$ = electric displacement vector, coul per m^2

ρ = charge density, coul per m^3

$K = K_r K_0$ = electric inductive capacity of medium

K_r = relative dielectric constant

$K_0 = 8.854 \times 10^{-12}$ farad per m

Since the curl of the vector field $\mathbf{E}$ vanishes, it is customary to let

$$\mathbf{E} = -\text{grad}\,\phi = -\nabla\phi \tag{4.4}$$

where ϕ is the electric potential.   Hence

$$\mathbf{D} = K\mathbf{E} = -K\nabla\phi \tag{4.5}$$

and, substituting this into Eq. (4.2), we have

$$\nabla \cdot \mathbf{D} = \nabla \cdot (-K\nabla\phi) = \rho \tag{4.6}$$

or

$$\nabla^2 \phi = -\frac{\rho}{K} \tag{4.7}$$

in a homogeneous medium.   This is Poisson's equation.   If the region under consideration is free of charges, $\rho = 0$ and Eq. (4.6) becomes Laplace's equation.

SURFACE CHARGE

Let us consider a solid conducting body as shown in Fig. 4.1.   If a charge Q is placed on the body, the electricity distributes itself so that the field inside the body is zero.   Hence the surface of the body is an equipotential surface, and by Eq. (4.2) we see that all the charge is on the surface of the body.   Also the electric field $\mathbf{E}$ has only a normal component to the surface of the body at the surface.

To determine the distribution of electricity per unit area, apply Gauss's theorem to a small disk-shaped volume as shown in the figure.   Since the

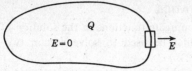

Fig. 4.1

only component of **E** is the one having the direction of the outward-drawn normal to the surface of the body, we have

$$\iint\limits_A \mathbf{D}\cdot d\mathbf{s} = \iint\limits_A Dn\,dA = \iiint (\nabla\cdot\mathbf{D})\,dv = \iiint \rho\,dv = Q \tag{4.8}$$

where A is the area of the disk and Q is the entire charge contained within the disk-shaped volume, and A has been assumed so small that D is sensibly constant throughout A.

Now if we let

$$\lim_{A\to 0} \frac{\displaystyle\iint\limits_A \mathbf{D}\cdot d\mathbf{s}}{A} = \lim_{A\to 0} \frac{Q}{A} = \sigma \tag{4.9}$$

where σ is the surface density, we have from (4.9)

$$\sigma = D_N \tag{4.10}$$

where D_N is the normal component of **D** at the surface of the body. Hence σ, the surface density of charge, may be determined by computing the absolute value of the vector **D** at the surface of the conductor.

In the usual electrostatic problem we are given the geometrical configuration of the charged conducting bodies and the total charge on each body (Fig. 4.2), and we are asked to find the charge density on the several conducting bodies and the electric field in the space outside the bodies.

Let Q_r be the total charge on the rth conducting body. This charge resides entirely on the surface of the rth conductor. The entire conducting body r is at one potential, and its surface is an equipotential surface at potential ϕ_r.

We must therefore solve Laplace's equation (4.7) subject to the condition that the potential ϕ should reduce to ϕ_1, ϕ_2, ϕ_r, etc., at the surface of the first,

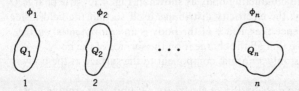

Fig. 4.2

second, and rth conducting body. In this chapter we concern ourselves with the solution of two-dimensional electrostatic problems.

Theoretically a two-dimensional electrostatic problem can never occur, since all conductors are finite. However, there are a great number of important cases in which the surfaces of the conducting bodies are cylindrical and can be generated by moving an infinite straight line parallel to some fixed straight line. If the lengths of the parallel cylindrical conductors are so great compared with the intervening spaces that the end effects are negligible, the problem becomes two-dimensional.

Since the real and imaginary parts of the analytic function $w(z)$ of the complex variable z satisfy Laplace's equation in two dimensions, it is natural that the use of conjugate functions should find application in the solution of two-dimensional electrostatic problems.

It is clear that if we are given a function $w(z)$ then we are assured that the real part u and the imaginary part v of $w(z)$ satisfy the two-dimensional Laplace equations

$$\frac{\partial^2 u}{\partial x^2} + \frac{\partial^2 u}{\partial y^2} = 0 \tag{4.11}$$

and

$$\frac{\partial^2 v}{\partial x^2} + \frac{\partial^2 v}{\partial y^2} = 0 \tag{4.12}$$

Hence either the function u or the function v may be taken as the electric potential function ϕ; if, for example, we let

$$\phi = u \tag{4.13}$$

then we have solved the electrostatic problem in which the family of curves

$$u = \text{const} \tag{4.14}$$

are the equipotential surfaces. To illustrate this, consider the function

$$w = A \ln z + c = u + jv \tag{4.15}$$

where A and c are real constants. Since

$$\ln z = \ln r + j\theta \tag{4.16}$$

where

$$r = \sqrt{x^2 + y^2} \qquad \tan \theta = \frac{y}{x} \tag{4.17}$$

we have

$$u = A \ln r + c \tag{4.18}$$

$$v = A\theta \tag{4.19}$$

If we take u for the electric potential function ϕ, we have

$$\phi = u = A \ln r + c \tag{4.20}$$

We may use this transformation to solve the electrostatic problem of two concentric circular cylinders as shown in Fig. 4.3. Since the curves $\phi = \text{const}$ are circles in the z plane, let us consider the problem of determining the electric field in the region between two concentric cylinders, one of radius a at potential ϕ_1, and the other of radius b at potential ϕ_2.

We thus have from Eq. (4.20)

$$\phi_1 = A \ln a + c \tag{4.21}$$

and

$$\phi_2 = A \ln b + c \tag{4.22}$$

Subtracting Eq. (4.21) from (4.22), we have

$$\phi_2 - \phi_1 = A \ln b - A \ln a = A \ln \frac{b}{a} \tag{4.23}$$

Hence

$$A = \frac{\phi_2 - \phi_1}{\ln (b/a)} \tag{4.24}$$

From (4.21), we have

$$c = \phi_1 - A \ln a = \phi_1 + \frac{\phi_1 - \phi_2}{\ln b - \ln a} \ln a$$

$$= \frac{\phi_2 \ln a - \phi_1 \ln b}{\ln a - \ln b} \tag{4.25}$$

Hence the potential in the region within the concentric cylindrical conductors is given by

$$\phi = \frac{\phi_2 - \phi_1}{\ln (b/a)} \ln r - \frac{\phi_2 \ln a - \phi_1 \ln b}{\ln (b/a)} \qquad 0 < a < b \tag{4.26}$$

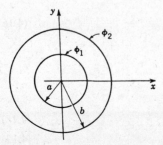

Fig. 4.3

The electric intensity $\mathbf{E}$ is given by

$$\mathbf{E} = -\nabla\phi = -\frac{\partial\phi}{\partial r} \tag{4.27}$$

Now, if q is the charge per unit length on the inner cylinder, we have

$$q = 2\pi a\sigma \tag{4.28}$$

where σ is the surface density of charge on the inner cylinder. Differentiating (4.20), we obtain

$$\mathbf{E} = -\frac{\partial\phi}{\partial r} = -\frac{A}{r} \tag{4.29}$$

If K is the dielectric constant of the medium within the cylinder, then we have

$$\mathbf{D} = K\mathbf{E} = -K\frac{\partial\phi}{\partial r} = -\frac{KA}{r} \tag{4.30}$$

The value of the normal component of the displacement vector at the surface of the inner cylinder gives the surface density σ, by Eq. (4.10). Hence

$$\sigma = -\frac{KA}{a} = +\frac{q}{2\pi a} \tag{4.31}$$

Hence

$$A = -\frac{q}{2\pi K} \tag{4.32}$$

The potential may then be expressed in the form

$$\phi = -\frac{q}{2\pi K}\ln r + c \tag{4.33}$$

In terms of the charge per unit length q of the inner cylinder the potentials of the inner and outer cylinders are

$$\phi_1 = -\frac{q}{2\pi K}\ln a + c \tag{4.34}$$

and

$$\phi_2 = -\frac{q}{2\pi K}\ln b + c \tag{4.35}$$

The *capacitance* per unit length of the capacitor formed from the two cylindrical conductors is defined to be

$$\frac{\text{Capacitance}}{\text{Length}} = \frac{q}{\phi_1 - \phi_2} = \frac{2\pi K}{\ln(b/a)} = c_0 \tag{4.36}$$

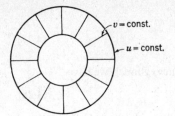

$v = \text{const.}$

$u = \text{const.}$

Fig. 4.4

This simple problem illustrates the use of the theory of functions in solving a potential problem. In this case we have chosen the real part u of the function $w = A \ln z + c$ to be the potential function ϕ. The curves $v = \text{const}$ are the lines of force. They are radial lines emanating from the center of the cylinders as shown in Fig. 4.4.

THE CHARGE ON THE SURFACE OF A CONDUCTOR

The charge on a strip of unit width between any two points P, Q on the surface of a conductor on which the potential $\phi = v$ is constant may be determined in a very simple manner.

As a consequence of Eqs. (4.10) and (4.5) we have, for the charge density σ on a conductor, the equation

$$\sigma = -K\left(\frac{\partial \phi}{\partial n}\right)_s = -K\left(\frac{\partial v}{\partial n}\right)_s \tag{4.37}$$

where $(\partial v/\partial n)_s$ denotes the normal derivative of v evaluated at the surface s, where v is constant. Now, by the Cauchy-Riemann equations and Eq. (3.21), we have

$$\frac{dw}{dz} = \frac{\partial v}{\partial y} + j\frac{\partial v}{\partial x} = \frac{\partial u}{\partial x} - j\frac{\partial u}{\partial y} \tag{4.38}$$

Hence

$$\left|\frac{dw}{dz}\right| = \frac{\partial v}{\partial n} = -\frac{\partial u}{\partial s} \tag{4.39}$$

where dn is an element of length *normal* to an equipotential line, $v = \phi = \text{const}$, and ds is an element of length *parallel* to an equipotential line. The sign is chosen so that if one faces in the direction of n the positive direction of s is to the left as shown in Fig. 4.5.

The total charge on a strip of unit width between any two points P, Q of the conductor is therefore

$$\int_P^Q \sigma \, ds = -K \int_P^Q \left(\frac{\partial v}{\partial n}\right) ds = K \int_P^Q \frac{\partial u}{\partial s} ds = K(u_Q - u_P) \tag{4.40}$$

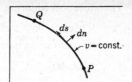

Fig. 4.5

If the surfaces $v_1 = $ const and $v_2 = $ const are closed surfaces and all the charge is situated on one side of one surface and on the opposite side of the other so that all lines of force $u = $ const in the region between the surfaces v_1 and v_2 pass from one surface to the other, then the surfaces form a capacitor. The total charge per unit width on either surface of this capacitor is

$$q = \oint \sigma \, ds = K \oint du = K[u] \tag{4.41}$$

where $[u]$ represents the increment in going once around the $v_1 = $ const curve. The potential difference is

$$|v_1 - v_2| = \varDelta v \tag{4.42}$$

The capacitance per unit length of the capacitor is

$$c_0 = \frac{q}{\varDelta v} = \frac{K[u]}{|v_2 - v_1|} \tag{4.43}$$

5   THE TRANSFORMATION $z = k \cosh w$

A very interesting and instructive transformation is

$$z = k \cosh w = k \cosh(u + jv) \tag{5.1}$$

where k is a real constant.   To study this transformation, we must determine the curves $u = $ const and $v = $ const.   Expanding $\cosh(u + jv)$ into its real and imaginary parts, we obtain

$$x + jy = k(\cosh u \cos v + j \sinh u \sin v) \tag{5.2}$$

Hence

$$\begin{aligned} x &= k \cosh u \cos v \\ y &= k \sinh u \sin v \end{aligned} \tag{5.3}$$

This may be written in the form

$$\begin{aligned} \cos v &= \frac{x}{k \cosh u} \\[2mm] \sin v &= \frac{y}{k \sinh u} \end{aligned} \tag{5.4}$$

Therefore, on squaring these equations and adding the results, we have

$$\frac{x^2}{k^2 \cosh^2 u} + \frac{y^2}{k^2 \sinh^2 u} = 1 \tag{5.5}$$

If we let

$$a = k \cosh u$$
$$b = k \sinh u \tag{5.6}$$

Eq. (5.5) may be written in the form

$$\frac{x^2}{a^2} + \frac{y^2}{b^2} = 1 \tag{5.7}$$

This is the equation of an ellipse with its center at the origin having a major axis of length $2a$ and a minor axis of length $2b$ as shown in Fig. 5.1.

The focal distance F is given by

$$F = \sqrt{a^2 - b^2} = \sqrt{k^2(\cosh^2 u - \sinh^2 u)} = k \tag{5.8}$$

Hence the curves $u = \text{const}$ are a family of *confocal ellipses* all having the focal distance k. If

$$u = 0 \qquad a = k \qquad b = 0 \tag{5.9}$$

This represents a degenerate ellipse and is a straight line extending over the range $-k \leqslant x \leqslant k$.

Now

$$\lim_{u \to \infty} \frac{b}{a} = \lim_{u \to \infty} \tanh u = 1 \tag{5.10}$$

Hence, as $u \to \infty$, the ellipses become more and more nearly circular. Therefore all the z plane is covered if u takes the range

$$0 \leqslant u \leqslant \infty \tag{5.11}$$

To obtain the curves $v = \text{const}$, we write Eqs. (5.4) in the form

$$\cosh u = \frac{x}{k \cos v} \qquad \sinh u = \frac{y}{k \sin v} \tag{5.12}$$

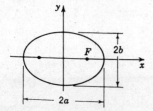

Fig. 5.1

Hence

$$\cosh^2 u - \sinh^2 u = \frac{x^2}{k^2 \cos^2 v} - \frac{y^2}{k^2 \sin^2 v} = 1 \qquad (5.13)$$

If we let

$$a' = k \cos v \qquad b' = k \sin v \qquad (5.14)$$

we have

$$\frac{x^2}{a'^2} - \frac{y^2}{b'^2} = 1 \qquad (5.15)$$

We thus see that the curves $v = $ const represent a family of hyperbolas, as shown in Fig. 5.2. The asymptotes of the hyperbolas are given by the equation

$$\frac{x}{a'} = \frac{y}{b'} \qquad (5.16)$$

The focal distance F is given by

$$F = \sqrt{a'^2 + b'^2} = k \sqrt{\cos^2 v + \sin^2 v} = k \qquad (5.17)$$

Hence the curves $v = $ const represent a confocal family of hyperbolas having the same foci as the family of ellipses $u = $ const.

The angle θ of the asymptote is given by

$$\tan \theta = \frac{y}{x} = \frac{b'}{a'} = \tan v \qquad (5.18)$$

If $v = 0$, we obtain a straight line starting from the focus F and extending to infinity along the x axis. This is the case of a degenerate hyperbola.

We thus see from the geometrical configuration of the curves $u = $ const and $v = $ const that with the proper choice of u or v to represent the potential function the transformation $z = k \cosh w$ gives the solution to the following problems:

1. The electric field around a charged elliptic conducting cylinder, as shown in Fig. 5.3. Here we let

$$\phi = u \qquad (5.19)$$

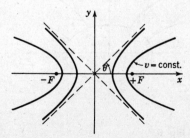

Fig. 5.2

The equipotential surfaces are now a family of confocal ellipses, and the lines of force are a family of confocal hyperbolas.

Consider a capacitor formed of a flat strip and an elliptic cylinder surrounding this strip, as shown in Fig. 5.4. Let the width of the flat strip be $2d$, where

$$d = \sqrt{a^2 - b^2} \tag{5.20}$$

The flat strip is then given by the degenerate ellipse $u = 0$.

Let it be required to find the capacitance c_0 of this capacitor. We determine c_0 by Eq. (4.43) modified to fit this choice of potential:

$$c_0 = \frac{K[v]}{|u_2 - u_1|} \tag{5.21}$$

In this case we have

$$u_1 = 0 \qquad \cosh u_2 = \frac{a}{k} = \frac{a}{d} \tag{5.22}$$

Hence

$$u_2 = \cosh^{-1} \frac{a}{d} \tag{5.23}$$

and

$$[v] = 2\pi \tag{5.24}$$

Hence

$$c_0 = \frac{2\pi K}{\cosh^{-1}(a/d)} \tag{5.25}$$

This is the capacitance per unit length of such a capacitor.

2. The electric field between two semi-infinite conducting plates with a slit separating them, as shown in Fig. 5.5. The two coplanar plates are degenerate hyperbolas.

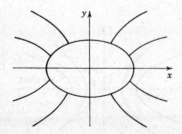

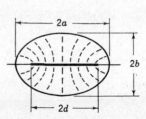

Fig. 5.3   **Fig. 5.4**

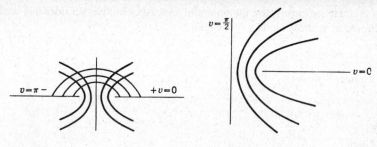

Fig. 5.5   Fig. 5.6

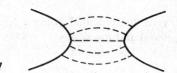

Fig. 5.7

3. The field between perpendicular semi-infinite plates separated by a gap, as shown in Fig. 5.6.
4. The field between two hyperbolic pole pieces, as shown in Fig. 5.7.

6   GENERAL POWERS OF z

Another instructive transformation is the transformation

$$w = Az^n = u + jv \tag{6.1}$$

where A is a real constant and n is a real number.

In order to obtain the curves $u = \text{const}$ and $v = \text{const}$, it is convenient to write z^n in the polar form

$$z^n = (r\,e^{j\theta})^n = r^n(\cos n\theta + j\sin n\theta) \tag{6.2}$$

Hence

$$u = Ar^n\cos n\theta$$
$$v = Ar^n\sin n\theta \tag{6.3}$$

where

$$r^2 = x^2 + y^2 \quad \text{and} \quad \tan\theta = \frac{y}{x} \tag{6.4}$$

If we now take v to be the potential function so that $\phi = v$, we see that

$$v = 0 \quad \begin{cases} \theta = 0 \\ n\theta = \pi \end{cases} \tag{6.5}$$

Hence in this case the potential v vanishes on the two sides of a wedge of angle ϕ, as shown in Fig. 6.1, where

$$n\phi = \pi \tag{6.6}$$

Hence, if we let

$$n = \frac{\pi}{\phi} \tag{6.7}$$

then

$$v = Ar^{\pi/\phi} \sin\frac{\pi\theta}{\phi} \tag{6.8}$$

gives the potential inside a conducting wedge whose sides are the surfaces $\theta = 0$ and $\theta = \phi$.

The function

$$u = Ar^{\pi/\phi} \cos\frac{\pi\theta}{\phi} \tag{6.9}$$

gives the lines of force in the region inside the wedge. To determine the surface density σ on the bottom plate, we have

$$\sigma = -K\left|\frac{dw}{dz}\right|_{\text{on }\theta=0} \tag{6.10}$$

Now

$$\frac{dw}{dz} = \frac{d}{dz}(Az^n) = nAz^{n-1}$$

$$= \frac{\pi}{\phi}Az^{(\pi/\phi-1)} \tag{6.11}$$

Hence

$$\sigma = \frac{-K\pi}{\phi}Ar^{(\pi/\phi-1)} \tag{6.12}$$

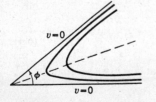

Fig. 6.1

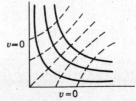

Fig. 6.2

As a special case let $\phi = \pi/2$. In this case we have the right-angle wedge shown in Fig. 6.2. We now have

$$v = Ar^2 \sin 2\theta = Ar^2 \times 2 \sin \theta \cos \theta$$

$$= 2Ar^2 \frac{y}{r} \frac{x}{r} = 2Axy \tag{6.13}$$

and

$$u = Ar^2 \cos 2\theta = Ar^2(\cos^2 \theta - \sin^2 \theta)$$

$$= A(x^2 - y^2) \tag{6.14}$$

We thus obtain two families of hyperbolas. The surface density on the lower plate is

$$\sigma = -2KAr \tag{6.15}$$

We thus see that the surface density is zero at the corner and increases linearly as we move away from the corner.

Another interesting case is the one for which $\phi = 2\pi$. This is the case where the wedge has been opened completely, as in Fig. 6.3. We now have

$$v = Ar^{\frac{1}{2}} \sin \frac{\theta}{2}$$
$$u = Ar^{\frac{1}{2}} \cos \frac{\theta}{2} \tag{6.16}$$

To obtain the rectangular form for the curves $u = $ const and $v = $ const in this case, it is more convenient to use (6.1) directly. In this case, $n = \frac{1}{2}$, and (6.1) becomes

$$w = Az^{\frac{1}{2}} = A\sqrt{z} \tag{6.17}$$

or

$$z = \left(\frac{w}{A}\right)^2 \tag{6.18}$$

For simplicity and with no loss of generality in studying the curves $u = $ const and $v = $ const let us take

$$A = \sqrt{2} \tag{6.19}$$

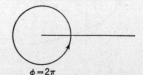

Fig. 6.3 $\phi = 2\pi$

The transformation then becomes

$$z = \frac{w^2}{2} = \frac{(u+jv)^2}{2} \qquad (6.20)$$

Hence

$$x + jy = \frac{u^2 - v^2 + 2juv}{2} \qquad (6.21)$$

or

$$x = \frac{u^2 - v^2}{2} \qquad (6.22)$$

$$y = uv$$

Eliminating u between these two equations, we get

$$y^2 = 2v^2 \left(x + \frac{v^2}{2} \right) \qquad (6.23)$$

These are the curves $v = \text{const}$. It is easy to show that this is a family of parabolas with their foci at the origin, as shown in Fig. 6.4. The parabolas intersect the x axis at

$$x = -\frac{v^2}{2} \qquad (6.24)$$

The parabola $v = 0$ degenerates into the positive side of the x axis. Eliminating v from Eqs. (6.22), we obtain

$$y^2 = 2u^2 \left(\frac{u^2}{2} - x \right) \qquad (6.25)$$

These are the curves $u = \text{const}$. They are a family of confocal parabolas with their foci at the origin and are orthogonal to the curves $v = \text{const}$, as shown in Fig. 6.5.

The curves $v = \text{const}$ give the equipotential lines in the region surrounding a semi-infinite charged plate, while the curves $u = \text{const}$ give the lines of force emanating from a semi-infinite charged plate.

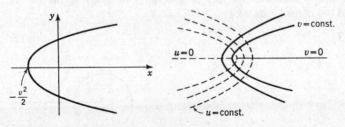

Fig. 6.4   **Fig. 6.5**

7 THE TRANSFORMATION $w = A \ln \dfrac{z-a}{z+a}$

We have seen that the transformation

$$w = -\frac{q}{2\pi K} \ln z = u + jv \tag{7.1}$$

gives the appropriate transformation to study the electric field in the region surrounding a charged circular cylinder with its center at the origin and having a charge q per unit length.   In this case the real part of the transformation, u, is given by

$$u = -\frac{q}{2\pi K} \ln r \tag{7.2}$$

and the imaginary part is given by

$$v = -\frac{q}{2\pi K} \theta \tag{7.3}$$

The family of curves $u = \text{const}$ and $v = \text{const}$ are given by Fig. 7.1. The lines $u = \text{const}$ are a family of concentric circles and represent the equipotentials.   The lines $v = \text{const}$ are a family of straight lines and represent the lines of force.   This transformation also represents the field of a line charge at the origin of charge q per unit length surrounded by a medium of dielectric constant K.

The transformation

$$w = -\frac{q}{2\pi K} \ln (z - a) \tag{7.4}$$

gives the appropriate transformation for a line charge situated at $z = a$, as may be seen by shifting the origin of the transformation (7.1) to $z = a$.

Let us now consider the field produced by a line charge of charge $+q$ per unit length situated at $z = a$ and another line charge of $-q$ units per unit

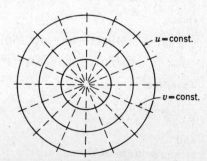

Fig. 7.1

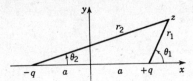

Fig. 7.2

length situated at $z = -a$, as shown in Fig. 7.2. The field produced by both line charges is equivalent to the superposition of the fields given by (7.4) and

$$w = + \frac{q}{2\pi K} \ln(z + a) \tag{7.5}$$

Hence

$$w = -\frac{q}{2\pi K}[\ln(z - a) - \ln(z + a)] = -\frac{q}{2\pi K}\ln\frac{z - a}{z + a} \tag{7.6}$$

represents the proper transformation to determine the field and equipotentials of the two line charges of Fig. 7.2.

Let

$$A = -\frac{q}{2\pi K} \tag{7.7}$$

We then have

$$u + jv = A \ln\frac{z - a}{z + a} \tag{7.8}$$

If we now let the distance from the points $z = a$ and $z = -a$ to the point z be r_1 and r_2, respectively, as shown in Fig. 7.2, we have

$$\begin{aligned} z - a &= r_1 e^{j\theta_1} \\ z + a &= r_2 e^{j\theta_2} \end{aligned} \tag{7.9}$$

where

$$\theta_1 = \arg(z - a) \qquad \theta_2 = \arg(z + a) \tag{7.10}$$

Hence

$$\begin{aligned} u + jv &= A[\ln(z - a) - \ln(z + a)] \\ &= A[\ln r_1 + j\theta_1 - \ln r_2 - j\theta_2] \end{aligned} \tag{7.11}$$

Therefore

$$u = A \ln\frac{r_1}{r_2} \tag{7.12}$$

and

$$v = A(\theta_1 - \theta_2) \tag{7.13}$$

These are the curves $u = \text{const}$ and $v = \text{const}$. Now from (7.12), we have

$$\ln\frac{r_1}{r_2} = \frac{u}{A} \tag{7.14}$$

Therefore

$$\frac{r_1}{r_2} = e^{u/A} \tag{7.15}$$

Hence

$$\left(\frac{r_1}{r_2}\right)^2 = \frac{(x-a)^2 + y^2}{(x+a)^2 + y^2} = e^{2u/A} = k \tag{7.16}$$

Therefore

$$y^2 + \left[x - \frac{a(1+k)}{1-k}\right]^2 = \frac{4a^2 k}{(1-k)^2} \tag{7.17}$$

We thus see that the curves $u = \text{const}$ are a family of circles with centers at

$$y = 0 \qquad x = \frac{a(1+k)}{1-k} \tag{7.18}$$

and radii

$$r = \frac{2a\sqrt{k}}{1-k} \tag{7.19}$$

These circles are shown in Fig. 7.3.

This transformation may be used to find the capacitance of two eccentric cylinders of equal radii R, whose centers are at a distance S apart, as shown in Fig. 7.4.

Since any of the circles $u = \text{const}$ of Fig. 7.3 may be replaced by a circular conductor at potential u, we must adjust the center of one of the $u = \text{const}$ circles to be at $x = S/2$. Hence by Eqs. (7.18), we have

$$\frac{S}{2} = \frac{a(1+k)}{1-k} \tag{7.20}$$

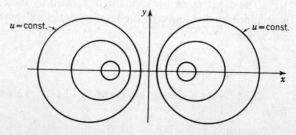

$u = \text{const.}$ $u = \text{const.}$

Fig. 7.3

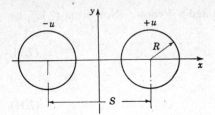

Fig. 7.4

The radius of the circle is fixed by Eq. (7.19) to be

$$R = \frac{2a\sqrt{k}}{1-k} \tag{7.21}$$

Eliminating a between Eqs. (7.20) and (7.21), we have

$$\frac{S}{R} = \frac{1+k}{\sqrt{k}} = \frac{1+e^{2u/A}}{e^{u/A}} \tag{7.22}$$

Therefore

$$\frac{S}{2R} = \frac{e^{u/A}+e^{-u/A}}{2} = \cosh\frac{u}{A} \tag{7.23}$$

By Eq. (7.7), we have

$$\cosh\left(-\frac{2\pi Ku}{q}\right) = \cosh\frac{2\pi Ku}{q} = \frac{S}{2R} \tag{7.24}$$

Hence

$$\frac{2\pi Ku}{q} = \cosh^{-1}\frac{S}{2R} \tag{7.25}$$

or

$$q = \frac{2\pi Ku}{\cosh^{-1}(S/2R)} \tag{7.26}$$

The capacitance per unit length formed by the conductors of Fig. 7.4 is obtained by dividing q by the total potential difference, which is obviously $2u$. Hence

$$C_0 = \frac{q}{2u} = \frac{K\pi}{\cosh^{-1}(S/2R)} \tag{7.27}$$

The curves $v = \text{const}$ are given by (7.13). This may be written in the form

$$\theta_1 - \theta_2 = \frac{v}{A} \tag{7.28}$$

Now

$$\theta_1 = \tan^{-1}\frac{y}{x-a} \qquad \theta_2 = \tan^{-1}\frac{y}{x+a} \tag{7.29}$$

Hence we have

$$\tan^{-1}\frac{y}{x-a} - \tan^{-1}\frac{y}{x+a} = \frac{v}{A} \tag{7.30}$$

We now use the trigonometric identity

$$\tan^{-1}M - \tan^{-1}N = \tan^{-1}\frac{M-N}{1+MN} \tag{7.31}$$

and let

$$M = \frac{y}{x-a} \qquad N = \frac{y}{x+a} \tag{7.32}$$

Then Eq. (7.30) becomes

$$\tan^{-1}\frac{2ya}{x^2 - a^2 + y^2} = \frac{v}{A} \tag{7.33}$$

and hence

$$\frac{2ya}{x^2 - a^2 + y^2} = \tan\frac{v}{A} \tag{7.34}$$

This is a family of circles whose centers are at

$$y = a\cot\frac{v}{A} \qquad x = 0 \tag{7.35}$$

and having radii given by

$$R = a\csc\frac{v}{A} \tag{7.36}$$

These circles are orthogonal to the circles $u = \text{const}$ and represent the lines of force in the electrical application (see Fig. 7.5).

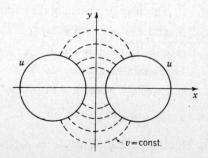

Fig. 7.5

8   DETERMINATION OF THE REQUIRED TRANSFORMATION WHEN THE BOUNDARY IS EXPRESSED IN PARAMETRIC FORM

In the last two sections we have studied transformations that enabled us to solve problems of physical interest. In general if we are given a two-dimensional potential problem arising from some physical investigation, we are required to find a transformation $w(z) = u + jv$ of such nature that either the curves $u = $ const or the curves $v = $ const should coincide with the boundary of the region. It is sometimes possible to express the desired equipotential boundary in parametric form.

Let

$$F(x,y) = 0 \tag{8.1}$$

be the equation of one of the desired equipotential boundaries. Let us suppose that this equation can be expressed in the parametric form

$$x = F_1(t) \qquad y = F_2(t) \tag{8.2}$$

where $F_1(t)$ and $F_2(t)$ are real functions of the real parameter t whose range of variation corresponds to the whole equipotential (8.1).

Then the transformation

$$z = F_1(kw) + jF_2(kw) \tag{8.3}$$

where k is a real constant, gives $v = 0$ over the surface (8.1). To see this, let $v = 0$ in (8.3); we then have

$$z = F_1(ku) + jF_2(ku) = x + jy \tag{8.4}$$

Hence

$$x = F_1(ku) \qquad y = F_2(ku) \tag{8.5}$$

These are exactly the parametric equations (8.2) with ku in the place of the parameter t.

This method enables one to find the proper transformations when the boundaries are confocal conics or various cycloidal curves. For example, consider the parabola with its focus at the origin

$$y^2 = 4a(x + a) \tag{8.6}$$

This may be written in the parametric form

$$(x + a) = at^2 \qquad y = 2at \tag{8.7}$$

Hence the transformation is

$$z = aw^2 - a + 2ajw = a(w + j)^2 \tag{8.8}$$

where we have placed $k = 1$ for simplicity. Therefore

$$w = -j + \left(\frac{z}{a}\right)^{\frac{1}{2}} \tag{8.9}$$

This transformation makes the parabola (8.6) the equipotential surface $v = 0$.

As another example, let us consider the ellipse

$$\frac{x^2}{a^2} + \frac{y^2}{b^2} = 1 \tag{8.10}$$

This may be expressed in the parametric form

$$x = a\cos t \qquad y = b\sin t \tag{8.11}$$

The required transformation is

$$z = a\cos w + jb\sin w \tag{8.12}$$

where we have again placed $k = 1$, for simplicity.   If we let

$$a = c\cosh\phi \qquad b = c\sinh\phi \tag{8.13}$$

then

$$\tanh\phi = \frac{b}{a} \tag{8.14}$$

and

$$c^2 = a^2 - b^2 \tag{8.15}$$

Then (8.12) may be written in the form

$$z = c(\cos w\cosh\phi + j\sin w\sinh\phi)$$
$$= c\cos(w + j\phi) = c\cos[u + j(v + \phi)] \tag{8.16}$$

As another example, suppose it is required to find the field on one side of a corrugated metal sheet whose equation is that of the cycloid

$$x = a(t - \sin t) \qquad y = a(1 - \cos t) \tag{8.17}$$

The required transformation is

$$z = a(w - \sin u) + aj(1 - \cos w)$$
$$= a(w + j - je^{-jw}) \tag{8.18}$$

9   SCHWARZ'S TRANSFORMATION

Schwarz[†] has shown how to obtain a transformation in which one equipotential is a linear polygon.   This is one of the most useful transformations.   It transforms the interior of a polygon in the z plane into the upper half of another plane, say the z_1 plane, in such a manner that the sides of the polygon in the z plane are transformed into the real axis of the z_1 plane.   Schwarz has shown that, given the required polygon, a certain differential equation may be written which when integrated gives directly the desired transformation.

[†] H. A. Schwarz, *Journal für Mathematik*, vol. 70, p. 5.105, 1869.

Consider the expression

$$\frac{dz}{dz_1} = A(z_1 - u_1)^{\phi_1}(z_1 - u_2)^{\phi_2} \cdots (z_1 - u_n)^{\phi_n} \tag{9.1}$$

where A is a complex constant; $u_1, u_2, \ldots, u_n$ and $\phi_1, \phi_2, \ldots, \phi_n$ are real numbers; and

$$u_n > u_{n-1} > \cdots u_2 > u_1 \tag{9.2}$$

Now since the argument of a product of complex numbers is equal to the sum of the arguments of the individual factors, we have

$$\arg\frac{dz}{dz_1} = \arg A + \phi_1 \arg(z_1 - u_1)$$
$$+ \phi_2 \arg(z_1 - u_2) + \cdots + \phi_n \arg(z_1 - u_n) \tag{9.3}$$

Let us now consider the z_1 plane of Fig. 9.1. The real numbers $u_1, u_2, \ldots, u_n$ are plotted on the real axis of the z_1 plane. If z_1 is a real number, then the number

$$N_r = z_1 - u_r \tag{9.4}$$

is positive if z_1 is greater than u_r and it is negative when z_1 is less than u_r. Hence

$$\arg(z_1 - u_r) = \begin{cases} 0 & \text{if } z_1 > u_r \\ \pi & \text{if } z_1 < u_r \end{cases} \tag{9.5}$$

Let us suppose that the point z_1 traverses the real axis of the z_1 plane from left to right, as shown by the arrow in Fig. 9.1. Let

$$\theta_r = \arg\frac{dz}{dz_1} \qquad \text{when } u_r < z_1 < u_{r+1} \tag{9.6}$$

Then, by (9.3) and (9.5), we have

$$\theta_r = \arg A + (\phi_{r+1} + \phi_{r+2} + \cdots + \phi_n)\,\pi \tag{9.7}$$

and

$$\theta_{r+1} = \arg A + (\phi_{r+2} + \phi_{r+3} + \cdots + \phi_n)\,\pi \tag{9.8}$$

Hence

$$\theta_{r+1} - \theta_r = -\pi\phi_{r+1} \tag{9.9}$$

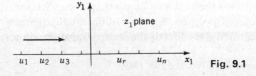

Fig. 9.1

Now

$$\arg\frac{dz}{dz_1} = \arg\frac{dx + j\,dy}{dx_1} = \tan^{-1}\frac{dy}{dx} \tag{9.10}$$

We see that this is the angle that the element dz in the z plane into which dz_1 is transformed by the transformation (9.1) makes with the real axis of the z plane. It is thus seen that as the point z_1 traverses the real axis of the z_1 plane the corresponding point z traverses a polygon in the z plane, as shown in Fig. 9.2.

When the point z_1 passes from the left of u_{r+1} to the right of u_{r+1} in the z_1 plane, then the direction of the point z in the z plane is suddenly changed by an angle of $-\pi\phi_{r+1}$, measured in the mathematically positive sense, as shown in Fig. 9.2.

If we imagine the broken line in Fig. 9.2 to form a closed polygon, then the angle α_{r+1} measured between two adjacent sides of the polygon is called an *interior angle*. We then have

$$\alpha_{r+1} - \pi\phi_{r+1} = \pi \tag{9.11}$$

Hence

$$\phi_{r+1} = \frac{\alpha_{r+1}}{\pi} - 1 \tag{9.12}$$

Substituting this into (9.1), we have

$$\frac{dz}{dz_1} = A(z_1 - u_1)^{\alpha_1/\pi - 1}(z_1 - u_2)^{\alpha_2/\pi - 1} \cdots (z_1 - u_n)^{\alpha_n/\pi - 1}$$

$$= A \prod_{r=1}^{r=n} (z_1 - u_r)^{\alpha_r/\pi - 1} \tag{9.13}$$

Integrating this expression with respect to z_1, we have

$$z = A \int \prod_{r=1}^{n} (z_1 - u_r)^{\alpha_r/\pi - 1}\,dz_1 + B \tag{9.14}$$

where B is an arbitrary constant.

This transformation transforms the real axis of the z_1 plane into a polygon in the z plane. The angles α_r are the interior angles of the polygon.

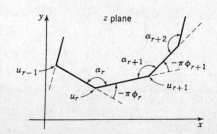

Fig. 9.2

The modulus of the constant A determines the size of the polygon, and the argument of the constant A determines the orientation of the polygon. The location of the polygon is determined by the constant B.

10 POLYGON WITH ONE ANGLE

As the simplest example of the Schwarz transformation (9.14), let us transform the real axis of the z_1 plane into an angle in the z plane, as shown in Fig. 10.1.

In this case the transformation reduces to

$$z = A \int (z_1 - u_1)^{\alpha/\pi - 1} \, dz_1 + B \tag{10.1}$$

For simplicity, let

$$u_1 = 0 \tag{10.2}$$

Therefore

$$z = A \int z_1^{\alpha/\pi - 1} \, dz_1 + B = K z_1^{\alpha/\pi} + B \qquad \alpha \neq 0 \tag{10.3}$$

where K is an arbitrary constant. If we let $B = 0$, we have

$$z = K z_1^{\alpha/\pi} \tag{10.4}$$

This transformation gives the correspondence shown between the two planes.

POLYGON WITH ANGLE ZERO

If we let the interior angle of the polygon, α, be equal to zero in (10.1), we have

$$z = A \int \frac{dz_1}{z_1} + B = A \ln z_1 + B \tag{10.5}$$

Let $B = 0$ and A be a real constant. We then obtain

$$z = A \ln z_1 \tag{10.6}$$

The correspondence between the z_1 and z_2 planes is shown in Fig. 10.2.

To see the correspondence between the planes, let

$$z_1 = r_1 e^{j\theta_1} \tag{10.7}$$

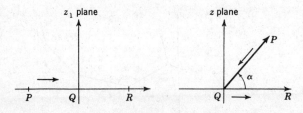

Fig. 10.1

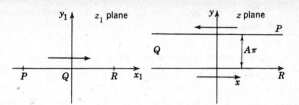

Fig. 10.2

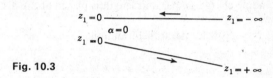

Fig. 10.3

Then

$$x + jy = A(\ln r_1 + j\theta_1) \tag{10.8}$$

Therefore

$$x = A \ln r_1 \qquad y = A\theta_1 \tag{10.9}$$

$\theta_1 = 0$ corresponds to the positive side of the real axis of the z_1 plane and to the entire real axis of the z plane. The origin in the z_1 plane corresponds to the point $x = -\infty$ in the z plane. The negative side of the real axis of the z_1 plane corresponds to the line $y = A\pi$ in the z plane. This transformation in the z plane is a limiting case of the transformation of Fig. 10.3.

11 SUCCESSIVE TRANSFORMATIONS

In solving two-dimensional potential problems, it is frequently convenient to use successive transformations.

Let

$$w = F_1(z_1) \tag{11.1}$$

and

$$z_1 = F_2(z) \tag{11.2}$$

Then, by the elimination of z_1 between (11.1) and (11.2), we obtain

$$w = F_3(z) \tag{11.3}$$

The relation (11.2) expresses a transformation from the z plane into a z_1 plane, while (11.1) expresses a further transformation from the z_1 plane into a w plane. Therefore the final transformation (11.3) may be regarded as the result of two successive transformations.

There are two uses of successive transformations that are of great importance.

a CONDUCTOR INFLUENCED BY A LINE CHARGE

We have seen in Sec. 7 that the transformation

$$w = -\frac{q}{2\pi K}\ln(z_1 - a) + \frac{q}{2\pi K}\ln(z_1 + a) = \frac{-q}{2\pi K}\ln\frac{z_1 - a}{z_1 + a} \qquad (11.4)$$

gives the solution when a line charge q per unit length is placed at $z_1 = a$ and a line charge $-q$ per unit length is placed at the distance $z_1 = -a$, as shown in Fig. 11.1.

Now let the transformation

$$z_1 = F(z) \qquad (11.5)$$

transform the real axis of the z_1 plane into the surface S and the point q into $z = z_0$, as shown in Fig. 11.2.   Then the transformation

$$w = -\frac{q}{2\pi K}\ln\frac{F(z) - F(z_0)}{F(z) + F(z_0)} \qquad (11.6)$$

gives the solution when a line charge q is placed at $z = z_0$ in the presence of the surface S that is at potential $u = 0$. The curves $u = $ const are the equipotentials, and the curves $v = $ const are the lines of force.

b CONDUCTORS AT DIFFERENT POTENTIALS

Consider the three planes of Fig. 11.3.   Let

$$z_1 = F(z) \qquad (11.7)$$

transform the surfaces S_1 and S_2 in the z plane into the real axis of the z_1 plane in such a manner that the surface S_2 is mapped into the negative part of the real axis of the z_1 plane and the surface S_1 into the positive part of the real axis of the z_1 plane.

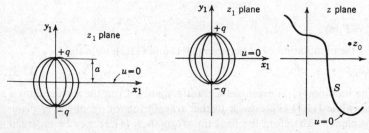

Fig. 11.1   **Fig. 11.2**

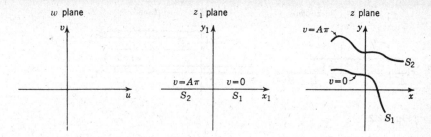

Fig. 11.3

The transformation

$$w = A \ln z_1 = u + jv \tag{11.8}$$

transforms the z_1 plane so that the positive part of the real axis corresponds to $v = 0$ and the negative part of the real axis corresponds to $v = A\pi$. Hence the transformation

$$w = A \ln F(z) \tag{11.9}$$

gives the solution for the problem in which the surface S_1 is at potential $v = 0$ and the surface S_2 is at potential $v = A\pi$. The $u = \text{const}$ curves give the lines of force.

12   THE PARALLEL-PLATE CAPACITOR; FLOW OUT OF A CHANNEL

As an illustration of the use of Schwarz's transformation and the above general principles, let us determine the proper transformation to determine the potential and field distribution of two semi-infinite conducting plates raised to different potentials as shown in Fig. 12.1.

By symmetry it is necessary only to solve the problem of a semi-infinite plane placed parallel to an infinite plane, as shown in Fig. 12.2. Since there are only two angles involved in the polygon in the z plane, we may write the Schwarz transformation equation (9.14) in the form

$$\frac{dz}{dz_1} = K(z_1 - u_1)^{\alpha_1/\pi - 1}(z_1 - u_2)^{\alpha_2/\pi - 1} \tag{12.1}$$

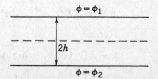

Fig. 12.1

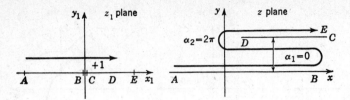

Fig. 12.2

where K is a constant. Since we are free to choose the numbers u_1 and u_2, we choose them so that the expression (12.1) is as simple as possible. The two interior angles of the polygon in the z plane are

$$\alpha_1 = 0 \quad \text{and} \quad \alpha_2 = 2\pi \tag{12.2}$$

Let us take

$$u_1 = 0 \quad \text{and} \quad u_2 = +1 \tag{12.3}$$

Then Eq. (12.1) becomes

$$\frac{dz}{dz_1} = Kz_1^{-1}(z_1 - 1) = \frac{K(z_1 - 1)}{z_1} = K\left(1 - \frac{1}{z_1}\right) \tag{12.4}$$

Integrating, we have

$$z = K(z_1 - \ln z_1) + C \tag{12.5}$$

where C is an arbitrary constant. Since the sides of the polygon in the z plane are horizontal and hence have zero slope, it is necessary for the constant K to be real.

BOUNDARY CONDITIONS

The arbitrary constants K and C must now be determined from the correspondence between the two planes.

1. As the point z_1 moves across the origin from B to C, the point z turns through a zero angle and the imaginary part of z is increased from 0 to h. Let

$$z_1 = r_1 e^{j\theta_1} \tag{12.6}$$

Therefore

$$x + jy = K(r_1 e^{j\theta_1} - \ln r_1 - j\theta_1) + C \tag{12.7}$$

Hence

$$\begin{aligned} x &= K(r_1 \cos\theta_1 - \ln r_1) + \operatorname{Re} C \\ y &= K(r_1 \sin\theta_1 - \theta_1) + \operatorname{Im} C \end{aligned} \tag{12.8}$$

where Re denotes the "real part of" and Im denotes the "imaginary part of."

Now

$$y = 0 \qquad \theta_1 = \pi$$
$$y = h \qquad \theta_1 = 0 \tag{12.9}$$

Hence

$$0 = K(-\pi) + \mathrm{Im}\, C$$
$$h = \mathrm{Im}\, C \tag{12.10}$$

$$K = \frac{h}{\pi} \tag{12.11}$$

2. Let the point $z_1 = 1$ correspond to the point $z = x_0 + jh$. The point $z_1 = 1$ is determined by

$$r_1 = 1 \qquad \theta_1 = 0 \tag{12.12}$$

Therefore

$$x_0 = \frac{h}{\pi} + \mathrm{Re}\, C \tag{12.13}$$

If we choose x_0 so that

$$x_0 = \frac{h}{\pi} \tag{12.14}$$

then we have

$$\mathrm{Re}\, C = 0 \tag{12.15}$$

With this choice of x_0 the constant C is a pure imaginary and from Eqs. (12.10) has the value

$$C = jh \tag{12.16}$$

The transformation (12.5) thus becomes

$$z = \frac{h}{\pi}(z_1 - \ln z_1) + jh$$

$$= \frac{h}{\pi}(z_1 - \ln z_1 + j\pi) \tag{12.17}$$

This transformation transforms the z_1 plane into the polygon in the z plane with the correspondence of points shown in Fig. 12.2.

TRANSFORMATION TO w PLANE

We now transform the z_1 plane to the w plane by the transformation

$$w = \ln z_1 = u + jv \tag{12.18}$$

Since

$$z_1 = r_1 e^{j\theta_1} \tag{12.19}$$

we have

$$u + jv = \ln r_1 + j\theta_1 \tag{12.20}$$

and

$$u = \ln r_1 \qquad v = \theta_1 \tag{12.21}$$

The correspondence between the two planes is shown in Fig. 12.3. By (12.18) we have

$$z_1 = e^w \tag{12.22}$$

If we now eliminate z_1 between Eqs. (12.17) and (12.22), we pass directly from the w plane to the z plane by the equation

$$z = \frac{h}{\pi}(e^w - w + j\pi) \tag{12.23}$$

The correspondence between the two planes is shown in Fig. 12.4. If we let v be the potential function, we have in the z plane a semi-infinite plane at potential zero parallel to an infinite plane at potential π. If we write (12.23) in the form

$$x + jy = \frac{h}{\pi}[e^u(\cos v + j \sin v) - u - jv + j\pi] \tag{12.24}$$

we have

$$x = \frac{h}{\pi}(e^u \cos v - u)$$

$$y = \frac{h}{\pi}(e^u \sin v - v + \pi) \tag{12.25}$$

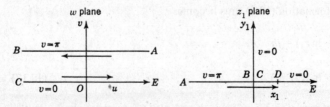

Fig. 12.3

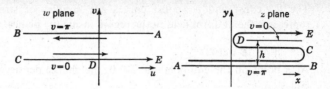

Fig. 12.4

Let $v = 0$. Then

$$x = \frac{h}{\pi}(e^u - u) \qquad y = h \tag{12.26}$$

For

$$u = 0 \qquad u = \infty \qquad u = -\infty$$

$$x = \frac{h}{\pi} \qquad x = \infty \qquad x = \infty \tag{12.27}$$

Negative values of u give the region inside the capacitor, and positive values of u give the region outside the capacitor. If u is large and positive, we have approximately

$$x \doteq \frac{h}{\pi}(e^u \cos v)$$

$$y \doteq \frac{h}{\pi}(e^u \sin v) \tag{12.28}$$

Hence

$$x^2 + y^2 \doteq \frac{h^2}{\pi^2}e^{2u} \qquad u \gg 1 \tag{12.29}$$

The $u = $ const curves are a family of circles, so that the lines of force are approximately circles at a large distance from the origin, as seen in Fig. 12.5. If u is large and negative, we have approximately

$$x \doteq -\frac{h}{\pi}u \qquad y \doteq \frac{h}{\pi}(\pi - v) \tag{12.30}$$

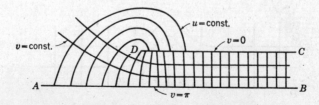

Fig. 12.5

The $u = \text{const}$ and $v = \text{const}$ curves are families of straight lines. This represents the uniform field in the region inside the planes.

THE CHARGE DENSITY

At a point on the equipotential $v = 0$, the charge density σ is given by the equation

$$\sigma = -K \left| \frac{dw}{dz} \right|_{v=0} = -\frac{K\pi}{h} \frac{1}{e^u - 1} \tag{12.31}$$

Inside the capacitor, u is large and negative, and we have approximately

$$\sigma \doteq -\frac{K\pi}{h} \tag{12.32}$$

At D, $u = 0$ and σ becomes infinite. On the upper part of the semi-infinite plane, u is large and positive, and we have

$$\sigma \doteq -\frac{K\pi}{he^u} \tag{12.33}$$

The surface density decreases as we move to the right of the point D.

This transformation also solves the hydrodynamical problem of the irrotational flow of fluid out of a long channel into a large reservoir.

13 THE EFFECT OF A WALL ON A UNIFORM FIELD

As another example of the use of the Schwarz transformation, let it be required to determine the effect produced by an indefinitely thin wall or ridge on a uniform field, as shown in Fig. 13.1.

The solution of this problem also gives the lines of flow of a fluid over an obstacle. We regard the line $PQRST$ as a polygon in the z plane and transform it into the real axis of the z_1 plane, as shown in Fig. 13.2.

Let us establish the following correspondence between the two planes:

$$P \to -\infty \qquad S \to u_3$$
$$Q \to u_1 \qquad T \to +\infty$$
$$R \to u_2$$

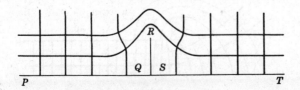

Fig. 13.1

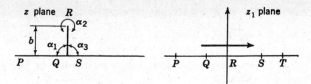

Fig. 13.2

From the figure we have

$$\alpha_1 = \frac{\pi}{2} \qquad \alpha_2 = 2\pi \qquad \alpha_3 = \frac{\pi}{2} \tag{13.1}$$

The Schwarz expression of Sec. 9 becomes

$$\frac{dz}{dz_1} = A(z_1 - u_1)^{\alpha_1/\pi - 1}(z_1 - u_2)^{\alpha_2/\pi - 1}(z_1 - u_3)^{\alpha_3/\pi - 1}$$

$$= A(z_1 - u_1)^{-\frac{1}{2}}(z_1 - u_2)(z_1 - u_3)^{-\frac{1}{2}} \tag{13.2}$$

For simplicity, let

$$u_1 = -a \qquad u_2 = 0 \qquad u_3 = +a \tag{13.3}$$

Then (13.2) becomes

$$\frac{dz}{dz_1} = \frac{Az_1}{\sqrt{z_1^2 - a^2}} \tag{13.4}$$

Integrating, we have

$$z = A \int \frac{z_1 \, dz_1}{\sqrt{z_1^2 - a^2}} + B = A \sqrt{z_1^2 - a^2} + B \tag{13.5}$$

BOUNDARY CONDITIONS

1. When $z_1 = \pm a$ let $z = 0$. Therefore

$$B = 0 \tag{13.6}$$

2. If b is the height of the wall, then, when $z = jb$, let $z_1 = 0$. Therefore

$$jb = A \sqrt{-a^2} = Aaj \tag{13.7}$$

Hence

$$A = \frac{b}{a} \tag{13.8}$$

The transformation

$$z = \frac{b}{a} \sqrt{z_1^2 - a^2} \tag{13.9}$$

or

$$z_1 = \frac{a}{b}\sqrt{z^2 + b^2} \qquad (13.10)$$

transforms the $y_1 = 0$ axis in the z_1 plane into the polygon $PQRST$ in the z plane. Now let

$$w = F_1 z = u + jv \qquad (13.11)$$

This transformation transforms the lines $v = \text{const}$ in the w plane into the lines

$$y_1 = \frac{v}{F_1} \qquad (13.12)$$

in the z_1 plane. Hence

$$w = F_1 z = \frac{F_1 a}{b}\sqrt{z^2 + b^2} \qquad (13.13)$$

in the transformation that transforms the line $v = 0$ into the polygon in the z plane. If the field is to be uniform at large distances from the wall, we must have

$$w = \frac{F_1 a}{b}\sqrt{z^2 + b^2} \doteq \frac{F_1 a}{b} z = Fz \qquad \text{for large } z \qquad (13.14)$$

Hence

$$F = \frac{F_1 a}{b} \qquad (13.15)$$

and the transformation may be written

$$w = F\sqrt{z^2 + b^2} = u + jv \qquad (13.16)$$

where v is the potential function, as shown in Fig. 13.3. Separating (13.16) into its real and imaginary parts, we have

$$u^2 - v^2 = F^2(x^2 - y^2) + F^2 b^2$$
$$uv = F^2 xy \qquad (13.17)$$

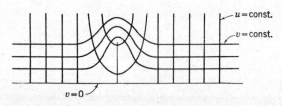

$u = \text{const.}$

$v = \text{const.}$

$v = 0$

Fig. 13.3

Eliminating u between these equations, we have

$$\frac{F^4 x^2 y^2}{v^2} - v^2 = F^2(x^2 - y^2) + F^2 b^2 \tag{13.18}$$

This is the family of equipotentials, and

$$u^2 - \frac{F^4 x^2 y^2}{u^2} = F^2(x^2 - y^2) + F^2 b^2 \tag{13.19}$$

are the lines of force.

In the hydrodynamical case $v = $ const gives lines of flow and $u = $ const gives constant-velocity potentials.

14   APPLICATION TO HYDRODYNAMICS

The theory of conjugate functions is of extreme usefulness in determining two-dimensional-flow distribution of moving fluids.   There is a very close analogy between the electrical problems discussed in the last few sections and problems in hydrodynamics.

Let **A** be the velocity vector field of an ideal nonviscous fluid.   If the flow is irrotational, then

$$\nabla \times \mathbf{A} = 0 \tag{14.1}$$

and, as we see in Appendix E, the velocity **A** is the gradient of a scalar, or

$$\mathbf{A} = -\nabla \phi \tag{14.2}$$

where ϕ is the *velocity potential*.   The lines of flow are everywhere normal to the surfaces $\phi = $ const.   If the fluid is incompressible, then the equation of continuity is

$$\nabla \cdot \mathbf{A} = 0 \tag{14.3}$$

so that

$$\nabla \cdot \mathbf{A} = \nabla(-\nabla \phi) = 0 \tag{14.4}$$

or

$$\nabla^2 \phi = 0 \tag{14.5}$$

so that, as in the case of the electrostatic potential, the velocity potential satisfies Laplace's equation in a region that contains no sources.

The Stream Function.   Let $\Phi(x, y)$ be a function of the two independent variables x and y.   Now the differential $d\Phi$ is given by

$$d\Phi = \frac{\partial \Phi}{\partial x} dx + \frac{\partial \Phi}{\partial y} dy \tag{14.6}$$

For two-dimensional motion the equation of continuity (14.3) reduces to

$$\frac{\partial A_x}{\partial x} + \frac{\partial A_y}{\partial y} = 0 \tag{14.7}$$

If we let

$$A_x = -\frac{\partial \Phi}{\partial y} \qquad A_y = \frac{\partial \Phi}{\partial x} \tag{14.8}$$

then (14.6) becomes

$$d\Phi = A_y \, dx - A_x \, dy \tag{14.9}$$

The condition that this expression be a perfect differential is

$$\frac{\partial A_y}{\partial y} = -\frac{\partial A_x}{\partial x} \tag{14.10}$$

This is exactly the equation of continuity (14.7). Hence from the velocity vector **A** we can determine a unique (to within an additive constant) function Φ, and the velocity components are given by (14.8). The function Φ is called the *stream function*. From Eq. (14.2) we have

$$A_x = -\frac{\partial \phi}{\partial x} = -\frac{\partial \Phi}{\partial y}$$
$$A_y = -\frac{\partial \phi}{\partial y} = \frac{\partial \Phi}{\partial x} \tag{14.11}$$

Hence

$$\frac{\partial \phi}{\partial x}\frac{\partial \Phi}{\partial x} + \frac{\partial \phi}{\partial y}\frac{\partial \Phi}{\partial y} = 0 \tag{14.12}$$

The two families $\phi = \text{const}$ and $\Phi = \text{const}$ intersect orthogonally. It is thus apparent that if we have

$$w = u + jv = w(z) \tag{14.13}$$

a function of the complex variable z, then, if we take u for the potential, v will be the stream function, or vice versa. When we consider fluid problems in which $w(z)$ represents the flow field, the function $w(z)$ is called the *complex potential*. For example if we consider the function

$$w = k \ln z = \phi + j\Phi \tag{14.14}$$

from Sec. 4 we have

$$\phi = k \ln r$$
$$\Phi = k\theta \tag{14.15}$$

where

$$z = r e^{j\theta} \tag{14.16}$$

Thus the stream lines are radial lines emanating from the origin, and the lines of constant ϕ are concentric circles about the origin. From hydrodynamics we know that a stream line is a line tangent everywhere to the velocity field; that is, it describes an infinitesimal flow tube. As a consequence we may state that the flow field represented by (14.14) is flow moving radially outward or toward the origin depending upon the sign of the constant k.

The flow velocity may be found from

$$A_r = -\frac{\partial \phi}{\partial r} = -\frac{k}{r}$$
$$A_\theta = -\frac{1}{r}\frac{\partial \phi}{\partial \theta} \tag{14.17}$$

From this equation it is readily seen that the flow varies inversely with the radius, and that when k is negative, the flow is outward from the origin, that is, *source flow* and when k is positive, the flow is inward, that is, *sink flow* (see Fig. 14.1). The constant k is termed the *source* or *sink strength*.

As a further example of fluid flow, consider the equation

$$w = -jk \ln z \tag{14.18}$$

From this relation we have

$$\phi = k\theta$$
$$\Phi = -k \ln r \tag{14.19}$$

This flow field represents flow in concentric circles about the origin. The radial and tangential velocities are

$$A_r = -\frac{\partial \phi}{\partial r} = 0$$
$$A_\theta = -\frac{1}{r}\frac{\partial \phi}{\partial \theta} = \frac{k}{r} \tag{14.20}$$

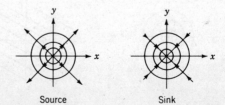

Fig. 14.1   Source   Sink

With a little imagination the reader may recognize this flow field as the mathematical representation of a filament vortex (cf. Fig. 14.2).

Another useful flow field is defined by

$$w = -A\, e^{-ja}\, z \qquad A, a\ \text{real} \tag{14.21}$$

This relation yields the following result for the potential and stream functions:

$$\phi = -A(x \cos a + y \sin a)$$
$$\phi = -A(-x \sin a + y \cos a) \tag{14.22}$$

The velocity components are

$$A_x = A \cos a$$
$$A_y = A \sin a \tag{14.23}$$

This flow field represents uniform (parallel) flow making the angle of attack a with the positive x axis and with velocity A (cf. Fig. 14.3).

The Joukowski transformation

$$\frac{w - ka}{w + ka} = \left(\frac{z - a}{z + a}\right)^k \tag{14.24}$$

is of importance in the practical problem of mapping an airplane wing profile on a nearly circular curve. The wing profile has a sharp point at the trailing edge. The angle

$$\phi = (2 - k)\pi \tag{14.25}$$

is the angle between the tangents to the upper and lower parts of the profile at this point. The circle that passes through the point $-a$ in the z plane so that it encloses the point $z = a$ and cuts the line joining $z = -a$ and $z = a + \delta$, where δ is small, is mapped by this transformation into a wing-shaped curve in the w plane.

In practical problems on the study of the flow of air around an airfoil, the desired transformation is one that maps the region *outside* a circle or a nearly circular curve. If we take the special case

$$a = 1 \qquad k = 2 \tag{14.26}$$

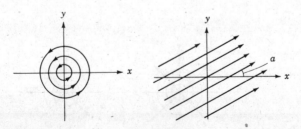

Fig. 14.2   **Fig. 14.3**

we have

$$\frac{w-2}{w+2} = \left(\frac{z-1}{z+1}\right)^2 \tag{14.27}$$

This transformation transforms a circle in the z plane passing through the point $z = -1$ and containing the point $z = 1$ into a wing-shaped curve in the w plane. This curve is known as *Joukowski's profile*.[†]

A useful quantity in two-dimensional hydrodynamics is the complex velocity[‡] which is defined by[§]

$$-\overline{\frac{dw}{dz}} = A_x + jA_y = q \tag{14.28}$$

From this equation it is easily seen that

$$A_x = -\mathrm{Re}\,\frac{dw}{dz}$$
$$A_y = \mathrm{Im}\,\frac{dw}{dz} \tag{14.29}$$

Hence, given the mapping function, (14.28) may be used to obtain the x and y components of the fluid velocity field. A useful application of (14.28) is to locate the stagnation points, that is, points at which the fluid has a zero velocity. These points are those for which (14.28) is equal to zero. A final note on (14.28) is to recognize that

$$\left|\frac{dw}{dz}\right| = (A_x^2 + A_y^2)^{\frac{1}{2}} \tag{14.30}$$

which is the total fluid velocity of the fluid field at a given point.

To illustrate the previous points we shall consider the following example in which the flow field is composed of a uniform flow parallel to the x axis and a source located at the origin. The total complex potential for this problem is

$$w = -Az - k\ln z \tag{14.31}$$

Applying (14.28) yields

$$-\frac{dw}{dz} = A + \frac{k}{z} \tag{14.32}$$

[†] Cf. Glauert, 1943 (see References).

[‡] Cf. Milne-Thomson, 1955 (see References).

[§] If the velocity vector field is defined by $\mathbf{A} = \phi$ rather than by (14.2), then (14.28) is replaced by

$$q = A_x + jA_y = \overline{\frac{dw}{dz}}$$

This equation implies that there is one stagnation point in the flow field located at $z = -k/A$. The two potential functions are

$$\phi = -Ax - \frac{k}{2}\ln(x^2 + y^2)$$

$$\Phi = -Ay - k\tan^{-1}\frac{y}{x}$$

(14.33)

Now, by using either (14.11) or (14.28), the velocity components are

$$A_x = A + \frac{kx}{x^2 + y^2}$$

$$A_y = \frac{ky}{x^2 + y^2}$$

(14.34)

The stream line that passes through the stagnation point is termed the *dividing stream line*. The equation of this line is found from (14.33) by noting that at the stagnation point $z = -k/A$, $\Phi = 0$. The desired equation is therefore

$$\Phi = -Ay - k\tan^{-1}\frac{y}{x} = 0$$

or

$$\frac{1}{x} = -\frac{1}{y}\tan\frac{Ay}{k}$$

(14.35)

The flow field of this system is shown in Fig. 14.4. Since a stream line is tangent to the flow field, no fluid flows across a stream line. This implies that any stream line may be considered a solid boundary. For our particular case we shall consider the dividing stream line as a solid boundary. From Fig. 14.4 one might visualize many possible flow fields represented by this complex potential. One possibility is the flow over a blunt body, and another, looking only at the upper half plane, might be the flow over a flat plane intersecting a smooth bluff.

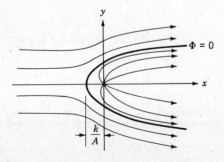

Fig. 14.4

From (14.35) it may be seen that the line $\Phi = 0$ approaches the asymptotes $y = \pm k\pi/A$. The total fluid velocity at any point in the fluid may be found from (14.30):

$$V = \left| \frac{dw}{dz} \right| = \left| -A - \frac{k}{z} \right|$$

$$= \left[A + \left(\frac{kx}{x^2 + y^2} \right)^2 + \left(\frac{ky}{x^2 + y^2} \right)^2 \right]^{\frac{1}{2}} \tag{14.36}$$

Substituting (14.35) yields the total velocity along the boundary curve $\Phi = 0$:

$$V = \left(A^2 + \frac{2Ak}{x} \cos^2 \frac{Ay}{k} + k^2 \sin^2 \frac{Ay}{k} \right)^{\frac{1}{2}} \tag{14.37}$$

As a concluding example let us investigate the transformation of an infinite row of sources along the y axis. It is assumed that the sources are all of equal strength and equally spaced at a distance a apart. The locations are

$$x = 0 \qquad y = 0, \pm a, \pm 2a, \ldots$$

From (14.14) we deduce that the complex potential is given by

$$w = \cdots -k \ln(z + ja) - k \ln z - k \ln(z - ja) - \cdots$$

$$= -k \sum_{n=-\infty}^{\infty} \ln(z - jna) \tag{14.38}$$

The equation may also be written in the form

$$w = -k \ln z(z + ja)(z - ja)(z + 2ja)(z - 2ja) \cdots$$

$$= -k \ln z \prod_{n=1}^{\infty} (z^2 + n^2 a^2)$$

$$= -k \ln z \prod_{n=1}^{\infty} \frac{1 + z^2}{n^2 a^2} - k \ln \prod_{n=1}^{\infty} n^2 a^2 \tag{14.39}$$

Since the last term in (14.39) is a constant,† it will be neglected, as previously discussed. To reiterate this argument we note that the velocity components are the items of interest in all problems and since they are given by the derivatives of the potential functions any additive constants are automatically eliminated. Accepting this step, we find

$$w = -k \ln z \prod_{n=1}^{\infty} \frac{1 + z^2}{n^2 a^2} \tag{14.40}$$

---

† This procedure may curl a few readers' hair since the constant being thrown away is in fact infinite! However, even if it is infinite, its derivative is zero and it therefore contributes nothing to the velocity components.

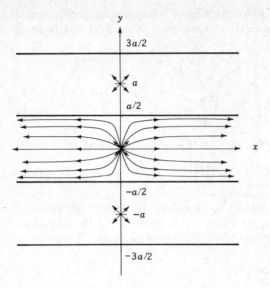

Fig. 14.5

The argument of the logarithm may be written as[†]

$$z \prod_{n=1}^{\infty} \frac{1 + z^2}{n^2 a^2} = \frac{a}{\pi} \sinh \pi \frac{z}{a} \tag{14.41}$$

Thus the complex potential for an infinite row of equidistant and equal strength sources is[‡]

$$w = -k \ln \left(\sinh \pi \frac{z}{a} \right) \tag{14.42}$$

The resulting flow field is indicated in Fig. 14.5 and may be interpreted as a representation of flow from a source midway between parallel walls or efflux from a small opening into a semi-infinite channel.

15 APPLICATION TO STEADY HEAT FLOW

As further illustration of the transformations previously discussed, two examples from steady-state heat conduction will be investigated in this final section.

First consider the problem of heat flow in an annular segment, as shown in Fig. 15.1. The problem is to find the temperature distribution $T(x,y)$

[†] Ryshik, 1963, No. 1.431.2, p. 37 (see References).

[‡] Again a constant term has been neglected.

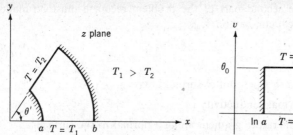

Fig. 15.1

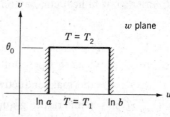

Fig. 15.2

inside the circular segment which is insulated on the two bounding radii a and b, while the surface bounded by the x axis is held at constant temperature T_1 and the surface bounded by the radial line θ' is held at a constant temperature T_2. It is also required to determine the total heat flux through the segment, assuming that the thermal conductivity is k.

To solve the first part of the problem the transformation $w = \ln z$ will be applied.

$$w = u + jv = \ln z = \ln r\, e^{j\theta}$$
$$= \ln r + j\theta \tag{15.1}$$

From (15.1)

$$u = \ln r = \tfrac{1}{2}\ln(x^2 + y^2)$$
$$v = \theta = \tan^{-1}\frac{y}{x} \tag{15.2}$$

Equations (15.2) can be seen to describe the mapping of the segment of Fig. 15.1 in the z plane into the region in the w plane, as shown in Fig. 15.2.

For the region of Fig. 15.2 the solution for the temperature distribution is

$$T(u,v) = T_1 - \frac{(T_1 - T_2)v}{\theta_o} \tag{15.3}$$

The solution in the w plane has been written down by inspection because the problem is simply one of parallel heat flow in a uniform medium. To obtain the solution in the z plane we substitute (15.2) into (15.3):

$$T(x,y) = T_1 - \frac{(T_1 - T_2)(\tan^{-1} y/x)}{\theta_o} \tag{15.4}$$

This solution indicates that the temperature is constant along any radial line, that is, $y/x = $ constant, and varies linearly with the angle θ for constant radius.

To solve the second part of the problem recourse must be made to Fourier's law of heat conduction:[†]

$$q = -k \frac{\partial T}{\partial n} \tag{15.5}$$

where q = heat flux per unit of surface area

k = thermal conductivity

$\partial T / \partial n$ = temperature gradient along a normal to the surface

The total heat flux into the segment may be found from either (15.3) or (15.4). For the purpose of illustration, the solution will be obtained from both equations. First consider (15.3). This equation indicates that the temperature distribution in the w plane segment is independent of the coordinate u. To evaluate the heat flux into the segment, (15.5) will be applied to the surface $v = 0$. In this case (15.5) becomes

$$q(u,v) = -k \frac{\partial T}{\partial v} \bigg|_{v=0} \tag{15.6}$$

Substituting (15.3) yields

$$q(u,v) = \frac{k(T_1 - T_2)}{\theta_o} \tag{15.7}$$

which represents the heat flux per unit area along the surface $v = 0$. To obtain the total heat flux into the segment, it will be assumed that it is one unit thick. Since q is constant, as indicated in (15.7), the total heat flux is

$$Q = qA = \frac{k(T_1 - T_2)(\ln b - \ln a)}{\theta_o}$$

$$= \frac{k(T_1 - T_2) \ln b/a}{\theta_o} \tag{15.8}$$

Now let us recompute this result in the z plane by using (15.4). The bounding surface along $y = 0$ will again be used. For this case (15.5) becomes

$$q(x,y) = -k \frac{\partial T}{\partial y} \bigg|_{y=0} \tag{15.9}$$

Applying this equation to (15.4) yields

$$q(x,y) = \frac{k(T_1 - T_2)}{\theta_o x} \tag{15.10}$$

This equation indicates that in the z plane q is a function of x along the surface $y = 0$. Thus in order to obtain the total heat flux, still assuming a unit

[†] Cf. Carslaw, 1945 (see References).

thickness, (15.10) must be integrated over the surface. Following this procedure gives

$$Q = \int_a^b q\,dx = \frac{k(T_1 - T_2)}{\theta_o} \int_a^b \frac{dx}{x}$$

$$= \frac{k(T_1 - T_2)\ln b/a}{\theta_o} \tag{15.11}$$

This result is identical to that given by (15.8); however, it should be noted that the result was obtained much more easily in the w plane.

Hence, by using a conformal transformation, the temperature distribution in the region of Fig. 15.1 has been obtained quite readily, and the evaluation of the total heat flux through the segment was also obtained.

The second example is to find the steady-state temperature distribution $T(x,y)$ in, and the total heat flux through, the region of Fig. 15.3.   This region may be considered as a slab of unit thickness bounded by the coordinate axes and the ellipse†

$$\frac{x^2}{a^2} + \frac{y^2}{a^2 - 1} = 1 \tag{15.12}$$

The elliptical surface and the segment from the origin to the point (1,0) are thermally insulated.   The surface from (1,0) to (a,0) is held at a constant temperature T_1, and the surface from (0,0) to $(0, \sqrt{a^2 - 1})$ is held at temperature T_2.

To solve this problem the following transformation will be employed:

$$z = \cos w \tag{15.13}$$

From (15.13) it is readily found that

$$x = \cos u \cosh v \tag{15.14}$$

$$y = -\sin u \sinh v \tag{15.15}$$

† The geometry has been selected here so as to make the point (1,0) a focus of the ellipse. This renders the problem much easier algebraically but does not imply that a more general problem cannot be treated by conformal transformations.

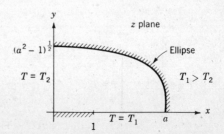

Fig. 15.3

from which

$$\frac{x^2}{\cosh^2 v} + \frac{y^2}{\sinh^2 v} = 1 \qquad (15.16)$$

$$\frac{x^2}{\cos^2 u} - \frac{y^2}{\sin^2 u} = 1 \qquad (15.17)$$

This transformation maps the region of Fig. 15.3 into the rectangular region of Fig. 15.4.

Again the transformed problem is one of simple linear heat flow; hence the temperature distribution is given by

$$T = T_1 - \frac{2(T_1 - T_2)u}{\pi} \qquad (15.18)$$

To return to the z plane we find from (15.17), after some algebraic manipulation, that

$$u = \cos^{-1}\left\{\frac{x^2 + y^2 + 1}{2} - \tfrac{1}{2}[(x^2 + y^2 + 1)^2 - 4x^2]^{\frac{1}{2}}\right\} \qquad (15.19)$$

hence $T(x,y)$ may be obtained by substituting (15.19) into (15.18).

The heat flux into the region may again be found from the temperature distribution in either the w or z plane. In this case it should be clear that it will be computed easier in the w plane. From (15.5), as applied to the face $u = 0$, gives

$$q = -k\frac{\partial T}{\partial u}\Big|_{u=0} \qquad (15.20)$$

Substituting (15.18) into (15.20) yields

$$q = k\frac{2(T_1 - T_2)}{\pi} \qquad (15.21)$$

Since the heat flux is constant, the total heat flux is simply, considering a unit thickness,

$$Q = 2k\frac{(T_1 - T_2)\cosh^{-1} a}{\pi} \qquad (15.22)$$

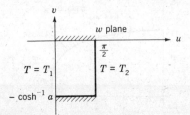

Fig. 15.4

PROBLEMS

1. Given the transformation $z = a \sin w$, where a is real. What are the curves $u = $ const and the curves $v = $ const? What possible electrical and hydrodynamical problems may be solved by this transformation?

2. Given the transformation $z = w + e^w$. Study this transformation in the range $-\pi \leqslant v \leqslant \pi$ for all values of u. Determine the curves $u = $ const and $v = $ const for very large and very small values of u. Draw a rough sketch of the lines $u = $ const and $v = $ const. What problems may be solved by this transformation?

3. Given the transformation $z = \tanh w$. Determine the curves $u = $ const and $v = $ const, and sketch them. What practical problems are solved by this transformation?

4. Using the theory of the complex variable, show that $R^n(\cos n\theta + \sin n\theta)$ and $1/R^n(\cos n\theta + \sin n\theta)$, where n is an integer, are solutions of the two-dimensional Laplace equation, where R and θ are polar coordinates in the xy plane.

5. Two parallel infinite planes at a distance a from each other are kept at zero potential under the influence of a line charge of strength q per unit length that is located between the planes and at a distance of b from the lower plane. Find the field and potential distribution in the region between the planes.

6. A cylinder $(x/a)^{2n} + (y/b)^{2n} = 1$ carries a charge of Q units per unit length. Show that the transformation that gives the field is

$$z = a \left(\cos \frac{nw}{2Q} \right)^{1/n} + jb \left(\sin \frac{nw}{2Q} \right)^{1/n}$$

7. Determine the transformation that gives the field inside a hollow cavity of rectangular shape that contains a line charge at a distance of a from the base whose width is b.

8. Determine the surface density on an earthed conducting plane (potential zero) under the influence of a line charge of strength q at a distance of a from the plane.

9. Write down the integral from which the transformation can be obtained which represents the field near the sharp edges of a slit beveled at $45°$.

10. Using the transformation

$$w = \frac{z + \sqrt{z^2 - b^2}}{2}$$

which gives the potential and stream functions when there is a slit of width $2b$ in a thin infinite-plane conducting sheet that forms one boundary of a uniform field of unit strength, find the field when a filament carrying a charge per unit length is placed in front of the slit.

11. Write down the integral that represents the field when a slit of uniform width $2b$ in an infinite sheet of thickness d forms one boundary of a uniform field.

12. For source flow show that the net volume outflow of a two-dimensional source of unit thickness (z direction) is $2\pi k$.

13. Bernoulli's equation relates the static pressure p to the total fluid velocity v at a point as follows:

$$p + \tfrac{1}{2}\rho v^2 = p_0$$

where ρ is the fluid density and p_0 the stagnation pressure or pressure at a point where the velocity is zero. Derive the pressure distribution for source flow.

14. Repeat Prob. 13 for uniform flow.

15. Repeat Prob. 13 for vortex flow.

16. Discuss the flow field represented by the transformation

$$w = k \ln \frac{z - a}{z + a}$$

where k and a are real.

17. Investigate the flow field represented by

$$w = \frac{A}{z}$$

18. Derive (14.28) from $w = \phi + j\Phi$ and (14.11).

19. In Fig. 14.4 find the transcendental equation whose solution yields the point on the solid at which the static pressure is 90 percent of the stagnation pressure p_0. (Assume $\rho = 1$ and refer to Prob. 13.)

20. Investigate the fluid flow represented by

$$w = -Az + k \ln \frac{z - a}{z + a}$$

21. Determine the points at which the following mapping functions cease to be conformal:

(a) $w = az^2 + bz + c$ (b) $w = z + \dfrac{1}{z}$ $z \neq 0$

(c) $w = e^z$ (d) $w = \sin z$

(e) $w = z^2 + \dfrac{1}{z^2}$ $z \neq 0$ (f) $w = e^{z^2}$

(Problems a, b, e, and f from Kreyszig, 1962)

22. The mapping $w = (az + b)/(cz + d)$ is called a *linear transformation*. Determine the condition that the constants a, b, c, and d must satisfy to guarantee that the mapping is conformal.

23. Find the images of the following straight lines and circles in the w plane as mapped by the transformation $w = 1/z$:

(a) $|z - 2| = 1$ (b) $|z - 2i| = 1$

(c) $x = 1$ (d) $y = x + 1$ (e) $|z - 1| = 1$

(Kreyszig, 1962)

24. Find an analytic function $w = f(z)$ which maps the region $x < 0$, $-\infty < y < \infty$ onto the interior of the circular region $|w| < 1$.

25. Show that the mapping of Prob. 24 transforms the y axis into the circle $|w| = 1$.

26. Find an analytic function $w = f(z)$ that maps the region $2 \leqslant y \leqslant x + 1$ onto the disk $|w| \leqslant 1$.

(Kreyszig, 1962)

27. Show that the transformation $w = z + 1/z$ maps the circle $r = c$ into the ellipse

$$u = \left(c + \frac{1}{c}\right)\cos\theta \qquad v = \left(c - \frac{1}{c}\right)\sin\theta$$

(Churchill, 1960)

28.   Apply the function $w = \ln z$ to solve for the steady-state temperature distribution in the circular wedge shown below:

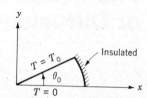

Fig. P 28

(Churchill, 1960)

29.   Apply the Schwartz-Christoffel transformation to derive two successive transformations to solve the Dirichlet problem

$$\frac{\partial^2 H}{\partial x^2} + \frac{\partial^2 H}{\partial y^2} = 0 \qquad \left(0 < x < \frac{\pi}{2} \qquad y > 0 \right)$$

$$H(0,y) = 1 \qquad H\left(\frac{\pi}{2}, y\right) = 0 \qquad H(x,0) = 0$$

(Churchill, 1960)

REFERENCES

1925. Jeans, J. H.: "The Mathematical Theory of Electricity and Magnetism," Cambridge University Press, New York.

1933. Walker, M.: "Conjugate Functions for Engineers," Oxford University Press, New York.

1938. Rothe, R. F., F. Ollendorf, and K. Pohlhausen: "Theory of Functions as Applied to Engineering Problems," M.I.T. Press, Cambridge, Mass.

1939. Smythe, W. R.: "Static and Dynamic Electricity," McGraw-Hill Book Company, New York.

1943. Glauert, H.: "The Elements of Aerofoil and Airscrew Theory," The Macmillan Company, New York.

1945. Carslaw, H. S.: "Mathematical Theory of Heat Conduction in Solids," Dover Publications, New York.

1960. Churchill, A. V.: "Complex Variables and Applications," McGraw-Hill Book Company, New York.

1962. Kreyszig, E.: "Advanced Engineering Mathematics," John Wiley & Sons, Inc., New York.

1963. Ryshik, I. M., and I. S. Gradstein: Tables of Series, Products, and Integrals," Veb Deutscher Verlag der Wissenschaften, Berlin.

1965. Milne-Thompson, L. M.: "Theoretical Hydrodynamics," 3d ed., The Macmillan Company, New York.

1966. Ostberg, D. R., and F. W. Perkins: "An Introduction to Linear Analysis," Addison-Wesley Publishing Company, Inc., Reading, Mass.

1966. Raven, F. H.: "Mathematics of Engineering Systems," McGraw-Hill Book Company, New York.

1966. Sokolnikoff, I. S., and R. M. Redheffer: "Mathematics of Physics and Modern Engineering," 2d ed., McGraw-Hill Book Company, New York.

1966. Wylie, C. R., Jr.: "Advanced Engineering Mathematics," 3d ed., McGraw-Hill Book Company, New York.

11
The Equation of Heat Conduction or Diffusion

1 INTRODUCTION

Perhaps next in importance to Laplace's equation in applied mathematics is the partial differential equation

$$\frac{\partial V}{\partial t} = h^2 \nabla^2 V \tag{1.1}$$

where h^2 is a constant and ∇^2 is the Laplacian operator. It is seen in Appendix E, Sec. 14, that this equation governs the distribution of temperature V in homogeneous solids. In Appendix E, Sec. 18, it is shown that as a consequence of Maxwell's electromagnetic equations the current-density vector $\mathbf{J}$ satisfies the equation

$$\nabla^2 \mathbf{J} = \mu\sigma \frac{\partial \mathbf{J}}{\partial t} \tag{1.2}$$

This equation has exactly the same form as Eq. (1.1) and is known in the electrical literature as the *skin-effect* equation. It may also be shown† that, if U is the concentration of a certain material in grams per cubic centimeter in a certain homogeneous medium of diffusivity constant K measured in square centimeters per second, U satisfies the equation

$$\nabla^2 U = \frac{1}{K} \frac{\partial U}{\partial t} \tag{1.3}$$

This equation is also of the form (1.1).

In the theory of the consolidation of soil,‡ it is shown that, if U is the excess hydrostatic pressure at any point at any time t and C_v is the coefficient of consolidation, U satisfies the equation

$$\nabla^2 U = \frac{1}{C_v} \frac{\partial U}{\partial t} \tag{1.4}$$

This is also of the form of a heat-flow equation.

In Chap. 12, Sec. 10, it is shown that the equations governing the propagation of potential e and current i along an electrical cable having a resistance of R ohms per unit length and a capacitance of C farads per unit length are given by

$$\frac{\partial^2 e}{\partial x^2} = RC \frac{\partial e}{\partial t} \tag{1.5}$$

$$\frac{\partial^2 i}{\partial x^2} = RC \frac{\partial i}{\partial t} \tag{1.6}$$

These equations are of the form (1.1) when V is a function of x only so that the Laplacian operator reduces to $\nabla^2 = \partial^2/\partial x^2$.

This chapter will be devoted to a discussion of some of the simpler methods of solution of Eq. (1.1) subject to certain initial and boundary conditions. We have already seen in Chap. 9 that when a steady state with respect to time has been reached, the term $\partial V/\partial t$ is absent in Eq. (1.1) and the problem reduces to a solution of Laplace's equation subject to the prescribed boundary conditions of the problem.

2   VARIABLE LINEAR FLOW

Let us suppose that we have a bar of length s and of uniform section, the diameter of which is small in comparison with the radius of curvature. Let us suppose that its surface is impervious to heat so that there is no radiation from the sides.

---

† A. B. Newman, The Drying of Porous Solids; Diffusion Calculation, *Transactions of the American Institute of Chemical Engineers*, vol. 27, p. 310, 1931.

‡ K. Terzaghi, "Erdbaumechanik," Johann Ambrosius Barth, Munich, 1925.

Let the initial temperature of the bar be given, and let its ends be kept at the constant temperature zero. If we take one end of the bar at the origin and we denote distances along the bar by x, we have

$$\frac{\partial v}{\partial t} = h^2 \frac{\partial^2 v}{\partial x^2} \tag{2.1}$$

as a special case of Eq. (1.1). The *boundary* conditions are

$$\begin{aligned} v = 0 \quad & \text{when } x = 0 \\ v = 0 \quad & \text{when } x = s \end{aligned} \Bigg\} \quad \text{for all values of } t \tag{2.2}$$

The *initial* condition is

$$v = F(x) \qquad \text{for } t = 0 \qquad v \neq \infty \qquad \text{for } t = \infty \tag{2.3}$$

To solve Eq. (2.1), let us try a solution of the form

$$v(x,t) = e^{mt} u(x) \tag{2.4}$$

where m is a constant and $u(x)$ is a function of x to be determined. On substituting this in (2.1) and dividing out the common factor e^{mt}, we have

$$mu = h^2 \frac{d^2 u}{dx^2} \tag{2.5}$$

or

$$\frac{d^2 u}{dx^2} = \frac{m}{h^2} u \tag{2.6}$$

If we let

$$a^2 = -\frac{m}{h^2} \tag{2.7}$$

Eq. (2.6) becomes

$$\frac{d^2 u}{dx^2} + a^2 u = 0 \tag{2.8}$$

The general solution of this equation is

$$u = A \sin ax + B \cos ax \tag{2.9}$$

Now u must satisfy the boundary conditions (2.2). The first condition, at $x = 0$, gives $B = 0$. In order for u to vanish at $x = s$, we must have

$$A \sin as = 0 \tag{2.10}$$

or

$$as = r\pi \qquad a = \frac{r\pi}{s} \qquad r = 0, 1, 2, 3, \ldots \tag{2.11}$$

for a nontrivial solution. To each value of r there corresponds a solution of the differential equation (2.6) of the form

$$u_r = A_r \sin \frac{r\pi x}{s} \tag{2.12}$$

where A_r is an arbitrary constant.

The possible values of the constant m are given by Eqs. (2.7) and (2.11) and may be written in the form

$$m_r = -\left(\frac{hr\pi}{s}\right)^2 \tag{2.13}$$

To each value of r there corresponds a solution of the differential equation (2.1) of the form

$$v_r = A_r e^{-(hr\pi/s)^2 t} \sin \frac{r\pi x}{s} \tag{2.14}$$

that satisfies the boundary conditions. By summing over all values of r, we construct the general solution

$$v = \sum_{r=1}^{\infty} A_r e^{-(hr\pi/s)^2 t} \sin \frac{r\pi x}{s} \tag{2.15}$$

To evaluate the arbitrary constants A_r, we place $t = 0$ in (2.15) and use the initial conditions (2.3). We thus have

$$F(x) = \sum_{r=1}^{r=\infty} A_r \sin \frac{r\pi x}{s} \tag{2.16}$$

We must therefore expand $F(x)$ in a half-range series of sines. We thus obtain

$$A_r = \frac{2}{s} \int_0^s F(x) \sin \frac{r\pi x}{s} \, dx \tag{2.17}$$

Equation (2.15) with these values of the constants A_r gives the solution of the problem.

If instead of the ends of the bar being kept at temperature zero they are impervious to heat, then the statement of the problem becomes

$$\left.\begin{aligned} \frac{\partial v}{\partial x} &= 0 \quad \text{at } x = 0 \\[2mm] \frac{\partial v}{\partial x} &= 0 \quad \text{at } x = s \end{aligned}\right\} \quad \text{for all values of } t \tag{2.18}$$

$$v = F(x) \quad \text{for } t = 0 \qquad v \neq \infty \qquad \text{at } t = \infty \tag{2.19}$$

In this case we have from Eq. (2.9)

$$\frac{du}{dx} = aA \cos ax - Ba \sin ax \tag{2.20}$$

If this is to vanish at $x = 0$ and $x = s$, we must have

$$A = 0 \qquad \sin as = 0 \tag{2.21}$$

Again we obtain

$$a = \frac{r\pi}{s} \qquad r = 0, 1, 2, \ldots \tag{2.22}$$

Continuing the same reasoning as before, we obtain the general solution

$$v = B_0 + \sum_{r=1}^{\infty} B_r e^{-(hr\pi/s)^2 t} \cos \frac{r\pi x}{s} \tag{2.23}$$

We must therefore expand $F(x)$ in a half-range cosine series. We thus obtain

$$B_0 = \frac{1}{s} \int_0^s F(x)\,dx \qquad B_r = \frac{2}{s} \int_0^s F(x) \cos \frac{r\pi x}{s}\,dx \tag{2.24}$$

It is interesting to note that when $t = \infty$ we have

$$v = B_0 = \frac{1}{s} \int_0^s F(x)\,dx \tag{2.25}$$

the average initial temperature of the bar. This result might of course have been inferred directly from the fact that no heat leaves the bar.

3 ELECTRICAL ANALOGY OF LINEAR HEAT FLOW

A very interesting and useful electrical analogy to linear heat flow may be deduced if we employ the following notation.

Consider a uniform bar or rod that is thermally insulated so that no heat escapes from its sides, as shown in Fig. 3.1. Let

$v(x,t) =$ temperature at point a distance x from one end of bar

$\quad \phi =$ heat flux or quantity of heat passing through cross section s of bar per unit area per unit time

$\quad R =$ thermal or heat resistance per unit length of material, that is, temperature drop per unit length when heat flux is unity ($1/R =$ thermal conductivity)

$\quad C =$ heat capacitance of material $=$ specific heat $\times$ density (number of heat units to raise block of unit area and unit length $1°$ in temperature)

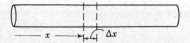

$\quad\quad x \longrightarrow \quad \Delta x \qquad\qquad$ **Fig. 3.1**

If $v(x + \varDelta x, t)$ is the temperature at a point at a distance of $x + \varDelta x$ from the end of the bar, we have

$$R\phi \, \varDelta x = v(x,t) - v(x + \varDelta x, t) \tag{3.1}$$

Using Taylor's expansion and letting $\varDelta x \to 0$ in the limit, we obtain

$$-\frac{\partial v}{\partial x} = R\phi \tag{3.2}$$

Now the heat flux $\phi(x,t)$ at the point x raises the temperature of a lamina of thickness $\varDelta x$ at the rate $\partial v / \partial t$, and a heat flux of magnitude $\phi(x + \varDelta x, t)$ emerges from the lamina of width $\varDelta x$. We thus have the equation

$$\phi(x,t) = C \varDelta x \frac{\partial v}{\partial t} + \phi(x + \varDelta x, t) \tag{3.3}$$

Again expanding $\phi(x + \varDelta x, t)$ by Taylor's expansion and letting $\varDelta x \to 0$, we obtain

$$-\frac{\partial \phi}{\partial x} = C \frac{\partial v}{\partial t} \tag{3.4}$$

Now, in Chap. 12, Sec. 10, it is shown that the flow of current and potential along an electric cable as shown in Fig. 3.2 having a resistance R and a capacitance C per unit length are governed by the equations

$$-\frac{\partial i}{\partial x} = C \frac{\partial e}{\partial t}$$

$$-\frac{\partial e}{\partial x} = Ri \tag{3.5}$$

Comparing Eqs. (3.2) and (3.4) with (3.5), we see that we have the following set of analogies:

$$\phi \text{ (heat flux)} \to i \text{ (current along cable)}$$

$$v \text{ (temperature)} \to e \text{ (electric potential)}$$

$$R \text{ (thermal resistivity)} \to R\left(\frac{\text{electrical resistance}}{\text{length}}\right)$$

$$C \text{ (heat capacitance)} \to C\left(\frac{\text{electrical capacitance}}{\text{length}}\right)$$

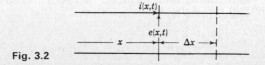

Fig. 3.2

On eliminating ϕ between Eqs. (3.2) and (3.4), we obtain

$$\frac{\partial^2 v}{\partial x^2} = CR \frac{\partial v}{\partial t} \tag{3.6}$$

or

$$\frac{\partial v}{\partial t} = \frac{1}{CR} \frac{\partial^2 v}{\partial x^2} \tag{3.7}$$

The quantity

$$h^2 = \frac{1}{CR} = \frac{\text{thermal conductivity}}{\text{density} \times \text{specific heat}} \tag{3.8}$$

is called the *diffusivity*, *diffusion coefficient*, or *thermometric conductivity*.

The above electrical analogy makes it possible to use the methods of electrical-circuit theory to solve problems in one-dimensional heat conduction.

4   LINEAR FLOW IN SEMI-INFINITE SOLID, TEMPERATURE ON FACE GIVEN AS SINUSOIDAL FUNCTION OF TIME

Let all space on the positive side of the yz plane be filled with a homogeneous solid of diffusivity h^2.   Let the temperature on the yz plane be given as a sinusoidal function of the time, and let it be the same for all values of y and z. It is required to find the temperature throughout the solid when the periodic state is established.

It is clear from symmetry that the temperature $v(x,t)$ is independent of y and z, and the conditions to be fulfilled are

$$\frac{\partial v}{\partial t} = h^2 \frac{\partial^2 v}{\partial x^2} \tag{4.1}$$

$$v = V_0 \sin \omega t \qquad \text{for } x = 0 \qquad v \neq \infty \qquad \text{at } x = \infty$$

$$= \text{Im}\,(V_0\,e^{j\omega t}) \qquad j = \sqrt{-1} \tag{4.2}$$

where Im stands for the "imaginary part of."

To solve (4.1), let us assume a solution of the form

$$v(x,t) = \text{Im}\,[u(x)\,e^{j\omega t}] \tag{4.3}$$

Discarding the Im symbol and substituting this in (4.1), we obtain

$$j\omega u = h^2 \frac{d^2 u}{dx^2} \tag{4.4}$$

where the common factor $e^{j\omega t}$ has been divided out.   Equation (4.4) may be written in the form

$$\frac{d^2 u}{dx^2} = \frac{j\omega}{h^2} u \tag{4.5}$$

If we let

$$a^2 = \frac{j\omega}{h^2} = \frac{\omega}{h^2} e^{j(\pi/2 + 2k\pi)} \qquad k = 0, 1, 2, \ldots \tag{4.6}$$

the solution of (4.5) may be written in the form

$$u = A e^{ax} + B e^{-ax} \tag{4.7}$$

where A and B are arbitrary constants. Now since a^2 is given by (4.6), the square root of a^2 is given by

$$a = \sqrt{\frac{\omega}{h^2}} e^{j(\pi/4)} \tag{4.8}$$

The positive root may be written in the form

$$a = \sqrt{\frac{\omega}{h^2}} e^{j(\pi/4)} = \sqrt{\frac{\omega}{2h^2}} (1 + j) \tag{4.9}$$

Now, at $x = \infty$, the temperature is finite, and hence we must have the constant A in (4.7) equal to zero. At $x = 0$ we have

$$u = B = V_0 \tag{4.10}$$

Hence the solution is

$$
\begin{aligned}
v(x,t) &= \mathrm{Im}\,(V_0 e^{-ax} e^{jwt}) \\
&= \mathrm{Im}\,(V_0 e^{-\sqrt{\omega/2h^2}\,x} e^{j(\omega t - \sqrt{\omega/2h^2}\,x)}) \\
&= V_0 e^{-\sqrt{\omega/2h^2}\,x} \sin\left(\omega t - \sqrt{\frac{\omega}{2h^2}}\,x\right)
\end{aligned}
\tag{4.11}
$$

The negative root leads to the same solution.

We thus see that the temperature decreases in magnitude and changes in phase as we proceed farther into the solid. This result explains why the daily variation of the temperature of the earth's surface cannot be traced to a depth of 3 or 4 ft, while the annual variation cannot be traced beyond a depth of 60 or 70 ft. The greater the frequency of the variation of the surface temperature, the more rapidly is the amplitude of the variation attenuated. This phenomenon has its counterpart in electromagnetic theory and is known in the electrical literature as the *skin effect*.

5 TWO-DIMENSIONAL HEAT CONDUCTION

To illustrate the solution of the two-dimensional diffusion equation, let us consider the following problem.

A thin rectangular plate (Fig. 5.1) whose surface is impervious to heat flow has at $t = 0$ an arbitrary distribution of temperature. Its four edges are

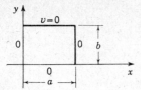

Fig. 5.1

kept at zero temperature. It is required to determine the subsequent temperature of the plate as t increases.

Let the plate extend from $x = 0$ to $x = a$ and from $y = 0$ to $y = b$. Expressing the problem mathematically, we must solve the equation

$$\frac{\partial v}{\partial t} = h^2 \left(\frac{\partial^2 v}{\partial x^2} + \frac{\partial^2 v}{\partial y^2} \right) \tag{5.1}$$

subject to the boundary conditions

$$
\begin{array}{cccc}
v = 0 & v = 0 & v = 0 & v = 0 \\
x = 0 & x = a & y = 0 & y = b \\
0 \leqslant y \leqslant b & 0 \leqslant y \leqslant b & 0 \leqslant x \leqslant a & 0 \leqslant x \leqslant a
\end{array} \tag{5.2}
$$

for all time t.   The initial conditions are

$$
\begin{aligned}
v = F(x, y) \quad & \text{for } 0 \leqslant x \leqslant a \quad 0 \leqslant y \leqslant b \quad \text{at } t = 0 \\
v = 0 \quad & \text{at } t = \infty
\end{aligned} \tag{5.3}
$$

To solve Eq. (5.1), assume a solution of the form

$$v = e^{-\theta t} F_1(x) F_2(y) \tag{5.4}$$

where F_1 is a function of x only and F_2 is a function of y only.   Substituting (5.4) in (5.1) and dividing out the common term $e^{-\theta t}$, we obtain

$$-\theta F_1 F_2 = h^2 (F_1'' F_2 + F_1 F_2'') \tag{5.5}$$

If we divide by $h^2 F_1 F_2$, we have

$$\frac{F_1''}{F_1} + \frac{F_2''}{F_2} = -\frac{\theta}{h^2} \tag{5.6}$$

This may be written in the form

$$\frac{F_1''}{F_1} + \frac{\theta}{h^2} = -\frac{F_2''}{F_2} = k^2 \tag{5.7}$$

We have now succeeded in separating the variables since the left-hand member of (5.7) is a function of x only and the right-hand member of (5.7) is a function of y only, and hence both members of (5.7) are equal to a constant which we have called k^2.

If we let

$$\frac{\theta}{h^2} - k^2 = q^2 \tag{5.8}$$

Eq. (5.7) separates into the two equations

$$\begin{aligned}F_1'' + q^2\, F_1 &= 0 \\ F_2'' + k^2\, F_2 &= 0\end{aligned} \tag{5.9}$$

These equations have the solutions

$$\begin{aligned}F_1 &= A_1 \sin qx + B_1 \cos qx \\ F_2 &= A_2 \sin ky + B_2 \cos ky\end{aligned} \tag{5.10}$$

where the A's and B's are arbitrary constants.   Now, to satisfy the boundary conditions (5.2), it is obvious that there cannot be any cosine terms present so that we must have $B_1 = B_2 = 0$.

Also we must have

$$\begin{aligned}\sin qa &= 0 \\ \sin kb &= 0\end{aligned} \tag{5.11}$$

Hence

$$\begin{aligned}q &= \frac{m\pi}{a} \qquad m = 0, 1, 2, \ldots \\ k &= \frac{n\pi}{b} \qquad n = 0, 1, 2, \ldots\end{aligned} \tag{5.12}$$

From (5.8) we find that

$$\theta_{mn} = h^2 \left[\left(\frac{m\pi}{a} \right)^2 + \left(\frac{n\pi}{b} \right)^2 \right] \tag{5.13}$$

Hence for each value of m and n we find a particular solution of (5.1) that satisfies the boundary conditions (5.2) of the form

$$v = B_{mn}\, e^{-\theta_{mn}t} \sin \frac{m\pi x}{a} \sin \frac{n\pi y}{b} \tag{5.14}$$

If we sum over all possible values of m and n, we construct the general solution

$$v = \sum_{m=1}^{\infty} \sum_{n=1}^{\infty} B_{mn}\, e^{-\theta_{mn}t} \sin \frac{m\pi x}{a} \sin \frac{n\pi y}{b} \tag{5.15}$$

where the quantities B_{mn} are *arbitrary* constants that must be determined from the initial conditions (5.3).   Placing $t = 0$ in (5.15) we must have

$$F(x,y) = \sum_{m=1}^{m=\infty} \sum_{n=1}^{n=\infty} B_{mn} \sin\frac{m\pi x}{a} \sin\frac{n\pi y}{b} \tag{5.16}$$

We must therefore expand the initial temperature function $F(x,y)$ into a double sine series.   To do this, let us multiply both sides of (5.16) by

$$\sin\frac{r\pi x}{a} \sin\frac{s\pi y}{b} dx\, dy$$

where r and s are integers and integrate from $x = 0$ to $x = a$ and $y = 0$ to $y = b$; because of the orthogonality properties of the sines all the terms in the summation vanish except the term for which $m = r$ and $n = s$, and we obtain the result

$$B_{rs} = \frac{4}{ab} \int_{x=0}^{x=a} \int_{y=0}^{y=b} F(x,y) \sin\frac{r\pi x}{a} \sin\frac{s\pi y}{b} dy\, dx \tag{5.17}$$

This determines the arbitrary constants of the general solution (5.15).

THE COOLING OF A HOT BRICK

By a simple extension of the above analysis it may be shown that, if we have a rectangular parallelepiped or brick whose sides are kept at zero temperature and whose internal temperature is given arbitrarily at $t = 0$ to be $F(x,y,z)$, then the subsequent temperature is given by

$$v = \sum_{m=1}^{\infty} \sum_{n=1}^{\infty} \sum_{r=1}^{\infty} B_{mnr}\, e^{-\theta_{mnr}t} \sin\frac{m\pi x}{a} \sin\frac{n\pi y}{b} \sin\frac{r\pi z}{c} \tag{5.18}$$

where

$$\theta_{mnr} = h^2\left[\left(\frac{m\pi}{a}\right)^2 + \left(\frac{n\pi}{b}\right)^2 + \left(\frac{r\pi}{c}\right)^2\right] \tag{5.19}$$

and

$$B_{mnr} = \frac{8}{abc} \int_{x=0}^{a} \int_{y=0}^{b} \int_{z=0}^{c} F(x,y,z) \sin\frac{m\pi x}{a} \sin\frac{n\pi y}{b} \sin\frac{r\pi z}{c} dx\, dy\, dz \tag{5.20}$$

where the brick is oriented with respect to a cartesian coordinate system, as shown in Fig. 5.2.

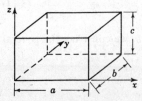

Fig. 5.2

6 TEMPERATURES IN AN INFINITE BAR

An illustration of the method of solution of the diffusion equation by the use of definite integrals is given by the following example:

Given an infinite bar of small cross section so insulated that there is no transfer of heat at the surface, we take the x axis along the bar. The temperature of the bar at $t = 0$ is given as an arbitrary function $F(x)$ of x. It is required to find the subsequent temperature.

Stated mathematically, we must solve the equation

$$\frac{\partial v}{\partial t} = h^2 \frac{\partial^2 v}{\partial x^2} \tag{6.1}$$

subject to the initial condition

$$v = F(x) \qquad \text{at } t = 0 \tag{6.2}$$

Let us assume a solution of (6.1) of the form

$$v = e^{-a^2 t} u(x) \tag{6.3}$$

Substituting this in (6.1) and dividing out the common factor $e^{-a^2 t}$, we obtain

$$-a^2 u = h^2 \frac{d^2 u}{dx^2} \tag{6.4}$$

or

$$\frac{d^2 u}{dx^2} + \frac{a^2}{h^2} u = 0 \tag{6.5}$$

If we let

$$\frac{a^2}{h^2} = \alpha^2 \tag{6.6}$$

a *particular* solution of (6.5) may be written in the form

$$u = \cos \alpha(x - \beta) \tag{6.7}$$

where β is an arbitrary constant. We thus obtain from (6.3) the following particular solution of Eq. (6.1),

$$v_1 = e^{-\alpha^2 h^2 t} \cos \alpha(x - \beta) \tag{6.8}$$

where α and β are arbitrary constants.

If we multiply (6.8) by $F(\beta)/\pi$, we obtain

$$v_2 = \frac{F(\beta)}{\pi} e^{-\alpha^2 h^2 t} \cos \alpha(x - \beta) \tag{6.9}$$

This is still a solution of (6.1). The integral of v_2 with respect to the parameters α and β in the form

$$v = \frac{1}{\pi} \int_{\alpha=0}^{\alpha=\infty} \int_{\beta=-\infty}^{\beta=+\infty} F(\beta) e^{-\alpha^2 h^2 t} \cos \alpha(x - \beta) \, d\alpha \, d\beta \tag{6.10}$$

is a solution of Eq. (6.1).   If we place $t = 0$ in (6.10), we obtain

$$v(x,0) = \frac{1}{\pi} \int_{\alpha=0}^{\alpha=\infty} \int_{\beta=-\infty}^{\beta=+\infty} F(\beta) \cos [\alpha(x - \beta)] \, d\alpha \, d\beta \tag{6.11}$$

But by Appendix C, Sec. 9, we see that this is the Fourier integral representation of $F(x)$.   Hence the solution of the problem is given formally by (6.10).   This may be simplified by using the well-known definite integral

$$\int_0^\infty e^{-y^2} \cos 2by \, dy = \frac{\sqrt{\pi}\, e^{-b^2}}{2} \tag{6.12}$$

which is found in most tables of integrals.   Using this integral, we transform (6.10) into

$$v(x,t) = \frac{1}{2h\sqrt{t\pi}} \int_{\beta=-\infty}^{\beta=+\infty} F(\beta) \, e^{-(x-\beta)^2/4h^2 t} \, d\beta \tag{6.13}$$

7   TEMPERATURES INSIDE A CIRCULAR PLATE

The use of cylindrical coordinates in the solution of a diffusion problem is illustrated by the following:

Consider a thin circular plate whose faces are impervious to heat flow and whose circular edge is kept at zero temperature.   At $t = 0$ the initial temperature of the plate is a function $F(r)$ of the distance from the center of the plate only.   It is required to find the subsequent temperature.   Let the radius of the plate be a.

It is obvious because of symmetry that the temperature v must be a function of r and t only.   Using cylindrical coordinates, we know that v must satisfy the equation

$$\frac{\partial v}{\partial t} = h^2 \nabla^2 v$$

$$= h^2 \left(\frac{\partial^2 v}{\partial r^2} + \frac{1}{r} \frac{\partial v}{\partial r} \right) \qquad 0 \leqslant r \leqslant a \tag{7.1}$$

The boundary condition is

$$v = 0 \qquad \text{at } r = a \tag{7.2}$$

The initial condition is

$$v = F(r) \qquad \text{at } t = 0 \tag{7.3}$$

To solve Eq. (7.1), assume the solution

$$v = e^{-mt} u(r) \tag{7.4}$$

where $u(r)$ is a function of r only. Substituting this into (7.1) and dividing the common factor e^{-mt}, we obtain

$$-mu = h^2 \left(\frac{d^2 u}{dr^2} + \frac{1}{r} \frac{du}{dr} \right) \tag{7.5}$$

Dividing by h^2 and multiplying by r, we obtain

$$r \frac{d^2 u}{dr^2} + \frac{du}{dr} + \frac{m}{h^2} ru = 0 \tag{7.6}$$

If we let

$$\frac{m}{h^2} = k^2 \tag{7.7}$$

we recognize that Eq. (7.6) has the same form as Eq. (9.1) of Appendix B with $n = 0$. Hence the general solution of (7.6) is

$$u = AJ_0(kr) + BY_0(kr) \tag{7.8}$$

where A and B are arbitrary constants. Since the temperature must remain finite at $r = 0$, the arbitrary constant B in (7.8) must be equal to zero since $Y_0(kr)$ goes to infinity at $r = 0$. We thus have the solution

$$u = AJ_0(kr) \tag{7.9}$$

However, the boundary of the plate $r = a$ is maintained at zero temperature for all values of time, and hence we must have

$$J_0(ka) = 0 \tag{7.10}$$

Hence only values of k are permissible that satisfy Eq. (7.10). Let these values be k_s ($s = 1, 2, 3, \ldots$). Equation (7.7) gives the following values for m:

$$m_s = (k_s h)^2 \tag{7.11}$$

A particular solution of (7.1) that satisfies the boundary condition is

$$v_s = A_s e^{-k_s^2 h^2 t} J_0(k_s r) \tag{7.12}$$

If we sum over all possible values of s, we construct the general solution

$$v = \sum_{s=1}^{\infty} A_s e^{-k_s^2 h^2 t} J_0(k_s r) \tag{7.13}$$

where the arbitrary constants A_s must be determined from the initial condition that, at $t = 0$, $v = F(r)$. Placing $t = 0$ in (7.13), we have

$$F(r) = \sum_{s=1}^{\infty} A_s J_0(k_s r) \tag{7.14}$$

We must therefore expand the arbitrary function $F(r)$ into a series of Bessel functions.   Using the results of Appendix B, Sec. 12, we have

$$A_s = \frac{2}{a^2[J_1(k_s a)]^2} \int_{r=0}^{r=a} rF(r)J_0(k_s r)\, dr \qquad s = 1, 2, \ldots \tag{7.15}$$

This determines the arbitrary constants in the solution (7.13).

8   SKIN EFFECT ON A PLANE SURFACE

A very interesting problem in electrodynamics is the one of determining the alternating-current distribution in homogeneous conducting mediums. Perhaps the simplest problem of this type is the following one.

Consider a semi-infinite conducting mass of metal of permeability μ and conductivity σ.   Let the equation of the surface of this mass be $z = 0$, and suppose that on this surface the current-density vector **i** has only an x component i_x which has the value

$$(i_x)_{z=0} = i_0 \sin \omega t \tag{8.1}$$

That is, on the surface of the semi-infinite mass we have a sheet of alternating current flowing in the x direction.   The problem is to determine the distribution of current density within the conducting mass.

In Appendix E, Sec. 18, it is shown that the current-density vector **i** satisfies the equation

$$\nabla^2 \mathbf{i} = \mu\sigma \frac{\partial \mathbf{i}}{\partial t} \tag{8.2}$$

in a region of conductivity σ and permeability μ.   This is a vector equation, and in the case we are considering we have

$$\mathbf{i} = (i_x, 0, 0) \tag{8.3}$$

That is, the current-density vector has only one component, the x component. Accordingly, in the case under consideration, we have

$$\nabla^2 i_x = \mu\sigma \frac{\partial i_x}{\partial t} = \frac{\partial^2 i_x}{\partial z^2} \tag{8.4}$$

This problem is mathematically identical to the heat-flow problem discussed in Sec. 4.   If we let

$$h^2 = \frac{1}{\mu\sigma} \tag{8.5}$$

and follow the analysis of Sec. 4, writing z instead of x and i_x instead of v_x, etc., we obtain

$$i_x = i_0 \, e^{-\sqrt{(\omega\mu\sigma/2)}z} \sin\left(\omega t - \sqrt{\frac{\omega\mu\sigma}{2}}\, z\right) \tag{8.6}$$

for the current density inside the mass of metal. We see that, because of the negative exponential factor, the current amplitude decreases as we penetrate the metal and is greatest at the surface of the mass. This phenomenon is the well-known *skin effect* of electrodynamics.

We may obtain the net current flowing per meter width by integrating

$$\int_0^\infty i_x \, dz = i_0 \int_0^\infty e^{-kz} \sin(\omega t - kz) \, dz \tag{8.7}$$

where

$$k = \sqrt{\frac{\omega\mu\sigma}{2}} \tag{8.8}$$

Using formulas Nos. 506 and 507 of Peirce's "Tables of Integrals," the integral (8.7) may be evaluated, and we obtain

$$\int_0^\infty i_z \, dz = \frac{i_0}{\sqrt{\omega\mu\sigma}} \sin\left(\omega t - \frac{\pi}{4}\right) \tag{8.9}$$

It may be noted that this value of current would be increased if all conducting matter below a certain depth were removed since for certain values of z the current is reversed.

The average power $\bar{P}$ converted to heat per square meter of surface is obtained by integrating

$$\bar{P} = \frac{\omega}{2\pi\sigma} \int_{z=0}^{z=\infty} \int_{t=0}^{2\pi/\omega} i_z^2 \, dt \, dz$$

$$= \frac{i_0^2}{2\sigma} \int_0^\infty e^{-2ks} \, dz$$

$$= \frac{i_0^2}{4\sigma k} \tag{8.10}$$

9   CURRENT DENSITY IN A WIRE

A very important problem in electrical engineering is the determination of the current density in a circular wire carrying alternating current.

Let us consider a long cylinder of radius a, oriented along the z axis of a cartesian coordinate system, as shown in Fig. 9.1. Let the material of the cylinder be homogeneous and have a permeability μ and conductivity σ. In this case the current-density vector **i** has only a z component, and Eq. (8.2) reduces to

$$\nabla^2 i_s = \mu\sigma \frac{\partial i_s}{\partial t} \tag{9.1}$$

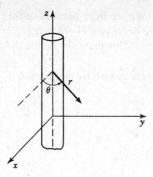

Fig. 9.1

Since we are considering the case of a wire carrying alternating current, the current density has the form

$$i_s = u(r)\cos \omega t$$

$$= \text{Re}\,[u(r)e^{j\omega t}] \tag{9.2}$$

where the symbol Re denotes the "real part of" and $u(r)$ is a function of r only. Suppressing the Re symbol and substituting this in (9.1), we obtain, on dividing out the common factor $e^{j\omega t}$,

$$\nabla^2 u = j\omega\mu\sigma u \tag{9.3}$$

Since u is a function of r only, we have (Appendix E, Sec. 12)

$$\nabla^2 u = \frac{1}{r}\frac{d}{dr}\left(r\frac{du}{dr}\right) \tag{9.4}$$

Hence Eq. (9.3) becomes

$$\frac{1}{r}\frac{d}{dr}\left(r\frac{du}{dr}\right) = j\omega\mu\sigma u \tag{9.5}$$

Differentiating, this becomes

$$\frac{d^2 u}{dr^2} + \frac{1}{r}\frac{du}{dr} - j\omega\mu\sigma u = 0 \tag{9.6}$$

If we let

$$k^2 = -j\omega\sigma \tag{9.7}$$

this may be written in the form

$$\frac{d^2 u}{dr^2} + \frac{1}{r}\frac{du}{dr} + k^2 u = 0 \tag{9.8}$$

Equation (9.8) is of the form of Eq. (9.1) of Appendix B with $n = 0$. Hence the solution of (9.8) is

$$u = AJ_0(kr) + BY_0(kr) \tag{9.9}$$

However, since the current density must remain finite at $r = 0$ and the function Y_0 goes to infinity for the zero value of its argument, we must have

$$B = 0 \tag{9.10}$$

The solution (9.9) reduces to

$$u = AJ_0(kr) \tag{9.11}$$

Now

$$k = \sqrt{-j\omega\mu\sigma} = j\sqrt{j}\sqrt{\omega\mu\sigma}$$
$$= j^{\frac{3}{2}}m \tag{9.12}$$

where

$$m = \sqrt{\omega\mu\sigma} \tag{9.13}$$

Hence (9.11) may be written in the form

$$u = AJ_0(j^{\frac{3}{2}}mr) \tag{9.14}$$

This may be written in terms of the ber and bei functions of Appendix B, Sec. 11, in the form

$$u = A[\text{ber}\,(mr) + j\,\text{bei}\,(mr)] \tag{9.15}$$

To determine the arbitrary constant A, place $r = 0$ in Eq. (9.14), and let u_0 be the value of the current-density amplitude at the center of the wire. Since $J_0(0) = 1$, we obtain

$$u_0 = AJ_0(0) = A \tag{9.16}$$

Substituting this in (9.15), we thus have

$$u = u_0[\text{ber}\,(mr) + j\,\text{bei}\,(mr)] \tag{9.17}$$

This expression may be written in the polar form

$$u = u_0\,M\,e^{j\theta} \tag{9.18}$$

where

$$M = \sqrt{\text{ber}^2\,(mr) + \text{bei}^2\,(mr)} \tag{9.19}$$

and

$$\tan\theta = \frac{\text{bei}\,(mr)}{\text{ber}\,(mr)} \tag{9.20}$$

We now obtain the instantaneous current density i_z by substituting (9.18) in (9.2). We thus have

$$i_z = \text{Re}\,(u_0\,M\,e^{j(\omega t + \theta)})$$

$$= u_0\,M\cos{(\omega t + \theta)} \tag{9.21}$$

TOTAL CURRENT IN WIRE

The total current $I(t)$ in the wire may be obtained by integrating the current density i_z throughout the cross section of the wire. We then have

$$I(t) = \int_{r=0}^{r=a} i_z \times 2\pi r\,dr$$

$$= \text{Re}\int_{r=0}^{r=a} 2\pi r u\,dr\,e^{j\omega t}$$

$$= \text{Re}\left[e^{j\omega t}\int_{r=0}^{r=a} u_0 J_0(j^{\frac{3}{2}}mr) \times 2\pi r\,dr\right] \tag{9.22}$$

To integrate this expression, we use expression (6.11) of Appendix B for $n = 1$. This equation may be written in the form

$$\int u J_0(u)\,du = u J_1(u) \tag{9.23}$$

Hence we have

$$I(t) = \text{Re}\left[\frac{2\pi u_0\,a J_1(j^{\frac{3}{2}}ma)}{jm\,\sqrt{j}}e^{j\omega t}\right]$$

$$= I_0\cos{(\omega t + \phi)} \tag{9.24}$$

where

$$I_0 = \left|\frac{2\pi u_0\,a J_1(j^{\frac{3}{2}}ma)}{jm\,\sqrt{j}}\right| \tag{9.25}$$

If we know the amplitude of the total current in the wire, we may compute u_0 from (9.25) and hence obtain i_z from (9.21).

10 GENERAL THEOREMS

In this section some general theorems useful in the solution of certain heat-flow and diffusion problems will be considered.

Theorem I If $u_1(x,t)$, $u_2(y,t)$, and $u_3(z,t)$ are solutions of the three linear heat-flow equations

$$\frac{\partial u_1}{\partial t} = h^2\frac{\partial^2 u_1}{\partial x^2} \tag{10.1}$$

$$\frac{\partial u_2}{\partial t} = h^2\frac{\partial^2 u_2}{\partial y^2} \tag{10.2}$$

$$\frac{\partial u_3}{\partial t} = h^2\frac{\partial^2 u_3}{\partial z^2} \tag{10.3}$$

then $u = u_1 u_2 u_3$ is necessarily a solution of the three-dimensional heat-flow equation

$$\frac{\partial u}{\partial t} = h^2 \left(\frac{\partial^2 u}{\partial x^2} + \frac{\partial^2 u}{\partial y^2} + \frac{\partial^2 u}{\partial z^2} \right) \tag{10.4}$$

Proof. To prove this theorem, substitute the function $u = u_1 u_2 u_3$ in (10.4). We then have

$$\frac{\partial}{\partial t}(u_1 u_2 u_3) = h^2 \left[\frac{\partial^2}{\partial x^2}(u_1 u_2 u_3) + \frac{\partial^2}{\partial y^2}(u_1 u_2 u_3) + \frac{\partial^2}{\partial z^2}(u_1 u_2 u_3) \right] \tag{10.5}$$

Differentiating, we have

$$u_2 u_3 \frac{\partial u_1}{\partial t} + u_1 u_3 \frac{\partial u_2}{\partial t} + u_1 u_2 \frac{\partial u_3}{\partial t}$$
$$= h^2 \left(u_2 u_3 \frac{\partial^2 u_1}{\partial x^2} + u_1 u_3 \frac{\partial^2 u_2}{\partial y^2} + u_1 u_2 \frac{\partial^2 u_3}{\partial z^2} \right) \tag{10.6}$$

If we now substitute Eqs. (10.1), (10.2), and (10.3) in (10.6), we obtain an identity. This proves the theorem.

This theorem may be used to facilitate the solution of the "brick" problem of Sec. 5.

Theorem II   If $u_1(r,t)$ is a solution of the symmetrical two-dimensional equation

$$\frac{\partial u_1}{\partial t} = h^2 \left(\frac{\partial^2 u_1}{\partial r^2} + \frac{1}{r} \frac{\partial u_1}{\partial r} \right) \tag{10.7}$$

and $u_2(z,t)$ is a solution of the linear heat-flow equation

$$\frac{\partial u^2}{\partial t} = h^2 \frac{\partial^2 u_2}{\partial z^2} \tag{10.8}$$

then $u = u_1 u_2$ is necessarily a solution of the three-dimensional heat-flow equation

$$\frac{\partial u}{\partial t} = h^2 \left(\frac{\partial^2 u}{\partial r^2} + \frac{1}{r} \frac{\partial u}{\partial r} + \frac{\partial^2 u}{\partial z^2} \right) \tag{10.9}$$

Proof. This is proved by substituting $u = u_1 u_2$ in Eq. (10.9). By the use of Eqs. (10.7) and (10.8) we again obtain an identity and this proves the proposition.

PROBLEMS

1. An infinite plate of thickness a and uniform material of permeability μ and conductivity σ carries alternating current of angular frequency ω in one direction. Show that the current density u_0 at the middle of the plate is related to the current density u_s at the surface of the plate by the equation

$$\frac{u_0}{u_s} = \frac{\sqrt{2}}{\sqrt{\cosh ma + \cos ma}}$$

where $m = \sqrt{\mu\sigma\omega/2}$.

2. An infinite iron plate is bounded by the parallel planes $x = h$ and $x = -h$. Wire is wound uniformly around the plate, and the layers of wire are parallel to the y axis. An alternating current is sent through the wire. This current produces a magnetic field $H_0 \cos \omega t$ parallel to the z axis at the surface of the plate. Show that the magnetic field H inside the plate at a distance of x from the center is given by

$$H = H_0 \left(\frac{\cosh 2mx + \cos 2mx}{\cosh 2mh + \cos 2mh} \right)^{\frac{1}{2}} \cos (\omega t + \phi)$$

where $m = \sqrt{\sigma\omega\mu/2}$ and ϕ is a phase angle. μ and σ are the permeability and the conductivity of the plate, respectively.

HINT: The magnetic field H satisfies the equation $\nabla^2 H = \mu\sigma(\partial H/\partial t)$ in the region inside the plate.

3. The rate at which heat is lost from a rod as a consequence of surface radiation into the surrounding air at constant temperature u_0 is proportional to the difference of temperature $u - u_0$ and to the surface area of the element. Show that this leads to an equation of the form

$$\frac{\partial u}{\partial t} = h^2 \frac{\partial^2 u}{\partial x^2} - k^2(u - u_0)$$

for the temperature of the rod.

4. If u_0 is zero in Prob. 3, show that in the electrical analogy discussed in Sec. 3 we must add a leakage-conductance term to take into account the radiation in the thermal problem.

5. An infinite bar has one end located at $x = 0$, and the other end extends to $x = \infty$. The temperature of the surrounding air is zero, and the end at $x = 0$ is kept at $u = u_0$. Find the steady-state distribution of temperature, taking into account radiation.

6. A thin bar of uniform section is bent into the form of a circular ring of large radius a. At one point P in the ring, let a steady temperature u_0 be maintained. Let heat be radiated from the ring to the air. Assume that the air is at zero temperature. Show that, when a steady state is established, the temperature of the ring is given by

$$u = u_0 \frac{\cosh k(x - \pi a)}{\cosh k\pi a}$$

where k^2 is the radiation coefficient as defined in Prob. 3 and x is a coordinate measured around the ring so that $x = 0$ at P. This arrangement is known as *Fourier's ring*.

7. It has been proposed to represent the rate of cooling of a surface by the empirical formula $e(v - v_0)^n$, where e is the emissivity of the surface and n has the value 1.2. Show that on this supposition the equation to be satisfied in a long thin rod cooling laterally is

$$\frac{\partial v}{\partial t} = h^2 \frac{\partial^2 v}{\partial x^2} - \frac{ep}{c\rho\sigma} (v - v_0)^n$$

where v is the temperature at a distance x measured along the rod from one end, v_0 is the temperature of the surrounding medium, h^2 is the diffusivity, ρ the density, p the perimeter, c the specific heat, and σ the area of cross section of the rod.

8. A bar of length s is heated so that its two ends are at temperature zero. If initially the temperature is given by $u = cx(s - x)/s^2$, find the temperature at time t.

9. If in the equation

$$\frac{\partial u}{\partial t} = h^2 \frac{\partial^2 u}{\partial x^2} - k(u - u_0)$$

we introduce the change in variable $u = u_0 + ve^{-kt}$, show that the equation in v is of the form

$$\frac{\partial v}{\partial t} = a^2 \frac{\partial^2 v}{\partial x^2}$$

10. A rectangular plate bounded by the lines $x = 0$, $y = 0$, $x = a$, $y = b$ has an initial distribution of temperature given by

$$v = A \sin \frac{\pi x}{a} \sin \frac{\pi y}{b}$$

The edges are kept at zero temperature, and the plane faces are impermeable to heat. Find the temperature at any point and time, and show that very close to any corner of the plate the lines of equal temperature and flow of heat are orthogonal systems of rectangular hyperbolas.

Show that the heat lost by the plate across the edges up to time t is

$$\frac{4mAab}{\pi^2}\left(1 - e^{-h^2\left(\frac{1}{a^2} + \frac{1}{b^2}\right)\pi^2 t}\right)$$

where m is the thermal capacity of the plate per unit area.

11. A conducting sphere initially at zero temperature has its surface kept at a constant temperature u_0 for a given time, after which it is kept at zero. Find the temperature at any time in the second stage.

12. A bar of uniform cross section is covered with impermeable varnish and extends from the point $x = 0$ to infinity. The bar being throughout at temperature zero, the extremity is brought at time $t = 0$ to temperature v_0 and kept at this temperature. Find the distribution of temperature in the bar at any subsequent time t, and verify that the solution gives the obvious solution at $t = \infty$.

13. Find the temperature distribution for the bar in Sec. 2 when $F(x) = v_0 = \text{const.}$ Assume that the ends of the bar are held at $0°$. Sketch the temperature distribution for $t = 0$, $t = (s/h\pi)^2$.

14. Find the temperature distribution for the bar in Sec. 2 when $F(x) = v_0 \sin \pi x/s$. Assume that the temperature of the ends of the bar is $0°$.

15. Find the solution to the heat conduction in a bar of length s when the temperature of the ends is given by

$$v(0,t) = v_0 \qquad v(s,t) = v_s$$

and the initial temperature distribution is

$$v(x,0) = F(x)$$

16. Find the temperature distribution for the problem in Sec. 5 when $F(x,y) = v_0 = \text{const.}$ Assume homogeneous boundary conditions.

17. Find the temperature distribution for the problem in Sec. 5 when $F(x,y) = xy/ab$, assuming homogeneous boundary conditions.

18. Under homogeneous boundary conditions solve the problem in Sec. 5 subject to the initial temperature distribution $F(x,y) = v_0 \sin \pi x/a$. Discuss any discrepancies that you come across.

19. Assuming nonhomogeneous boundary conditions for the problem in Sec. 5, how would you obtain the solution for an arbitrary initial temperature distribution $F(x,y)$?

20. Assuming constant initial temperature distribution and homogeneous boundary conditions, find the solution to the problem in Sec. 5.

21. Repeat Prob. 20, except assume that $F(x,y,z) = xyz/abc$.

22. Fill in the details of deriving Eq. (5.20).

REFERENCES

1893. Byerly, W. E.: "Fourier Series and Spherical Harmonics," Ginn and Company, Boston.

1939. Smythe, W. R.: "Static and Dynamic Electricity," McGraw-Hill Book Company, New York.

1941. Churchill, R. V.: "Fourier Series and Boundary Value Problems," McGraw-Hill Book Company, New York.

1945. Carslaw, H. S.: "Mathematical Theory of Heat Conduction in Solids," Dover Publications, Inc., New York.

1962. Kreyszig, E.: "Advanced Engineering Mathematics," John Wiley & Sons, Inc., New York.

1966. Ostberg, D. R., and F. W. Perkins: "An Introduction to Linear Analysis," Addison-Wesley Publishing Company, Inc., Reading, Mass.

1966. Raven, F. H.: "Mathematics of Engineering Systems," McGraw-Hill Book Company, New York.

1966. Sokolnikoff, I. S., and R. M. Redheffer: "Mathematics of Physics and Modern Engineering," 2d ed., McGraw-Hill Book Company, New York.

1966. Wylie, C. R., Jr.: "Advanced Engineering Mathematics," 3d ed., McGraw-Hill Book Company, New York.

12
The Wave Equation

1 INTRODUCTION

In this chapter solutions of the so-called "wave equation" in simple cases will be considered. One of the most fundamental and common phenomena that occur in nature is the phenomenon of wave motion. When a stone is dropped into a pond, the surface of the water is disturbed and waves of displacement travel radially outward; when a tuning fork or a bell is struck, sound waves are propagated from the source of sound. The electrical oscillations of a radio antenna generate electromagnetic waves that are propagated through space. These various physical phenomena have something in common. Energy is propagated with a finite velocity to distant points, and the wave disturbance travels through the medium that supports it without giving the medium any permanent displacement. We shall find in this chapter that whatever the nature of the wave phenomenon, whether it be the displacement of a tightly stretched string, the deflection of a stretched membrane, the propagation of currents and potentials along an electrical transmission line, or the propagation of electromagnetic waves in free space, these entities are

governed by a certain differential equation, the wave equation.   This equation
has the form

$$\nabla^2 u = \frac{\partial^2 u}{\partial x^2} + \frac{\partial^2 u}{\partial y^2} + \frac{\partial^2 u}{\partial z^2} = \frac{1}{c^2} \frac{\partial^2 u}{\partial t^2} \tag{1.1}$$

where c is a constant having dimensions of velocity, t is the time, x, y, and z
are the coordinates of a cartesian reference frame, and u is the entity under
consideration, whether it be a mechanical displacement, or the field components
of an electromagnetic wave, or the currents or potentials of an electrical
transmission line.

2   THE TRANSVERSE VIBRATIONS OF A STRETCHED STRING

The study of the oscillations of a tightly stretched string is a fundamental
problem in the theory of wave motion.   Consider a perfectly flexible string
that is stretched between two points by a constant tension T so great that
gravity may be neglected in comparison with it.   Let the string be uniform
and have a mass per unit length equal to m.   Let us take the undisturbed
position of the string to be the x axis and suppose that the motion is confined
to the xy plane.

Consider the motion of an element PQ of length ds, as shown in Fig. 2.1.
The net force in the y direction, F_y, is given by

$$F_y = T \sin \theta_2 - T \sin \theta_1 \tag{2.1}$$

Now, for small oscillations, we may write

$$\sin \theta_2 \doteq \tan \theta_2 = \left(\frac{\partial y}{\partial x}\right)_{x+dx} \tag{2.2}$$

$$\sin \theta_1 \doteq \tan \theta_1 = \left(\frac{\partial y}{\partial x}\right)_x \tag{2.3}$$

Hence we have

$$F_y = \left(T \frac{\partial y}{\partial x}\right)_{x+dx} - \left(T \frac{\partial y}{\partial x}\right)_x \tag{2.4}$$

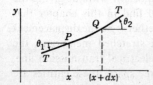

Fig. 2.1

By Taylor's expansion we have

$$\left(T\frac{\partial y}{\partial x}\right)_{x+dx} \doteq \left(T\frac{\partial y}{\partial x}\right)_x + \frac{\partial}{\partial x}\left(T\frac{\partial y}{\partial x}\right)_x dx \tag{2.5}$$

where we have neglected terms of order dx^2 and higher. Substituting this into (2.4), we get

$$F_y = \frac{\partial}{\partial x}\left(T\frac{\partial y}{\partial x}\right) dx \tag{2.6}$$

By Newton's law of motion we have

$$F_y = \frac{\partial}{\partial x}\left(T\frac{\partial y}{\partial x}\right) dx = m\,dx\left(\frac{\partial^2 y}{\partial t^2}\right) \tag{2.7}$$

where $m\,dx$ represents the mass of the section of string under consideration and where we have written dx for ds since the displacement is small. $\partial^2 y/\partial t^2$ is the acceleration of the section of string in the y direction. We thus have

$$\frac{\partial}{\partial x}\left(T\frac{\partial y}{\partial x}\right) = m\frac{\partial^2 y}{\partial t^2} \tag{2.8}$$

as the equation governing the small oscillations of a flexible string.

If the stretching force is constant throughout the string, we may write

$$\frac{\partial}{\partial x}\left(T\frac{\partial y}{\partial x}\right) = T\frac{\partial^2 y}{\partial x^2} = m\frac{\partial^2 y}{\partial t^2} \tag{2.9}$$

or

$$\frac{\partial^2 y}{\partial x^2} = \frac{1}{c^2}\frac{\partial^2 y}{\partial t^2}. \tag{2.10}$$

This is a special case of the general wave equation (1.1), where in this case we have

$$c = \sqrt{\frac{T}{m}} \tag{2.11}$$

It is easy to show that the constant c has dimensions of length/time and hence the same dimensions as a velocity.

3 D'ALEMBERT'S SOLUTION; WAVES ON STRINGS

We may obtain a general solution of (2.10) by a symbolic method in the following manner: If we write

$$D_x = \frac{\partial}{\partial x} \qquad D_x^2 = \frac{\partial^2}{\partial x^2} \tag{3.1}$$

then Eq. (2.10) may be written symbolically in the form

$$\frac{\partial^2 y}{\partial t^2} = c^2 \frac{\partial^2 y}{\partial x^2} = (cD_x)^2 y \tag{3.2}$$

Let us now treat Eq. (3.2) as an ordinary differential equation with constant coefficients of the form

$$\frac{d^2 y}{dt^2} = a^2 y \tag{3.3}$$

where

$$a = cD_x \tag{3.4}$$

The solution of (3.3) if a were a constant would be of the form

$$y = A_1 e^{at} + A_2 e^{-at} \tag{3.5}$$

where A_1 and A_2 are arbitrary constants. Thus Eq. (3.3) is formally satisfied by

$$y = e^{cD_x t} F_1(x) + e^{-cD_x t} F_2(x) \tag{3.6}$$

where, since the integration has been performed with respect to t, instead of the arbitrary constants A_1 and A_2 we have the arbitrary functions of x, $F_1(x)$ and $F_2(x)$. To interpret the solution, we use the symbolic form of Taylor's expansion, as written in Appendix C.

$$e^{hD_x} F(x) = F(x + h) \tag{3.7}$$

If we let

$$h = ct \tag{3.8}$$

we have

$$e^{cD_x t} F_1(x) = F_1(x + ct) \tag{3.9}$$

If we let

$$h = -ct \tag{3.10}$$

we have

$$e^{-cD_x t} F_2(x) = F_2(x - ct) \tag{3.11}$$

Hence, by (3.6), the general solution of the one-dimensional wave equation (2.10) is given by

$$y(x,t) = F_1(x + ct) + F_2(x - ct) \tag{3.12}$$

Now if $F_1(x + ct)$ is plotted as a function of x, it is exactly the same as $F_1(x)$ in shape, but every point on it is displaced a distance ct to the left of the corresponding point in $F_1(x)$. The function $F_1(x + ct)$ thus represents a wave of displacement of arbitrary shape traveling toward the left along the string with a velocity equal to c. In the same manner it may be seen that $F_2(x - ct)$ represents a wave of displacement traveling with a velocity c to the right along the string. The general solution is the sum of these two waves.

Consider a string having one end fixed at the origin, and let the other end be a great distance away at $x = s$. Suppose that a wave of arbitrary shape given by

$$y = F(ct + x) \tag{3.13}$$

is approaching the origin. At the origin the displacement must be zero for all time t; hence the reflected wave must have the form

$$y = -F(ct - x) \tag{3.14}$$

since the sum of this and the original expression is zero at $x = 0$ for all values of t. This shows that the transverse waves in a stretched string are inverted by reflection from a fixed end.

4 HARMONIC WAVES

Consider the expression

$$y = A \cos \frac{2\pi}{T}\left(t - \frac{x}{c}\right) = A \cos 2\pi \left(\frac{t}{T} - \frac{x}{\lambda}\right) \tag{4.1}$$

By plotting the expression as a function for x for successive values of t it may be shown to represent an infinite train of progressive harmonic waves. It is seen that, as t increases, the whole wave profile moves forward in the positive x direction with a velocity equal to c. The *wavelength*, that is, the distance between two successive crests at any time, is given by

$$\lambda = cT \tag{4.2}$$

The time that it takes a complete wave to pass a fixed point is called the *period* and is given by T. The *frequency* of the wave is denoted by f and is given by

$$f = \frac{1}{T} \tag{4.3}$$

The *wave number*, the number of waves that lie in a distance of 2π units, is denoted by k and is given by the equation

$$k = \frac{2\pi}{\lambda} \tag{4.4}$$

5   FOURIER-SERIES SOLUTION

The general solution of the wave equation for the vibrating string shows clearly the wave nature of the phenomenon but is not very well suited for certain types of physical investigations.

Consider a string fastened at $x = 0$ and $x = s$ to fixed supports, and let us suppose that at $t = 0$ we are given the displacement and velocity of every point of the string.   That is, we stipulate that

$$\left. \begin{aligned} y(x,t) &= y_0(x) \\ \frac{\partial y}{\partial t} &= v_0(x) \end{aligned} \right\} \quad \text{at } t = 0 \tag{5.1}$$

where $y_0(x)$ and $v_0(x)$ are the initial displacement and velocity of the string, respectively.

We then wish to determine the subsequent behavior of the string.   To do this, let us assume a solution of the form

$$y(x,t) = e^{j\omega t} v(x) \qquad j = \sqrt{-1} \tag{5.2}$$

where $v(x)$ is a function of x alone and ω is to be determined.   We thus have

$$\frac{\partial^2 y}{\partial t^2} = -\omega^2 e^{j\omega t} v \tag{5.3}$$

$$\frac{\partial^2 y}{dx^2} = e^{j\omega t} \frac{d^2 v}{\partial x^2} \tag{5.4}$$

On substituting these expressions into the wave equation

$$\frac{\partial^2 y}{\partial x^2} = \frac{1}{c^2} \frac{\partial^2 y}{\partial t^2} \tag{5.5}$$

it becomes, after some reductions,

$$\frac{d^2 v}{dx^2} + \left(\frac{\omega}{c}\right)^2 v = 0 \tag{5.6}$$

This ordinary differential equation has the solution

$$v = A \sin \frac{\omega}{c} x + B \cos \frac{\omega}{c} x \tag{5.7}$$

However, since the string is fastened at $x = 0$ and at $x = s$, we have the boundary conditions

$$v = 0 \quad \begin{cases} \text{at } x = 0 \\ \text{at } x = s \end{cases} \tag{5.8}$$

The first condition leads to

$$B = 0 \tag{5.9}$$

The second condition gives

$$0 = A \sin \frac{\omega s}{c} \tag{5.10}$$

Since we are looking for a nontrivial solution, $A \neq 0$; therefore

$$\sin \frac{\omega s}{c} = 0 \tag{5.11}$$

This transcendental equation leads to the possible values of ω. Therefore we have

$$\frac{\omega s}{c} = k\pi \qquad k = 0, 1, 2, \ldots \tag{5.12}$$

or

$$\omega_k = \frac{k\pi c}{s} \qquad k = 0, 1, 2, \ldots \tag{5.13}$$

where we have labeled ω_k the value corresponding to the particular value of k. For each value of k we may write in view of (5.7) and (5.9) the equation

$$v_k = A_k \sin \frac{k\pi x}{s} \tag{5.14}$$

As a consequence of (5.2) we have for every value of k the solution

$$y_k = e^{j\omega_k t} A_k \sin \frac{k\pi x}{s} \qquad A_k \text{ arbitrary} \tag{5.15}$$

This expression satisfies the wave equation (5.5) and the boundary conditions (5.8). Since the real and imaginary parts of (5.15) satisfy the wave equation and the equation is linear, we may write

$$y_k = \left(C_k \cos \frac{k\pi c t}{s} + D_k \sin \frac{k\pi c t}{s} \right) \sin \frac{k\pi x}{s} \tag{5.16}$$

where C_k and D_k are arbitrary constants.

By summing over all the values of k we can construct a general solution of the form

$$y = \sum_{k=1}^{\infty} y_k = \sum_{k=1}^{k=\infty} \left(C_k \cos \frac{k\pi c t}{s} + D_k \sin \frac{k\pi c t}{s} \right) \sin \frac{k\pi x}{s} \tag{5.17}$$

The arbitrary constants C_k and D_k must now be determined from the initial conditions (5.1). We thus have at $t = 0$

$$y_0(x) = \sum_{k=1}^{k=\infty} C_k \sin \frac{k\pi x}{s} \tag{5.18}$$

This requires the expansion of an arbitrary function, $y_0(x)$, the initial displacement into a series of sines. To obtain the typical coefficient C_r, we multiply both sides of (5.18) by the expression

$$\sin\frac{r\pi x}{s} \tag{5.19}$$

and integrate with respect to x from $x = 0$ to $x = s$. We then have

$$\int_0^s y_0(x)\sin\frac{r\pi x}{s}\,dx = \sum_{k=1}^{k=\infty} C_k \int_0^s \sin\frac{k\pi x}{s}\sin\frac{r\pi x}{s}\,dx \tag{5.20}$$

If we now make use of the result

$$\int_0^s \sin\frac{k\pi x}{s}\sin\frac{r\pi x}{s}\,dx = \begin{cases} 0 & \text{if } k \neq r \\ \dfrac{s}{2} & \text{if } k = r \end{cases} \tag{5.21}$$

then (5.20) reduces to

$$\int_0^s y_0(x)\sin\frac{r\pi x}{s}\,dx = \frac{s}{2}C_r \tag{5.22}$$

or

$$C_r = \frac{2}{s}\int_0^s y_0(x)\sin\frac{r\pi x}{s}\,dx \qquad r = 1, 2, 3, \ldots \tag{5.23}$$

This determines the arbitrary constants C_k in the general solution (5.17). To determine the arbitrary constants D_k, we make use of the second initial condition (5.1). We then have, on computing $\partial y/\partial t$ from (5.17) and putting $t = 0$,

$$v_0(x) = \sum_{k=1}^{k=\infty} \frac{k\pi c}{s} D_k \sin\frac{k\pi x}{s} \tag{5.24}$$

Proceeding in the same manner as before, we obtain

$$D_r = \frac{2}{r\pi c}\int_0^s v_0(x)\sin\frac{r\pi x}{s}\,dx \qquad r = 1, 2, \ldots \tag{5.25}$$

for the D_k coefficients.

From (5.23) and (5.25) we see that if the string has no initial velocity the D_k constants are all zero, while if the string has no initial displacement all the C_k coefficients are zero.

Each term of (5.17) represents a stationary wave; the wavelengths are given by $2s/n$, where n is any integer. The frequency, or number of periods per second, of the *fundamental* note is given by

$$F_1 = \frac{1}{2s}\sqrt{\frac{T}{m}} \tag{5.26}$$

and the frequency of the *harmonics*, as the other terms are called, is given by

$$F_n = \frac{n}{2s}\sqrt{\frac{T}{m}} \qquad n = 2, 3, \ldots \tag{5.27}$$

As an example of the application of Eqs. (5.23) and (5.25), let us consider the case of a string that is plucked at its midpoint as shown in Fig. 5.1. In this case, we have the initial condition

$$v_0(x) = 0 \tag{5.28}$$

Hence by (5.25) we have

$$D_r = 0 \qquad r = 1, 2, 3, \ldots \tag{5.29}$$

The initial displacement $y_0(x)$ is given by

$$y_0(x) = \frac{2hx}{s} \qquad 0 \leqslant x \leqslant \frac{s}{2}$$
$$\tag{5.30}$$
$$y_0(x) = \frac{2h}{s}(s - x) \qquad \frac{s}{2} \leqslant x \leqslant s$$

Substituting the analytical expression for the initial displacement into Eq. (5.22), we obtain

$$C_r = \frac{2}{s}\left[\int_0^s \frac{2hx}{s}\sin\frac{r\pi x}{s}\,dx + \int_{s/2}^s \frac{2h}{s}(s - x)\sin\frac{r\pi x}{s}\,dx\right] \tag{5.31}$$

On carrying out the integrations, we obtain

$$C_r = \frac{8h}{\pi^2 r^2}\sin\frac{\pi r}{2} \qquad \text{if } r \text{ is odd}$$
$$\tag{5.32}$$
$$C_r = 0 \qquad \text{if } r \text{ is even}$$

Substituting these results into (5.17), we obtain the following expression for the displacement of the string on being released from the initial position given by Fig. 5.1:

$$y = \frac{8h}{\pi^2}\left(\sin\frac{\pi x}{s}\cos\frac{\pi ct}{s} - \frac{1}{9}\sin\frac{3\pi x}{s}\cos\frac{3\pi ct}{s} + \cdots\right) \tag{5.33}$$

It is thus seen that no even harmonics are excited and that the second harmonic has one-ninth the amplitude of the fundamental. It is shown in treatises on sound that the energy emitted by an oscillating string as sound is

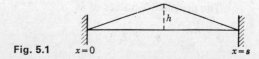

Fig. 5.1 $x = 0$ $x = s$

proportional to the square of the amplitude of the oscillation. It is thus evident that in this case the fundamental tone will appear much louder than the harmonic tone.

6 ORTHOGONAL FUNCTIONS

In the last section a general solution of the wave equation of the vibrating string was constructed by assuming a solution of the form

$$y(x,t) = e^{j\omega t} v(x) \tag{6.1}$$

This assumption, on substitution into the one-dimensional wave equation, led to the ordinary differential equation

$$\frac{d^2 v}{dx^2} + k^2 v = 0 \tag{6.2}$$

where k is given by

$$k = \frac{\omega}{c} \tag{6.3}$$

The general solution of (6.2) may be written in the form

$$v = A \sin kx + B \cos kx \tag{6.4}$$

Every solution of this form is perfectly acceptable as far as the differential equation is concerned, but it does not describe the behavior of the string. This solution permits the ends of the string to vibrate, whereas the physical condition of the problem requires these to be fixed. It is thus necessary to impose the following boundary conditions upon the solution (6.4):

$$v(0) = 0 \quad v(s) = 0 \tag{6.5}$$

The boundary conditions may, of course, be satisfied by placing $A = 0$ and $B = 0$, but this would lead to the trivial and unwanted solution $y = 0$ everywhere. Hence there is left only the arbitrary constant B for adjustment. It must be taken to be zero in order to satisfy the first condition of (6.5). However, the function

$$v = A \sin kx \tag{6.6}$$

will not satisfy the second condition of (6.5). The problem can be solved only if we are willing to prescribe only certain values to the undetermined constant k. If $\sin ks$ must be zero, then k must be 0, or π/s, $2\pi/s$, . . . , $n\pi/s$. The value $k = 0$ is rejected because it leads to the trivial solution $y = 0$.

The permissible values of k,

$$k = \frac{n\pi}{s} \quad n = 1, 2, 3, \ldots \tag{6.7}$$

are called *eigenvalues* in the modern literature of mathematical physics. To each eigenvalue there corresponds an *eigenfunction*

$$v_n = A_n \sin \frac{n\pi x}{s} \tag{6.8}$$

The eigenfunctions under consideration have two important properties: they are (1) orthogonality and (2) completeness.

Orthogonality is defined as

$$\int_0^s v_n(x) v_m(x)\, dx = 0 \qquad \text{if } n \neq m \tag{6.9}$$

The word comes originally from vector analysis, where two vectors **A** and **B** are said to be orthogonal if

$$\mathbf{A} \cdot \mathbf{B} = A_x B_x + A_y B_y + A_z B_z = 0 \tag{6.10}$$

In the same manner vectors in n dimensions having components A_i, B_i $(i = 1, 2, \ldots, n)$ are said to be orthogonal when

$$\sum_{i=1}^{i=n} A_i B_i = 0 \tag{6.11}$$

Imagine now a vector space of an infinite number of dimensions in which the components A_i and B_i become continuously distributed. Then i is no longer a denumerable index but a continuous variable x and the scalar product of (6.11) turns into

$$\int_0^s A(x) B(x)\, dx = 0 \tag{6.12}$$

In this case the functions A and B are said to be orthogonal. The idea of orthogonality is indefinite unless reference is made to a specific range of integration, which in the present case is from 0 to s. The fact that the eigenfunctions (6.8) are orthogonal may be verified at once since we have

$$\int_0^s v_n(x) v_m(x)\, dx = A_n A_m \int_0^s \sin \frac{n\pi x}{s} \sin \frac{m\pi x}{s}\, dx = 0 \qquad \text{if } n \neq m \tag{6.13}$$

If $n = m$, we have

$$\int_0^s v_n^2(x)\, dx = A_n^2 \int_0^s \sin^2 \frac{n\pi x}{s}\, dx = \frac{s}{2} A_n^2 \tag{6.14}$$

We now turn to the notion of *completeness*. A set of functions is said to be complete if an arbitrary function $f(x)$ satisfying the same boundary conditions as the functions of the set can be expanded as follows:

$$f(x) = \sum_{n=1}^{n=\infty} A_n v_n(x) \tag{6.15}$$

where the quantities A_n are constant coefficients. In the present discussion, Eq. (6.15) is equivalent to the theorem of Fourier discussed in Appendix C.

7 THE OSCILLATIONS OF A HANGING CHAIN

Let us consider the small coplanar oscillations of a uniform flexible string or chain hanging from a support under the action of gravity as shown in Fig. 7.1. We consider only small deviations y from the equilibrium position; x is measured from the free end of the chain. Let it be required to determine the position of the chain

$$y = y(x,t) \tag{7.1}$$

where at $t = 0$ we give the chain an arbitrary displacement

$$y = y_0(x) \tag{7.2}$$

In this case the tension T of the chain is variable, and hence Eq. (2.8) governs the displacement of the chain at any instant. Accordingly we have

$$\frac{\partial}{\partial x}\left(T\frac{\partial y}{\partial x}\right) = m\frac{\partial^2 y}{\partial t^2} \tag{7.3}$$

where m is the mass per unit length of the chain. In this case the tension T is given by

$$T = mgx \tag{7.4}$$

Hence we have

$$\frac{\partial}{\partial x}\left(mgx\frac{\partial y}{\partial x}\right) = m\frac{\partial^2 y}{\partial t^2} \tag{7.5}$$

Or, differentiating and dividing both members by the common factor m, we have

$$x\frac{\partial^2 y}{\partial x^2} + \frac{\partial y}{\partial x} = \frac{1}{g}\frac{\partial^2 y}{\partial t^2} \tag{7.6}$$

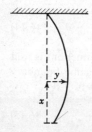

Fig. 7.1

As in the case of the tightly stretched string, let us assume

$$y(x,t) = e^{j\omega t}\, v(x) \tag{7.7}$$

Substituting this into (7.6), we obtain

$$x\frac{d^2 v}{dx^2} + \frac{dv}{dx} + \frac{\omega^2}{g} v = 0 \tag{7.8}$$

This equation resembles Bessel's differential equation [Appendix B, Eq. (9.6)]. The change in variable,

$$Z^2 = \frac{4\omega^2 x}{g} \tag{7.9}$$

reduces (7.8) to

$$Z^2\frac{d^2 v}{dZ^2} + Z\frac{dv}{dZ} + Z^2 v = 0 \tag{7.10}$$

whose general solution is

$$v = A J_0(Z) + B Y_0(Z) \tag{7.11}$$

In order to satisfy the condition that the displacement of the string y remain finite when $x = 0$, we must place

$$B = 0 \tag{7.12}$$

Accordingly, in terms of the original variable x, we have the solution

$$v = A J_0\left(2\omega \sqrt{\frac{x}{g}}\right) \tag{7.13}$$

for the function v.

So far, the value of ω is undetermined. In order to determine it, we make use of the boundary condition

$$v = 0 \qquad \text{at } x = s \tag{7.14}$$

This leads to the equation

$$0 = A J_0\left(2\omega \sqrt{\frac{s}{g}}\right) \tag{7.15}$$

Now, for a nontrivial solution, A cannot be equal to zero, and hence we have

$$J_0\left(2\omega \sqrt{\frac{s}{g}}\right) = 0 \tag{7.16}$$

If we let

$$u = 2\omega \sqrt{\frac{s}{g}} \tag{7.17}$$

we must find the roots of the equation

$$J_0(u) = 0 \tag{7.18}$$

If we consult a table of Bessel functions,† we find that the zeros of the Bessel function $J_0(u)$ are given by the values

2.405, 5.52, 8.654, 11.792, etc.

Accordingly the various possible values of ω are given by

$$\omega_1 = \frac{2.405}{2}\sqrt{\frac{g}{s}} \qquad \omega_2 = \frac{5.52}{2}\sqrt{\frac{g}{s}} \qquad \omega_3 = \frac{8.654}{2}\sqrt{\frac{g}{s}} \qquad \text{etc.} \tag{7.19}$$

To each value of ω we associate a characteristic function or eigenfunction v_n of the form

$$v_n = A_n J_0\left(2\omega_n\sqrt{\frac{x}{g}}\right) \tag{7.20}$$

Since the real and imaginary parts of the assumed solution (7.7) are solutions of the original differential equation, we can construct a general solution of (7.6) satisfying the boundary conditions by summing the particular solutions corresponding to the various possible values of n in the manner

$$y(x,t) = \sum_{n=1}^{n=\infty} J_0\left(2\omega_n\sqrt{\frac{x}{g}}\right)(A_n \cos \omega_n t + B_n \sin \omega_n t) \tag{7.21}$$

where the quantities A_n and B_n are arbitrary constants to be determined from the boundary conditions of the problem. In the case under consideration there is no initial velocity imparted to the chain; hence

$$\left(\frac{\partial y}{\partial t}\right)_{t=0} = 0 \tag{7.22}$$

This leads to the condition

$$B_n = 0 \tag{7.23}$$

At $t = 0$ we have

$$y_0(x) = \sum_{n=1}^{n=\infty} A_n J_0\left(2\omega_n\sqrt{\frac{x}{g}}\right) \tag{7.24}$$

That is, we must expand the arbitrary displacement $y_0(x)$ into a series of Bessel functions to zeroth order. To do this, we can make use of the results

---

† See, for example, N. W. McLachlan, "Bessel Functions for Engineers," Oxford University Press, New York, 1934.

(12.17) and (12.20) of Appendix B. It is there shown that an arbitrary function of $F(x)$ may be expanded in a series of the form

$$F(z) = \sum_{n=1}^{n=\infty} A_n J_0(u_n z) \tag{7.25}$$

where the quantities u_n are successive positive roots of the equation

$$J_n(u) = 0 \tag{7.26}$$

The coefficients A_n are then given by the equation

$$A_n = \frac{2}{J_1^2(u_n)} \int_0^1 z J_0(u_n z) F(z)\, dz \tag{7.27}$$

To make use of this result to obtain the coefficients of the expansion (7.24), it is necessary only to introduce the variable

$$z = \sqrt{\frac{x}{s}} \tag{7.28}$$

In view of (7.17) and (7.18), Eq. (7.24) becomes

$$y_0(x) = y_0(sz^2) = F(z) = \sum_{n=1}^{n=\infty} A_n J_0(u_n z) \tag{7.29}$$

This is in the form (7.25), and the arbitrary constants are determined by (7.27).

The determination of the possible frequencies and modes of oscillation of a hanging chain is of historical interest. It appears to have been the first instance where the various normal modes of a continuous system were determined by Daniel Bernoulli (1732).

8   THE VIBRATIONS OF A RECTANGULAR MEMBRANE

As another example leading to the solution of the wave equation, let us consider the oscillations of a flexible membrane. Let us suppose that the membrane has a density of m g per cm² and that it is pulled evenly around its edge with a tension of T dynes per cm length of edge. If the membrane is perfectly flexible, this tension will be distributed evenly throughout its area; that is, the material on opposite sides of any line segment dx is pulled apart with a force of $T dx$ dynes.

Let us call u the displacement of the membrane from its equilibrium position. It is a function of time and of the position on the membrane of the point in question. If we use rectangular coordinates to locate the point, u will be a function of x, y, and t. Let us consider an element $dx\, dy$ of the membrane shown in Fig. 8.1.

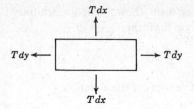

Fig. 8.1

If we refer to the analogous argument of Sec. 2 for the string, we see that the net force normal to the surface of the membrane due to the pair of tensions $T\,dy$ is given by

$$T\,dy\left[\left(\frac{\partial u}{\partial x}\right)_{x+dx} - \left(\frac{\partial u}{\partial x}\right)_x\right] = T\frac{\partial^2 u}{\partial x^2}dx\,dy \tag{8.1}$$

The net normal force due to the pair $T\,dx$ by the same reasoning is

$$T\,dx\left[\left(\frac{\partial u}{\partial y}\right)_{y+dy} - \left(\frac{\partial u}{\partial y}\right)_y\right] = T\frac{\partial^2 u}{\partial y^2}dx\,dy \tag{8.2}$$

The sum of these forces is the net force on the element and is equal to the mass of the element times its acceleration.   That is, we have

$$T\left(\frac{\partial^2 u}{\partial x^2} + \frac{\partial^2 u}{\partial y^2}\right)dx\,dy = m\frac{\partial^2 y}{\partial t^2}dx\,dy \tag{8.3}$$

Dividing out the common factors of both sides of this equation, we obtain

$$\frac{\partial^2 u}{\partial x^2} + \frac{\partial^2 u}{\partial y^2} = \frac{1}{c^2}\frac{\partial^2 u}{\partial t^2} \tag{8.4}$$

where

$$c = \sqrt{\frac{T}{m}} \tag{8.5}$$

Equation (8.4) is the wave equation for the membrane.

Let us consider the oscillations of the rectangular membrane of Fig. 8.2. Let us assume that

$$u = v(x,y)\,e^{j\omega t} \tag{8.6}$$

On substituting this assumption into (8.4) and canceling common factors, we obtain

$$\frac{\partial^2 v}{\partial x^2} + \frac{\partial^2 v}{\partial y^2} + k^2 v = 0 \tag{8.7}$$

where

$$k^2 = \left(\frac{\omega}{c}\right)^2 \tag{8.8}$$

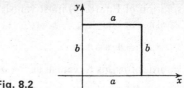

Fig. 8.2

Let us now assume

$$v(x,y) = F_1(x)\,F_2(y) \tag{8.9}$$

that is, that v is the product of two functions, a function only of x and the other one only of y. If we substitute this into (8.7), we obtain

$$F_2 \frac{d^2 F_1}{dx^2} + F_1 \frac{d^2 F_2}{dy^2} + k^2 F_1 F_2 = 0 \tag{8.10}$$

This equation may be written in the form

$$\frac{1}{F_1}\frac{d^2 F_1}{dx^2} = -\frac{1}{F_2}\frac{d^2 F_2}{dy^2} - k^2 \tag{8.11}$$

Now the first member of Eq. (8.11) is a function of x only, while the second member is a function of y only. Now a function of y cannot equal a function of x for all values of x and y if both functions really vary with x and y. Thus the only possible way for the equation to be true for both sides to be independent of both x and y is for it to be a constant. Let us call this constant $-m^2$; we then have

$$\frac{1}{F_1}\frac{d^2 F_1}{dx^2} = -\frac{1}{F_2}\frac{d^2 F_2}{dy^2} - k^2 = -m^2 \tag{8.12}$$

We thus obtain the two equations

$$\frac{d^2 F_1}{dx^2} + m^2 F_1 = 0 \tag{8.13}$$

and

$$\frac{d^2 F_2}{dy^2} + q^2 F_2 = 0 \tag{8.14}$$

where

$$q^2 = k^2 - m^2 \tag{8.15}$$

Now since the membrane is fastened at the edges $x = 0$, $x = a$, $y = 0$, $y = b$, the solution of Eq. (8.13) must vanish at $x = 0$ and $x = a$, while the solution of Eq. (8.14) must vanish at $y = 0$, $y = b$. A solution of (8.13) that vanishes at 0 is

$$F_1 = A_1 \sin mx \tag{8.16}$$

In order for this to vanish at $x = a$, we must have

$$ma = n\pi \qquad n = 1, 2, 3, \ldots \tag{8.17}$$

or

$$m = \frac{n\pi}{a} \tag{8.18}$$

A solution of Eq. (8.14) that vanishes at $y = 0$ is

$$F_2 = A_2 \sin qy \tag{8.19}$$

In order for this solution to vanish at

$$y = b \tag{8.20}$$

we must have

$$qb = r\pi \qquad r = 1, 2, 3, \ldots \tag{8.21}$$

Hence

$$q = \frac{r\pi}{b} \tag{8.22}$$

Now from (8.15) we have

$$k^2 = m^2 + q^2 = \pi^2 \left(\frac{n^2}{a^2} + \frac{r^2}{b^2} \right) \tag{8.23}$$

Hence the possible angular frequencies are given by

$$\omega_{nr} = c\pi \sqrt{\frac{n^2}{a^2} + \frac{r^2}{b^2}} \qquad \begin{cases} n = 1, 2, 3, \ldots \\ r = 1, 2, 3, \ldots \end{cases} \tag{8.24}$$

Substituting (8.16) and (8.19) into (8.9), we obtain

$$v(x, y) = A_1 A_2 \sin \frac{n\pi x}{a} \sin \frac{r\pi y}{b} \tag{8.25}$$

Since the indices r and n may take positive integral values and A_1 and A_2 are arbitrary, we may write instead of (8.25) the function $v(x, y)$ in the form

$$v_{nr} = A_{nr} \sin \frac{n\pi x}{a} \sin \frac{r\pi y}{b} \tag{8.26}$$

If the membrane is excited from rest with an initial displacement $u_0(x, y)$ and with *no* initial velocity, we retain the real part of (8.6) and construct a general solution by summing over all possible values of n and r. We thus obtain

$$u = \sum_{n=1}^{\infty} \sum_{r=1}^{\infty} A_{nr} \cos \omega_{nr} t \sin \frac{n\pi x}{a} \sin \frac{r\pi y}{b} \tag{8.27}$$

We are now confronted with the question of expanding the function $u_0(x,y)$ into the series

$$u_0(x,y) = \sum_{n=1}^{\infty} \sum_{r=1}^{\infty} A_{nr} \sin \frac{n\pi x}{a} \sin \frac{r\pi y}{b} \tag{8.28}$$

Let us first regard y as a constant. In that case $u_0(x,y)$ can be expanded in the form

$$u_0(x,y)_{y=\text{const}} = \sum_{n=1}^{\infty} A_n \sin \frac{n\pi x}{a} \tag{8.29}$$

where we have

$$A_n = \frac{2}{a} \int_0^a u_0(x,y) \sin \frac{n\pi x}{a} dx \tag{8.30}$$

After the integration with respect to x is performed and the limits substituted, regard A_n as a function of y and let it be expanded in terms of $\sin(r\pi y/b)$ by the series

$$A_n = \sum_{r=1}^{r=\infty} A_{nr} \sin \frac{r\pi y}{b} \tag{8.31}$$

where

$$A_{nr} = \frac{4}{ab} \int_0^b dy \int_0^a u_0(x,y) \sin \frac{n\pi x}{a} \sin \frac{r\pi y}{b} dx \tag{8.32}$$

This determines the arbitrary constants A_{nr}, and the subsequent motion is given by (8.27).

From (8.24) and (8.27) it is evident that if any mode is excited for which n or r is greater than unity we have nodal lines parallel to the edges. It also appears that, if the ratio a^2/b^2 is not equal to that of two integers, the frequencies are all distinct. However, if a^2/b^2 is commensurable, some of the periods coincide. The nodal lines may then assume a great variety of forms. The simplest example is the square membrane; in this case we have

$$\omega_{nr}^2 = \frac{c^2 \pi^2}{a^2} (n^2 + r^2) \tag{8.33}$$

for the squares of the angular frequencies.

9   THE VIBRATIONS OF A CIRCULAR MEMBRANE

In the case of the circular membrane we naturally have recourse to polar coordinates with the origin at the center. In this case the equation of motion deduced in Sec. 8 must be transformed from cartesian to polar coordinates. We may write the basic equation of motion of the membrane in the form

$$\nabla^2 u = \frac{1}{c^2}\frac{\partial^2 u}{\partial t^2} \tag{9.1}$$

where in this case ∇^2 is the Laplacian operator in two dimensions. To write this equation in polar coordinates, it is necessary only to write the invariant quantity $\nabla^2 u$ in polar coordinates. Using Eq. (12.20) of Appendix E, we have

$$\nabla^2 u = \frac{1}{r}\frac{\partial}{\partial r}\left(r\frac{\partial u}{\partial r}\right) + \frac{1}{r^2}\frac{\partial^2 u}{\partial \theta^2} \tag{9.2}$$

Accordingly the wave equation for the membrane becomes in this case

$$\frac{1}{r}\frac{\partial}{\partial r}\left(r\frac{\partial u}{\partial r}\right) + \frac{1}{r^2}\frac{\partial^2 u}{\partial \theta^2} = \frac{1}{c^2}\frac{\partial^2 u}{\partial t^2} \tag{9.3}$$

Let us consider the symmetrical case where the motion is started in a symmetrical manner about the origin so that the displacement u is a function only of r and t and independent of the angle θ.   This is the case of symmetrical oscillations about the origin.   Since in this case we have

$$u = u(r,t) \tag{9.4}$$

Eq. (9.3) becomes

$$\frac{1}{r}\frac{\partial}{\partial r}\left(r\frac{\partial u}{\partial r}\right) = \frac{1}{c^2}\frac{\partial^2 u}{\partial t^2} \tag{9.5}$$

Let us now study the symmetrical oscillations of a circular membrane or radius a that is given an initial displacement $u_0(r)$ at $t = 0$.   In this case we have to find a solution of (9.5) that vanishes at $r = a$ and has the initial value $u_0(r)$ at $t = 0$. As in the above examples, let us assume a solution of the form

$$u = v(r)e^{j\omega t} \tag{9.6}$$

Substituting this assumed form of the solution in (9.5), we obtain, on dividing out the factor $e^{j\omega t}$,

$$\frac{1}{r}\frac{d}{dr}\left(r\frac{dv}{dr}\right) + \left(\frac{\omega}{c}\right)^2 v = 0 \tag{9.7}$$

On carrying out the differentiation we have

$$r\frac{d^2 v}{dr^2} + \frac{dv}{dr} + \left(\frac{\omega}{c}\right)^2 vr = 0 \tag{9.8}$$

or

$$\frac{d^2 v}{dr^2} + \frac{1}{r}\frac{dv}{dr} + k^2 v = 0 \tag{9.9}$$

where

$$k = \frac{\omega}{c} \tag{9.10}$$

Equation (9.9) is an equation of the form discussed in Appendix B. Its general solution is

$$v = A J_0(kr) + B Y_0(kr) \tag{9.11}$$

where A and B are arbitrary constants and J_0 and Y_0 are the Bessel functions of the zeroth order and of the first and second kind, respectively. Since the amplitude of the oscillation is finite at the origin, we must have

$$B = 0 \tag{9.12}$$

We must now satisfy the condition that the amplitude of the oscillation must vanish at $r = a$. Hence we must have

$$0 = A J_0(ka) \tag{9.13}$$

For nontrivial solutions A cannot be equal to zero; hence we must have

$$0 = J_0(ka) \tag{9.14}$$

From a table of Bessel functions, we find that Eq. (9.14) is satisfied if ka has values 2.404, 5.520, 8.653, 11.791, 14.93, etc. Accordingly the fundamental angular frequency is given by

$$\omega_1 = \frac{2.404c}{a} \tag{9.15}$$

The other possible angular frequencies are given by the above numbers multiplied by c/a. It is thus seen that there are an infinite number of natural frequencies possible, but these frequencies are *not* multiples of each other as is the case of the vibrating string. By summing over all the possible values of ω we may construct a general solution of (9.5) that satisfies the boundary condition at the periphery of the membrane and the initial condition that at $t = 0$ the membrane has no initial velocity and an arbitrary initial displacement $u_0(r)$. We thus obtain

$$u = \sum_{n=1}^{n=\infty} A_n J_0\left(\frac{\omega_n}{c} r\right) \cos \omega_n t \tag{9.16}$$

Now, at $t = 0$, we have

$$u = u_0(r) \tag{9.17}$$

Accordingly we must expand an arbitrary function $u_0(r)$ in a series of the form

$$u_0(r) = \sum_{n=1}^{n=\infty} A_n J_0\left(\frac{\omega_n r}{c}\right) \tag{9.18}$$

To determine the arbitrary constants A_n, we may use Eq. (12.20) of Appendix B.   We thus obtain

$$A_n = \frac{2}{a^2 J_1^2[(\omega_n/c)a]} \int_0^a rJ_0\left(\omega_n \frac{r}{c}\right) u_0(r)\, dr \tag{9.19}$$

When these values of the constants A_n are substituted into (9.16), we obtain the subsequent displacement of the membrane.

10   THE TELEGRAPHIST'S, OR TRANSMISSION-LINE, EQUATIONS

In this section we shall consider the flow of electricity in a pair of linear conductors such as telephone wires or an electrical transmission line.

Consider a long, imperfectly insulated transmission line, as shown in Fig. 10.1.   Let us consider that an electric current is flowing from the source A to the receiving end B in the direction shown in the figure.   Let the distance measured along the length of the cable be denoted by x; then both the current i and the potential difference e between the two wires are functions of x and t. Denote the resistance per unit length of the wires by R, the conductance per unit length between the two wires by G, the capacitance per unit length of the two wires by C, and the inductance per unit length by L.

Now consider an element of the transmission line of length Δx.   If the electromotive force at the point x is $e(x,t)$, then, if we compute the potential drop along an element of length Δx, we have

$$e(x,t) = iR\,\Delta x + L\,\Delta x\,\frac{\partial i}{\partial t} + e(x + \Delta x,t) \tag{10.1}$$

However, by Taylor's expansion, we have

$$e(x + \Delta x,t) = e(x,t) + \frac{\partial e}{\partial x}\Delta x + \cdots \tag{10.2}$$

where we neglect terms of higher order.   Substituting this into (10.1), we have

$$-\frac{\partial e}{\partial x} = iR + L\frac{\partial i}{\partial t} \tag{10.3}$$

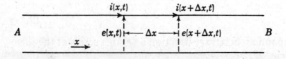

Fig. 10.1

If we denote by $i(x,t)$ the current entering a section of length Δx and by $i(x + \Delta x,t)$ the current leaving this element, we have

$$i(x,t) = eG\,\Delta x + \frac{\partial e}{\partial t} C\,\Delta x + i(x + \Delta x,t) \tag{10.4}$$

where $eG\,\Delta x$ is the current that leaks through the insulation and $(\partial e/\partial t)C\Delta x$ is the current involved in charging the capacitor formed by the proximity of the two conductors forming the transmission line. By Taylor's expansion we have

$$i(x + \Delta x,t) = i(x,t) + \frac{\partial i}{\partial x} \Delta x + \cdots \tag{10.5}$$

where we neglect higher-order terms. Substituting this into (10.4), we obtain

$$-\frac{\partial i}{\partial x} = eG + C\frac{\partial e}{\partial t} \tag{10.6}$$

Equations (10.3) and (10.6) are simultaneous partial differential equations for the potential difference and the current of the transmission line. To eliminate the potential difference, we take the partial derivative of (10.6) with respect to x. We then obtain

$$-\frac{\partial^2 i}{\partial x^2} = G\frac{\partial e}{\partial x} + C\frac{\partial}{\partial t}\left(\frac{\partial e}{\partial x}\right) \tag{10.7}$$

We then substitute the value of $\partial e/\partial x$ given by (10.3) into (10.7) and thus obtain

$$\frac{\partial^2 i}{\partial x^2} = LC\frac{\partial^2 i}{\partial t^2} + (LG + RC)\frac{\partial i}{\partial t} + RGi \tag{10.8}$$

In the same manner we have, on eliminating the current,

$$\frac{\partial^2 e}{\partial x^2} = LC\frac{\partial^2 e}{\partial t^2} + (LG + RC)\frac{\partial e}{\partial t} + RGe \tag{10.9}$$

Equations (10.8) and (10.9) are sometimes known as the *telephone equations* since they are used in discussing telephonic transmission.

In many applications to telegraph signaling the leakage G is small and the term for the effect of inductance L is negligible, so that we may place

$$G = L = 0 \tag{10.10}$$

The equations then take the simplified form

$$\frac{\partial^2 i}{\partial x^2} = RC\frac{\partial i}{\partial t} \tag{10.11}$$

and

$$\frac{\partial^2 e}{\partial x^2} = RC \frac{\partial e}{\partial t} \tag{10.12}$$

These are known as the *telegraph*, or *cable*, *equations*.

For high frequencies the terms in the time derivatives are large, and some qualitative properties of the solution may be found by neglecting the terms for the effect of leakage and resistance in comparison with them. On placing

$$G = R = 0 \tag{10.13}$$

the equations become

$$\frac{\partial^2 i}{\partial x^2} = LC \frac{\partial^2 i}{\partial t^2} \tag{10.14}$$

$$\frac{\partial^2 e}{\partial x^2} = LC \frac{\partial^2 e}{\partial t^2} \tag{10.15}$$

If we let

$$v = \frac{1}{\sqrt{LC}} \tag{10.16}$$

then we see that in this case both the current and the potential difference between the lines satisfy the one-dimensional wave equation

$$\frac{\partial^2 u}{\partial x^2} = \frac{1}{v^2} \frac{\partial^2 u}{\partial t^2} \tag{10.17}$$

This equation has the same form as that governing the displacement of the tightly stretched oscillating string. We thus see that a transmission line with negligible resistance and leakage propagates waves of current and potential with a velocity equal to $1/\sqrt{LC}$. The study of Eqs. (10.3) and (10.6) is fundamental in the theory of electrical power transmission and telephony.

THE DISTORTIONLESS TRANSMISSION LINE

A very interesting special case occurs when the parameters of the transmission line satisfy the relation

$$\frac{R}{L} = \frac{G}{C} \tag{10.18}$$

The amount of leakage indicated by this equation is reduced by increasing the inductance parameter L. A line having this relation between its parameters is called a *distortionless line*. This type of line is of considerable importance in telephony and telegraphy.

To analyze this special case, let

$$a^2 = RG \tag{10.19}$$

Now we have

$$RC = LG \tag{10.20}$$

Hence

$$R^2 C^2 = (LG)(RC) = (RG)(LC)$$
$$= \frac{a^2}{v^2} \tag{10.21}$$

But

$$LG + RC = 2RC = \frac{2a}{v} \tag{10.22}$$

Hence Eq. (10.9) in this case becomes

$$\frac{\partial^2 e}{\partial x^2} = \frac{1}{v^2}\frac{\partial^2 e}{\partial t^2} + \frac{2a}{v}\frac{\partial e}{\partial t} + a^2 e \tag{10.23}$$

Let us now introduce the variable $y(x,t)$ defined by the equation

$$e(x,t) = e^{-avt} y(x,t) \tag{10.24}$$

If we perform the indicated differentiation and substitute the results into (10.23), we obtain after some reductions

$$\frac{\partial^2 y}{\partial x^2} = \frac{1}{v^2}\frac{\partial^2 y}{\partial t^2} \tag{10.25}$$

This is a one-dimensional wave equation in y, and its general solution is

$$y = F(x - vt) + G(x + vt) \tag{10.26}$$

where F and G denote arbitrary functions of the arguments $x - vt$ and $x + vt$, respectively. Now

$$av = (RC)v^2 = \frac{R}{L} \tag{10.27}$$

Hence in view of (10.24) we have

$$e(x,t) = e^{-Rt/L}[F(x - vt) + G(x + vt)] \tag{10.28}$$

We thus see that the general solution of (10.9) in this case is given by the superposition of two arbitrary waves that travel without distortion to the left and to the right along the line with velocity of v but with an attenuation given by the factor $e^{-Rt/L}$.

LINE WITHOUT LEAKAGE

In most practical power-transmission lines the insulation between the line wires is so good that the leakage conductance coefficient G may be taken equal to zero.　In such a case the equations for the potential and current of the transmission line become

$$\frac{\partial^2 e}{\partial x^2} = LC \frac{\partial^2 e}{\partial t^2} + RC \frac{\partial e}{\partial t} \tag{10.29}$$

$$\frac{\partial^2 i}{\partial x^2} = LC \frac{\partial^2 i}{\partial t^2} + RC \frac{\partial i}{\partial t} \tag{10.30}$$

To solve (10.30), we let

$$i = y e^{-Rt/2L} \tag{10.31}$$

Making this substitution in (10.30), we obtain after some reductions

$$\frac{\partial^2 y}{\partial x^2} = \frac{1}{v^2} \frac{\partial^2 y}{\partial t^2} - \frac{a^2}{v^2} y \qquad a = \frac{R}{2L} \tag{10.32}$$

Now let

$$z = a \sqrt{t^2 - \frac{x^2}{v^2}} \tag{10.33}$$

We have

$$\frac{\partial^2 y}{\partial x^2} = \frac{d^2 y}{dz^2} \left(\frac{\partial z}{\partial x}\right)^2 + \frac{dy}{dz} \frac{\partial^2 z}{\partial x^2} \tag{10.34}$$

$$\frac{\partial^2 y}{\partial t^2} = \frac{d^2 y}{dz^2} \left(\frac{\partial z}{\partial t}\right)^2 + \frac{dy}{dz} \frac{\partial^2 z}{\partial t^2} \tag{10.35}$$

On substituting these expressions for $\partial^2 y/\partial x^2$ and $\partial^2 y/\partial t^2$ in (10.32), we obtain after some reductions

$$\frac{d^2 y}{dz^2} + \frac{1}{z} \frac{dy}{dz} - y = 0 \tag{10.36}$$

This equation is the modified Bessel differential equation of the zeroth order discussed in Appendix B, Sec. 10.　Its general solution is

$$y = AI_0(z) + BK_0(z) \tag{10.37}$$

where $I_0(z)$ and $K_0(z)$ are modified Bessel functions of the zeroth order of the first and second kind and A and B are arbitrary constants.　Since the function $K_0(z)$ does not remain finite at $z = 0$, we must have $B = 0$.　Hence the solution for the current is

$$i = e^{-Rt/2L} AI_0(z) \tag{10.38}$$

The constant A must be determined from a knowledge of the initial conditions of the transmission line.

THE STEADY-STATE SOLUTION

Let us consider the transmission line shown in Fig. 10.2.　In this case we consider the effect of impressing a harmonic potential difference of the form

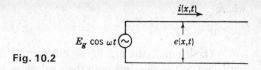

Fig. 10.2

$E_g \cos \omega t$ at one end of the line.   This case is of great importance in power transmission and in communication networks.

In this case we impress on the line at $x = 0$ a potential difference of the form

$$E_g \cos \omega t = \text{Re}\,(E_g\, e^{j\omega t}) \tag{10.39}$$

where Re denotes the "real part of."   To solve Eqs. (10.3) and (10.6) in this case, we assume a solution of the form

$$e(x,t) = \text{Re}\,[E(x)\,e^{j\omega t}]$$
$$i(x,t) = \text{Re}\,[I(x)\,e^{j\omega t}] \tag{10.40}$$

We now substitute these assumed forms of the solution in (10.3) and (10.6) and drop the Re sign, remembering that to get the instantaneous potential difference $e(x,t)$ and the instantaneous current $i(x,t)$ we must use Eqs. (10.40).   On substituting the above assumed form of the solution into (10.3) and (10.6) and dividing out the common factor $e^{j\omega t}$ we obtain

$$\frac{dE}{dx} + (R + j\omega L)I = 0$$
$$\frac{dI}{dx} + (G + j\omega C)E = 0 \tag{10.41}$$

It is convenient to introduce the notation

$$Z = R + j\omega L$$
$$Y = G + j\omega C \tag{10.42}$$

Then Eqs. (10.41) become

$$\frac{dE}{dx} + ZI = 0$$
$$\frac{dI}{dx} + YE = 0 \tag{10.43}$$

We may eliminate E and I by differentiating each equation with respect to x and substituting either dE/dx or dI/dx from the other.   We thus obtain

$$\frac{d^2 E}{dx^2} = a^2 E$$
$$\frac{d^2 I}{dx^2} = a^2 I \tag{10.44}$$

where

$$a^2 = ZY \tag{10.45}$$

These equations are linear differential equations of the second order with constant coefficients, and their solutions are

$$E(x) = A_1 e^{-ax} + A_2 e^{ax}$$
$$I(x) = B_1 e^{-ax} + B_2 e^{ax} \tag{10.46}$$

The evaluation of the arbitrary constants A_1, A_2, B_1, B_2 requires a knowledge of the terminal conditions of the transmission line. The evaluation of these arbitrary constants in the case of general terminal conditions is straightforward but rather tedious.†

To illustrate the general procedure, let us consider the case of a line short-circuited at $x = s$, as shown in Fig. 10.3. In this case, we have the boundary conditions

$$E(0) = E_g$$
$$E(s) = 0 \tag{10.47}$$

These boundary conditions lead to the two simultaneous equations in the arbitrary constants A_1 and A_2,

$$A_1 + A_2 = E_g$$
$$A_1 e^{-as} + A_2 e^{as} = 0 \tag{10.48}$$

Solving for A_1 and A_2 and substituting the results into (10.46), we obtain

$$E(x) = \frac{E_g}{e^{as} - e^{-as}} (e^{a(s-x)} - e^{-a(s-x)}) \tag{10.49}$$

This result may be written more concisely in terms of hyperbolic functions in the form

$$E(x) = \frac{E_g \sinh a(s - x)}{\sinh as} \tag{10.50}$$

† See, for example, E. A. Guillemin, "Communication Networks," chap. 2, John Wiley & Sons, Inc., New York, 1935.

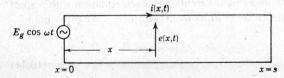

Fig. 10.3

We may obtain $I(x)$ directly from (10.43) in the form

$$I(x) = -\frac{1}{Z}\frac{dE}{dx}$$

$$= \frac{E_g a}{Z}\frac{\cosh a(s-x)}{\sinh as} \tag{10.51}$$

If we introduce the notation

$$Z_0 = \sqrt{\frac{Z}{Y}} \tag{10.52}$$

then (10.51) may be written in the form

$$I(x) = \frac{E_g}{Z_0}\frac{\cosh a(s-x)}{\sinh as} \tag{10.53}$$

Since a is in general complex, we see that $E(x)$ and $I(x)$ are complex functions of x. To obtain the instantaneous potential $e(x,t)$ and current $i(x,t)$, we must use Eqs. (10.40). An interesting physical significance of the solution may be obtained by considering an indefinitely long line. Since in general a is a complex number, we may write

$$a = a_1 + ja_2 \tag{10.54}$$

In this case the arbitrary constant A_2 in (10.46) must be equal to zero, since if it were not, the absolute value of $E(x)$ would increase indefinitely. Hence in this case we have

$$E(x) = E_g e^{-ax} = E_g e^{-a_1 x} e^{-ja_2 x} \tag{10.55}$$

since

$$E(0) = E_g$$

The instantaneous potential is then given by (10.40) in the form

$$e(x,t) = \text{Re}\,(E_g e^{-ax} e^{j\omega t})$$

$$= E_g e^{-a_1 x} \cos(\omega t - a_2 x) \tag{10.56}$$

That is, as we travel along the infinite line, the potential is diminished by the factor $e^{-a_1 x}$ and a phase difference $a_2 x$ is introduced. In this case we have

$$I(x) = \frac{E_g}{Z_0} e^{-ax} \tag{10.57}$$

and the instantaneous current is given by (10.40). If there is no dissipation of energy along the line, then $R = 0$, $G = 0$; hence

$$a = j\omega\sqrt{LC} \tag{10.58}$$

or

$$a_1 = 0 \qquad a_2 = \omega \sqrt{LC} \tag{10.59}$$

In this case there is no attenuation of potential or current along the line, but we have

$$e(x,t) = E_g \cos \omega(t - x\sqrt{LC}) \tag{10.60}$$

This represents a harmonic wave having a phase velocity of $1/\sqrt{LC}$. The current in this case is given by

$$i(x,t) = E_g \sqrt{\frac{C}{L}} \cos \omega(t - x\sqrt{LC}) \tag{10.61}$$

The study of more complicated boundary conditions may be undertaken in the same manner, beginning from Eqs. (10.46). An excellent discussion of the general solution will be found in the above-mentioned reference.

11 TIDAL WAVES IN A CANAL

Let us consider an incompressible liquid of uniform depth h. The waves which occur on the surface of a liquid may be divided into the following two types:

1. *Ripples:* These are waves whose wavelength is small, being of the order of $\frac{2}{3}$ in. These waves owe their propagation to the surface tension in the liquid-air surface and to gravity.
2. *Tidal waves:* These waves are characterized by having such a long wavelength that the effect of surface tension can be neglected. In the case of these waves the vertical displacement of the surface is small in comparison with the wavelength.

In this section the theory of tidal waves will be discussed. Consider Fig. 11.1. Let the bottom of the liquid be given by the plane $y = 0$. Let y be

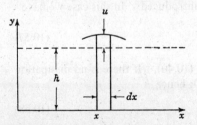

Fig. 11.1

measured so that the y axis points upward. Let the free surface in the xy plane be given by

$$y_s = h + u \tag{11.1}$$

where h is the depth of the liquid in the undisturbed state and u is the height of the wave above the undisturbed surface of the liquid.

Let the liquid be bounded in the z direction by fixed planes parallel to the xy plane. It will be assumed that these planes are smooth and the liquid slips along them without any frictional resistance.

As a consequence of this assumption the distance between these planes is of no consequence. However, for simplicity, it will be assumed that they are a unit distance apart; that is, the liquid is constrained to move in a uniform canal of unit width and depth h.

Let us use the following notation:

$$p_0 = \text{pressure at surface of liquid} \tag{11.2}$$

$$\rho = \text{density of liquid} \tag{11.3}$$

$$g = \text{gravitational constant} \tag{11.4}$$

If we neglect the vertical acceleration of the particles of the liquid, we may write the following expression for the pressure at the point x, y:

$$p(x,y) = p_0 + g\rho(y_s - y)$$
$$= p_0 + g\rho(h + u - y) \tag{11.5}$$

as a consequence of (11.1).

Now let us consider the vertical strip which is bounded by the planes x and $x + dx$ before the wave motion begins. The volume of this strip is $h\,dx$. An instant after the motion is started, let the elevation of the surface of the strip be $y_s = h + u$, and let the planes bounding it have moved to $x + v$ and $x + dx + v + (\partial v/\partial x)\,dx$. The volume of the strip is then $(h + u)[dx + (\partial v/\partial x)\,dx]$. Since the liquid is assumed incompressible, the volume of the strip does not vary with the time. Hence we have

$$h\,dx = (h + u)\left(dx + \frac{\partial v}{\partial x}\,dx\right) \tag{11.6}$$

or

$$u + h\frac{\partial v}{\partial x} = 0 \tag{11.7}$$

where the second term has been neglected since it is a quantity of higher order.

THE RESULTANT FORCE ON THE STRIP

Let the strip under consideration be divided into elements parallel to the bottom. Consider one of these elements of height dy. The force on this element is given by

$$p\, dy - \left(p + \frac{\partial p}{\partial x} dx\right) dy = -\frac{\partial p}{\partial x} dx\, dy \tag{11.8}$$

But from Eq. (11.5) we have

$$\frac{\partial p}{\partial x} = g\rho \frac{\partial u}{\partial x} \tag{11.9}$$

since u is the only function of x in (11.5). Hence the force on the element is given by

$$-\frac{\partial p}{\partial x} dx\, dy = -g\rho \frac{\partial u}{\partial x} dx\, dy \tag{11.10}$$

The force F_x on the whole strip is obtained by integrating the force on the element of height dy, and we have

$$F_x = -g\rho \frac{\partial u}{\partial x} dx \int_0^h dy = -g\rho h \frac{\partial u}{\partial x} dx \tag{11.11}$$

This is the rate of increase in momentum by Newton's second law of motion:

$$F_x = h\rho\, dx \frac{\partial^2 v}{\partial t^2} = -g\rho h \frac{\partial u}{\partial x} dx \tag{11.12}$$

However, differentiating (11.7), we have

$$\frac{\partial u}{\partial x} = -h \frac{\partial^2 v}{\partial x^2} \tag{11.13}$$

Substituting this into (11.12) and dividing by the common factor, we obtain

$$\frac{\partial^2 v}{\partial x^2} = \frac{1}{gh} \frac{\partial^2 v}{\partial t^2} \tag{11.14}$$

This is the one-dimensional wave equation discussed in Sec. 3. The velocity of propagation of long waves is therefore

$$c = \sqrt{gh} \tag{11.15}$$

Equation (11.14) may be used to study the pattern of long water waves in a rectangular trough whose length is much greater than its breadth and depth. The conditions at the ends of the trough are $v = 0$ at $x = 0$ and $v = 0$ at $x = s$, where s is the length of the trough.

This analysis is similar to that of the vibrating string of length s considered in Sec. 5. The *period* of the fundamental oscillation is $2s/\sqrt{gh}$. If the restriction that the long waves move only in one dimension is removed, then a similar analysis shows that the quantity v above satisfies the two-dimensional wave equation.

$$\frac{\partial^2 v}{\partial x^2} + \frac{\partial^2 v}{\partial y^2} = \frac{1}{c^2} \frac{\partial^2 v}{\partial t^2} \qquad \text{where } c = \sqrt{gh} \qquad (11.16)$$

12 SOUND WAVES IN A GAS

An important class of problems involving wave propagation is that of waves in a *compressible* fluid such as a gas. The case of sound waves is a particular example of this type of wave motion.

To derive the equation for the propagation of sound in a compressible gas, we recall (see Appendix E) that the general equation of continuity for a fluid is

$$\frac{\partial \rho}{\partial t} + \nabla \cdot (\rho \mathbf{v}) = 0 \qquad (12.1)$$

where ρ is the density of the fluid and $\mathbf{v}$ the vector velocity of the fluid defined by

$$\mathbf{v} = \mathbf{i} v_x + \mathbf{j} v_y + \mathbf{k} v_z \qquad (12.2)$$

where v_x, v_y, and v_z are the components of the velocity of the fluid in a cartesian frame of reference.

The Euler equation of motion in vector form is

$$\frac{\partial \mathbf{v}}{\partial t} + (\mathbf{v} \cdot \nabla) \mathbf{v} = \mathbf{F} - \frac{1}{\rho} \nabla p \qquad (12.3)$$

where p is the pressure of the fluid and $\mathbf{F}$ is the "body" force per unit mass.

These fundamental equations will now be applied to the case of wave motion in a perfect gas. The following assumptions will be made:

1. The motion is *irrotational* so that there exists a velocity potential ϕ defined by the equation

$$v = -\nabla \phi \qquad (12.4)$$

2. The velocities are so small that their squares and products can be neglected.
3. The body force $\mathbf{F}$ can be neglected.

Let us write

$$\rho = \rho_0 (1 + s) \qquad (12.4a)$$

where ρ_0 is the initial value of the density. The quantity s is called the *condensation*. Since the changes in density in a sound wave are small, the condensation s is small. It will be assumed that as a consequence of this the term $\nabla \cdot s$ can be neglected in the continuity equation.

Substituting (12.4) into (12.1), we obtain

$$\frac{\partial}{\partial t}[\rho_0(1+s)] + \nabla \cdot [\rho_0(1+s)\mathbf{v}] = 0 \tag{12.5}$$

By the assumption that s is small this equation becomes

$$\frac{\partial s}{\partial t} = -\nabla \cdot \mathbf{v} \tag{12.6}$$

In terms of the velocity potential of Eq. (12.4) this becomes

$$\frac{\partial s}{\partial t} = \nabla^2 \phi \tag{12.7}$$

Neglecting the squares and products of the velocities in Eq. (12.3) and writing the resulting vector equation in scalar form, we obtain

$$\frac{\partial}{\partial t}\frac{\partial \phi}{\partial x} = \frac{1}{\rho}\frac{\partial p}{\partial x}$$
$$\frac{\partial}{\partial t}\frac{\partial \phi}{\partial y} = \frac{1}{\rho}\frac{\partial p}{\partial y} \tag{12.8}$$
$$\frac{\partial}{\partial t}\frac{\partial \phi}{\partial z} = \frac{1}{\rho}\frac{\partial p}{\partial z}$$

If the first equation is multiplied by dx, the second one by dy, the third by dz and the results are added, we obtain

$$\frac{\partial}{\partial t}\left(\frac{\partial \phi}{\partial x}dx + \frac{\partial \phi}{\partial y}dy + \frac{\partial \phi}{\partial z}dz\right) = \frac{1}{\rho}\left(\frac{\partial p}{\partial x}dx + \frac{\partial p}{\partial y}dy + \frac{\partial p}{\partial z}dz\right) \tag{12.9}$$

or

$$\frac{\partial}{\partial t}(d\phi) = \frac{dp}{\rho} \tag{12.10}$$

Integrating, we obtain

$$\frac{\partial \phi}{\partial t} = \int \frac{dp}{\rho} = \text{const} \tag{12.11}$$

THE ISOTHERMAL ASSUMPTION (NEWTON'S EQUATION)

Isaac Newton assumed that the condensation and rarefaction process involved in the propagation of sound waves takes place in such a manner that the *temperature* of each element of mass is constant. In this case we may assume that Boyle's law holds, and we have

$$p = c\rho \tag{12.12}$$

where ρ is the density, p the pressure, and c a constant. Then we have

$$dp = c\,d\rho = c\rho_0\,ds \tag{12.13}$$

by Eq. (12.4a).

Consequently we have

$$\int \frac{dp}{\rho} = \int \frac{c\,ds}{1+s} = c\ln(1+s) \tag{12.14}$$

Substituting this into (12.11), we have

$$\frac{\partial \phi}{\partial t} = c\ln(1+s) = \text{const} \tag{12.15}$$

Differentiating with respect to t, we obtain

$$\frac{\partial^2 \phi}{\partial t^2} = \frac{c}{1+s}\frac{\partial s}{\partial t} = c\frac{\partial s}{\partial t} \tag{12.16}$$

since $s \ll 1$.

Eliminating $\partial s/\partial t$ between (12.16) and (12.7), we have

$$\nabla^2 \phi = \frac{1}{c}\frac{\partial^2 \phi}{\partial t^2} \tag{12.17}$$

This isothermal assumption gives the general wave equation (12.17). The velocity of propagation v_0 is

$$v_0 = \sqrt{c} = \sqrt{\frac{p}{\rho}} \tag{12.18}$$

This is the well-known expression derived by Newton for the velocity of sound in a gas. However, it does not agree with experiment since its value is too small.

THE ADIABATIC ASSUMPTION OF LAPLACE

Laplace showed that, in the case of sound waves, the condensation and rarefaction take place so rapidly that the heat produced has not time to disappear by conduction. Hence the *temperature* of each element of mass will not be constant, but its quantity of heat will remain constant. This process is not an isothermal one but an adiabatic one.

In this case, instead of Boyle's law (12.12) we have the adiabatic relation

$$p = c\rho^k \tag{12.19}$$

where c and k are constants. The constant k is given by the following relation:

$$k = \frac{\text{specific heat at constant pressure}}{\text{specific heat at constant volume}} \tag{12.20}$$

k is of the order of 1.41 for air, oxygen, hydrogen, and nitrogen.

As a consequence of Eqs. (12.19) and (12.4), we have

$$dp = ck\rho^{k-1} dp = c\rho_0^k k(1 + s)^{k-1} ds \tag{12.21}$$

We also have

$$\int \frac{dp}{\rho} = ck\rho_0^{k-1} \int (1 + s)^{k-2} ds = \frac{ck\rho_0^{k-1}}{k - 1} (1 + s)^{k-1} \tag{12.22}$$

Hence by Eq. (12.11)

$$\frac{\partial^2 \phi}{\partial t^2} = \frac{\partial}{\partial t} \int \frac{dp}{\rho} = ck\rho_0^{k-1}(1 + s)^{k-2} \frac{\partial s}{\partial t} \tag{12.23}$$

Since s is small, $(1 + s)^{k-2} \approx 1$ and

$$c\rho_0^{k-1} = \frac{p_0}{\rho_0} \tag{12.24}$$

as a consequence of (12.19).

Equation (12.23) then may be written in the form

$$\frac{\partial^2 \phi}{\partial t^2} = \frac{k\rho_0}{\rho_0} \frac{\partial s}{\partial t} \tag{12.25}$$

Using (12.7) to eliminate $\partial s/\partial t$, we obtain

$$\nabla^2 \phi = \frac{\rho_0}{k\rho_0} \frac{\partial^2 \phi}{\partial t^2} \tag{12.26}$$

and the velocity of propagation of the wave is

$$v_0 = \sqrt{\frac{kp_0}{\rho_0}} \tag{12.27}$$

This result agrees well with experiment. The velocity of sound in air at standard conditions (20°C, 760 mm Hg) is 34,400 cm per sec.

13 THE MAGNETIC VECTOR POTENTIAL

In this section a very important method of integrating the equations of electrodynamics will be discussed. This method of integration leads to an equation known in the literature as the *inhomogeneous wave equation*. The integration of this equation is discussed in Sec. 14.

The general equations of electrodynamics, or Maxwell's equations, are formulated in Appendix E, Sec. 16. They are

$$\nabla \times \mathbf{E} = -\frac{\partial \mathbf{B}}{\partial t} \tag{13.1}$$

$$\nabla \times \mathbf{H} = \mathbf{J} + \frac{\partial \mathbf{D}}{\partial t} \tag{13.2}$$

$$\nabla \cdot \mathbf{B} = 0 \tag{13.3}$$

$$\nabla \cdot \mathbf{D} = \rho \tag{13.4}$$

The significance of the various symbols is discussed in Appendix E.

In certain problems of electrodynamics the current-density distribution $\mathbf{J}$ and the charge density ρ are specified as given functions of space and time. It is then required to determine the electric intensity $\mathbf{E}$ and the magnetic intensity $\mathbf{H}$.

In addition to the above equations we have the following relations in a homogeneous isotropic medium:

$$\mathbf{D} = K\mathbf{E} \tag{13.5}$$

$$\mathbf{B} = \mu\mathbf{H} \tag{13.6}$$

$$\mathbf{J} = \sigma\mathbf{E} \tag{13.7}$$

In order to integrate the Maxwell equations, let us introduce a vector $\mathbf{A}$, defined by the following relation:

$$\mathbf{B} = \nabla \times \mathbf{A} \tag{13.8}$$

The vector $\mathbf{A}$ is called the *magnetic vector potential*.

Substituting (13.8) into (13.1), we obtain

$$\nabla \times \mathbf{E} = -\frac{\partial}{\partial t}(\nabla \times \mathbf{A}) = -\nabla \times \frac{\partial \mathbf{A}}{\partial t} \tag{13.9}$$

since the order of time differentiation and space differentiation may be interchanged.

Equation (13.9) may be written in the form

$$\nabla \times \left(\mathbf{E} + \frac{\partial \mathbf{A}}{\partial t}\right) = 0 \tag{13.10}$$

This shows that the vector $\mathbf{E} + \partial\mathbf{A}/\partial t$ is irrotational and may therefore be expressed as the gradient of a scalar point function in the form

$$\mathbf{E} + \frac{\partial \mathbf{A}}{\partial t} = -\nabla\phi \tag{13.11}$$

(ϕ is called the *scalar potential*) or

$$\mathbf{E} = -\frac{\partial \mathbf{A}}{\partial t} - \nabla\phi \tag{13.12}$$

If we multiply Eq. (13.2) by μ, we obtain

$$\nabla \times \mathbf{B} = \mu\mathbf{J} + \mu\frac{\partial \mathbf{D}}{\partial t} \tag{13.13}$$

in view of Eq. (13.6).

Now

$$\nabla \times \mathbf{B} = \nabla \times (\nabla \times \mathbf{A}) = \nabla(\nabla \cdot \mathbf{A}) - \nabla^2 \mathbf{A} \tag{13.14}$$

by Appendix E, Eq. (11.2).   If this result is substituted into (13.13), we have

$$\nabla(\nabla \cdot \mathbf{A}) - \nabla^2 \mathbf{A} = \mu \mathbf{J} + \mu K \frac{\partial \mathbf{E}}{\partial t} \tag{13.15}$$

Equation (13.12) may be differentiated with respect to t, and the result is

$$\frac{\partial \mathbf{E}}{\partial t} = -\frac{\partial^2 \mathbf{A}}{\partial t^2} - \nabla \frac{\partial \phi}{\partial t} \tag{13.16}$$

The quantity $\partial \mathbf{E}/\partial t$ may now be eliminated between Eqs. (13.16) and (13.15), and the resulting equation is

$$\nabla(\nabla \cdot \mathbf{A}) - \nabla^2 \mathbf{A} = \mu \mathbf{J} + \mu K \left(-\frac{\partial^2 \mathbf{A}}{\partial t^2} - \nabla \frac{\partial \phi}{\partial t} \right) \tag{13.17}$$

or

$$-\nabla^2 \mathbf{A} = \mu \mathbf{J} - \mu K \frac{\partial^2 \mathbf{A}}{\partial t^2} - \nabla \left(\nabla \cdot \mathbf{A} + \mu K \frac{\partial \phi}{\partial t} \right) \tag{13.18}$$

Now the curl of the magnetic vector potential $\mathbf{A}$ is specified by Eq. (13.8). The divergence of $\mathbf{A}$ has not been specified.

In order to determine the vector $\mathbf{A}$ uniquely, both its curl and its divergence must be specified.   Let

$$\nabla \cdot \mathbf{A} = -\mu K \frac{\partial \phi}{\partial t} \tag{13.19}$$

Then Eq. (13.18) may be written in the following form:

$$\nabla^2 \mathbf{A} - \mu K \frac{\partial^2 \mathbf{A}}{\partial t^2} = -\mu \mathbf{J} \tag{13.20}$$

Equation (13.4) may be written in the form

$$\nabla \cdot \mathbf{E} = \frac{\rho}{K} \tag{13.21}$$

However, $\mathbf{E}$ is given in terms of $\mathbf{A}$ and ϕ by Eq. (13.12).   Hence, if we substitute this value of $\mathbf{E}$ in (13.21), we obtain

$$\nabla \cdot \left(-\frac{\partial \mathbf{A}}{\partial t} - \nabla \phi \right) = \frac{\rho}{K} \tag{13.22}$$

or

$$-\frac{\partial}{\partial t}(\nabla \cdot \mathbf{A}) - \nabla^2 \phi = \frac{\rho}{K} \tag{13.23}$$

Eliminate $\nabla \cdot \mathbf{A}$ between Eqs. (13.19) and (13.23), and we have

$$\nabla^2 \phi - \mu K \frac{\partial^2 \phi}{\partial t^2} = -\frac{\rho}{K} \tag{13.24}$$

Let

$$c = \frac{1}{\sqrt{\mu K}} \tag{13.25}$$

Then Eqs. (13.20) and (13.24) may be written in the form

$$\nabla^2 \mathbf{A} - \frac{1}{c^2} \frac{\partial^2 \mathbf{A}}{\partial t^2} = -\mu \mathbf{J} \tag{13.26}$$

and

$$\nabla^2 \phi - \frac{1}{c^2} \frac{\partial^2 \phi}{\partial t^2} = -\frac{\rho}{K} \tag{13.27}$$

These two equations have the same form. In the literature of electro-dynamics they are known as the *inhomogeneous wave equations*, or Lorenz's equations. The quantity c is the velocity of propagation of electromagnetic waves. If the variation of the current density $\mathbf{J}$ and the charge density ρ is given in space and time, then Eqs. (13.26) and (13.27) may be solved for $\mathbf{A}$ and ϕ and the field vectors $\mathbf{B}$ and $\mathbf{E}$ determined by Eqs. (13.8) and (13.12). The solution of Eqs. (13.26) and (13.27) is discussed in the next section.

14   THE INHOMOGENEOUS WAVE EQUATION

In the last section it was demonstrated that both the magnetic vector potential $\mathbf{A}$ and the scalar potential ϕ are propagated in accordance with equations of the type

$$\nabla_u^2 - \frac{1}{c^2} \frac{\partial^2 u}{\partial t^2} = -f(x,y,z,t) \tag{14.1}$$

This is the *inhomogeneous wave equation*, or Lorenz's equation. $f(x,y,z,t)$ is a given function of the space variables and the time t.

Let it be required to solve this equation subject to the *initial conditions*

$$\left. \begin{aligned} u &= 0 \\ \frac{\partial u}{\partial t} &= 0 \end{aligned} \right\} \quad \text{at } t = 0 \tag{14.2}$$

In order to solve this equation, the method of the Laplace transform described in Chap. 4 may be used. Let us introduce the following two transforms,

$$Lu(x,y,z,t) = U(x,y,z,s) \tag{14.3}$$

and

$$Lf(x,y,z,t) = F(x,y,z,s) \tag{14.4}$$

That is, we introduce a Laplace transform with respect to the time variable. By Theorem (III) of Chap. 4, we have, using (14.2),

$$L\frac{\partial^2 u}{\partial t^2} \, s^2 \, U \tag{14.5}$$

Hence, by the Laplace transformation, Eq. (14.1) is transformed to

$$\nabla^2 U - \frac{s^2}{c^2} U = -F \tag{14.6}$$

Let

$$k^2 = -\frac{s^2}{c^2} \qquad k = \pm j\frac{s}{c} \tag{14.7}$$

Then Eq. (14.6) may be written in the form

$$\nabla^2 U + k^2 U + F = 0 \tag{14.8}$$

This equation is known in the literature as *Helmholtz's equation*.

Consider the equation

$$\nabla^2 U_0 + k^2 U_0 = 0 \tag{14.9}$$

A particular solution of this equation is

$$U_0 = \frac{e^{\pm jkr}}{r} \tag{14.10}$$

where r is the distance from the point where U_0 is calculated to another point. By the use of this particular solution the solution of the more general equation (14.8) may be constructed. It is

$$U(x,y,z,s) = \frac{1}{4\pi} \iiint \frac{F(x_1,y_1,z_1,s)\,e^{\pm jhr}}{r} \, dv \tag{14.11}$$

where x, y, z are the variables of integration and

$$r = \sqrt{(x - x_1)^2 + (y - y_1)^2 + (z - z_1)^2} \qquad dv = dx_1\,dy_1\,dz_1 \tag{14.12}$$

The fact that (14.11) satisfies (14.8) may be easily verified by differentiation. Let us substitute the value of k given by (14.7) into (14.11). We then obtain

$$U(x,y,z,s) = \frac{1}{4\pi} \iiint \frac{F(x_1,y_1,z_1,s)\,e^{\pm(s/c)r}\,dv}{r} \tag{14.13}$$

In order to obtain u, the solution of the inhomogeneous wave equation, we must take the inverse transform of $U(x,y,z,s)$ and obtain

$$u(x,y,z,t) = L^{-1}\,U(x,y,z,s) \tag{14.14}$$

To do this, we use Theorem (VII) of Chap. 4. This theorem states that, if $Lh(t) = g(s)$, then

$$L^{-1} e^{-as} g(s) = \begin{cases} 0 & t < a \\ h(t-a) & t > a \end{cases} \tag{14.15}$$

a is a constant. Applying this theorem to (14.13), we obtain

$$u(x,y,z,t) = \frac{1}{4\pi} \int \int \int \frac{F(x,y,z,t-r/c)\,dv}{r} \tag{14.16}$$

This indicates that the effects in the variation of $F(x_1,y_1,z_1,t)$ do not reach the point x, y, z until a retarded time $t - r/c$. If the plus sign in the exponential term of (14.13) were used, the inverse transform would lead to an anticipated term of the form $t + r/c$. These anticipated terms are of little importance in physical applications.

A more complete discussion of the solution of the inhomogeneous wave equation, taking into account more general initial conditions, is given by Pipes.†

Let us return now to the equations governing the propagation of the magnetic vector potential $\mathbf{A}$ and the scalar potential ϕ:

$$\nabla^2 \mathbf{A} - \frac{1}{c^2} \frac{\partial^2 \mathbf{A}}{\partial t^2} = -\mu \mathbf{J} \tag{14.17}$$

and

$$\nabla^2 \phi - \frac{1}{c^2} \frac{\partial^2 \phi}{\partial t^2} = -\frac{\rho}{k} \tag{14.18}$$

Comparing these equations with (14.1), we see that they have solutions of the form

$$\mathbf{A} = \frac{\mu}{4\pi} \int \int \int \frac{\mathbf{J}(x_1,y_1,z_1,t-r/c)\,dv}{r} \tag{14.19}$$

and

$$\phi = \frac{1}{4\pi k} \int \int \int \frac{\rho(x_1,y_1,z_1,t-r/c)\,dv}{r} \tag{14.20}$$

If the space and time distribution of the current density $\mathbf{J}$ and the charge density ρ is known in space and time, then the potentials $\mathbf{A}$ and ϕ may be computed. These potentials are called the *retarded potentials* of electrodynamics. They are of fundamental importance in the theory of electromagnetic waves.

† L. A. Pipes, Operational Solution of the Wave Equation, *Philosophical Magazine*, ser. 7, no. 26, pp. 333–340, September, 1938.

15 THE THEORY OF WAVEGUIDES

In this section the propagation of electromagnetic waves traveling in the longitudinal direction in a homogeneous isotropic medium that fills the interior of a metal tube of infinite length will be considered. The tube under consideration will be assumed to have a uniform cross section and will be assumed to be straight in the x direction.

It will be assumed that the conductivity of the metal of the tube is infinite. The equations governing the propagation of electromagnetic waves in the interior of tubes are of great importance in the field of communication engineering.†

Let us write the fundamental Maxwell equations in the form

$$\nabla \times \mathbf{E}_0 = -\mu \frac{\partial \mathbf{H}_0}{\partial t} \tag{15.1}$$

$$\nabla \times \mathbf{H}_0 = \sigma \mathbf{E}_0 + k \frac{\partial \mathbf{E}_0}{\partial t} \tag{15.2}$$

$$\nabla \cdot \mathbf{E}_0 = 0 \tag{15.3}$$

$$\nabla \cdot \mathbf{H}_0 = 0 \tag{15.4}$$

(See Appendix E, Sec. 16.)

These equations govern the distribution of the electric intensity $\mathbf{E}_0$ and the magnetic intensity $\mathbf{H}_0$ in a medium of conductivity σ, electric inductive capacity K, and magnetic inductive capacity μ. It is assumed that the medium is devoid of free charges.

In order to discuss the possible oscillations that may be propagated inside the waveguide, let us assume that the time dependence of the electric and magnetic fields $\mathbf{E}_0$ and $\mathbf{H}_0$ is of the form $e^{j\omega t}$. Let us take the y and z axes of a cartesian reference frame in the plane of the cross section of the waveguide, and let the x axis be taken in the longitudinal direction along the waveguide.

Let it be assumed that the fields $\mathbf{E}_0$ and $\mathbf{H}_0$ are of the form

$$\mathbf{E}_0 = \mathbf{E}\, e^{j(\omega t - ax)} \tag{15.5}$$

$$\mathbf{H}_0 = \mathbf{H}\, e^{j(\omega t - ax)} \qquad j = \sqrt{-1} \tag{15.6}$$

The frequency of the oscillations is $f = \omega/2\pi$. a is called the *propagation constant*. $\mathbf{E}$ and $\mathbf{H}$ are vectors of the form

$$\mathbf{E} = iE_x + jE_y + kE_z \tag{15.7}$$

and

$$\mathbf{H} = iH_x + jH_y + kH_z \tag{15.8}$$

† See L. J. Chu and W. L. Barrow, Electromagnetic Waves in Hollow Metal Tubes of Rectangular Cross Section, *Proceedings of the Institute of Radio Engineers*, vol. 26, p. 1520, 1938.

If we substitute Eqs. (15.5) and (15.6) into (15.1) and (15.2) and expand the resulting vector equations into cartesian components, we obtain

$$\frac{\partial E_z}{\partial y} - \frac{\partial E_y}{\partial z} = -j\omega\mu H_x$$

$$\frac{\partial E_x}{\partial z} + aE_z = -j\omega\mu H_y \qquad\qquad (15.9)$$

$$aE_y + \frac{\partial E_x}{\partial y} = j\omega\mu H_z$$

$$\frac{\partial H_z}{\partial y} - \frac{\partial H_y}{\partial z} = (\sigma + j\omega k) E_x$$

$$\frac{\partial H_x}{\partial z} + aH_z = (\sigma + j\omega k) E_y \qquad\qquad (15.10)$$

$$-aH_y - \frac{\partial H_x}{\partial y} = (\sigma + j\omega k) E_z$$

It will now be shown that there are two types of waves which satisfy these equations and which may exist independently. In the literature these waves are known as the TM, or E, waves and the TE, or H, waves.

a   TRANSVERSE ELECTRIC (TE), OR H, WAVES

These waves are characterized by the fact that

$$E_x = 0 \qquad \text{and} \qquad H_x \neq 0 \qquad\qquad (15.11)$$

That is, the *electric* field has no component in the direction of propagation, and the *magnetic* field has a component in the direction of propagation.

With the restriction that $E_x = 0$, Eqs. (15.9) and (15.10) become

$$\frac{\partial E_z}{\partial y} - \frac{\partial E_y}{\partial z} = -j\omega\mu H_x$$

$$aE_z = -j\omega\mu H_y \qquad\qquad (15.12)$$

$$aE_y = j\omega\mu H_z$$

and

$$\frac{\partial H_z}{\partial y} - \frac{\partial H_y}{\partial z} = 0$$

$$\frac{\partial H_x}{\partial z} + aH_z = (\sigma + j\omega k) E_y \qquad\qquad (15.13)$$

$$-aH_y - \frac{\partial H_x}{\partial y} = (\sigma + j\omega k) E_z$$

It will be left as an exercise for the reader to show that, if H_y, H_z, E_y, and E_z are eliminated between Eqs. (15.12) and (15.13), the following equation for H_x is obtained:

$$\frac{\partial^2 H_x}{\partial z^2} + \frac{\partial^2 H_x}{\partial y^2} = -[a^2 - (\sigma + j\omega k) j\omega\mu] H_z \qquad (15.14)$$

In the dielectric region inside the waveguide we have

$$\sigma \ll \omega k \qquad (15.15)$$

Hence (15.14) reduces to

$$\frac{\partial^2 H_x}{\partial z^2} + \frac{\partial^2 H_x}{\partial y^2} = -(a^2 + \omega^2 \mu k) H_x$$

$$= -k^2 H_x \qquad (15.16)$$

where

$$k^2 = a^2 + \omega^2 \mu k \qquad (15.17)$$

Equation (15.16) determines the magnetic intensity H_x subject to the boundary conditions. It may be shown by the use of (15.12) and (15.13) that

$$H_y = -\frac{a}{k^2} \frac{\partial H_x}{\partial y} \qquad (15.18)$$

$$H_z = -\frac{a}{k^2} \frac{\partial H_x}{\partial z} \qquad (15.19)$$

$$E_y = \frac{j\omega\mu}{a} H_z \qquad (15.20)$$

$$E_z = -\frac{j\omega\mu}{a} H_y \qquad (15.21)$$

This shows that, when H_x is known, then the vector **E** and the vector **H** are fully determined.

The boundary conditions If the surface of the metallic waveguide is a perfect conductor, then the tangential component of the **E** vector must vanish. In the case of a rectangular waveguide whose sides are parallel to the y and z axes we must have E_y and E_z equal to zero at the surface of the waveguide. From Eqs. (15.18) and (15.19) this means that $\partial H_x/\partial y$ and $\partial H_x/\partial z$ must vanish at the surface.

In the general case the fact that the electric lines of force must enter the perfectly conducting surface $\sigma = \infty$ normally implies that

$$\frac{\partial H_x}{\partial n} = 0 \qquad \text{at surface of waveguide} \qquad (15.21a)$$

n is the normal to the surface. In general coordinates, Eq. (15.16) may be written in the form

$$\nabla^2_{y,z} H_x + k^2 H_x = 0 \qquad \nabla^2_{y,z} = \frac{\partial^2}{\partial y^2} + \frac{\partial^2}{\partial z^2} \tag{15.22}$$

subject to the boundary condition (15.21). It will be noted that this equation has the same form as Eq. (8.7) for the vibrations of a membrane. However, the boundary condition is different.

The solution of Eqs. (15.22) subject to the boundary condition (15.21) determines the possible values of k. For given frequencies, therefore, the values of the propagation constant a are found from the equation

$$a = \sqrt{k^2 - \omega^2 \mu k} \tag{15.23}$$

Only imaginary values of a lead to possible wave propagation along the waveguide. If a is real, then the waves are rapidly attenuated as they move along the x direction of the waveguide.

b TRANSVERSE MAGNETIC (TM), OR E, WAVES

These TM, or E, waves are characterized by the fact that

$$H_x = 0 \qquad \text{and} \qquad E_x \neq 0 \tag{15.24}$$

If we place the restriction that $H_x = 0$ in Eqs. (15.9) and (15.10), we obtain

$$\frac{\partial E_z}{\partial y} - \frac{\partial E_y}{\partial z} = 0$$

$$\frac{\partial E_x}{\partial z} + a E_z = -j\omega\mu H_y \tag{15.25}$$

$$a E_y + \frac{\partial E_x}{\partial y} = j\omega\mu H_z$$

and

$$\frac{\partial H_z}{\partial y} - \frac{\partial H_y}{\partial z} = (\sigma + j\omega k) E_x$$

$$a H_z = (\sigma + j\omega k) E_y \tag{15.26}$$

$$-a H_y = (\sigma + j\omega k) E_z$$

It is possible to eliminate E_z, E_y, H_y, and H_z by the use of these equations and obtain the equation

$$\frac{\partial^2 E_x}{\partial y^2} + \frac{\partial^2 E_x}{\partial z^2} = -(a^2 + \omega^2 \mu k) E_x$$

$$= -k^2 E_x \qquad \text{in the dielectric} \tag{15.27}$$

In the general case, this may be written in the form

$$\nabla^2_{yz} E_x + k^2 E_x = 0 \tag{15.28}$$

The field **E** can have no tangential component at the surface of the perfectly conducting waveguide.   Hence we have the boundary condition

$$E_x = 0 \qquad \text{at surface of waveguide} \tag{15.29}$$

Equation (15.28) and the boundary condition (15.29) are precisely the same as the equation and the boundary condition of the vibrating membrane of Sec. 8.   This is the well-known analogy first noticed by Lord Rayleigh.†
The calculation of the possible values of k leads to the possible values of the propagation constant a.   The quantities E_z, E_y, H_y, and H_z are determined from E_x by the use of Eqs. (15.25) and (15.26).

PROBLEMS

1.   Find the form at time t of a vibrating string of length s, whose ends are fixed and which is initially displaced into an isosceles triangle.   The string is vibrating transversely, is under constant stretching force, and starts from rest.

2.   A transversely vibrating string of length s is stretched between two points A and B. The initial displacement of each point of the string is zero, and the initial velocity at a distance x from A is $kx(s - x)$.   Find the form of the string at any subsequent time.

3.   The differential equation governing the displacement of a viscously damped string is

$$\frac{\partial^2 y}{\partial t^2} = c^2 \frac{\partial^2 y}{\partial x^2} - 2k \frac{\partial y}{\partial t}$$

Find the general solution of this equation when the string has an initial displacement $y = y_0(x)$ and an initial velocity $\partial y/\partial t = v_0(x)$ at $t = 0$.

4.   A square membrane is made of material of density m g per cm² and is under a tension of T dynes per cm.   What must be the length of one side of the membrane in order that the fundamental frequency be F_0?   What will be the frequencies of the two lowest overtones?

5.   A rectangular membrane is struck at its center, starting from rest, in such a way that at $t = 0$ a small rectangular region about the center may be considered to have a velocity v_0 and the rest has no velocity.   Find the amplitudes of the various overtones.

6.   Write the differential equation governing the displacement of the *general* oscillations of a circular membrane.   Show that the allowed angular frequencies are determined by the equation

$$J_n \left(a \sqrt{\frac{m\omega^2}{T}} \right) = 0 \qquad n = 0, 1, 2, 3, \ldots$$

where J_n is the Bessel function of the first kind and nth order, a is the radius of the circular membrane, m the mass per unit area of the membrane, and T the tension per unit length.

† See Lord Rayleigh, *Philosophical Magazine*, ser. 5, no. 43, p. 125, 1897.

7. Show that the general solution of the differential equations governing the propagation of current and potential along a dissipationless transmission line $R = 0$, $G = 0$ is

$$e = f\left(x - \frac{t}{\sqrt{LC}}\right) + g\left(x + \frac{t}{\sqrt{LC}}\right)$$

$$i = \sqrt{\frac{C}{L}} f\left(x - \frac{t}{\sqrt{LC}}\right) - \sqrt{\frac{C}{L}} g\left(x + \frac{t}{\sqrt{LC}}\right)$$

where f and g are arbitrary functions.

These solutions may be interpreted as a combination of two waves, one moving to the left and one moving to the right each with velocity $1/\sqrt{LC}$.

8. Show that, if we place $G = 0$ and $L = 0$ in the transmission-line equations, we obtain an equation of the form

$$\frac{\partial^2 u}{\partial x^2} = h^2 \frac{\partial u}{\partial t}$$

This is the differential equation governing the distribution of temperature in the theory of one-dimensional heat conduction. Discuss the analogy between the transmission line in this case and the one-dimensional heat flow.

9. Show that the small longitudinal oscillations of a long rod satisfy the equation

$$\frac{\partial^2 u}{\partial t^2} = \frac{E}{m} \frac{\partial^2 u}{\partial x^2}$$

where u is the displacement of a point originally at a distance x from the end of the rod, E is the modulus of elasticity of the rod, and m is the density.

10. Find the natural frequencies of vibration of a long prismatic rod of length s, density m, and elastic modulus E, which is fixed at $x = 0$ and free at $x = s$.

11. Find the equation that determines the natural frequencies of vibration of the rod described above if the end $x = 0$ is fixed and the end $x = s$ is fastened to a mass M.

12. Starting with the differential equations of the dissipationless transmission line ($R = 0$, $G = 0$) and that governing the oscillations of a long prismatic bar, show that there exists a close analogy between the electrical and mechanical systems. What is the electrical analogue of a long rod oscillating freely whose end at $x = 0$ is fixed and whose other end at $x = s$ is free?

13. A string of length l is composed of two separate strings, each of different materials fastened together at the point a. The segment of string from 0 to the point a is one material, and the segment from a to l is the other. Formulate the equations describing this problem, along with the appropriate boundary conditions, and develop the solution in general terms, assuming that the initial position is given by $f(x)$ for $0 < x < l$. The string has no initial velocity.

HINT: Assume that the string vibrates as a whole at the same frequency.

14. Find the solution for the vibrations of a string whose initial velocity is zero and whose initial distribution is

(a) $f(x) = A(\sin x - \sin 2x)$ $0 \leqslant x \leqslant \pi$

(b)

(c) $f(x) = 0.1x(\pi - x)$ $0 \leqslant x \leqslant \pi$

(Kreyszig, 1962)

15. Solve the following partial differential equations by the method of separation of variables:

(a) $\dfrac{\partial u}{\partial x} - \dfrac{\partial u}{\partial y} = 0$

(b) $x\dfrac{\partial u}{\partial x} - y\dfrac{\partial u}{\partial y} = 0$

(c) $\dfrac{\partial u}{\partial x} + \dfrac{\partial u}{\partial y} - 2(x+y)\,u = 0$

(d) $x^2\dfrac{\partial^2 u}{\partial x\,\partial y} + 3y^2\,u = 0$

(Kreyszig, 1962)

16. Perform the indicated transformation of the independent variables and solve

(a) $\dfrac{\partial^2 u}{\partial x\,\partial y} - \dfrac{\partial^2 u}{\partial y^2} = 0 \qquad v = x \qquad z = x + y$

(b) $\dfrac{\partial^2 u}{\partial x^2} + \dfrac{\partial^2 u}{\partial x\,\partial y} - 2\dfrac{\partial^2 u}{\partial y^2} = 0 \qquad v = x + y \qquad z = 2x - y$

(Kreyszig, 1962)

17. Find the deflection of a unit square membrane if the initial velocity is zero and the initial deflection is

(a) $0.1xy(1-x)(1-y)$

(b) $k\sin \pi x \sin 2\pi y$

(Kreyszig, 1962)

18. Show that the forced vibrations of an elastic string under an external force $P(x,t)$ per unit length acting normal to the string are governed by the equation

$$\frac{\partial^2 u}{\partial t^2} = c^2\frac{\partial^2 u}{\partial x^2} + \frac{P}{\rho}$$

(Kreyszig, 1962)

REFERENCES

1935. Guillemin, E. A.: "Communication Networks," vol. II, John Wiley & Sons, Inc., New York.

1936. Morse, P. M.: "Vibration and Sound," McGraw-Hill Book Company, New York.

1937. Timoshenko, S.: "Vibration Problems in Engineering," D. Van Nostrand Company, Inc., Princeton, N.J.

1962. Kreyszig, E.: "Advanced Engineering Mathematics," John Wiley & Sons, Inc., New York.

1966. Ostberg, D. R., and F. W. Perkins: "An Introduction to Linear Analysis," Addison-Wesley Publishing Company, Inc., Reading, Mass.

1966. Raven, F. H.: "Mathematics of Engineering Systems," McGraw-Hill Book Company, New York.

1966. Sokolnikoff, I. S., and R. M. Redheffer: "Mathematics of Physics and Modern Engineering," 2d ed., McGraw-Hill Book Company, New York.

1966. Wylie, C. R., Jr.: "Advanced Engineering Mathematics," 3d ed., McGraw-Hill Book Company, New York.

13
Operational Methods in Applied Mathematics

1 INTRODUCTION

For a considerable time the use of symbolic or "operational" methods in mathematics has attracted a great deal of attention.† Leibnitz in 1695 noted that the symbol d of differentiation, now D in the symbolic method, has some of the properties of ordinary algebraic quantities. In the early part of the nineteenth century the separation of symbols of operation from their operands was studied by the French mathematicians Arbogast, Brisson, Servios, and others. From 1836 to 1860 the British school of mathematicians used the symbolic method in the differential and integral calculus, in the calculus of finite differences, and in the solution of differential and difference equations. These methods are still used in some elementary textbooks. However, these methods were frequently regarded as somewhat "shady" from the standpoint of rigorous mathematics because no explicit formulation was given of the conditions under which the use of these devices afforded correct results.

† Cf. Davis, 1936 (see References).

Heaviside's operational analysis of certain differential equations of importance in electrodynamics and electrical engineering, carried out in the years 1887 to 1898, was the last notable advance in the symbolic method. It is well known[†] that Heaviside's methods were not orthodox and that because of this the Royal Society of London refused to publish some of his papers. Several mathematicians and electrical engineers, notably Bromwich, Jeffreys, Wagner, Carson, and Levy, attempted to explain Heaviside's methods and place them on a firm mathematical foundation. The activity stimulated by Heaviside's methods has been aptly summarized by the eminent British mathematician E. T. Whittaker[‡] in the following statement:

> Looking back on the controversy after thirty years, we should now place the operational calculus with Poincarè's discovery of automorphic functions and Ricci's discovery of the tensor calculus as the three most important mathematical advances of the last quarter of the nineteenth century. Applications, extensions, and justification of it constitute a considerable part of the mathematical activity of today.

It is now generally recognized that Heaviside's operational caculus and its extensions may be based on the Laplace transform. However, it is not very well known that Oliver Heaviside himself clearly pointed this out.[§] It is interesting to note that the Laplace transform was known long before Heaviside used it. Euler and Gauss both used transformations resembling the transform introduced by Laplace in 1779.[||] Cauchy showed its relation to the Fourier transform, which is a special case, and Abel derived many of its properties. In recent years many investigators have contributed to bring the subject into its present form, and there are now several standard texts on the subject (cf. references at end of this chapter).

2 INTEGRAL TRANSFORMS

In discussing operational methods it is difficult not to cover the broad area of integral transforms as a subtopic. In fact, it frequently occurs today that when operational methods are mentioned, the reference is to integral transforms. The reason for this occurrence is that integral transforms are in fact a subclass of operational methods and perhaps even the largest. The most extensively used integral transform at the present time is the Laplace transform (cf. Chap. 4). Besides this, however, there are several other transforms in existence, the most common of which are the Fourier,[¶] Hankel,

[†] Cf. Davis, 1936 (see References).
[‡] See Whittaker, 1928 (see References).
[§] Heaviside, 1950 (see References).
[||] Laplace, 1779 (see References).
[¶] There are actually three forms of the Fourier transform.

and Mellin transforms. The formal definitions of these four important transform pairs are given in Eqs. (2.1) through (2.6) below.†

LAPLACE

$$L[f(t)] = \int_0^\infty f(t) e^{-st} dt = g(s)$$

$$L^{-1}[g(s)] = \frac{1}{2\pi j} \int_{c-j\infty}^{c+j\infty} g(s) e^{st} ds = f(t)$$

(2.1)

FOURIER: COMPLEX

$$F[f(t)] = \int_{-\infty}^\infty f(t) e^{-jwt} dt = g(w)$$

$$F^{-1}[g(w)] = \frac{1}{2\pi} \int_{-\infty}^\infty g(w) e^{jwt} dw = f(t)$$

(2.2)

FOURIER: SINE

$$F_s[f(t)] = \int_0^\infty f(t) \sin wt \, dt = g(w)$$

$$F_s^{-1}[g(w)] = \frac{2}{\pi} \int_0^\infty g(w) \sin wt \, dw = f(t)$$

(2.3)

FOURIER: COSINE

$$F_c[f(t)] = \int_0^\infty f(t) \cos wt \, dt = g(w)$$

$$F_c^{-1}[g(w)] = \frac{2}{\pi} \int_0^\infty g(w) \cos wt \, dw = f(t)$$

(2.4)

HANKEL

$$H_n[f(t)] = \int_0^\infty f(t) t J_n(st) \, dt = g(s)$$

$$H_n^{-1}[g(s)] = \int_0^\infty g(s) s J_n(st) \, ds = f(t)$$

(2.5)

This transform is useful for solving problems involving cylindrical symmetry, that is, Bessel's differential equation, and the value of n is prescribed by the particular equation being solved. For example,

$$H_n\left(\ddot{f} + \frac{1}{t}\dot{f} - \left(\frac{n}{t}\right)^2 f\right) = -s^2 g(s)$$

provided $f(t)$ and $tf(t)$ both vanish as t approaches 0 or ∞.

† The symbol t will be used to denote the independent variable since it most frequently is time; however, it should be clearly understood that it represents any independent variable.

MELLIN

$$M[f(t)] = \int_0^\infty f(t)\, t^{s-1}\, dt = g(s)$$

$$M^{-1}[g(s)] = \frac{1}{2\pi j} \int_{c-j\infty}^{c+j\infty} g(s)\, t^{-s}\, ds = f(t) \tag{2.6}$$

For multidimensional problems these transforms may be applied in various combinations, as will be illustrated in a few examples later in this chapter. The remainder of this chapter will be devoted to several example problems illustrating the use of operational methods and integral transforms. For brevity, only infinite transforms will be discussed, and the reader is referred to the references† for thorough discussions of finite transforms.

3 APPLICATION OF THE OPERATIONAL CALCULUS TO THE SOLUTION OF PARTIAL DIFFERENTIAL EQUATIONS

We have seen in Chap. 4 the manner in which the Laplace transform may be used to obtain the solution of ordinary differential equations with constant coefficients.

The basic principle of the method is to introduce a Laplacian transformation with respect to the independent variable and in this manner obtain an algebraic equation for the transform of the dependent variable. The determination of the inverse transform, either by a calculation of the complex Bromwich integral or by a consultation of a table of transforms, then supplies the solution to the problem.

The use of the Laplace transform in the solution of linear partial differential equations with constant coefficients will now be illustrated by some examples.

a THE DISSIPATIONLESS TRANSMISSION LINE

As an example of the transform method of solution of partial differential equations, let us consider the following problem. Consider the short-circuited electrical transmission line of Fig. 3.1. Let the line have a length of l, and let it be assumed that the conductors of the line are perfect and that the insulation between the conductors is perfect. In this case, the series resistance

† Such as Sneddon, 1951, or Tranter, 1956.

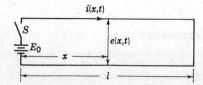

Fig. 3.1

R of the line is zero and the leakage conductance G is also zero. Let L be the inductance per unit length of the line and C the capacitance per unit length. The equations governing the distribution of potential and current along the line are (see Chap. 12, Sec. 10)

$$-\frac{\partial e}{\partial x} = L\frac{\partial i}{\partial t} \tag{3.1}$$

$$-\frac{\partial i}{\partial x} = C\frac{\partial e}{\partial t} \tag{3.2}$$

At $t = 0$ the switch S is closed, and it is desired to determine the subsequent distribution of current and potential along the line.

In order to solve this problem by the Laplacian-transform method, let us introduce the following transforms with respect to the independent variable t:

$$\mathscr{L}e(x,t) = E(x,s) \tag{3.3}$$

$$\mathscr{L}i(x,t) = I(x,s) \tag{3.4}$$

where $\mathscr{L}$ denotes the Laplace transform.

Since the initial current and potential distributions of the line are assumed to be zero, we have

$$\mathscr{L}\frac{\partial i}{\partial t} = sI \tag{3.5}$$

$$\mathscr{L}\frac{\partial e}{\partial t} = sE \tag{3.6}$$

Now, since in the transformations (3.3) and (3.4) x is a parameter, it follows that

$$\mathscr{L}\frac{\partial e}{\partial x} = \frac{dE}{dx} \tag{3.7}$$

and

$$\mathscr{L}\frac{\partial i}{\partial x} = \frac{dI}{dx} \tag{3.8}$$

Therefore Eqs. (3.1) and (3.2) transform into

$$-\frac{dE}{dx} = LsI \tag{3.9}$$

$$-\frac{dI}{dx} = CsE \tag{3.10}$$

Eliminating I from these two equations, we obtain

$$\frac{d^2 E}{dx^2} = LCs^2 E \tag{3.11}$$

Eliminating E, we have

$$\frac{d^2 I}{dx^2} = LCs^2 I \tag{3.12}$$

If we let

$$v = \frac{1}{\sqrt{LC}} \tag{3.13}$$

Eq. (3.11) becomes

$$\frac{d^2 E}{dx^2} = \frac{s^2}{v^2} E \tag{3.14}$$

This is an ordinary differential equation with constant coefficients, and its solution is

$$E = A_1 e^{-sx/v} + A_2 e^{+sx/v} \tag{3.15}$$

where A_1 and A_2 are arbitrary constants.

In order to determine A_1 and A_2, we use the boundary conditions

$$\begin{array}{cc} E = E_0 & E = 0 \\ x = 0 & x = l \end{array} \tag{3.16}$$

Using these conditions, we obtain the two simultaneous linear equations

$$\begin{aligned} E_0 &= A_1 + A_2 \\ 0 &= A_1 e^{-ls/v} + A_2 e^{ls/v} \end{aligned} \tag{3.17}$$

Hence

$$A_1 = \frac{\begin{vmatrix} E_0 & 1 \\ 0 & e^{ls/v} \end{vmatrix}}{\begin{vmatrix} 1 & 1 \\ e^{-ls/v} & e^{ls/v} \end{vmatrix}} = \frac{E_0 e^{ls/v}}{e^{ls/v} - e^{-ls/v}} \tag{3.18}$$

and

$$A_2 = \frac{\begin{vmatrix} 1 & E_0 \\ e^{-ls/v} & 0 \end{vmatrix}}{e^{ls/v} - e^{-ls/v}} = \frac{-E_0 e^{-ls/v}}{e^{ls/v} - e^{-ls/v}} \tag{3.19}$$

Hence, with these values of the constants A_1 and A_2, (3.15) becomes

$$E(x,s) = \frac{E_0}{e^{ls/v} - e^{-ls/v}} \left(e^{-(s/v)(x-l)} - e^{+(s/v)(x-l)} \right) \tag{3.20}$$

Waves along an infinite line. Before obtaining the inverse transform of (3.20), let us consider the solution for $l \to \infty$ so that the line is of infinite length. In this case we have

$$\lim_{l \to \infty} E(x,s) = E_0 e^{-sx/v} \tag{3.21}$$

To calculate the inverse transform of this we make use of the following theorem: If

$$L^{-1} g(s) = h(t) \tag{3.22}$$

then

$$L^{-1} e^{-ks} g(s) = \begin{cases} 0 & t < k \\ h(t-k) & t > k \end{cases} \qquad k > 0$$

Hence we have

$$L^{-1} E_0 e^{-sx/v} = \begin{cases} 0 & t < \dfrac{x}{v} \\ E_0 & t > \dfrac{x}{v} \end{cases} \tag{3.23}$$

This represents a potential wave of amplitude E_0 traveling with velocity v to the right end, as shown in Fig. 3.2.

Now, in order to determine the inverse transform of (3.20), let us multiply the numerator and denominator of (3.20) by $e^{-ls/v}$, and we then obtain

$$E(x,s) = \frac{E_0}{1 - e^{-2ls/v}} [e^{-sx/v} - e^{-(s/v)(2l-x)}] \tag{3.24}$$

Now we have

$$\frac{1}{1 - e^{-2ls/v}} = 1 + e^{-2ls/v} + e^{-4ls/v} + e^{-6ls/v} + \cdots \tag{3.25}$$

Hence we may write (3.21) in the form

$$E(x,s) = E_0(e^{-sx/v} + e^{-(s/v)(x+2l)} + e^{-(s/v)(x+4l)} + \cdots)$$
$$- E_0(e^{-(s/v)(2l-x)} + e^{-(s/v)(4l-x)} + e^{-(s/v)(6l-x)} + \cdots) \tag{3.26}$$

We may interpret the various terms as traveling waves. The first group of terms represent waves traveling to the right with a velocity v, while the second group of terms represent waves that have been reflected at the short-circuited end and are traveling to the left with a velocity of v.

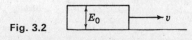

Fig. 3.2

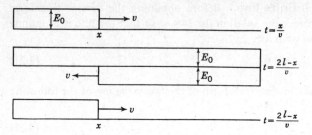

Fig. 3.3

The course of events is interpreted graphically in Fig. 3.3. The system of reflected waves keeps the end at $x = l$ at potential zero and the end $x = 0$ at potential E_0. To obtain the current waves, we may use Eq. (3.9) and write it in the form

$$I = -\frac{1}{Ls}\frac{dE}{dx} \tag{3.27}$$

Differentiating (3.26), we have

$$\frac{dE}{dx} = -\frac{sE_0}{v}(e^{-sx/v} + e^{-(s/v)(x+2l)} + e^{-(s/v)(x+4l)} + \cdots)$$

$$-\frac{sE_0}{v}(e^{-(s/v)(2l-x)} + e^{-(s/v)(4l-x)} + e^{-(s/v)(6l-x)} + \cdots) \tag{3.28}$$

Since

$$Lv = L\frac{1}{\sqrt{LC}} = \sqrt{\frac{L}{C}} = Z_0 \tag{3.29}$$

the characteristic impedance of the line, we have from (3.27) and (3.28)

$$I(x,s) = \frac{E_0}{Z_0}(e^{-sx/v} + e^{-(s/v)(2l-x)} + e^{-(s/v)(x+2l)} + e^{-(s/v)(4l-x)} + \cdots)$$

$$\tag{3.30}$$

It is interesting to compute the current at the short-circuited end $x = l$. Placing $x = l$ in (3.30), we have

$$I(l,s) = \frac{E_0}{Z_0}(e^{-ls/v} + e^{-(s/v)l} + e^{-(s/v)3l} + e^{-(s/v)3l} + \cdots)$$

$$= \frac{2E_0}{Z_0}(e^{-ls/v} + e^{-3ls/v} + e^{-3ls/v} + \cdots) = \mathscr{L}i(l,t) \tag{3.31}$$

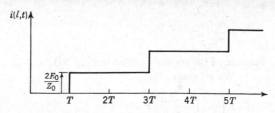

Fig. 3.4

The inverse transform of (3.13) is given graphically in Fig. 3.4, where

$$T = \frac{l}{v} \tag{3.32}$$

so that the current continues to increase in finite jumps of magnitude $2E_0/Z_0$.

b   LINEAR FLOW OF HEAT IN A SEMI-INFINITE SOLID

As another example of the use of the Laplacian transformation in the solution of partial differential equations, let us consider the determination of the flow of heat into a semi-infinite solid, $x > 0$ (Fig. 3.5).

Initially the solid is at temperature $v = 0$, and at $t = 0$ the boundary $x = 0$ is raised to a temperature v_0. In order to determine the subsequent distribution of temperature in the solid, we must solve the heat-flow equation (Chap. 11)

$$\frac{\partial v}{\partial t} = h^2 \frac{\partial^2 v}{\partial x^2} \tag{3.33}$$

where h^2 is the thermometric conductivity of the solid and v the temperature, subject to the boundary conditions

$$\begin{array}{lll} v = v_0 & x = 0 & t > 0 \\ v = 0 & x > 0 & t = 0 \end{array} \tag{3.34}$$

To do this, let us introduce the Laplacian transform

$$Lv(x,t) = u(x,s) \tag{3.35}$$

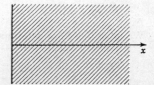

Fig. 3.5

Hence

$$L\frac{\partial v}{\partial t} = su \tag{3.36}$$

because of the second condition of (3.34).

We also have

$$L\left(\frac{\partial^2 v}{\partial x^2}\right) = \frac{d^2 u}{dx^2} \tag{3.37}$$

Hence Eq. (3.33) transforms to

$$su = h^2 \frac{d^2 u}{dx^2} \tag{3.38}$$

or

$$\frac{d^2 u}{dx^2} = \frac{s}{h^2} u \tag{3.39}$$

This is a linear equation with constant coefficients in the independent variable x. The general solution of (3.39) is

$$u = A e^{-\sqrt{-s/h^2}\, x} + B e^{+\sqrt{s/h^2}\, x} \tag{3.40}$$

Since the temperature cannot be infinite for infinite values of x, we must have

$$B = 0 \tag{3.41}$$

Hence the solution reduces to

$$u = A e^{-\sqrt{s/h^2}\, x} \tag{3.42}$$

Now, since $v = v_0$ at $x = 0$, we must have

$$A = \frac{v_0}{s} \tag{3.43}$$

Therefore

$$u = \frac{v_0 e^{-\sqrt{s/h^2}\, x}}{s} = Lv \tag{3.44}$$

To determine the inverse Laplacian transform of this expression, let

$$a = \frac{x}{h} \tag{3.45}$$

We must therefore determine

$$L^{-1}\frac{v_0 e^{-a\sqrt{s}}}{s} = v(x,t) \tag{3.46}$$

To do this, let us expand $e^{-a\sqrt{s}}$ into a Maclaurin series of the form

$$e^m = 1 + \frac{m}{1!} + \frac{m^2}{2!} + \frac{m^3}{3!} + \cdots \tag{3.47}$$

Hence

$$\frac{e^{-a\sqrt{s}}}{s} = \left[1 - as^{\frac{1}{2}} + \frac{a^2 s}{2!} - \frac{a^3 s^{\frac{3}{2}}}{3!} + \frac{a^4 s^2}{4!} - \frac{a^5 s^{\frac{5}{2}}}{5!} + \cdots\right] \tag{3.48}$$

We now compute the inverse transforms of each individual term by the fundamental relation

$$L^{-1} s^{-n-1} = \frac{t^n}{\Gamma(n+1)} \qquad t > 0 \tag{3.49}$$

The inverse transforms of all positive powers of s are zero for $t > 0$. Hence

$$L^{-1} \frac{e^{-a\sqrt{s}}}{s} = 1 - \frac{2}{\sqrt{\pi}}\left[\frac{a}{2\sqrt{t}} - \frac{1}{3}\left(\frac{a}{2\sqrt{t}}\right)^3 + \frac{1}{2! \times 5}\left(\frac{a}{2\sqrt{t}}\right)^5 - \cdots\right] \tag{3.50}$$

Now in Appendix B, Sec. 8, we consider the function

$$\mathrm{erf}(w) = \frac{2}{\sqrt{\pi}} \int_0^w e^{-\theta^2}\, d\theta \tag{3.51}$$

where $\mathrm{erf}(w)$ is the well-known *error function*. If we expand $e^{-\theta^2}$ in (3.51) and integrate term by term, we obtain

$$\mathrm{erf}(w) = \frac{2}{\sqrt{\pi}}\left(w - \frac{w^3}{3 \times 1!} + \frac{w^5}{5 \times 2!} - \frac{w^7}{7 \times 3!} + \cdots\right) \tag{3.52}$$

If we now let

$$w = \frac{a}{2\sqrt{t}} \tag{3.53}$$

we have, on comparing (3.50) and (3.52),

$$L^{-1} \frac{e^{-a\sqrt{s}}}{s} = 1 - \mathrm{erf}\left(\frac{a}{2\sqrt{t}}\right) \tag{3.54}$$

Hence, in view of (3.44), the temperature in the solid is given by

$$v(x,t) = v_0\left[1 - \mathrm{erf}\left(\frac{x}{2h\sqrt{t}}\right)\right] \tag{3.55}$$

This equation shows how the temperature diffuses into the solid. The time required to produce a given rise of temperature at a distance x from the surface is proportional to x^2. We notice the absence of wave phenomena in the increase in temperature.

c THE OSCILLATIONS OF A BAR

As a further example of the method, let us consider the following problem. A uniform bar of length l is at rest in its equilibrium position with the end $x = 0$ fixed as shown in Fig. 3.6.   At $t = 0$ a constant force F_0 per unit area is applied at the free end.   It is required to determine the subsequent state of motion of the bar.

If F_x is the stress at the point x of the bar, $u(x,t)$ the displacement, ρ the density of the bar, and E the Young's modulus, we have

$$F_x = E \frac{\partial u}{\partial x} \tag{3.56}$$

and the equation of motion is

$$\frac{\partial F_x}{\partial x} = \rho \frac{\partial^2 u}{\partial t^2} = E \frac{\partial^2 u}{\partial x^2} \tag{3.57}$$

Hence we have

$$\frac{\partial^2 u}{\partial x^2} = \frac{1}{c^2} \frac{\partial^2 u}{\partial t^2} \tag{3.58}$$

where

$$c = \sqrt{\frac{E}{\rho}} \tag{3.59}$$

Equation (3.58) determines the longitudinal displacement of the bar. The constant c is the velocity of sound in the bar.   The initial conditions of the problem are

$$\left. \begin{array}{c} u(x,t) = 0 \\ \dfrac{\partial u}{\partial t} = 0 \end{array} \right\} \quad t = 0 \quad\quad 0 < x < l \tag{3.60}$$

The boundary conditions are

$$\begin{array}{cc} u = 0 & \dfrac{\partial u}{\partial x} = \dfrac{F_0}{E} \\[2mm] x = 0 & x = l \\[1mm] t > 0 & t > 0 \end{array} \tag{3.61}$$

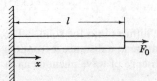

Fig. 3.6

To solve Eq. (3.58) by the Laplace-transform method, let

$$Lu(x,t) = v(x,s) \tag{3.62}$$

Then (3.58) transforms to

$$\frac{d^2 v}{dx^2} = \left(\frac{s}{c}\right)^2 v \tag{3.63}$$

The general solution of this equation is

$$v = A e^{-sx/c} + B e^{+sx/c} \tag{3.64}$$

The boundary conditions of (3.64) are

$$
\begin{array}{ccc}
v = 0 & \dfrac{dv}{dx} = \dfrac{F_0}{Es} & \\
x = 0 & x = l &
\end{array}
\tag{3.65}
$$

Making use of the boundary conditions, we obtain

$$
\begin{aligned}
v &= \frac{F_0 c}{Es^2} \frac{\sinh(sx/c)}{\cosh(ls/c)} \\
&= \frac{F_0 c}{Es^2} \frac{e^{sx/c} - e^{-sx/c}}{e^{ls/c} + e^{-ls/c}} \\
&= \frac{F_0 c}{Es^2} \frac{1 - e^{-2sx/c}}{1 + e^{-2ls/c}} e^{(s/c)(x-l)}
\end{aligned}
\tag{3.66}
$$

Now we have

$$\frac{1}{1 + e^{-2sl/c}} = 1 - e^{-2sl/c} + e^{-4sl/c} - e^{-6sl/c} + \cdots \tag{3.67}$$

Hence we may write (3.66) in the form

$$v = \frac{F_0 c}{Es^2}[1 - e^{(-2s/c)x} - e^{-2ls/c} + e^{-2s(l+x)/c} + \cdots] \tag{3.68}$$

Let us determine the motion of the end $x = l$. Placing $x = l$ in (3.68), we have

$$v(l,s) = \frac{F_0 c}{Es^2}(1 - 2e^{-2sl/c} + 2c^{-4sl/c} - 2e^{-6sl/c} + \cdots) \tag{3.69}$$

The inverse transform of (3.69) has the value

$$
\begin{aligned}
u(l,t) &= \frac{F_0 c}{E} t & 0 < t < \frac{2l}{c} \\
u(l,t) &= \frac{F_0 c}{E} t - \frac{2F_0 c(t - 2l/c)}{E} & \frac{2l}{c} < t < \frac{4l}{c} \quad \text{etc.}
\end{aligned}
\tag{3.70}
$$

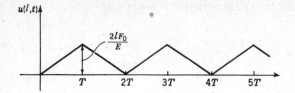

Fig. 3.7

This is given graphically in Fig. 3.7, where

$$T = \frac{2l}{c} \tag{3.71}$$

It is thus seen that the end moves by jerks as shown in Fig. 3.7.

The above examples illustrate the general procedure by which the solution of linear partial differential equations may be effected by the use of the Laplacian transformation. For an extended discussion of the use of the method in this connection the reader may consult the references at the end of this chapter.

4   EVALUATION OF INTEGRALS

The evaluation of certain definite integrals may be effected simply by the use of the Laplacian transformation. In this section we shall consider certain examples that illustrate the general procedure. Many integrals may be evaluated by the use of the following theorem:

If

$$Lh(t) = g(s) \tag{4.1}$$

then

$$\int_0^\infty g(s)\,ds = \int_0^\infty \frac{h(t)\,dt}{t} \tag{4.2}$$

This theorem may be proved in the following manner: By hypothesis we have

$$g(s) = \int_0^\infty e^{-st} h(t)\,dt \tag{4.3}$$

Therefore

$$\int_0^\infty g(s)\,ds = \int_0^\infty \left[\int_0^\infty e^{-st} h(t)\,dt \right] ds$$
$$= \int_0^\infty \int_0^\infty e^{-st} h(t)\,ds\,dt \tag{4.4}$$

provided that it is permissible to reverse the order of integration. But we have

$$\int_0^\infty e^{-st}\, ds = \frac{1}{t} \tag{4.5}$$

Hence we obtain

$$\int_0^\infty g(s)\, ds = \int_0^\infty \frac{h(t)\, dt}{t} \tag{4.6}$$

As an example of the use of this theorem, let it be required to evaluate the integral

$$\int_0^\infty \frac{\sin at\, dt}{t} \qquad \text{where } a > 0 \tag{4.7}$$

We have from the table of transforms (Appendix A)

$$L \sin at = \frac{a}{s^2 + a^2} \tag{4.8}$$

Hence, by (4.6), we have

$$\int_0^\infty \frac{\sin at\, dt}{t} = \int_0^\infty \frac{a\, ds}{s^2 + a^2} = \tan^{-1} \frac{s}{a} \Big|_0^\infty$$

$$= \frac{\pi}{2} \tag{4.9}$$

As another example, consider the integral

$$\int_0^\infty \frac{e^{-at} - e^{-bt}}{t}\, dt \tag{4.10}$$

From the table of transforms (Appendix A) we have

$$L e^{-at} = s + a \qquad L e^{-bt} = s + b \tag{4.11}$$

Hence by the use of (4.6) we have

$$\int_0^\infty \frac{(e^{-at} - e^{-bt})\, dt}{t} = \int_0^\infty \left(\frac{ds}{s+a} - \frac{ds}{s+b} \right) = \ln \frac{s+a}{s+b} \Big|_0^\infty = \ln \frac{b}{a} \tag{4.12}$$

Another method of evaluating definite integrals operationally depends on the introduction of a parameter in the integrand. As an example, consider the integral

$$I = \int_0^\infty e^{-x^2}\, dx \tag{4.13}$$

As seen in Appendix F, Sec. 11, this integral is usually evaluated by a trick. Now let us consider the transform

$$L e^{-tx^2} = s + x^2 \tag{4.14}$$

Hence we have

$$L \int_0^\infty e^{-tx^2}\, dx = \int_0^\infty \frac{dx}{s+x^2} = \frac{1}{\sqrt{s}} \tan^{-1} \frac{x}{\sqrt{s}} \Big|_0^\infty$$

$$= \frac{\pi}{2} \frac{1}{\sqrt{s}} \tag{4.15}$$

Therefore

$$\int_0^\infty e^{-tx^2}\, dx = L^{-1} \frac{\pi}{2} \frac{1}{\sqrt{s}} \tag{4.16}$$

But

$$L^{-1} \frac{1}{\sqrt{s}} = \frac{1}{\sqrt{\pi t}} \tag{4.17}$$

Hence we have

$$\int_0^\infty e^{-tx^2}\, dx = \frac{\pi}{2} \frac{1}{\sqrt{\pi t}} = \frac{1}{2}\sqrt{\frac{\pi}{t}} \tag{4.18}$$

Placing $t = 1$, we finally obtain

$$\int_0^\infty e^{-x^2}\, dx = \frac{\sqrt{\pi}}{2} \tag{4.19}$$

As another example, consider the integral

$$\int_0^\infty \frac{\cos x\, dx}{1+x^2} \tag{4.20}$$

In this case we make use of the transform

$$L \cos tx = \frac{s}{s^2 + x^2} \tag{4.21}$$

Hence we have

$$L \int_0^\infty \frac{\cos tx\, dx}{1+x^2} = \int_0^\infty \frac{s\, dx}{(s^2+x^2)(1+x^2)}$$

$$= \frac{s}{s^2-1} \int_0^\infty \left(\frac{1}{x^2+1} - \frac{1}{s^2+x^2} \right) dx$$

$$= \frac{\pi}{2} s + 1 \tag{4.22}$$

Therefore we have

$$\int_0^\infty \frac{\cos tx\, dx}{1+x^2} = L^{-1} \frac{\pi}{2} s + 1 = \frac{\pi}{2} e^{-t} \tag{4.23}$$

If we now place $t = 1$, we finally obtain

$$\int_0^\infty \frac{\cos x \, dx}{1 + x^2} = \frac{\pi}{2e} \tag{4.24}$$

These examples are typical and illustrate the general procedure.

A critical examination of the validity of the above procedure reveals that it is dependent on the possibility of reversing the order of integration of integrals involving infinite limits.†

5   APPLICATIONS OF THE LAPLACE TRANSFORM TO THE SOLUTION OF LINEAR INTEGRAL EQUATIONS

An integral equation is one in which the unknown function appears under the sign of integration.   The importance of integral equations in mathematical applications to physical problems lies in the fact that it is usually possible to reformulate a differential equation together with its boundary conditions as a single integral equation.   In this section certain types of integral equations that can be solved by the use of the Laplace-transform theory will be discussed.

a   VOLTERRA'S INTEGRAL EQUATION OF THE SECOND KIND WITH A DIFFERENCE KERNEL

Consider the equation

$$F(t) = G(t) - \int_0^t K(t - u) \, G(u) \, du \tag{5.1}$$

where $F(t)$ and $K(t - u)$ are *known* functions and the function $G(t)$ is not known.   This equation is known in the literature as *Volterra's integral equation of the second kind with the difference kernel* $K(t - u)$.   The unknown function $G(t)$, and hence the solution of (5.1), may be readily found by the use of the Laplace-transform theory.   In order to solve (5.1), introduce the following Laplace transforms:

$$LG(t) = g(s) \tag{5.2}$$

$$LF(t) = f(s) \tag{5.3}$$

$$LK(t) = k(s) \tag{5.4}$$

Then, by the Faltung theorem (Theorem V), we have

$$L \int_0^t K(t - u) \, G(u) \, du = g(s) k(s) \tag{5.5}$$

Hence the Laplace transform of the integral equation (5.1) is

$$f(s) = g(s) - k(s) g(s) \tag{5.6}$$

† A discussion of this will be found in H. S. Carslaw, "Fourier Series and Integrals," Dover Publications, Inc., New York, 1951.

This equation may be solved for $g(s)$ in the form

$$g(s) = \frac{f(s)}{1 - k(s)} \tag{5.7}$$

The solution of (5.1) is therefore

$$G(t) = L^{-1} g(s) = L^{-1} \frac{f(s)}{1 - k(s)} \tag{5.8}$$

As a simple example of this method of solution, let it be required to solve the following equation:

$$a \sin t = G(t) - \int_0^t \sin (t - u) G(u) \, du \tag{5.9}$$

This is a special case of (5.1) with

$$LF(t) = La \sin t = \frac{a}{s^2 + 1} = f(s) \tag{5.10}$$

$$LK(t) = L \sin t = \frac{1}{s^2 + 1} = k(s) \tag{5.11}$$

Therefore by (5.8) we have

$$G(t) = L^{-1} \frac{a}{s^2 + 1} \frac{s^2 + 1}{s^2} = L^{-1} \frac{a}{s^2} = at \tag{5.12}$$

b   THE LIOUVILLE-NEUMANN SERIES SOLUTION

The well-known series expansion of Liouville and von Neumann for the solution of Volterra's integral equation (5.1) may be readily obtained by expanding (5.7) in the following form:

$$g(s) = \frac{f(s)}{1 - k(s)} = \sum_{n=0}^{\infty} [k(s)]^n f(s) = \sum_{n=0}^{\infty} g_n(s) \tag{5.13}$$

The functions $g_n(s)$ clearly fulfill the following recurrence relation:

$$g_n(s) = k(s) g_{n-1}(s) \qquad g_0(s) = f(s) \tag{5.14}$$

If we let

$$L^{-1} g_n(s) = G_n(t) \tag{5.15}$$

it is evident from (5.14) and the Faltung theorem that

$$G_0(t) = F(t) \qquad \text{and} \qquad G_n(t) = \int_0^t K_n(t - u) F(u) \, du$$
$$n = 1, 2, 3, \ldots \tag{5.16}$$

where the functions $K_n(t)$ are given by

$$K_n(t) = L^{-1} [k(s)]^n \qquad n = 1, 2, 3, \ldots \tag{5.17}$$

These functions are known as the *iterated kernels*. If the above processes converge, the solution of the integral equation (5.1) is given by

$$G(t) = \sum_{n=0}^{\infty} G_n(t) \qquad\qquad (5.18)$$

c VOLTERRA'S INTEGRAL EQUATION OF THE FIRST KIND WITH A DIFFERENCE KERNEL

Volterra's integral equation of the first kind with a difference kernel is a special case of (5.1) and is usually written in the following form:

$$F(t) = \int_0^t K(t - u) \, G(u) \, du \qquad\qquad (5.19)$$

The kernel $K(t - u)$ and the function $F(t)$ are given known functions, and it is required to determine the unknown function $G(t)$. If the Laplace transforms $f(s) = LF(t)$ and $k(s) = LK(t)$ are introduced, it is easy to show, by the same reasoning as that used in Sec. 5a, that the solution of (5.19) is given by

$$G(t) = L^{-1} \frac{f(s)}{k(s)} \qquad\qquad (5.20)$$

In many cases of practical importance it is possible to factor the right-hand member of (5.20) and to compute the required inverse Laplace transform. An alternative form of solution may be obtained if we write

$$\frac{f(s)}{k(s)} = f(s) \cdot \frac{1}{k(s)} \qquad\qquad (5.21$$

and let

$$H(t) = L^{-1} \frac{1}{k(s)} \qquad\qquad (5.22)$$

In terms of $H(t)$ the solution of (5.19) may be written in the following form:

$$G(t) = \int_0^t H(t - u) \, F(u) \, du \qquad\qquad (5.23)$$

The function $H(t)$ is called the *reciprocal kernel* of $K(t)$.

d ABEL'S INTEGRAL EQUATION; THE TAUTOCHRONE

The integral equation

$$F(t) = \int_0^t \frac{G(u) \, du}{(t - u)^n} \qquad 0 < n < 1 \qquad\qquad (5.24)$$

is a special case of (5.19) and is the classical equation of Abel. In order to obtain the solution of Abel's equation by (5.20), we have

$$LK(t) = Lt^{-n} = s^{n-1} \, \Gamma(1 - n) = k(s) \qquad 0 < n < 1 \qquad\qquad (5.25)$$

by a table of Laplace transforms. Hence by (5.20) the solution of (5.24) is given by

$$G(t) = L^{-1}\frac{f(s)}{k(s)} = L^{-1}\frac{f(s)}{s^{n-1}\,\Gamma(1-n)} \tag{5.26}$$

By the use of a table of Laplace transforms (Appendix A) we obtain the following result:

$$L^{-1}\frac{1}{s^n\,\Gamma(1-n)} = \frac{1}{\Gamma(1-n)\,\Gamma(n)\,t^{1-n}} \tag{5.27}$$

A fundamental result of the theory of gamma functions (see Appendix B) is that

$$\Gamma(1-n)\,\Gamma(n) = \frac{\pi}{\sin n\pi} \qquad 0 < n < 1 \tag{5.28}$$

Therefore

$$L^{-1}\frac{1}{s^n\,\Gamma(1-n)} = \frac{\sin n\pi}{\pi t^{1-n}} \tag{5.29}$$

Therefore, by an application of the Faltung theorem, we obtain

$$L^{-1}\frac{f(s)}{s^{n-1}\,\Gamma(1-n)} = \frac{\sin \pi n}{\pi}\frac{d}{dt}\int_0^t \frac{F(u)\,du}{(t-u)^{1-n}} = G(t) \tag{5.30}$$

The solution of Abel's integral equation (5.24) is therefore given by

$$G(t) = \frac{\sin n\pi}{\pi}\frac{d}{dt}\int_0^t \frac{F(u)\,du}{(t-u)^{1-n}} \qquad 0 < n < 1 \tag{5.31}$$

The tautochrone   When $n = \frac{1}{2}$, Eq. (5.31) gives the solution of the classical problem of the tautochrone (the curve of equal descent). The general form of this famous problem is to determine the path of a particle of mass m under the influence of the earth's gravitational field so that its time of fall from P (at the variable height y) to the origin at $y = 0$ shall be some given function $T(y)$ of y (see Fig. 5.1).

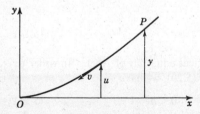

Fig. 5.1

The time of fall of the particle from the point P to the origin O is given by the equation

$$T(y) = \int_0^P \frac{ds}{v} \tag{5.32}$$

where v is the velocity of the particle at some intermediate point of height u and ds is the line element of the required curve. Since the particle is assumed to move without friction, the velocity v satisfies the equation of conservation of energy:

$$\tfrac{1}{2}mv^2 = mg(y - u) \tag{5.33}$$

or

$$v = \sqrt{2g(y - u)} \tag{5.34}$$

The line element ds of the required curve may be written in the form

$$ds = \left(\frac{ds}{du}\right) du \tag{5.35}$$

If (5.33) and (5.35) are substituted into (5.32), the result may be written in the form

$$T(y) = \frac{1}{\sqrt{2g}} \int_0^y \frac{(ds/du)\, du}{\sqrt{y - u}} \tag{5.36}$$

This is an Abel integral equation for the function ds/dy of the form (5.24). By (5.31) the solution of (5.36) is seen to be

$$\frac{ds}{dy} = \frac{\sqrt{2g}}{\pi} \frac{d}{dy} \int_0^y \frac{T(u)\, du}{\sqrt{y - u}} \tag{5.37}$$

If (5.37) is integrated with respect to y and the arc length of the required curve is measured from the origin ($y = 0$), we obtain

$$s(y) = \frac{\sqrt{2g}}{\pi} \int_0^y \frac{T(u)\, du}{\sqrt{y - u}} \tag{5.38}$$

If the function $T(y)$ is specified, (5.38) enables the actual path to be determined.

In the problem of the tautochrone the time of falling is given to be independent of the height, so that

$$T(y) = T_0 \qquad \text{a given constant} \tag{5.39}$$

In this case the integration of (5.38) may be performed, and the result is

$$s(y) = \frac{2T_0}{\pi} \sqrt{2gy} \tag{5.40}$$

Therefore

$$\frac{ds}{dy} = \frac{T_0}{\pi}\sqrt{\frac{2g}{y}} = \sqrt{1 + \left(\frac{dx}{dy}\right)^2} \tag{5.41}$$

The differential equation of the required curve is therefore

$$1 + \left(\frac{dx}{dy}\right)^2 = \frac{2gT_0^2}{\pi^2 y} = \frac{a}{y} \tag{5.42}$$

where

$$a = \frac{2gT_0^2}{\pi^2} \tag{5.43}$$

Equation (5.42) may be written in the form

$$dx = \sqrt{\frac{a-y}{y}}\, dy \tag{5.44}$$

The integration of (5.44) may be easily effected by the following trigonometric substitution:

$$y = a\sin^2\frac{\theta}{2} \tag{5.45}$$

With this substitution (5.43) gives

$$dx = a\cos^2\frac{\theta}{2}\, d\theta = \frac{a}{2}(1 + \cos\theta)\, d\theta \tag{5.46}$$

Since the curve must pass through the origin, $x = 0$ when $y = 0$. Therefore the parametric equations of the tautochrone are

$$x = \frac{a}{2}(\theta + \sin\theta) \qquad y = \frac{a}{2}(1 - \cos\theta) \qquad a = \frac{2gT_0^2}{\pi^2} \tag{5.47}$$

These are the equations of a cycloid generated by the motion of a point on a circle of radius $a/2$ as the circle rolls along the lower side of the line $y = a$.

e INTEGRAL EQUATIONS OF THE FIRST KIND REDUCIBLE TO EQUATIONS WITH A DIFFERENCE KERNEL

Integral equations of the type

$$F(t) = \int_0^t K(tu)\, G(u)\, du \tag{5.48}$$

and

$$F(t) = \int_0^t K\left(\frac{t}{u}\right) G(u)\, du \tag{5.49}$$

sometimes occur in the analysis of physical problems. In these equations the kernel is a function either of the product tu or of the quotient t/u of the two variables t and u.   It is possible to transform these integral equations into equations with a kernel depending on the difference $x - y$ by the substitutions $t = e^x$ and $u = e^{-y}$ in (5.48) and $t = e^x$ and $u = e^y$ in (5.49).   The transformed equations are then of the type (5.15).

f   THE SOLUTION OF LINEAR TIME-VARYING ELECTRICAL CIRCUITS BY THE USE OF INTEGRAL EQUATIONS

Electrical circuits whose parameters vary with the time alone are of considerable technical importance.   The response of circuits of this class may sometimes be obtained by formulating the circuit equations in terms of linear integral equations.

To illustrate the use of integral equations in electrical-circuit problems of this type, consider the electrical circuit of Fig. 5.2.   This circuit consists of the constant resistance, inductance, and elastance parameters R, L, S, in series with a time-varying resistance $R_0(t)$ and an electromotive force $e(t)$. Both $R_0(t)$ and $e(t)$ are known functions of the time t, and it is desired to determine the current $i(t)$ as a function of the time.   By Kirchhoff's second law the differential equation that expresses the potential balance of the circuit is

$$\frac{L_0\,di}{dt} + Ri + S \int i\,dt + R_0(t)\,i(t) = e(t) \tag{5.50}$$

This is the linear integrodifferential equation satisfied by the current of the circuit.

In order to transform (5.50) into an integral equation, introduce the following Laplace transforms:

$$Li(t) = I(s) \qquad Le(t) = E(s) \tag{5.51}$$

By the use of these transforms (5.50) is transformed into the equation

$$\left(L_0 s + R + \frac{S}{s}\right) I(s) + L[R_0(t)\,i(t)] = E(s) \tag{5.52}$$

where it has been assumed that the initial current and the initial charge of the circuit are both zero at $t = 0$.   Let $Z(s)$, the operational impedance of the constant part of the circuit, be given by

$$Z(s) = L_0 s + R + \frac{S}{s} \tag{5.53}$$

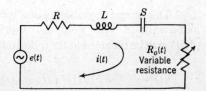

Fig. 5.2

and let

$$L[R_0(t)\,i(t)] = G(s) \tag{5.54}$$

With this notation (5.52) may be written in the form

$$I(s) = \frac{E(s)}{Z(s)} - \frac{G(s)}{Z(s)} \tag{5.55}$$

Now let

$$H(s) = \frac{1}{Z(s)} = \frac{s}{L_0 s^2 + Rs + S} \tag{5.56}$$

and

$$h(t) = L^{-1} H(s) = L^{-1} \frac{s}{L_0 s^2 + Rs + S} \tag{5.57}$$

In terms of the transform $H(s)$, (5.55) may be written in the following form:

$$I(s) = H(s) E(s) - H(s) G(s) \tag{5.58}$$

By the Faltung theorem the inverse transform of (5.58) may be written in the following form:

$$i(t) = \int_0^t h(t-u)\,e(u)\,du - \int_0^t h(t-u)\,R_0(u)\,i(u)\,du \tag{5.59}$$

This is the integral equation satisfied by the current $i(t)$ of the circuit of Fig. 5.2. Let

$$i_0(t) = \int_0^t h(t-u)\,e(u)\,du \tag{5.60}$$

This is the current that would flow in the circuit if the variable resistance $R_0(t)$ were placed equal to zero. $i_0(t)$ thus represents the current that would flow in a series circuit having the constant parameters R, L, and S under the influence of the electromotive force $e(t)$. The integral equation (5.59) may therefore be written in the form

$$i(t) = i_0(t) - \int_0^t h(t-u)\,e(u)\,du \tag{5.61}$$

This integral equation is of a type that has been studied by the eminent mathematician Vito Volterra. Volterra has shown that the solution of (5.61) may be obtained in the following form:

$$i(t) = \sum_{n=0}^{\infty} (-1)^n i_n(t) = i_0(t) - i_1(t) + i_2(t) - i_3(t) + \cdots \tag{5.62}$$

where the sequence of functions $i_k(t)$ is given by the equations

$$i_0(t) = \int_0^t h(t-u)\, e(u)\, du$$

$$i_1(t) = \int_0^t h(t-u)\, R_0(u)\, i_0(u)\, du$$

$$i_2(t) = \int_0^t h(t-u)\, R_0(u)\, i_1(u)\, du \qquad (5.63)$$

$$\cdots\cdots\cdots\cdots\cdots\cdots\cdots$$

$$i_{n+1}(t) = \int_0^t h(t-u)\, R_0(u)\, i_n(u)\, du$$

It is easy to show that the series (5.62) is absolutely convergent and satisfies the integral equation (5.61). For purposes of computation, it is convenient to express the sequence of functions (5.63) in terms of the following Laplace transforms:

$$i_0(t) = L^{-1}\frac{E(s)}{Z(s)}$$

$$i_1(t) = L^{-1}\frac{L[R_0(t)\, i_0(t)]}{Z(s)}$$

$$i_2(t) = L^{-1}\frac{L[R_0(t)\, i_1(t)]}{Z(s)} \qquad (5.64)$$

$$\cdots\cdots\cdots\cdots\cdots\cdots$$

$$i_{n+1}(t) = L^{-1}\frac{L[R_0(t)\, i_n(t)]}{Z(s)}$$

In practice, it is usually found that the sequence $i_k(t)$ may be more simply computed by Eqs. (5.64) than by (5.63) because a table of suitable Laplace transforms greatly assists in effecting the required integrations. In the usual applications it is necessary only to compute a few terms of the series (5.62) to obtain the required accuracy for the solution of practical circuit problems.

6 SOLUTION OF ORDINARY DIFFERENTIAL EQUATIONS WITH VARIABLE COEFFICIENTS

It is sometimes possible, by the introduction of a Laplacian transformation, to transform certain linear differential equations with variable coefficients to other equations that may be integrated readily. By such a procedure we are able to obtain the Laplace transforms of the solutions of the original equations. Then by determining the inverses of these transforms we find the solutions of the original equations with variable coefficients.

The procedure will be illustrated by the following example. Let us consider the Bessel differential equation of the nth order (discussed in Appendix B):

$$\frac{d^2 z}{du^2} + \frac{1}{u}\frac{dz}{du} + \left(1 - \frac{n^2}{u^2}\right)z = 0 \qquad (6.1)$$

In order to simplify the coefficients of the equation, let us change the variables by the equations

$$z = u^{-n} y \quad \text{and} \quad u^2 = 4t \qquad (6.2)$$

These changes in variable transform the equation into the form

$$t\frac{d^2 y}{dt^2} + (1 - n)\frac{dy}{dt} + y = 0 \qquad (6.3)$$

To transform this equation, let

$$Y(s) = \int_0^\infty e^{-st} y(t)\, dt \qquad (6.4)$$

that is,

$$Ly(t) = Y(s) \qquad (6.5)$$

Now, by an integration by parts, we have

$$\int_0^\infty e^{-st}\frac{dy}{dt}\, dt = s\, Y(s) - y(0) \qquad (6.6)$$

$$\int_0^\infty e^{-st}\frac{d^2 y}{dt^2}\, dt = s^2\, Y(s) - sy(0) - y'(0) \qquad (6.7)$$

Differentiating (6.7) with respect to s, we obtain

$$\frac{d}{ds}\int_0^\infty e^{-st}\frac{d^2 y}{dt^2}\, dt = -\int_0^\infty e^{-st}\, t\frac{d^2 y}{dt^2}\, dt$$

$$= \frac{d}{ds}[s^2\, Y(s) - sy(0) - y'(0)] \qquad (6.8)$$

We thus have the result

$$\int_0^\infty e^{-st}\, t\frac{d^2 y}{dt^2}\, dt = y(0) - \frac{d}{ds}(s^2\, Y) \qquad (6.9)$$

Now since

$$y(t) = u^n z = 2^n\, t^{n/2} z \qquad (6.10)$$

we see that

$$y(0) = 0 \qquad (6.11)$$

Hence Eq. (6.3) transforms into

$$\frac{d}{ds}(s^2 Y) + (n-1)sY - Y = 0 \qquad (6.12)$$

Therefore

$$sY + s^2 Y' + nsY - Y = 0 \qquad (6.13)$$

or

$$s\frac{d(sY)}{ds} + \left(n - \frac{1}{s}\right)sY = 0 \qquad (6.14)$$

If we now let

$$w(s) = sY \qquad (6.15)$$

we may write (6.14) in the form

$$s\frac{dw}{ds} + \left(n - \frac{1}{s}\right)w = 0 \qquad (6.16)$$

Hence

$$\frac{dw}{w} = -\left(n - \frac{1}{s}\right)\frac{ds}{s} \qquad (6.17)$$

or

$$\ln w = -n\ln s - \frac{1}{s} + c \qquad (6.18)$$

where c is an arbitrary constant. Therefore,

$$y = Ks^{-(n+1)}e^{-1/s} \qquad (6.19)$$

where K is a new arbitrary constant. Now, by (6.5), we have

$$y(t) = L^{-1}y(s) = L^{-1}Ks^{-(n+1)}e^{-1/s} \qquad (6.20)$$

or

$$2^n t^{n/2}(u) = L^{-1}Ks^{-(n+1)}e^{-1/s} \qquad (6.21)$$

Now a solution of (6.1) that is finite at $u = 0$ is

$$z = J_n(u) = J_n(2\sqrt{t}) \qquad (6.22)$$

Hence we have from (6.21)

$$Lt^{n/2}J_n(2\sqrt{t}) = \frac{K}{2^n}s^{-(n+1)}e^{-1/s} \qquad (6.23)$$

where K must be properly chosen.

Expanding the right-hand member of (6.23) in inverse powers of s, calculating the inverse transforms of the individual terms of the expansion, it can be seen by comparing with the series expansion of $t^{n/2} J_n(2\sqrt{t})$ that K must have the value

$$K = 2^n \tag{6.24}$$

Hence we have the result

$$Lt^{n/2} J_n(2\sqrt{t}) = s^{-(n+1)} e^{-1/s} \tag{6.25}$$

If we now place $n = 0$, we obtain

$$LJ_0(2\sqrt{t}) = \frac{e^{-1/s}}{s} \tag{6.26}$$

If we now make use of the theorem that if

$$Lh(t) = g(s) \tag{6.27}$$

then

$$Lh(kt) = \frac{1}{k} g\left(\frac{s}{k}\right) \tag{6.28}$$

where k is a constant whose absolute value is greater than zero, we have from (6.26)

$$LJ_0(2\sqrt{kt}) = \frac{e^{-k/s}}{s} \tag{6.29}$$

If we now let

$$k = -j \tag{6.30}$$

we obtain

$$J_0(2\sqrt{-jt}) = \mathrm{ber}\,(2\sqrt{t}) + j\,\mathrm{bei}\,(2\sqrt{t})$$

$$= L^{-1} \frac{e^{j/s}}{s} \tag{6.31}$$

Placing

$$k = +j \tag{6.32}$$

in (6.29), we have

$$J_0(2\sqrt{jt}) = \mathrm{ber}\,(2\sqrt{t}) - j\,\mathrm{bei}\,(2\sqrt{t})$$

$$= L^{-1} \frac{e^{-j/s}}{s} \tag{6.33}$$

Adding Eqs. (6.31) and (6.33), we obtain

$$L^{-1} \frac{1}{s} \cos\frac{1}{s} = \mathrm{ber}\,(2\sqrt{t}) \tag{6.34}$$

Subtracting (6.33) from (6.31), we have

$$L^{-1} \frac{1}{s} \sin \frac{1}{s} = \text{bei} (2\sqrt{t})$$ (6.35)

These two transform pairs may be used to obtain many interesting properties of these functions. The procedure followed in this example on Bessel's equation may be carried out in the case of several other linear differential equations with variable coefficients and the properties of their solutions thus studied.

7 THE SUMMATION OF FOURIER SERIES BY THE LAPLACE TRANSFORM

In the mathematical analysis of problems in electrical-circuit theory, mechanical vibrations, heat conduction, and other branches of physics and engineering one frequently arrives at solutions that are expressed in the form of infinite Fourier series. For purposes of calculation it is more convenient to have the solution of these problems in closed form. It is sometimes possible to sum and interpret certain Fourier series by the use of the Laplace-transform theory. In this section a method for doing this will be given.

The method to be described is based on the fact that certain hyperbolic functions can be expressed as infinite series of exponential functions. With the notation

$$A_k = e^{-kas}$$ (7.1)

the following expansions are known:

$$\tanh as = 1 - 2A_2 + 2A_4 - 2A_6 + \cdots$$ (7.2)

$$\coth as = 1 + 2A_2 + 2A_4 + 2A_6 + \cdots$$ (7.3)

$$\frac{1}{\sinh as} = 2(A_1 + A_3 + A_5 + A_7 + \cdots)$$ (7.4)

$$\frac{1}{\cosh as} = 2(A_1 - A_3 + A_5 - A_7 + \cdots)$$ (7.5)

where a is a positive real number and s is a number whose real part is positive. The following rational-fraction expansions of these functions are also known:†

$$\tanh as = \frac{2}{as} \sum_{n=1}^{\infty} \frac{1}{1 + \{[(2n-1\,\pi]/2as\}^2}$$ (7.6)

---

† E. T. Whittaker and G. N. Watson, "A Course in Modern Analysis," pp. 134–136, Cambridge University Press, New York, 1927.

$$\coth as = \frac{1}{as}\left[1 + \sum_{n=1}^{\infty} \frac{2}{1 + (n\pi/as)^2}\right] \tag{7.7}$$

$$\frac{1}{\cosh as} = \frac{1}{a^2 s^2} \sum_{n=1}^{\infty} \frac{(-1)(2n-1)\pi}{1 + (n\pi/2as)^2} \tag{7.8}$$

$$\frac{1}{\sinh as} = \frac{1}{as}\left[1 + \sum_{n=1}^{\infty} (-1)^n \frac{2}{1 + (n\pi/as)^2}\right] \tag{7.9}$$

It is possible to expand other combinations of hyperbolic functions in a similar manner. However, the above expansions are sufficient to illustrate the general principles of the method to be described.

Example 1 In certain investigations of physical problems the following Fourier series is obtained:

$$F_1(t) = 4\left(\frac{\sin t}{1} + \frac{\sin 3t}{3} + \frac{\sin 5t}{5} + \cdots\right)$$

$$= 4 \sum_{n=1}^{\infty} \frac{\sin(2n-1)t}{2n-1} \qquad t > 0 \tag{7.10}$$

The Laplace transform of the function $F_1(t)$ represented by the series (7.10) may be obtained by taking the Laplace transform of each term of the series. If this is done, one obtains

$$LF_1(t) = \frac{4}{s^2}\left[\frac{1}{1 + (1/s)^2} + \frac{1}{1 + (3/s)^2} + \frac{1}{1 + (5/s)^2} + \cdots\right] \tag{7.11}$$

We notice that if in (7.6) we place

$$a = \frac{\pi}{2} \tag{7.12}$$

then, by comparing the expansions (7.6) and (7.11), we have

$$LF_1(t) = \frac{1}{s} 2a \tanh as = \frac{\pi}{s} \tanh \frac{\pi s}{2} \tag{7.13}$$

The right-hand member of (7.13) may now be expanded by (7.2) in the form

$$LF_1(t) = \frac{\pi}{s}(1 - 2e^{-\pi s} + 2e^{-2\pi s} - 2e^{-3\pi s} + \cdots) \tag{7.14}$$

This is the Laplace transform of the Fourier series (7.10).

The inverse transform of (7.13) may be obtained by the use of Theorem VII of Chap. 4, Sec. 3. By the use of this theorem it is seen that the inverse transform of the first term of (7.13) is a step function of height π formed at $t = 0$. The inverse transform of the second term is a step function of height -2π formed at $t = \pi$. The inverse transform of the third term is a step function of height $+2\pi$ formed at $t = 2$, and so on. The sum of all these step functions is illustrated by Fig. 7.1. The graph of this figure represents the function defined by the Fourier series (7.10).

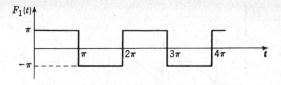

Fig. 7.1

Example 2   As a second example, let it be required to interpret the following Fourier series that occurs in certain electrical and heat-conduction problems:

$$F_2(t) = \frac{2}{\pi} \sum_{n=1}^{\infty} \frac{1}{n} \sin \frac{n\pi t}{l} \tag{7.15}$$

If the Laplace transform of $\sin(n\pi t/l)$ is taken from a table of transforms, the Laplace transform of (7.15) is seen to be

$$LF_2(t) = \frac{1}{s^2 l} \sum_{n=1}^{\infty} \frac{2}{1 + (n\pi/ls)^2} \tag{7.16}$$

If this series is now compared with (7.7), it is seen that we have

$$LF_2(t) = \frac{1}{s} \coth sl - \frac{1}{ls^2} \tag{7.17}$$

By the use of (7.3) this may be written in the following form:

$$LF_2(t) = \frac{1}{s}\left(1 - \frac{1}{sl} + 2e^{-2sl} + 2e^{4sl} + 2e^{-6sl} + \cdots \right) \tag{7.18}$$

To obtain the inverse transform of (7.18) and hence the original function $F_2(t)$, we note that the inverse transform of the first term is a unit step function formed at $t = 0$, the inverse transform of the term $-1/s^2 l$ is $-t/l$, the inverse transform of $(2/s)e^{-2sl}$ is a step function of height 2 that begins at $t = 2l$, etc. The sum of the various inverse transforms of (7.18) gives the function $F_2(t)$. The result is given graphically in Fig. 7.2.

Example 3   As another example, let it be required to obtain the graph of the function represented by the following Fourier series:

$$F_3(t) = \sum_{n=1}^{\infty} (-1)^{n-1} \frac{\sin nt}{n} \tag{7.19}$$

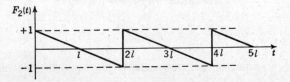

Fig. 7.2

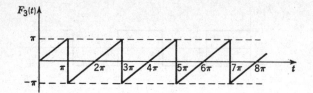

Fig. 7.3

If the Laplace transform of (7.19) is computed, its similarity to (7.19) will be noted and the result may be transformed by (7.4) into the following form:

$$LF_3(t) = \frac{1}{s}\left(\frac{1}{s} - 2e^{-s} - 2e^{-3s} - 2e^{-5s} - \cdots\right) \tag{7.20}$$

The sum of the various inverse transforms gives the function $F_3(t)$. This function is given graphically in Fig. 7.3.

Example 4　A row of impulses　An interesting series which arises in the analysis of the propagation of charge along a transmission line will now be considered. This series represents a row of impulses and the function it represents is important in the analysis of the disturbances created when lightning hits a transmission line. The series to be considered is the following one:

$$F_4(x) = \frac{2}{a}\sum_{n=1}^{\infty}\sin\frac{n\pi x}{a}\sin\frac{n\pi b}{a} \tag{7.21}$$

To study this series more easily, the product of sines may be transformed into two cosines and written in the following form:

$$F_4(x) = \frac{1}{a}\sum_{n=1}^{\infty}\cos\frac{n\pi(x-b)}{a} - \frac{1}{a}\sum_{n=1}^{\infty}\cos\frac{n\pi(x+b)}{a} \tag{7.22}$$

In order to study this series by the Laplace-transform method, substitute $t = x - b$ in the first series of (7.22). This series may be written in the form

$$\phi(t) = \frac{1}{a}\sum_{n=1}^{\infty}\cos\frac{n\pi t}{a} \tag{7.23}$$

The Laplace transform of this series is

$$L\phi(t) = \frac{1}{sa}\sum_{n=1}^{\infty}1 + (n\pi/as)^2 \tag{7.24}$$

By comparing (7.24) and (7.7) it can be seen that (7.24) may be written in the form

$$L\phi(t) = \frac{2}{as}(as\coth as - 1) = 2(1 + 2e^{-2as} + 2e^{-4as} + 2e^{-6as} + \cdots) - \frac{2}{as} \tag{7.25}$$

The last result is obtained by means of the expansion (7.3). The inverse transform of (7.25) may be calculated in terms of the Dirac impulse function $\delta(t)$. The result is

$$\phi(t) = 2[\delta(t) + 2\delta(t - 2a) + 2\delta(t - 4a) + 2\delta(t - 6a) + \cdots] - \frac{2}{a}$$

$$= 2[\delta(x - b) + 2\delta(x - b - 2a) + 2\delta(x - b - 4a)$$
$$+ 2\delta(x - b - 6a) + \cdots] - \frac{2}{a} \quad (7.26)$$

In the same manner the second series of (7.22) may be computed in terms of impulse functions by the substitution $t = x + b$, and it can be shown that this series represents a row of negative impulses at $x = -b$, $x = 2a - b$, etc. The impulse form of $F_4(x)$ is therefore given by

$$F_4(x) = 2[\delta(x - b) + 2\delta(x - b - 2a) + 2\delta(x - b - 4a)$$
$$+ 2\delta(x - b - 6a) + \cdots]$$
$$- 2[\delta(x + b) + 2\delta(x + b - 2a) + 2\delta(x + b - 4a)$$
$$+ 2\delta(x + b - 6a) + \cdots] \quad (7.27)$$

The function $F_4(x)$ is therefore zero for all values of x except at the places $x = b$, $b - 2a$, $b - 4a$, . . . and $x = -b$, $x = -b - 2a$, $x = -b - 4a$, . . . , where it has values of positive and negative impulses of the Dirac type.

GENERAL CONSIDERATIONS

The general Fourier series may be written in the form

$$F(t) = A_0 + \sum_{n=1}^{\infty} A_n \cos nwt + \sum_{n=1}^{\infty} B_n \sin nwt \quad (7.28)$$

where the coefficients A_n and B_n are known functions of n. The Laplace transform of (7.28) is

$$LF(t) = \frac{A_0}{s} + \frac{1}{s} \sum_{n=1}^{\infty} \frac{A_n}{1 + (nw/s)^2} + \frac{1}{s^2} \sum_{n=1}^{\infty} \frac{wB_n}{[1 + (nw/s)^2]} \quad (7.29)$$

The success of the method of summation described above depends upon the possibility of effecting the summations of (7.29) by the use of (7.6) to (7.9) or by formulas derivable from them. The interpretation of the Fourier series can then be carried out by the use of (7.2) to (7.5) or combinations of these expansions. In many cases, however, the summation of the fractions (7.29) may present insurmountable difficulties, and, in such cases, this method of interpretation is *not* applicable. An interesting extension of the above method will be found in a paper by Spiegel.†

8 THE DEFLECTION OF A LOADED CORD

Perhaps the simplest problem encountered in the theory of structures is the determination of the deflection of a cord stretched between supports under various conditions of loading.

† M. R. Spiegel, the Dirac Delta-function and the Summation of Fourier Series, *Journal of Applied Physics*, vol. 23, no. 8, pp. 906–909, August, 1952.

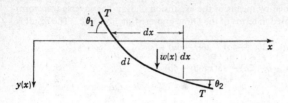

Fig. 8.1

Consider a section dl of a perfectly flexible cord as shown in Fig. 8.1. Let $y(x)$ be the deflection curve of the cord, $w(x)$ be the load per unit length, and T be the tension of the cord.　Then the equilibrium condition obtained by equating the net force on the segment in the y direction is

$$T\sin\theta_1 - T\sin\theta_2 = w(x)\,dx \tag{8.1}$$

Equating the forces in the x direction, we have

$$T\cos\theta_1 = T\cos\theta_2 = H \tag{8.2}$$

where H is the horizontal component of the tension T and is constant for the span.　Dividing (8.1) by (8.2), we have

$$\tan\theta_1 - \tan\theta_2 = \frac{w(x)\,dx}{H} \tag{8.3}$$

But we have

$$\tan\theta_1 = \left(\frac{dy}{dx}\right)_x \qquad \tan\theta_2 = \left(\frac{dy}{dx}\right)_{x+dx} \tag{8.4}$$

Hence, by Taylor's expansion, we have

$$\tan\theta_2 = \left(\frac{dy}{dx}\right)_x + \left(\frac{d^2 y}{dx^2}\right)_x dx + \text{higher-order terms in } dx \tag{8.5}$$

Substituting this into (8.3), we obtain

$$\frac{d^2 y}{dx^2} = -\frac{w(x)}{H} \tag{8.6}$$

This is the fundamental differential equation governing the deflection of a cord under the influence of a load $w(x)$ per unit length and a horizontal component of tension H.

Equation (8.6) may be solved most conveniently by the Laplace-transform, or operational, method.　To do this, let us introduce the transforms

$$Ly(x) = Y(s)$$
$$Lw(x) = W(s) \tag{8.7}$$

We then have

$$L\left(\frac{d^2 y}{dx^2}\right) = s^2\, Y - s y_0 - y_1 \tag{8.8}$$

where

$$y_0 = y(0) \qquad y_1 = \left(\frac{dy}{dx}\right)_{x=0} \tag{8.9}$$

Hence Eq. (8.6) is transformed into

$$s^2\, Y - s y_0 - y_1 = -\frac{W(s)}{H} \tag{8.10}$$

a UNIFORM LOAD

Let us first consider the case of a uniform load w_0 per unit length.

$$w(x) = w_0 \tag{8.11}$$

Let the cord be suspended from the points $x = 0$ and $x = l$ that are at equal heights, as shown in Fig. 8.2.   In this case we have

$$W(s) = \frac{w_0}{s} \qquad y_0 = 0 \qquad y(l) = 0 \tag{8.12}$$

Hence (8.10) becomes

$$Y = \frac{y_1}{s^2} - \frac{w_0}{H s^3} = L y(x) \tag{8.13}$$

To determine $y(x)$, we use the table of transforms (Appendix A) and we have

$$y(x) = L^{-1}\, Y(s) = y_1\, x - \frac{w_0\, x^2}{2H} \tag{8.14}$$

To determine y_1, we use the condition $y(l) = 0$ and obtain

$$y_1 = \frac{w_0\, l}{2H} \tag{8.15}$$

Substituting this into (8.14), we obtain

$$y = \frac{w_0}{2H}(lx - x^2) \tag{8.16}$$

for the deflection of the uniformly loaded cord.

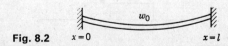

Fig. 8.2 $x = 0$ w_0 $x = l$

b UNIFORM LOAD EXTENDING OVER PART OF THE SPAN

The power of the operational method is demonstrated in the solution of the case of Fig. 8.3.   In this case we have

$$W(s) = \int_{x_1}^{x_2} e^{-sx} w_0 \, dx = \frac{w_0}{s}(e^{-x_1 s} - e^{-x_2 s}) \tag{8.17}$$

for the transform of the unit load w_0 extending from x_1 to x_2.   In this case, Eq. (8.10) becomes

$$Y(s) = \frac{y_1}{s^2} - \frac{w_0}{H} \frac{e^{-x_1 s} - e^{-x_2 s}}{s^3} = Ly(x) \tag{8.18}$$

By the use of the table of transforms and the rule that if

$$L^{-1} g(s) = h(x) \tag{8.19}$$

then

$$L^{-1} e^{-kp} g(s) = \begin{cases} 0 & x < k \\ h(x-k) & x > k \end{cases} \tag{8.20}$$

where $k > 0$, it is seen that we have

$$y(x) = \begin{cases} y_1 x & 0 < x < x_1 \\ y_1 x - \dfrac{w_0}{2H}(x - x_1)^2 & x_1 < x < x_2 \\ y_1 x - \dfrac{w_0}{2H}(x - x_1)^2 + \dfrac{w_0}{2H}(x - x_2)^2 & x_2 < x < l \end{cases} \tag{8.21}$$

To determine y_1, we use $y(l) = 0$, and we find from (8.21) that

$$y_1 = \frac{w_0}{2Hl}[(l - x_1)^2 - (l - x_2)^2] \tag{8.22}$$

Substituting this into (8.21), we obtain the deflection curve.

The advantage of the operational method over the classical method is that we are able to write one differential equation for the entire span, while in the classical procedure it is necessary to write several equations depending on the discontinuous nature of the load and then evaluate the constants by using the condition that the deflection is continuous.

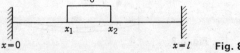

$x = 0$ $x = l$ **Fig. 8.3**

c THE EFFECT OF A CONCENTRATED LOAD

Let us compute the deflection of a flexible cord fixed at both ends and support-ing a concentrated load P_0, as shown in Fig. 8.4.   The concentrated load is located at the point $x = a$.   In order to solve this problem operationally, we must compute the transform of the concentrated load.   This may be done by considering the concentrated load P_0 as the limit of a distributed load w_0 distributed over a very small region, as shown in Fig. 8.5.

The transform of such a load is

$$W(s) = \int_a^{a+\delta} w_0 e^{-sx}\, dx = \frac{w_0}{s}(e^{-sa} - e^{-s(a+\delta)})$$

$$= \frac{w_0}{s} e^{-sa}(1 - e^{-s\delta}) \tag{8.23}$$

Now we take the limit $\delta \to 0$ and $w_0 \to \infty$ in such a way that

$$\lim_{\substack{w_0 \to \infty \\ \delta \to 0}} w_0\, \delta = P_0 \tag{8.24}$$

The exponential function $e^{-s\delta}$ may be expanded in the form

$$e^{-s\delta} = 1 - s\delta + \frac{s^2 \delta^2}{2!} - \frac{s^3 \delta^3}{3!} + \cdots \tag{8.25}$$

Substituting this into (8.23), we have

$$\lim_{\substack{\delta \to 0 \\ w_0 \to \infty}} W(s) = \lim_{\substack{\delta \to 0 \\ w_0 \to \infty}} e^{-sa} w_0\, \delta s = P_0 e^{-sa} \tag{8.26}$$

This is the transform of the concentrated load situated at the point $x = a$.   Substituting this into Eq. (8.10), we have

$$Y(s) = \frac{y_1}{s^2} - \frac{P_0 e^{-as}}{H\, s^2} = Ly(x) \tag{8.27}$$

Computing the inverse transform of this expression by the table of transforms, we obtain

$$y(x) = \begin{cases} y_1 x & 0 < x < a \\ y_1 x - \dfrac{P_0(x - a)}{H} & a < x < l \end{cases} \tag{8.28}$$

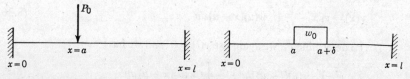

Fig. 8.4   **Fig. 8.5**

The constant y_1 may be determined by the condition $y(l) = 0$, and it is

$$y_1 = \frac{P_0}{Hl}(l - a) \tag{8.29}$$

Substituting this value of y_1 into (8.28), we obtain the following equations for the deflection $y(x)$:

$$y(x) = \begin{cases} \dfrac{P_0}{H} x \left(1 - \dfrac{a}{l}\right) & 0 < x < a \\[2ex] \dfrac{P_0}{H} a \left(1 - \dfrac{x}{l}\right) & a < x < l \end{cases} \tag{8.30}$$

d THE EFFECT OF AN ARBITRARY LOAD

A useful relation may be easily obtained giving the deflection of an arbitrary load by the use of the Faltung theorem established in Chap. 4. The theorem states that if

$$L^{-1} g_1(s) = h_1(x)$$
$$L^{-1} g_2(s) = h_2(x) \tag{8.31}$$

then

$$L^{-1} g_1 g_2 = \int_0^x h_1(u) h_2(x - u)\, du$$
$$= \int_0^x h_2(u) h_1(x - u)\, du \tag{8.32}$$

Returning to Eq. (8.10) with the initial deflection $y_0 = 0$, we have

$$Y(s) = \frac{y_1}{s^2} - \frac{W(s)}{Hs^2} = Ly(x) \tag{8.33}$$

Now since

$$L^{-1} \frac{1}{s^2} = x \qquad L^{-1} W(s) = w(x) \tag{8.34}$$

we have, by the Faltung theorem,

$$L^{-1} \frac{W(s)}{Hs^3} = L^{-1} \frac{W(s)}{Hs^2} \int_0^x \frac{w(u)(x - u)\, du}{H} \tag{8.35}$$

Hence, from (8.33), we have

$$y(x) = y_1 x - \frac{1}{H} \int_0^x w(u)(x - u)\, du \tag{8.36}$$

To determine y_1, we again use $y(l) = 0$ and we have

$$y = \frac{x}{Hl} \int_0^l w(u)(l - u)\, du - \frac{1}{H} \int_0^z w(u)(x - u)\, du \tag{8.37}$$

This gives the deflection of the cord in the span from $x = 0$ to $x = l$ due to the influence of a general load $w(x)$.

9 STRETCHED CORD WITH ELASTIC SUPPORT

Let us assume that the vertical deflection of the cord is restrained by a large number of springs such that their effect can be considered as a distributed restoring force per unit length equal to ky, where k is a measure of the *spring constant* of the support and y is the deflection. In this case we must add the amount $-ky$ to the vertical load of Eq. (8.6), and we obtain the differential equation

$$H\frac{d^2 y}{dx^2} - ky = -w(x) \tag{9.1}$$

for the deflection $y(x)$. To solve this equation operationally, we again introduce the transforms

$$Ly(x) = Y(s) \tag{9.2}$$

$$Lw(x) = W(s) \tag{9.3}$$

and Eq. (9.1) transforms to

$$(s^2 - c^2)\,Y = sy_0 + y_1 - \frac{W(s)}{H} \tag{9.4}$$

where

$$c^2 = \frac{k}{H} \tag{9.5}$$

a CORD FIXED AT ITS ENDS, CONCENTRATED LOAD

Let us consider the case shown in Fig. 9.1. In this case the transform of the concentrated load is by (8.26) $e^{-as}P_0$. Since $y_0 = 0$, Eq. (9.4) becomes

$$Y(s) = \frac{y_1}{s^2 - c^2} - \frac{P_0}{H}e^{-as}s^2 - c^2 = Ly(x) \tag{9.6}$$

Consulting the table of transforms (Appendix A), we find the inverse transform to be

$$y(x) = \begin{cases} y_1\dfrac{\sinh cx}{c} & 0 < x < a \\[2mm] y_1\dfrac{\sinh cx}{c} - \dfrac{P_0}{Hc}\sinh c(x - a) & a < x < l \end{cases} \tag{9.7}$$

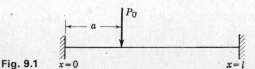

Fig. 9.1 $x=0$ $\qquad\qquad\qquad\qquad\qquad\qquad x=l$

If we use the boundary condition $y(l) = 0$, we obtain

$$y_1 = \frac{P_0}{H} \frac{\sinh c(l-a)}{\sinh cl} \tag{9.8}$$

Substituting this value of y_1 into (9.7), we obtain the following deflection of the cord:

$$y(x) = \frac{P_0}{Hc} \frac{\sinh c(l-a)\sinh cx}{\sinh lc} \qquad 0 < x < a \tag{9.9}$$

$$y(x) = \frac{P_0}{Hc} \frac{\sinh c(l-a)}{\sinh cl} \sinh cx - \frac{P_0}{Hc} \sinh c(x-a) \qquad a < x < l \tag{9.10}$$

b INFINITE CORD ELASTICALLY SUPPORTED, CONCENTRATED LOAD

The deflection of an infinitely long elastically supported cord under the influence of a concentrated load P_0 may be obtained as a special case of (9.10).

In order to obtain the deflection in this case, let

$$z = x - a \qquad \text{and} \qquad a = \frac{l}{2} \tag{9.11}$$

Making these substitutions in (9.10), we obtain

$$y(z+a) = y_1(z) = \frac{P_0}{Hc} \frac{\sinh(cl/2)\sinh[(cl/2) + cz]}{\sinh cl}$$

$$- \frac{P_0}{Hc} \sinh cz \qquad -0\frac{l}{2} < z < \frac{l}{2} \tag{9.12}$$

Now

$$\lim_{l\to\infty} y_1(z) = \frac{P_0}{2Hc} e^{-cz} = \frac{P_0}{2\sqrt{Hk}} e^{-cz} \qquad -0\frac{l}{2} < z < \infty \tag{9.13}$$

This is the required deflection where z is measured from the point of application of the load.

c ELASTICALLY SUPPORTED CORD WITH UNIFORM LOAD

Let us consider the case of Fig. 9.2. In this case the cord is loaded with a uniform load w_0 per unit length. The load extends from $x = a$ to $x = b$.

For this case we have

$$Lw(x) = \frac{w_0}{s} (e^{-as} - e^{-bs}) \tag{9.14}$$

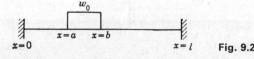

Fig. 9.2

Then the general equation (9.4) becomes

$$Y(s) = y_1 s^2 - c^2 - \frac{w_0}{Hs} \frac{e^{-as} - e^{-bs}}{s^2 - c^2} = Ly(x) \tag{9.15}$$

Consulting the table of transforms (Appendix A), we obtain

$$y(x) = y_1 \frac{\sinh cx}{c} \qquad\qquad\qquad 0 < x < a$$

$$y(x) = y_1 \frac{\sinh cx}{c} - \frac{w_0}{Hc^2} [\cosh c(x - a) - 1] \qquad a < x < b$$

$$y(x) = y_1 \frac{\sinh cx}{c} - \frac{w_0}{Hc^2} \cosh c(x - a) \tag{9.16}$$

$$\qquad\qquad\qquad + \frac{w_0}{Hc^2} \cosh c(x - b) \qquad b < x < l$$

Using the boundary condition $y(l) = 0$, we obtain the following value for the constant y_1:

$$y_1 = \frac{w_0}{Hc \sinh cl} [\cosh c(l - a) - \cosh c(l - b)] \tag{9.17}$$

Substituting this value of y_1 into (9.16) gives the required deflection.

d ELASTICALLY SUPPORTED CORD, GENERAL LOADING

The transform of the deflection of an elastically supported cord under the influence of a general load $w(x)$ is given by Eq. (9.4) in the form

$$Y(s) = \frac{y_1}{s^2 - c^2} - \frac{1}{H} \frac{W(s)}{s^2 - c^2} = Ly(x) \tag{9.18}$$

where

$$Lw(x) = W(s) \tag{9.19}$$

Now, by the Faltung theorem, we have

$$L^{-1} \frac{W(s)}{s^2 - c^2} = \frac{1}{c} \int_0^x w(u) \sinh [c(x - u)] \, du \tag{9.20}$$

Hence the inverse transform of (9.18) is given by

$$y(x) = y_1 \frac{\sinh cx}{c} - \frac{1}{Hc} \int_0^x w(u) \sinh [c(x - u)] \, du \tag{9.21}$$

Using the boundary condition $y(l) = 0$, we find the following value of y_1:

$$y_1 = \frac{1}{H \sinh cl} \int_0^l w(u) \sinh [c(l - u)] \, du \tag{9.22}$$

Substituting this value of y_1 into (9.21) gives the required deflection produced by the general loading $w(x)$.

10   THE DEFLECTION OF BEAMS BY TRANSVERSE FORCES

Consider a uniform straight beam supported, as shown in Fig. 10.1.   Let us measure the deflection $y(x)$ of the beam at any point x downward.   Then it is shown in works on elasticity† that $y(x)$ satisfies the following differential equation:

$$EI\frac{d^4 y}{dx^4} = w(x) \tag{10.1}$$

where E is Young's modulus of elasticity of the material of the beam and I is the moment of inertia of the cross section of the beam with respect to a line passing through the center of gravity of the cross section and perpendicular to the x axis and to the vertical direction y.   The quantity EI is called the *flexural rigidity* of the beam. $w(x)$ is the load per unit length of the beam.

There also exist the two following relations:

$$F(x) = -EI\frac{d^3 y}{dx^3} \tag{10.2}$$

and

$$M(x) = -EI\frac{d^2 y}{dx^2} \tag{10.3}$$

where $F(x)$ is the *shear force* and $M(x)$ is the *bending moment*.   The deflection $y(x)$ is measured positive downward, the load per unit length $w(x)$ is measured positive downward, the shear force $F(x)$ is measured positive upward, and the bending moment $M(x)$ positive clockwise.

To solve Eq. (10.1), let us introduce the transforms

$$Ly(x) = Y(s) \tag{10.4}$$
$$Lw(x) = W(s) \tag{10.5}$$

Hence, since

$$L\frac{d^4 y}{dx^4} = s^4 Y - s^3 y_0 - s^2 y_1 - s y_2 - y_3 \tag{10.6}$$

---

† R. V. Southwell, "An Introduction to the Theory of Elasticity for Engineers and Physicists," Oxford University Press, New York, 1936.

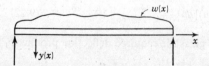

Fig. 10.1

where

$$y_r = \left(\frac{d^r y}{dx^r}\right) \qquad \text{evaluated at } x = 0 \qquad\qquad (10.7)$$

Eq. (10.1) transforms into

$$Y(s) = \frac{W(s)}{EIs^4} + \frac{y_0}{s} + \frac{y_1}{s^2} + \frac{y_2}{s^3} + \frac{y_3}{s^4} \qquad\qquad (10.8)$$

a UNIFORM BEAM CLAMPED HORIZONTALLY AT BOTH ENDS UNDER THE INFLUENCE OF A UNIFORMLY DISTRIBUTED LOAD

Consider the deflection of the beam of Fig. 10.2. In this case, the deflection and slope at $x = 0$ are both equal to zero; hence

$$y_0 = y_1 = 0 \qquad\qquad (10.9)$$

The transform of the uniform load w_0 is given by

$$W(s) = \frac{w_0}{s} \qquad\qquad (10.10)$$

Hence, in this case, (10.8) reduces to

$$Y(s) = \frac{w_0}{EIs^5} + \frac{y^2}{s^3} + \frac{y_3}{s^4} = Ly(x) \qquad\qquad (10.11)$$

The inverse transform of this gives

$$y(x) = \frac{w_0 x^4}{EI\,4!} + y_2 \frac{x^2}{2!} + y_3 \frac{x^3}{3!} \qquad\qquad (10.12)$$

The constants y_2 and y_3 are determined by the conditions

$$y = \frac{dy}{dx} = 0 \qquad \text{at } x = l \qquad\qquad (10.13)$$

These conditions give

$$y_2 = +\frac{w_0 l^2}{12EI} \qquad y_3 = -\frac{w_0 l}{2EI} \qquad\qquad (10.14)$$

Substituting these values into (10.12), we obtain the following equation for the deflection:

$$y(x) = \frac{w_0 x^2 (l - x)^2}{24EI} \qquad\qquad (10.15)$$

Fig. 10.2 $x=0$ $x=l$

b   UNIFORM BEAM CLAMPED HORIZONTALLY AT BOTH ENDS AND CARRYING A CONCENTRATED LOAD

Consider the problem of determining the deflection of the beam shown in Fig. 10.3. In this case the transform of the concentrated load is

$$W(s) = e^{-as} P_0 \tag{10.16}$$

Since the deflection and slope are zero at $x = 0$, Eq. (10.8) in this case reduces to

$$Y(s) = \frac{P_0 \, e^{-as}}{EI \, s^4} + \frac{y_2}{s^3} + \frac{y_3}{s^4} = Ly(x) \tag{10.17}$$

The inverse transform of this is

$$y(x) = y_2 \frac{x^2}{2!} + y_3 \frac{x^3}{3!} \qquad\qquad 0 < x < a$$
$$\tag{10.18}$$
$$y(x) = \frac{P_0}{6EI}(x - a)^3 + y_2 \frac{x^2}{2} + y_3 \frac{x^3}{6} \qquad a < x < l$$

The conditions that the deflection and slope must vanish at $x = l$ enable y_1 and y_2 to be determined.   Inserting these values of y_2 and y_3 into (10.18), we obtain the following equations for the deflection:

$$y(x) = \frac{P_0}{6EIl^3} x^2(l - a)^2[3al - (l + 2a)x] \qquad 0 < x < a$$
$$\tag{10.19}$$
$$y(x) = \frac{P_0 a^2}{6EIl^3} (a - l)^2(3l - 2a - al) \qquad a < x < l$$

c   UNIFORM BEAM CLAMPED HORIZONTALLY AT ONE END AND FREE AT THE OTHER AND CARRYING A CONCENTRATED LOAD

Let us determine the deflection of the beam shown in Fig. 10.4.   In this case, Eq. (10.8) becomes

$$Y(s) = \frac{P_0 \, e^{-as}}{EIs^4} + \frac{y_2}{s^3} + \frac{y_3}{s^4} \tag{10.20}$$

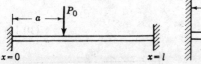

Fig. 10.3

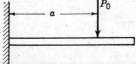

Fig. 10.4

The inverse transform of $Y(s)$ is

$$y(x) = y_2 \frac{x_2}{2} + y_3 \frac{x^3}{6} \qquad\qquad 0 < x < a$$

$$y(x) = \frac{P_0}{6EI}(x-a)^3 + y_2 \frac{x^2}{2!} + y_3 \frac{x^3}{6} \qquad a < x < l \tag{10.21}$$

In this case, since the end $x = l$ is free, it follows that the bending moment and shear force at that point must vanish. We therefore have by (10.2) and (10.3)

$$\frac{d^2 y}{dx^2} = \frac{d^3 y}{dx^3} = 0 \qquad \text{at } x = l \tag{10.22}$$

Applying these conditions to (10.21), we obtain

$$y_2 = \frac{aP_0}{EI}$$

$$y_3 = -\frac{P_0}{EI} \tag{10.23}$$

Substituting these values into (10.21), we obtain the following equations for the deflection:

$$y(x) = \frac{P_0 x^2}{EI}\left(\frac{a}{2} - \frac{x}{6}\right) \qquad 0 < x < a$$

$$y(x) = \frac{P_0 a^2}{EI}\left(\frac{x}{2} - \frac{a}{6}\right) \qquad a < x < l \tag{10.24}$$

11 DEFLECTION OF BEAMS ON AN ELASTIC FOUNDATION

Let us assume that a uniform beam is attached to a rigid base by means of a uniform elastic medium, as shown in Fig. 11.1. The action of the elastic medium may be taken into account by introducing a restoring force $-ky$ acting in a direction opposite to that of the load $w(x)$. In this case Eq. (10.1) becomes

$$EI \frac{d^4 y}{dx^4} + ky = w(x) \tag{11.1}$$

The constant k is called the *modulus of the foundation*.

Fig. 11.1

Let us divide Eq. (11.1) by EI. We then obtain

$$\frac{d^4 y}{dx^4} + \frac{k}{EI} y = \frac{w(x)}{EI} \tag{11.2}$$

If we now let

$$a = \left(\frac{k}{4EI}\right)^{\frac{1}{4}} \tag{11.3}$$

Eq. (11.2) may be written in the form

$$\frac{d^4 y}{dx^4} + 4a^4 y = \frac{w(x)}{EI} \tag{11.4}$$

To solve this equation operationally, we write

$$Ly(x) = Y(s) \tag{11.5}$$
$$Lw(x) = W(s) \tag{11.6}$$

Equation (11.4) is then transformed to

$$(s^4 + 4a^4)\, Y = \frac{W(s)}{EI} + s^3 y_0 + s^2 y_1 + s y_2 + y_3 \tag{11.7}$$

a BEAM ON ELASTIC FOUNDATION AND CLAMPED HORIZONTALLY AT BOTH ENDS UNDER THE INFLUENCE OF A CONCENTRATED LOAD

Let us consider the deflection of the beam shown in Fig. 11.2. In this case, the conditions that the slope and deflection at $x = 0$ must vanish lead to

$$y_1 = y_0 = 0 \tag{11.8}$$

The transform of the concentrated load P_0 acting at $x = b$ is

$$W(s) = P_0 e^{-sb} \tag{11.9}$$

Hence, in this case, Eq. (11.7) becomes

$$Y(s) = \frac{P_0}{EI}\frac{e^{-sb}}{s^4 + 4a^4} + y_2 \frac{s}{s^4 + 4a^4} + y_3 \frac{1}{s^4 + 4a^4} \tag{11.10}$$

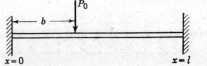

$x = 0$ $\qquad$ $x = l$ $\qquad$ **Fig. 11.2**

The transform of dy/dx is given by

$$L\left(\frac{dy}{dx}\right) = s\,Y(s)$$

$$= \frac{P_0}{EI}\frac{s\,e^{-sb}}{s^4 + 4a^4} + y_2\frac{s^2}{s^4 + 4a^4} + y_3\frac{s}{s^4 + 4a^4} \tag{11.11}$$

We have the following inverse transforms:

$$L^{-1}(s^4 + 4a^4) = \frac{1}{4a^3}(\sin ax \cosh ax - \cos ax \sinh ax)$$

$$= \phi_1(x) \tag{11.12}$$

$$L^{-1}\frac{s}{s^4 + 4a^4} = \frac{1}{2a^2}(\sin ax \sinh ax) = \phi_2(x) \tag{11.13}$$

$$L^{-1}\frac{s^2}{s^4 + 4a^4} = \frac{1}{2a}(\sin ax \cosh ax + \cos ax \sinh ax) = \phi_3(x) \tag{11.14}$$

$$L^{-1}\frac{s^3}{s^4 + 4a^4} = \cos ax \cosh ax = \phi_4(x) \tag{11.15}$$

In terms of these functions the inverse of $Y(s)$ in Eq. (11.10) is

$$
\begin{aligned}
y &= y_2\,\phi_2(x) + y_3\,\phi_1(x) & 0 < x < b \\
y &= \frac{P_0}{EI}\phi_1(x - b) + y_2\,\phi_2(x) + y_3\,\phi_1(x) & b < x < l
\end{aligned}
\tag{11.16}
$$

The constants y_2 and y_3 may be found from the condition that the deflection and slope must vanish at $x = l$. From (11.16) we thus obtain

$$0 = \frac{P_0}{EI}\phi_1(l - b) + y_2\,\phi_2(l) + y_3\,\phi_1(l) \tag{11.17}$$

From the transform of (11.11) we obtain

$$0 = \frac{P_0}{EI}\phi_2(l - b) + y_2\,\phi_3(l) + y_3\,\phi_2(l) \tag{11.18}$$

These two equations may be solved for the constants y_2 and y_3. This gives

$$y_2 = \frac{P_0}{EI}\frac{\phi_1(l)\,\phi_2(l - b) - \phi_2(l)\,\phi_1(l - b)}{\phi_2^2(l) - \phi_3(l)\,\phi_1(l)} \tag{11.19}$$

$$y_3 = \frac{P_0}{EI}\frac{\phi_2(l)\,\phi_2(l - b) - \phi_3(l)\,\phi_1(l - b)}{\phi_1(l)\,\phi_2(l) - \phi_2^2(l)} \tag{11.20}$$

The deflection is obtained by substituting these values of y_2 and y_3 into (11.16).

b   BEAM ON ELASTIC FOUNDATION AND CLAMPED HORIZONTALLY AT BOTH ENDS UNDER THE INFLUENCE OF A GENERAL LOAD

In this case, since the deflection and slope vanish at $x = 0$, Eq. (11.7) becomes

$$Y(s) = \frac{W(s)}{EI(s^4 + 4a^4)} + \frac{sy_2}{s^4 + 4a^4} + \frac{y_3}{s^4 + 4a^4} \tag{11.21}$$

The inverse transform of the slope is given by

$$L\left(\frac{dy}{dx}\right) = \frac{sW(s)}{EI(s^4 + 4a^4)} + \frac{s^2 y_2}{s^4 + 4a^4} + \frac{sy_3}{s^4 + 4a^4} \tag{11.22}$$

Now, by the Faltung theorem, we have

$$L^{-1}\frac{W(s)}{s^4 + 4a^4} = \int_0^x w(u)\,\phi_1(x - u)\,du \tag{11.23}$$

$$L^{-1}\frac{W(s)}{s^4 + 4a^4} = \int_0^x w(u)\,\phi_2(x - u)\,du \tag{11.24}$$

Hence the inverses of (11.21) and (11.22) give the following equations for the deflection and slope:

$$y = \frac{1}{EI}\int_0^x w(u)\,\phi_1(x - u)\,du + y_2\,\phi_2(x) + y_3\,\phi_1(x) \tag{11.25}$$

$$\frac{dx}{dy} = \frac{1}{EI}\int_0^x w(u)\,\phi_2(x - u)\,du + y_2\,\phi_3(u) + y_3\,\phi_2(x) \tag{11.26}$$

Now, if we let

$$A = \frac{1}{EI}\int_0^l w(u)\,\phi_1(l - u)\,du \tag{11.27}$$

$$B = \frac{1}{EI}\int_0^l w(u)\,\phi_2(l - u)\,du \tag{11.28}$$

and make use of the fact that the slope and deflection both vanish at $x = l$, we obtain the following values for the constants y_2 and y_3:

$$y_2 = \frac{B\phi_1(l) - A\phi_2(l)}{\phi_2^2(l) - \phi_1(l)\,\phi_3(l)} \tag{11.29}$$

$$y_3 = \frac{B\phi_2(l) - A\phi_3(l)}{\phi_1(l)\,\phi_3(l) - \phi_2^2(l)} \tag{11.30}$$

If these values of y_2 and y_3 are substituted into (11.25), the value of the deflection is obtained.

12 BUCKLING OF A UNIFORM COLUMN UNDER AXIAL LOAD

Consider a column hinged at the point $x = l$ and supported at $x = 0$ in such a way that lateral deflection is prevented but free rotation is allowed as shown in Fig. 12.1.   The column is under the influence of an axial load considered positive when acting in a downward direction to cause a compression.   Let $y(x)$ be the deflection of the column.   The bending moment at any point x is given by

$$EI\frac{d^2 y}{dx^2} = -M \tag{12.1}$$

The bending moment at any point x due to the force P is given by

$$M = Py \tag{12.2}$$

Hence, substituting this into (12.1), we have

$$EI\frac{d^2 y}{dx^2} + Py = 0 \tag{12.3}$$

If we let

$$a^2 = \frac{P}{EI} \tag{12.4}$$

we obtain

$$\frac{d^2 y}{dx^2} + a^2 y = 0 \tag{12.5}$$

The general solution of this equation is

$$y = A \cos ax + B \sin ax \tag{12.6}$$

where A and B are arbitrary constants.   In order to satisfy the boundary condition $y(0) = 0$, we must have

$$A = 0 \tag{12.7}$$

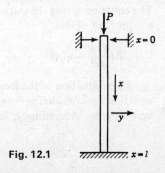

Fig. 12.1

so that the solution reduces to

$$y = B \sin ax \tag{12.8}$$

To satisfy the condition $y(l) = 0$, we must have

$$\sin al = 0 \tag{12.9}$$

or

$$al = k\pi \qquad k = 0, 1, 2, 3, \ldots \tag{12.10}$$

To each value of k, there corresponds a solution

$$y_k = B_k \sin \frac{k\pi x}{l} \tag{12.11}$$

where the B_k's are arbitrary constants. These deflections are called the *modes of buckling.* To each mode there corresponds a load

$$P_k = k^2 \pi^2 \frac{EI}{l^2} \tag{12.12}$$

These loads are called the *critical loads.* For each of these loads the corresponding mode of buckling represents an equilibrium position with an arbitrary amplitude. The first critical load is obtained by placing $k = 1$ in (12.12), so that we have

$$P_1 = \pi^2 \frac{EI}{l^2} \tag{12.13}$$

This equation is known as *Euler's formula.* It gives the upper limit for the stability of the undeflected equilibrium position of the column.

13   THE VIBRATION OF BEAMS

To find the equation of motion for the free vibrations of a uniform beam, we use D'Alembert's principle that states that any dynamical problem may be treated as a static problem by the addition of appropriate inertia forces. The equation giving the static deflection of a beam $y(x)$ under the influence of a static load $w(x)$ is

$$EI \frac{d^4 y}{dx^4} = w(x) \tag{13.1}$$

The equation of the free vibration of a beam may be obtained from this equation by considering $-m(\partial^2 y/\partial t^2)$ as the load, where m is the mass per unit length. Accordingly, letting

$$w(x) = -m \frac{\partial^2 y}{\partial t^2} \tag{13.2}$$

in (13.1), we obtain

$$EI\frac{\partial^4 y}{\partial x^4} = -m\frac{\partial^2 y}{\partial t^2} \tag{13.3}$$

where partial differentiation symbols must be used because the deflection y is now a function of the two *independent* variables x and t.

Let us investigate oscillations of the harmonic type. To do this, we let

$$y(x,t) = v(x)\sin \omega t \tag{13.4}$$

If we substitute (13.4) into (13.3), we obtain the following ordinary differential equation for the variable $v(x)$:

$$EI\frac{d^4 v}{dx^4} = m\omega^2 v \tag{13.5}$$

If we let

$$k^4 = \frac{m\omega^2}{EI} \tag{13.6}$$

Eq. (13.5) may be written in the form

$$\frac{d^4 v}{dx^4} = k^4 v \tag{13.7}$$

This equation has solutions of the exponential type

$$v = c\,e^{\theta x} \tag{13.8}$$

where c is an arbitrary constant. Substituting this into (13.7), we find that the possible values of θ are given by

$$\theta^4 = k^4 \tag{13.9}$$

or

$$\theta^2 = \pm k^2 \tag{13.10}$$

and hence

$$\theta = \pm k \text{ or } \pm jk \qquad j = \sqrt{-1} \tag{13.11}$$

Therefore, the general solution of (13.7) is given by

$$v = c_1 e^{kx} + c_2 e^{-kx} + c_3 e^{jkx} + c_4 e^{-jkx} \tag{13.12}$$

where the c_r quantities are *arbitrary* constants.

It is convenient to express the solution of (13.7) in terms of hyperbolic and trigonometric functions instead of exponential functions in the form

$$v(x) = A\cos kx + B\sin kx + C\cosh kx + D\sinh kx \tag{13.13}$$

where A, B, C, D are arbitrary constants.

a NATURAL FREQUENCIES OF A CANTILEVER

Let us determine the natural frequencies and modes of oscillation of the cantilever beam shown in Fig. 13.1. To determine the natural frequencies of the cantilever beam, we have the boundary conditions

(a) $\left.\begin{array}{l} v = 0 \\ \dfrac{dv}{dx} = 0 \end{array}\right\}$ $x = 0$ fixed end

(b) $\dfrac{d^2 v}{dx^2} = 0$ at $x = l$ bending moment $= 0$

(c) $\dfrac{d^3 v}{dx^3} = 0$ at $x = l$ shear force $= 0$

Imposing these conditions on the general solution (13.13), we have

$$A + C = 0 \qquad B + D = 0 \tag{13.14}$$

Hence we have

$$A = -C \qquad B = -D \tag{13.15}$$
$$C(\cosh kl + \cos kl) + D(\sinh kl + \sin kl) = 0 \tag{13.16}$$
$$C(\sinh kl - \sin kl) + D(\cosh kl + \cos kl) = 0 \tag{13.17}$$

In order that these homogeneous linear equations in C and D may have a nontrivial solution, it is necessary that

$$\begin{vmatrix} \cosh kl + \cos kl & \sinh kl + \sin kl \\ \sinh kl - \sin kl & \cosh kl + \cos kl \end{vmatrix} = 0 \tag{13.18}$$

or

$$1 + \cosh kl \cos kl = 0 \tag{13.19}$$

Therefore

$$\cos kl = -\frac{1}{\cosh kl} \tag{13.20}$$

This equation may be solved graphically by letting $kl = Z$ and plotting $\cos Z$ and $-1/\cosh Z$, as shown in Fig. 13.2. If the Z_r are the abscissas of the

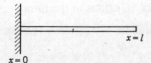

$x = 0$ **Fig. 13.1**

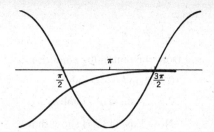

Fig. 13.2

points of intersection of these curves, then they are the solutions of Eq. (13.20).   The first few roots of (13.20) are given by

$$Z_1 = 1.8751 \qquad Z_2 = 4.694 \qquad Z_3 = 7.854$$
$$Z_4 = 10.996 \qquad Z_5 = 14.13, \text{ etc.}$$

(13.21)

Each root Z_r fixes a value k_r of k by the equation

$$k_r = \frac{Z_r}{l}$$

(13.22)

By (13.6) each value k_r of k fixes a value of the possible natural frequency ω by the equation

$$\omega_r = k_r^2 \sqrt{\frac{EI}{m}} = \left(\frac{Z_r}{l}\right)^2 \sqrt{\frac{EI}{m}}$$

(13.23)

The lowest natural angular frequency is given by

$$\omega_1 = \left(\frac{1.875}{l}\right)^2 \sqrt{\frac{EI}{m}}$$

(13.24)

b   NATURAL FREQUENCIES OF A HINGED BEAM

Let us consider the determination of the natural frequencies of oscillation of a beam hinged at both ends $x = 0$ and $x = l$.   The boundary conditions for this case are

$(a) \quad \begin{matrix} v = 0 & v = 0 \\ x = 0 & x = l \end{matrix}$

$(b) \quad \begin{matrix} \dfrac{d^2 v}{dx^2} = 0 & \dfrac{d^2 v}{dx^2} = 0 \\ x = 0 & x = l \end{matrix}$

Imposing two of these conditions on Eq. (13.13), we obtain the following equations:

$$0 = A + C$$
$$0 = -A + C$$

(13.25)

Hence

$$A = C = 0 \tag{13.26}$$

and from the remaining conditions

$$0 = B \sin kl + D \sinh kl$$
$$0 = -B \sin kl + D \sinh kl \tag{13.27}$$

Now Eqs. (13.27) show that

$$B \sin kl = 0 \tag{13.28}$$

and

$$D \sinh kl = 0 \tag{13.29}$$

Now, since $\sinh kl$ cannot be zero for real values of its argument, other than $kl = 0$, it follows that $D = 0$. Then, for a nontrivial solution $B \neq 0$, we must have

$$\sin kl = 0 \tag{13.30}$$

Hence

$$kl = r\pi \qquad r = 0, 1, 2, 3, \ldots \tag{13.31}$$

and to each value of r there corresponds a value k_r of k, given by

$$k_r = \frac{r\pi}{l} \tag{13.32}$$

By (13.6) the natural frequencies are given by

$$\omega_r = \left(\frac{r\pi}{l}\right)^2 \sqrt{\frac{EI}{m}} \qquad r = 1, 2, 3, 4, 5, \ldots \tag{13.33}$$

The natural modes are given by

$$v_r(x) = B_r \sin \frac{r\pi x}{l} \qquad r = 1, 2, 3, \ldots \tag{13.34}$$

c   BEAM CLAMPED AT ONE END AND CARRYING A MASS AT THE FREE END

As a more complicated example of the general method, let us consider the system of Fig. 13.3. This represents a cantilever beam supporting a mass M at its free end. In this case we must adjust the general solution (13.13) to the boundary conditions:

$$(a) \quad \left.\begin{array}{l} v = 0 \\ \dfrac{dv}{dx} = 0 \end{array}\right\} \quad x = 0 \qquad \text{clamped end}$$

$$(b) \quad \frac{d^2 v}{dx^2} = 0 \quad \text{at } x = l \qquad \text{zero bending moment}$$

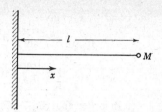

Fig. 13.3

The fourth condition is that at the free end the shearing force is the force due to the inertia of the mass M.   Now the shearing force F is given by

$$F = -EI\frac{\partial^3 y}{\partial x^3} \tag{13.35}$$

in terms of the deflection $y(x,t)$.   The inertia force of the mass is

$$-M\left(\frac{\partial^2 y}{\partial t^2}\right)_{x=l}$$

Hence we have

$$EI\left(\frac{\partial^3 y}{\partial x^3}\right)_{x=l} = M\left(\frac{\partial^2 y}{\partial t^2}\right)_{x=l} \tag{13.36}$$

To obtain the boundary condition in terms of the variable $v(x)$, we use the fact that

$$y(x,t) = v(x)\sin \omega t \tag{13.37}$$

Hence, substituting this into (13.36) and dividing both sides by the common factor $\sin \omega t$, we obtain

(c) $$EI\left(\frac{d^3 v}{dx^3}\right)_{x=l} = -(M\omega^2 v)_{x=l}$$

This is the required boundary condition.

We now impose these boundary conditions on the general solution (13.13).   Condition (*a*) gives

$$\begin{aligned} 0 &= A + C \\ 0 &= B + D \end{aligned} \tag{13.38}$$

Therefore we may write

$$v(x) = A\,(\cos kx - \cosh kx) + B\,(\sin kx - \sinh kx) \tag{13.39}$$

Condition (*b*) gives

$$-A\,(\cos kl + \cosh kl) - B\,(\sin kl + \sinh kl) = 0 \tag{13.40}$$

and condition (c) gives

$$-k^3[A(\sinh kl - \sin kl) + B(\cosh kl + \cos kl)]$$

$$= + \frac{M\omega^2}{EI}[A(\cosh kl - \cos kl) + B(\sinh kl - \sin kl)] \quad (13.41)$$

In order that Eqs. (13.40) and (13.41) may have solutions other than the trivial solution $A = B = 0$, the determinant of their coefficients must vanish. If we let

$$\phi = \frac{M}{ml} = \frac{M}{M_B} \tag{13.42}$$

so that ϕ is the ratio of the mass M to the mass of the beam M_B, and also let

$$Z = kl \tag{13.43}$$

we obtain, after some reductions, the equation

$$\frac{1 + \cosh Z \cos Z}{\cosh Z \sin Z - \sinh Z \cos Z} = \phi Z \tag{13.44}$$

This equation can be solved by plotting the curve

$$y_1 = \frac{1 + \cosh Z \cos Z}{\cosh Z \sin Z - \sinh Z \cos Z}$$

$$= \frac{\cos Z + \operatorname{sech} Z}{\sin Z - \cos Z \tanh Z} \tag{13.45}$$

and the straight line

$$y_2 = \phi Z \tag{13.46}$$

and finding the values of Z at the intersections.

If $\phi = 1$, a graph gives the following approximate values of Z for the intersection of y_1 and y_2:

$$Z = 1.238, 4.045, \frac{9\pi}{4}, \frac{13\pi}{4}, \ldots \tag{13.47}$$

If we call the rth root of (13.45) Z_r, then we have

$$\omega_r = \left(\frac{Z_r}{l}\right)^2 \sqrt{\frac{EI}{m}} \tag{13.48}$$

for the natural angular frequencies of the oscillations of the system.

If $M \gg M_B$ so that the supported mass is much larger than the mass of the beam, then $\phi \gg 1$. In this case the angular frequency is very small, and Z is small. Now, when Z is small, we can use the following approximations:

$$\cos Z \doteq 1 - \frac{Z^2}{2} \qquad \cosh Z \doteq 1 + \frac{Z^2}{2}$$
$$\sin Z \doteq Z - \frac{Z^3}{6} \qquad \sinh Z \doteq Z + \frac{Z^3}{6} \tag{13.49}$$

Using these approximations, Eq. (13.44) becomes

$$1 + 1 - \frac{Z^4}{4} = \phi Z^2 \left[1 + \frac{Z^2}{3} - \left(1 - \frac{Z^2}{3} \right) \right] \tag{13.50}$$

or

$$2 - \frac{Z^4}{4} = \frac{2}{3} \phi Z^4 \tag{13.51}$$

Neglecting $Z^4/4$ in comparison with 2, we have

$$Z_1 = \left(\frac{3}{\phi} \right)^{\frac{1}{4}} \tag{13.52}$$

In this case we have from (13.48)

$$\omega_1 = \frac{1}{l^2} \sqrt{\frac{3EI}{m\phi}} = \frac{1}{l^2} \sqrt{\frac{3EIM_B}{mM}} = \sqrt{\frac{3EI}{l^3 M}} \tag{13.53}$$

for the fundamental frequency in the case that the supported mass is much greater than the mass of the beam. This result may be computed more simply by computing the *effective spring constant* of the beam by impressing a unit force at the free end and computing the resulting deflection.

14 EXAMPLES OF FOURIER TRANSFORMS

As a first example of the use of Fourier transforms† let us return to the example of Sec. 3b. This problem was one of linear heat flow in a semi-infinite medium. The governing equation and boundary conditions are

$$\frac{\partial v}{\partial t} = h^2 \frac{\partial^2 v}{\partial x^2} \tag{14.1}$$

$$v = v_0 \qquad \text{at } x = 0 \qquad t > 0$$
$$v = 0 \qquad \text{at } x = 0 \qquad t = 0 \tag{14.2}$$

† Tranter, 1956, p. 34 (see References).

Instead of using a Laplace transform on the time variable, as in Sec. 4, we will apply a Fourier sine transform to the space coordinate x; that is,

$$\bar{v}(w,t) = F_s[v(x,t)]$$

$$= \int_0^\infty v(x,t) \sin wx \, dx \tag{14.3}$$

To apply the Fourier sine transform to Eq. (14.1), simply multiply through by $\sin(wx)$ and integrate from 0 to ∞:

$$\int_0^\infty \frac{\partial v}{\partial t} \sin wx \, dx = h^2 \int_0^\infty \frac{\partial^2 v}{\partial x^2} \sin wx \, dx \tag{14.4}$$

Since the integral is over x, the left-hand side of (14.4) reduces to

$$\int_0^\infty \frac{\partial v}{\partial t} \sin wx \, dx = \frac{\partial}{\partial t} \int_0^\infty v \sin wx \, dx$$

$$= \frac{\partial \bar{v}}{\partial t} \tag{14.5}$$

To decompose the second partial with respect to x, the right-hand side of (14.4) must be integrated by parts twice. The result is

$$F_s\left[\frac{\partial^2 v}{\partial x^2}\right] = \frac{\partial v}{\partial x} \sin wx \Big|_0^\infty - wv \cos wx \Big|_0^\infty - w^2 \bar{v} \tag{14.6}$$

To evaluate the first two terms on the right-hand side of (14.6), it is argued on physical grounds that both v and $\partial v/\partial x$ vanish at $x = \infty$ for all time, hence the contribution from the upper limits vanish. The evaluation of the lower limits is straightforward and gives as a final result:

$$F_s\left[\frac{\partial^2 v}{\partial x^2}\right] = wv_0 - w^2 \bar{v} \tag{14.7}$$

Therefore the complete transform of (14.1) is

$$\frac{\partial \bar{v}}{\partial t} + h^2 w^2 \bar{v} = h^2 wv_0 \tag{14.8}$$

This equation is a linear first-order differential equation which is easily solved by any one of several classical methods to yield

$$\bar{v}(w,t) = \frac{v_0}{w}(1 - e^{-h^2 w^2 t}) \tag{14.9}$$

The solution in the real domain is obtained by employing the inversion integral given in (2.3):

$$
\begin{aligned}
v(x,t) &= F_s^{-1}[\bar{v}(w,t)] \\
&= \frac{2v_0}{\pi} \int_0^\infty (1 - e^{-h^2 w^2 t}) \sin wx \, \frac{dw}{w} \\
&= \frac{2v_0}{\pi} \int_0^\infty \sin wx \, \frac{dw}{w} - \frac{2v_0}{\pi} \int_0^\infty e^{-w^2 h^2 t} \sin wx \, \frac{dw}{w}
\end{aligned}
\tag{14.10}
$$

It is well known that the value of the first integral is $\pi/2$.† The second integral is also a well-known relation:‡

$$
\int_0^\infty e^{-h^2 w^2 t} \sin wx \, \frac{dw}{w} = \frac{\pi}{2} \operatorname{erf} \frac{x}{2ht^{\frac{1}{2}}}
\tag{14.11}
$$

Thus the solution for the temperature distribution is

$$
\begin{aligned}
v(x,t) &= v_0 \left(1 - \operatorname{erf} \frac{x}{2ht^{\frac{1}{2}}}\right) \\
&= v_0 \operatorname{erfc} \frac{x}{2ht^{\frac{1}{2}}}
\end{aligned}
\tag{14.12}
$$

which agrees with Eq. (4.55).

As another example, consider the vibration of a string of infinite length:

$$
\frac{\partial^2 u}{\partial x^2} = \frac{1}{c^2} \frac{\partial^2 u}{\partial t^2}
\tag{14.13}
$$

$$
u(x,0) = f(x) \qquad \frac{\partial u}{\partial t}(x,0) = 0
\tag{14.14}
$$

To solve this problem, the complex Fourier transform will be used; that is,

$$
\bar{u}(w,t) = F[u(x,t)] = \int_{-\infty}^\infty u(x,t) e^{-jwx} \, dx
\tag{14.15}
$$

Applying this transform to (14.13) and requiring u and $\partial u/\partial x$ to vanish as $x \to \infty$ yields§

$$
-w^2 \bar{u} = \frac{1}{c^2} \frac{d^2 \bar{u}}{dt^2}
\tag{14.16}
$$

† Cf. E. Janke and F. Emde, "Tables of Functions," p. 3, Dover Publications, Inc., New York, 1945.

‡ Cf. H. B. Dwight, "Tables of Integrals and Other Mathematical Data," 4th ed., no. 861.22, The Macmillan Company, New York, 1961.

§ Ordinary derivatives have been used since w may be considered constant.

or

$$\frac{d^2 \bar{u}}{dt^2} + w^2 c^2 \bar{u} = 0 \qquad (14.17)$$

The general solution of this equation is

$$\bar{u} = A(w) \sin wct + B(w) \cos wct \qquad (14.18)$$

Before the initial conditions can be applied, they must be transformed as well:

$$\bar{u}(w,0) = F[u_0(x)] = \int_{-\infty}^{\infty} f(x) e^{-jwx} dx = \bar{f}(x)$$

$$\frac{\partial u}{\partial t}(w,0) = 0 \qquad (14.19)$$

Imposing the initial conditions on (14.18) yields the desired transform solution:

$$\bar{u} = \bar{f}(w) \cos wct \qquad (14.20)$$

Applying the inverse transform defined in (2.2) gives

$$u(x,t) = \frac{1}{2\pi} \int_{-\infty}^{\infty} \bar{f}(w) \cos wct \, e^{jwx} dw \qquad (14.21)$$

Making use of Euler's expansion for the cosine and rearranging terms, (14.21) may be written as

$$u(x,t) = \frac{1}{4\pi} \int_{-\infty}^{\infty} [e^{jw(x+ct)} + e^{jw(x-ct)}] \bar{f}(w) \, dw$$

$$= \frac{1}{4\pi} \int_{-\infty}^{\infty} e^{jw(x+ct)} \bar{f}(w) \, dw + \frac{1}{4\pi} \int_{-\infty}^{\infty} e^{jw(x-ct)} \bar{f}(w) \, dw \qquad (14.22)$$

Examining the inverse of the first boundary condition (14.19), we see that

$$f(x) = \frac{1}{2\pi} \int_{-\infty}^{\infty} \bar{f}(w) e^{jwx} dw \qquad (14.23)$$

From this equation it is easy to deduce the following relation:

$$f(x \pm ct) = \frac{1}{2\pi} \int_{-\infty}^{\infty} \bar{f}(w) e^{jw(x \pm ct)} dw \qquad (14.24)$$

Comparing this relation with (14.22) permits us to express the solution as follows:

$$u(x,t) = \tfrac{1}{2} f(x + ct) + \tfrac{1}{2} f(x - ct) \qquad (14.25)$$

which is exactly D'Alembert's solution of Chap. 12, Eq. (3.12).

15 MELLIN AND HANKEL TRANSFORMS

Consider the infinite wedge† shown in Fig. 15.1 lying in the region $r > 0$, $0 \leqslant \theta \leqslant \alpha$.   It is desired to determine the steady-state temperature distribution $v(r,\theta)$ within the wedge subject to the following conditions:

$$v = u(r) - u(r - a) \qquad \theta = \alpha$$

$$\frac{\partial v}{\partial \theta} = 0 \qquad \theta = 0 \tag{15.1}$$

where $u(r)$ denotes the Heaviside unit step function.   Since only the steady-state solution is desired, the governing equation is Laplace's differential equation in cylindrical coordinates:

$$\frac{r^2 \partial^2 v}{\partial r^2} + \frac{r \partial v}{\partial r} + \frac{\partial^2 v}{\partial \theta^2} = 0 \tag{15.2}$$

To solve this problem we shall make use of the Mellin transform:

$$\bar{v}(s,\theta) = M\left[v(r,\theta)\right] = \int_0^\infty v(r,\theta)\, r^{s-1}\, dr \tag{15.3}$$

The transforms of each of the three terms in (15.2) become‡

$$M\left[\frac{r \partial v}{\partial r}\right] = -s\bar{v}$$

$$M\left[\frac{r^2 \partial^2 v}{\partial r^2}\right] = s(s+1)\,\bar{v} \tag{15.4}$$

$$M\left[\frac{\partial^2 v}{\partial \theta^2}\right] = \frac{d^2 \bar{v}}{d\theta^2}$$

Hence, upon transforming (15.2), we find

$$\frac{d^2 \bar{v}}{d\theta^2} + s^2 \bar{v} = 0 \tag{15.5}$$

† This problem is modified from one in Tranter, 1956, no. 5, p. 58 (see References).

‡ It must be assumed that $v(r,\theta) \approx 1/r^{s+1}$ as $r \to \infty$.

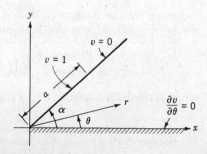

Fig. 15.1

The general solution is

$$\bar{v} = A(s)\sin s\theta + B(s)\cos s\theta \tag{15.6}$$

To evaluate the coefficients, the Mellin transform of the boundary conditions must be computed. From the second boundary condition of (15.1)

$$\left.\frac{\partial \bar{v}}{\partial \theta}\right|_{\theta=0} = 0 \tag{15.7}$$

and from the first

$$\bar{v}(s,\alpha) = \int_0^\infty [u(r) - u(r-a)]\, r^{s-1}\, dr$$

$$= \int_0^a r^{s-1}\, dr = \frac{a^s}{s} \tag{15.8}$$

Applying (15.7) and (15.8) to (15.6) results in the following transformed solution:

$$\bar{v}(s,\theta) = \frac{a^s \cos s\theta}{s \cos s\alpha} \tag{15.9}$$

Using the inversion relation of (2.6), we find

$$v(r,\theta) = \frac{1}{2\pi j} \int_{c-j\infty}^{c+j\infty} \frac{\cos s\theta}{\cos s\alpha} \left(\frac{a}{r}\right)^s \frac{ds}{s} \tag{15.10}$$

This integral may be evaluated by integrating along the imaginary axis from $-\infty$ to $+\infty$, except for a small semicircle on the right side of the origin which avoids crossing the simple pole there (cf. Fig. 15.2). Because the singularity at the origin is a simple pole, the integral over the small semicircle is just one-half the residue at $s = 0$.† If Γ represents the small semicircle, then

$$\frac{1}{2\pi j} \int_\Gamma \frac{\cos s\theta}{\cos s\alpha} \left(\frac{a}{r}\right)^s \frac{ds}{s} = \frac{1}{2} \tag{15.11}$$

Since the path of integration lies on the imaginary axis, we then may set $s = jw$, $ds = j\,dw$. Using this substitution and (15.11), (15.10) becomes

$$v(r,\theta) = \frac{1}{2} + \frac{1}{\pi} \int_0^\infty \frac{\cosh w\theta}{\cosh w\alpha} \sin\left(w \ln\frac{a}{r}\right) \frac{dw}{w} \tag{15.12}$$

which is the desired solution to the problem.

---

† A very useful fact to remember is that an integral over a small semicircular contour centered at a *simple pole* is always equal to half the residue of the simple pole.

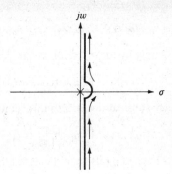

Fig. 15.2

To illustrate Hankel transforms we shall consider the vibrations of a circular membrane of infinite radius. The equation of motion for this system is

$$\nabla^2 u = \frac{1}{c^2} \frac{\partial^2 u}{\partial t^2} \tag{15.13}$$

where $u(r,\theta,t)$ represents the displacement of the membrane from the plane of its equilibrium position. If we investigate only symmetrical vibrations such that the motion is independent of θ, that is, $u(r,\theta,t) = u(r,t)$, then (15.13) becomes

$$\frac{\partial^2 u}{\partial r^2} + \frac{1}{r} \frac{\partial u}{\partial r} = \frac{1}{c^2} \frac{\partial^2 u}{\partial t^2} \tag{15.14}$$

The initial conditions are assumed to be

$$u(r,0) = u_0(r) \qquad \frac{\partial u}{\partial t}(r,0) = 0 \tag{15.15}$$

Because of the polar geometry of this problem, a Hankel transform is implied. If we let

$$\bar{u}(s,t) = H_0[u(r,t)] \tag{15.16}$$

then the Hankel transform† of (15.14) is

$$\frac{d^2 \bar{u}}{dt^2} + c^2 s^2 \bar{u} = 0 \tag{15.17}$$

The solution of this equation subject to the initial conditions of (15.15) is

$$\bar{u}(s,t) = u_0(s) \cos cst \tag{15.18}$$

† Cf. note on Hankel transform following Eq. (2.5).

where

$$u_0(s) = H_0[u_0(r)] = \int_0^\infty u_0(r)J_0(rs)r\,dr \tag{15.19}$$

Applying the inversion integral as given by (2.5), the result is

$$u(r,t) = \int_0^\infty u_0(s)\cos cst\,J_0(sr)s\,ds \tag{15.20}$$

By substituting (15.19) this solution may be rewritten in the following way:

$$u(r,t) = \int_0^\infty u_0(\xi)\left[\int_0^\infty \cos cst\,J_0(\xi s)J_0(rs)s\,ds\right]\xi\,d\xi \tag{15.21}$$

These integrals are difficult to integrate, and the reader is referred to Sneddon†
for a further discussion.

16   REPEATED USE OF TRANSFORMS

In multidimensional problems it is sometimes convenient to apply integral transforms to more than one independent variable. Several interesting examples of this may be found in the literature.‡   Because of the linearity of integral transforms, the same one may be applied repeatedly, or combinations may be used.   Linearity allows the order of application to be rearranged as desired; that is, one might solve a problem by Laplace transforming first and then applying a Fourier transform.   If order were important, the solution would have to be found by inverse Fourier transforming first before inverse Laplace transforming.   However, due to the linearity of these transform operators, the solution may be found by inverting the transformed solution in any order.

To illustrate the use of multiple transforms, let us consider the first example of Sec. 14.   This problem was to determine the temperature distribution inside a semi-infinite medium subjected to the conditions given in (14.2). The governing equation is given by (14.1).   To solve this problem both Laplace and Fourier transforms will be employed.   Applying the Laplace transform to the time variable first, we find that the transform of (14.1) is

$$h^2\frac{\partial^2 \bar{v}}{\partial x^2} - s\bar{v} = 0 \tag{16.1}$$

where

$$L[v(x,t)] = \bar{v}(x,s) \tag{16.2}$$

If we now introduce the Fourier sine transform

$$F_s[\bar{v}(x,s)] = \bar{\bar{v}}(w,s) \tag{16.3}$$

† Sneddon, 1951, p. 125ff. (see References).
‡ See Webster, 1955, and Estrin, 1951 (see References).

and follow the same procedure of Sec. 14, the transform of (16.1) is

$$h^2(w\bar{v}_0 - w^2\,\bar{\bar{v}}) - s\bar{\bar{v}} = 0 \tag{16.4}$$

where

$$\bar{v}_0 = L(v_0) = \frac{v_0}{s} \tag{16.5}$$

From (16.5) we have

$$\bar{\bar{v}} = \frac{wv_0 h^2}{s(s + w^2 h^2)} \tag{16.6}$$

To invert this solution, we will proceed in the reverse order. The "natural order" would be to inverse Fourier transform first and then inverse Laplace transform. The inverse Laplace transform of (16.6) is readily determined from entry No. 2.1 of the table of transforms (Appendix A).

$$\bar{v} = \frac{v_0}{w}(1 - e^{-w^2 h^2 t}) \tag{16.7}$$

This result is identical to (14.9), from which it should be evident that upon applying the inverse Fourier sine transform, the final solution is

$$v(x,t) = v_0 \operatorname{erfc} \frac{x}{2ht^{\frac{1}{2}}} \tag{16.8}$$

which is the previously obtained solution.

As a second example let us consider a slight modification of the foregoing problem. Instead of considering a constant surface temperature for the semi-infinite solid, consider that it is subjected to a sinusoidally varying temperature described by

$$v(0,t) = v_m \sin at \qquad x = 0 \qquad t > 0 \tag{16.9}$$

The remaining boundary condition is assumed to remain the same. The solution to this problem is identical to the previous one up to (16.5). At this point the new boundary condition must be inserted; that is,

$$\bar{v}_0 = L(v_m \sin at) = \frac{av_m}{s^2 + a^2} \tag{16.10}$$

Inserting this result into (16.5) and solving for $\bar{\bar{v}}$ yields

$$\bar{\bar{v}} = \frac{awv_m}{(s^2 + a^2)(w^2 + s/h^2)} \tag{16.11}$$

To invert this solution we will first find the inverse Fourier sine transform. From the definition of the inverse transform, we have

$$\bar{v} = \frac{2av_m}{\pi(s^2 + a^2)} \int_0^\infty \frac{w \sin wx \, dw}{(w^2 + s/h^2)} \tag{16.12}$$

To evaluate this integral it is convenient to introduce a new variable of integration $p = wx$. Performing the change of variable gives

$$\bar{v} = \frac{2av_m}{\pi(s^2 + a^2)} \int_0^\infty \frac{p \sin p \, dp}{(p^2 + sx^2/h^2)} \tag{16.13}$$

The above integral† is tabulated in many reference tables, and the result is

$$\bar{v} = \frac{av_m \, e^{-s^{\frac{1}{2}}x/h}}{s^2 + a^2} \tag{16.14}$$

To accomplish the inverse Laplace transform of this result, the convolution theorem will be employed. To do this we shall rewrite (16.14) as follows:

$$\bar{v} = v_m \bar{v}_1(s) \bar{v}_2(s) \tag{16.15}$$

where

$$\bar{v}_1 = \frac{a}{s^2 + a^2} \qquad \bar{v}_2 = e^{-s^{\frac{1}{2}}x/h} \tag{16.16}$$

The inversion of these two functions is accomplished by using Nos. 2.4 and R.2 of the table of transforms (Appendix A):

$$v_1 = \sin at \qquad v_2 = \frac{x \, e^{-x^2/4h^2 t}}{2h\pi^{\frac{1}{2}} t^{\frac{3}{2}}} \tag{16.17}$$

Employing the convolution theorem, the final result is

$$v(x,t) = \frac{v_m x}{2h\pi^{\frac{1}{2}}} \int_0^t \sin at' \, e^{-x^2/4h^2(t-t')} \frac{dt'}{(t-t')^{\frac{3}{2}}} \tag{16.18}$$

This integral may be placed in a slightly more convenient form by introducing the following change of variables:‡

$$t' = t - \frac{x^2}{4h^2 \xi^2} \tag{16.19}$$

thus

$$v = \frac{2v_m}{\pi^{\frac{1}{2}}} \int_{x/2ht^{\frac{1}{2}}}^\infty \sin\left[a\left(t - \frac{x^2}{4h^2 \xi^2}\right)\right] e^{-\xi^2} d\xi \tag{16.20}$$

which agrees with the result for a general input function obtained by Sneddon.‡

17 GREEN'S FUNCTIONS

In addition to the foregoing transform methods of solving boundary value problems, there is another method which is known as the *method of Green's functions*. Actually, a Green's function is a solution for a system

---

† Cf. Chap. 1, Prob. 14.

‡ Cf. Sneddon, 1951, p. 171 (see References).

excited by a point-source type of forcing function. Once the Green's function is known, the solution for any arbitrary forcing function may be expressed in the form of a convolution-type integral. Referring to the impulse responses discussed in Chap. 8, it may be stated that Green's functions and impulse responses are identical, except that the former are involved in the solutions to boundary value problems, whereas the latter are involved in solutions to initial value problems.

For the sake of brevity, we shall consider only one-dimensional systems in this section. The interested reader is referred to any of the several outstanding texts which cover the subject in greater detail.†

The approach followed in this section for determining Green's functions will be to employ the Laplace transform, which the authors feel is a very convenient tool for this purpose. However, it should be noted that any other method of solving the differential equations involved is an equally valid approach. For example, Dettman‡ uses the method of variation of parameters for the problem which we will consider as our first example.

When studying the vibrations of a string subject to a distributed forcing function, the following boundary value problem arises.§

$$\frac{d^2 y}{dx^2} + k^2 y = -f(x) \tag{17.1}$$

with the boundary conditions

$$y(0) = y(a) = 0 \tag{17.2}$$

To solve (17.1) for the Green's function by means of the Laplace transform, we replace $f(x)$ with an impulse function that occurs at an arbitrary point x'; that is, (17.1) becomes

$$\frac{d^2 y}{dx^2} + k^2 y = -\delta(x - x') \tag{17.3}$$

In solving (17.3) by means of Laplace transforms, it should be noted that both $y(0)$ and $y'(0)$ are required. To circumvent this we leave $y'(0)$ as a parameter which is evaluated by applying the boundary condition $y(a) = 0$ to the inverted solution.

To proceed, let

$$\bar{y} = L[y(x)]$$

and hence

$$L[y''(x)] = s^2 \bar{y} - y'(0)$$

† Cf. Courant and Hilbert, 1953; Dettman, 1962; or Kreider et al., 1966 (see References).

‡ Dettman, 1962 (see References).

§ *Ibid.*, p. 190.

since $y(0) = 0$.   Now the Laplace transform of (17.3) is

$$s^2 \bar{y} - y'(0) + k^2 \bar{y} = -e^{-x's} \tag{17.4}$$

and the transformed solution is

$$\bar{y} = \frac{y'(0)}{s^2 + k^2} - \frac{e^{-x's}}{s^2 + k^2} \tag{17.5}$$

From No. 2.4 of the table of transforms (Appendix A) the inverse transform is found to be

$$y(x,x') = \frac{y'(0)\sin kx}{k} - \frac{\sin k(x - x')\,u(x - x')}{k} \tag{17.6}$$

where $u(x)$ is the Heaviside unit step function.   We may now eliminate the unknown $y'(0)$ in the manner described above.   Applying $y(a) = 0$ yields

$$0 = \frac{y'(0)\sin ka}{k} - \frac{\sin k(a - x')}{k} \tag{17.7}$$

and therefore

$$y'(0) = \frac{\sin k(a - x')}{\sin ka} \tag{17.8}$$

Hence the desired solution is

$$y(x,x') = \frac{\sin kx \sin k(a - x')}{k \sin ka} - \frac{\sin k(x - x')\,u(x - x')}{k} \tag{17.9}$$

The solution given by (17.9) is called the *Green's function of* (17.1) and physically represents the displacement of the point x when a unit load has been applied at the point x'.   Rather than express the answer to the above problem as (17.9), it is more common to write it in two separate expressions as follows:

$$y(x,x') = \frac{\sin kx' \sin k(a - x)}{k \sin ka} \qquad 0 \leqslant x' \leqslant x \tag{17.10}$$

$$y(x,x') = \frac{\sin kx \sin k(a - x')}{k \sin ka} \qquad x \leqslant x' \leqslant a \tag{17.11}$$

Now, if we are given an arbitrary forcing function as in (17.1), the answer is written as

$$y(x) = \int_0^a f(x')\,y(x,x')\,dx' \tag{17.12}$$

where it should be noted that the integral must be divided into two parts, one ranging from 0 to x and the other from x to a.

A few readers may have some concern over the source of equation (17.12).   The following heuristic argument on physical grounds is presented as an attempt to satisfy any questions that may arise. Also, any of the references listed at the end of this chapter which deal with this subject in greater detail may be consulted to obtain more formalized derivations.

It was previously stated that $y(x,x')$ represents the displacement of the string, assuming that this is the physical problem which leads to (17.1), at point x due to a unit load applied at point x'. Now the term $f(x')y(x,x')dx'$ is simply the incremental displacement at x due to the actual load intensity at x'. The integral over x' from 0 to a sums the contributions of the entire distributed load to yield the total displacement at the point x.

To further illustrate the use of Green's functions, consider the problem of the deflection of a cantilever beam, as shown in Fig. 17.1. The deflection of the beam from the unloaded position is measured as positive downward and is denoted by $y(x)$. The basic beam relations are

$$y(x) = \text{deflection}$$

$$\theta(x) = \frac{dy}{dx} = \text{slope}$$

$$M(x) = EI\frac{d^2 y}{dx^2} = \text{bending moment}$$

$$F(x) = EI\frac{d^3 y}{dx^3} = \text{shear force}$$

$$E = \text{Young's modulus of elasticity}$$

$$I = \text{moment of inertia}$$

The product EI is commonly termed the *flexural rigidity* of the beam, and we shall assume that it is a constant in our following examples. If the applied load is denoted as $w(x)$ and is positive in the downward direction, then the beam equation for static deflections is

$$\frac{dF(x)}{dx} = -w(x) \tag{17.13}$$

or, in terms of the deflection, this equation becomes

$$EI\frac{d^4 y}{dx^4} = w(x) \tag{17.14}$$

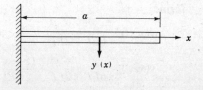

Fig. 17.1

Now, in order to determine the Green's function for the beam, we consider an impulse load applied at an arbitrary point x'; that is, $w(x) = \delta(x - x')$. Hence the equation for the Green's function becomes

$$EI\frac{d^4 y}{dx^4} = \delta(x - x') \tag{17.15}$$

Laplace transforming this equation, we have

$$EI[s^4 \bar{y} - s^3 y(0) - s^2 y'(0) - sy''(0) - y'''(0)] = e^{-sx'} \tag{17.16}$$

To proceed we must specify the boundary conditions appropriate for the cantilever beam, as shown in Fig. 17.1. Specifically these conditions are that the deflection and slope must vanish at $x = 0$, while the moment and shear must vanish at $x = a$; that is,

$$y(0) = 0 \qquad y'(0) = 0 \qquad y''(a) = 0 \qquad y'''(a) = 0$$

Once again we are faced with the fact that we do not know values for $y''(0)$ and $y'''(0)$, which necessitates leaving them as parameters to be eliminated from the inverted solution by application of the conditions at $x = a$.

Applying the two known conditions to (17.16) and solving for the transform solution gives

$$\bar{y} = \frac{e^{-sx'}}{s^4 EI} + \frac{y'''(0)}{s^4 EI} + \frac{y''(0)}{s^3 EI} \tag{17.17}$$

Inverting this equation yields

$$y(x,x') = \frac{y'''(0)\, x^3}{6EI} + \frac{y''(0)\, x^2}{2EI} + \frac{(x - x')^3 u(x - x')}{6EI} \tag{17.18}$$

Now we eliminate $y''(0)$ and $y'''(0)$ by applying the boundary conditions at a which gives the desired Green's function:

$$y(x,x') = \frac{-x^3}{6EI} + \frac{x'\, x^2}{2EI} + \frac{(x - x')^3 u(x - x')}{6EI} \tag{17.19}$$

The form of the Green's function as given in (17.19) may be unfamiliar to some readers because of the use of the Heaviside step function. The following two equations are equivalent to (17.19) but are in the more familiar form:

$$y(x,x') = \frac{-x^3}{6EI} + \frac{x'\, x^2}{2EI} \qquad x \leqslant x' \leqslant a \tag{17.20}$$

$$y(x,x') = \frac{-x^3}{6EI} + \frac{x'\, x^2}{2EI} + \frac{(x - x')^3}{6EI} \qquad 0 \leqslant x' \leqslant x \tag{17.21}$$

Using this Green's function, we may now write the solution to the deflection of a cantilever beam subjected to an arbitrary loading distribution $w(x)$:

$$y(x) = \int_0^a w(x')\, y(x,x')\, dx' \tag{17.22}$$

To illustrate the use of this solution let us consider the problem of the determination of the static deflection of the cantilever beam subjected to a constant distributed load:

$$w(x) = w_0\, u(x) \tag{17.23}$$

Substituting this relation along with (17.19) into (17.22), we have

$$y(x) = \frac{w_0}{6EI}\left[-x^3 \int_0^a dx' + 3x^2 \int_0^a x'\, dx' + \int_0^x (x-x')^3\, dx'\right] \tag{17.24}$$

Performing the indicated integrations and combining terms yields the desired solution:

$$y(x) = \frac{w_0\, x^2}{24EI}(6a^2 - 4ax + x^2) \tag{17.25}$$

As another example, consider the case when the beam carries a load, as shown in Fig. 17.2. Using the Heaviside unit step function, the distributed load may be represented as

$$w(x) = w_0\, u(x-a) \tag{17.26}$$

The reader might question the above representation and prefer the following:

$$w(x) = w_0[u(x-a) - u(x-b)] \tag{17.27}$$

in which the second term is included to terminate the load at the end of the beam. This term is unnecessary since the beam ends at $x = b$. This particular point was neglected in the previous example, and the correct answer was obtained as can be easily verified.

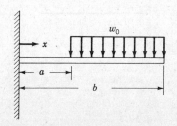

Fig. 17.2

Substituting (17.26) and (17.19) into (17.22) gives

$$y(x) = \frac{w_0}{6EI}\left[-x^3 \int_a^b dx' + 3x^2 \int_a^b x' \, dx' \right.$$
$$\left. + \int_0^b u(x'-a)u(x-x')(x-x')^3 \, dx'\right] \quad (17.28)$$

Because of the two step functions in the last integral, the result must be divided into two separate answers, depending upon whether $x \leqslant a$ or $x \geqslant a$. The reason for this may be seen in Fig. 17.3. It is seen that the product $u(x'-a)u(x-x')$ is 0 for all x' over the interval $0 \leqslant x' \leqslant b$ as long as $x < a$. However, from Fig. 17.3b we see that for $x > a$, the product $u(x'-a)u(x-x')$ is 0 everywhere over the interval $0 \leqslant x' \leqslant b$, except over $a \leqslant x' \leqslant x$ where the product is 1.

As a result of this consideration, we see that the solution as given by (17.28) must be written as follows:

$$y(x) = \frac{w_0}{6EI}\left[-x^3 \int_a^b dx' + 3x^2 \int_a^b x' \, dx'\right] \qquad x < a \qquad (17.29)$$

$$y(x) = \frac{w_0}{6EI}\left[-x^3 \int_a^b dx' + 3x^2 \int_a^b x' \, dx' + \int_a^x (x-x')^3 \, dx'\right] \qquad x > a$$
$$(17.30)$$

Evaluating these integrals gives the final result:

$$y(x) = \frac{w_0 x^2(b-a)}{12EI}[3(b^2-a^2)-2x] \qquad x \leqslant a \qquad (17.31)$$

$$y(x) = \frac{w_0}{24EI}[6x^2(b^2-a^2)-4x^3(b-a)+(x-a)^4] \qquad x \geqslant a \qquad (17.32)$$

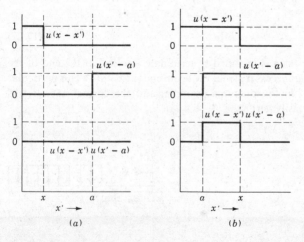

Fig. 17.3

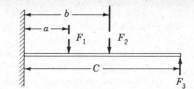

Fig. 17.4

As a concluding example, consider the problem in Fig. 17.4.   The loading function for this problem may be represented as follows:

$$w(x) = F_1 \delta(x - a) + F_2 \delta(x - b) - F_3 \delta(x - c) \qquad (17.33)$$

where $\delta(x) = $ Dirac delta function.

Substituting (17.33) and (17.19) into (17.22) and evaluating the first two integrals yields

$$y(x) = \frac{1}{6EI} \Big\{ F_1 x^2(3a - x) + F_2 x^2(3b - x)$$
$$- F_3 x^2(3c - x) + \int_0^c [F_1 \delta(x' - a)$$
$$+ F_2 \delta(x' - b) - F_3 \delta(x' - c)] u(x - x')(x - x')^3 \, dx' \Big\} \qquad (17.34)$$

The evaluation of the last integral gives three solutions, each valid in each of the different regions of the beam.   The reader should note that this remaining integral gives the value of 0 for $0 \leqslant x \leqslant a$, $F_1(x - a)^3$ for $a \leqslant x \leqslant b$, and $F_1(x - a)^3 + F_2(x - b)^3$ for $b \leqslant x \leqslant c$.   Combining all these results gives the final solution to the stated problem:

$$y(x) = \frac{1}{6EI} [F_1 x^2(3a - x) + F_2 x^2(3b - x) - F_3 x^2(3c - x)]$$
$$0 \leqslant x \leqslant a \quad (17.35)$$

$$y(x) = \frac{1}{6EI} [F_1 a^2(3x - a) + F_2 x^2(3b - x) - F_3 x^2(3c - x)]$$
$$a \leqslant x \leqslant b \quad (17.36)$$

$$y(x) = \frac{1}{6EI} [F_1 a^2(3x - a) + F_2 b^2(3x - b) - F_3 x^2(3c - x)]$$
$$b \leqslant x \leqslant c \quad (17.37)$$

This last example illustrates the convenience of using impulse functions to represent point loads as well as the Green's function.

PROBLEMS

1.   A flexible string is held under a horizontal component of tension H and extends between $x = 0$ and $x = l$.   It is loaded by a uniformly distributed load w_0 per unit length, extending from $x = a$ to $x = b$.   Find the location and magnitude of the maximum deflection.

2. Find the deflection of the string of Prob. 1 if two concentrated loads P_a and P_b act on it at $x = a$ and $x = b$.

3. A string extending from $x = 0$ to $x = l$ is under a horizontal component of tension H and rests on an elastic foundation of spring constant k. A load P_a at $x = a$ and a load P_b at $x = b$ act on the string. Determine the deflection of the string.

4. Determine the deflection of a cantilever beam of flexural rigidity EI under the influence of a uniform load w_0 per unit length extending from $x = a$ to $x = b$. The beam is clamped at $x = 0$ and extends to $x = l$.

5. An infinite beam of flexural rigidity EI rests on an elastic foundation. The modulus of the foundation is k. The beam extends from $x = -\infty$ to $x = +\infty$, and at $x = 0$ there is applied a concentrated load P_0. Determine the deflection of the beam.

6. Consider a beam of length l clamped at $x = 0$ and $x = l$. Let $v_i(x)$ be a mode of oscillation corresponding to an angular frequency ω_i and let $v_j(x)$ be a mode corresponding to an angular frequency ω_j. Show that

$$\int_0^l v_i(x) v_j(x) \, dx = 0 \qquad \text{if } i \neq j$$

7. Determine the natural frequencies of oscillation of a uniform beam with free ends.

8. Determine the natural frequencies of oscillation of a uniform beam with a built-in end at $x = 0$ and simply supported at $x = l$.

9. Derive the equation giving the natural frequencies of a beam clamped at $x = 0$ and $x = l$ resting on an elastic foundation of modulus equal to k.

10. Show that the solution of the differential equation $d^2 y/dt^2 + a^2 y = F(t)$ subject to the initial conditions

$$\left. \begin{array}{l} y = 0 \\ y' = 0 \end{array} \right\} \qquad \text{at } t = 0$$

is

$$y = \frac{1}{a} \int_0^t F(u) \sin a(t - u) \, du$$

11. Solve the integral equation

$$F(t) = a \sin bt + c \int_0^t \sin b(t - u) F(u) \, du \qquad b > c > 0$$

12. A particle of mass m can perform small oscillations about a position of equilibrium under a restoring force mn^2 times the displacement. It is started from rest by a constant force F_0 which acts for a time T and then ceases. Show that the amplitude of the subsequent oscillation is $2F_0/mn^2 \sin(nT/2)$.

13. Three flywheels A, B, C of moments of inertia $3I$, $4I$, $3I$, respectively, are connected by equal shafts of stiffness k and negligible moment of inertia. At $t = 0$, when the system is at rest and unstrained, A is suddenly given an angular velocity w. Show that the subsequent angular velocity of c is

$$\frac{w}{10} \left(3 - 5 \cos nt + 2 \cos nt \sqrt{\frac{5}{2}} \right)$$

where $n^2 = k/3I$.

14. An electrical transmission line of resistance R, capacitance C, inductance L, and leakage conductance G per unit length is short-circuited at the end $x = l$, and at $t = 0$ a constant electromotive force is impressed on the line at the end $x = 0$. If the parameters of the line satisfy the relation $RC = LG$, determine the subsequent potential and current distribution along the line.

15. A uniform prismatic bar of density ρ and Young's modulus E is at rest with its end $x = l$ free to move. At $t = 0$ a constant force F_0 is applied to the rod at $x = 0$. Find the subsequent displacement of the rod.

16. A uniform prismatic bar is hung vertically, and its lower end is clamped so that its displacement is zero at all points. At $t = 0$ it is released so that it hangs from its upper point. Find the subsequent motion of the rod.

17. An electrical cable of length l is short-circuited at $x = l$. The cable has a capacitance C and resistance R per unit length. Initially the current and potential along the cable are zero. At $t = 0$ a constant electromotive force E_0 is applied to the cable at $x = 0$. Find the subsequent current and potential distribution along the cable.

18. Show that the transform of the full-wave rectification curve of a sine wave $|\sin t|$ is given by

$$L|\sin t| = \frac{1}{s^2 + 1} \coth \frac{\pi s}{2}$$

19. A uniform rod has its sides thermally insulated and is initially at a temperature v_0. At time zero the end $x = 0$ is cooled to temperature zero and afterward maintained at that temperature. The end $x = s$ is kept at temperature v_0. Find the variation in temperature at other points of the rod.

20. Show that $LSi(t) = 1/s \cot^{-1} s$, where $Si(t)$ is the *sine integral function* defined by

$$Si(t) = \int_0^t \frac{\sin u}{u} du$$

21. Show that $LEi(-t) = -\ln(1 + s)/s$, where $Ei(u)$ is the *exponential integral* defined by

$$Ei(u) = \int_{-\infty}^u \frac{e^x dx}{x}$$

Evaluate the following integrals operationally:

22. $\displaystyle\int_0^\infty \frac{\sin tx \, dx}{\sqrt{x}} = \sqrt{\frac{\pi}{2t}} \qquad t > 0.$

23. $\displaystyle\int_{-\infty}^{+\infty} \frac{x \sin tx}{a^2 + x^2} dx = \pi e^{-at} \qquad a > 0 \qquad t > 0.$

24. $\displaystyle\int_0^\infty \frac{e^{-tx^2}}{1 + x^2} dx = \frac{\pi}{2} e^t \operatorname{erf}(\sqrt{t}).$

25. Solve Prob. 2 of Chap. 12 by applying a Fourier transform to the time variable.

26. Solve Prob. 2 of Chap. 12 using the Laplace transform.

27. Solve Prob. 3 of Chap. 12 by Laplace transforms.

28. By means of a Fourier transform of the space variable, solve for the conduction of heat in a semi-infinite solid subject to the following boundary and initial conditions:

$$v(0,t) = v_0 \sin wt \qquad v(x,0) = 0$$

29. Solve Prob. 28 by applying a Laplace transform to the time variable.

30. Discuss the problems involved in attempting to solve Prob. 28 by applying a Laplace transform to the space variable.

31. Solve for the temperature distribution in a semi-infinite solid by means of a suitable transform subject to the following boundary conditions:

$$v(0,t) = v_0[H(t) - H(t-a)] \qquad v(x,0) = 0$$

where $H(t)$ is the Heaviside unit step function.

32. In the example in Sec. 2b, what is the heat flux into the solid at the surface $x = 0$?

33. A particle of mass m moves on a vertical x axis under two forces: the force of gravity and a resistance proportional to the velocity. If the axis is taken positive downward, the equation of motion is

$$\frac{md^2 x}{dt^2} = mg - \frac{kdx}{dt}$$

Show that its solution under the initial conditions

$$x(0) = 0 \qquad \frac{dx}{dt}(0) = v_0$$

is

$$x(t) = (bv_0 - g)(1 - e^{-bt}) + \frac{bgt}{b^2}$$

where $b = k/m$. Also discuss the motion.

(Churchill, 1958)

34. Solve the following boundary value problems and verify the results:

(a) $\dfrac{\partial y}{\partial x} + x \dfrac{\partial y}{\partial t} = 0 \qquad y(x,0) = 0 \qquad y(0,t) = t$

(b) $\dfrac{\partial^2 u}{\partial x^2} - 2 \dfrac{\partial^2 u}{\partial x \, \partial t} + \dfrac{\partial^2 u}{\partial t^2} = 0$

$u(x,0) = 0 \qquad \dfrac{\partial u}{\partial t}(x,0) = 0 \qquad u(0,t) = 0$

$u(1,t) = f(t)$

(c) $\dfrac{\partial y}{\partial x} + 2x \dfrac{\partial y}{dt} = 2x \qquad y(x,0) = y(0,t) = 1$

(Churchill, 1958)

35. An unstrained semi-infinite elastic bar is moving lengthwise with velocity $-v_0$ when the end $x = 0$ is suddenly brought to rest, the other end remaining free. Set up and solve the boundary value problem for the longitudinal displacements $u(x,t)$. Also find the force per unit area exerted by the support at $x = 0$ upon the end of the bar.

(Churchill, 1958)

36. A semi-infinite solid $x < 0$ has a thermal diffusivity k_1 and thermal conductivity K_1. It is in contact along the plane $x = 0$ with the semi-infinite solid $x > 0$ in which these quantities are k_2 and K_2. If the initial temperature of the first solid is a constant T_0 and that of the second solid is 0 at $t = 0$, find the temperature at a time t in the second solid.

(Tranter, 1956)

37. Find the steady-state temperature in the semi-infinite annulus bounded by the plane $z = 0$ and the cylindrical surfaces $r = a$ and $r = b$. The boundary conditions are that the surfaces $z = 0$ and $r = a$ are maintained at $0°$, and the surface $r = b$ is kept at the arbitrary temperature $f(z)$. Solve this problem by means of a Fourier sine transform.

(Tranter, 1956)

38. Use the complex form of the Fourier transform to show that

$$V = \frac{1}{2\pi^{\frac{1}{2}} t^{\frac{1}{2}}} \int_{-\infty}^{\infty} f(u) e^{-(x-u)^2/4t} \, du$$

is the solution to the boundary value problem

$$\frac{\partial V}{\partial t} = \frac{\partial^2 V}{\partial t^2} \qquad -\infty < x < \infty \qquad t > 0$$

$$V = f(x) \qquad \text{when } t = 0$$

(Tranter, 1956)

39. The magnetic potential ϕ for a circular disk of radius a and strength w, magnetized parallel to its axis, satisfies Laplace's equation, is equal to $2\pi w$ on the disk itself, and vanishes at exterior points in the plane of the disk. If the disk lies in the plane $z = 0$, show that the potential at the point (r,z), $z > 0$, is

$$\phi = 2\pi a w \int_0^{\infty} e^{-pz} J_0(rp) J_1(ap) \, dp$$

(Tranter, 1956)

40. The faces of the infinite wedge $-\alpha < \theta < \alpha$, $r > 0$, are maintained at unit temperature for a distance a measured from the apex and at $0°$ elsewhere. Use a Mellin transform to show that the steady temperature at the point (r,θ) is

$$\frac{1}{2} + \frac{1}{\pi} \int_0^{\infty} \sin p \ln \frac{a}{r} \cosh p\theta \operatorname{sech} p\alpha \frac{dp}{p}$$

(Tranter, 1956)

41. In Chap. 12, Sec. 12, it was shown that infinitesimal waves propagating in an ideal fluid may be described by a velocity potential which satisfies Laplace's equation. Utilize the Hankel transform to solve for the velocity potential in a semi-infinite medium bounded by the plane $x = 0$ and driven by a piston of radius a located at $x = 0$ which is oscillating harmonically (see Fig. P 40).

The problem is to solve Laplace's equation in cylindrical polar coordinates subject to the boundary conditions

$$\frac{\partial \phi}{\partial x} = jwA e^{jwt} \qquad \text{on } x = 0 \qquad 0 \leqslant r \leqslant a$$

$$\frac{\partial \phi}{\partial x} = 0 \qquad \text{on } x = 0 \qquad r > a$$

$\phi = $ velocity potential

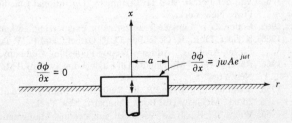

Fig. P 40

If the pressure exerted on the piston is given by

$$p = -\rho \left. \frac{\partial \phi}{\partial t} \right|_{x=0}$$

find the total force exerted on the piston by the fluid.

(Miles, 1961)

REFERENCES

1779. Laplace, P. S.: Sur les suites, oeuvres complètes, *Mémoires de l'académie des sciences*, vol. 10, pp. 1–89.

1928. Whittaker, E. T.: Oliver Heaviside, *Bulletin of the Calcutta Mathematical Society*, vol. 20, p. 199.

1936. Davis, H. T.: "The Theory of Linear Operators," The Principia Press, Bloomington, Ind.

1937. Timoshenko, S.: "Vibration Problems in Engineering," D. Van Nostrand Company, Inc., Princeton, N.J.

1940. Von Kármán, T., and M. A. Biot: "Mathematical Methods in Engineering," McGraw-Hill Book Company, New York.

1943. Pipes, L. A.: Application of the Operational Calculus to the Theory of Structures, *Journal of Applied Physics*, September, pp. 486–495.

1950. Heaviside, O: "Electromagnetic Theory," Dover Publications, Inc., New York.

1951. Sneddon, I. N.: "Fourier Transforms," McGraw-Hill Book Company, New York.

1953. Carslaw, H. S., and J. C. Jaeger: "Operational Methods in Applied Mathematics," Oxford University Press, 2d ed., London.

1953. Courant, R., and D. Hilbert: "Methods in Mathematical Physics," vol. 1 and 2, Interscience Publishers, Inc., New York.

1955. Webster, A. G.: "Partial Differential Equations of Mathematical Physics," 2d ed., Dover Publications, Inc., New York.

1956. Tranter, C. J.: "Integral Transforms in Mathematical Physics," John Wiley & Sons, Inc., New York.

1958. Churchill, R. V.: "Operational Mathematics," 2d ed., McGraw-Hill Book Company, New York.

1961. Lanczos, C.: "Linear Differential Operators," D. Van Nostrand Company, Inc., New York.

1961. Miles, J. W.: Integral Transforms, in E. F. Beckenbach (ed.), "Modern Mathematics for the Engineer," 2d ser., McGraw-Hill Book Company, New York.

1962. Dettman, J. W.: "Mathematical Methods in Physics and Engineering," McGraw-Hill Book Company, New York.

1964. Van der Pol, B., and H. Bremmer: "Operational Calculus," Cambridge University Press, New York.

1965. Fodor, G.: "Laplace Transforms in Engineering," Akadémiai Kiadó, Budapest.

1966. Kreider, D. L., R. G. Kuller, D. R. Ostberg, and F. W. Perkins: "An introduction to Linear Analysis," Addison-Wesley Publishing Company, Inc., Reading, Mass.

1966. Raven, F. H.: "Mathematics of Engineering Systems," McGraw-Hill Book Company, New York.

1966. Sokolnikoff, I. S., and R. M. Redheffer: "Mathematics of Physics and Engineering," 2d ed., McGraw-Hill Book Company, New York.

1966. Wiley, C. R., Jr.: "Advanced Engineering Mathematics," 3d ed., McGraw-Hill Book Company, New York.

14
Approximate Methods in Applied Mathematics

1 INTRODUCTION

This chapter is included to provide the reader with a brief introduction to some of the approximate techniques commonly employed in applied mathematics.†

The need for approximate methods arises from many diverse circumstances. One such case is experimental work where data is accumulated and there is no a priori knowledge of an analytical representation of the physical system; at most there is only a general idea of a mathematical model with undefined parameters. In order to obtain an analytical representation or a satisfactory evaluation of the parameters, curve-fitting techniques of various types are employed.

Another situation in which approximate methods must be resorted to is that in which the mathematical model of the system is known but exact solutions cannot be obtained for various reasons. In these situations when solutions must be obtained, the only recourse is to resort to some approximate

† All numerical computations presented in this chapter were made on an IBM/1130 computer located at the University of Redlands, Redlands, Calif.

method. It should be pointed out that this chapter deals with approximate methods of solving mathematical models of physical systems which are themselves only approximate descriptions of physical systems. The problem the analyst must always concern himself with, regardless of whether he is using exact or approximate solutions, is that he must be always on guard to be sure that his basic model is consistent with the physical problem he is dealing with. In other words, he must always check to see that the underlying postulates of the mathematical model are indeed applicable to the problem at hand.

As an illustration, let us consider fluid flow. If we postulate inviscid and incompressible flow, it may be shown that the fluid potential functions satisfy Laplace's equation (see Appendix E, Sec. 13). With Laplace's equation as the mathematical model, it may be further shown that the drag coefficient varies inversely with the square root of 1 minus the mach number. This result predicts that the drag will become infinite when the speed of sound is approached. We know this is not true; however, many people were quite certain of this in the not too far distant past. The fallacy lay in the fact that the postulates are no longer valid at speeds near the speed of sound. Those investigators who knew the very close fit of this model with experimental data for low velocities and failed to check the postulates against the physical problem for near-sonic conditions were greatly disappointed.

In reading this chapter the reader should also remember that the material presented is only an introduction to a few of the many approximate techniques that have been developed to date. Those desiring further information should consult the references at the end of this chapter.

2 THE METHOD OF LEAST SQUARES

Quite frequently in experimental work there arises a need to generate a curve to fit a set of given data points whose trend might be linear, quadratic, or of higher order. The *method of least squares* is a method of computing a curve in such a way that it minimizes the error of the fit at the data points.

To begin, let us assume that we have a set of n data points (x_i, y_i) through which we desire to pass a straight line. This line is to represent the "best fit" in the least-squares sense. As we know, a straight line may be expressed in the form

$$y = a_0 + a_1 x \tag{2.1}$$

The formulas we are about to derive will be a set of simultaneous equations for the arbitrary parameters a_0 and a_1.

To assist in visualizing the process, the reader may refer to Fig. 2.1, which depicts the data points as well as the line to be fitted. Also shown are other quantities to be defined later.

Unless the data fall in a straight line, the general curve will not pass through all of the data points. For convenience, let us consider the kth point. The ordinate of the point is given as y_k. The ordinate at x_k as given

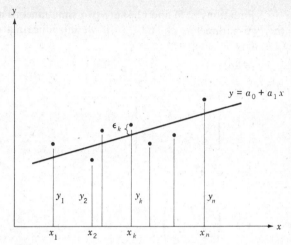

Fig. 2.1

by the general curve is $a_0 + a_1 x_k$. The difference between these two values is the error of fit at the kth point:

$$\epsilon_k = y_k - (a_0 + a_1 x_k)$$

If we now sum the squares of all the errors at the data points, we will obtain the total error of fit:

$$E = \sum_{k=1}^{n} \epsilon_k^2 = \sum_{k=1}^{n} [y_k - (a_0 + a_1 x_k)]^2 \qquad (2.2)$$

The errors have been squared to eliminate possible cancellation.

The total error is a function of how well the line "fits" the data points: that is, it is a function of the parameters a_0 and a_1 which control the positioning of the line in xy space. It should be fairly evident that the "best" fit is that position which minimizes the total error. To find this position we consider E as a function of a_0 and a_1 and require that its derivatives vanish. Performing this operation, we have

$$\frac{\partial E}{\partial a_0} = -2 \sum_{k=1}^{n} [y_k - (a_0 + a_1 x_k)] = 0 \qquad (2.3)$$

$$\frac{\partial E}{\partial a_1} = -2 \sum_{k=1}^{n} [y_k - (a_0 + a_1 x_k)] x_k = 0 \qquad (2.4)$$

Dividing each of these equations by -2 and rearranging terms yields

$$na_0 + a_1 \sum_{k=1}^{n} x_k = \sum_{k=1}^{n} y_k \qquad (2.5)$$

$$a_0 \sum_{k=1}^{n} x_k + a_1 \sum_{k=1}^{n} x_k^2 = \sum_{k=1}^{n} x_k y_k \qquad (2.6)$$

Equations (2.5) and (2.6) are two simultaneous algebraic equations for the two parameters a_0 and a_1. If we introduce matrix notation, (2.5) and (2.6) may be combined to give

$$[A](a) = (b) \tag{2.7}$$

where

$$[A] = \begin{bmatrix} n & \sum_{k=1}^{n} x_k \\ \sum_{k=1}^{n} x_k & \sum_{k=1}^{n} x_k^2 \end{bmatrix}$$

$$(a) = \begin{pmatrix} a_0 \\ a_1 \end{pmatrix}$$

$$(b) = \begin{pmatrix} \sum_{k=1}^{n} y_k \\ \sum_{k=1}^{n} x_k y_k \end{pmatrix}$$

The solution to (2.7) is

$$(a) = [A]^{-1}(b) \tag{2.8}$$

Thus, knowing the data points, all the terms in $[A]$ and (b) may be computed, and hence (a) is determined.

This is the method of least squares, and it may be extended to fit a polynomial of any order. In the general case, one may desire to fit an mth-order curve to the n given data points (x_k, y_k); that is,

$$y = a_0 + a_1 x + a_2 x^2 + \cdots + a_m x^m \tag{2.9}$$

Following the same derivation procedure, it may be shown that the general result is again (2.7) where

$$A = \begin{bmatrix} n & \sum_{k=1}^{n} x_k & \sum_{k=1}^{n} x_k^2 & \cdots & \sum_{k=1}^{n} x_k^m \\ \sum_{k=1}^{n} x_k & \sum_{k=1}^{n} x_k^2 & \sum_{k=1}^{n} x_k^3 & \cdots & \sum_{k=1}^{n} x_k^{m+1} \\ \cdots & \cdots & \cdots & \cdots & \cdots \\ \sum_{k=1}^{n} x_k^m & \sum_{k=1}^{n} x_k^{m+1} & \sum_{k=1}^{n} x_k^{m+2} & \cdots & \sum_{k=1}^{n} x_k^{2m} \end{bmatrix} \tag{2.10}$$

$$(a) = \begin{pmatrix} a_0 \\ a_1 \\ \vdots \\ a_m \end{pmatrix} \qquad (b) = \begin{pmatrix} \sum_{k=1}^{n} y_k \\ \sum_{k=1}^{n} y_k x_k \\ \vdots \\ \sum_{k=1}^{n} y_k x_k^m \end{pmatrix}$$

Thus, given a set of data points for which it is desired to fit a curve with minimum error, one only needs to compute the elements of the matrices $[A]$ and (b).

As an example let us consider the following set of data:

x	0.00	1.00	2.00	3.00	4.00
y	0.99	0.03	−1.02	−1.94	−3.04

If we examine this data, we may observe that it follows a somewhat linear relation. From this observation we seek to determine the coefficients of

$$y = a_0 + a_1 x \tag{2.11}$$

by the method of least squares. Following the indicated procedure, we find that

$$a_0 = 1.00999 \qquad a_1 = -1.00299 \tag{2.12}$$

It is of interest to note what happens when a higher-order polynomial is fitted to the above data. Again if we apply the method of least squares to find the coefficients for

$$y = a_0 + a_1 x + a_2 x^2 \tag{2.13}$$

we find that

$$a_0 = 0.98856 \qquad a_1 = -0.96013 \qquad a_2 = -0.01071 \tag{2.14}$$

Going one step further and attempting to fit

$$y = a_0 + a_1 x + a_2 x^2 + a_3 x^3 \tag{2.15}$$

yields the result

$$a_0 = 0.99756 \qquad a_1 = -1.02463$$
$$a_2 = 0.03428 \qquad a_3 = -0.00749 \tag{2.16}$$

If we compare the results for the coefficients in each of these three cases, we can readily see that the coefficients of the second- and third-order terms are small compared to the constant and linear coefficients. This shows that the linear character of the data strongly influences the results. The implication is that even though a higher-order polynomial is fitted to a set of data, the method of least squares adjusts the coefficients of the dominate terms to reflect the dominate trend of the data.

From the above derivation, it should be clear that any set of n data points may be fitted exactly with an $n - $ 1st-order polynomial. There are, however, a few precautions that the user should be aware of. One of these is that when the number of data points becomes large, the equations for the coefficients may

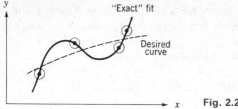

Fig. 2.2

become ill-conditioned.† When this situation arises, the procedure is to fit a series of orthogonal polynomials such as Legendre's or Chebyshev's.‡

Another problem arises when one attempts to fit a set of data exactly with a polynomial of order one less than the number of given points. In this case the curve is forced to pass exactly through every data point, and hence the total error is zero. However, the resulting curve may be far from the desired result. A simple example of this is illustrated in Fig. 2.2, where a third-order polynomial has been fitted through four data points. These points seem to indicate a linear trend, but the curve does not respond.

The advantage of using a larger number of data points than the order of the polynomial is that the curve tends to smooth out the data; that is, it is not sensitive to high-order fluctuations.

3 MATRIX FORMULATION OF THE METHOD OF LEAST SQUARES

An interesting alternative derivation of the method of least squares may be developed from the use of matrices without the requirement of taking partial derivatives to obtain the minimum error. This method of formulation is convenient because of the fact that most digital computing facilities have built-in subroutines for handling matrix manipulations. However, if a computer facility is available, it would be wise to first check to see if a least-squares routine is supplied before looking into the matrix subroutines.

To present this formulation, let us consider a situation in which we have a set of n data points (x_i, y_i) for which it is desired to determine the coefficients of

$$y = a_0 + a_1 x + a_2 x^2 \tag{3.1}$$

for a "best" fit. We begin by writing this equation for every data point; that is,

$$
\begin{aligned}
y_1 &= a_0 + a_1 x_1 + a_2 x_1^2 \\
y_2 &= a_0 + a_1 x_2 + a_2 x_2^2 \\
&\cdots\cdots\cdots\cdots \\
y_n &= a_0 + a_1 x_n + a_2 x_n^2
\end{aligned}
\tag{3.2}
$$

† The term *ill-conditioned* is used to denote the condition that the determinant of coefficients becomes very small.

‡ For a more complete discussion, the reader is referred to Kuo, 1965 (see References).

Obviously all of these equations would be satisfied only when all the data points fall exactly on the desired curve. However, satisfied or not, let us consider the set of Eqs. (3.2) and rewrite them in matrix notation. If we define the following matrices

$$[C] = \begin{bmatrix} 1 & 1 & 1 & \cdots & 1 \\ x_1 & x_2 & x_3 & \cdots & x_n \\ x_1^2 & x_2^2 & x_3^2 & \cdots & x_n^2 \end{bmatrix} \tag{3.3}$$

$$(y) = \begin{Bmatrix} y_1 \\ \vdots \\ y_n \end{Bmatrix} \qquad (a) = \begin{Bmatrix} a_0 \\ a_1 \\ a_2 \end{Bmatrix}$$

Eq. (3.2) may be written in terms of the single matrix equation

$$[C]'\{a\} = \{y\} \tag{3.4}$$

At first glance one might jump at the conclusion that (3.4) could be solved directly by premultiplying through by the inverse of the matrix $[C]'$. This operation is not possible since $[C]'$ is not a square matrix. Only in the special case when there are only three data points can this be done; that is, only when the number of data points is one greater than the order of the degree of the polynomial to be fitted. We can, however, reduce (3.4) to a square system by premultiplying through by $[C]$:

$$[C][C]'\{a\} = [C]\{y\} \tag{3.5}$$

From (3.5) it can be seen that the order of the system is 3×3. If we introduce the notation

$$[D] = [C][C]'$$

the solution to (3.5) may be expressed as

$$\{a\} = [D]^{-1}[C]\{y\} \tag{3.6}$$

If the matrix $[D]$ is examined carefully along with $[C]\{y\}$, it should be seen that the former is exactly the matrix $[A]$ and the latter the matrix $\{b\}$ from the previous section [see Eq. (2.7)]. Thus, the procedure followed here has generated the equations for a least-squares fit. For large problems this approach is recommended because the use of matrices greatly increases the efficiency of the process as a result of the systematic notation.

4 NUMERICAL INTEGRATION OF FIRST-ORDER DIFFERENTIAL EQUATIONS

It happens very frequently in practical applications that one encounters a differential equation describing some physical system which cannot be readily solved by analytical methods. The reason is that it is usually either too complex or nonlinear and cannot be solved. At this point the only recourse left to the analyst is to integrate the equation numerically. With the advent of high-speed digital computers, this procedure has become a straightforward task.

The methods available for numerical integration are numerous, and we shall only attempt to illustrate a few of the classical ones. However, the reader should not forget the fact that an analytical solution is usually more desirable whenever it can be obtained.

As an introduction to the numerical integration of a differential equation, let us consider the ordinary first-order equation

$$y' = f(x, y) \tag{4.1}$$

Since we are going to integrate this equation numerically, we must have a specific initial condition given:

$$y(x_0) = y_0 \tag{4.2}$$

Assuming that we are seeking continuous solutions, we may apply Taylor's expansion to obtain a relation which gives a value for y at a point $x + \Delta x$ as follows:

$$y(x + \Delta x) = y(x) + y' \Delta x + \tfrac{1}{2} y'' \Delta x^2 + \cdots \tag{4.3}$$

If Δx is small, we may neglect all the terms of order Δx^2 and higher to obtain

$$y(x + \Delta x) = y(x) + y' \Delta x \tag{4.4}$$

In further considerations we will replace the incremental step Δx with the term h which is called the *step size*.

To apply this result to obtain a solution to (4.1), we may proceed as follows. From (4.2) we know a value of y at a specified point x_0. Using (4.4) a value of y can be computed at a neighboring point a distance h away:

$$y_1 = y_0 + hf(x_0, y_0) \tag{4.5}$$

where y' has been replaced by (4.1). This process of projecting from a known point to a neighboring point is illustrated in Fig. 4.1.

Now, knowing y_1 we may compute another value for y a step away from y_1, or two steps away from y_0:

$$y_2 = y_1 + hf(x_1, y_1) \tag{4.6}$$

By continuing this procedure we may extend the solution stepwise, with the general relation

$$y_{n+1} = y_n + hf(x_n, y_n) \tag{4.7}$$

Equation (4.7) is known as the *Euler-Cauchy formula*. It essentially is a second-order formula since terms of order h^2 and higher were neglected in its derivation.

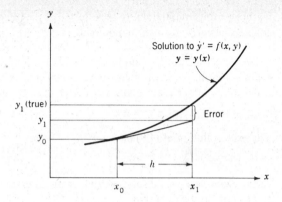

Fig. 4.1

It should be evident that the accuracy of this method is highly dependent upon the step size h. Even with a modern digital computer, a very small step size can result in long run times to integrate (4.1) to a desired final value with small error. To improve this situation, several higher-order methods have been devised so that acceptable accuracies may be obtained with reasonable step sizes and run times.

One of the most widely used higher-order methods is that of Runge-Kutta, which is a fourth-order system. Those readers who are interested in the derivation of this procedure are referred to the excellent book by Kuo.† The basic equations for this method are

$$y_{n+1} = y_n + \frac{h}{6}(k_0 + 2k_1 + 2k_2 + k_3) \tag{4.8}$$

where $k_0 = f(x_n, y_n)$

$$k_1 = f\left(x_n + \frac{h}{2}, y_n + k_0\frac{h}{2}\right)$$

$$k_2 = f\left(x_n + \frac{h}{2}, y_n + k_1\frac{h}{2}\right)$$

$$k_3 = f(x_n + h, y_n + k_2 h)$$

This procedure essentially yields a more accurate estimate of the average slope. As a result, the Runge-Kutta method is sensitive to the higher-order derivatives and tracts the function closer. One of the advantages is that at each computation point the error can be predicted for the next step. This

† Kuo, 1965, chap. 7 (see References).

feature permits the step size to be adjusted as the integration proceeds so that a constant accuracy can be maintained.

To illustrate both the Euler-Cauchy and the Runge-Kutta methods, we will consider a simple example so that the results may be compared with the exact solution. The problem to be considered is the following linear first-order differential equation and associated initial condition:

$$y' = y \qquad y(0) = 1 \qquad (4.9)$$

The exact solution of (4.9) is

$$y = e^x \qquad (4.10)$$

In the example we shall solve (4.9) over the interval 0 to 1. Comparing (4.9) to (4.1), it can be seen that

$$f(x,y) = y \qquad (4.11)$$

and hence the general Euler-Cauchy formula (4.7) becomes

$$y_{n+1} = (1 + h)y_n \qquad (4.12)$$

Also, the constants in the Runge-Kutta equations are

$$k_0 = y_n$$

$$k_1 = \left(1 + \frac{h}{2}\right) y_n$$

$$k_2 = \left(1 + \frac{h}{2} + \frac{h^2}{4}\right) y_n$$

$$k_3 = \left(1 + h + \frac{h^2}{2} + \frac{h^3}{4}\right) y_n$$

In order to obtain a fair comparison of these two methods, it is suggested† that the step size in the Euler-Cauchy formula be set approximately four times smaller than the step size in the Runge-Kutta procedure. The reason for this method of comparison lies in the fact that four terms must be computed in the Runge-Kutta method, as opposed to one in the Euler-Cauchy method. With the inclusion of this modification, both methods will require approximately the same amount of computer time. Thus the comparison is based on equal computations or computer time as opposed to equal step size.

In this specific example, the step size for the Euler-Cauchy method has been set at 0.025 and at 0.1 for the Runge-Kutta method. Figure 4.2 presents the tabulated results of both of these methods along with the exact solution. The absolute discrepancies have also been included for easier comparison.

† This suggestion was made by Robert D. Engel, Assistant Professor of Engineering Science at the University of Redlands, Redlands, Calif.

COMPARISON OF EULER-CAUCHY AND RUNGE-KUTTA METHODS OF NUMERICAL INTEGRATION

x	Exact	Euler-Cauchy ($h = 0.025$)		Runge-Kutta ($h = 0.1$)	
		Solution	Absolute error	Solution	Absolute error
0.0	1.000000	1.000000	0.000000	1.000000	0.000000
0.1	1.105170	1.103813	0.001357	1.105170	0.000000
0.2	1.221402	1.218403	0.002999	1.221402	0.000000
0.3	1.349858	1.344889	0.004969	1.349858	0.000000
0.4	1.491824	1.484506	0.007318	1.491823	0.000001
0.5	1.648721	1.638616	0.010105	1.648720	0.000001
0.6	1.822118	1.808726	0.013392	1.822117	0.000001
0.7	2.013752	1.996495	0.017257	2.013750	0.000002
0.8	2.225540	2.203157	0.022383	2.225538	0.000002
0.9	2.459603	2.432535	0.027068	2.459599	0.000004
1.0	2.718281	2.685064	0.033217	2.718277	0.000004

Fig. 4.2

It should be of interest to note that the Runge-Kutta solution is far more accurate even with the modified step size discussed above.

One of the great values of these numerical methods is that the operation is identical whether the differential equation is linear or nonlinear. Also, almost every computer facility now has several standard subroutines based on one or more of the standard higher-order methods which provide for fast and efficient solution of differential equations with a minimum of programming effort.

5 NUMERICAL INTEGRATION OF HIGHER-ORDER DIFFERENTIAL EQUATIONS

The methods of Sec. 4 may be extended to apply to equations of the second order or higher as the case may be. However, rather than have a modified method for equations of different order, the most common approach at present is to reduce higher-order differential equations to a set of simultaneous first-order equations. The reason for this is that systems of simultaneous first-order equations can be handled with nearly the same degree of ease as a single equation on a digital computer.

To illustrate this approach, let us consider the following second-order equation and its associated initial conditions:

$$y'' + y = 0 \qquad y(0) = 0 \qquad y'(0) = 1 \qquad (5.1)$$

It may be seen that the exact solution to this problem is $y = \sin x$, which will be used for checking the numerical results.

To reduce (5.1) to a first-order system, we introduce the variable

$$z = y' \tag{5.2}$$

Substituting this variable into (5.1) gives

$$z' + y = 0 \tag{5.3}$$

Thus, the set of simultaneous first-order differential equations representing (5.1) and its initial conditions is

$$
\begin{aligned}
y' &= z & y(0) &= 0 \\
z' &= -y & z(0) &= 1
\end{aligned}
\tag{5.4}
$$

If we convert (5.4) into the standard form for the Euler-Cauchy method, we have

$$
\begin{aligned}
y_{n+1} &= y_n + hz_n \\
z_{n+1} &= z_n - hy_n
\end{aligned}
\tag{5.5}
$$

where h is the step size. These equations have been solved for a step size of 0.025, and the results are tabulated in Fig. 5.1. Also shown are the results obtained by using the Runge-Kutta technique with a step size of 0.1.

As previously stated, numerical methods are particularly useful due to the existence of modern high-speed digital computers. Perhaps the greatest value is realized when one must obtain solutions to nonlinear differential equations for which exact solutions do not exist in closed form. When handling problems of this type, one must be careful to maintain a check on the growth of errors which usually implies the need for smaller step sizes and

		Euler-Cauchy ($h = 0.025$)		Runge-Kutta ($h = 0.1$)	
x	Exact	Solution	Absolute error	Solution	Absolute error
0.0	0.0000000	0.0000000	0.0000000	0.0000000	0.0000000
0.1	0.0998334	0.0998438	0.0000104	0.0998333	0.0000001
0.2	0.1986693	0.1986898	0.0000205	0.1986691	0.0000002
0.3	0.2955202	0.2955505	0.0000303	0.2955198	0.0000004
0.4	0.3894183	0.3894579	0.0000396	0.3894178	0.0000005
0.5	0.4794255	0.4794738	0.0000483	0.4794249	0.0000006
0.6	0.5646424	0.5646985	0.0000561	0.5646418	0.0000006
0.7	0.6442176	0.6442807	0.0000631	0.6442168	0.0000008
0.8	0.7173560	0.7174252	0.0000692	0.7173551	0.0000009
0.9	0.7833269	0.7834008	0.0000739	0.7833257	0.0000012
1.0	0.8414709	0.8415485	0.0000776	0.8414697	0.0000012

Fig. 5.1

hence longer computation times. Another advantage is that the procedure is exactly the same for linear and nonlinear equations.

6   MONTE CARLO METHOD

Sometimes the basic nature of a problem makes it difficult to formulate for numerical solutions in the previously described manner. In cases such as these the Monte Carlo method can sometimes be applied to obtain approximate solutions. Perhaps one of the greatest areas of application of this method is in the solution of boundary value problems. In these applications the name *random walk* is common terminology.

Basically, Monte Carlo is a probabilistic method for which the high-speed digital computer is ideally suited. To illustrate this method we will begin with a simple example.† Let us set out to determine the area under the curve $y = x^3$ over the interval $0 \leqslant x \leqslant 1$. The answer is $\frac{1}{4}$; however, we will attempt to obtain this answer by the Monte Carlo procedure. Figure 6.1 shows the curve and the desired area.

Basically the method proceeds as follows: We may consider the desired area to lie entirely within the square bounded by the axes and the lines $x = 1$, $y = 1$. The actual area is bounded by the x axis, the line $x = 1$, and the curve itself. The method is to arbitrarily select points lying within the square in a random fashion. Once an arbitrary point has been selected, a decision is made as to whether or not it falls within the area of interest. If it does not, then it is simply counted in the total count. If it does fall within the area, then it is counted as a "hit" and also in the total count. As the number of points becomes large, the ratio of total hits to the total count will converge to the actual area. The result is actually a fraction of the total "target" area,

† This example and the following one were worked out in detail by Bruce Sinclair, an undergraduate engineering major at the University of Redlands, Redlands, Calif.

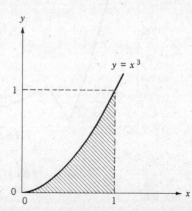

Fig. 6.1

so that if this area is known, then the desired area is simply the resulting fraction times the total area.   In our problem the total area is one; hence the resulting fraction should converge to the answer of $\frac{1}{4}$.

The digital computer program for this problem used the following pattern.   Using a standard random number generator, a random number between 0 and 1 was selected.   This number was assumed to be the abscissa of the first point.   Another random number was selected which was interpreted to be the ordinate of the random point.   The point was considered to fall within the desired area if the random ordinate was less than the value for the ordinate computed for the random abscissa using the equation of the curve.

If the random point happens to fall exactly on the curve, then some policy must be established to decide how the point is to be counted.   The alternatives are to consider such a point as a hit or a miss or some fraction of a hit.   In this example problem, the policy was set to consider these points as one-half a hit.   Actually if the number of trials is large, then any decision policy will have little consequence on the result.

The digital computer repeats the above process several thousand times so that a large number of points is obtained.   In the example problem an approximate value for the area was printed out every 500 trials for a total of 5,000 samples.   Figure 6.2 shows the results in graphical form.   It is probably of interest to know that the time required to run this problem on an IBM/1130 computer was something less than three seconds, excluding printing time.

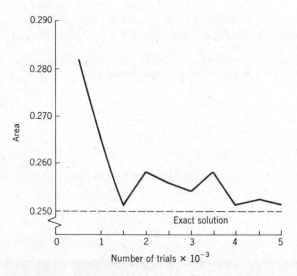

Fig. 6.2

The second example to be presented in this section illustrates the application of the Monte Carlo method to a steady-state boundary value problem. Consider a homogeneous uniform plate of length l and width w. The sides are insulated, and one end is held at 0°F while the other is at 250°F. The problem is to determine the steady-state temperature distribution. This problem is two-dimensional since variations normal to the plate will be neglected. The analytical solution for this problem is easily determined to be a linear temperature distribution along the bar. It may seem that we are killing a moth with a cannon in this problem, but it is useful in illustrating the method. Specifically the procedure to be used is commonly known as a *random walk method*.

Before we continue with the example, a few comments should be made about the method in general. At the discretion of the investigators, the region comprising the problem is subdivided into a square mesh, as indicated in Fig. 6.3. Rather than show the rectangular region representing the plate problem, an arbitrarily shaped region has been illustrated. Each intersection point of the mesh inside the region of interest is a point at which the temperature may be determined by the random walk method. From each point there are four possible paths or steps to a neighboring point. It should be evident that the finer the mesh, the more detailed the resulting temperature distribution will be. However, as will become apparent, the finer the mesh, the longer the computation time required to solve the problem. As a result the investigator is faced with the ever-present problem of deciding between increased accuracy and computation time.

Once the mesh is established, the procedure is to select one of the mesh nodes as a starting point. One of the four possible steps from that node is selected in a random manner which moves the point to the new position. From the new position another random step is made in any of the four possible directions. This process is repeated many times, and the path from the starting point will wander randomly over the region, thus the name *random walk*. The random walk continues until the path meets a boundary of the

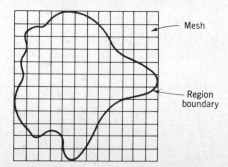

Fig. 6.3

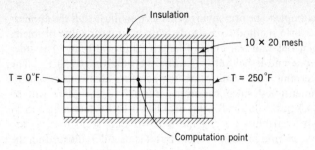

Fig. 6.4

region.   If the boundary condition at this point is one of an insulated wall, then the point is reflected back into the region and the walk continues.   If the boundary condition prescribes a fixed temperature, then this temperature is recorded as a score and another walk is started from the same initial point. The walk is also terminated whenever another point is hit for which the temperature has already been calculated.   The temperature of this point is included in the calculations as a score in the same way as the temperature of boundaries.

After a large number of walks have been made from a single starting point, the temperature is computed by dividing the total score by the total

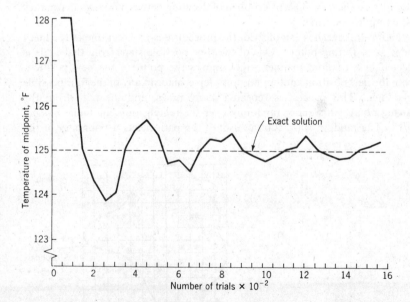

Fig. 6.5

number of walks. This method is repeated for each mesh point until the entire region has been covered.

Returning to the example described earlier, the problem was set up using a 10×20 mesh to describe the plate. The problem is shown schematically in Fig. 6.4. Using the random walk method, the temperature at the central mesh point was computed. The results are shown in Fig. 6.5, which presents the computed temperature for every 100 random walks, so that the convergence of the method may be visualized.

One of the primary difficulties with this method is the excessive time required for the computations. However, a new approach has been recently devised which reduces the computer run times by nearly an order of magnitude. This method is the *floating random walk*† procedure and appears to be an extremely valuable innovation.

7 APPROXIMATE SOLUTIONS OF DIFFERENTIAL EQUATIONS

Throughout the history of applied mathematics, approximate methods for solving differential equations have received considerable attention. At present most of the widely known procedures have been unified into one standardized procedure termed the *method of weighted residuals.*‡ Essentially the above-mentioned methods are based on a procedure of selecting a series of standard functions with arbitrary coefficients. This series is used as an approximate solution to the differential equation. These solutions are then forced to satisfy the differential equation by any one of several methods which are based on minimizing the error of the "fit." This procedure leads to a set of equations for the arbitrary or undetermined coefficients.

With regard to boundary conditions, the approximate solution should preferably satisfy them completely; however, this is not an absolute requirement. It should be clear that if all the boundary conditions are satisfied, the approximate solution will give a more reasonable fit. One convenient method of selecting functions which satisfy the boundary conditions of a problem is to choose them so that each one satisfies one or more of the conditions and is homogeneous at the other boundary points. When all of these functions are combined to form the complete approximate solution, all of the boundary conditions may be satisfied. Another common approach is to use the boundary conditions themselves to generate additional equations for the arbitrary coefficients.

---

† A. Haji-Sheikh and E. M. Sparrow, The Floating Random Walk and Its Applications to Monte Carlo Solutions of Heat Equations, *SIAM Journal of Applied Mathematics*, vol. 14, no. 2, pp. 370–389, March, 1966.

‡ An excellent review of this procedure is given in B. A. Finlayson and L. E. Scriven, The Method of Weighted Residuals—A Review, *Applied Mechanics Reviews*, vol. 19, no. 9, September, 1966.

8 RAYLEIGH'S METHOD OF CALCULATING NATURAL FREQUENCIES

Lord Rayleigh[†] has given an approximate method for finding the lowest natural frequency of a vibrating system. The method depends upon the energy considerations of the oscillating system.

From mechanics it is known that, if a conservative system (one that does not undergo loss of energy throughout its oscillation) is vibrating freely, then when the system is at its maximum displacement the kinetic energy is instantaneously zero, since the system is at rest at this instant. At this same instant the potential of the system is at its maximum value. This is evident since the potential energy is the work done against the elastic restoring forces, and this is clearly a maximum at the maximum displacement.

In the same manner, when the system passes through its mean position, the kinetic energy is a maximum and the potential energy is zero. By realizing these facts it is possible to compute the natural frequencies of conservative systems.

To illustrate the general procedure, consider the simple mass-and-spring system shown in Fig. 8.1. Let V be the potential energy and T be the kinetic energy of the system at any instant. We then have

$$V = \int_0^z Kx\,dx = K\frac{x^2}{2} \tag{8.1}$$

since the spring force is given by Kx, where K is the spring constant. The kinetic energy is

$$T = \frac{1}{2}M\left(\frac{dx}{dt}\right)^2 \tag{8.2}$$

Let the system be executing free harmonic vibrations of the type

$$x = a\sin \omega t \tag{8.3}$$

Now at the maximum displacement

$$\sin \omega t = \pm 1 \qquad \cos \omega t = 0 \tag{8.4}$$

[†] Lord Rayleigh, "Theory of Sound," 2d ed., vol. I, pp. 111, 287, Dover Publications, Inc., New York, 1945.

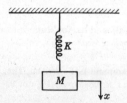

Fig. 8.1

Hence, from (8.1) and (8.2), we have

$$V_M = \tfrac{1}{2}Ka^2 \tag{8.5}$$

where V_M denotes the maximum potential energy.

Now, when the system is passing through its equilibrium position, we have

$$\sin \omega t = 0 \qquad \cos \omega t = \pm 1 \tag{8.6}$$

Hence we have

$$T_M = \tfrac{1}{2}Ma^2 \omega^2 \tag{8.7}$$

Now the total energy in the system is the sum of the kinetic and potential energies. In the absence of damping forces this total remains constant (conservative system). It is thus evident that the maximum kinetic energy is equal to the maximum potential energy, and hence

$$T_M = V_M \tag{8.8}$$

or

$$\tfrac{1}{2}Ka^2 = \tfrac{1}{2}Ma^2 \omega^2 \tag{8.9}$$

We thus find

$$\omega = \sqrt{\frac{K}{M}} \tag{8.10}$$

for the natural frequency of the harmonic oscillations. The above illustrative example shows the essence of Rayleigh's method of computing natural frequencies. The method is particularly useful in determining the lowest natural frequency of continuous systems in the absence of frictional forces.

The method will now be applied to determine the natural frequencies of uniform beams.

KINETIC ENERGY OF AN OSCILLATING BEAM

Consider a section of length dx of a uniform beam having a mass m per unit length. The kinetic energy of this element of beam is given by

$$dT = \tfrac{1}{2}m\left(\frac{\partial y}{\partial t}\right)^2 dx \tag{8.11}$$

since $\partial y/\partial t$ is the velocity of the element and $m\,dx$ is its mass. The entire kinetic energy is given by

$$T = \tfrac{1}{2}m \int_0^s \left(\frac{\partial y}{\partial t}\right)^2 dx \tag{8.12}$$

where s is the length of the beam.

THE POTENTIAL ENERGY OF AN OSCILLATING BEAM

Consider an element of beam of length dx as shown in Fig. 8.2.

If the left-hand end of the section is fixed, the bending moment M turns the right-hand end through the angle ϕ and the bending moment is proportional to ϕ; that is, we have

$$M(\phi) = k\phi \tag{8.13}$$

where k is a constant of proportionality. If now the section is bent so that the right-hand end is turned through an angle θ, the amount of work done is given by

$$dV = \int_{\phi=0}^{\phi=\theta} M(\phi)\, d\phi = \int_{\phi=0}^{\phi=\theta} k\phi\, d\phi = \frac{k\theta^2}{2} = \frac{M(\theta)\,\theta}{2} \tag{8.14}$$

Now the slope of the displacement curve at the left end of the section is $\partial y/\partial x$, and the slope at the right-hand end is

$$\left(\frac{\partial y}{\partial x}\right)_{x+dx} = \left(\frac{\partial y}{\partial x}\right)_x + \left(\frac{\partial^2 y}{\partial x^2}\right)_x dx + \cdots \tag{8.15}$$

by Taylor's expansion in the neighborhood of x.

Hence, neglecting higher-order terms, we have

$$\theta = -\frac{\partial^2 y}{\partial x^2}dx \tag{8.16}$$

Substituting this into (8.14), we have

$$dV = -\frac{M}{2}\frac{\partial^2 y}{\partial x^2}dx \tag{8.17}$$

for the potential energy of the section of the beam. However, we have

$$M = -EI\frac{\partial^2 y}{\partial x^2} \tag{8.18}$$

Hence, substituting this into (8.17), we obtain

$$dV = \frac{EI}{2}\left(\frac{\partial^2 y}{\partial x^2}\right)^2 dx \tag{8.19}$$

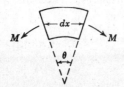

Fig. 8.2

The potential energy of the whole beam is

$$V = \frac{EI}{2} \int_0^s \left(\frac{\partial^2 y}{\partial x^2}\right)^2 dx \tag{8.20}$$

If now the beam is executing harmonic vibrations of the type

$$y(x,t) = v(x) \sin \omega t \tag{8.21}$$

we have, on substituting this into (8.12),

$$T_M = \tfrac{1}{2} m\omega^2 \int_0^s v^2(x)\, dx \tag{8.22}$$

for the *maximum* kinetic energy, and, substituting into (8.20), we obtain

$$V_M = \frac{EI}{2} \int_0^s \left(\frac{d^2 v}{dx^2}\right)^2 dx \tag{8.23}$$

for the *maximum* potential energy.

The natural angular frequency ω can be found if the deformation curve $v(x)$ is known by equating the two expressions (8.22) and (8.23). The procedure is to guess a certain function $v(x)$ that satisfies the boundary conditions at the ends of the beam and then equate expressions (8.22) and (8.23) to determine ω.

As an example of the general procedure, consider the system of Fig. 8.3. This system consists of a heavy uniform beam simply supported at each end and carrying a concentrated mass M at the center of the span. The boundary conditions at the ends are

$$\left.\begin{array}{c} y = 0 \\ \dfrac{\partial^2 y}{\partial x^2} = 0 \end{array}\right\} \quad x = 0 \qquad \left.\begin{array}{c} y = 0 \\ \dfrac{\partial^2 y}{\partial x^2} = 0 \end{array}\right\} \quad x = s \tag{8.24}$$

This expresses the fact that the displacement and bending moments at each end are zero. To determine the fundamental frequency, let us assume the curve

$$v(x) = A \sin \frac{\pi x}{s} \tag{8.25}$$

where A is the maximum deflection at the center. This assumed curve satisfies the boundary conditions and approximates the first mode of oscillation.

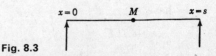

Fig. 8.3

From (8.22) the maximum kinetic energy of the beam is

$$T_M = \frac{m\omega^2}{2} \int_0^s A^2 \sin^2 \frac{\pi x}{s} \, dx = \frac{m\omega^2 A^2 s}{4} = \frac{M_B \omega^2 A^2}{4} \tag{8.26}$$

where

$$M_B = ms \tag{8.27}$$

is the mass of the beam.   The maximum kinetic energy of the central load is

$$T'_M = \tfrac{1}{2} M \omega^2 A^2 \tag{8.28}$$

Hence, for the whole system, the total maximum kinetic energy is

$$\sum T_M = \tfrac{1}{2}\omega^2 A^2 \left(\frac{M_B}{2} + M \right) \tag{8.29}$$

From (8.23) the maximum potential energy of the system is

$$V_M = \frac{EI}{2} \int_0^s \left(\frac{\partial^2 v}{\partial x^2} \right)^2 dx = \frac{EI A^2 \pi^4}{4s^4} \int_0^s 2 \sin^2 \frac{\pi x}{s} \, dx \tag{8.30}$$

We have

$$\int_0^s 2 \sin^2 \frac{\pi x}{s} \, dx = s \tag{8.31}$$

Hence

$$V_M = \frac{EI A^2 \pi^4}{4s^3} \tag{8.32}$$

Equating expressions (8.29) and (8.32), we obtain

$$\omega^2 = \frac{48.7 EI}{s^3 (M_B/2 + M)} \tag{8.33}$$

This gives the fundamental angular frequency ω.   There are two interesting special cases:

1. If the central load is zero, then $M = 0$ and we have

$$\omega^2 = \frac{97.4 EI}{M_B s^3} \tag{8.34}$$

This turns out to be the exact expression which would be obtained by the use of the differential equations of Chap. 13, Sec. 12.

2. If the beam is light compared with the central load, then $M_B \ll M$ and we have

$$\omega^2 = \frac{48.7 EI}{M s^3} \tag{8.35}$$

Rayleigh's method is very useful in the computation of the lowest natural frequencies of systems having distributed mass and elasticity. The success of the method depends on the fact that a large error in the assumed mode $v(x)$ produces a small error in the frequency ω. If it happens that we choose $v(x)$ to be one of the true modes, then Rayleigh's method will give the exact value for ω. In general it may be shown that Rayleigh's method gives values of the fundamental frequency ω that are somewhat greater than the true values.

9 THE COLLOCATION METHOD†

Consider the general linear differential equation of motion for a one-dimensional‡ continuous system, that is,

$$L(u) - \frac{\partial^2 u}{\partial t^2} = 0 \tag{9.1}$$

where $\quad L = $ linear differential spatial operator

$\quad u(x,t) = $ dependent variable

If we were considering the transverse motion of a string, the operator L would be

$$L = c^2 \frac{\partial^2}{\partial x^2} \tag{9.2}$$

whereas if we wanted to describe the transverse oscillations of a beam, we would have

$$L = \frac{1}{m(x)} \frac{\partial^2}{\partial x^2} EI(x) \frac{\partial^2}{\partial x^2} \tag{9.3}$$

To simplify this discussion we will limit ourselves to harmonic solutions of (9.1) only; that is, we are assuming that

$$u(x,t) = y(x) \sin (\omega t + \psi) \tag{9.4}$$

Thus, with this assumption, Eq. (9.1) reduces to

$$L(y) + \omega^2 y = 0 \tag{9.5}$$

† The material in this section and in Sec. 10 was organized with the assistance of Michael Gollong, a graduate of the Mathematics Department of the University of Redlands, Redlands, Calif.

‡ This method is equally applicable to two- and three-dimensional systems.

Even with this reduction, Eq. (9.5) is difficult to solve exactly in many two-dimensional cases. That is particularly true in those cases where the spatial region is nonseparable.†

To proceed with the collocation method, a solution to (9.5) is assumed as follows:

$$y = \sum_{i=1}^{N} a_i \phi_i(x) \qquad (9.6)$$

where the a_i are arbitrary constants and the $\phi_i(x)$ are known functions selected by the user which should be constructed in dimensionless form. It is desirable that these functions satisfy the boundary conditions imposed on (9.1). At this point (9.6) is substituted into (9.5) to yield

$$\sum_{i=1}^{N} a_i L[\phi_i(x)] + \omega^2 \sum_{i=1}^{N} a_i \phi_i(x) = E(a_i, x) \qquad (9.7)$$

where E is the error resulting from the fact that the assumed solution is only an approximate solution. The collocation procedure is to force the error to be zero at N points x_k lying within the region over which the solution is desired. These points are usually equally spaced, and the result of this procedure is a system of N equations for the N arbitrary coefficients a_i. If the exact solution were used, the error would be identically zero over the entire interval.

Thus if we set (9.7) to zero at the N points x_k, we have

$$\sum_{i=1}^{N} a_i L[\phi_i(x_k)] + \omega^2 \sum_{i=1}^{N} a_i \phi_i(x_k) = 0 \qquad (9.8)$$

If we utilize matrix notation, we may write (9.8) in the form

$$[A](a) + \omega^2 [B](a) = (0) \qquad (9.9)$$

where $[A_{ij}] = [L\phi_j(x_i)]$

$\quad\quad [B_{ij}] = [\phi_j(x_i)]$

$\quad\quad\quad (a) =$ vector of unknown coefficients, a_k

By matrix manipulations, we can write (9.9) in the form

$$[c](a) = \lambda(a) \qquad (9.10)$$

where $[c] = [A]^{-1}[B]$

$\quad\quad \lambda = -1/\omega^2$

---

† For an excellent example of applying approximate techniques to nonseparable domains, see Daniel Dicker, Solutions of Heat Conduction Problems with Nonseparable Domains, *Journal of Applied Mechanics*, vol. 30, no. 1, December, 1963.

which indicates that the problem has been reduced to a standard eigenvalue problem. In terms of the physical system, each eigenvalue of (9.10) represents the reciprocal of a natural frequency, and each eigenvector represents a mode shape. We have set up the eigenvalue problem of (9.10) so that the largest eigenvalue will correspond to the lowest natural or fundamental frequency. This formulation was chosen since most iteration procedures for determining eigenvalues converge to the one with the largest magnitude first (see Chap. 3, Sec. 34, or Chap. 6, Sec. 7). It is also usually true that as the smaller eigenvalues are computed, many routines lose accuracy. Since an analyst is usually concerned with the fundamental and first few overtones, the problem should be structured as above so that these values are determined first.

As for eigenvectors, we mentioned that they are associated with the modes of the various vibrations. In the case of a vibrating-string problem, the largest eigenvalue should yield an approximate value for the fundamental frequency; the corresponding eigenvector should yield the coefficients for the ϕ_i which will combine to approximate the fundamental mode which is a sine function.

To illustrate this method, consider a vibrating string of length l and L is given by (9.2). The problem to be solved is

$$c^2 \frac{\partial^2 u}{\partial x^2} - \frac{\partial^2 u}{\partial t^2} = 0 \tag{9.11}$$

where $u(x,t) = $ displacement from equilibrium.

Equation (9.11) is reduced to

$$c^2 \frac{d^2 y}{dx^2} + \omega^2 y = 0 \tag{9.12}$$

by the substitution

$$u(x,t) = y(x)\sin(\omega t + \psi) \tag{9.13}$$

which assumes harmonic vibrations. The boundary conditions are for fixed ends; that is,

$$u(0,t) = u(l,t) = 0$$

Applying these to (9.13) gives

$$y(0) = y(l) = 0 \tag{9.14}$$

At this point we assume an approximate function of the form

$$y(x) = \sum_{i=1}^{2} a_i \phi_i(x) \tag{9.15}$$

with

$$\phi_1 = \frac{x(l-x)}{l^2} \qquad \phi_2 = \frac{x^2(l-x)}{l^3} \tag{9.16}$$

both of which satisfy (9.14). The functions have been chosen in dimensionless form to conform with the basic requirements of this method. Following the collocation procedure, Eq. (9.15) is substituted into (9.12) to obtain the error term

$$E(a_i,x) = c^2 \sum_{i=1}^{2} a_i \phi_i'' + \omega^2 \sum_{i=1}^{2} a_i \phi_i \qquad (9.17)$$

Since we have used two functions to form the approximate solution, two points in the interval $0 \leqslant x \leqslant l$ must be chosen at which E will be set to zero. For convenience the points $x_1 = l/3$ and $x_2 = 2l/3$ will be used. Evaluating the coefficient matrices in (9.9) gives

$$[A] = -\frac{2c^2}{l^2}\begin{bmatrix} 1 & 0 \\ 1 & 1 \end{bmatrix} \qquad [B] = \frac{2}{27}\begin{bmatrix} 3 & 1 \\ 3 & 2 \end{bmatrix} \qquad (a) = \begin{pmatrix} a_1 \\ a_2 \end{pmatrix}$$

At this point we may write (9.9) as

$$-\frac{2c^2}{l^2}\begin{bmatrix} 1 & 0 \\ 1 & 1 \end{bmatrix}(a) + \frac{2\omega^2}{27}\begin{bmatrix} 3 & 1 \\ 3 & 2 \end{bmatrix}(a) = (0) \qquad (9.18)$$

Rearranging terms, (9.18) becomes

$$\begin{bmatrix} 3 & 1 \\ 3 & 2 \end{bmatrix}(a) = \frac{27c^2}{\omega^2 l^2}\begin{bmatrix} 1 & 0 \\ 1 & 1 \end{bmatrix}(a) \qquad (9.19)$$

Now if we let

$$\lambda = \frac{27c^2}{\omega^2 l^2} \qquad (9.20)$$

and premultiply by $[A]^{-1}$, we have the eigenvalue problem

$$\begin{bmatrix} 3 & 1 \\ 0 & 1 \end{bmatrix}(a) = \lambda(a) \qquad (9.21)$$

The eigenvalues and associated eigenvectors are

$$\lambda_1 = 3 \qquad (a)_1 = \begin{pmatrix} 1 \\ 0 \end{pmatrix}$$

$$\lambda_2 = 1 \qquad (a)_2 = \begin{pmatrix} 1 \\ -2 \end{pmatrix}$$

From (9.20) we see that

$$\omega_1 = \frac{3c}{l}$$

$$y_1(x) = \frac{a' x(l - x)}{l^2} \qquad (9.22)$$

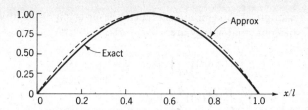

Fig. 9.1

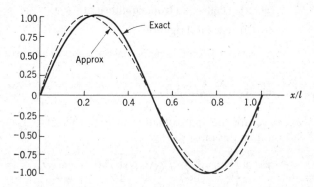

Fig. 9.2

where a' = arbitrary amplitude for the fundamental

$\omega_2 = 3\sqrt{3}\, c/l$

$y_2(x) = a''[x(l - x)/l^2 - 2x^2(l - x)/l^3]$

$\qquad = a''\, x(l - x)(l - 2x)/l^3$

a'' = arbitrary amplitude for the first harmonic

From Chap. 12, Sec. 5, the exact fundamental frequency is

$$\omega_1(\text{exact}) = \frac{\pi c}{l}$$

which indicates an error of 4.5 percent in the natural frequency as predicted by the approximate solution (9.16). Figures 9.1 and 9.2 show the approximate and exact solutions in dimensionless form for the fundamental and first harmonic. As shown by these figures, the two-term approximation yields a very reasonable fit to the exact solution.

10   THE METHOD OF RAYLEIGH-RITZ

This approximate technique is generally approached from energy considerations. For purposes of illustration we will concern ourselves with the specific example of the vibrations of a cantilever beam of length l.   The total kinetic and potential energy are

$$T = \tfrac{1}{2} \int_0^l m(x) \left(\frac{\partial u}{\partial t} \right)^2 dx$$

$$V = \tfrac{1}{2} \int_0^l EI(x) \left(\frac{\partial^2 u}{\partial x^2} \right)^2 dx$$

(10.1)

where $m(x)$ = mass per unit length

$\quad\quad u(x,t)$ = deflection from equilibrium

$\quad\quad EI(x)$ = flexural rigidity

As in the collocation method, we assume an approximate solution in the form

$$u(x,t) = \sum_{k=1}^n \phi_k(x) q_k(t)$$

(10.2)

where ϕ_k are the approximate functions and q_k are general time functions. Substituting this into (10.1) gives

$$T = \tfrac{1}{2} \int_0^l m(x) \left[\sum_{k=1}^n \phi_k(x) \dot{q}_k(t) \right]^2 dx$$

$$V = \tfrac{1}{2} \int_0^l EI(x) \left[\sum_{k=1}^n \phi_k''(x) q_k(t) \right]^2 dx$$

(10.3)

Now, to obtain the equation of motion, we make use of Lagrange's equation in the form

$$\frac{d}{dt}\left(\frac{\partial T}{\partial \dot{q}_r} \right) - \frac{\partial T}{\partial q_r} + \frac{\partial V}{\partial q_r} = 0$$

(10.4)

From (10.3) we see that

$$\frac{\partial T}{\partial q_r} = 0 \quad\quad r = 1, 2, \ldots, n$$

$$\frac{\partial T}{\partial \dot{q}_r} = \int_0^l m(x) \phi_r(x) \sum_{k=1}^n \phi_k(x) \dot{q}_k(t) \, dx$$

(10.5)

Likewise from (10.3) we find

$$\frac{\partial V}{\partial q_r} = \int_0^l EI(x) \phi_r''(x) \sum_{k=1}^n \phi_k''(x) q_k(t) \, dx$$

(10.6)

Computing the first term for (10.4) from (10.5) gives

$$\frac{d}{dt}\left(\frac{\partial T}{\partial \dot{q}_r} \right) = \int_0^l m(x) \phi_r(x) \sum_{k=1}^n \phi_k(x) \ddot{q}_k(t) \, dx$$

(10.7)

Combining all these results into (10.4) yields

$$\sum_{k=1}^{n} \ddot{q}_k(t) \int_0^l m(x)\,\phi_r(x)\,\phi_k(x)\,dx$$
$$+ \sum_{k=1}^{n} q_k(t) \int_0^l EI(x)\,\phi_r''(x)\,\phi_k''(x)\,dx = 0 \qquad (10.8)$$

If we now restrict ourselves to periodic time functions

$$q_k(t) = a_k \sin(\omega t + \psi) \qquad (10.9)$$

then (10.8) reduces to

$$\sum_{k=1}^{n} a_k \int_0^l EI(x)\,\phi_r''(x)\,\phi_k''(x)\,dx$$
$$- \omega^2 \sum_{k=1}^{n} a_r \int_0^l m(x)\,\phi_r(x)\,\phi_k(x)\,dx = 0 \qquad (10.10)$$

The system of equations given by (10.10) represents n linear algebraic equations in the n unknowns a_k. If we now introduce the following matrices

$$[A] = [A_{ij}] = \left[\int_0^l EI(x)\,\phi_i''(x)\,\phi_j''(x)\,dx \right]$$
$$[B] = [B_{ij}] = \left[\int_0^l m(x)\,\phi_i(x)\,\phi_j(x)\,dx \right] \qquad (10.11)$$
$$\{a\} = \{a_k\}$$

the system of (10.10) becomes

$$[A]\{a\} - \omega^2 [B]\{a\} = \{0\} \qquad (10.12)$$

which is a standard eigenvalue problem of the form (9.10).

As an example of this method, consider a cantilever beam for which the mass per unit length and the flexural rigidity are constant. If the beam is mounted at $x = 0$ with $x = l$ being the free end, the boundary conditions are

$$u(0,t) = 0 \qquad \text{(no deflection at support)}$$

$$\frac{\partial u}{\partial x}(0,t) = 0 \qquad \text{(built-in end)}$$

$$\frac{\partial^2 u}{\partial x^2}(l,t) = 0 \qquad \text{(no moment at free end)}$$

$$\frac{\partial^3 u}{\partial x^3}(l,t) = 0 \qquad \text{(no shear at free end)}$$

For an approximate solution, let us use a two-term expansion for (10.2) in which

$$\phi_1 = \left(\frac{x}{l}\right)^2 \qquad \phi_2 = \left(\frac{x}{l}\right)^3 \qquad (10.13)$$

It should be noted that these functions do not satisfy all the boundary conditions; however, this is not an absolute requirement. Substituting these functions into (10.11), we have

$$[A] = \frac{2EI}{l^3}\begin{bmatrix} 2 & 3 \\ 3 & 6 \end{bmatrix} \qquad [B] = \frac{ml}{210}\begin{bmatrix} 42 & 35 \\ 35 & 30 \end{bmatrix} \qquad (10.14)$$

In the form of the standard eigenvalue problem (9.10), our problem reduces to

$$[C]\{a\} = \lambda\{a\} \qquad (10.15)$$

where

$$[C] = \begin{bmatrix} 147 & 120 \\ -56 & -45 \end{bmatrix} \qquad \lambda = \frac{1{,}260EI}{\omega^2 ml^4} \qquad (10.16)$$

Using standard procedures to find the eigenvalues and eigenvectors yields

$$\lambda_1 = 100.1 \qquad \lambda_2 = 1.1$$
$$\{a\}_1 = \begin{Bmatrix} -2.55 \\ 1.00 \end{Bmatrix} \qquad \{a\}_2 = \begin{Bmatrix} 1.00 \\ -1.22 \end{Bmatrix} \qquad (10.17)$$

From the largest eigenvalue we find that the fundamental frequency is

$$\omega_0 \text{ (approx.)} = \frac{3.55}{l^2}\left(\frac{EI}{m}\right)^{\frac{1}{4}} \qquad (10.18)$$

Comparing this with the exact solution from Chap. 13, Sec. 7, indicates an error of 0.85 percent. Substituting into (10.2) gives the modal solution

$$u(x,t) \, A\left[-2.55\left(\frac{x}{l}\right)^2 + 1.00\left(\frac{x}{l}\right)^3\right]\sin(\omega_0 t + \psi) \qquad (10.19)$$

where $A, \psi = $ arbitrary constants.

11 GALERKIN'S METHOD

This method was developed by B. Galerkin in 1915[†] and is perhaps the most widely used approximate method today. The procedure is similar to the collocation method, and we shall use the same approach to illustrate its derivation.

Again, for convenience, we shall assume only one spatial dimension since the extension to more dimensions is a straightforward procedure. We assume that the system under consideration may be described by the equation

$$L(u) - \frac{\partial^2 u}{\partial t^2} = 0 \qquad (11.1)$$

[†] Cf. Galerkin, 1915 (see References).

where L and u are the same as for (9.1).   This equation reduces to

$$Ly + \omega^2 y = 0 \tag{11.2}$$

under the assumption of harmonic solutions [cf. (9.4)].   We continue along the same lines and construct an approximate solution

$$y = \sum_{k=1}^{n} a_k \phi_k(x) \tag{11.3}$$

Substituting this into (11.2) yields the error

$$E(x,a_k) = \sum_{k=1}^{n} a_k L\phi_k + \omega^2 \sum_{k=1}^{n} a_k \phi_k \tag{11.4}$$

At this point the procedure differs from the collocation method in that the Galerkin method requires the error to be orthogonal with all the approximating functions ϕ_k.   Specifically this requirement is

$$\int_a^b \phi_i(x) E(x,a_k) \, dx = 0 \qquad i = 1, 2, \ldots, n \tag{11.5}$$

where a and b are the limits of the interval of x over which the problem is defined.   Equation (11.5) represents n linear algebraic equations for the n unknowns a_k.

Substituting (11.4) into (11.5) and rearranging terms gives

$$\sum_{k=1}^{n} a_k \int_a^b \phi_i L\phi_k \, dx + \omega^2 \sum_{k=1}^{n} a_k \int_a^b \phi_i \phi_k \, dx = 0 \qquad i = 1, 2, \ldots, n \tag{11.6}$$

Introducing matrix notation

$$[A_{ik}] = \left[\int_a^b \phi_i L\phi_k \, dx \right]$$

$$[B_{ik}] = \left[\int_a^b \phi_i \phi_k \, dx \right]$$

$$\{a\} = \{a_k\}$$

we have

$$[A]\{a\} + \omega^2[B]\{a\} = \{0\} \tag{11.7}$$

which again may be reduced to the eigenvalue problem

$$[C]\{a\} = \lambda\{a\} \tag{11.8}$$

where $[C]$ and λ are defined in (9.10).

To illustrate this method, let us consider the same problem used in Sec. 9. In that problem we were attempting to solve the wave equation for the vibrating string of length l. In this problem

$$L = c^2 \frac{\partial^2}{\partial x^2}$$

and it was assumed that

$$\phi_1 = \frac{x(l-x)}{l^2}.$$

$$\phi_2 = \frac{x^2(l-x)}{l^3}$$

(11.9)

The interval for x is 0 to l. Evaluating the coefficient matrices, we have

$$[A] = -\frac{c^2}{30l} \begin{bmatrix} 10 & 5 \\ 5 & 4 \end{bmatrix}$$

$$[B] = \frac{l}{420} \begin{bmatrix} 14 & 7 \\ 7 & 4 \end{bmatrix}$$

and converting into the form of (11.8) we have

$$[C] = \begin{bmatrix} 21 & 8 \\ 0 & 5 \end{bmatrix}$$

$$\lambda = \frac{210c^2}{\omega^2 l^2}$$

(11.10)

The eigenvalues and their associated eigenvectors for this problem are

$$\lambda_1 = 21 \qquad \lambda_2 = 5$$

$$\{a\}_1 = \begin{Bmatrix} 1 \\ 0 \end{Bmatrix} \qquad \{a\}_2 = \begin{Bmatrix} 1 \\ -2 \end{Bmatrix}$$

(11.11)

At this point it is interesting to note that the eigenvectors are identical with those obtained in the collocation method. From the largest eigenvalue, we obtain the following approximation for the fundamental frequency:

$$\omega_0 \text{ (approx.)} = \sqrt{10} \frac{c}{l} \approx 3.16 \frac{c}{l}$$

(11.12)

indicating an error of only 0.6 percent, which is nearly an order-of-magnitude improvement over the result using the collocation method. The error in the first harmonic is 3.0 percent.

PROBLEMS

1. Repeat the steps of Sec. 2 to obtain the formula for a least-squares fit for a second-order equation.

2. Using the least-squares method, find the coefficients for the best-fit second-order equation for the following data:

x	0	1	2	3	4
y	0.1	0.8	4.2	9.2	15.6

3. Integrate $y' = -0.5y$, $y(0) = 1$ from 0 to 2 by the Euler-Cauchy method using a step size of 0.5.

4. Repeat Prob. 3 with $h = 0.2$.

5. Assume that you have access to a digital computer that has a random number generator which will generate random numbers over the interval 0 to 1. Discuss how you would set up the following problems for solution by the Monte Carlo method:

$$(a) \int_0^\pi \sin x \, dx \qquad (b) \int_2^3 x^2 \, dx \qquad (c) \int_0^2 x^3 \, dx$$

6. A string of length s under a tension t carries a mass M at its midpoint. Using Rayleigh's method, determine the fundamental frequency of oscillation. (HINT: Obtain the total potential and kinetic energies, etc.)

7. A uniform beam is built in at one end and carries a mass M at its other end. Use Rayleigh's method to determine the fundamental angular frequency of oscillation.

8. A uniform beam is hinged at $x = 0$ and is elastically supported by a spring whose constant is K at $x = 5$. Apply Rayleigh's method to determine the fundamental frequency.

9. Find an approximate value for the fundamental frequency of a vibrating string using the collocation method. Use the same two approximating functions introduced in the text as well as the third function $\phi_3 = x(l - x)^2/l^3$. For the collocating points use $\frac{1}{4}, \frac{1}{2}, \frac{3}{4}$.

10. Solve Prob. 9 by using Galerkin's method and the same approximating functions.

11. Solve the sample beam problem of Sec. 10 by the collocation method.

12. Solve Prob. 11 using Galerkin's method.

13. Use the Rayleigh-Ritz method to solve for the fundamental frequency of the vibrating string of Sec. 9 with the same approximating functions. Compare your answer with those obtained by the collocation and Galerkin methods.

REFERENCES

1915. Galerkin, B. G.: Sterzhni i plastiny. Ryady v nekotorykh voprosakh uprogogo ravnovesiya sterzhnei i plastin (Rods and Plates. Series Occurring in Some Problems of Elastic Equilibrium of Rods and Plates), *Tekh. Petrograd*, vol. 19, pp. 897–908; Translation 63-18924, Clearinghouse for Federal Scientific and Technical Information.

1933. Temple, G., and W. G. Bickley: "Rayleigh's Principle," Oxford University Press, New York.

1937. Timoshenko, S.: "Vibration Problems in Engineering," D. Van Nostrand Company, Inc., Princeton, N.J.

1938. Duncan, W. J.: "The Principles of Galerkin's Method," Gt. Brit. Aero. Res. Council Rept. and Memo. No. 1848; reprinted in Gt. Brit. Air Ministry Aero Res. Comm. Tech. Rept., vol. 2, pp. 589–612.

1951. "Monte Carlo Method," National Bureau of Standards Applied Mathematics Series, vol. 12.

1956. Brown, G. W.: Monte Carlo Method, in E. F. Bechenbach (ed.), "Modern Mathematics for the Engineer," McGraw-Hill Book Company, New York, pp. 279–306.

1956. Sokolnikoff, I. S.: "Mathematical Theory of Elasticity," 2d ed., chap. 7, McGraw-Hill Book Company, New York.

1965. Kuo, S. S.: "Numerical Methods and Computers," Addison-Wesley Publishing Company, Inc., Reading, Mass.

1966. Finlayson, B. A., and L. E. Scriven: The Method of Weighted Residuals—A Review, *Applied Mechanics Reviews*, vol. 19, no. 9, September, pp. 735 ff.

15
The Analysis of Nonlinear Oscillatory Systems

1 INTRODUCTION

The mathematical analysis of many of the oscillatory phenomena that occur in nature and in technology leads to the solution of nonlinear differential equations. Such systems as a pendulum executing large oscillations, the flow of electrical current in a circuit consisting of a capacitance in series with an iron-cored inductance coil, the free oscillations of a regenerative triode oscillator, the motion of a mass restrained by a spring and undergoing dry or solid friction are typical examples of systems whose analysis leads to nonlinear differential equations.

The equations of these systems are well known in astronomy and have been studied by such investigators as Liapounoff, Linstedt, and especially H. Poincaré. A number of Russian scientists have contributed greatly to the solution of these nonlinear problems. Two of the most active Russian workers in this field are N. Kryloff and N. Bogoliuboff. In the field of nonlinear electrical oscillations van der Pol has made some notable contributions.

In this chapter some of the simpler analytical methods that have been devised to solve nonlinear oscillatory problems will be considered. The interested reader will find the references at the end of this chapter of value.

2   OSCILLATOR DAMPED BY SOLID FRICTION

Perhaps the simplest nonlinear vibrating system is the oscillator of Fig. 2.1. This system consists of a mass m connected to rigid supports by means of springs of total effective spring constant k.   The mass is supposed to slide on a rough surface and hence experience a constant frictional force F, which is directed in a direction opposite to the velocity.   Let the motion start in such a manner that

$$\left.\begin{aligned} x &= a \\ \dot{x} &= 0 \end{aligned}\right\} \quad \text{at } t = 0 \qquad\qquad (2.1)$$

That is, the mass is pulled out a distance a from its equilibrium position and released at $t = 0$.   The equation of motion is, by Newton's law,

$$m\ddot{x} = kx = \pm F \qquad\qquad (2.2)$$

If we let

$$\omega^2 = \frac{k}{m} \qquad A = \frac{F}{m} \qquad\qquad (2.3)$$

then

$$\ddot{x} + \omega^2 x = +A \qquad 0 < t < \frac{\pi}{\omega} \qquad\qquad (2.4)$$

when the force is acting in the positive x direction. It is expressly understood that this is true provided that the initial position $x = a$ is such that

$$\omega^2 a > A \qquad\qquad (2.5)$$

so that the spring force is greater than the frictional force.   This means that the initial displacement lies outside the so-called "dead region"

$$x = \pm \frac{A}{\omega^2} \qquad\qquad (2.6)$$

If motion is possible and the first swing to the left is of such magnitude that it is still outside the dead region, then

$$\ddot{x} + \omega^2 x = -A \qquad \frac{\pi}{\omega} < t < \frac{2\pi}{\omega} \qquad\qquad (2.7)$$

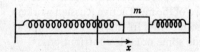

Fig. 2.1

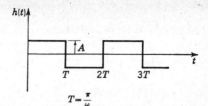

Fig. 2.2

$T = \dfrac{\tau}{\omega}$

We can thus write the equation of motion in the form

$$\ddot{x} + \omega^2 x = h(t) \tag{2.8}$$

where $h(t)$ is a discontinuous function of the meander type whose graph is given by Fig. 2.2.   This is provided that $\dot{x} \neq 0$, that is, that the motion continues, and that successive swings are of such amplitude that they lie outside the dead region.

Equation (2.8) may be solved by the Laplacian-transform, or operational, method.   To do this, let

$$Lx(t) = y(s)$$
$$Lh(t) = g(s) = A(1 - 2e^{-Ts} + 2e^{-2Ts} - 2e^{-3Ts} + \cdots) \tag{2.9}$$

Now, as a consequence of the initial conditions (2.1), we have

$$L\ddot{x} = s^2 y - sa \tag{2.10}$$

Hence Eq. (2.8) transforms to

$$(s^2 + \omega^2)y = sa + A(1 - 2e^{-Ts} + 2e^{-2Ts} - 2e^{-3Ts} + \cdots) \tag{2.11}$$

Therefore we have

$$y = \frac{sa}{s^2 + \omega^2} + \frac{A}{s^2 + \omega^2}(1 - 2e^{-Ts} + 2e^{-2Ts} - 2e^{-3Ts} + \cdots) \tag{2.12}$$

Now, from the table of transforms (Appendix A) we have

$$L^{-1}\frac{s}{s^2 + \omega^2} = \cos \omega t$$
$$L^{-1}\frac{1}{s(s^2 + \omega^2)} = \frac{1 - \cos \omega t}{\omega^2} \tag{2.13}$$

Hence we have, on calculating the inverse transforms of (2.12),

$$x = a\cos \omega t + \frac{A}{\omega^2}(1 - \cos \omega t) \qquad 0 < t < T \tag{2.14}$$

$$x = a\cos \omega t + \frac{A}{\omega^2}(1 - \cos \omega t) - \frac{2A}{\omega^2}[1 - \cos \omega(t - T)]$$
$$\text{for } T < t < 2T \tag{2.15}$$

The solution for greater values of t may be written down in a similar manner. It is noted that at

$$t = T \qquad x = -a + \frac{2A}{\omega^2} \qquad\qquad\qquad (2.16)$$

$$t = 2T \qquad x = a - \frac{4A}{\omega^2} \qquad \text{etc.} \qquad\qquad (2.17)$$

That is, each successive swing is $2A/\omega^2$ shorter than the preceding one. The solution should be continued until one of the swings lies inside the dead region $x = \pm A/\omega^2$. In such a case the motion stops.

3 THE FREE OSCILLATIONS OF A PENDULUM

As another simple example of a nonlinear dynamical system, let us discuss the motion of the simple pendulum of Fig. 3.1. Figure 3.1 represents a particle of mass m suspended from a fixed point O by a massless inextensible rod of length s, free to oscillate in the plane of the paper under the influence of gravity. If the pendulum is displaced by an angle θ from its position of equilibrium as shown in the figure, then the potential energy of the system V is the work done against gravity to lift the mass of the pendulum a distance h, given by

$$h = s(1 - \cos\theta) \qquad\qquad\qquad (3.1)$$

Hence the potential energy is

$$V = mgs(1 - \cos\theta) \qquad\qquad\qquad (3.2)$$

The kinetic energy of motion of the pendulum is given by

$$T = \tfrac{1}{2}mv^2 \qquad\qquad\qquad (3.3)$$

where v is the linear velocity of the pendulum. In terms of the angle θ we have

$$v = s\frac{d\theta}{dt} = s\dot\theta \qquad\qquad\qquad (3.4)$$

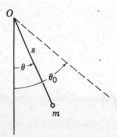

Fig. 3.1

Hence

$$T = \tfrac{1}{2}ms^2 \dot{\theta}^2 \tag{3.5}$$

Now, since there is no loss of energy from the system, we must have

$$T + V = C \tag{3.6}$$

where C is a constant representing the total energy. Hence, substituting (3.2) and (3.5) into (3.6), we have

$$mgs(1 - \cos \theta) + \tfrac{1}{2}ms^2 \dot{\theta}^2 = C \tag{3.7}$$

To determine the constant C, let us assume that $\theta = \theta_0$ is the maximum amplitude of swing so that

$$\dot{\theta} = 0 \quad \text{when } \theta = \theta_0 \tag{3.8}$$

Substituting this condition into (3.7), we have

$$C = mgs(1 - \cos \theta_0) \tag{3.9}$$

Hence we have

$$g(1 - \cos \theta) + \frac{s}{2}\dot{\theta}^2 = g(1 - \cos \theta_0) \tag{3.10}$$

or

$$\dot{\theta}^2 = \frac{2g}{s}(\cos \theta - \cos \theta_0) \tag{3.11}$$

To obtain the equation of motion, we differentiate (3.11) with respect to time and obtain

$$2\dot{\theta}\ddot{\theta} = -\frac{2g}{s}\sin \theta \dot{\theta} \tag{3.12}$$

or

$$\frac{d^2\theta}{dt^2} + \frac{g}{s}\sin \theta = 0 \tag{3.13}$$

This is the equation of motion, and Eq. (3.11) is its first integral.

Equation (3.13) is a nonlinear differential equation because of the presence of the trigonometric function $\sin \theta$. In the theory of *small* oscillations of a pendulum we expand $\sin \theta$ into a Maclaurin series of the form

$$\sin \theta = \theta - \frac{\theta^3}{3!} + \frac{\theta^5}{5!} - \frac{\theta^7}{7!} + \cdots \tag{3.14}$$

If θ is always small, the approximation

$$\sin \theta \doteq \theta \tag{3.15}$$

is a very good one, and in this case (3.13) becomes

$$\frac{d^2\theta}{dt^2} + \frac{g}{s}\theta = 0 \tag{3.16}$$

This is a linear equation with constant coefficients whose general solution is of the form

$$\theta = A\sin(\omega t + \phi) \tag{3.17}$$

where

$$\omega = \sqrt{\frac{g}{s}} \tag{3.18}$$

and A and ϕ are arbitrary constants that depend on the initial conditions of the motion. The period of the motion P_0 is given by

$$P_0 = \frac{2\pi}{\omega} = 2\pi\sqrt{\frac{g}{s}} \tag{3.19}$$

We thus see that the period is independent of the amplitude of oscillation in this case. By measuring s and P_0 the acceleration g due to gravity may be determined by (3.19) to a high degree of accuracy.

To determine the period of oscillation for large amplitudes, we return to Eq. (3.11) and write it in the form

$$\frac{d\theta}{dt} = \sqrt{\frac{2g}{s}}\sqrt{\cos\theta - \cos\theta_0} \tag{3.20}$$

or

$$dt = \sqrt{\frac{s}{2g}}\frac{d\theta}{\sqrt{\cos\theta - \cos\theta_0}} \tag{3.21}$$

Now the period P_{θ_0} is twice the time taken by the pendulum to swing from $\theta = -\theta_0$ to $\theta = \theta_0$. Therefore

$$P_{\theta_0} = 2\sqrt{\frac{s}{2g}}\int_{-\theta_0}^{\theta_0}\frac{d\theta}{\sqrt{\cos\theta - \cos\theta_0}} \tag{3.22}$$

If we use the trigonometric identities

$$\cos\theta = 1 - 2\sin^2\frac{\theta}{2} \qquad \cos\theta_0 = 1 - 2\sin^2\frac{\theta_0}{2} \tag{3.23}$$

then we may write (3.22) in the form

$$P_{\theta_0} = \sqrt{\frac{s}{g}}\int_{-\theta_0}^{+\theta_0}\frac{d\theta}{\sqrt{\sin^2(\theta_0/2) - \sin^2(\theta/2)}} \tag{3.24}$$

Let us now introduce the two new variables k and ϕ by the equations

$$k = \sin \frac{\theta_0}{2} \tag{3.25}$$

and

$$\sin \frac{\theta}{2} = k \sin \phi \tag{3.26}$$

Hence

$$d\theta = \frac{2k \cos \phi \, d\phi}{\cos(\theta/2)} = \frac{2k \cos \phi \, d\phi}{\sqrt{1 - k^2 \sin^2 \phi}} \tag{3.27}$$

In terms of these two new variables the integral of (3.24) becomes

$$P_{\theta_0} = 4 \sqrt{\frac{s}{g}} \int_0^{\pi/2} \frac{d\phi}{\sqrt{1 - k^2 \sin^2 \phi}} \tag{3.28}$$

Now the integral

$$K(k) = \int_0^{\pi/2} \frac{d\phi}{\sqrt{1 - k^2 \sin^2 \phi}} \tag{3.29}$$

is called the complete elliptic integral of the first kind and is tabulated.[1]   We therefore may write

$$P_{\theta_0} = 4 \sqrt{\frac{s}{g}} K(k) \tag{3.30}$$

Equation (3.25) determines k in terms of the maximum angle of swing θ_0. From the tables of the function $K(k)$ and Eq. (3.30), we may determine the period P_{θ_0}.   We thus find that, for $\theta_0 = 60°$, we have

$$P_{60°} = 1.07 P_0 \tag{3.31}$$

and, for $\theta_0 = 2°$, we have

$$P_{2°} = 1.000076 P_0 \tag{3.32}$$

where P_0 is given by (3.19).

We thus see that the period of a pendulum depends on its amplitude of oscillation.

4   RESTORING FORCE A GENERAL FUNCTION OF THE DISPLACEMENT

Let us consider the motion of a particle of mass m.   Let the mass be subjected to a restoring force $F(x)$ of an elastic nature tending to restore the mass m to

[1] See, for example, B. O. Peirce, "A Short Table of Integrals," p. 121, Ginn & Company, Boston, 1929.

the position of equilibrium $x = 0$. The equation of motion of this system is, by Newton's law,

$$m\frac{d^2 x}{dt^2} + F(x) = 0 \tag{4.1}$$

If we let

$$v = \frac{dx}{dt} \tag{4.2}$$

be the velocity, then we have

$$\frac{d^2 x}{dt^2} = \frac{dv}{dt} = \frac{dv}{dx}\frac{dx}{dt} = v\frac{dv}{dx} \tag{4.3}$$

Hence Eq. (4.1) may be written in the form

$$mv\,dv + F(x)\,dx = 0 \tag{4.4}$$

Now the potential energy of the system is, at any instant,

$$V(x) = \int_0^x F(u)\,du \tag{4.5}$$

The kinetic energy is

$$T = \tfrac{1}{2}mv^2 \tag{4.6}$$

Let us suppose that the motion is started in such a way that the mass is pulled out a certain distance $x = a$ and then released. We then have

$$\begin{aligned} x &= a \\ \dot{x} &= v = 0 \end{aligned} \tag{4.7}$$

In this case the *total energy E* imparted to the system is

$$E = \int_0^a F(u)\,du = V(a) \tag{4.8}$$

Now, integrating Eq. (4.4), we obtain

$$\frac{mv^2}{2} + \int_0^x F(u)\,du = C \tag{4.9}$$

where C is a constant of integration. To determine C, we use the conditions (4.7), and we have

$$\int_0^a F(u)\,du = C = E \tag{4.10}$$

Equation (4.9) may therefore be written in the form

$$\frac{mv^2}{2} = E - V(x) \tag{4.11}$$

or

$$v^2 = \frac{2(E-V)}{m} = \left(\frac{dx}{dt}\right)^2 \tag{4.12}$$

Hence

$$\frac{dx}{dt} = \sqrt{\frac{2}{m}} \sqrt{E-V} \tag{4.13}$$

or

$$t = \sqrt{\frac{m}{2}} \int \frac{dx}{\sqrt{E-V(x)}} + t_0 \tag{4.14}$$

where t_0 is an arbitrary constant.   If we measure t so that

$$t = 0 \qquad \text{at } x = 0 \tag{4.15}$$

then we have from Eq. (4.14)

$$t = \sqrt{\frac{m}{2}} \int_0^x \frac{du}{\sqrt{E-V(u)}} = N(x) \tag{4.16}$$

If it is possible to integrate this expression, we obtain

$$t = N(x) \tag{4.17}$$

or, inversely,

$$x = F(t) \tag{4.18}$$

In some cases it is possible to integrate the expression (4.16) explicitly. Usually the integration leads to elliptic integrals, and Eq. (4.18) leads to elliptic functions.

The period of the motion is given by the equation

$$P = 4\sqrt{\frac{m}{2}} \int_0^a \frac{du}{\sqrt{E-V(u)}} \tag{4.19}$$

A special case of this general theory is the motion of a pendulum with large amplitude discussed in Sec. 3.   In the case of the pendulum the kinetic energy is

$$T = \tfrac{1}{2}ms^2\,\dot{\theta}^2 \tag{4.20}$$

The potential energy is

$$V(\theta) = mgs(1 - \cos\theta) \tag{4.21}$$

The total energy is

$$E = mgs(1 - \cos\theta_0) \tag{4.22}$$

Substituting these expressions into (4.16), we have

$$t = s \sqrt{\frac{m}{2}} \int_0^\theta \frac{du}{\sqrt{mgs(\cos u - \cos \theta_0)}} \qquad (4.23)$$

or

$$t = \sqrt{\frac{s}{2g}} \int_0^\theta \frac{du}{\sqrt{\cos u - \cos \theta_0}} \qquad (4.24)$$

The period of the pendulum P is given by

$$P = 4 \sqrt{\frac{s}{2g}} \int_0^{\theta_0} \frac{du}{\sqrt{\cos u - \cos \theta_0}} \qquad (4.25)$$

Equation (4.19) enables one to calculate the natural frequency of a conservative oscillatory system whose restoring force is nonlinear. If the integration cannot be performed, numerical or graphical integration may be resorted to in order to obtain the period P.

5 AN OPERATIONAL ANALYSIS OF NONLINEAR DYNAMICAL SYSTEMS

A powerful method of determining the free oscillations of certain nonlinear systems will be given in this section. The method presented here is an operational adaptation of the one developed by Linstedt and Liapounoff.[1]

The method may be illustrated by a consideration of a mechanical oscillating system consisting of a mass attached to a spring. The equation of the free vibration of such a system is

$$\frac{m d^2 x}{dt^2} + F(x) = 0 \qquad (5.1)$$

where $m d^2 x/dt^2$ is the inertia force of the mass, $F(x)$ is the spring force, and x is measured from the position of equilibrium of the mass when the spring is not stressed.   Let us consider the symmetrical case where

$$F(x) = kx + bx^3 \qquad (5.2)$$

Hence (5.1) becomes

$$\frac{m d^2 x}{dt^2} + kx + bx^3 = 0 \qquad (5.3)$$

or

$$\frac{d^2 x}{dt^2} + \omega^2 x + \alpha x^3 = 0 \qquad (5.4)$$

[1] See A. M. Kryloff, *Bulletin of the Russian Academy of Sciences*, no. 1, p. 1, 1933.

where

$$\omega = \left(\frac{k}{m}\right)^{\frac{1}{2}} \tag{5.5}$$

$$\alpha = \frac{b}{m} \tag{5.6}$$

Equation (5.4) occurs in the theory of nonlinear vibrating systems and certain types of nonlinear electrical systems and serves to illustrate the general method of analysis. Let us consider the solution of Eq. (5.4) subject to the initial conditions

$$\left. \begin{array}{l} x = a \\ \dot{x} = 0 \end{array} \right\} \quad \text{at } t = 0 \tag{5.7}$$

That is, the mass is displaced a distance a and allowed to oscillate freely. We are interested in studying the subsequent behavior of the motion.

Let us now multiply Eq. (5.4) by $e^{-st} dt$ and integrate from 0 to ∞. We thus obtain

$$\int_0^\infty e^{-st}\left(\frac{d^2 x}{dt^2} + \omega^2 x + \alpha x^3\right) dt = 0 \tag{5.8}$$

Let us use the notation of Chap. 14 and write

$$Lx(t) = y(s) \tag{5.9}$$

Now, by an integration by parts, it may easily be shown that

$$\frac{L d^2 x}{dt^2} = s^2 y - sx_0 - \dot{x}_0 \tag{5.10}$$

where $\dot{x}_0$ and x_0 are the initial velocity and displacement of the particle at $t = 0$.

In view of the initial conditions (5.7) and the transform (5.10), Eq. (5.8) may be written in the form

$$(s^2 + \omega^2) y = sa - \alpha Lx^3 \tag{5.11}$$

Now let

$$x = x_0 + \alpha x_1 + \alpha^2 x_2 + \alpha^3 x_3 + \cdots \tag{5.12}$$

$$\omega^2 = \omega_0^2 + c_1 \alpha + c_2 \alpha^2 + c_3 \alpha^3 + \cdots \tag{5.13}$$

$$y = y_0 + \alpha y_1 + \alpha^2 y_2 + \alpha^3 y_3 + \cdots \tag{5.14}$$

$$y_i(s) = Lx_i(t) \tag{5.15}$$

In these expressions the quantities $x_r(t)$ are functions of time to be determined, and ω_0 is the frequency, which will be determined later. The c_i quantities are constants which are chosen to eliminate resonance conditions in a manner that will become clear as we proceed. The functions $y_r(s)$ are the Laplace transforms of the functions $x_r(t)$.

In most nonlinear dynamical systems the quantity α is small compared with ω^2, and the series (5.12) may be shown to converge. In the following discussion let us limit our calculations by omitting all the terms containing α to a power higher than the third. Substituting the above expressions into (5.11), we obtain

$$s^2(y_0 + \alpha y_1 + \alpha^2 y_2 + \alpha^3 y_3)$$
$$+ (\omega_0^2 + c_1\alpha + c_2\alpha^2 + c_3\alpha^3)(y_0 + \alpha y_1 + \alpha^2 y_2 + \alpha^3 y_3)$$
$$= sa - L(x_0 + \alpha x_1 + \alpha^2 x_2 + \alpha^3 x_3)^3 \quad (5.16)$$

If we now neglect all terms containing α to powers higher than the third, we obtain

$$(s^2 y_0 + \omega_0^2 y_0) + \alpha(s^2 y_1 + \omega_0^2 y_1 + c_1 y_0 + Lx_0^3)$$
$$+ \alpha^2(s^2 y_2 + \omega_0^2 y_2 + c_2 y_0 + c_1 y_1 + L3x_0^2 x_1)$$
$$+ \alpha^3[s^2 y_3 + \omega_0^2 y_3 + c_3 y_0 + c_2 y_1 + c_1 y_2 + L(3x_0^2 x_2 + 3x_0 x_1^2)]$$
$$= s^2 a \quad (5.17)$$

This equation must hold for any value of the quantity α. This means that each factor for each of the three powers of α must be zero. Hence Eq. (5.17) splits up into the following system of equations:

$$s^2 y_0 + \omega_0^2 y_0 = s^2 a \quad (5.18)$$

$$s^2 y_1 + \omega_0^2 y_1 = -c_1 y_0 - Lx_0^3 \quad (5.19)$$

$$s^2 y_2 + \omega_0^2 y_2 = -c_2 y_0 - c_1 y_1 - L3x_0^3 x_1 \quad (5.20)$$

$$s^2 y_3 + \omega_0^2 y_3 = -c_3 y_0 - c_2 y_1 - c_1 y_2 - L(3x_0^2 x_2 + 3x_0 x_1^2) \quad (5.21)$$

Using the notation

$$T(\phi) = \frac{1}{s^2 + \phi^2} \quad (5.22)$$

Eq. (5.18) may be written in the form

$$y_0 = saT(\omega_0) = Lx_0 \quad (5.23)$$

From the table of transforms (see p. 617), we have

$$L^{-1} saT(\omega_0) = a\cos\omega_0 t = x_0 \quad (5.24)$$

This represents the first approximation to the solution of Eq. (5.4) subject to the initial conditions (5.7). The transform of the second approximation as given by (5.19) may be written in the form

$$y_1 = -c_1 y_0 T(\omega_0) - T(\omega_0) Lx_0^3 \quad (5.25)$$

From the table of transforms we have

$$Lx_0^3 = La^3 \cos^3 \omega_0 t = \frac{a^3}{4} [3sT(\omega_0) + sT(3\omega_0)] \tag{5.26}$$

Substituting (5.23) and (5.26) into (5.25), we obtain

$$y_1 = -sT^2(\omega_0)\left(c_1 a + \frac{3a^3}{4}\right) - \frac{a^3}{4} sT(\omega_0) T(3\omega_0) \tag{5.27}$$

Now, from the table of transforms, it is seen that

$$L^{-1} sT^2(\omega_0) = \frac{t}{2\omega_0} \sin \omega_0 t \tag{5.28}$$

Hence the first term of the right-hand member of (5.27) corresponds to a condition of resonance. We may eliminate this condition of resonance by placing the coefficient of this term equal to zero. Then

$$c_1 a + \frac{3a^3}{4} = 0 \tag{5.29}$$

or

$$c_1 = -\frac{3a^2}{4} \tag{5.30}$$

This determines the constant c_1. With the resonance condition eliminated, (5.27) reduces to

$$y_1 = -\frac{a^3}{4} sT(\omega_0) T(3\omega_0) \tag{5.31}$$

Making use of the table of transforms, we obtain the second approximation

$$x_1 = L^{-1} y_1 = L^{-1}\left[-\frac{a^3}{4} sT(\omega_0) \ T(3\omega_0)\right]$$

$$= \frac{a^3}{32\omega_0^2} (\cos 3\omega_0 t - \cos \omega_0 t) \tag{5.32}$$

If we limit our calculations to the second approximation, we obtain from (5.12), (5.24), and (5.32)

$$x = a \cos \omega_0 t + \frac{\alpha a^3}{32\omega_0^2} (\cos 3\omega_0 t - \cos \omega_0 t) \tag{5.33}$$

The angular frequency is obtained by substituting the value of c_1 given by (5.30) into (5.13). This gives

$$\omega_0^2 = \omega^2 + \tfrac{3}{4} a^2 \alpha \tag{5.34}$$

From this we see that the presence of the term x^3 in the equation introduces a higher harmonic term $\cos 3\omega_0 t$ and the fundamental frequency is not constant but depends on the amplitude a and increases with a provided that the quantity α is positive.

The third approximation is obtained by substituting the above values of y_0, y_1, x_0, and x_1 into (5.20). This gives

$$y_2 = -c_2 saT^2(\omega_0) + \frac{c_1 a^3}{4} T(\omega_0) sT(\omega_0) T(3\omega_0) - T(\omega_0) L(3x_0^2 x_1)$$

$$(5.35)$$

We must now compute

$$L(3x_0^2 x_1) = L\left(\frac{3a^5}{32\omega_0^2} \cos^2 \omega_0 t \cos 3\omega_0 t - \cos^3 \omega_0 t\right) \qquad (5.36)$$

By using the table of transforms we easily obtain

$$L(3x_0^2 x_1) = \frac{3a^5 s}{(4 \times 32) \omega_0^2} [T(5\omega_0) + T(3\omega_0) + T(3\omega_0) - 2T(\omega_0)] \qquad (5.37)$$

Substituting this value of $L(3x_0^2 x_1)$ and making use of the identity

$$T(a) T(b) = \frac{1}{b^2 - a^2} [T(a) - T(b)] \qquad (5.38)$$

we write (5.35) in the form

$$y_2 = sT^2(\omega_0)\left(-c_2 a + \frac{c_1 a^3}{32\omega_0^2} + \frac{3a^5}{64\omega_0^2}\right)$$
$$+ sT(\omega_0) T(3\omega_0)\left[-\frac{c_1 a^3}{32\omega_0^2} - \frac{3a^5}{(4 \times 32) \omega_0^2}\right]$$
$$- sT(\omega_0) T(5\omega_0) \frac{3a^5}{(4 \times 32) \omega_0^2} \qquad (5.39)$$

To eliminate the condition of resonance, we equate the coefficient of the $sT^2(\omega_0)$ term to zero, substituting the value of c_1 into the coefficient. We thus obtain

$$c_2 = \frac{3a^4}{128\omega_0^2} \qquad (5.40)$$

On substituting the value of c_1 into the second member of (5.39), we see that this term vanishes and we have

$$y_2 = -\frac{3a^5}{128\omega_0^2} sT(\omega_0) T(5\omega_0) \qquad (5.41)$$

Using the table of transforms to obtain the inverse transform of y_2, we have the third approximation

$$x_2 = \frac{a^5}{1{,}024\omega_0^4}(\cos 5\omega_0 t - \cos \omega_0 t) \tag{5.42}$$

From (5.12) we thus have the third approximation

$$x = a \cos \omega_0 t + \frac{\alpha a^3}{32\omega_0^2}(\cos 3\omega_0 t - \cos \omega_0 t)$$
$$+ \frac{\alpha^2 a^5}{1{,}024\omega_0^4}(\cos 5\omega_0 t - \cos \omega_0 t) \tag{5.43}$$

where now the fundamental frequency is given by (5.13) as

$$\omega_0^2 = \omega^2 + \tfrac{3}{4}a^2\alpha - \frac{3a^4\alpha^2}{128\omega_0^2} \tag{5.44}$$

The fourth approximation is obtained in the same manner from (5.21). We compute

$$L(3x_0^2 x_2 + 3x_0 x_1^2) = \frac{3a^7 s}{(4 \times 1{,}024)\omega_0^4}[2T(7\omega_0) + T(5\omega_0) - 3T(3\omega_0)] \tag{5.45}$$

by using the table of transforms. Substituting the values of the quantities y_0, y_1, and y_2 as given above into Eq. (5.21) and making use of the relation (5.45), we obtain

$$y_3 = -c_3 saT^2(\omega_0) - \frac{c_2 T(\omega_0) a^3 s}{32\omega_0^2}[T(3\omega_0) - T(\omega_0)]$$
$$- \frac{c_1 T(\omega_0) a^5 s}{1{,}024\omega_0^4}[T(5\omega_0) - T(\omega_0)]$$
$$- \frac{3a^7 s T(w_0)}{(4 \times 1{,}024)\omega_0^4}[2T(7\omega_0) + T(5\omega_0) - 3T(3\omega_0)] \tag{5.46}$$

The condition for no resonance leads to

$$c_3 a = \frac{c_2 a^3}{32\omega_0^2} + \frac{c_1 a^5}{1{,}024\omega_0^4} \tag{5.47}$$

Substituting the values of c_1 and c_2 given by (5.30) and (5.40) into (5.47), we obtain

$$c_3 = 0 \tag{5.48}$$

Suppressing the resonance terms, Eq. (5.46) reduces to

$$y_3 = sT(\omega_0) T(3\omega_0) \frac{3a^7}{2{,}048\omega_0^4} + sT(\omega_0) T(7\omega_0)\left(-\frac{6a^7}{4{,}096\omega_0^4}\right) \tag{5.49}$$

Computing the inverse transform by the use of the table of transforms, we obtain

$$x_3 = \frac{a^7}{32{,}768\omega_0^6}(5\cos\omega_0 t - 3\cos 3\omega_0 t + \cos 7\omega_0 t) \tag{5.50}$$

Substituting x_0, x_1, x_2, x_3 into (5.12), we obtain the fourth approximation

$$x = a\cos\omega_0 t + \frac{\alpha a^3}{32\omega_0^2}(\cos 3\omega_0 t - \cos\omega_0 t)$$

$$+ \frac{\alpha^2 a^5}{1{,}024\omega_0^4}(\cos 5\omega_0 t - \cos\omega_0 t)$$

$$+ \frac{\alpha^3 a^7}{32{,}768\omega_0^6}(5\cos\omega_0 t - 3\cos 3\omega_0 t + \cos 7\omega_0 t) \tag{5.51}$$

To this approximation the fundamental frequency ω_0 is given by (5.13) and is

$$\omega_0^2 = \omega^2 + \tfrac{3}{4}\alpha a^2 - \tfrac{3}{128}\alpha^2 \frac{a^4}{\omega_0^2} \tag{5.52}$$

Now, in all the calculations, terms to higher power than the third have been omitted. We may simplify Eq. (5.52) by substituting on the right-hand side the value of ω_0^2 given by (5.34). We thus obtain

$$\omega_0^2 = \omega^2 + \tfrac{3}{4}\alpha a^2 - \tfrac{3}{128}\alpha^2\left(\frac{a^4}{\omega_0^2 + \tfrac{3}{4}\alpha a^2}\right) \tag{5.53}$$

Expanding the term in parentheses in powers of α and retaining powers of α only up to the third, we have

$$\omega_0^2 = \omega^2 + \tfrac{3}{4}\alpha a^2 - \frac{3\alpha^2 a^4}{128\,\omega^2} + \frac{9\alpha^3 a^6}{512\omega^4} \tag{5.54}$$

Further approximations may be carried out by the same general procedure.

a THE VIBRATIONS OF A PENDULUM

The above analysis may be applied to the study of a theoretical pendulum. The equation of motion of such a pendulum is

$$\ddot{\theta} + \frac{g}{l}\sin\theta = 0 \tag{5.55}$$

where l is the length of the pendulum and g is the gravitational constant. If we develop $\sin\theta$ in a power series in θ and retain only the first two terms of the series, we obtain

$$\ddot{\theta} + \frac{g}{l}\theta - \frac{g}{6l}\theta^3 = 0 \tag{5.56}$$

If we stipulate the initial condition that, at $t = 0$, $\theta = \theta_0$, we may make use of the preceding analysis by letting

$$\theta_0 = a \tag{5.57}$$

$$\omega^2 = \frac{g}{l} \tag{5.58}$$

$$\alpha = -\frac{g}{6l} \tag{5.59}$$

Using the first approximation for the angular frequency as given by (5.34), we have

$$\omega_0^2 = \frac{g}{l} - \frac{g\theta_0^2}{8l} \tag{5.60}$$

To this approximation the period of the oscillations is given by

$$T = \frac{2\pi}{\omega_0} = \frac{2\pi}{[(g/l)(1 - \theta_0^2/8)]^{\frac{1}{2}}} \tag{5.61}$$

For small amplitude θ_0 the radical may be expanded in powers of θ_0, and we may write

$$T = 2\pi \left(\frac{l}{g}\right)^{\frac{1}{2}} \left(1 + \frac{\theta_0^2}{16}\right) \tag{5.62}$$

This formula gives excellent results for small amplitudes of oscillation.

b NONLINEAR ELECTRICAL CIRCUIT

As another example of the method, let us consider the free oscillations of an electrical circuit consisting of an inductance and a capacitor in series. The equation of such a circuit may be written in the form

$$\frac{d\phi}{dt} + \frac{Q}{C} = 0 \tag{5.63}$$

where ϕ is the total flux linking the circuit, Q is the charge on the capacitor, and C is the capacitance of the system. For simplicity we assume that the circuit is devoid of resistance. If the inductance of the system consists of a coil of wire wound on an iron core, we have in general

$$i = F(\phi) \tag{5.64}$$

where i is the current flowing in the circuit. The functional dependence between the current in the circuit and the flux ϕ depends on the functional variation of the flux density B of the material of the inductance coil with the magnetizing force, which is proportional to the current i. This function is many-valued in general. However, it may be taken to be single-valued at large magnetizing forces for Nicalloy, Permalloy, and low-loss steels.

Several analytical expressions have been suggested to express approximately the relation between the magnetizing current i and the total flux ϕ in a ferromagnetic material.   Some of the expressions are

$$i = \sum_{n=1}^{m} a_n \phi^n \tag{5.65}$$

$$i = A \sinh \phi \tag{5.66}$$

$$i = A_0 + \sum_{n=1}^{\infty} A_n \cos n\phi + \sum_{n=1}^{\infty} B_n \sin n\phi \tag{5.67}$$

and other forms may be derived.   In ordinary circuit theory the linear relation

$$i = \frac{1}{L} \phi \tag{5.68}$$

is assumed.   The coefficient L is then the inductance of the system.   In our analysis we shall assume that the relation between i and ϕ is given by

$$i = \frac{1}{L} \phi + \frac{1}{b} \phi^3 \tag{5.69}$$

where the coefficient $1/b$ is a measure of the nonlinearity of the system.

This expression may be regarded as the first two terms of the power series of the relation (5.66) and by a proper choice of the coefficients may be made to approximate the actual i-ϕ curve of a material quite accurately. If we differentiate Eq. (5.63), we obtain

$$\frac{d^2 \phi}{dt^2} + \frac{i}{C} = 0 \tag{5.70}$$

in view of the relation

$$i = \frac{dQ}{dt} \tag{5.71}$$

Substituting the value of i given by (5.69) into (5.70) yields

$$\frac{d^2 \phi}{dt^2} + \frac{1}{LC} \phi + \frac{1}{Cb} \phi^3 = 0 \tag{5.72}$$

If we now let

$$\omega^2 = \frac{1}{LC} \tag{5.73}$$

$$\alpha = \frac{1}{bC} \tag{5.74}$$

$$\phi = x \tag{5.75}$$

then Eq. (5.72) is identical with Eq. (5.4).

Let us now consider the case in which at $t = 0$ there is no current flowing in the system and the capacitor has an initial charge of Q_0. In view of Eqs. (5.63) and (5.69) we therefore have the following initial conditions of the independent variable x:

$$\left. \begin{array}{l} x = 0 \\ \dfrac{dx}{dt} = -\dfrac{Q_0}{C} = v_0 \end{array} \right\} \quad \text{at } t = 0 \tag{5.76}$$

In this case, therefore, we have to solve Eq. (5.4) subject to the initial conditions (5.76). As before, we let y be the transform of x. The transformed equation is in this case

$$(s^2 + \omega^2) y = v_0 - Lx^3 \tag{5.77}$$

To treat this case, we again establish the sequence (5.16) to (5.21), and by equating like powers of α we obtain the set of equations

$$s^2 y_0 + \omega_0^2 y_0 = v_0 \tag{5.78}$$

$$s^2 y_1 + \omega_0^2 y_1 = -c_1 y_0 - Lx_0^3 \tag{5.79}$$

$$s^2 y_2 + \omega_0^2 y_2 = -c_2 y_0 - c_1 y_1 - L(3x_0^2 x_1) \tag{5.80}$$

$$s^2 y_3 + \omega_0^2 y_3 = -c_3 y_0 - c_2 y_1 - c_1 y_2 - L(3x_0^2 x_2 + 3x_0 x_1^2) \tag{5.81}$$

This set of equations is the same as the set (5.18) to (5.21) with the exception of the first one because of the different boundary conditions. The first approximation is obtained from (5.78) by writing

$$y_0 = sv_0 T(\omega_0) = Lx_0 \tag{5.82}$$

From the table of transforms this gives the first approximation

$$x_0 = \frac{v_0}{\omega_0} \sin \omega_0 t \tag{5.83}$$

We now write (5.79) in the form

$$y_1 = T(\omega_0)(-c_1 y_0 - Lx_0^3) \tag{5.84}$$

From the table of transforms we have

$$Lx_0 = \frac{3v_0^3}{4\omega_0^2}[T(\omega_0) - T(3\omega_0)] \tag{5.85}$$

Substituting (5.82) and (5.85) into (5.84), we obtain

$$y_1 = T^2(\omega_0)\left(-c_1 v_0 - \frac{3v_0^3}{4\omega_0^2}\right) + \frac{3v_0^3}{4\omega_0^2} \tag{5.86}$$

$$T(\omega_0) T(3\omega_0) = Lx_1$$

The term $T^2(\omega_0)$ leads to resonance, as may be seen from the table of transforms; hence its coefficient must be set equal to zero.   This yields

$$c_1 = -\frac{3v_0^2}{4\omega_0^2} \tag{5.87}$$

By means of the table of transforms we then obtain

$$x_1 = L^{-1} y_1 = \frac{3v_0^3}{32\omega_0^5} (\sin \omega_0 t - \sin 3\omega_0 t) \tag{5.88}$$

To this approximation we obtain the angular frequency ω_0 from (5.13) as

$$\omega_0^2 = \omega^2 + \frac{3v_0^2 \alpha}{4\omega^2} \tag{5.89}$$

The second approximation is therefore

$$x = \frac{v_0 \sin \omega_0 t}{\omega_0} + \frac{3v_0^3 \alpha}{32\omega_0^5}\left(\sin \omega_0 t - \frac{\sin 3\omega_0 t}{3}\right) \tag{5.90}$$

The variation of the charge on the capacitor may be obtained from (5.63) in the form

$$Q = -C\frac{dx}{dt} = Q_0 \cos \omega_0 t + \frac{3Q_0^3}{32\omega_0^4 bc^3}(\cos \omega_0 t - \cos 3\omega_0 t) \tag{5.91}$$

In view of (5.89) the natural frequency of the system may be written in the form

$$f = \frac{1}{2\pi}\left(\frac{1}{LC} + \frac{3}{4}\frac{Q_0^2 L^2}{bC}\right)^{\frac{1}{2}} \tag{5.92}$$

to the second approximation.   We see that the frequency varies with the initial charge on the capacitor Q_0.

c OSCILLATOR NONLINEARLY DAMPED

As another application of the method in the study of free oscillations, let us apply it to the study of a freely vibrating system whose damping force is proportional to the square of the velocity.   The equation of motion of this system is given by

$$\ddot{x} + \omega^2 x \pm \alpha\dot{x}^2 = 0 \tag{5.93}$$

The minus sign must be taken when the velocity is in the direction of the negative x axis and the plus sign when the velocity is in the direction of the positive x axis.   Let us assume the initial conditions

$$\left.\begin{array}{l} x = a \\ \dot{x} = 0 \end{array}\right\} \quad \text{at } t = 0 \tag{5.94}$$

We thus have to consider the equation

$$\ddot{x} + \omega^2 x - \alpha \dot{x}^2 = 0 \tag{5.95}$$

for the first half of the oscillation. As before, we set up the sequence (5.12) to (5.15), and if we limit our calculations to the terms containing α^2, we obtain the following set of equations:

$$s^2 y_0 + \omega_0^2 y_0 = s^2 a \tag{5.96}$$

$$s^2 y_1 + \omega_0^2 y_1 = -c_1 y_0 + L \dot{x}_0^2 \tag{5.97}$$

$$s^2 y_2 + \omega_0^2 y_2 = -c_1 y_1 - c_2 y_0 + L(2\dot{x}_0 \dot{x}_1) \tag{5.98}$$

From (5.96) we obtain

$$y_0 = s^2 a T(\omega_0) = La \cos \omega_0 t = L x_0 \tag{5.99}$$

Substituting this into (5.96), we get

$$y_1 = -c_1 s^2 a T^2(\omega_0) + \frac{\omega_0^2 a^2}{2} [T(\omega_0) - s^2 T(\omega_0) T(2\omega_0)] \tag{5.100}$$

In order to eliminate resonance, we must choose

$$c_1 = 0 \tag{5.101}$$

Hence

$$y_1 = \frac{\omega_0^2 a^2}{2} \left[T(\omega_0) - \frac{s^2}{3\omega_0^2} T(\omega_0) + \frac{s^2}{3\omega_0^2} T(2\omega_0) \right] \tag{5.102}$$

By means of the table of transforms we obtain

$$x_1 = L^{-1} y_1 = \frac{a^2}{2} - \tfrac{2}{3}a^2 \cos \omega_0 t + \frac{a^2}{6} \cos 2\omega_0 t \tag{5.103}$$

Substituting the various transforms involved into (5.98) gives

$$y_2 = s^2 T^2(\omega_0) \left(-c_2 a + \frac{2a^3 \omega_0^2}{6} \right) - \frac{2a^3 \omega_0^2 s^2}{6} T(\omega_0) T(3\omega_0)$$

$$+ \frac{4a^3 \omega_0^2}{6} s^2 T(\omega_0) T(2\omega_0) - \frac{4a^3 \omega_0^2}{6} T(\omega_0) \tag{5.104}$$

The condition that there be no resonance in this case gives

$$c_2 = \frac{a^2 \omega_0^2}{3} \tag{5.105}$$

From the table of transforms we obtain

$$x_2 = L^{-1} y_2 = -\tfrac{2}{3}a^3 + \frac{a^3}{72} (61 \cos \omega_0 t - 16 \cos 2\omega_0 t + 3 \cos 3\omega_0 t) \tag{5.106}$$

Substituting the above values of x_0, x_1, and x_2 into (5.12), we obtain

$$x = a \cos \omega_0 t + \frac{\alpha a^2}{6} (3 - 4 \cos \omega_0 t + \cos 2\omega_0 t)$$

$$- \frac{\alpha^2 a^3}{72} (48 - 61 \cos \omega_0 t + 16 \cos 2\omega_0 t - 3 \cos 3\omega_0 t) \quad (5.107)$$

Substituting the above values of c_1 and c_2 into (5.13) gives

$$\omega^2 = \omega_0^2 + \frac{\omega_0^2 a^2 \alpha^2}{3} \quad (5.108)$$

or

$$\omega_0 = \frac{\omega}{(1 + \frac{1}{3} a^2 \alpha^2)^{\frac{1}{2}}} \quad (5.109)$$

If we let $T_1/2$ be the time required for the oscillation to reach its maximum negative displacement to the left, we have

$$\frac{T_1}{2} = \frac{\pi}{\omega_0} = \frac{\pi (1 + 1 a^2 \alpha^2)^{\frac{1}{2}}}{\omega} \doteq \frac{\pi}{\omega} (1 + \frac{1}{6} a^2 \alpha^2) \quad (5.110)$$

To determine the maximum negative displacement $+a_1$, we place $\omega_0 t = \pi$ in (5.107), and we obtain

$$+a_1 = -a + \frac{4}{3} \alpha a^2 - \frac{16}{9} \alpha^2 a^3 \quad (5.111)$$

Now, to determine the motion for the next half cycle, we must solve the equation

$$\ddot{x} + \omega^2 x + \alpha \dot{x}^2 = 0 \quad (5.112)$$

subject to the initial conditions

$$\left. \begin{array}{l} x = -a_1 \\ \dot{x} = a \end{array} \right\} \quad \text{at } t = 0 \quad (5.113)$$

However, if we make the change in variable

$$x = -y \quad (5.114)$$

we see that we must then solve the equation

$$y + \omega^2 y - \alpha \dot{y}^2 = 0 \quad (5.115)$$

subject to the conditions

$$\left. \begin{array}{l} y = a_1 \\ \dot{y} = 0 \end{array} \right\} \quad \text{at } t = 0 \quad (5.116)$$

Table of transforms

$$\text{Let } T(\omega) = \frac{1}{s^2 + \omega^2} \qquad \mathscr{L}f(t) = \int_0^\infty e^{-st} F(t)\, dt$$

1. $L \sin \omega t = \omega T(\omega)$

2. $L \sin^2 \omega t = \frac{1}{2}[1 - sT(2\omega)]$

3. $L \sin^3 \omega t = \frac{1}{4}[3\omega T(\omega) - 3\omega T(3\omega)]$

4. $L \cos \omega t = sT(\omega)$

5. $L \cos^2 \omega t = \dfrac{1}{2s}[1 + sT(2\omega)]$

6. $L \cos^3 \omega t = \frac{1}{4}[3sT(\omega) + sT(3\omega)]$

7. $L \sin At \cos Bt = \frac{1}{2}[(A + B)T(A + B) + (A - B)T(A - B)]$

8. $L \cos At \cos Bt = \dfrac{s}{2}[T(A + B) + T(A - B)]$

9. $L \sin At \sin Bt = \dfrac{s}{2}[T(A - B) - T(A + B)]$

10. $L \sin At \sin Bt \sin ct = \dfrac{1}{4}[(A + B - c)T(A + B - c)$
$$+ (B + c - A)T(B + c - A) + (c + A - B)T(c + A - B)$$
$$- (A + B + c)T(A + B + c)]$$

11. $L \sin At \cos Bt \cos ct = \dfrac{1}{4}[(A + B - c)T(A + B - c)$
$$- (B + c - A)T(B + c - A) + (c + A - B)T(c + A - B)$$
$$+ (A + B + c)T(A + B + c)]$$

12. $L \sin At \sin Bt \cos ct = \dfrac{s}{4}[T(B + c - A) - T(A + B - c)$
$$+ T(c + A - B) - T(A + B + c)]$$

13. $L \cos At \cos Bt \cos ct = \dfrac{s}{4}[T(A + B - c) + T(B + c - A)$
$$+ T(c + A - B) + T(A + B + c)]$$

14. $L^{-1} sT(a)T(b) = \dfrac{\cos at - \cos bt}{b^2 - a^2}$

15. $L^{-1} sT^2(\omega) = \dfrac{t \sin \omega t}{2\omega}$

16. $L^{-1} T^2(\omega) = \dfrac{\sin \omega t}{2\omega^3} - \dfrac{t \cos \omega t}{2\omega^2}$

17. $L^{-1} \dfrac{1}{s} T(\omega) = \dfrac{1 - \cos \omega t}{\omega^2}$

18. $T(a)T(b) = \dfrac{1}{b^2 - a^2}[T(a) - T(b)]$

However, this is the same equation whose solution we have just obtained. The only difference is that we have a_1 instead of a in the initial conditions. We thus obtain

$$\frac{T_2}{2} \doteq \frac{\pi}{\omega}(1 + \tfrac{1}{6}a_1^2 \alpha^2) \tag{5.117}$$

for the time required for the second half cycle and

$$a_2 = -a_1 + \tfrac{4}{3}\alpha a_1^2 - \tfrac{16}{9}\alpha^2 a_1^3 \tag{5.118}$$

for the displacement of the system at the end of the second half cycle. It is thus evident that the oscillation has a gradually decreasing amplitude. Its amplitude after the $(n + 1)$st half cycle is given by

$$a_{n+1} = -a_n + \tfrac{4}{3}\alpha a_n^2 - \tfrac{16}{9}\alpha^2 a_n^3 \tag{5.119}$$

6 FORCED VIBRATIONS OF NONLINEAR SYSTEMS

In the last section the *free vibrations* of oscillating systems with nonlinear restoring forces were considered. In this section the *forced* oscillations of systems with nonlinear restoring forces will be considered.

As a typical example of such a system, let us consider the mechanical case of an oscillator of mass m acted upon by a nonlinear elastic restoring force $F(x)$ and by a periodic external force $F_0 \cos \omega t$. The equation of motion of such a system is

$$m\frac{d^2 x}{dt^2} + F(x) = F_0 \cos \omega t \tag{6.1}$$

Let us first discuss the case in which the restoring force is symmetric, that is, has equal magnitude at corresponding points on both sides of the position of equilibrium or position of rest. In this case only *odd* powers may occur in the law of force. Otherwise we have an unsymmetrical law of force and hence an unsymmetric vibration. This is expressed mathematically by the condition

$$F(-x) = F(x) \tag{6.2}$$

Since the methods of analysis and the qualitative results do not depend greatly upon the special form of $F(x)$, we shall choose the following form for the restoring force $F(x)$,

$$F(x) = kx - \delta x^3 \tag{6.3}$$

where $k > 0$.

If $\delta > 0$, it is said that the restoring force corresponds to a *soft* spring, while if $\delta < 0$, the restoring force is said to correspond to a *hard* spring. In the case that $\delta > 0$ the restoring force decreases with the amplitude of oscillation as in the case of a pendulum.   In this case the *natural frequency* decreases with increasing amplitude.

Inserting (6.3) into (6.1), we have the equation of motion

$$m\frac{d^2 x}{dt^2} + kx - \delta x^3 = F_0 \cos \omega t \tag{6.4}$$

This equation is known in the literature as *Duffing's equation* [7]. [1]

Experiments performed on dynamical systems whose equations of motion are of the form (6.4) show that as the time t increases, the motion of the system becomes periodic after some transient motions have died out.   The period of the resulting oscillations is found to have a fundamental frequency of $\omega/2\pi$ and may therefore be represented by a Fourier series in multiples of ω.

The amplitude of the steady state (as $t \to \infty$) may be calculated by the following approximate method.   As a first approximation let us assume

$$x_1 = a \cos \omega t \tag{6.5}$$

where the amplitude a is to be determined.   If we substitute this expression for x in (6.4) and make use of the trigonometric identity

$$\cos^3 \omega t = \tfrac{1}{4}\cos 3\omega t + \tfrac{3}{4}\cos \omega t \tag{6.6}$$

we obtain the equation

$$-ma\omega^2 + ka - \tfrac{3}{4}\delta a^3 - \frac{\delta}{4}a^3 \cos 3\omega t = F_0 \cos \omega t \tag{6.7}$$

If the fundamental vibration is to satisfy Eq. (6.4), we must have

$$\frac{3}{4}\frac{\delta}{m}a^3 + (\omega^2 - \omega_0^2)a + \frac{F_0}{m} = 0 \tag{6.8}$$

where

$$\omega_0 = \sqrt{\frac{k}{m}} \tag{6.9}$$

This is the natural angular frequency of the system in the absence of the nonlinear term.

Equation (6.8) determines the amplitude of the oscillation.   If we divide Eq. (6.8) by ω_0^2, we obtain

$$\frac{3\delta a^3}{4m\omega_0^2} = \left(1 - \frac{\omega^2}{\omega_0^2}\right)a - \frac{F_0}{\omega_0^2 m} \tag{6.10}$$

The roots of this cubic equation in a may be obtained graphically by constructing a y-a coordinate system, as shown in Fig. 6.1.   This figure represents the cubical parabola

$$y = \frac{3\delta a^3}{4m\omega_0^2} \tag{6.11}$$

---

[1] Bracketed numbers are keyed to the References at the end of this chapter.

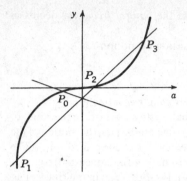

Fig. 6.1

and the straight line

$$y = \left(1 - \frac{\omega^2}{\omega_0^2}\right) a - \frac{F_0}{\omega_0^2 m} \tag{6.12}$$

The possible values of a are the abscissa of the points of intersection of these curves.

If ω is large, the slope of the straight line is negative and there is only one point of intersection P_0. There is also only one point of intersection for $\omega = \omega_0$. If now ω decreases, the straight line rotates until it intersects the cubical parabola at three points P_1, P_2, and P_3. The abscissa of these points corresponds to three possible amplitudes. The amplitude-versus-frequency curve has the form shown in Fig. 6.2.

A more precise analysis[1] shows that if we approach from the low-frequency side, the amplitude corresponding to the lower branch is the stable one. As ω increases, we arrive at the limiting point G. As a continues to increase beyond this point, only the upper branch yields a real point of intersection in Fig. 6.1. It is then seen that with a continuous increase of ω, the amplitude a will suddenly jump from the lower branch to the upper one at G.

---

[1] E. V. Appleton, On the Anomalous Behaviour of a Galvanometer, *Philosophical Magazine*, ser. 6, vol. 47, p. 609, 1924.

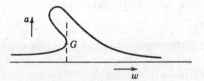

Fig. 6.2

These discontinuities or jumps in amplitude are frequently observed in nonlinear vibration processes both electrical and mechanical.[1]

THE HIGHER APPROXIMATIONS

It will now be shown how the next approximations of the motion are obtained. If Eq. (6.4) is solved for d^2x/dt^2 and if the first approximation (6.5) is substituted in the right-hand member for x, we obtain

$$\frac{d^2 x}{dt^2} = \frac{F_0}{m} \cos \omega t - \frac{ka}{m} \cos \omega t + \frac{3}{4} \frac{\delta a^3}{m} \cos \omega t + \frac{\delta a^3}{4m} \cos 3\omega t \tag{6.13}$$

Making use of (6.8), this reduces to

$$\frac{d^2 x}{dt^2} = -\omega^2 a \cos \omega t + \frac{\delta a^3}{4m} \cos 3\omega t \tag{6.14}$$

Integration gives

$$x_2 = a \cos \omega t - \frac{\delta a^3}{36m\omega^2} \cos 3\omega t = a \cos \omega t - \frac{\delta a^3 \omega_0^2}{36k\omega^2} \cos 3\omega t \tag{6.15}$$

This second approximation may then be substituted into (6.4) to obtain the third approximation. In this manner any number of terms of the Fourier-series solution may be obtained. The investigation of the convergence of the process shows that the series obtained converges if δ is small.

THE CASE OF AN UNSYMMETRIC RESTORING FORCE

If we add a quadratic term to the elastic restoring force so that

$$F(x) = kx + \delta x^2 \tag{6.16}$$

then the vibration becomes unsymmetric since changing the sign of x does not change that of the quadratic term, and hence the restoring force has different values at two points that are symmetric with respect to the origin. The equation of motion is now

$$m\frac{d^2 x}{dt^2} + kx + \delta x^2 = F_0 \cos \omega t \tag{6.17}$$

In this case we assume

$$x_1 = a \cos \omega t + b \tag{6.18}$$

as the first approximation. The constant b is introduced to allow for the lack of symmetry. We insert this approximation into (6.17) and determine a and b in such a way that the constant term and the fundamental vibration satisfy the differential equation.

---

[1] See O. Martienssen, Ueber Neue Resonanzer-scheinungen in Wechselstromkreisen, *Physikalische Zeitung*, vol. 11, p. 448, 1910.

Using the trigonometric identity

$$\cos^2 \omega t = \frac{1 + \cos 2\omega t}{2} \tag{6.19}$$

we obtain the two equations

$$b^2 + \frac{k}{\delta} b + \frac{a^2}{2} = 0 \tag{6.20}$$

and

$$a(\omega^2 - \omega_0^2) - \frac{2\delta ab}{m} + \frac{F_0}{m} = 0 \tag{6.21}$$

If δ is small, we have from (6.20)

$$b = -\frac{\delta a^2}{2k} \tag{6.22}$$

If we now substitute this into (6.21), we have

$$\frac{\delta^2}{km} a^3 + a(\omega^2 - \omega_0^2) + \frac{F_0}{m} = 0 \tag{6.23}$$

This is a cubic equation for the amplitude a. It may be solved graphically, and it is found that under certain conditions it has three roots so that the "jump" phenomenon occurs here as in the case of the symmetrical vibrations. The higher approximations are obtained in the same manner as in the symmetrical case.

SUBHARMONIC RESPONSE

Periodic solutions of the Duffing equation (6.4) have been considered. These solutions have a fundamental period $P = 2\pi/\omega$ equal to the period of the external exciting force. Experiments show that permanent oscillations with a frequency of $\frac{1}{2}, \frac{1}{3}, \ldots, 1/n$ of that of the applied force can occur in nonlinear systems. This phenomenon is called *subharmonic resonance*.

It is known that in linear systems having damping the permanent oscillations of the system have a frequency exactly equal to that of the exciting force, and hence subharmonic resonance is impossible in linear systems. In nonlinear systems, however, even with damping present, the phenomenon of subharmonic resonance is exhibited.

The usual explanation offered of the phenomenon of subharmonic resonance is that the oscillations of a nonlinear system contain higher harmonics in profusion. It is therefore possible that an external force with a frequency the same as one of the higher harmonics may be able to sustain and excite harmonics of lower frequency. This of course requires certain conditions to be true of the system. The mathematical discussion of the problem of subharmonic resonance is a matter of some difficulty.   An interesting discussion of the question will be found in the works of Kármán and Stoker given in the references at the end of this chapter.

As an example of a typical investigation of the possibility of subharmonic response, consider the following nonlinear differential equation:

$$\frac{d^2 x}{dt^2} + w_0^2 x + bx^3 = \frac{F_0}{m} \cos wt \tag{6.24}$$

Assume that a possible periodic solution of this equation has the following form:

$$x = A_0 \cos \frac{wt}{3} \tag{6.25}$$

If this assumed form of the solution is substituted into (6.24), the result may be written in the following form:

$$\left(w_0^2 A_0 - \frac{w^2}{9} A_0 + \tfrac{3}{4}bA_0^3\right) \cos \frac{wt}{3} + \frac{b}{4} A_0^3 \cos wt = \frac{F_0}{m} \cos wt \tag{6.26}$$

For (6.25) to satisfy (6.24) we must therefore have

$$w_0^2 A_0 - w^2 \frac{A_0}{9} + 3b \frac{A_0^3}{4} = 0 \tag{6.27}$$

and

$$\frac{b}{4} A_0^3 = \frac{F_0}{m} \tag{6.28}$$

Hence an oscillation of the type (6.25) with an amplitude of

$$A_0 = \left(\frac{4F_0}{mb}\right)^{\frac{1}{3}} \tag{6.29}$$

is possible provided the angular frequency of the forcing function satisfies the equation

$$w^2 = 9w_0^2 + \tfrac{27}{4}b\left(\frac{4F_0}{mb}\right)^{\frac{2}{3}} \tag{6.30}$$

The *stability* of this solution requires a separate investigation [14].

7 FORCED OSCILLATIONS WITH DAMPING

In many practical problems, particularly in the field of electrical-circuit theory, approximate solutions of nonlinear differential equations must be determined in which the forcing function contains a constant plus a harmonic term and the circuit contains damping.   The following equation is typical of this class of differential equations:

$$L\frac{di}{dt} + Ri + S_0 q + aq^3 = E_0 + E_m \sin(wt + \theta) \tag{7.1}$$

This equation governs the oscillations of an electrical circuit consisting of an inductance L and a resistance R, both of which are linear, in series with a nonlinear capacitor. S_0 is the constant *initial elastance* of this capacitor, and a is a constant which is the measure of the departure of the capacitor from linearity. In practical circuits a is a small positive number. The functions i and q are the current of the circuit and the charge separation on the plates of the capacitor, respectively. The functions are related by the differential equation

$$i = \frac{dq}{dt} \tag{7.2}$$

In order to obtain an approximate steady-state solution by Duffing's method, a periodic solution of the following form is assumed for (7.1),

$$i = I_m \sin wt \tag{7.3}$$

$$q = Q_0 - \frac{I_m}{w} \cos wt \tag{7.4}$$

where Q_0 is the constant charge accumulation on the plates of the capacitor and I_m is the maximum amplitude of the alternating current of the circuit. Substitute these expressions into (7.1), and let

$$F(t) = \left(L\frac{di}{dt} + Ri + S_0 q + aq^3\right) i = I_m \sin wt \qquad q = Q_0 - \frac{I_m}{w} \cos wt$$

$$= \left(S_0 Q_0 + aQ_0^3 + \frac{3aQ_0 I_m^2}{2w^2}\right) + RI_m \sin wt$$

$$+ \left(wLI_m - S_0\frac{I_m}{w} - \frac{3aI_m^3}{4w^3} - 3aQ_0^2\frac{I_m}{w}\right) \cos wt$$

$$+ \frac{3Q_0 aI_m^2}{2w^2} \cos 2wt - \frac{aI_m^3}{4w^3} \cos 3wt \tag{7.5}$$

With this notation Eq. (7.1) becomes

$$F(t) = E_0 + E_m \cos \theta \sin wt + E_m \sin \theta \cos wt \tag{7.6}$$

The undetermined coefficients I_m and Q_0 and the phase angle θ may be adjusted to make (7.3) and (7.4) an approximate solution of (7.1) by equating the constant term and the coefficients of $\sin wt$ and $\cos wt$ to the right-hand member of (7.6). This procedure leads to the following three equations:

$$S_0 Q_0 + aQ_0\left(Q_0^2 + \frac{3I_m^2}{2w^2}\right) = E_0 \tag{7.7}$$

$$RI_m = E_m \cos \theta \tag{7.8}$$

$$I_m\left(wL - \frac{S_0}{w}\right) - 3a\frac{I_m}{w}\left(Q_0^2 + \frac{I_m^2}{4w^2}\right) = E_m \sin \theta \tag{7.9}$$

The above equations are three simultaneous equations for the determination of the unknowns I_m, Q_0, and the phase angle θ. If numerical values for the parameters involved are available, a solution of these equations may be effected by a graphical procedure similar to that outlined in Sec. 6. In practical applications, however, considerable simplification may be introduced because the parameter a is a very small number. Since a is small, the second term of (7.7) may be neglected and the following approximate value for Q_0 obtained,

$$Q_0 = \frac{E_0}{S_0} = C_0 E_0 \qquad (7.10)$$

where $C_0 = 1/S_0$ is the *initial capacitance* of the nonlinear capacitor. If the direct potential E_0 is large in comparison with the maximum value of the harmonic potential so that $E_0 \gg E_m$, then, for the frequencies ordinarily used in practice, we have

$$Q_0^2 \gg \frac{I_m^2}{4w^2} \qquad (7.11)$$

If the term $I_m^2/4$ is neglected in (7.9), this equation may be written in the form

$$I_m \left[wL - \frac{1}{w}(S_0 + 3aC_0^2 E_0^2) \right] = E_m \sin \theta \qquad (7.12)$$

Hence, if (7.8) and (7.12) are squared and the results added, we obtain

$$I_m^2 \left\{ R^2 + \left[wL - \frac{1}{w}(S_0 + 3aC_0^2 E_0^2) \right]^2 \right\} = E_m^2 \qquad (7.13)$$

and the amplitude of the alternating current of the system is given by

$$I_m = \frac{E_m}{\sqrt{R^2 + [wL - (1/w)(S_0 + 3aC_0^2 E_0^2)]^2}} \qquad (7.14)$$

It is thus apparent that the effect of the direct potential is to increase the *effective elastance* of the system by an amount $3aC_0^2 E_0^2$. The tangent of the phase angle θ may be determined by the division of (7.12) by (7.8). This procedure yields

$$\tan \theta = \frac{wL - (1/w)(S_0 + 3aC_0^2 E_0^2)}{R} \qquad (7.15)$$

The principal harmonics of the current i may be obtained by writing (7.1) in the form

$$Ri = E_0 + E_m \sin(wt + \theta) - \left(L\frac{di}{dt} + S_0 q + aq^3 \right) \qquad (7.16)$$

and substituting (7.3) and (7.4) into the right-hand member of (7.16). If this is done and Eqs. (7.7), (7.8), and (7.9) used to simplify the resulting expression, the result yields the following value for the current i:

$$i = I_m \sin wt + \frac{aI_m^3}{4w^3 R} \cos 3wt - \frac{3aQ_0 I_m^2}{2w^2 R} \cos 2wt \qquad (7.17)$$

Higher-order harmonics may be obtained by substituting (7.17) into the right-hand member of (7.16) and repeating the procedure. It may be noted that, if $a = 0$, the system becomes a linear one and the denominator of (7.14) reduces to the ordinary impedance of a linear series circuit.

8 SOLUTION OF NONLINEAR DIFFERENTIAL EQUATIONS BY INTEGRAL EQUATIONS

A method for solving nonlinear differential equations based on the theory of nonlinear integral equations will now be described. The general features of the method are clearly demonstrated by applying it to a series electrical circuit containing a nonlinear element. Such a circuit is depicted in Fig. 8.1.

This circuit consists of a linear component whose operational impedance is $Z(D)$, where $D = d/dt$, the time derivative operator, in series with a nonlinear component. It will be assumed that the potential drop across the nonlinear component is a function $F(i)$ of the current that flows through it. The differential equation that expresses the potential balance of the circuit may be written in the following symbolic form:

$$Z(D)i(t) + F[i(t)] = e(t) \qquad (8.1)$$

where $e(t)$ is the applied potential and $i(t)$ is the circuit current.

In order to formulate Eq. (8.1) as an integral equation, let the Laplace transform be introduced:

$$\mathscr{L}i(t) = \int_0^\infty e^{-st} i(t)\, dt = I(s) \qquad (8.2)$$

$$\mathscr{L}e(t) = E(s) \qquad (8.3)$$

$$\mathscr{L}F[i(t)] = G(s) \qquad (8.4)$$

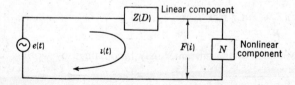

Fig. 8.1

Let it be assumed that initially the circuit is at rest so that, at $t = 0$, $i(0) = 0$ and $q(0) = 0$. That is, the initial current i and the charge separation q of the circuit are both zero at $t = 0$. With these initial conditions the Laplace transform of Eq. (8.1) is given by

$$Z(s) I(s) + G(s) = E(s) \tag{8.5}$$

or

$$I(s) = \frac{E(s)}{Z(s)} - \frac{G(s)}{Z(s)} \tag{8.6}$$

For simplicity, introduce the notation

$$H(s) = \frac{1}{Z(s)} \tag{8.7}$$

With this notation (8.6) may be written in the following form:

$$I(s) = E(s) H(s) - G(s) H(s) \tag{8.8}$$

Let the inverse transform of $H(s)$ be $h(t)$; that is,

$$L^{-1} H(s) = L^{-1} \frac{1}{Z(s)} = h(t) \tag{8.9}$$

The current of the circuit $i(t)$ may be obtained by taking the inverse Laplace transform of Eq. (8.8) in the form

$$i(t) = L^{-1} E(s) H(s) - L^{-1} G(s) H(s) \tag{8.10}$$

If the Faltung, or superposition, theorem (see Chap. 4, Sec. 3) is applied to the right-hand member of Eq. (8.10), the following result is obtained:

$$i(t) = \int_0^t h(t - u) e(u) \, du - \int_0^t h(t - u) F[i(u)] \, du \tag{8.11}$$

This is a nonlinear integral equation of the type studied by T. Lalesco.[1]

If $F(i) = 0$, so that the nonlinear component of the circuit is not present, then Eq. (8.11) reduces to

$$i_0(t) = \int_0^t h(t - u) e(u) \, du \tag{8.12}$$

This is the current that would flow in the circuit of Fig. 8.1 in the absence of the nonlinear component. Thus it is seen that Eq. (8.11) may be written in the form

$$i(t) = i_0(t) - \int_0^t h(t - u) F[i(u)] \, du \tag{8.12a}$$

[1] See T. Lalesco, "Introduction à la théorie des équations intégrales," Gauthier-Villars & Cie, Paris, 1912.

Lalesco has shown that the limit of the infinite sequence of functions

$$i_0(t) = \int_0^t h(t-u)\,e(u)\,du$$

$$i_1(t) = i_0(t) - \int_0^t h(t-u)\,F[i_0(u)]\,du$$

$$i_2(t) = i_0(t) - \int_0^t h(t-u)\,F[i_1(u)]\,du \qquad (8.13)$$

. . . .

$$i_{n+1}(t) = i_0(t) - \int_0^t h(t-u)\,F[i_n(u)]\,du$$

is the solution of the integral equation (8.11).   That is,

$$i(t) = \lim_{n\to\infty} i_n(t) \qquad (8.14)$$

is the desired solution.

The investigation of the convergence of the sequence (8.13) is given in the treatise of Lalesco.   The results of this investigation indicate that the sequence converges rapidly in applications of this method of solution to special cases arising from physical problems.

SOLUTION OF THE INTEGRAL EQUATION BY AN OPERATIONAL PROCEDURE

A convenient operational procedure that has been found useful in the calculation of the sequence of functions $i_n(t)$ of (8.13) is based on the inverse Laplace transform of (8.6), which may be written in the form

$$i(t) = L^{-1}\,I(s) = L^{-1}\frac{E(s)}{Z(s)} - L^{-1}\frac{LF[i(t)]}{Z(s)} \qquad (8.15)$$

The current $i_0(t)$ is given by

$$i_0(t) = L^{-1}\frac{E(s)}{Z(s)} \qquad (8.16)$$

Hence (8.15) may be written in the form

$$i(t) = i_0(t) - L^{-1}\frac{LF[i(t)]}{Z(s)} \qquad (8.17)$$

The sequence of functions $i_1(t),\ i_2(t),\ \ldots,\ i_n(t)$ of (8.13) may be conveniently expressed by the following equations:

$$i_1(t) = i_0(t) - L^{-1}\frac{LF[i_0(t)]}{Z(s)}$$

$$i_2(t) = i_0(t) - L^{-1}\frac{LF[i_1(t)]}{Z(s)} \qquad (8.18)$$

. . . .

$$i_{n+1}(t) = i_0(t) - L^{-1}\frac{LF[i_n(t)]}{Z(s)}$$

In practice, it is usually found that the sequence $i_1(t)$, $i_2(t)$, . . . may be more simply computed by (8.18) than by (8.13) because a table of suitable Laplace transforms greatly assists in effecting the required integrations [16].

As a simple example of the method, let it be required to solve the following nonlinear differential equation:

$$L\frac{di}{dt} + Ri - k\frac{d}{dt}i^3 = E \tag{8.19}$$

This equation occurs in the study of the transient analysis of a nonlinear inductor with constant excitation.   In this special case we have

$$Z(D) = LD + R \qquad F(i) = k\frac{d}{dt}i^3 \qquad D = \frac{d}{dt} \tag{8.20}$$

Hence

$$H(p) = \frac{1}{Z(s)} = \frac{1}{Ls + R} \qquad h(t) = L^{-1}H(s) = \frac{e^{-bt}}{L} \qquad b = \frac{R}{L} \tag{8.21}$$

The integral equation satisfied by the current in the circuit is in this case

$$i(t) = i_0(t) - \frac{k}{L}\int_0^t e^{b(u-t)} Di^3(u)\,du \tag{8.22}$$

where

$$i_0(t) = \frac{L^{-1}E}{Ls + R} = \frac{E(1 - e^{-bt})}{R} \tag{8.23}$$

The operational form of the current (8.17) is

$$i(t) = \frac{E}{R}(1 - e^{-bt}) - kL^{-1}\frac{L(Di^3)}{Ls + R} \tag{8.24}$$

The second approximation for the current is given by

$$i_1(t) = \frac{E}{R}(1 - e^{-bt}) - kL^{-1}\frac{L(Di_0^3)}{Ls + R} \tag{8.25}$$

Carrying out the indicated operations with the aid of a table of Laplace transforms, the following result is obtained:

$$i_1(t) = \frac{E}{R}(1 - e^{-bt}) - \frac{kE^3}{LR^3}(3bt\,e^{-bt} - \tfrac{9}{2}e^{-bt} + 6e^{-2bt} - \tfrac{3}{2}e^{-3bt}) \tag{8.26}$$

The higher approximations may be obtained by computing more functions of the sequence (8.18).   If the coefficient of the nonlinear term is small, the sequence converges rapidly and the second or third approximations usually give the accuracy required.

9   THE METHOD OF KRYLOFF AND BOGOLIUBOFF

The mathematical analysis of many important technical problems leads to differential equations of the following general form:

$$\ddot{x} + \mu f(x,\dot{x}) + w^2 x = 0 \tag{9.1}$$

where the parameter μ is a small constant and $f(x,\dot{x})$ is a given function of x and its derivative.

Equations of this type are called *quasi-linear differential equations*. The Russian scientists Kryloff and Bogoliuboff have devised a method for the approximate solution of equations of this type [6].   This method is similar in principle to the familiar *method of variation of parameters*, introduced by Lagrange in 1774.[1]   The Kryloff-Bogoliuboff method is based on the fact that, if $\mu = 0$ in (9.1), this equation becomes the equation of simple harmonic motion

$$\ddot{x} + w^2 x = 0 \tag{9.2}$$

whose solution is

$$x = a \sin(wt + \phi) \tag{9.3}$$

where a and ϕ are arbitrary constants.   The derivative of (9.3) is

$$\dot{x} = wa \cos(wt + \phi) \tag{9.4}$$

Kryloff and Bogoliuboff attempt to find a solution of (9.1) that preserves the form (9.3) and has a derivative of the form (9.4) but in which a and ϕ are now functions of the time $a(t)$ and $\phi(t)$, so that the solution of (9.1) now has the form

$$x(t) = a(t) \sin[wt + \phi(t)] \tag{9.5}$$

If (9.5) is differentiated, the following result is obtained:

$$\dot{x}(t) = aw \cos(wt + \phi) + \dot{a} \sin(wt + \phi) + a\dot{\phi} \cos(wt + \phi) \tag{9.6}$$

For $\dot{x}$ to retain the form (9.4), the sum of the last two terms of (9.6) must be equal to zero, so that

$$\dot{a} \sin(wt + \phi) + a\dot{\phi} \cos(wt + \phi) = 0 \tag{9.7}$$

This ensures that the derivative of x will have the form

$$\dot{x} = wa(t) \cos[wt + \phi(t)] \tag{9.8}$$

If (9.8) is differentiated, the result is

$$\ddot{x} = -w^2 a \sin(wt + \phi) + w\dot{a} \cos(wt + \phi) - wa\dot{\phi} \sin(wt + \phi) \tag{9.9}$$

---

[1] See E. L. Ince, "Ordinary Differential Equations," pp. 122–123, Longmans, Green & Co., Ltd., London, 1927.

Introduce the notation $\theta = wt + \phi$ and substitute (9.9) into (9.1). The result of this operation gives

$$w\dot{a}\cos\theta - wa\dot{\phi}\sin\theta = -\mu f(a\sin\theta, wa\cos\theta) \tag{9.10}$$

Equations (9.7) and (9.10) may be written in the following form:

$$\frac{da}{dt}\sin\theta + a\frac{d\phi}{dt}\cos\theta = 0 \tag{9.11}$$

$$\frac{da}{dt}\cos\theta - a\frac{d\phi}{dt}\sin\theta = \frac{-\mu}{w}f(a\sin\theta, wa\cos\theta) \tag{9.12}$$

These two equations may be solved for da/dt and $d\phi/dt$ simultaneously, with the following results:

$$\frac{da}{dt} = -\frac{\mu}{w}f(a\sin\theta, wa\cos\theta)\cos\theta \tag{9.13}$$

$$\frac{d\phi}{dt} = \frac{\mu}{wa}f(a\sin\theta, wa\cos\theta)\sin\theta \tag{9.14}$$

These equations give the variations of the amplitude a and the phase ϕ with the time t. So far in the analysis no approximations have been made, and therefore (9.13) and (9.14) are *exact* equations.

THE KRYLOFF AND BOGOLIUBOFF APPROXIMATION

Since da/dt and $d\phi/dt$ are proportional to the small quantity μ, it follows that these derivatives are small and that therefore $a(t)$ and $\phi(t)$ are slowly varying functions of the time t. Therefore, in the time $T = 2\pi/w$, $\theta = wt + \phi$ will increase approximately by 2π, while a and ϕ have not changed appreciably. In order to compute da/dt and $d\phi/dt$, Kryloff and Bogoliuboff replace the right-hand members of (9.13) and (9.14), as a first approximation, by their *average values* over a range of 2π in θ. The amplitude a is *regarded as a constant* in taking the average. This procedure leads to the following equations:

$$\frac{da}{dt} = -\frac{\mu}{2\pi w}\int_0^{2\pi} f(a\sin\theta, aw\cos\theta)\cos\theta\, d\theta \tag{9.15}$$

$$\frac{d\phi}{dt} = \frac{\mu}{2\pi wa}\int_0^{2\pi} f(a\sin\theta, aw\cos\theta)\sin\theta\, d\theta \tag{9.16}$$

The exact equations (9.13) and (9.14) are thus replaced by the approximate ones (9.15) and (9.16). The approximations involved in these equations are physically reasonable. A justification of these approximations and methods of higher approximations will be found in the book by Kryloff and Bogoliuboff. The general manner in which various terms in the function $f(x,\dot{x})$ affect the amplitude $a(t)$ and the phase $\phi(t)$ of the oscillation may be deduced from (9.15) and (9.16). If $f(x,\dot{x})$ contains only powers of x, such as x^2, x^3, . . . , the amplitude a is not affected because the integral in (9.15) vanishes. These terms affect the function $\phi(t)$, since the integral in (9.16) does not vanish. If, on the other hand, $f(x,\dot{x})$ contains terms of the form $\dot{x}$, $\dot{x}^2$, $\dot{x}^3$, etc., these terms affect the amplitude a but not the phase function $\phi(t)$.

10 APPLICATIONS OF THE METHOD OF KRYLOFF AND BOGOLIUBOFF

In this section several applications of the Kryloff-Bogoliuboff procedure to effect the approximate solutions of typical quasi-linear differential equations of technical importance will be given.

a THE ANHARMONIC OSCILLATOR

Let the differential equation of the anharmonic oscillator of Sec. 5 be written in the following form,

$$\frac{d^2 x}{dt^2} + w^2 x + \mu x^3 = 0 \tag{10.1}$$

where μ is a small parameter. This equation is of the quasi-linear form (9.1) with

$$\mu f(x,\dot{x}) = \mu x^3 \tag{10.2}$$

Hence (9.15) and (9.16) in this case are given by

$$\frac{da}{dt} = -\frac{\mu}{2\pi w} \int_0^{2\pi} a^3 \sin^3 \theta \cos \theta \, d\theta = 0 \tag{10.3}$$

and

$$\frac{d\phi}{dt} = \frac{\mu}{2\pi wa} \int_0^{2\pi} a^3 \sin^4 \theta \, d\theta = \frac{3a^2 \mu}{8w} \tag{10.4}$$

Therefore, since $da/dt = 0$, the amplitude does not change with time. The phase function $\phi(t)$ can be determined by integrating (10.4) and thus obtaining

$$\phi = 3\mu \frac{a^2 t}{8w} + \phi_0 \qquad \phi_0 = \text{an arbitrary constant} \tag{10.5}$$

Hence the solution (9.5) is, in this case,

$$x = a \sin\left[wt \left(1 + 3\frac{\mu a^2}{8w^2} \right) + \phi_0 \right] \tag{10.6}$$

The period of the oscillations T is given by

$$T = \frac{2\pi}{w} \left(1 + 3\frac{\mu a^2}{8w^2} \right)^{-1} \doteq \frac{2\pi}{w} \left(1 - 3\frac{\mu a^2}{8w^2} \right) \qquad \text{for } \mu \ll 1 \tag{10.7}$$

Therefore, if μ is *positive*, the period *decreases* with increasing amplitude, and if μ is *negative*, the period *increases* with increasing amplitude.

b THE QUASI-LINEAR VAN DER POL EQUATION

The differential equation

$$\frac{d^2 x}{dt^2} - \mu(1 - x^2)\frac{dx}{dt} + x = 0 \qquad 0 < \mu \ll 1 \tag{10.8}$$

is a very important equation in the theory of vacuum-tube oscillators and is known in the literature as *van der Pol's equation*.[1] This equation is of the form (9.1) with

$$\mu f(x,\dot{x}) = \mu(x^2 - 1)\frac{dx}{dt} \qquad w^2 = 1 \tag{10.9}$$

In this case (9.15) and (9.16) reduce to

$$\frac{da}{dt} = \frac{\mu a}{2\pi}\int_0^{2\pi}(1 - a^2\sin^2\theta)\cos^2\theta\, d\theta = \frac{\mu a}{2}\left(1 - \frac{a^2}{4}\right) \tag{10.10}$$

and

$$\frac{d\phi}{dt} = \frac{\mu}{2\pi}\int_0^{2\pi}(a^2\sin^2\theta - 1)\cos\theta\sin\theta\, d\theta = 0 \tag{10.11}$$

From (10.10) it can be seen that, after the amplitude has reached a steady state so that $da/dt = 0$, then $a = 2$. In order to determine the growth of the amplitude a, we may write (10.10) in the form

$$\frac{da}{a(1 - a^2/4)} = \frac{\mu\, dt}{2} \tag{10.12}$$

If this differential equation is integrated with the stipulation that $a = a_0$ when $t = 0$, then, after some algebraic reductions, the result may be written in the following form:

$$a = \frac{a_0 e^{\mu t/2}}{\sqrt{1 + (a_0^2/4)(e^{\mu t} - 1)}} \tag{10.13}$$

This solution indicates that, if $a_0 = 0$, then $a = 0$ and no oscillation will develop. However, if $a_0 > 0$, no matter how small, the amplitude of the oscillation will develop so that

$$\lim_{t\to\infty} a(t) = 2 \tag{10.14}$$

---

[1] See B. van der Pol, Nonlinear Theory of Electric Oscillations, *Proceedings of the Institute of Radio Engineers*, vol. 22, p. 1051, 1934.

If $a_0 > 2$, then the amplitude will *decay* so that (10.14) is satisfied. Since by (10.11) $d\phi/dt = 0$, we have

$$\phi = \phi_0 \tag{10.15}$$

and the solution (9.5) becomes

$$x(t) = \frac{a_0 \, e^{\mu t/2} \sin(t + \phi_0)}{\sqrt{1 + (a_0^2/4)(e^{\mu t} - 1)}} \tag{10.16}$$

When the parameter μ in the van der Pol equation is large, $\mu \gg 1$, the solutions of the equation are known as *relaxation oscillations*. These oscillations are discussed in Sec. 13.

c NEGATIVELY DAMPED OSCILLATOR

In a fundamental paper Lord Rayleigh[1] considered the motion of an electrically driven tuning fork by studying the following equation,

$$\frac{d^2 x}{dt^2} - 2k \frac{dx}{dt} + c\dot{x}^3 + w^2 x = 0 \tag{10.17}$$

where k and c are small positive constants. The term $-2k\dot{x}$ represents an effective negative resistance which is equivalent to a driving force. In this case, we have

$$\mu f(x,\dot{x}) = -2k\dot{x} + c\dot{x}^3 \tag{10.18}$$

Equations (9.15) and (9.16) now reduce to

$$\frac{da}{dt} = \frac{2k}{2\pi w} \int_0^{2\pi} aw \cos^2\theta \, d\theta - \frac{c}{2\pi w} \int_0^{2\pi} a^3 w^3 \cos^4\theta \, d\theta$$
$$= a(k - \tfrac{3}{8}cw^2 a^2) \tag{10.19}$$

$$\frac{d\phi}{dt} = -\frac{2k}{2\pi wa} \int_0^{2\pi} aw \cos\theta \sin\theta \, d\theta + \frac{c}{2\pi wa} \int_0^{2\pi} a^3 w^3 \cos^3\theta \sin\theta \, d\theta$$
$$= 0 \tag{10.20}$$

As a consequence of (10.19), it may be seen that, when a steady state has been reached so that $da/dt = 0$, we must have

$$k - \tfrac{3}{8}cw^2 a^2 = 0 \tag{10.21}$$

or

$$a = \frac{2}{w} \left(\frac{2k}{3c}\right)^{\frac{1}{2}} = a_s \tag{10.22}$$

[1] See Lord Rayleigh, On Maintained Vibrations, *Philosophical Magazine*, vol. 15, p. 229, 1883.

The equation for the amplitude (10.19) may be written in the following form:

$$\frac{da}{ak(1 - g^2 a^2)} = dt \qquad \text{where } g^2 = \frac{3cw^2}{8k} \tag{10.23}$$

If this differential equation is solved subject to the condition that $a = a_0$ at $t = 0$, then, after some algebraic reductions, the result may be written in the form

$$a(t) = \frac{a_0 e^{kt}}{\sqrt{1 + g^2 a_0^2(e^{2kt} - 1)}} \tag{10.24}$$

From this it may be seen that, if $a_0 = 0$, there is no motion. However, if the motion is started so that $a_0 \neq 0$, the amplitude attains a final steady-state value of a_s given by (10.22). Since $d\phi/dt = 0$ by (10.20), we have $\phi = \phi_0$, where ϕ_0 is arbitrary. The motion is therefore given by

$$x(t) = \frac{a_0 e^{kt} \sin(wt + \phi_0)}{\sqrt{1 + g^2 a_0^2(e^{2kt} - 1)}} \tag{10.25}$$

Additional examples of applications of this method will be found in the book by Kryloff and Bogoliuboff.

11 TOPOLOGICAL METHODS: AUTONOMOUS SYSTEMS

If the nonlinear differential equation under consideration is of the second order and does not contain the independent variable t *explicitly*, much information concerning the properties of the solution may be obtained by a geometrical procedure. Dynamical systems whose equations of motion do not contain the time explicitly are called *autonomous* systems. The equations of motions of second-order autonomous systems usually have the following form:

$$\ddot{x} + \phi(\dot{x}) + f(x) = 0 \tag{11.1}$$

If the velocity $y = \dot{x}$ of the system is introduced as another dependent variable, the differential equation (11.1) may be reduced to the following first-order system of equations:

$$\frac{dy}{dt} = -[\phi(y) + f(x)] \tag{11.2}$$

$$\frac{dx}{dt} = y \tag{11.3}$$

This is a special case of the more general system

$$\frac{dy}{dt} = Q(x,y) \tag{11.4}$$

$$\frac{dx}{dt} = P(x,y) \tag{11.5}$$

where $Q(x,y)$ and $P(x,y)$ are functions of x and y. In order to study Eq. (11.1) topologically, the quantities x and y can be considered as cartesian coordinates in a certain (x,y) plane called the *phase plane*. If (11.2) is divided by (11.3), the following equation is obtained:

$$\frac{dy}{dx} = -\frac{[\phi(y)+f(x)]}{y} \tag{11.6}$$

This differential equation specifies a definite curve in the phase plane. This curve is called the *phase trajectory*, or the *trajectory*, of the differential equation (11.1). The equation

$$\frac{dy}{dx} = \frac{Q(x,y)}{P(x,y)} \tag{11.7}$$

specifies the phase trajectory of the system (11.4). A point (x_0,y_0) of the phase plane for which

$$P(x_0,y_0) = Q(x_0,y_0) = 0 \tag{11.8}$$

is called a *singular point*. The points of the phase plane that *do not* have the property (11.8) are called *ordinary points*. It is convenient to regard the derivatives (11.4) and (11.5) as the y and x components of the velocity of a point in the phase plane. This point is called the *representative* point of the system, and its component velocities are given by

$$v_x = \frac{dx}{dt} = P(x,y) \qquad v_y = \frac{dy}{dt} = Q(x,y) \tag{11.9}$$

At a singular point we have $v_x = 0$ and $v_y = 0$, so that such a point may be regarded as a position of equilibrium with zero velocity. For purposes of illustration, it is convenient to obtain the phase trajectory of the differential equation of a linear harmonic oscillator having inertia m and elastic spring constant k. The differential equation of such an oscillator is

$$m\frac{d^2 x}{dt^2} + kx = 0 \qquad \frac{dx}{dt} = y \tag{11.10}$$

In this case Eq. (11.6) takes the following form:

$$\frac{dy}{dx} = -\frac{kx}{my} \qquad \text{or} \qquad my\,dy + kx\,dx = 0 \tag{11.11}$$

If this equation is integrated, the result is

$$\frac{my^2}{2} + \frac{kx^2}{2} = K \tag{11.12}$$

The arbitrary constant K can be determined by requiring that $x = x_0$ at $y = y_0$. This fixes the total energy E of the oscillator, and K is given by

$$K = \frac{my_0^2}{2} + \frac{kx_0^2}{2} = E \tag{11.13}$$

Hence (11.12) may be written in the form

$$my^2 + kx^2 = 2E \tag{11.14}$$

or

$$\frac{x^2}{2E/k} + \frac{y^2}{2E/m} = 1 \tag{11.15}$$

This equation has the form

$$\frac{x^2}{a^2} + \frac{y^2}{b^2} = 1 \tag{11.16}$$

the well-known equation for an ellipse with its center at the origin and a semimajor axis $a = (2E/k)^{\frac{1}{2}}$ and a semiminor axis $b = (2E/m)^{\frac{1}{2}}$ as shown in Fig. 11.1.   The phase trajectory of the differential equation (11.10) is a definite ellipse for a definite amount of total energy E.   For different values of total energy (11.11) fixes an entire family of ellipses.   The size of the ellipses depends on the total energy E of the system; large values of E determine large ellipses. The equation $y = \dot{x}$ indicates that the representative point P traverses the phase trajectory in a *clockwise* direction.   If $x = x_0$, $y = y_0$ at $t = 0$, the time integral of the differential equation (11.10) is

$$\left.\begin{array}{l} x = \dfrac{y_0}{w} \sin wt + x_0 \cos wt \\[2mm] y = y_0 \cos wt - wx_0 \sin wt \end{array}\right\} \qquad w = \left(\frac{k}{m}\right)^{\frac{1}{2}} \tag{11.17}$$

These are the parametric equations of the ellipse (11.15).   It can be seen from the periodicity of the solution (11.17) that the representative point completes one revolution of the ellipse in time

$$T = 2\pi \left(\frac{m}{k}\right)^{\frac{1}{2}} = \frac{2\pi}{w}$$

The time required for the representative point to perform one revolution may be obtained also by writing $dx/dt = y$ or $dt = dx/y$ and integrating this expression in the form

$$T = \oint \frac{dx}{y} = 4 \int_0^a \frac{dx}{\sqrt{(2E/m - w^2 x)\,dx}} = \frac{2\pi}{w} \tag{11.18}$$

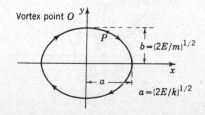

Vortex point O

$b = (2E/m)^{1/2}$

$a = (2E/k)^{1/2}$

Fig. 11.1

The significance of the circular arrow in (11.18) is that the line integral must be taken about a closed path. The center of the ellipse, $x = 0$, $y = 0$, is a singular point of Eq. (11.10). The possible phase trajectories are ellipses that enclose this point. A singular point of this type is called a *vortex point*. It is noted that all the phase trajectories enclose the vortex point and none approaches it. The vortex point is a position of stable equilibrium of the system.

APERIODIC MOTION

The equation of a mass being repulsed from the origin by a force proportional to its distance from the origin is

$$m\frac{d^2 x}{dt^2} = cx \quad \text{or} \quad \frac{d^2 x}{dt^2} = a^2 x \quad a = \left(\frac{c}{m}\right)^{\frac{1}{2}} \tag{11.19}$$

The equation for the phase trajectories is in this case

$$\frac{dy}{dx} = a^2 \frac{x}{y} \tag{11.20}$$

If this equation is integrated, the result is

$$y^2 - a^2 x^2 = h \qquad h = \text{an arbitrary constant} \tag{11.21}$$

For various values of the constant h, (11.21) is the equation for a family of *hyperbolas*. If it is required that $x = x_0$ and $y = y_0$ at $t = 0$, the solution of the differential equation (11.19) takes the following form:

$$x = x_0 \cosh at + \frac{y_0}{a} \sinh at \tag{11.22}$$

$$y = ax_0 \sinh at + y_0 \cosh at$$

These are the parametric equations of the hyperbolic phase trajectories (11.21), shown in Fig. 11.2. The family of hyperbolas (11.21) has two asymptotes $y = \pm ax$. The origin O is a singular point of a special type called a *saddle point*. In this case there are two special trajectories that pass through the saddle point. The motion of the representative point along the trajectories

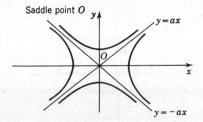

Fig. 11.2

that approach the saddle point is asymptotic with the time t. The trajectories in the neighborhood of a saddle point represent the possible motions that occur in the neighborhood of a point of *unstable equilibrium*.

THE MOTION OF A DAMPED OSCILLATOR

The differential equation of a linear oscillator with damping is

$$\frac{d^2 x}{dt^2} + 2h\frac{dx}{dt} + w_0^2 x = 0 \qquad y = \dot{x} \tag{11.23}$$

The equation for the phase trajectories of this oscillator is

$$\frac{dy}{dx} = -\frac{2hy + w_0^2 x}{y} \tag{11.24}$$

If $h^2 < w_0^2$, the oscillator performs damped oscillations. Let

$$w = (w_0^2 - h^2)^{\frac{1}{2}} \tag{11.25}$$

The general solution of (11.23) is

$$x = Ke^{-ht}\cos(wt + \theta)$$
$$y = \dot{x} = -Ke^{-ht}[h\cos(wt + \theta) + w\sin(wt + \theta)] \tag{11.26}$$

where K and θ are arbitrary constants.

These equations are the parametric equations of a family of spirals, one of which is depicted in Fig. 11.3. The origin O is a singular point of a type called a *focal point*. The representative point spirals around the phase plane and approaches the focal point at the origin in an asymptotic manner. The focal point is a point of *stable* equilibrium. Since as the representative point approaches the origin it spirals an infinite number of times about it, it is clear that it does not approach O with a *definite direction*.

THE MOTION OF AN OVERDAMPED OSCILLATOR

If the frictional term of the damped oscillator is large, the parameter h of (11.23) satisfies the inequality $h^2 > w_0^2$ and the motion is no longer oscillatory. Let

$$g = (h^2 - w_0^2)^{\frac{1}{2}} \tag{11.27}$$

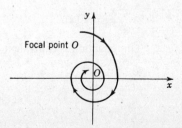

Focal point O

Fig. 11.3

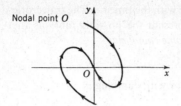

Nodal point O

Fig. 11.4

In this case the general solution of (11.23) may be expressed in the form

$$x = A e^{-ht} \cosh (bt + \theta)$$

$$y = \dot{x} = A e^{-ht} [b \sinh (bt + \theta) - h \cosh (bt + \theta)]$$

(11.28)

where A and θ are arbitrary constants.

These are the parametric equations of the phase trajectories. A typical trajectory is depicted in Fig. 11.4. The origin, $x = 0$, $y = 0$, is a singular point called a *nodal point*. The phase trajectories are now such that the representative point moves toward the nodal point with a *definite* direction. The four common types of singular points are summarized in Table 1:

Table 1   Basic types of singular points

Name	Type of motion	Stability	Approach
Vortex point	Oscillatory	Stable	None
Saddle point	Aperiodic	Unstable	Only along asymptotes
Focal point	Damped oscillatory	Stable	With no definite direction
Nodal point	Aperiodic	Stable	With a definite direction

12   NONLINEAR CONSERVATIVE SYSTEMS

The motions of nonlinear conservative systems that have one degree of freedom may be effectively studied by topological methods. As an example of the application of this method to nonlinear systems of this type, consider the motion of a mass m that is attracted toward a fixed point, the origin, by a nonlinear restoring force $F(x)$. The equation of motion of such a mass is

$$m \frac{d^2 x}{dt^2} = -F(x)$$

(12.1)

This equation is equivalent to the first-order differential equations

$$\frac{dy}{dt} = -\frac{F(x)}{m}$$

(12.2)

$$\frac{dx}{dt} = y$$

(12.3)

If (12.2) is divided by (12.3), we obtain the following equation for the phase trajectory of the motion of the nonlinear oscillator:

$$\frac{dy}{dx} = -\frac{F(x)}{my} \qquad (12.4)$$

If the variables are separated and the resulting equation is integrated, the result may be written in the form

$$\frac{m}{2}y^2 + \int_{x_0}^{x} F(x)\,dx = C \qquad (12.5)$$

The arbitrary constant may be evaluated by requiring that, at $t = 0$, $x = x_0$ and $y = y_0$. If this is done, (12.5) becomes

$$\frac{m}{2}y^2 + \int_{x_0}^{x} F(x)\,dx = \frac{m}{2}y_0^2 \qquad (12.6)$$

The quantity $(m/2)y^2$ is the *kinetic* energy of the system, and

$$V(x) = \int_0^x F(x)\,dx \qquad (12.7)$$

is the *potential* energy of the system. In terms of $V(x)$, (12.6) may be written in the form

$$\frac{m}{2}y^2 + V(x) = \frac{m}{2}y_0^2 + \int_0^{x_0} F(x)\,dx = E \qquad (12.8)$$

where E is the *total energy* of the system. Hence (12.8) may be solved for the velocity y in the form

$$y = \sqrt{\frac{2}{m}[E - V(x)]} \qquad (12.9)$$

This equation can be given an interesting topological interpretation. In order to fix the ideas involved, let

$$F(x) = k\sin x \qquad (12.10)$$

where k is a positive constant, to take a special case. The potential energy is in this case given by

$$V(x) = \int_0^x k\sin(x)\,dx = k(1 - \cos x) = 2k\sin^2\frac{x}{2} \qquad (12.11)$$

In order to study the topology of the motion, it is convenient to plot the potential energy $V(x)$ against x. Directly below this diagram the phase-plane trajectory is plotted as shown in Fig. 12.1. The upper part of Fig. 12.1 is a plot of $V(x) = 2k\sin^2(x/2)$ against x. The three horizontal lines on this diagram represent three distinct values of *total energy* E_1, E_s, and E_2. The

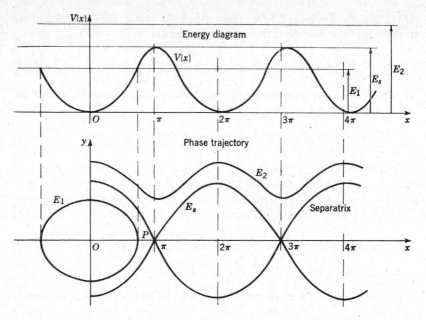

Fig. 12.1

lower diagram is a plot of three different phase trajectories corresponding to the three values of the total energy given above.   The phase trajectories are obtained by plotting $y = \sqrt{(2/m)[E - V(x)]}$ for the three given values of the total energy E.   It may be noted that if $k/m = g/s$ then Fig. 12.1 represents the phase trajectories of the equation of the simple pendulum discussed in Sec. 3.

The physical significance of the phase trajectories of Fig. 12.1 can be clearly visualized by regarding it as the phase-plane plot of the simple pendulum.   The phase trajectory for E_1 represents the oscillations of a pendulum having less energy than is necessary for it to go "over the top." The phase trajectory corresponding to E_s is the one in which the system has just enough total energy to take it to the top at $x = 2\pi$.   This phase trajectory is called the *separatrix* because it separates the motions of one type from those of an entirely different type.   The point P is a *saddle point;* this is a point at which the potential energy $V(x)$ is a *maximum* and is a point of *unstable equilibrium*.   This point corresponds to the inverted position of the pendulum and has entirely different properties than the point O, which is a *vortex point* and corresponds to a position of *stable equilibrium*, the bottom position of the pendulum.

If the pendulum is given a total energy E_2 which exceeds the critical energy E_s to take it "over the top," then the phase trajectory is a curve that continues with x and the motion is no longer periodic. In this case the pendulum continues to revolve about its point of support with *maximum* angular velocity at $x = 0$. 2π, 4π, etc., and *minimum* angular velocity at $x = \pi$, 3π, 5π, etc., when the pendulum passes the top position. If the motion is periodic so that the phase trajectory is a closed path, the period T of the oscillations may be obtained by the equation

$$T = \oint \frac{dx}{y} = \frac{m}{2} \oint \frac{dx}{\sqrt{E - V(x)}} \qquad (12.12)$$

where the line integral (12.12) must be taken around the closed trajectory.

The simple discussion given above illustrates the power of the topological aspects of the phase trajectories to give general qualitative information on the behavior of autonomous systems. If the system being studied contains damping of a negative or positive type, the phase trajectories may be rather complicated but a great deal of information may be obtained by their construction.

13 RELAXATION OSCILLATIONS

A great many periodic oscillatory processes in nature take place in such a manner that external energy is supplied to the system over part of a period and dissipated within the system in another part of the period. The name *relaxation oscillations* has been given to periodic processes of this type. As a simple example of a relaxation oscillation, consider the electrical circuit depicted in Fig. 13.1.

This circuit consists of a battery of potential E connected in series with a resistor R and a capacitor C. A diode tube D is connected across the capacitor. This tube has the property that, when the potential across its plates exceeds a value E_0, it suddenly becomes conducting and discharges the capacitor completely. After this has happened, the capacitor begins to charge again and the procedure repeats itself.

The charge on the capacitor, q, satisfies the differential equation

$$R\frac{dq}{dt} + \frac{q}{C} = E \qquad (13.1)$$

If we assume $q = 0$ at $t = 0$, the solution of this equation is

$$q = CE(1 - e^{-t/RC}) \qquad (13.2)$$

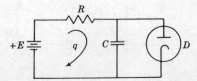

Fig. 13.1

The potential $e(t)$ across the plates of the capacitor is

$$E(t) = \frac{q}{C} = E(1 - e^{-t/RC}) \tag{13.3}$$

If we let T be the time required for the potential $e(t)$ to attain the breakdown potential of the diode, we have

$$E_0 = E(1 - e^{-t/RC}) \tag{13.4}$$

or

$$T = RC \ln \frac{E}{E - E_0} \tag{13.5}$$

At this value of time the capacitor is discharged, and the process repeats itself with a period of T. The graph of the periodic variation of the charge on the capacitor q is given by Fig. 13.2. It is thus seen that the period T of the oscillations of the diode oscillator of Fig. 13.1 is not determined by inertia and elastance parameters but by some form of *relaxation time*. There are many types of relaxation times, such as the time necessary for a current to build up in a system containing self-inductance, a diffusion time, heating-up time, etc. It is typical in the case of relaxation oscillations for *aperiodic* phenomena to repeat themselves periodically. This automatic periodic reoccurrence of the aperiodic phenomena is closely related to the presence of some form of energy source, such as the battery of the diode oscillator.

Another interesting example of a relaxation oscillation is the phenomenon involved in the action of the aeolian harp. In this case the oscillation of the system is by the formation of air eddies that develop alternately on one side and the other of the harp string. As each eddy is formed, it diffuses away and allows another eddy to be formed. The time period of the oscillation of the harp string is determined by a diffusion or relaxation time and does not depend on the natural period of the string when it is performing sinusoidal oscillations.

In a fundamental paper van der Pol[1] has enumerated many other instances of relaxation oscillations, such as a pneumatic hammer, the scratching noise of a knife on a plate, the waving of a flag in the wind, the

[1] See B. van der Pol, Relaxation Oscillations, *Philosophical Magazine*, vol. 2, p. 928, 1926.

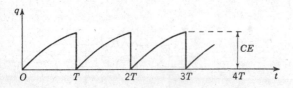

Fig. 13.2

humming noise of a water tap, the squeaking of a door, the oscillations of a steam engine with a flywheel that is too small, the multivibrator, the periodic sparks of a Whimshurst machine, the periodic reoccurrence of epidemics and economic crises, the fluctuation of populations of animal species, the motion of sleeping flowers, the periodic reoccurrence of showers behind a depression, the shivering from cold, and the beating of the heart.

By suitable changes in variables van der Pol has shown that the mathematical analyses of many relaxation phenomena lead to a study of the following differential equation:

$$\frac{d^2 x}{dt^2} - \mu(1 - x^2)\frac{dx}{dt} + x = 0 \tag{13.6}$$

This equation has been the subject of considerable investigation and is known as *van der Pol's equation*. The approximate solution of this equation has been studied by the method of Kryloff and Bogoliuboff in Sec. 10. It was there shown that, for *small* values of the parameter μ, the solution is a sustained *sinusoidal* oscillation. These sinusoidal oscillations exist for $0 < \mu \ll 1$, and in this case (13.6) is a quasi-linear differential equation.

It is of interest to consider the nature of the solution of (13.6) for *large* values of μ so that $\mu \gg 1$. The physical significance of (13.6) is facilitated if it is regarded as the equation of motion of an oscillator of unit mass and unit spring constant and variable damping. If μ is large, the initial damping is large and *negative*; this makes the system *highly aperiodic*. This aperiodic system with large negative resistance tends to run away from zero without any oscillations whatever. However, as soon as the momentary value of x^2 becomes larger than unity, the negative character of the resistance ceases to exist and the resistance becomes positive. This makes the amplitude x decrease aperiodically. It is thus seen that for large values of μ the system has the tendency initially to jump away from $x = 0$ to a positive value of x, decreasing afterward gradually, and suddenly to jump to a negative value of x, etc. The variation of x with time is thus a relaxation oscillation. This qualitative discussion will now be examined topologically.

14 PHASE TRAJECTORIES OF THE VAN DER POL EQUATION

The parametric equations of the phase trajectories of the van der Pol equation (13.6) are

$$\frac{dy}{dt} = \mu(1 - x^2)y - x \tag{14.1}$$

$$\frac{dx}{dt} = y \tag{14.2}$$

If (14.1) is divided by (14.2), the equation of the phase trajectory is obtained. This equation has the form

$$\frac{dy}{dx} = \mu(1 - x^2) - \frac{x}{y} \qquad (14.3)$$

In his paper[1] van der Pol plotted the phase-trajectory curves for three different values of the parameter μ ($\mu = 0.1$, $\mu = 1.0$, $\mu = 10.0$). The phase trajectories corresponding to these values of μ are reproduced in Figs. 14.1, 14.2, and 14.3.

It can be seen by examining the figures of the phase trajectories that, no matter where the motion begins, the trajectories converge to closed curves. These closed curves represent periodic motions with constant amplitude. For small values of μ, for example, $\mu = 0.1$, the closed curve is only slightly different from a circle. However, as μ is increased, the closed curves differ quite radically from circles and thus represent periodic motions that are distinctly nonharmonic.

[1] See B. van der Pol, Relaxation Oscillations, *Philosophical Magazine*, vol. 2, p. 928, 1926.

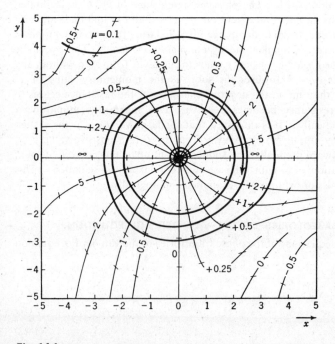

Fig. 14.1

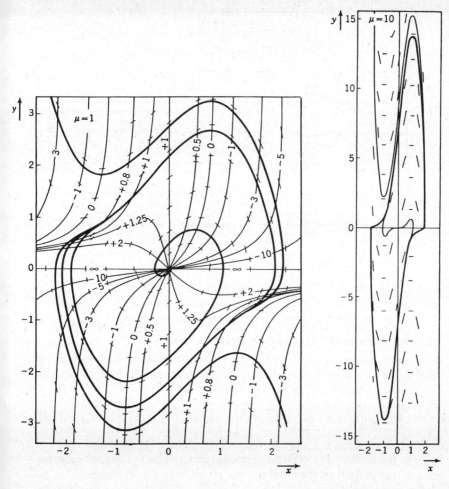

Fig. 14.2   **Fig. 14.3**

Figure 14.3 is the phase trajectory of a relaxation oscillation ($\mu = 10$), and in this case the motion is made up of sudden transitions between deflections of opposite sign. The amplitudes of the oscillations corresponding to the various values of μ are given in Fig. 14.4. In this figure the manner in which the oscillations build up for $\mu = 0.1$, 1.0, and 10.0 is given.

The final values of the oscillations for $\mu = 0.1$, 1.0, and 10.0 are also clearly illustrated by these graphs. The graph of the oscillation for $\mu = 10.0$ is a typical relaxation oscillation; it differs radically from the sinusoidal oscillation for $\mu = 0.1$.

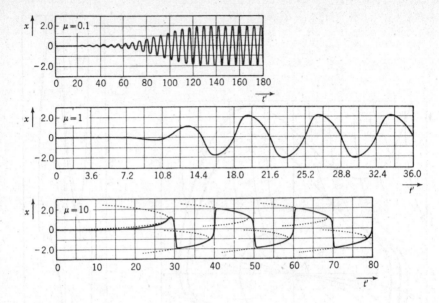

Fig. 14.4

15 THE PERIOD OF RELAXATION OSCILLATIONS

An approximate expression for the period of relaxation oscillations may be obtained by a graphical procedure.   For this purpose it is more convenient to consider the following differential equation:

$$\frac{d^2 x}{dt^2} + \mu\left(\frac{\dot{x}^3}{3} - \dot{x}\right) + x = 0 \tag{15.1}$$

This equation is known in the literature as the *Rayleigh equation*.   It may be shown that, if (15.1) is differentiated with respect to t and the substitution $u = dx/dt$ is made, the result is

$$\frac{d^2 u}{dt^2} + \mu(u^2 - 1)\frac{du}{dt} + u = 0 \tag{15.2}$$

Hence (15.1) is seen to be equivalent to the van der Pol equation (15.2), and will have oscillations of the same period.

The parametric equations of the phase trajectories of the Rayleigh equation are

$$\frac{dy}{dt} = \mu\left(y - \frac{y^3}{3}\right) - x \tag{15.3}$$

$$\frac{dx}{dt} = y \tag{15.4}$$

The equation of the phase trajectories is obtained by dividing (15.3) by (15.4). It is

$$\frac{dy}{dx} = \frac{\mu(y - y^3/3) - x}{y} \tag{15.5}$$

It is convenient to introduce the variable $z = x/\mu$ and write (15.5) in the form

$$\frac{dy}{dz} = \frac{\mu^2(y - y^3/3 - x)}{y} \tag{15.6}$$

and to consider the form taken by the periodic phase trajectory in the yz plane given by (15.6). If the parameter μ is *large*, it is seen from (15.6) that the slope $dy/dz = \pm M$, where M is a large number unless

$$z = y - \frac{y^3}{3} \tag{15.7}$$

and in this case $dy/dz = 0$ and the phase trajectory is horizontal.

The nature of the phase trajectories of (15.6) for large values of μ ($\mu \gg 1$) may be explained by an examination of Fig. 15.1. In this diagram the curve C_0 is a graph of $z = y - y^3/3$. Let it be supposed that the representative point of a phase trajectory begins to move from P. Since (15.6) gives a large negative value for dy/dz, the phase trajectory will be nearly vertical and the point P will move down in a vertical direction until it reaches C_0. On C_0 the dy/dz is zero, but since the field direction at all points other than those on C_0 is nearly vertical, the point P will follow the curve C_0 until B is reached. At B the representative point will move vertically upward until C_0 is reached at C. After reaching C the representative point will follow C_0 until the point D is reached, and it will then move downward and meet C_0 again at A.

By this type of reasoning, one can see that, for large values of μ, it does not matter where the motion starts and that eventually the representative

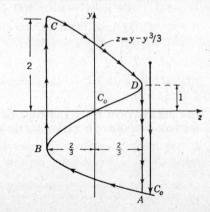

Fig. 15.1

point moves along the closed trajectory $ABCDA$ and the motion is periodic. To compute the period, the formula

$$T = \oint \frac{dx}{y} = \mu \oint \frac{dz}{y} \tag{15.8}$$

may be used. In order to evaluate this line integral, it may be noticed that the vertical-straight-line portions of the cycle $ABCDA$ make no contribution to T. Because of symmetry, (15.8) may be evaluated by integrating from C to D and taking twice the value of this integral to obtain T. From C to D we have $z = y - y^3/3$ or $dz = 1 - y^2$. Hence

$$T = \mu \oint \frac{dz}{y} = 2\mu \int_C^D \frac{dz}{y}$$

$$= 2\mu \int_{y=2}^{y=1} \left(\frac{1}{y} - y \right) dy = 2\mu \left(\ln y - \frac{y^2}{2} \right) \Big|_2^1$$

$$= 2\mu \left(\frac{y^3}{2} - \ln y \right) \Big|_1^2 = \mu(3 - 2\ln 2) = 1.612\mu \tag{15.9}$$

This is the approximate formula for the period of the relaxation oscillations for *very large* values of μ given by van der Pol. For very small values of μ we have seen that the period is independent of μ and given by 2π. The value $T = 1.612\mu$ is based on the assumption that μ is very large, and it gives a good estimate of the length of the period for moderately large values of μ. However, even for the case $\mu = 10$, Eq. (15.9) gives results that are about 20 percent too low. A better approximation to the period has been given by Shohat,[1] whose result is

$$T = \frac{2\pi}{S(\mu)} \tag{15.10}$$

where

$$S(\mu) = \frac{1}{1 + \mu} + \frac{\mu}{(1 + \mu)^2} + \frac{15\mu^2}{16(1 + \mu)^3} + \frac{13\mu^3}{16(1 + \mu)^4} + \cdots$$

It is interesting to note that for very large values of μ (15.9) gives

$$\lim_{\mu \to \infty} \left(\frac{2\pi\mu}{T} \right) = 3.89 \cdots$$

while (15.10) gives 3.75 for the same expression.

16   RELAXATION OSCILLATIONS OF A MOTOR-GENERATOR COMBINATION

A very interesting and spectacular relaxation oscillator occurs in electrical engineering. This oscillator consists of a separately excited direct-current

---

[1] J. Shohat, On van der Pol's and Related Nonlinear Differential Equations, *Journal of Applied Physics*, vol. 15, no. 7, pp. 568–574, July, 1944.

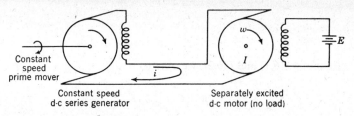

Fig. 16.1

motor operating with no load, whose current is supplied by a direct-current generator that has a series field as shown in Fig. 16.1.

If the generator of this combination is driven at constant speed by a prime mover and the circuit parameters are adjusted properly, it is found that the rotation of the motor will reverse periodically in such a manner that its angular velocity of rotation w will execute *relaxation oscillations*. In order to explain the action of this oscillator, consider the potential E_g of the generator. Since the machine is driven at *constant* speed by the prime mover, E_g is proportional to the strength of the magnetic field of the generator. Because the generator has a *series* field, E_g is given approximately by

$$E_g = C_1 i - C_2 i^3 \tag{16.1}$$

where C_1 and C_2 are positive constants. The cubic term in (16.1) represents the influence of the saturation of the iron of the generator field. The generated potential also satisfies the following equation:

$$E_g = L\frac{di}{dt} + Ri + C_3 w \tag{16.2}$$

The first term of (16.2) represents the inductive potential drop in the circuit of Fig. 16.1. i is the generator current, and L is the effective inductance of the field coils of the generator. Ri is the resistive drop of the circuit, and the term $C_3 w$ represents the counterelectromotive force of the separately excited direct-current motor. w is the angular velocity of the motor, and C_3 is a positive constant. The power supplied to the motor is

$$P_m = iC_3 w \tag{16.3}$$

or the product of the current and the counterelectromotive force. If I is the moment of inertia of the rotating parts of the motor, its kinetic energy of rotation is $Iw^2/2$. Since it is assumed that the motor is running with no load, the power being furnished to it is expended in changing its kinetic energy, so that

$$P_m = C_3 wi = \frac{d}{dt}\left(\frac{Iw^2}{2}\right) = Iw\frac{dw}{dt} \tag{16.4}$$

or

$$\frac{dw}{dt} = \frac{C_3 i}{I} \tag{16.5}$$

Equating (16.1) and (16.2), we obtain

$$E_g = L\frac{di}{dt} + Ri + C_3 w = C_1 i - C_2 i^3 \tag{16.6}$$

If (16.6) is differentiated with respect to t and dw/dt eliminated by (16.5), the result is

$$L\frac{d^2 i}{dt^2} - (C_1 - R - 3C_2 i^2)\frac{di}{dt} + \frac{C_3^2 i}{I} = 0 \tag{16.7}$$

Let the following parameters be introduced:

$$w_n = C_3(LI)^{-\frac{1}{2}} \tag{16.8}$$

and

$$\mu = \frac{(C_1 - R)(I/L)^{\frac{1}{2}}}{C_3} \tag{16.9}$$

Then, if the following changes in variables are made in (16.7),

$$x = i\left(\frac{3C_2}{C_1 - R}\right)^{\frac{1}{2}} \tag{16.10}$$

and

$$\tau = w_n t \tag{16.11}$$

this equation is transformed into the van der Pol equation

$$\frac{d^2 x}{d\tau^2} - \mu(1 - x^2)\frac{dx}{d\tau} + x = 0 \tag{16.12}$$

In the usual installation the value of μ given by (16.9) is much larger than unity, and the value (15.9) may be used to estimate the period of the oscillation. If this is done, the period of the oscillation governed by (16.7) is seen to be given by

$$T = \frac{1.612(C_1 - R)I}{C_3^2} \tag{16.13}$$

so that the period of the reversal of the current, and hence the period of the angular velocity $w(t)$, is proportional to the inertia I of the motor and *not* to the *square root* of the inertia, as would be the case if the oscillations were harmonic.

THE REVERSION METHOD FOR SOLVING NONLINEAR DIFFERENTIAL EQUATIONS†

17   INTRODUCTION

It is the purpose of this discussion to present a method that is applicable to the solution of nonlinear differential equations of a certain class.   The method under consideration is based on an algebraic procedure that reduces nonlinear differential equations to a system of linear differential equations.   When the methods of the Laplace-transform theory are applied to the solution of the resulting system of differential equations, the method saves a great deal of numerical effort.   This method was suggested to the author by A. C. Sim, formerly of the British Thomson-Houston Company, Ltd.   Other procedures leading to the same general ideas have been published.[1-3]

The author has had occasion to use the reversion method in the study of certain physical problems whose mathematical formulation leads to nonlinear differential equations.   It has been found that the method indicates the complexity of the solution before much labor has been expended.   In general, the solution of the nonlinear differential equations is expressed in the form of a combination of polynomials and exponential functions.

18   GENERAL DESCRIPTION OF THE METHOD

Let the nonlinear differential equation whose solution is desired be written in the form

$$a_1 y + a_2 y^2 + \cdots + a_5 y^5 + \cdots = k\phi(t) \tag{18.1}$$

where t is the independent variable; y, the desired dependent variable; k, a constant; $\phi(t)$, a given function of t.   The a_i coefficients are, in general, operators, and are functions of the operator, D, where

$$D = \frac{d}{dt} \quad \text{and} \quad a_1 \neq 0 \tag{18.2}$$

Assume a solution of the form

$$y = A_1 k + A_2 k^2 + A_3 k^3 + \cdots \tag{18.3}$$

† Reprinted from the *Journal of Applied Physics*, vol. 23, no. 2, pp. 202–207, February, 1952. Copyright 1952 by the American Institute of Physics.
[1] L. A. Pipes, *Journal of the Acoustical Society*, vol. 10, p. 29, July, 1938.
[2] L. A. Pipes, *Journal of Applied Physics*, vol. 13, p. 117, 1942.
[3] C. S. Roys, *Proceedings of the National Electronics Conference*, p. 663, 1947.

The A_i coefficients may be determined by substituting (18.3) into (18.1) and equating coefficients of equal powers of k. If this is done, the following equations for the coefficients, A_1, A_2, A_3, etc., are obtained:

$$A_1 = \frac{\phi(t)}{a_1} \tag{18.4}$$

$$A_2 = -\frac{a_2 A_1^2}{a_1} \tag{18.5}$$

$$A_3 = -\frac{1}{a_1}[2a_2 A_1 A_2 + a_3 A_1^3] \tag{18.6}$$

$$A_4 = -\frac{1}{a_1}[a_2(A_2^2 + 2A_1 A_3) + 3a_3 A_1^2 A_2 + a_4 A_1^4] \tag{18.7}$$

$$A_5 = -\frac{1}{a_1}[2a_2(A_1 A_4 + A_2 A_3) + 3a_3(A_1 A_2^2 + A_1^2 A_3)$$
$$+ 4a_4 A_1^3 A_2 + a_5 A_1^5] \tag{18.8}$$

$$A_6 = -\frac{1}{a_1}[a_2(A_3^2 + 2A_1 A_5 + 2A_2 A_4) + a_3(A_2^3 + 3A_1^2 A_4 + 6A_1 A_2 A_3)$$
$$+ 2a_4(2A_1^2 A_2^2 + 2A_1^3 A_3) + 5a_5 A_1^4 A_2 + a_6 A_1^6] \tag{18.9}$$

$$A_7 = -\frac{1}{a_1}[2a_2(A_1 A_6 + A_2 A_5 + A_3 A_4)$$
$$+ 3a_3(A_1^2 A_5 + A_2^2 A_3 + A_1 A_3^2 + 2A_1 A_2 A_4)$$
$$+ 4a_4(A_1 A_2^3 + A_1^3 A_4 + 3A_1^2 A_2 A_3) + 5a_5(2A_1^3 A_2^2 + A_1^4 A_3)$$
$$+ 6a_6 A_1^5 A_2 + a_7 A_1^7] \tag{18.10}$$

Additional coefficients may be found in Van Orstrand's paper,[1] in which a list of the first thirteen coefficients is given.

19 EXAMPLES ILLUSTRATING THE METHOD

The general procedure of the method of revision may best be illustrated by applying it to several special cases.

Example I   Consider the differential equations

$$\frac{dy}{dt} - y^2 = 1 \qquad \text{with the initial condition that } y = 0 \text{ at } t = 0 \tag{19.1}$$

If this is compared with the general equation, (18.1), it is seen that we must have

$$a_1 = \frac{d}{dt} = D \tag{19.2}$$

[1] C. E. Van Orstrand, *Philosophical Magazine*, vol. 19, p. 366, 1910.

and

$$a_2 = -1 \qquad k\phi(t) = 1 \qquad k = 1 \tag{19.3}$$

Hence from (18.4) we see that

$$A_1 = \frac{1}{a_1} \qquad \text{or} \qquad DA_1 = 1 \tag{19.4}$$

This is a differential equation that determines the function A_1.

Now

$$\frac{dA_1}{dt} = 1 \qquad \text{or} \qquad A_1 = t + \text{const} \tag{19.5}$$

In the method of reversion, the initial conditions are impressed on the function A_1 as if it were the entire solution. Hence the initial condition $y = 0$ at $t = 0$ leads to

$$A_1 = t \tag{19.6}$$

The function A_2 is determined from Eq. (18.5) and is given by

$$A_2 = -\frac{a_2 A_1^2}{a_1} \tag{19.7}$$

or

$$\frac{dA_2}{dt} = t^2 \tag{19.8}$$

This integrates into

$$A_2 = \frac{t^3}{3} \tag{19.9}$$

The constant of integration is zero because the subsequent functions $A_2, A_3, \ldots$, etc., are determined by the condition that they have zero initial values. Equation (18.6) enables the function A_3 to be determined. This is given by

$$A_3 = -\frac{1}{a_1} - \frac{2t^4}{3} \tag{19.10}$$

or

$$\frac{dA_3}{dt} = \frac{2t^5}{3} \tag{19.11}$$

Hence

$$A_3 = \frac{2t^5}{15} \tag{19.12}$$

Similarly, the other terms, $A_4, A_5, \ldots$, may be determined. The solution is given by (18.3), and in this case the result

$$y = t + \frac{t^3}{3} + \frac{2t^5}{15} + \cdots \tag{19.13}$$

is obtained.

The original differential equations (19.1) may be written in the form

$$\frac{dy}{dt} = 1 + y^2 \qquad \text{or} \qquad \frac{dy}{1 + y^2} = dt \tag{19.14}$$

Hence

$$\tan -1y = t \tag{19.15}$$

or

$$y = \tan t \qquad \text{when } y = 0 \text{ at } t = 0 \tag{19.16}$$

The series (19.13) may be recognized as the Taylor's series expansion of the function $\tan t$.

Example II Let it be required to solve the following differential equation:

$$\frac{dy}{dt} + \alpha y^2 = t \tag{19.17}$$

with the initial condition that $y = y_0$ at $t = 0$. The reversion procedure is simplified if the following change of variable is made:

$$v = y - y_0 \qquad \text{or} \qquad y = v + y_0 \tag{19.18}$$

This simplifies the initial condition in the variable v, giving $v = 0$ at $t = 0$. This change in variable transforms Eq. (19.1) into

$$\frac{dv}{dt} + \alpha(v^2 + 2vy_0 + y_0^2) = t \tag{19.19}$$

or

$$\frac{dv}{dt} + 2y_0\,\alpha v + \alpha v^2 = t - \alpha y_0^2 \tag{19.20}$$

If this is compared with (18.1), it is seen that

$$a_1 = \frac{d}{dt} + 2y_0\,\alpha \tag{19.21}$$

$$a_2 = \alpha \tag{19.22}$$

$$k = 1 \tag{19.23}$$

and

$$\phi(t) = (t - \alpha y_0^2) \tag{19.24}$$

From (18.4) the differential equation for A_1 is obtained in the form

$$a_1 A_1 = t - \alpha y_0^2 \tag{19.25}$$

or

$$\left(\frac{d}{dt} + 2y_0\,\alpha\right) A_1 = t - \alpha y_0^2 \tag{19.26}$$

Since the initial conditions on v are that $v = 0$ at $t = 0$, it is necessary that $A_1 = 0$ at $t = 0$.

Equation (19.26) may most easily be solved by the use of the method of the Laplace transform. If

$$LA_1 = \bar{A}_1 \tag{19.27}$$

and the Laplace transform of both members of the Eq. (19.26) is taken, the equation

$$(s + 2y_0\,\alpha)\,\bar{A}_1 = \frac{1}{s^2} - \frac{\alpha y_0^2}{s} \tag{19.28}$$

is obtained. Hence

$$\bar{A}_1 = \frac{1}{s^2(s + 2y_0\,\alpha)} - \frac{\alpha y_0^2}{s(s + 2y_0\,\alpha)} \tag{19.29}$$

The inverse transform of this equation is

$$A_1 = \frac{t}{2y_0\,\alpha} - \frac{1}{4y_0^2\,\alpha^2} + \frac{\epsilon^{-2y_0\alpha t}}{4y_0^2\,\alpha^2} - \alpha y_0^2 \frac{1 - \epsilon^{-2y_0\alpha t}}{2y_0\,\alpha} \tag{19.30}$$

The function A_2 is now determined from (18.5) and satisfies the differential equation

$$\left(\frac{d}{dt} + 2y_0\,\alpha\right) A_2 = \alpha A_1^2 = -\alpha\left[\frac{t}{2y_0\,\alpha} + \frac{1}{4y_0^2\,\alpha^2}(\epsilon - 1^{-2y_0\alpha t}) - \frac{y_0}{2}(1 - \epsilon^{-2y_0\alpha t})\right]^2 \tag{19.31}$$

This equation may be solved by the Laplace-transform method, although the procedure is somewhat tedious. The higher-order terms, A_3, A_4, etc., may be obtained in a similar manner. The solution is then given by

$$v = A_1 + A_2 + A_3 + \cdots \tag{19.32}$$

or

$$y = y_0 + A_1 + A_2 + A_3 + \cdots \tag{19.33}$$

Example III The equation

$$\frac{dy}{dt} + \alpha y^2 = t \tag{19.34}$$

discussed previously may be solved in another form without introducing a change in variable. The equation as it stands gives the following values:

$$a_1 = \frac{d}{dt} \tag{19.35}$$

$$a_2 = \alpha \tag{19.36}$$

$$k = 1 \tag{19.37}$$

$$\phi(t) = t \tag{19.38}$$

Hence the equation for A_1 is

$$\frac{dA_1}{dt} = t \tag{19.39}$$

The solution of this equation subject to $A_1 = y_0$ at $t = 0$ is

$$A_1 = \frac{t^2}{2} + y_0 \tag{19.40}$$

The equation for A_2 is

$$\frac{dA_2}{dt} = -\alpha \left(\frac{t^2}{2} + y_0 \right)^2 \tag{19.41}$$

or

$$A_2 = -\alpha \left(\frac{t^5}{20} + \frac{y_0 t^3}{3} + y_0^2 t \right) \tag{19.42}$$

where use has been made of the condition that $y = y_0$ at $t = 0$. The higher-order terms may be easily computed and lead to polynomials in t. The solution is of the form

$$y = \left(\frac{t^2}{2} + y_0 \right) - \alpha \left(\frac{t^5}{20} + \frac{y_0 t^3}{3} + y_0^2 t \right) + \cdots \tag{19.43}$$

and has a different form from the one obtained in Example II above.

Example IV   The differential equation of an anharmonic oscillator is

$$\frac{d^2 y}{dt^2} + \omega_0^2 y + \alpha y^2 = B \cos \omega t \tag{19.44}$$

where $\omega_0 \neq \omega$. Let it be required to solve this differential equation subject to the condition that $y = \dot{y} = 0$ at $t = 0$.

In this case we have

$$a_1 = \frac{d^2}{dt^2} + \omega_0^2 \tag{19.45}$$

$$a_2 = \alpha \tag{19.46}$$

$$k = 1 \tag{19.47}$$

$$\phi(t) = B \cos \omega t \tag{19.48}$$

The differential equation for A_1 is, in this case,

$$\left(\frac{d^2}{dt^2} + \omega_0^2 \right) A_1 = B \cos \omega t \tag{19.49}$$

If the Laplace transform of both sides of this equation is taken, we obtain

$$(s^2 + \omega_0^2) \bar{A}_1 = \frac{Bs}{s^2 + \omega^2} \tag{19.50}$$

where $\bar{A}_1$ is the Laplace transform of A_1 and use has been made of the fact that $A_1 = \dot{A}_1 = 0$ at $t = 0$. Hence we have

$$\bar{A}_1 = B \frac{s}{(s^2 + \omega_0^2)(s^2 + \omega^2)} \tag{19.51}$$

and the inverse Laplace transform from a table of transforms gives

$$A_1 = \frac{B}{(\omega^2 - \omega_0^2)} (\cos \omega_0 t - \cos \omega t) \tag{19.52}$$

under the hypothesis that $\omega_0 \neq \omega$.

The differential equation for A_2 is

$$\left(\frac{d^2}{dt^2} + \omega_0^2\right) A_2 = -\alpha A_1^2 \tag{19.53}$$

Now

$$A_1^2 = \frac{B^2}{(\omega^2 - \omega_0^2)^2} (\cos^2 \omega_0 t + \cos^2 \omega t - 2 \cos \omega_0 t \cos \omega t) \tag{19.54}$$

In order to compute the Laplace transform of A_1^2 and to save writing, let us write

$$T(\omega) = \frac{1}{s^2 + \omega^2} \tag{19.55}$$

We then have

$$L \cos^2 \omega_0 t = \frac{1}{2s} [1 + s^2 T(2\omega_0)] \tag{19.56}$$

$$L \cos^2 \omega t = \frac{1}{2s} [1 + s^2 T(2\omega)] \tag{19.57}$$

$$L \cos \omega_0 t \cos \omega t = \frac{s}{2} [T(\omega_0 + \omega) + T(\omega_0 - \omega)] \tag{19.58}$$

Hence on taking the Laplace transform of Eq. (19.53), we obtain

$$\bar{A}_2 = -\frac{\alpha}{s^2 + \omega_0^2} \frac{B^2}{\omega^2 - \omega_0^2} 2 \left[1 + \frac{s}{2} T(2\omega_0) + \frac{s}{2} T(2\omega) \right.$$
$$\left. - sT(\omega_0 + \omega) - s_1 T(\omega_0 - \omega) \right] \tag{19.59}$$

or

$$\bar{A}_2 = \frac{-\alpha B^2}{(\omega^2 - \omega_0^2)^2} \left[T(\omega_0) + \frac{s}{2} T(\omega_0) T(2\omega_0) + \frac{s}{2} T(\omega_0) T(2\omega) \right.$$
$$\left. - sT(\omega_0) T(\omega_0 + \omega) - sT(\omega_0) T(\omega_0 - \omega) \right] \tag{19.60}$$

To compute the inverse transform of $\bar{A}_2$ use may be made of the equations

$$L^{-1} sT(a) T(b) = \frac{\cos at - \cos bt}{b^2 - a^2} \tag{19.61}$$

$$L^{-1} \frac{1}{s} T(a) = \frac{1 - \cos at}{a^2} \tag{19.62}$$

Hence, on taking the inverse transform of $\bar{A}_2$, we obtain

$$A_2 = \frac{-\alpha B^2}{\omega^2 - \omega_0^2} 2 \left[\frac{1 - \cos \omega_0 t}{\omega_0^2} + \frac{1}{2} \frac{(\cos \omega_0 t - \cos 2\omega_0 t)}{3\omega_0^2} + \frac{1}{2} \frac{(\cos \omega_0 t - \cos 2\omega t)}{4\omega^2 - \omega_0^2} \right.$$
$$\left. - \frac{\cos \omega_0 t - \cos (\omega_0 + \omega) t}{\omega(2\omega_0 + \omega)} - \frac{\cos \omega_0 t - \cos (\omega_0 - \omega) t}{\omega(\omega - 2\omega_0)} \right] \tag{19.63}$$

The solution of (19.44) is therefore given by

$$y = A_1(t) + A_2(t) + \cdots$$

Equation (19.44) was first proposed by Helmholtz to explain the theory of combination tones of music.

Example V   Series electric circuit with nonlinear inductor   An example of some
practical importance may be obtained from electrical-circuit theory. Consider the electrical circuit illustrated in Fig. 19.1. The differential equation that governs the flow of current in the electric circuit of Fig. 19.1 is

$$\frac{Ldi}{dt} + Ri + \frac{d\phi}{dt} = E \qquad t > 0 \tag{19.64}$$

where the switch is supposed closed at $t = 0$ so that $i = 0$ at $t = 0$.

ϕ is the total flux linkage of the nonlinear inductor and is in general a complicated function of i, depending on the nature of the $B - H$ curve of the material of the inductor. A typical analytical expression that is sometimes assumed for the functional relation between ϕ and i is

$$\phi = \tanh i = i - \frac{i^3}{3} + \frac{2i^5}{15} - \cdots \tag{19.65}$$

or

$$\phi = \alpha_0 i - \alpha i^3 \tag{19.66}$$

Let ϕ be represented by (19.66). Then the Eq. (19.64) may be written in the form

$$\frac{Ldi}{dt} + Ri + \frac{d}{dt}(\alpha_0 i - \alpha i^3) = E \tag{19.67}$$

or in terms of the operator $D = d/dt$, we have

$$(L + \alpha_0) Di + Ri - \alpha Di^3 = E \tag{19.68}$$

or

$$Di + \frac{R}{L + \alpha_0} i - \frac{\alpha}{L + \alpha_0} Di^3 = \frac{E}{L + \alpha_0} \tag{19.69}$$

Now let

$$b = \frac{R}{L + \alpha_0} \tag{19.70}$$

Hence (19.69) may be written in the form

$$Di + bi - \frac{b\alpha Di^3}{R} = \frac{Eb}{R} \tag{19.71}$$

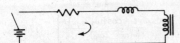

Fig. 19.1

In this equation, we have

$$a_1 = D + b \tag{19.72}$$

$$a_2 = 0 \tag{19.73}$$

$$a_3 = -\frac{b\alpha D}{R} \tag{19.74}$$

$$k = 1 \tag{19.75}$$

$$\phi(t) = \frac{Eb}{R} \tag{19.76}$$

The differential equation for A_1 is

$$\left(\frac{d}{dt} + b\right) A_1 = \frac{Eb}{R} \tag{19.77}$$

with the initial condition that $A_1 = 0$ at $t = 0$. The Laplace transform of this equation is

$$\bar{A}_1 = \frac{Eb}{Rs(s + b)} \tag{19.78}$$

The inverse Laplace transform is

$$A_1 = \frac{E}{R(1 - \epsilon^{-bt})} \tag{19.79}$$

The function A_2 is given by

$$A_2 = -\frac{a_2}{a_1} A_1^2 = 0 \qquad \text{since } a_2 = 0 \tag{19.80}$$

The function A_3 is given by

$$A_3 = -\frac{a_3 A_1^3}{a_1} \tag{19.81}$$

hence the differential equation for A_3 is given by

$$\left(\frac{d}{dt} + b\right) A_3 = \frac{b\alpha d A_1^3}{R\,dt} \tag{19.82}$$

$$A_1^3 = \frac{E^3}{R^3(1 - \epsilon^{-bt})^3}$$

$$= \frac{E^3}{R^3(1 - 3\,\epsilon^{-bt} + 3\,\epsilon^{-2bt} - \epsilon^{-3bt})} \tag{19.83}$$

If this is substituted into (19.82) and the Laplace transform of both members of the equation is taken, we obtain

$$(s + b)\bar{A}_3 = \frac{b\alpha}{R}\,s\,\frac{E^3}{R^3}\left(1 - \frac{3s}{s + b} + \frac{3s}{s + 2b} - \frac{s}{s + 3b}\right) \tag{19.84}$$

Hence

$$\bar{A}_3 = \frac{b\alpha E^3}{R^4}\left[\frac{s}{s + b} - \frac{3s^2}{(s + b)^2} + \frac{3s^2}{(s + b)(s + 2b)} - \frac{s^2}{(s + b)(s + 3b)}\right] \tag{19.85}$$

The inverse Laplace transform of $\bar{A}_3$ may be easily obtained from a table of Laplace transforms, and the result is

$$A_3 = \frac{b\alpha E^3}{R^4}\left[\epsilon^{-bt} - 3\epsilon^{-bt}(1 - bt) + 3\frac{b\epsilon^{-bt} - 2b\epsilon^{-2bt}}{-b} - \frac{b\epsilon^{-bt} - 3b\epsilon^{-3bt}}{-2b}\right] \qquad (19.86)$$

Hence the current in the circuit is given by the equation

$$i = A_1 + A_3 + \cdots = \frac{E}{R}(1 - \epsilon^{-bt}) + \frac{b\alpha E^3}{R^4}\left[\epsilon^{-bt} - 3\epsilon^{-bt}(1 - bt)\right.$$
$$\left. -3(\epsilon^{-bt} - 2\epsilon^{-2bt}) + \frac{\epsilon^{-bt} - 3\epsilon^{-bt}}{2}\right] + \cdots \qquad (19.87)$$

Example VI Periodic electromotive force applied to a nonlinear inductor
Consider the electric circuit of Fig. 19.2. Let the periodic electromotive force be applied at $t = 0$. The differential equation satisfied by the current for $t > 0$ is

$$\frac{L di}{dt} + Ri + \frac{d\phi}{dt} = E\sin wt \qquad (19.88)$$

If the nonlinear relation between the flux ϕ and the current is of the ferromagnetic core, it is represented by the idealized relation

$$\phi = c_1 i - c_3 i^3 \qquad (19.89)$$

where c_1 and c_3 are constants determined from the saturation curve of the ferromagnetic material of the core, then Eq. (19.88) may be written in the form

$$\frac{L di}{dt} + Ri + \frac{d}{dt}(c_1 i + c_3 i^3) = E\sin wt \qquad (19.90)$$

Let the notation

$$b = \frac{R}{L + c_1} \qquad (19.91)$$

be introduced. Equation (19.90) may be written in the form

$$Di + bi - \frac{bc_3 Di^3}{R} = \frac{Eb\sin wt}{R} \qquad (19.92)$$

where $D = d/dt$. If this equation is compared with (18.1) it is seen that

$$a_1 = D + b \qquad a_2 = 0 \qquad a_3 = -\frac{bc_3 D}{R} \qquad (19.93)$$

and all the other a's are zero. In this case, Eq. (18.4) is

$$a_1 A_1 = (D + b) A_1 = \frac{Eb\cos wt}{R} \qquad (19.94)$$

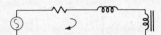

Fig. 19.2

To solve the differential equation (19.94) for A_1, let the Laplace transform of A_1, $\bar{A}_1$, be introduced and the Laplace transforms of both members of (19.94) be taken. The transformed equation is

$$(s + b)\bar{A}_1 = \frac{Eb}{R} \frac{w}{(s^2 + w^2)} \tag{19.95}$$

where use has been made of the initial condition that

$$A'(0) = A(0) = 0 \tag{19.96}$$

Hence $\bar{A}_1$ is given by

$$\bar{A}_1 = \frac{wEb}{R} \frac{1}{(s + b)(s^2 + w^2)} \tag{19.97}$$

From a table of Laplace transforms the following inverse transform is obtained:

$$A_1 = \frac{Eb}{R}\left[\frac{w\,e^{-bt}}{Z^2} + \frac{\sin(wt - \theta)}{Z}\right] \tag{19.98}$$

where

$$Z = (b^2 + w^2)^{\frac{1}{2}} \quad \text{and} \quad \theta = \tan^{-1}\frac{w}{b} \tag{19.99}$$

To continue the solution, it is now necessary to compute the sequence of functions, A_2, A_3, etc. Equation (18.5) reduces in this case to

$$A_2 = -\frac{a_2 A_1^2}{a_1} = 0 \quad \text{since } a_2 = 0 \tag{19.100}$$

A_3 is given by (18.6) in the form

$$A_3 = -\frac{a_3 A_1^3}{a_1} \tag{19.101}$$

or in this case,

$$(D + b)A_3 = \frac{Eb^4 c_3 D}{R^4 Z}\left[\frac{w\,e^{-bt}}{Z} + \sin(wt - \theta)\right]^3 \tag{19.102}$$

In order to solve Eq. (19.102), let

$$L\left[\frac{w\,e^{-bt}}{Z} + \sin(wt - \theta)\right]^3 = \phi(s) \tag{19.103}$$

Then if $\bar{A}_3$ is the transform of A_3, the Laplace transform of the differential equation (19.102) subject to the initial conditions $A_3'(0) = A(0) = 0$ is

$$\bar{A}_3 = \frac{E^3 c_3}{Z(L + c_1)^4} \frac{\phi(s)}{s + b} \tag{19.104}$$

Now since

$$L^{-1}\frac{1}{s + b} = e^{-bt} \tag{19.105}$$

then the Faltung theorem of the Laplace-transform theory leads to the following inverse transform of (19.104):

$$A_3 = \frac{E^3 c_3}{Z(L + c_1)^4} \frac{d}{dt} \int_0^t e^{-b(t-u)} \left[\frac{w\,e^{-bu}}{Z} + \sin(wu - \theta) \right]^3 du \qquad (19.106)$$

If the next approximation of the solution is desired, it can be seen from (18.7) and (18.8) that

$$A_4 = 0 \qquad a_1 A_5 = -3a_3 A_1^2 A_3 \qquad (19.107)$$

The current in the circuit is therefore given by

$$i = A_1 + A_3 + A_5 + \cdots = \frac{E}{Z(L + c_1)} \left[\frac{w\,e^{-bt}}{Z} + \sin(wt - \theta) \right]$$

$$+ \frac{E^3 c_3}{Z(L + c_1)^4} \frac{d}{dt} \int_0^t e^{-b(t-u)} \left[\frac{w\,e^{-bu}}{Z} + \sin(wu - \theta) \right]^3 du \qquad (19.108)$$

Example VII   General circuit in series with a nonlinear inductor   The analytical procedure described above may be applied to the case of a general circuit in series with a nonlinear inductor, as shown in Fig. 19.3.

If the circuit of Fig. 19.3 has an operational impedance $Z(D)$, the differential equation for the current is

$$Z(D)i + \frac{d\phi}{dt} = e(t) \qquad D = \frac{d}{dt} \qquad (19.109)$$

where $e(t)$ is the applied electromotive force as a function of time. Let it be assumed the switch is closed at $t = 0$ so that $i = 0$ at $t = 0$. In general, the flux ϕ is an odd function of the current i that for analytical purposes may be approximated by a series of the following form:

$$\phi = c_1 i + c_3 i^3 + c_5 i^5 + \cdots$$

$$= \sum_{n\text{ odd}} c_n i^n \qquad \text{where the } c\text{'s are constants} \qquad (19.110)$$

If this is substituted into (19.109), we have

$$Z(D)i + \sum c_n D i^n = e(t) \qquad (19.111)$$

or

$$Z(D)i + c_1 Di + c_n Di^n = e(t) \qquad (19.112)$$

Hence, comparing this with Eq. (18.1), we obtain

$$a_1(D) = Z(D) + c_1 D \qquad (19.113)$$

$$a_n = \begin{cases} c_n & n = 3, 5, 7, \ldots \\ 0 & n = 2, 4, 6, 8, \ldots \end{cases} \qquad (19.114)$$

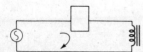

Fig. 19.3

In this case, Eq. (18.4) may be written in the form

$$[Z(D) + c_1 D] A_1 = c(t) \tag{19.115}$$

If $\bar{e}(s)$ is the Laplace transform of the electromotive force $e(t)$, and $\bar{A}(s)$ the transform of $A(t)$, then the Laplace transform of the differential equation (19.115) is

$$\bar{A}_1(s) = \frac{\bar{e}(s)}{Z(s) + c_1 s} = \frac{\bar{e}(s)}{a_1(s)} \tag{19.116}$$

Hence

$$A_1(t) = L^{-1} \bar{A}(s) \tag{19.117}$$

Since $a_2 = 0$, $A_2 = 0$. The Laplace transform of A_3 is given by

$$\bar{A}_3 = -\frac{a_3(s)}{a_1(s)} = -\frac{c_3 \, sLA_1^3}{Z(s) + c_1 s} \tag{19.118}$$

$A_4 = 0$ because a_2 and a_4 are equal to zero. The Laplace transform of A_5 is given by

$$\bar{A}_5 = -\frac{1}{a_1(s)} [3a_3(s) LA_1^2 LA_3 + a_5(s) LA_1^5] \tag{19.119}$$

Further terms may be obtained in a similar manner, and the current in the circuit is given by

$$i = A_1 + A_3 + A_5 + \cdots \tag{19.120}$$

20 CONCLUSION

From the preceding examples, the power of the method of reversion can be appreciated. This method is closely associated with the method of Picard for solving differential equations. A study of the convergence of the process indicates that convergence is rapid if the nonlinearities involved are small, and in such cases it is only necessary to compute a few terms of the series of A's to obtain an accurate solution of the differential equation under consideration. The economy of labor effected by the use of the reversion method may be seen by comparing it with other methods of nonlinear mechanics.[1]

FORCED OSCILLATIONS OF NONLINEAR CIRCUITS†

21 INTRODUCTION

There has been considerable activity in recent years in the development of analytical methods for the solution of nonlinear electrical-circuit problems of

---

[1] See, for example, E. G. Keller, "Mathematics of Modern Engineering," vol. II, pp. 294–301, John Wiley & Sons, Inc., New York, 1942, where problems similar to those given previously are treated.

† Paper 54-297, recommended by the AIEE Basic Sciences Committee and approved by the AIEE Committee on Technical Operations for presentation at the AIEE Summer and Pacific General Meeting, Los Angeles, Calif., June 21–25, 1954. Manuscript submitted March 15, 1954; made available for printing April 27, 1954.

practical importance [15–26]. These investigations are based on the assumption that the forcing function applied to the circuit is a sinusoid. There exist many practical circuits, such as those involving saturable reactors, saturable capacitors, magnetic and dielectric amplifiers, etc., in which the forcing functions involved contain constant biasing potentials in series with sinusoids. Since the principle of superposition does not apply to nonlinear systems, the required solutions for the circuit response under the action of a forcing function containing a constant plus a sinusoid cannot be readily obtained from the response of the circuit to a forcing function containing a sinusoid alone. It is therefore necessary to develop entirely new solutions for these more complicated forcing functions.

The method presented in this paper is an extension of the method of undetermined coefficients which is a standard method for solving linear problems to the nonlinear case. The method is an approximate one but gives useful information regarding the phase and amplitude of the forced periodic oscillations. No abstract general theory will be presented, but the method will be illustrated by applying it to typical nonlinear circuits of practical importance.

22 FORCED OSCILLATIONS OF A NONLINEAR INDUCTOR

An important technical problem is the computation of the amplitude and phase of the fundamental and the harmonic content of the steady-state current in a circuit of the type depicted by Fig. 22.1.

This circuit contains a harmonic potential $E_m \sin(\omega t + \theta)$ and a bias potential E_0 in series with a linear resistor and a nonlinear inductor. The nonlinear inductor consists of a coil of N turns wound on a magnetic core having a cross-sectional area A and a mean length s.

The differential equation that determines the circuit current i is

$$Ri + N\dot{\phi} = E_0 + E_m \sin(\omega t + \theta) \tag{22.1}$$

where ϕ is the magnetic flux in the core of the nonlinear inductor. The flux ϕ is related to the current i by Ampere's law which, expressed in suitable units, is

$$H = \frac{Ni}{s} \tag{22.2}$$

where H = magnetic intensity of the core

s = mean length of the magnetic path in the core

For many practical purposes, the magnetization curve of the core material of the inductor may be represented by the following third-degree polynomial:

$$B = \mu_0 H - kH^3 \tag{22.3}$$

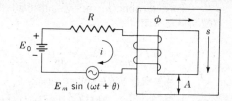

Fig. 22.1

$E_m \sin(\omega t + \theta)$

In this expression B is the magnetic induction and k is a constant determined empirically by adjusting Eq. (22.3) to fit the actual magnetization curve of the material of the core. μ_0 is the initial permeability of the core material. It is defined by the equation

$$\mu_0 = \left(\frac{dB}{dH}\right)_{H=0} \tag{22.4}$$

If A is the mean cross-sectional area of the inductor core, the flux ϕ may be expressed in the form

$$\phi = AB = A(\mu_0 - kH^3) = A\frac{N}{s}i\mu_0 - kA\left(\frac{Ni}{s}\right)^3 \tag{22.5}$$

Hence

$$N\phi = \frac{AN^2\mu_0}{s}i - \frac{kaN^4}{s^3}i^3 \tag{22.6}$$

It is convenient to write this expression in the following form:

$$N\phi = L_0 i - bi^3 \tag{22.7}$$

where

$$L_0 = \mu_0\frac{N^2 A}{s} \qquad b = \frac{kAN^4}{s^3} \tag{22.8}$$

L_0 is the initial inductance of the nonlinear inductor. If $N\phi$ as given by Eq. (22.7) is substituted into (22.1), the result is

$$L_0\frac{di}{dt} + Ri - b\frac{d}{dt}i^3 = E_0 + E_m\sin(\omega t + \theta) \tag{22.9}$$

To determine the steady-state response of the nonlinear inductor, it is necessary to determine a periodic solution of Eq. (22.9). The method of undetermined coefficients suggests that a periodic solution of the following form be assumed for the current:

$$i_0 = I_0 + I_m\sin\omega t \tag{22.10}$$

where I_0 = undetermined dc component of the steady-state current

I_m = the undetermined amplitude of the fundamental of the steady-state alternating current of the circuit

To determine I_0 and I_m, substitute Eq. (22.10) into the left member of (22.9) and write

$$F(t) = L_0 \frac{di_0}{dt} + Ri_0 - b \frac{d}{dt} i_0^3$$

$$= RI_0 + RI_m \sin \omega t + (\omega L_0 I_m - \tfrac{3}{4} b\omega I_m^3 \omega) \cos \omega t$$

$$- 3bI_0^2 I_m \omega - (3\omega I_0 I_m^2 b) \sin 2\omega t + \tfrac{3}{4} bI_m^3 \omega \cos 3\omega t \qquad (22.11)$$

$F(t)$ represents the potential drop that would exist across the circuit elements if the current flowing through the circuit had the form of Eq. (22.10). Since the impressed potential of the circuit has the form

$$E(t) = E_0 + E_m \sin (\omega t + \theta) = E_0 + E_4 \sin \omega t + E_2 \cos \omega t \qquad (22.12)$$

where

$$E_1 = E_m \cos \theta \qquad E_2 = E_m \sin \theta \qquad (22.13)$$

it is evident that $F(t) \neq E(t)$ and that Eq. (22.10) cannot be adjusted [1] to give the exact solution of the differential equation (22.9). However, an approximate solution of practical utility may be obtained by requiring that the constant term, the sine term, and the cosine term of $F(t)$ be made equal to $E(t)$. This stipulation leads to the following three equations:

$$RI_0 = E_0 \qquad (22.14)$$

$$RI_m = E_1 = E_m \cos \theta \qquad (22.15)$$

$$\omega L_0 I_m - 3b\omega I_m \frac{I_m^2}{4 + I_0^2} = E_2 = E_m \sin \theta \qquad (22.16)$$

These three simultaneous equations serve to determine the unknown amplitudes I_0 and I_m and the phase angle θ between the applied harmonic potential and the fundamental of the resulting alternating current of the circuit. Equation (22.14) gives the following value for the direct component of the current:

$$I_0 = \frac{E_0}{R} \qquad (22.17)$$

The amplitude of the alternating current I_m may be obtained by squaring Eqs. (22.15) and (22.16) and adding the result. This procedure gives

$$I_m^2 [R^2 + \omega^2 (L_0 - \tfrac{3}{4} bI_m^2 - 3bI_0^2)] = E_m^2 \qquad (22.18)$$

This is a cubic equation in I_m^2 and can be solved by a graphical construction for a given frequency ω. In general Eq. (22.18) will have either one or three real roots for the amplitude I_m. The possibility of different amplitudes may lead to "jump phenomena" in special cases.

If the bias potential is large so that the direct current I_0 is also large, it may be assumed that

$$\frac{I_m^2}{4} \ll I_0^2 \tag{22.19}$$

If the term $I_m^2/4$ is neglected in Eq. (22.18), this equation can be solved for the amplitude I_m directly.   The result is as follows:

$$I_m = \frac{E_m}{\sqrt{R^2 + \omega^2(L_0 - 3bI_0^2)^2}} \tag{22.20}$$

or

$$I_m = \frac{E_m}{\sqrt{R^2 + \omega^2[L_0 - 3(b/R^2)E_0^2]^2}} \tag{22.21}$$

Equation (22.21) shows that the effect of increasing the bias potential is to decrease the effective inductance of the circuit and hence to increase the amplitude of the alternating current.   This indicates that by changing the magnitude of the biasing potential it is possible to effect a considerable change in the amplitude I_m of the alternating current of the circuit.

The tangent of the phase angle θ of the alternating current of the non-linear inductor may be obtained by means of Eqs. (22.15) and (22.16) in the form

$$\tan \theta = \frac{E_2}{E_1} = \frac{\omega}{R}L_0 - \frac{3bE_0^2}{R^2} \tag{22.22}$$

To this degree of approximation, it is seen that the nonlinear inductor circuit behaves as if it were a linear circuit that has a resistance R and an inductive reactance $X_L = \omega(L_0 - 3bE_0^2/R^2)$.

DETERMINATION OF PRINCIPAL HARMONICS

To determine the amplitudes of the principal harmonics of the circuit, write the circuit differential Eq. (22.9) in the following form:

$$Ri = (E_0 + E_1 \sin \omega t + E_2 \cos \omega t) - \left(L_0 \frac{di}{dt} - b\frac{d}{dt}i^3\right) \tag{22.23}$$

and substitute $i = I_0 + I_m \sin \omega t$ into the right member of Eq. (22.23).   If this is done and relations (22.14), (22.15), and (22.16) are used, the resulting equation may be solved for i. The result is

$$i = I_0 + I_m \sin \omega t + 3\frac{\omega}{R}I_0 I_m^2 b \sin 2\omega t - \frac{3b}{4R}I_m^3 \omega \cos 3\omega t \tag{22.24}$$

If the higher harmonics are desired, Eq. (22.24) may be substituted into (22.23) and the resulting equation solved for i.   The convergence of this procedure requires a separate investigation.   If b is a small quantity, it may be shown to converge.

23   OSCILLATIONS OF A SATURABLE REACTOR

As a second example of the general method, consider the saturable reactor circuit of Fig. 23.1. This circuit consists of an iron core on which are wound two separate windings. The control winding contains N_0 turns and the output winding N turns. In series with the control windings are a resistance R_0 and a direct potential source E_0. The output winding contains a load resistance R in series with a harmonic potential $E_m \sin(\omega t + \theta)$. By varying the magnitude of the control potential E_0 of the circuit, it is possible to vary the amplitude of the output current.

An application of Kirchhoff's voltage law to the control and output circuits yields the following equations:

$$R_0 i_0 + N_0 \dot{\phi} = E_0 \tag{23.1}$$

$$Ri + N\dot{\phi} = E_m \sin(\omega t + \theta) \tag{23.2}$$

where ϕ is the magnetic flux in the iron core. In suitable units, Ampere's circuital law applied to the magnetic circuit of the core leads to the equation

$$N_0 i_0 + Ni = sH \tag{23.3}$$

where s is the mean length of the magnetic path of the core. The magnetization curve of the core material may be expressed by the empirical relation

$$H = \frac{\phi}{\mu_0 A} + c\phi^3 \tag{23.4}$$

where H = magnetic intensity

      ϕ = core flux

      μ_0 = initial permeability of the core material

      c = an empirical constant

Equations (23.1) and (23.2) may be solved for the currents i_0 and i in the form

$$i_0 = \frac{E_0}{R_0} - \frac{N_0}{R_0}\dot{\phi} \tag{23.5}$$

$$i = \frac{E_m}{R} \sin(\omega t + \theta) - \frac{N}{R}\dot{\phi} \tag{23.6}$$

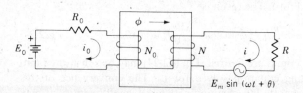

$E_m \sin(\omega t + \theta)$

Fig. 23.1

If H, i_0, and i are substituted from Eqs. (23.4), (23.5), and (23.6) into Eq. (23.3), the result is

$$N_0\left(\frac{E_0}{R_0} - \frac{N_0}{R_0}\dot{\phi}\right) + N\left[\frac{E_m}{R}\sin(\omega t + \theta) - \frac{N}{R}\dot{\phi}\right] = s\left(\frac{\phi}{\mu_0 A} + c\phi^3\right) \qquad (23.7)$$

This equation may be put into the more convenient form

$$T\dot{\phi} + \phi + K\phi^3 = A + B\sin(\omega t + \theta) \qquad (23.8)$$

where

$$T = \frac{L_0}{R_0} + \frac{L}{R} \qquad K = \mu_0 Ac \qquad A = \frac{L_0 E_0}{N_0 R_0} \qquad B = \frac{L_0 E_m}{N R} \qquad (23.9)$$

L_0 and L are two inductances given by

$$L_0 = \mu_0 \frac{N^2}{s} A \qquad L = \mu_0 \frac{N^2}{s} A \qquad (23.10)$$

To obtain the steady-state solution of Eq. (23.8) by the method of undetermined coefficients, assume a solution of the form

$$\phi = \phi_0 + \phi_m \sin \omega t \qquad (23.11)$$

Let

$$\begin{aligned}
F(t) &= (T\dot{\phi} + \phi + K\phi^3)_{\phi = \phi_0 + \phi m \sin \omega t} \\
&= (\phi_0 + K\phi_0^2 + \tfrac{3}{2}\phi_0 \phi_m^2) + \left[\phi_m + 3K\phi_m\left(\frac{\phi}{4}m^2 + \phi_0^2\right)\right]\sin \omega t \\
&\quad + T\omega\phi_m \cos \omega t - \frac{K}{4}\phi_m^3 \sin 3\omega t - \frac{3K\phi_0}{2}\phi_m^2 \cos 2\omega t \qquad (23.12)
\end{aligned}$$

If the constant term of $F(t)$ and the coefficients of the sine and cosine terms are equated to the corresponding terms of the right member of Eq. (23.8), the following three equations are obtained:

$$\phi_0 + K\phi_0(\phi_0^2 + \tfrac{3}{2}\phi_m^2) = A \qquad (23.13)$$
$$\phi_m + 3K\phi_m(\phi_m^2 + \phi_0^2) = B\cos\theta \qquad (23.14)$$
$$T\omega\phi_m = B\sin\theta \qquad (23.15)$$

These are three simultaneous equations to determine the unknowns ϕ_0, ϕ_m, and the phase angle θ. If K is small, the second term of Eq. (23.13) may be neglected and the following approximate value for ϕ_0 obtained:

$$\phi_0 = A = \frac{L_0 E_0}{N_0 R_0} \qquad (23.16)$$

If the control potential E_0 is large so that $(\phi/4)m^2 \ll \phi_0^2$, the term $(\phi_m^2)/4$ may be neglected in the second member of Eq. (23.14) and this equation written in the form

$$\phi_m(1 + 3K\phi_0^2) = B\cos\theta \tag{23.17}$$

The sum of the squares of Eqs. (23.15) and (23.17) is

$$\phi_m^2[T^2\omega^2 + (1 + 3K\phi_0^2)^2] = B^2 \tag{23.18}$$

or

$$\phi_m = \frac{B}{[T^2\omega^2 + (1 + 3K\phi_0^2)^2]^{\frac{1}{2}}} \tag{23.19}$$

The tangent of the phase angle θ may be determined by dividing Eq. (23.15) by (23.17). This procedure gives

$$\tan(\theta) = \frac{T\omega}{1 + 3K\phi_0^2} \tag{23.20}$$

If Eqs. (23.16) and (23.19) are substituted into Eq. (23.11), the following value for the core flux ϕ is obtained:

$$\phi = \frac{L_0}{N_0}\frac{E_0}{R_0} + LE_m\frac{\sin\omega t}{NR}[T^2\omega^2 + (1 + 3K\phi_0^2)^2]^{\frac{1}{2}} \tag{23.21}$$

This expression gives the approximate value of the core flux to terms of fundamental order

DETERMINATION OF PRINCIPAL HARMONICS OF CORE FLUX

To determine the principal harmonics of the core flux, write Eq. (23.8) as

$$\phi = A + B\sin(\omega t + \theta) - (T\dot{\phi} + K\phi^3) \tag{23.22}$$

If $\phi = \phi_0 + \phi_m\sin(\omega t + \theta)$ is substituted into the right member of Eqs. (23.22) and (23.13), Eqs. (23.14) and (23.15) are used to simplify the resulting expression and the following equation is obtained:

$$\phi = \phi_0 + \phi_m\sin\omega t + \frac{K}{4}\phi_m^3\sin 3\omega t + \tfrac{3}{2}K\phi_0\phi_m^2\cos 2\omega t \tag{23.23}$$

The control current i_0 may now be determined by Eq. (23.5) in the form

$$i_0 = \frac{E_0}{R_0} - \frac{N_0}{R_0}\dot{\phi}$$

$$= \frac{E_0}{R_0} - \frac{N_0}{R_0}\omega\phi_m(\cos\omega t + \tfrac{3}{4}K\phi_m^2\cos 3\omega t - 3K\phi_0\phi_m\sin 2\omega t) \tag{23.24}$$

The output current i can be determined by the use of Eq. (23.6) and expressed in the following form:

$$i = \frac{E_m}{R} \sin(\omega t + \theta) - \frac{N}{R} \dot{\phi}$$

$$= \frac{E_m}{R} \sin(\omega t + \theta) - \frac{N}{R} \omega \phi_m (\cos \omega t + \tfrac{3}{4} K \phi_m^2 \cos 3\omega t$$

$$- 3K\phi_0 \phi_m \sin 2\omega t) \quad (23.25)$$

The effect of the control potential E_0 in influencing the amplitude of the output current is clearly apparent if ϕ_0 as given by Eq. (23.16) is substituted into Eq. (23.19) and ϕ_m written in the following form:

$$\phi_m = \frac{LE_m}{NR} \left[T^2 \omega^2 + \left(1 + \frac{3KL_0^2 E_0^2}{N_0^2 R_0^2} \right) \right]^{-\frac{1}{2}} \quad (23.26)$$

It is evident from Eq. (23.26) that an increase in the control potential E_0 produces a decrease in ϕ_m and this in turn produces a change in the output current i, as is apparent from Eq. (23.25).

24   FORCED OSCILLATIONS OF A NONLINEAR CAPACITOR

As a third example of the general procedure, consider the circuit of Fig. 24.1. This circuit consists of a linear resistor and inductor in series with a nonlinear capacitor. The circuit is energized by a harmonic potential $E_m \sin(\omega t + \theta)$ and a direct biasing potential E_0. It will be assumed that the saturation curve of the nonlinear capacitor may be approximated by the following cubic polynomial [9, 10]:

$$E_c = S_0 q + aq^3 \quad (24.1)$$

where E_c = potential across the plates of the nonlinear capacitor

q = charge separation on the plates of the capacitor

S_0 = initial elastance of the capacitor

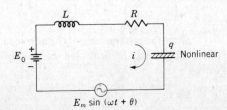

Fig. 24.1

The coefficient a is a positive constant that depends on the characteristics of the nonlinear dielectric of the capacitor. The capacitor charge q and the circuit current i satisfy the differential equation

$$L\frac{di}{dt} + Ri + S_0 q + aq^3 = E_0 + E_m \sin(\omega t + \theta) \qquad (24.2)$$

To obtain the steady-state periodic solution, the method of undetermined coefficients suggests that a periodic solution of the following form be assumed for the current and charge of the circuit:

$$i = I_m \sin \omega t \qquad (24.3)$$

$$q = Q_0 - \frac{I_m}{\omega} \cos \omega t \qquad (24.4)$$

Q_0 is the constant charge accumulation on the plates of the capacitor and I_m is the maximum amplitude of the alternating current of the circuit. We now substitute Eqs. (24.3) and (24.4) into the left member of Eq. (24.2) and let

$$F(t) = L\left(\frac{di}{dt} + Ri + S_0 q + aq^3\right)_{i=I_m \sin \omega t}$$

$$= \left(S_0 Q_0 + aQ_0^3 + \frac{3aQ_0 I_m^2}{2\omega^2}\right) + [RI_m]\sin \omega t$$

$$+ \left(\omega L I_m - S_0 \frac{I_m}{\omega} - \frac{3aI_m^3}{4\omega^3} - \frac{3aQ_0^2 I_m}{\omega}\right)\cos \omega t$$

$$- \frac{aI_m^3}{4\omega^3}\cos 3\omega t + \frac{3aQ_0 I_m^2 \cos 2\omega t}{2\omega^2} \qquad (24.5)$$

Following the general procedure, we now equate the coefficients of the constant, $\sin \omega t$ and $\cos \omega t$ terms of $F(t)$ and the corresponding terms of

$$E(t) = E_0 + E_m \sin(\omega t + \theta) = E_0 + E_m \cos \theta \sin \omega t$$
$$+ E_m \sin \theta \cos \omega t \qquad (24.6)$$

and obtain the following three equations:

$$S_0 Q_0 + aQ_0\left(Q_0^2 + \frac{3I_m^2}{2\omega^2}\right) = E_0 \qquad (24.7)$$

$$RI_m = E_m \cos \theta \qquad (24.8)$$

$$I_m\left(\omega L - \frac{S_0}{\omega}\right) - 3\frac{a}{\omega}I_m\left(Q_0^2 + \frac{I_m^2}{4\omega^2}\right) = E_m \sin \theta \qquad (24.9)$$

Since the parameter a of the nonlinear capacitor is small, the second term of Eq. (24.7) may be neglected and the following approximate value obtained for Q_0:

$$Q_0 = \frac{E_0}{S_0} = C_0 E_0 \qquad (24.10)$$

where $C_0 = 1/S_0$ is the initial capacitance.   If the bias potential E_0 is large in comparison with the maximum value of the applied harmonic potential so that $E_0 \gg E_m$, then for the usual frequencies used in practice we have

$$Q_0^2 \gg \frac{I_m^2}{4\omega^2} \tag{24.11}$$

If the term $I_m^2/4\omega^2$ is neglected in Eq. (24.9), this equation may be written in the form

$$I_m\left[\omega L - \frac{1}{\omega}(S_0 + 3aC_0^2 E_0^2)\right] = E_m \sin\theta \tag{24.12}$$

Hence if Eqs. (24.8) and (24.12) are squared and the results added, we obtain the following equation:

$$I_m^2\left\{R^2 + \left[\omega L - \frac{1}{\omega}(S_0 + 3aC_0^2 E_0^2)\right]^2\right\} = E_m^2 \tag{24.13}$$

The amplitude of the alternating current of the circuit is therefore given by

$$I_m = \frac{E_m}{\left\{R^2 + \left[\omega L - \frac{1}{\omega}(S_0 + 3aC_0^2 E_0^2)\right]^2\right\}^{\frac{1}{2}}} \tag{24.14}$$

It is thus apparent that the effect of the bias potential is to increase the effective elastance of the system by an amount $3aC_0^2 E_0^2$.

The tangent of the phase angle θ may be obtained by the division of Eq. (24.12) by (24.2).   The result is

$$\tan\theta = \frac{\omega L - \frac{1}{\omega}(S_0 + 3aC_0^2 E_0^2)}{R} \tag{24.15}$$

It may be noticed that if the circuit of Fig. 24.1 does not contain inductance, then $L = 0$ and Eq. (24.15) reduces to

$$I_m = \frac{E_m}{[R^2 + (S_0 + 3aC_0^2 E_0^2)^2/\omega^2]^{\frac{1}{2}}} \tag{24.16}$$

The effect of the bias potential in increasing the effective impedance of the circuit is clearly evident in Eq. (24.16).   It is thus apparent that an increase of the magnitude of the bias potential produces a corresponding decrease in the amplitude of the alternating current.   This principle is the basis of the operation of dielectric amplifiers.

DETERMINATION OF PRINCIPAL HARMONICS

The amplitudes of the principal harmonics of the circuit may be determined by writing Eq. (24.2) in the following form:

$$S_0 q = [E_0 + E_m \sin(\omega t + \theta)] - \left(L\frac{di}{dt} + Ri + aq^3\right) \tag{24.17}$$

and substituting Eqs. (24.3) and (24.4) into the right member of Eq. (24.17).

If Eqs. (24.7), (24.8), and (24.9) are used to simplify the resulting expression, it can be written in the form

$$q = Q_0 - \frac{I_m}{\omega}\cos\omega t + \frac{aI_m^3 \cos 3\omega t}{4\omega^3 S_0} - \frac{3aQ_0 I_m^2 \cos 2\omega t}{2\omega^2 S_0} \tag{24.18}$$

The circuit current is then given by

$$i = \frac{dq}{dt} = I_m \sin\omega t - \frac{3aC_0 I_m^3 \sin 3\omega t}{4\omega^2} + \frac{3aQ_0 I_m^2 \sin 2\omega t}{\omega S_0} \tag{24.19}$$

The current therefore contains a second- and third-harmonic current component.

25 STEADY-STATE OSCILLATIONS OF A SERIES-CONNECTED MAGNETIC AMPLIFIER

In recent years, magnetic amplifiers have been used more and more extensively as reliable substitutes for vacuum-tube amplifiers and many analytical papers explaining their operation and design have been published [27].

Because of the difficulties involved in the solution of the nonlinear differential equations that govern the response of magnetic amplifiers, the published analyses of these devices contain various simplifying assumptions and approximations [28]. The method of undetermined coefficients gives useful information when applied to the study of the response of magnetic amplifiers. To illustrate the general principles involved, the method will be applied to the study of the steady-state response of the symmetrical series-connected magnetic amplifier depicted in Fig. 25.1.

The basic equations of the circuit of Fig. 25.1 are obtained by an application of Kirchhoff's voltage laws and Ampere's circuital law for the electric and magnetic loops of the amplifier. These equations are

$$2Ri_1 + N(\dot\phi_a + \dot\phi_b) = E_0 \tag{25.1}$$

$$2Ri_2 + N(\dot\phi_a - \dot\phi_b) = E_m \sin(\omega t + \theta) \tag{25.2}$$

$$N(i_1 + i_2) = sH_a \tag{25.3}$$

$$N(i_1 - i_2) = sH_b \tag{25.4}$$

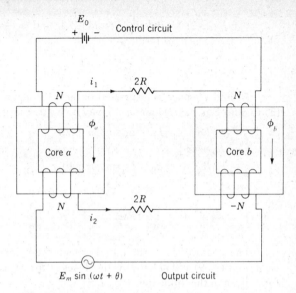

Fig. 25.1

where H_a, H_b = magnetic intensities of the cores a and b

ϕ_a, ϕ_b = magnetic fluxes in the corresponding cores

For simplicity, it is assumed that the amplifier under consideration is symmetric, so that all the windings have an equal number of turns N and equal ohmic resistances R; s is the mean length of the magnetic path of the two cores, and A their cross-sectional area. It will be assumed that the saturation curve of the core material is such that the relation between the magnetic intensity and the magnetic flux of the cores may be expressed by the following equations:

$$H_a = \frac{\phi_a}{\mu_0 A} + c\phi^3 \tag{25.5}$$

$$H_b = \frac{\phi_b}{\mu_0 A} + c\phi^3 \tag{25.6}$$

where μ_0 = initial permeability of the core material

c = an empirical constant

If Eqs. (25.1) and (25.2) are added and the sum $(i_1 + i_2)$ is expressed in terms of H_a by Eq. (25.3), the result is

$$\dot{\phi}_a + \frac{RsH_a}{N^2} = \frac{E_0 + E_m \sin{(\omega t + \theta)}}{2N} \tag{25.7}$$

If Eq. (25.2) is subtracted from (25.1) and $(i_1 - i_2)$ is expressed in terms of H_b by Eq. (25.4), the result is

$$\dot{\phi}_b = \frac{RsH_b}{N^2} = \frac{E_0 - E_m \sin(\omega t + \theta)}{2N} \tag{25.8}$$

H_a and H_b may be expressed in terms of ϕ_a and ϕ_b by Eqs. (25.5) and (25.6), and Eqs. (25.7) and (25.8) may be written in the following form:

$$\dot{\phi}_a + \frac{R\phi_a}{L_0 + k\phi_a^3} = \frac{E_0 + E_m \sin(\omega t + \theta)}{2N} \tag{25.9}$$

$$\dot{\phi}_b + \frac{R\phi_b}{L_0 + k\phi_b^3} = \frac{E_0 - E_m \sin(\omega t + \theta)}{2N} \tag{25.10}$$

where

$$L_0 = \frac{\mu_0 N^2 A}{s} \qquad k = \frac{Rsc}{N^2} \tag{25.11}$$

The quantity L_0 is the initial inductance of one of the core windings. Equations (25.9) and (25.10) for the fluxes of the cores are both of the general form given in the following:

$$\dot{\phi} + \frac{R\phi}{L_0} + k\phi^3 = \frac{E_0 + E_m \sin(\omega t + \theta)}{2N} \tag{25.12}$$

To obtain the approximate steady-state solution of this equation by the method of undetermined coefficients, assume a solution of the form

$$\phi = \phi_0 + \phi_m \sin \omega t \tag{25.13}$$

If this expression for ϕ is substituted into the left member of Eq. (25.12), the result is

$$\begin{aligned}
F(t) &= \left(\frac{\dot{\phi} + R\phi}{L_0 + k\phi^3} \right)_{\phi = \phi_0 + \phi_m \sin \omega t} \\
&= \frac{R}{L_0} \phi_0 + k\phi_0 \left(\phi_0^2 + \frac{\phi_m^2}{2} \right) + \left[\frac{R}{L_0} \phi_m + 3k\phi_m \left(\frac{\phi_m^2}{4} + \phi_0^2 \right) \right] \sin \omega t \\
&\quad + (\omega \phi_m) \cos \omega t - \frac{k}{2} \phi_0 \phi_m^2 \cos 2\omega t - \frac{k}{4} \phi_m^3 \sin 3\omega t \tag{25.14}
\end{aligned}$$

The constant term and the coefficients of $\sin(\omega t)$ and $\cos(\omega t)$ are now equated to the corresponding terms of the right member of Eq. (25.12). This procedure leads to the following three equations:

$$\frac{R}{L_0} \phi_0 + k\phi_0 \left(\phi_0^2 + \frac{\phi_m^2}{2} \right) = \frac{E_0}{2N} \tag{25.15}$$

$$\frac{R}{L_0} \phi_m + 3k\phi_m \left(\frac{\phi_m^2}{4} + \phi_0^2 \right) = \frac{E_m \cos \theta}{2N} \tag{25.16}$$

$$\omega \phi_m = \frac{E_m \sin \theta}{2N} \tag{25.17}$$

The system of algebraic equations, (25.15), (25.16), and (25.17), determines the unknown quantities ϕ_0, ϕ_m and the phase angle θ. The solution of this system of equations presents formidable difficulties, but an approximate solution may be effected by realizing that the constant $k = Rsc/N^2$ is a small quantity in cases of practical importance. If the term containing k in Eq. (25.15) is neglected in comparison with the first term, then this equation may be solved for ϕ_0 with the following result:

$$\phi_0 = \frac{L_0 E_0}{2NR} \tag{25.18}$$

Now, for sufficiently large values of the bias potential E_0, the assumption $\phi_m^2 \ll \phi_0^2$ may be made, and Eq. (25.16) may be written in the form

$$\phi_m\left(\frac{R}{L_0} + 3k\phi_0\right) = \frac{E_m \cos\theta}{2N} \tag{25.19}$$

If Eqs. (25.17) and (25.18) are squared and the results added, the following equation is obtained:

$$\phi_m^2\left[\omega^2 + \left(3k\phi_0^2 + \frac{R}{L_0}\right)^2\right] = \frac{E_m^2}{4N^2} \tag{25.20}$$

or

$$\phi_m = \frac{E_m}{2N}\left[\omega^2 + \left(\frac{R}{L_0} + 3k\phi_0^2\right)^2\right]^{\frac{1}{2}} \tag{25.21}$$

Hence the approximate solution of Eq. (25.12) is given by

$$\phi = \frac{L_0 E_0}{2NR} + \frac{E_m \sin\omega t}{2N}\left[\omega^2 + \left(\frac{R}{L_0} + 3k\phi_0^2\right)^2\right]^{\frac{1}{2}} \tag{25.22}$$

The tangent of the phase angle θ is obtained by dividing Eq. (25.17) by (25.19):

$$\tan(\theta) = \frac{\omega}{R/L_0 + 3k\phi_0^2} \tag{25.23}$$

PRINCIPAL HARMONICS OF CORE FLUX

To compute the principal harmonics of the core flux, write Eq. (25.12) in the following form:

$$\frac{R\phi}{L_0} = \frac{E_0 + E_m \sin(\omega t + \theta)}{2N} - (\dot{\phi} + k\phi^3) \tag{25.24}$$

and substitute $\phi = \phi_0 + \phi_m \sin(\omega t)$ in the right member of Eq. (25.24). If this is done and Eqs. (25.15), (25.16), and (25.17) are used to simplify the result, the following expression for ϕ is obtained:

$$\phi = \phi_0 + \phi_m \sin(\omega t) + \frac{kL_0 \phi_0 \phi_m^2}{2R}\cos 2\omega t + \frac{k}{4R}L_0 \phi_m^3 \sin 3\omega t \tag{25.25}$$

To this degree of approximation, the core flux is seen to have second- and third-harmonic components. Comparing Eqs. (25.9) and (25.10) for the determination of ϕ_a and ϕ_b with Eq. (25.12), it is seen that the fluxes ϕ_a and ϕ_b are, in the periodic steady state:

$$\phi_a = \phi_0 + \phi_m \sin \omega t + A_2 \cos 2\omega t + B_3 \sin 3\omega t \tag{25.26}$$

$$\phi_b = \phi_0 - \phi_m \sin \omega t + A_2 \cos 2\omega t - B_3 \sin 3\omega t \tag{25.27}$$

where

$$A_2 = \frac{kL_0 \phi_0 \phi_m^2}{2R} \qquad B_3 = \frac{kL_0 \phi_m^3}{4R} \tag{25.28}$$

To determine the control current i_1 of the magnetic amplifier, Eq. (25.1) may be solved for i_1 in the form

$$i_1 = \frac{E_0}{2R} - \frac{N(\dot{\phi}_a + \dot{\phi}_b)}{2R} \tag{25.29}$$

If Eqs. (25.26) and (25.27) are differentiated and the results substituted into (25.29), we obtain

$$i_1 = \frac{E_0}{2R} + \frac{2N\omega A_2 \sin 2\omega t}{R} \tag{25.30}$$

and the control current is seen to have a small second-harmonic component of the output potential. The output current i_2 may be obtained by solving Eq. (25.2) in the form

$$i_2 = \frac{E_m \sin(\omega t + \theta)}{2R} + \frac{N(\dot{\phi}_b - \dot{\phi}_a)}{2R} \tag{25.31}$$

If the time derivatives of Eqs. (25.26) and (25.27) are substituted into (25.31), the result is

$$i_2 = \frac{E_m \sin(\omega t + \theta)}{2R} - \frac{\omega N(\phi_m \cos \omega t + 3B_3 \cos 3\omega t)}{R} \tag{25.32}$$

and the output current is seen to contain a third-harmonic component of the applied potential. The effect of the control potential on the output current may be clearly seen if Eq. (25.18) is substituted into (25.21) and ϕ_m written in the following form:

$$\phi_m = \frac{E_m}{2N[\omega^2 + (R/L_0 + 3kL_0^2 E_0^2/4N^2 R^2)^2]^{\frac{1}{2}}} \tag{25.33}$$

26 CONCLUSIONS

The analysis presented in this section is an attempt to apply the classical method of undetermined coefficients which has proved so useful in the analysis of linear systems so as to obtain the approximate solutions of the basic differential equations of typical nonlinear electrical circuits of importance in engineering.

The examples of the application of the method indicate that useful practical results may be obtained by this procedure in a very simple manner. The method appears to be particularly well adapted to the study of electrical circuits that contain biasing control potentials. The effect of the control potentials on the current amplitudes involved and the amplitudes of the principal harmonics that are present in the circuits may be easily determined by the method of undetermined coefficients.

MATRIX SOLUTION OF EQUATIONS OF THE MATHIEU-HILL TYPE†

27 INTRODUCTION

The mathematical analysis of a considerable variety of physical problems leads to a formulation involving a differential equation that may be reduced to the following general form:

$$\frac{d^2 x}{dt^2} + F(t)\, x = 0 \tag{27.1}$$

where $F(t)$ is a single-valued *periodic* function of fundamental period T that may be represented by a general Fourier series of the form

$$F(t) = A_0 + \sum_1^\infty A_n \cos nwt + \sum_1^\infty B_n \sin nwt \tag{27.2}$$

where $w = 2\pi/T$.

If the Fourier series expansion for $F(t)$ degenerates into the simple form

$$F(t) = A_0 + A_1 \cos wt \tag{27.3}$$

then Eq. (27.1) is known as *Mathieu's* differential equation. If the function $F(t)$ is a periodic function of the general form (27.2), the equation is known as *Hill's* differential equation.

The general properties of the solution of Eq. (27.1) have been extensively discussed in the mathematical literature.[1-3] A survey of the nature of the

† Reprinted from *Journal of Applied Physics*, vol. 24, no. 7, pp. 902–910, July, 1953. Copyright 1953 by the American Institute of Physics.

[1] N. W. McLachlan, "Theory and Application of Mathieu Functions," Oxford University Press, London, 1947. Contains 226 references to the international literature.

[2] E. T. Whittaker and G. N. Watson, "Modern Analysis," pp. 404–428, Cambridge University Press, Cambridge, 1927.

[3] L. Brillouin, *Quarterly Journal of Applied Mathematics*, vol. 6, pp. 167–178, 1948; vol. 7, pp. 363–380, 1950.

analyses arising from practical applications indicates that they may be divided into two main categories. In the first category are problems in which an equation of the type (27.1) arises as a consequence of the separation of variables of a boundary value problem. In this case the appropriate solution is required to be a periodic function. In the second category are found problems that may be regarded as initial value problems in which an equation of the type (27.1) is involved. In this case the solutions are *not* restricted to periodic ones. The following physical problems are of the second type:

1. The production of electric currents by periodically varying the parameters of electrical circuits. This phenomenon is commonly called *parametric excitation* in the current literature.[1-4]
2. The vibrations of the side rods of locomotives.[5]
3. The propagation of waves in stratified media.[6, 7]
4. The propagation of electric currents in filter circuits and other periodic electric structures.[7, 8]
5. The theory of modulation in wireless.[9]
6. The theory of super-regeneration.[10]
7. The theory of the stability of an inverted pendulum by periodic movements of its point of support.[11]
8. The theory of astronomic perturbations.[12, 13]
9. The quantum theory of metals.[14]
10. The theory of the stability of the solutions of certain nonlinear differential equations.[15]

[1] M. Brillouin, *Eclairage Electrique*, vol. 11, pp. 49–59, 1897.

[2] A. Erdelyi, *Annales Physika*, vol. 19, p. 585, 1934.

[3] Barrow, Smith, and Baumann, *Journal of the Franklin Institute*, vol. 221, no. 3, p. 403, 1936.

[4] N. Minorsky, *Journal of the Franklin Institute*, vol. 240, no. 1, pp. 25–46, 1945.

[5] S. Timoshenko, "Vibration Problems in Engineering," pp. 167–181, D. Van Nostrand Company, Inc., New York, 1937.

[6] Lord Rayleigh, *Philosophical Magazine*, vol. 24, pp. 145–159, 1887.

[7] L. Brillouin, "Wave Propagation in Periodic Structures," McGraw-Hill Book Company, New York, 1946.

[8] G. A. Campbell, *Bell System Technical Journal*, vol. 1, pp. 1–32, 1922.

[9] J. R. Carson, *Proceedings of the Institute of Radio Engineers*, vol. 10, p. 62, 1922.

[10] A. Erdelyi, *Annales Physika*, vol. 23, p. 21, 1935.

[11] A. Stephenson, *Proceedings of the Manchester Philosophical Society*, vol. 52, no. 8, 1908.

[12] G. W. Hill, *Acta mathematica*, vol. 8, no. 1, 1886.

[13] J. H. Poincaré, "Les Methodes Nouvelles de la Mécanique Céleste," vol. II, Paris, 1893.

[14] F. Seitz, "Modern Theory of Solids," chap. 8, McGraw-Hill Book Company, New York, 1940.

[15] N. W. McLachlan, "Ordinary Nonlinear Differential Equations in Engineering and Physical Science," pp. 188–190, Oxford University Press, London, 1950.

    This is by no means an exhaustive list of the types of physical problems whose mathematical analyses involve equations of the type (27.1) directly, but it serves to indicate the importance of this equation in mathematical physics. The determination of the *periodic* solutions of equations of the type (27.1) has been the subject of considerable investigation.[1] The purpose of this discussion is to present a method for the solution of equations of the type (27.1) subject to prescribed initial conditions when the solutions are *not* required to be periodic. This method is based on an application of Sylvester's theorem of matrix algebra[2] to the solution of the problem. The method presented reduces the solution of Eq. (27.1) subject to given initial conditions to that of computing powers of matrices and is applicable to the solutions of the physical problems enumerated above.

28 THE USE OF MATRIX ALGEBRA IN SOLVING HILL'S EQUATION

A general method for the determination of the solution of (27.1) in terms of given initial values of $x(t)$ and $\dot{x}(t) = v(t)$ by means of the theory of matrices will now be presented. It has been found that in solving problems of the initial value type, the matrix method has certain advantages over methods that make use of the conventional Floquet theory of Hill's equation.[1, 3]

    Let $u_1(t)$ and $u_2(t)$ be two linearly independent solutions of the equation

$$\frac{d^2 x}{dt^2} + F(t) x = 0 \tag{28.1}$$

in the fundamental interval $0 \leqslant t \leqslant T$. The value of $x(t)$ and its first derivative $\dot{x}(t) = v(t)$ may be expressed in the following form:

$$\left.\begin{array}{l} x(t) = A_1 u_1(t) + A_2 u_2(t) \\ v(t) = A_1 \dot{u}_1(t) + A_2 \dot{u}_2(t) \end{array}\right\} \quad \text{for } 0 \leqslant t \leqslant T \tag{28.2}$$

It is convenient to write (28.2) in the matrix form

$$\begin{bmatrix} x(t) \\ v(t) \end{bmatrix} = \begin{bmatrix} x \\ v \end{bmatrix}_t = \begin{bmatrix} u_{11}(t) & u_{12}(t) \\ u_{21}(t) & u_{22}(t) \end{bmatrix} \begin{bmatrix} A_1 \\ A_2 \end{bmatrix} \tag{28.3}$$

A_1 and A_2 are *arbitrary* constants, and

$$\begin{bmatrix} u_{11}(t) & u_{12}(t) \\ u_{21}(t) & u_{22}(t) \end{bmatrix} = \begin{bmatrix} u_1(t) & u_2(t) \\ \dot{u}_1(t) & \dot{u}_2(t) \end{bmatrix} \tag{28.4}$$

[1] McLachlan, *op. cit.*, and Whittaker and Watson, *op. cit.*

[2] R. A. Frazer, W. J. Duncan, and A. R. Collar, "Elementary Matrices," pp. 64–85, Cambridge University Press, Cambridge, 1938.

[3] G. Floquet, *Annales L'Ecole Normale Supérieure*, vol. 12, pp. 47–88, 1883.

The Wronskian [1,2] of the two solutions $u_1(t)$ and $u_2(t)$ is a *constant* in the fundamental interval $0 \leqslant t \leqslant T$ and is given by the determinant

$$W_0 = \begin{vmatrix} u_{11}(t) & u_{12}(t) \\ u_{21}(t) & u_{22}(t) \end{vmatrix} = u_1(t)\dot{u}_2(t) - \dot{u}_1(t)u_2(t) \tag{28.5}$$

Since the two solutions $u_1(t)$ and $u_2(t)$ are linearly independent, $W_0 \neq 0$ and the matrix (28.4) is nonsingular and has the following inverse:

$$\begin{bmatrix} u_{11}(t) & u_{12}(t) \\ u_{21}(t) & u_{22}(t) \end{bmatrix}^{-1} = \frac{1}{W_0}\begin{bmatrix} u_{22}(t) & -u_{12}(t) \\ -u_{21}(t) & u_{11}(t) \end{bmatrix} \tag{28.6}$$

Let x_0 and v_0 be the *given* initial values of $x(t)$ and $v(t)$ at $t = 0$, and introduce the following notation:

$$\begin{bmatrix} x \\ v \end{bmatrix}_0 = \begin{bmatrix} x_0 \\ v_0 \end{bmatrix} \tag{28.7}$$

Hence as a consequence of (28.3) we have

$$\begin{bmatrix} x_0 \\ v_0 \end{bmatrix} = \begin{bmatrix} u_{11}(0) & u_{12}(0) \\ u_{21}(0) & u_{22}(0) \end{bmatrix}\begin{bmatrix} A_1 \\ A_2 \end{bmatrix} \tag{28.8}$$

or

$$\begin{bmatrix} A_1 \\ A_2 \end{bmatrix} = \begin{bmatrix} u_{11}(0) & u_{12}(0) \\ u_{21}(0) & u_{22}(0) \end{bmatrix}^{-1}\begin{bmatrix} x_0 \\ v_0 \end{bmatrix}$$

$$= \frac{1}{W_0}\begin{bmatrix} u_{22}(0) & -u_{12}(0) \\ -u_{21}(0) & u_{11}(0) \end{bmatrix}\begin{bmatrix} x_0 \\ v_0 \end{bmatrix} \tag{28.9}$$

This determines the column of arbitrary constants in terms of the given initial conditions. If (28.9) is substituted into (28.3), the result may be written in the form

$$\begin{bmatrix} x \\ v \end{bmatrix}_t = \frac{1}{W_0}\begin{bmatrix} u_{11}(t) & u_{12}(t) \\ u_{21}(t) & u_{22}(t) \end{bmatrix}\begin{bmatrix} u_{22}(0) & -u_{12}(0) \\ -u_{21}(0) & u_{11}(0) \end{bmatrix}\begin{bmatrix} x_0 \\ v_0 \end{bmatrix} \tag{28.10}$$

CASCADING THE SOLUTION BY THE USE OF MATRICES
The extension of the solution beyond the fundamental period $0 \leqslant t \leqslant T$ may be effectively carried out by matrix multiplication. If a change of variable of the form

$$\tau = t - nT \qquad \text{where } 0 \leqslant \tau \leqslant T$$
$$n = 0, 1, 2, 3, \ldots \tag{28.11}$$

is introduced in the Hill equation (28.1), the equation is transformed to

$$\frac{d^2 x}{d\tau^2} + F(\tau)x = 0 \tag{28.12}$$

[1] L. A. Pipes, *Philosophical Magazine*, vol. 30, pp. 370–395, 1940.
[2] E. Meissner, *Schweiz. Bauz.* vol. 72, p. 95, 1918.

as a consequence of the periodicity of $F(t)$ since

$$F(\tau + nT) = F(\tau) \tag{28.13}$$

Hence Hill's equation is *invariant* to the change in variable (28.11). It is thus apparent that a solution of the form (28.10) may be obtained in each interval $0 \leqslant \tau \leqslant T$. The *final* values of x and v at the end of one interval of the variation of $F(t)$ are the *initial* values of x and v in the following interval.

At the end of the fundamental period $t = T$, (28.10) becomes

$$\begin{bmatrix} x \\ v \end{bmatrix}_T = \frac{1}{W_0} \begin{bmatrix} u_{11}(T) & u_{12}(T) \\ u_{21}(T) & u_{22}(T) \end{bmatrix} \begin{bmatrix} u_{22}(0) & -u_{12}(0) \\ -u_{21}(0) & u_{11}(0) \end{bmatrix} \begin{bmatrix} x_0 \\ v_0 \end{bmatrix} \tag{28.14}$$

Let

$$\begin{aligned}
[M] &= \begin{bmatrix} A & B \\ C & D \end{bmatrix} = \frac{1}{W_0} \begin{bmatrix} u_{11}(T) & u_{12}(T) \\ u_{21}(T) & u_{22}(T) \end{bmatrix} \begin{bmatrix} u_{22}(0) & -u_{12}(0) \\ -u_{21}(0) & u_{11}(0) \end{bmatrix} \\
&= \begin{bmatrix} [u_{11}(T)u_{22}(0) - u_{12}(T)u_{21}(0)] & [u_{12}(T)u_{11}(0) - u_{11}(T)u_{12}(0)] \\ [u_{21}(T)u_{22}(0) - u_{22}(T)u_{21}(0)] & [u_{22}(T)u_{11}(0) - u_{21}(T)u_{12}(0)] \end{bmatrix}
\end{aligned} \tag{28.15}$$

Then by the use of this notation, (28.14) may be written in the following form:

$$\begin{bmatrix} x \\ v \end{bmatrix}_T = [M] \begin{bmatrix} x_0 \\ v_0 \end{bmatrix} = \begin{bmatrix} A & B \\ C & D \end{bmatrix} \begin{bmatrix} x \\ v \end{bmatrix}_0 \tag{28.16}$$

Since the Wronskian W_0 of the two solutions $u_1(t)$ and $u_2(t)$ is a constant, it may be seen that the determinant of the matrix $[M]$ has the following value:

$$|M| = \begin{vmatrix} A & B \\ C & D \end{vmatrix} = (AD - BC) = 1 \tag{28.17}$$

At the end of the second period of the variation of $F(t)$, we have

$$\begin{aligned}
\begin{bmatrix} x \\ v \end{bmatrix}_{2T} &= \begin{bmatrix} A & B \\ C & D \end{bmatrix} \begin{bmatrix} x \\ v \end{bmatrix}_T \\
&= \begin{bmatrix} A & B \\ C & D \end{bmatrix}^2 \begin{bmatrix} x_0 \\ v_0 \end{bmatrix} = [M]^2 \begin{bmatrix} x \\ v \end{bmatrix}_0
\end{aligned} \tag{28.18}$$

Similarly, it may be seen that at the end of n periods, we have

$$\begin{bmatrix} x \\ v \end{bmatrix}_{nT} = \begin{bmatrix} A & B \\ C & D \end{bmatrix}^n \begin{bmatrix} x_0 \\ v_0 \end{bmatrix} = [M]^n \begin{bmatrix} x \\ v \end{bmatrix}_0 \tag{28.19}$$

The solution within the $(n + 1)$th interval is given by

$$\begin{bmatrix} x \\ v \end{bmatrix}_{nT+\tau} = \begin{bmatrix} u_{11}(\tau) & u_{12}(\tau) \\ u_{21}(\tau) & u_{22}(\tau) \end{bmatrix} \begin{bmatrix} u_{22}(0) & -u_{12}(0) \\ -u_{21}(0) & u_{11}(0) \end{bmatrix} \begin{bmatrix} x \\ v \end{bmatrix}_{nT} \tag{28.20}$$

where $0 \leqslant \tau \leqslant T$.

Equation (28.20) is the solution of Hill's equation at any time $t > 0$ in terms of the *initial* conditions and two linearly independent solutions of Hill's equation in the fundamental interval $0 \leqslant t \leqslant T$.

COMPUTATION OF INTEGRAL POWERS OF [M]

The integral powers of the matrix $[M]$ required in Eq. (28.19) may be most simply obtained by the use of *Sylvester's theorem*.[1]　This important theorem states that if the characteristic or latent roots r_1 and r_2 of the matrix $[M]$ are *distinct* and $P[M]$ is any *polynomial* of $[M]$, then

$$P[M] = P(r_i)[H_0(r_1)] + P(r_2)[H_0(r_2)] \tag{28.21}$$

where

$$[H_0(r_i)] = \frac{[F(r_i)]}{(d\Delta/dr)_{r=r_i}} \tag{28.22}$$

$[F(r)]$ is the *adjoint* of the characteristic matrix of the matrix $[M]$ given by No. 9 of Table 2.　$\Delta(r)$ is the characteristic function of the matrix $[M]$ as given by No. 5 of Table 2; it has the form

$$\Delta(r) = r^2 - r(A + D) + 1 \tag{28.23}$$

Hence

$$\frac{d\Delta(r)}{dr} = 2r - (A + D) \tag{28.24}$$

In order to apply Sylvester's theorem to the computation of $[M]^n$, let the polynomial of (28.21) be given by $P[M] = [M]^n$.　If $A + D \neq \pm 2$, the two latent roots of $[M]$ are distinct, and entry No. 10 of Table 2 readily follows from Sylvester's theorem.　In case that $A + D = 2$, both latent roots become equal to unity and No. 12 of Table 2 gives the powers of $[M]$ as a consequence of a confluent form of Sylvester's theorem.　If $A + D = -2$, both latent roots of $[M]$ become equal to -1.　In this case entry No. 13 gives the proper form for the powers of $[M]$.　In the important case of distinct latent roots and symmetry of $[M]$ so that $A = D$, the proper form for the powers of $[M]$ is given by No. 11 of Table 2.　Entries No. 10 to No. 13 of Table 2 cover all possible special cases that can arise in practice.　These forms are useful in the theory of four-terminal networks.[2]　The general theory will now be applied to the solution of some representative special cases.

---

[1] Frazer et al., *op. cit.*

[2] Pipes, *op. cit.*

Table 2 Fundamental relations involving the matrix $[M]$

1 The matrix $[M]$	$[M] = \begin{bmatrix} A & B \\ C & D \end{bmatrix}$
2 The determinant of $[M]$	$\|M\| = AD - BC = 1$
3 The inverse of $[M]$	$[M]^{-1} = \begin{bmatrix} D & -B \\ -C & A \end{bmatrix}$
4 The characteristic matrix of $[M]$	$[f(r)] = \begin{bmatrix} (r - A) & -B \\ -C & (r - D) \end{bmatrix}$
5 The characteristic function of $[M]$	$\Delta(r) = $ determinant of $[f(r)]$ $= r^2 - r(A + D) + 1$
6 The characteristic equation of $[M]$	$\Delta(r) = r^2 - r(A + D) + 1 = 0$
7 The characteristic or latent roots of $[M]$ (three cases)	Roots of $\Delta(r) = 0$ If $A + D \neq 2$, $r_1 = e^a$, $r_2 = e^{-a}$ where $\cosh a = \dfrac{(A + D)}{2}$ If $A + D = +2$, $r_1 = r_2 = +1$ If $A + D = -2$, $r_1 = r_2 = -1$
8 The inverse of the characteristic matrix of $[M]$	$[f(r)]^{-1} = \dfrac{1}{\Delta(r)} \begin{bmatrix} (r - D) & B \\ C & (r - A) \end{bmatrix}$
9 The adjoint of the characteristic matrix of $[M]$	$[F(r)] = (r)[f(r)]^{-1}$ $= \begin{bmatrix} (r - D) & B \\ C & (r - A) \end{bmatrix}$
10 Integral powers of $[M]$ (distinct latent roots) $A + D \neq \pm 2$ $n = 0, 1, 2, 3, \ldots$ $\cosh a = \dfrac{(A + D)}{2}$	$[M]^n = \dfrac{1}{S_1} \begin{bmatrix} S_{n+1} - DS_n & BS_n \\ CS_n & S_{n+1} - AS_n \end{bmatrix}$ where $S_n = \sinh an$
11 Integral powers of $[M]$ symmetric case $A = D$ (distinct latent roots) $A = D \neq \pm 1$ $\cosh a = A$	$[M]^n = \begin{bmatrix} \cosh an & Z_0 \sinh an \\ \dfrac{\sinh an}{Z_0} & \cosh an \end{bmatrix}$ $Z_0 = \left(\dfrac{B}{C}\right)^{\frac{1}{2}}$ $\quad n = 0, 1, 2, 3, \ldots$
12 Integral powers of $[M]$ (equal latent roots) $(A + D) = +2$ $r_1 = r_2 = +1$	$[M]^n = \begin{bmatrix} n(A - 1) + 1 & nB \\ nC & n(D - 1) + 1 \end{bmatrix}$ $n = 0, 1, 2, 3, \ldots$
13 Integral powers of $[M]$ (equal latent roots) $(A + D) = -2$ $r_1 = r_2 = -1$ $n = 0, 1, 2, 3, \ldots$	$[M]^n = nE_{n-1} \begin{bmatrix} (A + 1) & B \\ C & (D + 1) \end{bmatrix} + E_n I$ $E_n = e^{jn\pi}$ $I = \begin{bmatrix} 1 & 0 \\ 0 & 1 \end{bmatrix} = $ unit matrix

29 THE SOLUTION OF THE HILL-MEISSNER EQUATION

The Hill equation (27.1) in the special case for which the function $F(t)$ has the form of the rectangular ripple of Fig. 29.1 was used by Meissner[1] in his analysis of the vibrations of the driving rods of locomotives. The stability of the solution of this equation has been studied by van der Pol and Strutt.[2]

Let T be the fundamental period of the rectangular ripple, and let the ripple vary in height from h_1 to h_2. Hill's equation now reduces to

$$\frac{d^2 x}{dt^2} + h_1 x = 0 \qquad 0 \leqslant t \leqslant \frac{T}{2}$$
$$\frac{d^2 x}{dt^2} + h_2 x = 0 \qquad \frac{T}{2} \leqslant t \leqslant T \tag{29.1}$$

If we let

$$w_1 = \sqrt{h_1} \qquad \text{and} \qquad w_2 = \sqrt{h_2} \tag{29.2}$$

it can be seen that the solution of (29.1) in the range $0 \leqslant t \leqslant T/2$ may be written in the matrix form

$$\begin{bmatrix} x \\ v \end{bmatrix}_t = \begin{bmatrix} \cos w_1 t & \sin \dfrac{w_1 t}{w_1} \\ -w_1 \sin w_1 t & \cos w_1 t \end{bmatrix} \begin{bmatrix} x_0 \\ v_0 \end{bmatrix} \qquad 0 \leqslant t \leqslant \frac{T}{2} \tag{29.3}$$

where x_0 and v_0 are the initial values of x and $\dot{x}$ at $t = 0$ and the column on the left gives the values of x and $\dot{x}$ at t. The values of x and v at $t = T/2$ are given by

$$\begin{bmatrix} x \\ v \end{bmatrix}_{T/2} = \begin{bmatrix} \cos \theta_1 & \dfrac{\sin \theta_1}{w_1} \\ -w_1 \sin \theta_1 & \cos \theta_1 \end{bmatrix} \begin{bmatrix} x_0 \\ v_0 \end{bmatrix} \tag{29.4}$$

where

$$\theta_1 = w_1 \frac{T}{2} = \frac{T}{2} \sqrt{h_1} \tag{29.5}$$

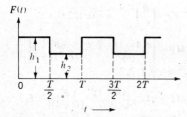

Fig. 29.1

[1] Meissner, *op. cit.*
[2] B. van der Pol and M. J. O. Strutt, *Philosophical Magazine*, vol. 5, p. 18, 1928.

Since the values of x and v at $t = T/2$ are the *initial* values of x and v in the next interval, $T/2 \leqslant t \leqslant T$, we have

$$\begin{bmatrix} x \\ v \end{bmatrix}_T = \begin{bmatrix} \cos\theta_2 & \dfrac{\sin\theta_2}{w_2} \\ -w_2\sin\theta_2 & \cos\theta_2 \end{bmatrix} \begin{bmatrix} \cos\theta_1 & \dfrac{\sin\theta_1}{w_1} \\ -w_1\sin\theta_1 & \cos\theta_1 \end{bmatrix} \begin{bmatrix} x_0 \\ v_0 \end{bmatrix} \tag{29.6}$$

where

$$\theta_2 = w_2\frac{T}{2} = \frac{T}{2}\sqrt{h_2} \tag{29.7}$$

If the indicated multiplication in (29.6) is performed, the result may be written in the form

$$\begin{bmatrix} x \\ v \end{bmatrix}_T = \begin{bmatrix} A & B \\ C & D \end{bmatrix} \begin{bmatrix} x_0 \\ v_0 \end{bmatrix} = [M]\begin{bmatrix} x_0 \\ v_0 \end{bmatrix} \tag{29.8}$$

In this case the matrix $[M]$ has the form

$$[M] = \begin{vmatrix} A & B \\ C & D \end{vmatrix}$$

$$= \begin{bmatrix} \left(M_1 M_2 - N_1 N_2 \dfrac{w_1}{w_2}\right) & \left(M_2 \dfrac{N_1}{w_1} + M_1 \dfrac{N_2}{w_2}\right) \\ -(M_1 N_2 w_2 + M_2 N_1 w_1) & \left(M_1 M_2 - N_1 N_2 \dfrac{w_2}{w_1}\right) \end{bmatrix} \tag{29.9}$$

where the notation

$$M_i = \cos\theta_i \quad\text{and}\quad N_i = \sin\theta_i \tag{29.10}$$

has been introduced.

The values of x and v at the end of n complete periods of the rectangular tipple of Fig. 29.1 are given by

$$\begin{bmatrix} x \\ v \end{bmatrix}_{nT} = \begin{bmatrix} A & B \\ C & D \end{bmatrix} \begin{bmatrix} x_0 \\ v_0 \end{bmatrix} = [M]^n \begin{bmatrix} x_0 \\ v_0 \end{bmatrix} \tag{29.11}$$

If numerical values are given the matrix $[M]^n$ may be computed by No. 10 of Table 2 or its alternative forms.

THE STABILITY OF THE SOLUTION

The nature of the solution depends on the form taken by the powers of the matrix $[M]$. If

$$(A + D) \neq \pm 2 \tag{29.12}$$

the matrix $[M]$ has two *distinct* latent roots and $[M]^n$ is given by No. 10 of Table 2. Now

$$\cosh a = \frac{(A + D)}{2} = \cos\theta_1\cos\theta_2 - \frac{\sin\theta_1}{2}\sin\theta_2\frac{\theta_1^2 + \theta_2^2}{\theta_1\theta_2} \tag{29.13}$$

as can be seen directly from (29.9). There are two separate cases to be considered. If

$$\left|\frac{A+D}{2}\right| > 1 \tag{29.14}$$

then $\cosh(a) > 1$, and a must have a positive real part. In this case (29.11) may be written with the aid of No. 10 of Table 2 in the following form:

$$\begin{bmatrix} x \\ v \end{bmatrix}_{nT} = \frac{1}{\sinh a} \begin{bmatrix} \begin{bmatrix} \sinh a(n+1) \\ -D\sinh an \end{bmatrix} & B\sinh an \\ C\sinh an & \begin{bmatrix} \sinh a(n+1) \\ -A\sinh an \end{bmatrix} \end{bmatrix} \begin{bmatrix} x_0 \\ v_0 \end{bmatrix} \tag{29.15}$$

Since the real part of the number a is positive, it can be seen that x and v will *increase* with t. In physical applications this leads to *unstable* solutions. If

$$\left|\frac{A+D}{2}\right| < 1$$

then

$$a = jb \tag{29.16}$$

where

$$\cos b = \frac{A+D}{2} \tag{29.17}$$

If the value $a = jb$ is substituted into (29.15), the result is

$$\begin{bmatrix} x \\ v \end{bmatrix}_{nT} = \frac{1}{\sin b} \begin{bmatrix} \begin{bmatrix} \sin b(n+1) \\ -D\sin bn \end{bmatrix} & B\sin bn \\ C\sin bn & \begin{bmatrix} \sin b(n+1) \\ -A\sin bn \end{bmatrix} \end{bmatrix} \begin{bmatrix} x_0 \\ v_0 \end{bmatrix} \tag{29.18}$$

It can be seen that in this case the values of x and v *oscillate* with increasing t. This case leads to *stable* solutions in physical applications. The dividing line between stable and unstable solutions leads to the case for which

$$\left|\frac{A+D}{2}\right| = 1 \tag{29.19}$$

In this unique case the matrix $[M]$ has equal latent roots. If $(A + D) = 2$, the latent roots are $r_1 = r_2 = 1$ and the proper form for $[M]^n$ is given by No. 12 of Table 2. If on the other hand, $(A + D) = -2$, the latent roots of $[M]$ are $r_1 = r_2 = -1$, and the proper form for $[M]^n$ is given by No. 13 of Table 2. In general, the condition (29.19) leads to instability in the physical applications.

The question of the stability of the solution may be summarized by the following expressions:

$$\left| \frac{A+D}{2} \right| > 1 \qquad \text{unstable}$$

$$\left| \frac{A+D}{2} \right| < 1 \qquad \text{stable} \tag{29.20}$$

$$\left| \frac{A+D}{2} \right| = 1 \qquad \text{unstable}$$

These are the conclusions reached by van der Pol and Strutt[1] in a different manner by the use of the Floquet theory of Hill's equation. The present analysis indicates the intimate connection between the criterion for stability and the nature of the latent roots of $[M]$.

30   SOLUTION OF HILL'S EQUATION IF $F(t)$ IS A SUM OF STEP FUNCTIONS

The matrix method of solution is particularly effective in solving Hill's equation in cases when the function $F(t)$ may be represented as a sum of step function in the interval $0 \leqslant t \leqslant T$, as shown in Fig. 30.1.

Let the function $F(t)$ be composed of m step functions each of length T_0 and heights $h_1, h_2, \ldots, h_m$, where

$$mT_0 = T \tag{30.1}$$

Let the following notation be introduced:

$$w_k = \sqrt{h_k} \qquad \theta_k = T_0 \sqrt{h_k} \tag{30.2}$$

$$[M]_k = \begin{bmatrix} P_k & Q_k \\ R_k & P_k \end{bmatrix} = \begin{bmatrix} \cos \theta_k & \dfrac{\sin \theta_k}{w_k} \\ -w_k \sin \theta_k & \cos \theta_k \end{bmatrix} \tag{30.3}$$

[1] B. van der Pol and M. J. O. Strutt, *Philosophical Magazine*, vol. 5, p. 18, 1928.

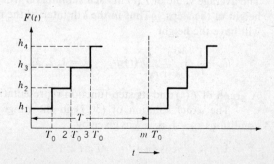

Fig. 30.1

If we let

$$\begin{bmatrix} A & B \\ C & D \end{bmatrix} = [M]_m [M]_{m-1} \cdots [M]_2 [M]_1 = [M] \tag{30.4}$$

it can be seen by reasoning similar to that used in Sec. 29 that at the end of the fundamental interval $T = mT_0$ the solution is given by

$$\begin{bmatrix} x \\ v \end{bmatrix}_T = \begin{bmatrix} A & B \\ C & D \end{bmatrix} \begin{bmatrix} x_0 \\ v_0 \end{bmatrix} \tag{30.5}$$

The solution at the end of n intervals when $t = nT$ is given by

$$\begin{bmatrix} x \\ v \end{bmatrix}_{nT} = \begin{bmatrix} A & B \\ C & D \end{bmatrix}^n \begin{bmatrix} x_0 \\ v_0 \end{bmatrix} \tag{30.6}$$

The stability of the solution is then determined by (29.20).

This method can be used to study the stability and to obtain the approximate solution of equations of the Hill type by representing the given function $F(t)$ throughout the fundamental interval by a sum of step functions as in the method of mean coefficients.[1]   To illustrate the general procedure, consider a Hill's equation of the form

$$\frac{d^2 x}{dt^2} + F(t) x = 0 \tag{30.7}$$

where the function $F(t)$ has the form

$$F(t) = 16\pi^2 e^{-2t} - \tfrac{1}{4} \qquad 0 \leqslant t \leqslant 1 \tag{30.8}$$

in the fundamental interval $0 \leqslant t \leqslant 1$ and is a *periodic* function of fundamental period $T = 1$.

In order to represent the function $F(t)$ by step functions, divide the fundamental interval $T = 1$ into ten equal subintervals of length $T_0 = \tfrac{1}{10}$. The average value of $F(t)$ in each subinterval will be taken to be equal to the height of the steps.   Thus in the kth interval the representative step function will have the height

$$h_k = \frac{1}{T_0} \int_{(k-1)T_0}^{kT_0} F(t)\, dt \qquad k = 1, 2, 3, \ldots \tag{30.9}$$

A graph of $F(t)$ and its step-function representation is given in Fig. 30.2.

The *exact* solution of (30.7) in the range $0 \leqslant t \leqslant 1$, with the initial conditions $x_0 = 1, v_0 = \tfrac{1}{2}$, is given by

$$x(t) = e^{t/2} \cos 4\pi\, e^{-t} \qquad 0 \leqslant t \leqslant 1 \tag{30.10}$$

[1] Fraser et al., *op. cit.*

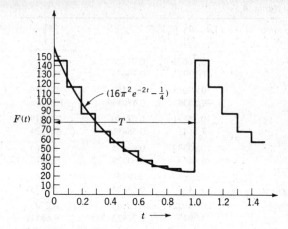

Fig. 30.2

The values of h_k, w_k, and the elements of the matrices $[M]_k$ are tabulated in Table 3.   The matrices $[M]_k$ and the values of x_k and v_k of the solution at the end of the kth subinterval as a result of the initial conditions $x_0 = 1$ and $v_0 = \frac{1}{2}$ are given in Table 4.   The matrix $[M]$ is obtained by the multiplication of the chain of matrices in the form

$$[M] = [M]_{10}[M]_9[M]_8 \cdots [M]_1 = \begin{bmatrix} A & B \\ C & D \end{bmatrix}$$

$$= \begin{bmatrix} 0.6504 & -1.8412 \\ 0.1956 & -3.9488 \end{bmatrix} \tag{30.11}$$

As a consequence of (30.11), we have

$$\frac{A + D}{2} = -1.6492 = \cosh a \tag{30.12}$$

Table 3

Step	h_k	w_k	P_k	Q_k	R_k
1	142.88	11.953	0.36673	0.07783	−11.1202
2	116.93	10.813	0.47015	0.08162	−9.5438
3	95.69	9.782	0.55851	0.08480	−8.1142
4	78.30	8.849	0.63339	0.08745	−6.8473
5	64.06	8.004	0.69644	0.08966	−5.7437
6	52.41	7.239	0.74922	0.09149	−4.7945
7	42.86	6.547	0.79326	0.09301	−3.9862
8	35.04	5.920	0.82983	0.09426	−3.3032
9	28.64	5.352	0.86016	0.09529	−2.7298
10	23.40	4.838	0.88523	0.09614	−2.2506

Table 4

Step k	$\begin{bmatrix} A & B \\ C & D \end{bmatrix} = [M]_k$		$\begin{bmatrix} x \\ v \end{bmatrix}$
1	0.36673	0.07783	$1.0 \quad x_0$
	−11.1202	0.36673	$0.5 = v_0$
2	0.47015	0.08162	0.4056
	−9.5438	0.47015	−10.9368
3	0.55851	0.08480	−0.7019
	−8.1142	0.55851	−9.0134
4	0.63339	0.08745	−1.1564
	−6.8473	0.63339	0.6617
5	0.69644	0.08966	−0.6746
	−5.7437	0.69644	8.3373
6	0.74922	0.09149	0.2777
	−4.7945	0.74922	9.6811
7	0.79326	0.09301	1.0938
	−3.9862	0.79326	5.9218
8	0.82983	0.09426	1.4184
	−3.3032	0.82983	0.3374
9	0.86016	0.09529	1.2088
	−2.7298	0.86016	−4.4053
10	0.88523	0.09614	0.6200
	−2.2506	0.88523	−7.0890

The stability criterion (29.20) indicates that the solution is *unstable*. A table of hyperbolic functions gives the following value for a:

$$a = 1.085 + j\pi \tag{30.13}$$

The solution at the end of n periods of the function $F(t)$ is given by (29.15) with the value given by (30.13) for a.

31 A CLASS OF HILL'S EQUATIONS WITH EXPONENTIAL VARIATION

If Hill's equation has the form

$$\frac{d^2 x}{dt^2} + (k^2 e^{-2t} - n^2) x = 0 \qquad 0 \leqslant t \leqslant T \tag{31.1}$$

throughout the fundamental interval $0 \leqslant t \leqslant T$, it may be solved in this interval by means of Bessel functions[1] in the form

$$x(t) = A_1 J_n(k e^{-t}) + A_2 Y_n(k e^{-t}) \qquad 0 \leqslant t \leqslant T \tag{31.2}$$

[1] E. Kamke, "Differentialgleichungen Lösungsmethoden und Lösungen," p. 442, Chelsea Publishing Company, New York, 1948.

where J_n and Y_n are Bessel functions of the first and second kinds of order n^1 and the A's are arbitrary constants. If we let

$$z = k e^{-t} \tag{31.3}$$

and compute the derivative

$$\frac{d}{dt} J_n(z) = \frac{d}{dz} J_n(z) \frac{dz}{dt} = -z J_n'(z) \tag{31.4}$$

it is evident that we may write $x(t)$ and $\dot{x}(t) = v$ in the convenient matrix form

$$\begin{bmatrix} x \\ v \end{bmatrix}_t = \begin{bmatrix} J_n(z) & Y_n(z) \\ -z J_n'(z) & -z Y_n'(z) \end{bmatrix} \begin{bmatrix} A_1 \\ A_2 \end{bmatrix} \tag{31.5}$$

The Wronskian, Eq. (28.5), now takes the form

$$W_0 = -z[J_n(z) Y_n'(z) - J_n'(z) Y_n(z)] \tag{31.6}$$

From the theory of Bessel functions[1] we have the following result:

$$J_n(z) Y_n'(z) - J_n'(z) Y_n(z) = \frac{2}{\pi z} \tag{31.7}$$

If this is substituted into (31.6), we obtain

$$W_0 = -\frac{2}{\pi} \tag{31.8}$$

The matrix $[M]$ of (28.15) now becomes

$$[M] = \begin{bmatrix} A & B \\ C & D \end{bmatrix}$$

$$= -\frac{\pi}{2} \begin{bmatrix} J_n(k e^{-T}) & Y_n(k e^{-T}) \\ -k e^{-T} J_n'(k e^{-T}) & -k e^{-T} Y_n'(k e^{-T}) \end{bmatrix}$$

$$\begin{bmatrix} -k Y_n'(k) & -Y_n(k) \\ +k J_n'(k) & +J_n(k) \end{bmatrix} \tag{31.9}$$

The conditions for stability are given by (29.20) and the solution after n periods is given by

$$\begin{bmatrix} x \\ v \end{bmatrix}_{nT} = \begin{bmatrix} A & B \\ C & D \end{bmatrix}^n \begin{bmatrix} x_0 \\ v_0 \end{bmatrix} \tag{31.10}$$

If $k = 4\pi$ and $n = \frac{1}{2}$, Eq. (31.1) reduces to the one discussed in Sec. 30 by another method.

---

[1] N. W. McLachlan, "Bessel Functions for Engineers," Oxford University Press, London, 1941.

32   HILL'S EQUATION WITH A SAWTOOTH VARIATION

As a final example of the general method, consider the solution of Hill's equation (27.1) when the function $F(t)$ has the form of a sawtooth wave as shown in Fig. 32.1.   The function $F(t)$ has the form

$$F(t) = at + b \qquad 0 \leqslant t \leqslant T \tag{32.1}$$

in the fundamental interval and repeats the variation at intervals of T.   Hill's equation now takes the form

$$\frac{d^2 x}{dt^2} + (at + b)x = 0 \qquad 0 \leqslant t \leqslant T \tag{32.2}$$

in the fundamental interval.   Let

$$z = at + b \tag{32.3}$$

This change of variable transforms (32.2) into the form

$$\frac{d^2 x}{dz^2} + \frac{z}{a^2}x = 0 \tag{32.4}$$

Equation (32.4) is a form of Stokes's equation.   The linearly independent solutions of this equation have been tabulated.[1]   The solution of (32.4) may also be expressed in terms of Bessel functions of order one-third[2] in the following form:

$$x = (z)^{\frac{1}{2}}[A_1 J_{\frac{1}{3}}(kz^{\frac{3}{2}}) + A_2 Y_{\frac{1}{3}}(kz^{\frac{3}{2}})] \tag{32.5}$$

where the A's are arbitrary constants and

$$k = \frac{2}{3a} \tag{32.6}$$

[1] "Tables of the Modified Hankel Functions of Order One-Third and of Their Derivatives," Harvard University Press, Cambridge, Mass., 1945.

[2] Kamke, *op. cit.*

$F(t)$

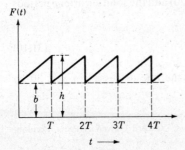

Fig. 32.1

In the notation of Sec. 28 we have

$$u_1(t) = (z)^{\frac{1}{2}} J_{\frac{1}{3}}(kz^{\frac{3}{2}}) \qquad u_2(t) = (z)^{\frac{1}{2}} Y_{\frac{1}{3}}(kz^{\frac{3}{2}}) \tag{32.7}$$

By the use of the recurrence relation of the theory of Bessel functions [1]

$$\frac{d}{dy} J_n(y) = -\frac{n}{y} J_n(y) + J_{n-1}(y) \tag{32.8}$$

the following derivatives may be calculated:

$$\frac{du_1}{dt} = z J_{-\frac{2}{3}}(kz^{\frac{3}{2}}) \qquad \frac{du_2}{dt} = z Y_{-\frac{2}{3}}(kz^{\frac{3}{2}}) \tag{32.9}$$

Hence the solution of (32.2) in the fundamental interval $0 \leqslant t \leqslant T$ may be expressed in the following form:

$$\begin{bmatrix} x \\ v \end{bmatrix}_t = \begin{bmatrix} (z)^{\frac{1}{2}} J_{\frac{1}{3}}(kz^{\frac{3}{2}}) & (z)^{\frac{1}{2}} Y_{\frac{1}{3}}(kz^{\frac{3}{2}}) \\ z J_{-\frac{2}{3}}(kz^{\frac{3}{2}}) & z Y_{-\frac{2}{3}}(kz^{\frac{3}{2}}) \end{bmatrix} \begin{bmatrix} A_1 \\ A_2 \end{bmatrix} \tag{32.10}$$

Let the following notation be introduced:

$$\theta = kb^{\frac{3}{2}} = \frac{2b^{\frac{3}{2}}}{3a} \tag{32.11}$$

$$\phi = kh^{\frac{3}{2}} = \frac{2h^{\frac{3}{2}}}{3a} \tag{32.12}$$

The Wronskian W_0 of (28.5) in this case is given by

$$W_0 = \begin{vmatrix} (b)^{\frac{1}{2}} J_{\frac{1}{3}}(\theta) & (b)^{\frac{1}{2}} Y_{\frac{1}{3}}(\theta) \\ b J_{-\frac{2}{3}}(\theta) & b Y_{-\frac{2}{3}}(\theta) \end{vmatrix}$$
$$= b^{\frac{3}{2}} [J_{\frac{1}{3}}(\theta) Y_{-\frac{2}{3}}(\theta) - J_{-\frac{2}{3}}(\theta) Y_{\frac{1}{3}}(\theta)] \tag{32.13}$$

The matrix $[M]$ of (28.15) is now

$$[M] = \frac{1}{W_0} \begin{bmatrix} (h)^{\frac{1}{2}} J_{\frac{1}{3}}(\phi) & (h)_{\frac{1}{2}} Y_{\frac{1}{3}}(\phi) \\ h J_{-\frac{2}{3}}(\phi) & h Y_{-\frac{2}{3}}(\phi) \end{bmatrix}$$

$$\begin{bmatrix} b Y_{-\frac{2}{3}}(\theta) & (b)^{\frac{1}{2}} Y_{\frac{1}{3}}(\theta) \\ -b J_{-\frac{2}{3}}(\theta) & (b)^{\frac{1}{2}} J_{\frac{1}{3}}(\theta) \end{bmatrix} = \begin{bmatrix} A & B \\ C & D \end{bmatrix} \tag{32.14}$$

Hence

$$A = \frac{b(h)^{\frac{3}{2}}}{W_0} [J_{\frac{1}{3}}(\phi) Y_{-\frac{2}{3}}(\theta) - J_{-\frac{2}{3}}(\theta) Y_{\frac{1}{3}}(\phi)]$$

$$D = \frac{h(b)^{\frac{3}{2}}}{W_0} [J_{\frac{1}{3}}(\theta) Y_{-\frac{2}{3}}(\phi) - J_{-\frac{2}{3}}(\phi) Y_{\frac{1}{3}}(\theta)]$$

$$\tag{32.15}$$

$$B = \frac{(hb)^{\frac{3}{2}}}{W_0} [J_{\frac{1}{3}}(\theta) Y_{\frac{1}{3}}(\phi) - J_{\frac{1}{3}}(\phi) Y_{\frac{1}{3}}(\theta)]$$

$$C = \frac{hb}{W_0} [J_{-\frac{2}{3}}(\phi) Y_{-\frac{2}{3}}(\theta) - J_{-\frac{2}{3}}(\theta) Y_{-\frac{2}{3}}(\phi)]$$

---

[1] N. W. McLachlan, "Bessel Functions for Engineers," Oxford University Press, London, 1941.

The stability conditions for the solution are given by (29.20) and the magnitude of the solution after n periods is given by (28.19) in terms of the initial conditions.

ASYMPTOTIC APPROXIMATIONS

If b is large so that

$$(b^3)^{\frac{1}{4}} \gg \frac{3a}{2} \tag{32.16}$$

then terms of the order of $1/\theta^{\frac{3}{2}}$ and $1/\phi^{\frac{3}{2}}$ can be neglected and the various Bessel functions involved may be represented by the *dominant term* of their asymptotic expansions. If this is done, we have

$$J_{\frac{1}{3}}(w) = \left(\frac{2}{\pi w}\right)^{\frac{1}{2}} \cos\left(w - \frac{\pi}{4} - \frac{\pi}{6}\right)$$

$$Y_{\frac{1}{3}}(w) = \left(\frac{2}{\pi w}\right)^{\frac{1}{2}} \sin\left(w - \frac{\pi}{4} - \frac{\pi}{6}\right)$$

$$J_{-\frac{1}{3}}(w) = \left(\frac{2}{\pi w}\right)^{\frac{1}{2}} \cos\left(w - \frac{\pi}{4} + \frac{\pi}{3}\right) \tag{32.17}$$

$$Y_{-\frac{1}{3}}(w) = \left(\frac{2}{\pi w}\right)^{\frac{1}{2}} \sin\left(w - \frac{\pi}{4} + \frac{\pi}{3}\right)$$

If the Bessel functions in (32.13) and (32.14) are represented by their dominant terms (32.17), the following expressions are obtained for W_0 and $[M]$:

$$W_0 = \frac{3a}{\pi} \tag{32.18}$$

and

$$[M] = \frac{2}{3a}\left(\frac{bh}{\theta\phi}\right)^{\frac{1}{4}}\begin{bmatrix} (b)^{\frac{1}{4}}\cos(\theta - \phi) & \sin(\phi - \theta) \\ (bh)^{\frac{1}{4}}\sin(\theta - \phi) & (h)^{\frac{1}{4}}\cos(\theta - \phi) \end{bmatrix} \tag{32.19}$$

In this case we have

$$\frac{(A + D)}{2} = \frac{\cos(\theta - \phi)}{3a}[b(h)^{\frac{1}{4}} + h(b)^{\frac{1}{4}}] \tag{32.20}$$

and the stability criterion (29.20) may be easily applied.

33 CONCLUSION

The natural manner by which equations of the Mathieu-Hill type may be solved in terms of given initial conditions by the use of matrix algebra has been demonstrated. The powerful matrix notation and the use of Sylvester's theorem to raise the matrices involved to the required powers greatly facilitate the solution of equations of this type.

THE ANALYSIS OF TIME-VARYING ELECTRICAL CIRCUITS†

34 INTRODUCTION

The general theory of electrical circuits whose component parameters are linear and constant has been extensively developed and is well understood. In recent years, considerable attention and effort has been directed to the analysis and performance of circuits whose parameters vary with the time. Many of the most important and interesting problems of circuit theory involve variable parameters. For example, the equivalent circuits of the microphone transmitter, the condenser microphone, the induction generator, the super-regenerator, and many other practical devices contain parameters that are time-varying. Many systems in mechanical and acoustical engineering in which the compliance or the inertia parameters vary with the time lead to the same mathematical formulation of their behavior as do the time-varying circuit problems.

The analyses of electrical and mechanical systems involving time-varying parameters have been undertaken by many investigators using mathematical techniques. It is the purpose of this discussion to present four distinct mathematical methods that have been found useful for the analysis of such systems in the hope that an expository account of these techniques will prove useful to workers in this field of investigation.

35 THE CLASSICAL THEORY OF DIFFERENTIAL EQUATIONS

The differential equations that determine the current or charge distribution of several time-variable circuits of practical importance are first-order linear equations of the following form:

$$\frac{dy}{dt} + P(t)y = F(t) \tag{35.1}$$

where $P(t)$ and $F(t)$ are given functions of the time t. The initial conditions of these circuits are usually of such form that the dependent variable $y(t)$ has the following initial value:

$$y(0) = 0 \tag{35.2}$$

The general solution of (35.1), subject to the initial condition (35.2), is

$$y(t) = \exp\left[-\int P(t)\,dt\right] \int_0^t \exp\left[\int P(t)\,dt\right] F(t)\,dt \tag{35.3}$$

† Reprinted from the *Transactions of the IRE Professional Group on Circuit Theory*, vol. CT-2, no. 1, March, 1955.

THE CARBON MICROPHONE

This general solution of (35.1) is very useful in the solution of many time-variable circuits of practical importance.   As an example of the use of (35.3), consider the circuit of the carbon microphone shown in Fig. 35.1.[1] The equivalent circuit of this device consists of a constant potential source E in series with a constant resistance R, a constant inductance L, and a time-varying resistance $r \sin(\theta t)$.   The output terminals of the carbon microphone appear across the inductance of the circuit.

The differential equation satisfied by the current of the circuit of Fig. 35.1 is

$$\frac{di}{dt} + \frac{1}{L}(R + r \sin \theta t) i = \frac{E}{L} \tag{35.4}$$

This differential equation has the same form as (35.1) with $y = i$ and $F(t) = E/L$. In this case, the integral $\int P(t)\,dt$ is given by

$$\int P(t)\,dt = \frac{1}{L}\int (R + r \sin \theta t)\,dt = bt - a \cos \theta t \tag{35.5}$$

where $b = R/L$, $a = r/\theta L$.

If the initial conditions of the circuit are $i(0) = 0$, the current is given by (35.3) in the following form:

$$i(t) = e^{-bt + a \cos \theta t} \frac{E}{L} \int_0^t e^{bt - a \cos \theta t}\,dt \tag{35.6}$$

The harmonic components of the current may be obtained by the well-known result in the theory of Bessel functions:[2]

$$e^{a \cos \theta t} = I_0(a) + 2 \sum_{n=1}^{\infty} I_n(a) \cos n\theta t \tag{35.7}$$

where $I_n(a)$ is the modified Bessel function of the first kind of order n.   These functions have been extensively tabulated.[3]

[1] E. A. Guillemin, "Communication Networks," vol. I, pp. 403–416, John Wiley & Sons, Inc., New York, 1931.

[2] W. G. Bickley, "Bessel Functions and Formulae," pp. 15–17, Cambridge University Press, London, 1953.

[3] E. Jahnke and F. Emde, "Tables of Functions," pp. 137–138, Dover Publications, Inc., New York, 1943.

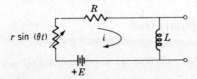

Fig. 35.1

The modified Bessel functions have the following property:

$$I_n(-a) = I_n(a) \qquad \text{for } n = 0, 2, 4, 6, \ldots$$
$$I_n(-a) = -I_n(a) \qquad \text{for } n = 1, 3, 5, 7, \ldots \tag{35.8}$$

In order to obtain the various modulation products of the current from the general solution (35.6), it is necessary to evaluate the following integral:

$$M = \int_0^t e^{bt - a \cos \theta t} \, dt$$
$$= \int_0^t e^{bt} \left[I_0(a) + 2 \sum_1^\infty I_n(-a) \cos n\theta t \right] dt \tag{35.9}$$

This integral may be evaluated by the use of the following well-known result:

$$\int e^{bt} \cos n\theta t \, dt = K_n e^{bt} \cos (n\theta t - \phi_n) \tag{35.10}$$

where

$$K_n = [b^2 + (n\theta)^2]^{-\frac{1}{2}} \qquad \tan \phi_n = \frac{n\theta}{b} \tag{35.11}$$

By the use of this result, the following expression for M may be obtained:

$$M = e^{bt} \left[\frac{I_0}{b}(a) + 2 \sum_1^\infty I_n(-a) K_n (\cos n\theta t - \cos \phi_n) \right] - \frac{I_0}{b}(a) \tag{35.12}$$

If these results are substituted into (35.6), the circuit current $i(t)$ may be expressed in the following form:

$$i(t) = \frac{E}{L} \left[I_0(a) + 2 \sum_{k=1}^\infty I_k(a) \cos k\theta t \right]$$
$$\times \left[\frac{I_0}{b}(a) + 2 \sum_{n=1}^\infty I_n(-a) K_n (\cos n\theta t - \cos \phi_n) \right]$$
$$- \frac{E I_0}{Lb}(a) e^{-bt} \left[I_0(a) + 2 \sum_{k=1}^\infty I_k(a) \cos k\theta t \right] \tag{35.13}$$

This expression gives the steady-state and transient current of the microphone circuit. The modulation products of the steady-state solution may be readily computed by the use of this result.

THE CONDENSER TRANSMITTER

Another circuit of practical importance whose analysis leads to a differential equation of the form (35.1) is the equivalent circuit of the condenser transmitter.[1] The equivalent circuit of this device is shown in Fig. 35.2. This circuit consists of a variable elastance parameter of the form $S(t) = S_0(1 + k \sin \theta t)$, where $k < 1$ in series with a constant potential source E and a constant resistance R. The output voltage of this device is taken from the

[1] E. A. Guillemin, *op. cit.*

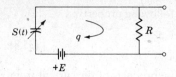

+E   Fig. 35.2

terminals of the resistance R.   The differential equation satisfied by the mesh charge q of this circuit is

$$\frac{dq}{dt} + \frac{S_0}{R}(1 + k \sin \theta t)q = \frac{E}{R} \tag{35.14}$$

This equation is of the form (35.1).   If the initial condition of the circuit is $q(0) = 0$, the mesh charge is given by (35.3) in the form

$$q(t) = \frac{E}{R} e^{-mt+n \cos \theta t} \int_0^t e^{mt-n \cos \theta t} \, dt \tag{35.15}$$

where $m = S_0/R$, $n = kS_0/\theta R$.

The modulation products may be obtained by using (35.7) and performing the indicated integration.   The response of other time-varying circuits whose equations are of first order may be obtained in a similar manner.

36   MATRIX METHODS IN THE ANALYSIS
OF TIME-VARIABLE CIRCUITS

The response and stability of linear time-varying circuits that contain parameters whose magnitudes vary in a periodic manner with the time may be obtained by a matrix multiplication technique.[1]   This method of analysis is applicable when the differential equation governing the charge or current distribution of the circuit can be written as a Mathieu-Hill differential equation having the form

$$\frac{d^2 x}{dt^2} + F(t) x = 0 \tag{36.1}$$

in which $F(t)$ is a given periodic function of the time t and suitable initial conditions are prescribed.

To illustrate this general mathematical technique, consider the circuit illustrated by Fig. 36.1.   This circuit consists of a constant resistance and inductance parameters R and L in series with a time-varying capacitance whose elastance parameter is $S(t)$.   It is desired to determine the behavior of the circulating charge $q(t)$ when the periodic variation of $S(t)$ and the initial circuit charge q_0 and current i_0 of the circuit are prescribed at $t = 0$.

[1] L. A. Pipes, Matrix Analysis of Linear Time-varying Circuits, *Journal of Applied Physics*, vol. 25, pp. 1179–1185, September, 1954.

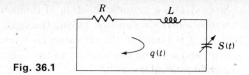

Fig. 36.1

Since there is a possibility of energy being furnished to the circuit by the agent that is causing the time variation of the capacitor, this circuit is said to be parametrically excited.[1]

The differential equation that determines the mesh charge $q(t)$ of this circuit is

$$\frac{L\,d^2q}{dt^2} + \frac{R\,dq}{dt} + S(t)q = 0 \tag{36.2}$$

It will be assumed that the elastance $S(t)$ is a periodic function of the time of fundamental period T so that

$$S(t+T) = S(t) \tag{36.3}$$

The initial conditions of the circuit will be assumed to be given by the equations

$$i(0) = i_0 \quad \text{and} \quad q(0) = q_0 \tag{36.4}$$

To simplify (36.2) it is convenient to introduce the new variable $x(t)$ by the following substitution:

$$q(t) = e^{-bt}x(t) \quad \text{where } b = \frac{R}{2L} \tag{36.5}$$

In terms of $x(t)$, (36.2) is transformed into

$$\frac{d^2x}{dt^2} + F(t) = 0 \tag{36.6}$$

where the function $F(t)$ is given by

$$F(t) = \frac{S}{L}(t) - b^2 \tag{36.7}$$

The initial conditions (36.4) may be expressed as initial conditions on $x(t)$ by the following equations:

$$x(0) = q_0 = x_1 \qquad \dot{x}(0) = (i_0 + bq_0) = x_2 \tag{36.8}$$

---

[1] N. Minorsky, "Introduction to Nonlinear Mechanics," pp. 362–369, J. W. Edwards, Publisher, Incorporated, Ann Arbor, Mich., 1947.

Since $S(t)$ is a periodic function of fundamental period T, the function $F(t)$ is also a periodic function having the same period. The interval from $t = 0$ to $t = T$ is therefore the fundamental interval of the variation of $F(t)$. Let $u_1(t)$ and $u_2(t)$ be two linearly independent solutions of (36.6) in the fundamental interval $0 \leqslant t \leqslant T$.

The values of $x(t) = x_1$, and its first derivative $\dot{x}(t) = x_2$ may be expressed in the following form:

$$\left.\begin{array}{l} x_1(t) = K_1 u_1(t) + K_2 u_2(t) \\ x_2(t) = K_1 \dot{u}_1(t) + K_2 \dot{u}_2(t) \end{array}\right\} \quad \text{for } 0 \leqslant t \leqslant T \tag{36.9}$$

where K_1 and K_2 are arbitrary constants. It is convenient to write (36.9) as the single matrix equation

$$\begin{bmatrix} x_1(t) \\ x_2(t) \end{bmatrix} = \begin{bmatrix} u_1(t) & u_2(t) \\ \dot{u}_1(t) & \dot{u}_2(t) \end{bmatrix} \begin{bmatrix} K_1 \\ K_2 \end{bmatrix} \tag{36.10}$$

To simplify the writing, the following notation will be introduced:

$$[x(t)] = \begin{bmatrix} x_1(t) \\ x_2(t) \end{bmatrix} \qquad [K] = \begin{bmatrix} K_1 \\ K_2 \end{bmatrix}$$

$$[u(t)] = \begin{bmatrix} u_1(t) & u_2(t) \\ \dot{u}_1(t) & \dot{u}_2(t) \end{bmatrix} = \begin{bmatrix} u_{11}(t) & u_{12}(t) \\ u_{21}(t) & u_{22}(t) \end{bmatrix} \tag{36.11}$$

In terms of this notation, (36.10) may be expressed in the following concise form:

$$[x(t)] = [u(t)][K] \tag{36.12}$$

The Wronskian [1] W_0 of the two solutions $u_1(t)$ and $u_2(t)$ is a *constant* in the fundamental interval $0 \leqslant t \leqslant T$ and is the following determinant:

$$W_0 = \det [u(t)] = u_1(t) \dot{u}_2(t) - \dot{u}_1(t) u_2(t) \tag{36.13}$$

Since the two solutions $u_1(t)$ and $u_2(t)$ are linearly independent, $W_0 \neq 0$. The matrix $[u(t)]$ is therefore nonsingular and has the following inverse:

$$[u(t)]^{-1} = \frac{1}{W_0} \begin{bmatrix} u_{22}(t) & -u_{12}(t) \\ -u_{21}(t) & u_{11}(t) \end{bmatrix} \tag{36.14}$$

at $t = 0$, (36.12) becomes

$$[x(0)] = [u(0)][K] \tag{36.15}$$

hence

$$[K] = [u(0)]^{-1}[x(0)] \tag{36.16}$$

---

[1] L. R. Ford, "Differential Equations," pp. 115–117, McGraw-Hill Book Company, New York, 1933.

This equation determines the arbitrary constants in terms of the initial conditions.   If this value of $[K]$ is now substituted into (36.12) the result is

$$[x(t)] = [u(t)][u(0)]^{-1}[x(0)] \tag{36.17}$$

At the end of the fundamental interval $t = T$, the solution is given in terms of the initial conditions by the following equation:

$$[x(T)] = [u(T)][u(0)]^{-1}[x(0)] \tag{36.18}$$

or

$$[x(T)] = [M][x(0)] \tag{36.19}$$

where

$$[M] = [u(T)][u(0)]^{-1} = \begin{bmatrix} A & B \\ C & D \end{bmatrix} \tag{36.20}$$

CASCADING THE SOLUTION BY THE USE OF MATRICES

It is easy to see that as a consequence of the periodicity of $F(t)$ expressed by the equation

$$F(t + T) = F(t) \tag{36.21}$$

(36.6) is invariant to the change of variable

$$t = \tau + nT \quad \text{where } 0 \leqslant \tau \leqslant T \quad n = 0, 1, 2, 3, \ldots \tag{36.22}$$

It is thus apparent that a solution of the form (36.17) may be obtained in each interval $0 \leqslant \tau \leqslant T$.   The *final* values of x_1 and x_2 at the end of one interval of the variation of $F(t)$ are the *initial* values of x_1 and x_2 in the following interval.   Therefore the solution at the instant $t = 2T$ is

$$[x(2T)] = [M][x(T)] = [M]^2[x(0)] \tag{36.23}$$

At the end of n fundamental periods of $F(t)$, $t = nT$, the solution is

$$[x(nT)] = [M]^n[x(0)] \quad n = 0, 1, 2, 3, 4, \ldots \tag{36.24}$$

The solution at the time $t = nT + \tau$ is

$$[x(nT + \tau)] = [u(\tau)][u(0)]^{-1}[x(nT)] \quad 0 \leqslant \tau \leqslant T$$
$$n = 0, 1, 2, 3, \ldots \tag{36.25}$$

The solution of (36.6) is given by (36.25).   The practical use of this solution depends on having an efficient technique for the computation of positive integral powers of the matrix $[M]$.   These required powers of $[M]$ are easily obtained by the use of Sylvester's theorem of matrix algebra.[1] Results are summarized in Table 5.

[1] R. A. Frazer, W. J. Duncan, and A. R. Collar, pp. 79–80, "Elementary Matrices," Cambridge University Press, London, 1938.

SQUARE-WAVE VARIATION OF ELASTANCE

To illustrate the general method of solution, consider the case in which the elastance parameter $S(t)$ of the circuit of Fig. 36.1 varies in the manner illustrated by Fig. 36.2. In this case the function $F(t)$ has the following values:

$$F(t) = \begin{cases} F_1 = \dfrac{S_1}{L - b^2} & 0 \leqslant t \leqslant \dfrac{T}{2} \\[2ex] F_2 = \dfrac{S_2}{L - b^2} & \dfrac{T}{2} \leqslant t \leqslant T \end{cases} \tag{36.26}$$

It is easy to show that in this case (36.18) takes the following form:

$$\begin{bmatrix} x_1(t) \\ x_2(t) \end{bmatrix} = \begin{bmatrix} \cos\theta_2 & \dfrac{\sin\theta_2}{w_2} \\ -w_2\sin\theta_2 & \cos\theta_2 \end{bmatrix} \begin{bmatrix} \cos\theta_1 & \dfrac{\sin\theta_1}{w_1} \\ -w_1\sin\theta_1 & \cos\theta_1 \end{bmatrix} \begin{bmatrix} x_1(0) \\ x_2(0) \end{bmatrix} \tag{36.27}$$

where

$$w_i = (F_i)^{\frac{1}{2}} \qquad \theta_i = \frac{T}{2}(F_i)^{\frac{1}{2}} \qquad i = 1, 2$$

The matrix $[M]$ of (36.20) now takes the form

$$\begin{aligned} [M] &= \begin{bmatrix} \cos\theta_2 & \dfrac{\sin}{w_2}\theta_2 \\ -w_2\sin\theta_2 & \cos\theta_2 \end{bmatrix} \begin{bmatrix} \cos\theta_1 & \dfrac{\sin}{w_1}\theta_1 \\ -w_1\sin\theta_1 & \cos\theta_1 \end{bmatrix} \\[1ex] &= \begin{bmatrix} A & B \\ C & D \end{bmatrix} \end{aligned} \tag{36.28}$$

If the indicated matrix multiplication in (36.28) is performed, the following values are obtained for the elements A, B, C, D of the matrix $[M]$:

$$\begin{aligned} A &= \cos\theta_1\cos\theta_2 - \sin\theta_1 \frac{\sin\theta_2\, w_1}{w_2} \\[1ex] B &= \cos\theta_2 \frac{\sin\theta_1}{w_1} - \cos\theta_1 \frac{\sin\theta_2}{w_2} \\[1ex] C &= (w_2\cos\theta_1\sin\theta_2 + w_1\cos\theta_2\sin\theta_1) \\[1ex] D &= \cos\theta_1\cos\theta_2 - \sin\theta_1 \frac{\sin\theta_2\, w_2}{w_1} \end{aligned} \tag{36.29}$$

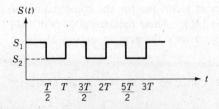

Fig. 36.2

Table 5 Positive integral powers of $[M] = \begin{Vmatrix} A & B \\ C & D \end{Vmatrix}$

$(A + D) \neq \pm 2$

1    $\cosh a = \dfrac{A + D}{2}$ $[M]^n = \dfrac{1}{S_1} \begin{bmatrix} (S_{n+1} - DS_n) & (BS_n) \\ (CS_n) & (S_{n+1} - AS_n) \end{bmatrix}$

                   $S_n = \sinh an$ $n = 0, 1, 2, 3, 4, \ldots$

$A = D \neq \pm 1$

2    $\cosh a = A$ $[M]^n = \begin{bmatrix} \cosh an & Z_0 \sinh an \\ \dfrac{\sinh an}{Z_0} & \cosh an \end{bmatrix}$

            $Z_0 = \left[\dfrac{B}{C} \right]^{\frac{1}{2}}$ $n = 0, 1, 2, 3, 4, \ldots$

3    $(A + D) = +2$ $[M]^n = \begin{bmatrix} n(A - 1) + 1 & nB \\ nC & n(D - 1) + 1 \end{bmatrix}$

4    $(A + D) = -2$ $[M]^n = nE_{n-1} \begin{bmatrix} A + 1 & B \\ C & D + 1 \end{bmatrix} + E_n I$

            $E_n = e^{jn\pi}$ $I = \begin{bmatrix} 1 & 0 \\ 0 & 1 \end{bmatrix}$

            $n = 0, 1, 2, 3, 4, \ldots$

The values of x and $\dot{x}$ after the end of n complete periods the square wave $S(t)$ are given by

$$[x(nT)] = [M]^n[x(0)] \tag{36.30}$$

The magnitudes of the circulating charge and the current of the circuit at $t = nT$ may be obtained by the use of (36.5) and are given by

$$q(nT) = e^{-bnT} x_1(nT) \qquad n = 0, 1, 2, 3, 4, \ldots$$
$$i(nT) = e^{-bnT}[x_2(nT) - bx_1(nT)] \tag{36.31}$$

The powers of $[M]^n$ of the matrix $[M]$ may be easily computed by the use of Table 5.

STABILITY OF THE CIRCUIT

The nature of the charge and current oscillations of the circuit depend on the form taken by the high powers of the matrix $[M]$. In the usual case, $A + D \neq 2$

and the matrix $[M]^n$ is given by the first entry of Table 5. In this case, the circuit charge at $t = nT$ is given by (36.31) in the form

$$q(nT) = \frac{e^{-bnT}}{S_1} [x_1(0)(S_{n+1} - DS_n) + x_2(0) BS_n] \qquad (36.32)$$

where $S_n = \sinh(an)$, and $x_1(0)$ and $x_2(0)$ are given in terms of the initial conditions of the circuit by (36.8). An examination of (36.32) shows that for large values of n, the charge $q(nT)$ will grow indefinitely with n if

$$a > bT \qquad (36.33)$$

and the solution will be *unstable*. If, on the other hand,

$$a < bT \qquad (36.34)$$

$q(nT)$ will eventually be damped out by the resistance and the circuit will be *stable*. If $a = bT$, steady oscillations of the circuits can be maintained by the energy source that is varying the elastance.

Since

$$\cosh a = \frac{A + D}{2} = \cos \theta_1 \cos \theta_2 - \sin \theta_1 \frac{\sin \theta_2 (\theta_1^2 + \theta_2^2)}{2\theta_1 \theta_2}$$

$$= k \qquad (36.35)$$

it can be seen that the oscillations of the charge and the current of the circuit will grow indefinitely with time and the circuit therefore will be unstable if the circuit parameters are such that

$$\cosh^{-1} k > bT \qquad (36.36)$$

The special cases in which $A + D = \pm 2$ lead to the forms for $[M]^n$ given in the third and fourth entries of Table 5 and may be treated similarly.

The matrix method is applicable when a solution of (36.6) in the interval $0 \leqslant t \leqslant T$ is readily obtained and is particularly useful if the circuit has a parameter that undergoes square-wave or sawtooth variations. Additional examples of this method of solution are available.[1,2]

37 APPROXIMATE SOLUTION OF TIME-VARIABLE CIRCUIT PROBLEMS BY THE USE OF THE BWK APPROXIMATION

A great many time-variable circuits that occur in practice have the property that the variable parameters involved exhibit only small variations about a

---

[1] L. A. Pipes, Matrix Solution of Equations of the Mathieu-Hill Type, *Journal of Applied Physics*, vol. 24, pp. 902–910, July, 1953.

[2] L. Brillouin, "Wave Propagation in Periodic Structures," pp. 125–127, Dover Publications, Inc., New York, 1953.

large average value. The fundamental differential equations of such circuits can usually be transformed so that they take the following general form:

$$\frac{d^2 y}{dt^2} + G^2(t)y = F(t) \tag{37.1}$$

where $F(t)$ involves the driving potential of the circuit. If no external driving potentials are applied to the circuit, $F(t) = 0$, and (37.1) reduces to the homogeneous equation

$$\frac{d^2 y}{dt^2} + G^2(t)y = 0 \tag{37.2}$$

If $G^2(t)$ is a real positive function that exhibits small variations about a large average value, it can be shown that [1,2]

$$y(t) = [G(t)]^{-\frac{1}{2}}[A \cos \phi(t) + B \sin \phi(t)] \tag{37.3}$$

where A and B are arbitrary constants and $\phi(t)$ is given by

$$\phi(t) = \int G(t)\,dt \tag{37.4}$$

is a very good approximate solution of (37.2). More precisely, it can be shown [1,2] that if the variations of $G(t)$ are such that

$$|G^2(t)| \gg \left| \frac{G''}{2G} - 3\left(\frac{G'}{2G}\right)^2 \right| \tag{37.5}$$

in the range in t under consideration, $y(t)$ is an excellent approximation to the solution of (37.2). This approximation is known in the literature of mathematical physics as the *Brillouin-Wentzel-Kramers (BWK) approximation*. It has been used extensively in problems of quantum mechanics.[3]

INVERSE CAPACITANCE MODULATION

As a first example of the use of the BWK approximation in the analysis of time-varying circuits, consider the circuit of Fig. 37.1. This is a series circuit that consists of a constant inductance L in series with a time-varying capacitance $C(t)$ whose magnitude is

$$C(t) = \frac{C_0}{1 - 2h \cos 2\theta t} \tag{37.6}$$

[1] L. Brillouin, A Practical Method for Solving Hill's Equation, *Quarterly of Applied Mathematics*, vol. 6, pp. 167–178, July, 1948.

[2] L. Brillouin, The B.W.K. Approximation and Hill's Equation, *Quarterly of Applied Mathematics*, vol. 7, pp. 363–380, January, 1950.

[3] E. C. Kemble, "The Fundamental Principles of Quantum Mechanics," pp. 90–112, McGraw-Hill Book Company, New York, 1937.

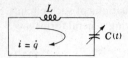

Fig. 37.1

The differential equation that is satisfied by the circulating charge $q(t)$ of the circuit is

$$\frac{d^2q}{dt^2} + w_0^2(1 - 2h\cos 2\theta t)q = 0 \qquad w_0^2 = \frac{1}{LC_0} \tag{37.7}$$

This is the differential equation of the equivalent circuit of a condenser microphone in which a sound wave of angular frequency 2θ is caused to vary the distance d between the two plates of a capacitor so that $d = d_0(1 - 2h\cos 2\theta t)$ where d_0 is the average distance of separation of the plates of the capacitor. Typical values of the various parameters in the case of radio-frequency modulation are

$$\begin{aligned} w_0 &= (LC_0)^{-\frac{1}{2}} = 2\pi(5 \times 10^{17}) \text{ sec}^{-1} \\ 2\theta &= 2\pi(5 \times 10^3) \text{ sec}^{-1} \qquad h = 2 \times 10^{-4} \end{aligned} \tag{37.8}$$

It is therefore justifiable to neglect terms of the order h^2 and higher powers of h. Equation (37.7) is a special case of (37.1) with

$$G(t) = w_0(1 - 2h\cos 2\theta t)^{\frac{1}{2}} \doteq w_0(1 - h\cos 2\theta t) \tag{37.9}$$

provided that terms of h^2 and higher are neglected. Since in this case $G(t)$ performs small amplitude oscillations about the large average value w_0, the inequality (37.5) is satisfied. The function $\phi(t)$ of (37.4) now takes the following form:

$$\begin{aligned} \phi(t) &= \int G(t)\,dt = w_0 \int (1 - h\cos 2\theta t)\,dt \\ &= w_0\left(t - h\frac{\sin 2\theta t}{2\theta}\right) \end{aligned} \tag{37.10}$$

Therefore the BWK solution of (37.7) is

$$\begin{aligned} q(t) = w_0(1 - h\cos 2\theta t)^{-\frac{1}{2}} A \cos [w_0 t - c\sin 2\theta t \\ + B\sin (w_0 t - c\sin 2\theta t)] \end{aligned} \tag{37.11}$$

where A and B are arbitrary constants and $c = w_0 h/2\theta$.

In order to determine the various harmonic components of the charge oscillations of the circuit, these expansions involving Bessel functions may be used:[1]

$$\begin{aligned} \cos (x\sin \phi) &= J_0(x) + 2[J_2(x)\cos 2\phi + J_4(x)\cos 4\phi + \cdots] \\ \sin (x\sin \phi) &= 2[J_1(x)\sin \phi + J_3(x)\sin 3\phi + \cdots] \end{aligned} \tag{37.12}$$

---

[1] W. G. Bickley, "Bessel Functions and Formulae," pp. 15–17, Cambridge University Press, London, 1953.

where the $J_n(x)$ functions are Bessel functions of the first kind and nth order. If the quantity h in the square-root term of (37.11) is neglected, and Eqs. (37.12) are used, then, after certain algebraic reductions, (37.11) may be written in the form

$$q(t) = K_0 \sum_{n=1}^{\infty} J_n(c) \cos\left[(w_0 - 2n\theta)\, t - b\right] \tag{37.13}$$

where K_0 and b are arbitrary constants. This solution of Eq. (37.7) was obtained by J. R. Carson by an entirely different procedure.[1]

FORCED OSCILLATIONS OF A CIRCUIT WITH VARIABLE ELASTANCE

As a second example of the use of the BWK procedure, consider the circuit of Fig. 37.2. This circuit consists of a series connection of a resistance R, a constant inductance L, and a time-varying elastance $S(t)$ in series with a periodic electromotive force $E(t) = E_m \sin(wt)$. Let it be assumed that the variation of the elastance is of the form

$$S(t) = S_0(1 - 2h \cos 2\theta t) \tag{37.14}$$

where $h^2 \ll 1$.

S_0 is the large average value of the elastance, and h is a small amplitude parameter. The differential equation satisfied by the circulating charge of the circuit of Fig. 37.2 is

$$\frac{L\, d^2 q}{dt^2} + \frac{R\, dq}{dt} + S_0(1 - 2h \cos 2\theta t) q = E_m \sin wt \tag{37.15}$$

To simplify the form of (37.15), introduce the notation

$$b = \frac{R}{2L} \qquad w_0 = \left(\frac{S_0}{L}\right)^{\frac{1}{2}} \tag{37.16}$$

and the two variables $x(t)$ and $y(t)$ defined by the following equations:

$$x = \theta t \qquad y(t) = e^{bt} q(t) \tag{37.17}$$

In terms of the new variables x and y, (37.15) takes the form

$$\frac{d^2 y}{dx^2} + \frac{1}{\theta^2}\left[(w_0^2 - b^2) - 2hw_0^2 \cos 2x\right] y = \frac{E_m}{L\theta} e^{bx/\theta} \sin \frac{wx}{\theta} \tag{37.18}$$

[1] J. R. Carson, Notes on the Theory of Frequency Modulation, *Proceedings of the Institute of Radio Engineers*, vol. 10, pp. 57–64, February, 1922.

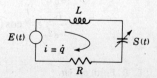

$E(t)$ $i = \dot{q}$ L $S(t)$ R

Fig. 37.2

If the resistance of the circuit R is small, then $b^2 \ll w_0^2$ and the term b^2 may be neglected in (37.18). The complementary function of the differential equation (37.18) is the solution of the equation

$$\frac{d^2 y}{dx^2} + \left(\frac{w_0}{\theta}\right)^2 (1 - 2h \cos 2x) y = 0 \tag{37.19}$$

This equation has the same *form* as (37.2). If the quantity h is so small that it can be neglected in comparison with unity, the BWK solution of (37.19) can be written in the form

$$y(x) = A_1 y_1(x) + A_2 y_2(x) \tag{37.20}$$

where A_1 and A_2 are arbitrary constants and the functions $y_1(x)$ and $y_2(x)$ are given by

$$y_1(x) = \cos [ax - c \sin (2x)] \qquad a = \frac{w_0}{\theta}$$

$$\tag{37.21}$$

$$y_2(x) = \sin [ax - c \sin (2x)] \qquad c = \frac{w_0 h}{2\theta}$$

The general solution of (37.18) for the forced oscillations of the circuit may now be obtained by the *method of variation of parameters*[1] in the form:

$$y(x) = A_1 y_1(x) + A_2 y_2(x) + y_2 \int \frac{F}{W}(x) y_1 \, dx - y_1 \int \frac{F}{W}(x) y_2 \, dx \tag{37.22}$$

where $F(x)$ is the right member of (37.18) given by

$$F(x) = \frac{E_m}{L\theta} e^{bx/\theta} \sin \frac{wx}{\theta} \tag{37.23}$$

and W is the Wronskian of the solutions of (37.19) whose value is

$$W = y_1 y_2' - y_1' y_2 = \frac{w_0}{\theta} = a \tag{37.24}$$

for the case in which $h \ll 1$. The terms containing the arbitrary constants in (37.22) constitute the complementary function of (37.19) and the terms involving the integrals give the particular integral of the equation.

The approximate general solution of (37.15) may now be obtained by expressing (37.22) in terms of the original variables q and t. It has the form

$$q(t) = A_1 e^{-bt} \cos (w_0 t - c \sin 2\theta t) + A_2 e^{-bt} \sin (w_0 t - c \sin 2\theta t)$$

$$+ \frac{E_m}{w_0 L} e^{-bt} \sin (w_0 t - c \sin 2\theta t) \int dt \, e^{bt} \sin wt \cos (w_0 t - c \sin 2\theta t)$$

$$- \frac{E_m}{w_0 L} e^{-bt} \cos (w_0 t - c \sin 2\theta t)$$

$$\times \int e^{bt} \sin wt \sin (w_0 t - c \sin 2\theta t) \, dt \tag{37.25}$$

---

[1] L. R. Ford, *op. cit.*

The arbitrary constants A_1 and A_2 of the transient part of the solution may be evaluated if the initial conditions of the circuit at $t = 0$ are given. The harmonic content of the steady-state solution may be obtained by expanding the integrands as a series of harmonic terms by means of Eqs. (37.12) and integrating term by term. Further examples of the BWK method applied to the solution of the differential equations of time-varying circuits have been found.[1]

38 THE USE OF LAPLACE TRANSFORMS AND INTEGRAL EQUATIONS IN THE SOLUTION OF TIME-VARIABLE CIRCUIT PROBLEMS

The Laplace-transform technique which has proved so useful in the analysis of linear circuits with constant parameters may be used to obtain useful results in the analysis of time-variable circuit problems. In order to illustrate the use of the Laplace-transform method in the analysis of time-varying circuits, consider the circuit of Fig. 38.1. This circuit is a series combination of a constant component whose operational impedance is $Z(D)$, $D = d/dt$, in series with a time-varying resistance $r(t)$ and a potential source $e(t)$.

The differential equation that expresses the potential balance of this circuit may be written in the following operational form:

$$Z(D) i(t) + r(t) i(t) = e(t) \qquad D = \frac{d}{dt} \tag{38.1}$$

In order to formulate this equation as an integral equation let the following Laplace transforms be introduced:

$$L_t i(t) = s e^{-st} i(t) \, dt = I(p)$$

$$L_t = \text{"Laplace transform of"} \tag{38.2}$$

$$L_t e(t) = E(s) \tag{38.3}$$

$$L_t[r(t) i(t)] = G(s) \tag{38.4}$$

---

[1] L. A. Pipes, Analysis of Linear Time-varying Circuits by the Brillouin-Wentzel-Kramers Method, *Transactions of the AIEE, Communication and Electronics*, no. 11, pp. 93–96, March, 1954.

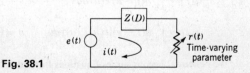

Fig. 38.1

Let it be assumed that initially the circuit is at rest so that at $t = 0$, $i(0) = 0$, and $q(0) = 0$. That is, the initial current $i(t)$ and the charge $q(t)$ of the circuit are both zero at $t = 0$. With these initial conditions, the Laplace transform of (38.1) is given by

$$Z(s)I(s) + G(s) = E(s) \tag{38.5}$$

or

$$I(s) = \frac{E(s)}{Z(s)} - \frac{G(s)}{Z(s)} \tag{38.6}$$

For simplicity, introduce the notation

$$H(s) = \frac{1}{Z(s)} \tag{38.7}$$

With this notation (38.6) may be written in the following form:

$$I(s) = E(s)\,H(s) - G(s)\,H(s) \tag{38.8}$$

Let the inverse Laplace transform of $H(s)$ be $h(t)$, so that

$$h(t) = L_t^{-1}\,H(s) = L_t^{-1}\,\frac{1}{Z(s)} \tag{38.9}$$

The current of the circuit $i(t)$ is therefore obtained by taking the inverse transform of $I(s)$ in the form

$$i(t) = L_t^{-1}\,I(s) = L_t^{-1}\,E(s)\,H(s) - L_t^{-1}\,G(s)\,H(s) \tag{38.10}$$

If the Faltung or convolution theorem of the Laplace-transform theory is applied to the right-hand member of (38.10), the following result is obtained:

$$i(t) = \int_0^t h(t-u)\,e(u)\,du - \int_0^t h(t-u)\,r(u)\,i(u)\,du \tag{38.11}$$

This is an integral equation of the Volterra type. Volterra[1] has shown that the solution of (38.11) is given by the following series of functions:

$$i(t) = i_0(t) - i_1(t) + i_2(t) - i_3(t) + \cdots \tag{38.12}$$

where

$$i_0(t) = \int_0^t h(t-u)\,e(u)\,du$$

$$i_{n+1} = \int_0^t h(t-u)[r(u)\,i_n(u)]\,du \qquad n = 0, 1, 2, 3, 4, \ldots \tag{38.13}$$

[1] V. Volterra, "Leçons sur les Équations Intégrales," Gauthier-Villars, Paris, France, 1913.

The series (38.12) may be shown to be a rapidly convergent one. The $i_0(t)$ term represents the current in the circuit without the presence of the variable resistance element; the $i_1(t)$ term represents the first correction term caused by the resistance variation, etc.

OPERATIONAL PROCEDURE

In the computation of the various functions $i_n(t)$ of the series (38.12), it is frequently simpler to use the following equations:

$$i_0(t) = \frac{L_t^{-1} E(s)}{Z(s)} \qquad i_n(t) = L_t^{-1} I_n(s) \qquad n = 1, 2, 3, 4, \ldots$$

$$I_{n+1}(s) = L_t\left[\frac{r(t)\, i_n(t)}{Z(s)}\right] \qquad n = 0, 1, 2, 3, 4, \ldots \tag{38.14}$$

As a special example consider a circuit whose constant part consists of a constant inductance L in series with a constant resistance R. In this case, $Z(s)$ has the form

$$Z(s) = Ls + R \tag{38.15}$$

Let the applied potential be $E_m \sin(wt)$ and the variable resistance have the form $r(t) = r_0 \sin(\theta t)$ with $r < R$. Therefore,

$$E(s) = L_t E_m \sin wt = \frac{wE_m}{s^2 + w^2} \tag{38.16}$$

$$H(s) = \frac{1}{Z(s)} = \frac{1}{L(s+a)} \qquad a = \frac{R}{L} \tag{38.17}$$

$$h(t) = L_t^{-1} H(s) = \frac{e^{-at}}{L} \tag{38.18}$$

The integral equation (38.11) now takes the form

$$i(t) = \frac{E_m}{L} e^{-at} \int_0^t e^{au} \sin wu\, du - r_0 e^{-at} \int_0^t e^{au} \sin \theta u\, i(u)\, du \tag{38.19}$$

The solution of this integral equation is

$$i(t) = i_0(t) - i_1(t) + i_2(t) - i_3(t) + \cdots \tag{38.20}$$

where

$$i_0(t) = \frac{E_m}{L} e^{-at} \int_0^t e^{au} \sin wu\, du$$

$$= \frac{E_m}{L} \frac{e^{-at}}{(a^2 + w^2)} + E_m \frac{\sin(wt - \phi)}{(R^2 + w^2 L^2)^{\frac{1}{2}}} \tag{38.21}$$

$$\tan \phi = \frac{wL}{R}$$

$$\tag{38.22}$$

$$i_{n+1}(t) = r_0 e^{-at} \int_0^t e^{au} \sin \theta u\, i_n(u)\, du \qquad n = 0, 1, 2, 3, \ldots$$

The transient and steady-state oscillations of the circuit may therefore be computed to the degree of accuracy desired.

CIRCUIT WITH VARIABLE INDUCTANCE

If the time-varying circuit contains a variable inductance $L(t)$ in series with a constant part of operational impedance $Z(D)$, the circuit equation takes the following form:

$$Z(D)i(t) + \frac{d}{dt}[L(t)i(t)] = e(t) \tag{38.23}$$

If the circuit is at rest at $t = 0$, this equation can be transformed into the following Volterra-type integral equation:

$$i(t) = \int_0^t h(t-u)e(u)\,du - \int_0^t h(t-u)\frac{d}{du}[L(u)i(u)]\,du \tag{38.24}$$

where $h(t) = L_t^{-1}s/sZ(s)$.

The solution of (38.24) has the form (38.20) with

$$i_0(t) = \int_0^t h(t-u)e(u)\,du$$

$$i_{n+1} = \int_0^t h(t-u)\frac{d}{du}[L(u)i_n(u)]\,du \qquad n = 0, 1, 2, 3, \ldots \tag{38.25}$$

The case involving a variable capacitance may be treated in a similar manner. Further applications of the Laplace transform and the integral equation technique in the analysis of time-varying systems are available.[1-4]

39 CONCLUSION

The present state of the mathematical analysis of circuits with time-varying parameters may be compared with the status of alternating-current networks before the practical development and use of complex impedances and symbolic methods of solution. Although methods for obtaining the transient and steady-state solutions of these networks were available, yet the use of alternating-current networks by engineers and physicists proceeded in a halting manner until the mathematical methods for the analysis of problems involving these networks had been interpreted and systematized so that they could be readily applied to circuits in common practical usage.

[1] J. R. Carson, Theory and Calculation of Variable Electric Systems, *Physics Review*, vol. 17, pp. 116–134, February, 1921.

[2] J. R. Carson, "Electric Circuit Theory and the Operational Calculus," pp. 159–173, McGraw-Hill Book Company, New York, 1926.

[3] V. Bush, "Operational Circuit Analysis," pp. 338–358, John Wiley & Sons, Inc., New York, 1927.

[4] L. A. Pipes, Operational and Matrix Methods in Linear Variable Networks, *Philosophical Magazine*, ser. 7, vol. 25, pp. 585–600, April, 1938.

The purpose of this discussion has been to catalogue some of the analytical methods that have been found useful in the analysis of time-varying systems and to interpret and systematize them for practical use. Unless existing methods and methods yet to be developed for the solution of problems involving these circuits are interpreted and systematized for practical use, the attempt to utilize time-varying circuits in practical work will be limited and severely circumscribed.

ANALYSIS OF LINEAR TIME-VARYING CIRCUITS BY THE BRILLOUIN-WENTZEL-KRAMERS METHOD†

40 INTRODUCTION

The general behavior of electrical circuits whose parameters are constants has been extensively studied and is well understood. In recent years, a great deal of attention has been paid to the general theory and performance of circuits whose parameters are functions of the time, especially when they vary periodically [30–38]. Examples of linear time-varying circuits of practical importance occur in the theory of electrical communications. Frequency modulation, for instance, utilizes variations of capacitance or, to a lesser degree, inductance. The microphone transmitter contains a variable resistance whose value is varied by some source of energy outside the circuit, and the capacitor microphone contains a variable capacitance. The introduction of superregeneration has especially stimulated an interest in circuits that contain periodic variation of a resistance parameter [39].

The mathematical analysis of the performance of electrical circuits whose parameters are periodic functions of the time leads generally to the solution of Mathieu's and Hill's differential equations. These equations have the following general form

$$\frac{d^2 y}{dx^2} + G^2(x)y = H(x) \tag{40.1}$$

where $G^2(x)$ is a periodic function of x and $H(x)$ is a continuous function of x. Despite the vast amount of literature on the theory of Mathieu functions [40], it is still a formidable task to obtain rigorous solutions of Eq. (40.1) in cases of practical importance, and approximate methods must be employed.

† A reprint from *Communication and Electronics*, published by American Institute of Electrical Engineers. Copyright 1954 and reprinted by permission of the copyright owner. The Institute assumes no responsibility for statements and opinions made by contributors.

41 BWK APPROXIMATION

The great majority of linear time-varying circuits that occur in communications engineering have the property of the variable parameters involved exhibiting only small variations about a large average value.   In such cases a very good approximate solution of Eq. (40.1) can be obtained by a method used by Brillouin, Wentzel, and Kramers to solve equations of the type

$$\frac{d^2y}{dx^2} + G^2(x)\,y = 0 \tag{41.1}$$

in connection with problems of wave mechanics [41, 42].   Although the BWK method was used by Brillouin, Wentzel, and Kramers to solve problems in wave mechanics in 1926, the procedure is a very old one and the first indication of its use is found in the collected papers of the eminent mathematician Liouville, published in 1837.

To determine the type of approximation involved in the BWK procedure, consider the following function:

$$y = \frac{1}{\sqrt{G(x)}} [C_1 \, \epsilon^{j\phi(x)} + C_2 \, \epsilon^{-j\phi(x)}] \tag{41.2}$$

where ϵ is the base of the natural logarithms, C_1 and C_2 are arbitrary constants, and $j = \sqrt{-1}$.   The function $\phi(x)$ is given by

$$\phi(x) = \int G(x)\,dx \tag{41.3}$$

By direct differentiation of Eq. (41.2), it can be shown that this function satisfies the following differential equation:

$$\frac{d^2 y}{dx^2} + \left[G_2 + \frac{G''}{2G} - \frac{3}{4}\left(\frac{G'}{G}\right)^2 \right] y = 0 \tag{41.4}$$

If $G^2(x)$ is a periodic function that exhibits only small variations about a large average value, it can be shown that

$$|G^2(x)| \gg \left| \left[\frac{G''}{2G} - 3\left(\frac{G'}{2G}\right)^2 \right] \right| \tag{41.5}$$

throughout the range in x, and hence that Eq. (41.2) is an approximate solution of Eq. (41.1).

In the case where $G^2(x)$ is a positive periodic function of x, Eq. (41.2) may be written in the alternative form

$$y = \frac{1}{\sqrt{G(x)}} [A \cos \phi(x) + B \sin \phi(x)] \tag{41.6}$$

where A and B are arbitrary constants, and $\phi(x)$ is given by (41.3).   Equation (41.6) is the usual BWK approximate solution of Eq. (41.1).

42 CAPACITANCE MODULATION

As a first example of the use of the general BWK procedure, consider the circuit of Fig. 42.1. This is a series circuit that consists of a constant inductance L in series with a capacitance $C(t)$ that varies with time in such a manner that

$$C(t) = C_0(1 + 2h \cos 2pt) \qquad (42.1)$$

where C_0 and h are constant parameters with $h^2 \ll 1$. If q denotes the charge on the capacitance, the differential equation of the circuit is

$$\frac{d^2 q}{dt^2} + \frac{q}{LC_0(1 + 2h \cos 2pt)} = 0 \qquad (42.2)$$

Typical values of the various parameters in the case of radio-frequency modulation are

$$\omega_0 - \frac{1}{\sqrt{LC_0}} = 2\pi(5 \times 10^7) \ \text{sec}^{-1} \qquad (42.3)$$

$$2p = 2\pi(5 \times 10^3) \ \text{sec}^{-1} \qquad (42.4)$$

$$h = 2 \times 10^{-4} \qquad (42.5)$$

It is therefore justifiable to neglect terms containing h^2 and higher and write

$$\frac{1}{1 + 2h \cos 2pt} = (1 - 2h \cos 2pt) \qquad (42.6)$$

Consequently Eq. (42.2) may be written in the following form:

$$\frac{d^2 q}{dt^2} + \omega_0^2(1 - 2h \cos 2pt)q = 0 \qquad (42.7)$$

If the new variables, $y = q$ and $x = pt$, are introduced, Eq. (42.7) takes the following form:

$$\frac{d^2 y}{dx^2} + \left(\frac{\omega_0}{p}\right)^2 (1 - 2h \cos 2x) y = 0 \qquad (42.8)$$

This equation is a special case of Eq. (41.1) with

$$G(x) = \frac{\omega_0}{p} \sqrt{1 - 2h \cos 2x} = \frac{\omega_0}{p} 1 - h \cos 2x \qquad (42.9)$$

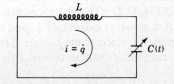

Fig. 42.1

provided that terms of order h^2 and higher are neglected. Since $\omega_0/p = 2 \times 10^4$ and $h = 2 \times 10^{-4}$, it can be seen that Eq. (41.5) is satisfied and an accurate approximate solution of Eq. (42.8) is given by Eq. (41.6). In this case we have

$$\phi(x) = \int G(x)\,dx = \frac{\omega_0}{p}\left(x - \frac{h}{2}\sin 2x\right) \tag{42.10}$$

Therefore the BWK solution of Eq. (42.8) is

$$y = \left[\frac{\omega_0}{p}(1 - h\cos 2x)\right]^{-\frac{1}{2}}[A\cos(ax - c\sin 2x) + B\sin(ax - c\sin 2x)] \tag{42.11}$$

where A and B are arbitrary constants and $a = \omega_0/p$ and $c = \omega_0 h/2p$.

To determine the various harmonic components, the following expansions involving Bessel functions [43] may be used:

$$\cos(a\sin b) = J_0(a) + 2[J_2(a)\cos 2b + J_4(a)\cos 4b + \cdots]$$

$$\sin(a\sin b) = 2[J_1(a)\sin b + J_3(a)\sin 3b + \cdots] \tag{42.12}$$

where the $J_n(a)$ functions are Bessel functions of the first kind and the nth order. With the aid of Eq. (42.12), (42.11) may be written in the following form after certain algebraic reductions:

$$y = q = K\left[\frac{\omega_0}{p}(1 - h\cos 2pt)\right]^{-\frac{1}{2}} \sum_{n=-\infty}^{+\infty} J_n\frac{\omega_0 h}{2p}\cos[(\omega_0 - 2np)t - \theta] \tag{42.13}$$

where K and θ are arbitrary constants. This is the BWK solution of Eq. (42.7) for the circuit charge. If the quantity h in the square-root term of Eq. (42.13) is neglected in comparison with unity, this equation may be written in the form

$$q = K_0 \sum_{n=-\infty}^{+\infty} J_n\frac{\omega_0 h}{2p}\cos[(\omega_0 - 2np)t - \theta] \tag{42.14}$$

where K_0 and θ are arbitrary constants. Equation (42.14) is a less accurate solution of (42.6) and was obtained by Carson [44] by a different procedure.

43 FORCED OSCILLATIONS OF A CIRCUIT WITH VARIABLE ELASTANCE

As a second example of the use of the BWK procedure, consider the circuit of Fig. 43.1. This circuit consists of a series connection of a constant resistance R, a constant inductance L, and a time-varying elastance $S(t)$ in series with a periodic electromotive force $E(t) = E_m \sin \omega t$. Let it be assumed that the variation of the elastance is of the form

$$S(t) = S_0(1 - 2h\cos 2pt) \qquad \text{where } h^2 \ll 1 \tag{43.1}$$

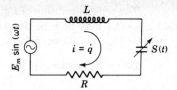

Fig. 43.1

S_0 is the constant part of the elastance, and h is a small amplitude parameter. The differential equation satisfied by the charge q on the elastance is

$$\frac{L\,d^2q}{dt^2} + \frac{R\,dq}{dt} + S_0(1 - 2h\cos 2pt)q = E_m \sin \omega \tag{43.2}$$

Let

$$b = \frac{R}{2L} \qquad \omega_0 = \sqrt{\frac{S_0}{L}} \tag{43.3}$$

and introduce the new variable $y(t)$ by means of the transformation

$$q(t) = \epsilon^{-bt} y(t) \tag{43.4}$$

In terms of these parameters, Eq. (43.2) may be written in the form

$$\frac{d^2y}{dt^2} + [(\omega_0^2 - b^2) - 2h\omega_0^2 \cos 2pt]y = \frac{E_m}{L}\epsilon^{bt} \sin \omega t \tag{43.5}$$

Let $x = pt$; then Eq. (43.5) becomes

$$\frac{d^2y}{dx^2} + \frac{1}{p^2}[(\omega_0^2 - b^2) - 2h\omega_0^2 \cos 2x]y = \frac{E_m}{Lp}\epsilon^{bx/p} \sin \frac{\omega x}{p} \tag{43.6}$$

If the resistance R of the circuit is small, then $b^2 \ll \omega_0^2$ and the terms b^2 may be neglected in Eq. (43.6). The complementary function of the differential equation (43.6) is the solution of the equation

$$\frac{d^2y}{dx^2} + \frac{\omega_0^2}{p^2}(1 - 2h\cos 2x)y = 0 \tag{43.7}$$

This is Eq. (42.9), and its BWK solution is given by Eq. (42.11). If h is small, it can be neglected in comparison with unity in the square-root term of Eq. (42.11), and the general solution of (43.7) may be written in the form

$$y(x) = A_1 y_1(x) + A_2 y_2(x) \tag{43.8}$$

where A_1 and A_2 are arbitrary constants and functions $y_1(x)$ and $y_2(x)$ are given by

$$
\begin{aligned}
y_1(x) &= \cos (ax - c\sin 2x) \\
y_2(x) &= \sin (ax - c\sin 2x)
\end{aligned}
\qquad
\begin{cases}
a = \dfrac{\omega_0}{p} \\[2mm]
c = \dfrac{\omega_0 h}{2p}
\end{cases}
\tag{43.9}
$$

The general solution of Eq. (43.6) for the forced oscillations of the circuit of Fig. 43.1 may now be obtained by the method of the variation of parameters [16] in the following form:

$$y(x) = A_1 y_1(x) + A_2 y_2(x) + y_2(x) \int \frac{R}{W}(x) y_1(x) \, dx$$

$$- y_1(x) \int R(x) y_2(x) \, dx \quad (43.10)$$

where $R(x)$ is the right member of Eq. (43.6) given by

$$R(x) = \frac{(E_m/Lp^2) \, \epsilon^{bx}}{p \sin \omega x/p} \quad (43.11)$$

and W is the Wronskian of the solutions, Eq. (43.12), whose value is

$$W = (y_1 y_2' - y_2 y_1') = \frac{\omega_0}{p} = a \quad (43.12)$$

for the case in which $h \ll 1$. The terms containing the arbitrary constants in Eq. (43.10) constitute the complementary function of (43.6), and the terms involving the indefinite integrals give the particular integral of the equation. The approximate general solution of Eq. (43.2) may now be obtained by expressing Eqs. (43.4) and (43.12) in terms of the original variables. It has the following form:

$$q(t) = A_1 \, \epsilon^{-bt} \cos(\omega_0 t - c \sin 2pt) + A_2 \, \epsilon^{-bt} \sin(\omega_0 t - c \sin 2pt)$$

$$+ \frac{E_m}{\omega_0 L} \, \epsilon^{-bt} \sin(\omega_0 t - c \sin 2pt) \int \epsilon^{bt} \sin \omega t \cos(\omega_0 t - c \sin 2pt)$$

$$\times \, dt - \frac{E_m}{\omega_0 L} \, \epsilon^{-bt} \cos \omega_0 t - c \sin 2pt)$$

$$\times \int \epsilon^{bt} \sin \omega t \sin(\omega_0 t - c \sin 2pt) \, dt \quad (43.13)$$

The arbitrary constants A_1 and A_2 of the transient part of the solution may be evaluated if the initial conditions of the circuit at $t = 0$ are given. The harmonic content of the steady-state solution may be obtained by expanding the integrand as a series of harmonic terms by means of Eq. (42.12) and integrating term by term.

44 SERIES CIRCUIT WITH PERIODICALLY VARYING RESISTANCE

In recent years, the introduction of superregeneration has stimulated an interest in the study of circuits that involve a periodic variation of the resistance parameter [7]. Consider the circuit of Fig. 44.1. This circuit consists of a constant inductance and capacitance in series with a periodically varying resistance $R(t)$. It is assumed that a harmonic forcing potential $E(t) = E_m \sin(\omega t)$ is impressed on the circuit.

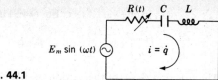

$E_m \sin(\omega t)$ $i = \dot{q}$

Fig. 44.1

The circulating charge of this circuit satisfies the following differential equation

$$\frac{d^2 q}{dt^2} + \frac{R}{L}(t)\frac{dq}{dt} + \frac{q}{LC} = \frac{E_m}{L}\sin \omega t \tag{44.1}$$

Let the resistance parameter have the following periodic variation:

$$R(t) = R_0(1 + k \cos pt) \qquad \text{where } k^2 \ll 1 \tag{44.2}$$

Let

$$u(t) = \frac{1}{2L}\int R(t)\,dt = b\left(t + \frac{k}{p}\sin pt\right) \tag{44.3}$$

with

$$b = \frac{R_0}{2L} \tag{44.4}$$

and introduce the transformation

$$q(t) = \epsilon^{-u(t)} y(t) \tag{44.5}$$

This transformation transforms Eq. (44.1) into the following equation:

$$\frac{d^2 y}{dt^2} + \left(\frac{1}{LC} - \frac{R^2}{4L^2} - \frac{R'}{2L}\right)y = \frac{E_m}{L}\epsilon^{u(t)}\sin \omega t \tag{44.6}$$

If Eq. (44.2) is substituted into Eq. (44.6) and terms of the order k^2 are neglected, the result is

$$\frac{d^2 y}{dt^2} + [\omega_0^2 + kb(p \sin pt - 2b \cos pt)]y = \frac{E_m}{L}\epsilon^{u(t)}\sin \omega t \tag{44.7}$$

where

$$\dot{\omega}_0 = \sqrt{\frac{1}{LC} - b^2} \tag{44.8}$$

If b is small, $p \gg 2b$ and the $\cos(pt)$ term in Eq. (44.7) may be neglected. The complementary function of Eq. (44.7) is then the solution of

$$\frac{d^2 y}{dt^2} + \omega_0^2\left(1 + \frac{kbp}{\omega_0^2}\sin pt\right)y = 0 \tag{44.9}$$

Let $x = pt$. In the variable x, Eq. (44.9) is transformed into

$$\frac{d^2 y}{dx^2} + \left(\frac{\omega_0}{p}\right)^2 \left(1 + \frac{kbp}{\omega_0^2} \sin x\right) y = 0 \tag{44.10}$$

The function $G(x)$ (of the section on BWK approximation) is now given by

$$G(x) = \frac{\omega_0}{p} \sqrt{1 + \frac{kbp}{\omega_0^2} \sin x} = \left(\frac{\omega_0}{p}\right) + m \sin x \tag{44.11}$$

where

$$m = \frac{kb}{2\omega_0} \tag{44.12}$$

The function $\phi(x)$ is in this case given by

$$\phi(x) = \int G(x)\, dx = \frac{\omega_0 x}{p} - m \cos x \tag{44.13}$$

Two linearly independent BWK solutions of Eq. (44.10) are

$$y_1(x) = \cos\left(\frac{\omega_0 x}{p} - m \cos x\right)$$

$$y_2(x) = \sin\left(\frac{\omega_0 x}{p} - m \cos x\right) \tag{44.14}$$

By the method described in Sec. 43 and after some algebraic reductions, the following approximate general solution of Eq. (44.1) may be obtained:

$$q(t) = A_1 \epsilon^{-u(t)} \cos(\omega_0 t > m \cos pt) + A_2 \epsilon^{-u(t)} \sin(\omega_0 t - m \cos pt)$$

$$+ \frac{E_m}{\omega_0 L} \epsilon^{-u(t)} \sin(\omega_0 t - m \cos pt) \int \epsilon^{u(t)} \sin \omega t \cos \omega_0 t - m \cos pt)$$

$$\times dt - \frac{E_m}{\omega_0 L} \epsilon^{-u(t)} \cos(\omega_0 t - m \cos pt) \int \epsilon^{u(t)} \sin \omega t$$

$$\times \sin(\omega_0 t - m \cos pt)\, dt \tag{44.15}$$

The quantities A_1 and A_2 are the arbitrary constants of the transient part of the solution, and the terms involving the integrals give the steady-state response of the system to the BWK order of approximation.

45 SERIES CIRCUIT WITH PERIODICALLY VARYING INDUCTANCE

As a final example of the BWK method, consider the circuit of Fig. 45.1. This circuit consists of a constant resistance and capacitance in series with a periodically varying inductance parameter. The circuit is energized by a periodic electromotive force $E(t) = E_m \sin \omega t$.

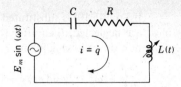

Fig. 45.1

The differential equation satisfied by the charge on the capacitance is [8]

$$\frac{d\phi}{dt} + \frac{R\,dq}{dt} + \frac{q}{C} = E_m \sin \omega t \tag{45.1}$$

where ϕ is the total magnetic flux linking the circuit, R is the resistance, and C the capacitance of the circuit. The flux ϕ is related to the inductance of the circuit by the fundamental equation

$$\phi = Li = \frac{L\,dq}{dt} \tag{45.2}$$

Hence if Eq. (45.2) is substituted into (45.1), the following equation is obtained:

$$\frac{d}{dt}\frac{L\,dq}{dt} + \frac{R\,dq}{dt} + \frac{q}{C} = E_m \sin \omega t \tag{45.3}$$

If the indicated differentiation is carried out, then

$$\frac{L\,d^2q}{dt^2} + \left(R + \frac{dL}{dt}\right)q + \frac{q}{C} = E_m \sin \omega t \tag{45.4}$$

This equation may be written in the following form:

$$\frac{d^2q}{dt^2} + 2P(t)\frac{dq}{dt} + \frac{q}{LC} = \frac{E_m}{L}\sin \omega t \tag{45.5}$$

where

$$P(t) = \frac{1}{2L}\left(R + \frac{dL}{dt}\right) \tag{45.6}$$

Let

$$v(t) = \int P(t)\,dt \tag{45.7}$$

$$q(t) = \epsilon^{-v(t)} y(t) \tag{45.8}$$

then Eq. (45.5) is transformed into

$$\frac{d^2y}{dt^2} + \left(\frac{1}{LC} - \frac{dP}{dt} - P^2\right)y = \frac{E_m}{L}\epsilon^{v(t)}\sin \omega t \tag{45.9}$$

Let it now be assumed that the variable inductance has the following form:

$$L(t) = L_0(1 + a\sin pt) \qquad \text{where } a^2 \ll 1 \tag{45.10}$$

If Eq. (45.10) is substituted into Eqs. (45.6), (45.7), and (45.9) and terms in a^2 are neglected, then after some algebraic reductions Eq. (45.9) may be written in the following form:

$$\frac{d^2 y}{dt^2} + (\omega_0^2 + 2g \sin pt)\, y = \frac{E_m}{L}\, \epsilon^{v(t)} \sin \omega t \tag{45.11}$$

where

$$\omega_0 = \sqrt{\frac{1}{L_0 C} - \frac{R^2}{4L_0^2}} \tag{45.12}$$

$$g = \frac{a}{4}\left(p - \frac{2}{L_0 C} + \frac{R^2}{L_0^2}\right) \tag{45.13}$$

$$v(t) = \int P(t)\, dt = \frac{Rt}{2L_0} + \frac{a}{2}\left(\sin pt + \frac{R}{L_0 p} \cos pt\right) \tag{45.14}$$

By the transformation $x = pt$, Eq. (45.11) may be transformed into the form

$$\frac{d^2 y}{dx^2} + \left[\left(\frac{\omega_0}{p}\right)^2 + 2g\,\frac{\sin x}{p^2}\right] y = E_m \frac{\epsilon^{v(x/p)}}{Lp^2} \sin \frac{\omega x}{p} \tag{45.15}$$

By the BWK method discussed in Sec. 41, the following complementary function of Eq. (45.15) y_c may be obtained:

$$y_c = A_1 \cos\left(\frac{\omega_0 x}{p} - z \cos x\right) + A_2 \sin\left(\frac{\omega_0 x}{p} - z \cos x\right) \tag{45.16}$$

where

$$z = \frac{g}{\omega_0 p} \tag{45.17}$$

A_1 and A_2 in Eq. (45.16) are arbitrary constants. The method of the variation of parameters, discussed in Sec. 43, gives the following approximate general solution of Eq. (45.4) with the inductance variation equation (45.10):

$$q(t) = A_1 \epsilon^{-v(t)} \cos(\omega_0 t - z \cos pt) + A_2 \sin(\omega_0 t - z \cos pt)\, \epsilon^{-v(t)}$$

$$+ \frac{E_m}{\omega_0 L} \epsilon^{-v(t)} \sin(\omega_0 t - z \cos pt) \int \epsilon^{v(t)} \sin \omega t \cos(\omega_0 t - z \cos pt)$$

$$- \frac{E_m}{\omega_0 L} \epsilon^{-v(t)} \sin(\omega_0 t - z \cos pt)$$

$$\times \int \epsilon^{v(t)} \cos \omega t \cos(\omega_0 t - z \cos pt) \tag{45.18}$$

The transient part of the solution contains the arbitrary constants, and the integral terms contain the steady-state response. The harmonic content of the steady-state response may be obtained by using the Bessel function expansions (42.12) and term by term integration.

46 CONCLUSION

The analysis presented in this discussion is an attempt to apply the method of the BWK approximation, which has yielded very useful and interesting results in problems of wave mechanics to the analysis of linear circuits whose parameters vary periodically with the time. A survey of the literature of circuit analysis reveals that this powerful method of analysis has been neglected by the electrical engineers, and it is hoped that this presentation will given an impetus to its future use for the solution of problems involving linear time-varying parameters.

PROBLEMS

1. Given the second-order linear differential equation,

$$\frac{d^2 x}{dt^2} + F^2(x)x = 0$$

where $F(x)$ is a slowly varying positive function of x. Show that the substitution $x = e^{j\phi(x)}$ leads to the following nonlinear differential equation of the Ricatti type:

$$j\frac{d^2\phi}{dx^2} - \phi^2 + F^2 = 0 \qquad j = (-1)^{\frac{1}{2}}$$

Obtain an approximate solution of the Ricatti equation, and show that

$$x = [F(x)]^{-\frac{1}{2}}[A\cos\phi(x) + B\sin\phi(x)]$$

where $\phi(x) = \int F(x)dx$ and A and B are arbitrary constants, is the approximate solution of the original equation (Brillouin-Wentzel-Kramers approximation).

2. A body of mass m falls through a medium that opposes its fall with a force proportional to the square of its velocity so that its equation of motion is

$$m\frac{d^2 x}{dt^2} = mg - Kv^2 \qquad v = \dot{x}$$

where x is the distance of fall and v is the velocity. If, at $t = 0$, $x = 0$, and $v = 0$, find the velocity of the body and the distance it has fallen at a time t. Plot the phase trajectory of the motion.

3. Solve Bernoulli's equation

$$\frac{dx}{dt} + P(t)x = Q(t)x^n$$

subject to the initial condition $x = x_0$ at $t = 0$.

4. Given the differential equation

$$\frac{d^2 x}{dt^2} + w_0^3 x + hx^2 = A\cos w_1 t + B\cos w_2 t$$

where w_0, h, A, B, w_1, and w_2 are constants. Devise a method of successive approximations for the solution of this equation. Let the initial conditions be $x = x_0$ and $\dot{x} = 0$ at $t = 0$. (This equation[1] is the basis of Helmholtz's *theory of combinations of tones*.)

[1] See L. A. Pipes, An Operational Treatment of Nonlinear Dynamical Systems, *Journal of the Acoustical Society of America*, vol. 10, pp. 29–31, July, 1938.

5. A particle oscillates in a straight line under the action of a central force tending to a fixed point O on the straight line and varying as the distance from O. If the motion takes place in a medium that resists the motion with a force proportional to the square of the velocity of the particle, find the relation between the amplitudes of any two successive arcs on each side of O. Plot the phase trajectory of the motion.

6. The nonlinear resistive material thyrite has a voltage-current characteristic of the form $V = Bi^b$, where V is the potential difference applied across the thyrite resistor, i the current, B a constant that depends on the length and cross section of the resistor. The parameter b has the value 0.28. An electrical circuit is composed of a linear inductor L and a thyrite resistor in series with a harmonic electromotive force. The equation of the circuit is

$$L\frac{di}{dt} + Bi^b = E_m \sin wt \qquad b = 0.28$$

Obtain an approximate expression for the steady-state periodic current of this circuit.

7. Show that the solution of the differential equation

$$\frac{d^2x}{dt^2} + k e^x = 0$$

with the initial conditions $dx/dt = 0$, $x = x_0$, at $t = 0$, is

$$x = x_0 - 2\ln(\cosh at) \qquad \text{where } a = \left(\frac{k e^{x_0}}{2}\right)^{\frac{1}{2}}$$

This is the equation of motion of a particle undergoing a restoring force which increases exponentially with the displacement.

8. Show that $x = 2b^{-\frac{1}{2}}\sin wt$ is the *exact* solution of the differential equation

$$\frac{d^2x}{dt^2} - 2aw(1 - bx^2)\frac{dx}{dt} + w^2 x = -\frac{4aw^2}{\sqrt{b}}\sin 3wt$$

This is a special case of a phenomenon called *frequency demultiplication* by van der Pol.[1]

9. Show that, if the motor of the motor-generator oscillator of Fig. 16.1 is replaced by a large capacitor and the rest of the system is left unchanged, the current will execute relaxation oscillations.

10. Discuss the oscillations of the height $h(t)$ of the level of water in the tank of Fig. P.10. Does $h(t)$ execute relaxation oscillations? Plot the phase trajectory of $h(t)$.

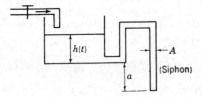

Fig. P 10

11. Given the differential equation

$$\frac{d^2x}{dt^2} + c\frac{dx}{dt} + w_0^2 x - hx^3 = F\sin wt \qquad h \ll 1$$

[1] B. van der Pol, Frequency Demultiplication, *Nature*, vol. 120, p. 363, 1927.

Determine a solution of this equation of the form

$$x = A(t)\sin wt + B(t)\cos wt$$

where $A(t)$ and $B(t)$ are slowly varying functions of t. Assume that $dA/dt \ll wA$, $dB/dt \ll wB$, and $d^2 A/dt^2 = d^2 B/dt^2 = 0$.

12. Solve the *linear* equation

$$\frac{d^2 x}{dt^2} + \mu \frac{dx}{dt} + w^2 x = 0$$

for small values of μ by the method of Kryloff and Bogoliuboff.

13. Show that the differential equation

$$\frac{d^2 x}{dt^2} + k^2 \sin x = 0$$

has the approximate solution $x = x_0 \sin wt$, where the angular frequency w is given by $w = k[2J_1(x_0)/x_0]^{\frac{1}{2}}$. $J_1(x_0)$ is the Bessel function of the first kind and first order. If $k = (g/L)^{\frac{1}{2}}$, the above equation governs the motion of a simple pendulum of length L.

14. The forced oscillations of a simple pendulum are governed by the equation

$$\frac{d^2 \theta}{dt^2} + k^2 \sin \theta = F\sin wt \qquad k = \left(\frac{g}{L}\right)^{\frac{1}{2}}$$

Show that the first approximation for the periodic motion is given by $\theta = A\sin wt$, where the amplitude A satisfies the equation $A(1 - w^2/k^2) - F/k^2 = A - 2J_1(A)$, where $J_1(A)$ is the Bessel function of the first kind and first order.

15. A particle of mass m moves along the x axis under the action of a restraining force $F(x) = 1/x^2 - 1/x$. Show that, if the motion of the mass is started with a small amount of energy E_0, the motion will be oscillatory. If, however, the motion is started with a large amount of energy E, show that the motion is *not* periodic and extends to infinity. Determine the critical energy that forms the dividing line between the two cases. Draw the phase trajectories for the oscillatory and nonoscillatory cases.

16. A pendulum (Froude's pendulum) is suspended from a dry horizontal shaft, of circular cross section, by a ring. It is well known that, if the shaft rotates with a suitable constant angular velocity w_0, the pendulum oscillates with a gradually increasing amplitude which reaches a final ultimate value. Write the differential equation of motion of this pendulum, and use it to explain the above phenomenon.

    (HINT: Assume the frictional force at the suspension to be a function of $w_0 - \theta$, the difference between the angular velocity of the shaft and the pendulum.)

17. Devise a graphical method for the solution of the equation

$$m\frac{d^2 x}{dt^2} + F(x) = 0$$

subject to the initial conditions $x = a$ and $\dot{x} = 0$ at $t = 0$.[1]

18. The differential equation of an oscillator whose motion is retarded by viscous and Coulomb friction and is being driven by a harmonic force is

$$m\frac{d^2 x}{dt^2} + c\frac{dx}{dt} + kx + Fsgn(\dot{x}) = F_0 \sin(wt + \theta)$$

[1] See Lord Kelvin, On Graphic Solutions of Dynamical Problems, *Philosophical Magazine*, vol. 34, pp. 332–348, 1892.

where

$$sgn(\dot{x}) = \begin{cases} +1 & \text{if } \dot{x} \geqslant 0 \\ -1 & \text{if } \dot{x} < 0 \end{cases}$$

The term $Fsgn(\dot{x})$ represents the effect of dry friction.   Obtain an approximate expression for the periodic-steady-state solution of this equation.[1]

REFERENCES

Books and reports

1. McLachlan, N. W.: "Ordinary Nonlinear Differential Equations in Engineering and Physical Science," 2d ed., Oxford University Press, New York, 1956.   (This book contains an excellent discussion of the better-known analytical procedures and an extensive bibliography of 211 titles covering the international literature to 1950.)
2. Stoker, J. J.: "Nonlinear Vibrations," Interscience Publishers, Inc., New York, 1950. (Contains a clear exposition of the topological methods for solving autonomous systems and a fairly complete discussion of stability.)
3. Andronow, A. A., and C. E. Chaikin: "Theory of Oscillations," Princeton University Press, Princeton, N.J., 1949.   (Contains a clear and interesting exposition of the topological properties of many autonomous systems.)
4. Minorsky, N.: "Introduction to Nonlinear Mechanics," J. W. Edwards, Publisher, Inc., Ann Arbor, Mich., 1947.   (An extensive survey of the international literature to 1947; a clear discussion of topological methods, analytical methods, nonlinear resonance, and relaxation oscillations.)
5. von Kármán, T.: The Engineer Grapples with Nonlinear Problems, *Bulletin of the American Mathematical Society*, vol. 46, no. 8, pp. 615–683, 1940.   (An excellent exposition of self-excited nonlinear oscillations, subharmonic resonance, large-deflection theory, elasticity with finite deformations, fluid jets, waves and finite amplitudes, viscous fluids, and compressible fluids; contains excellent bibliographies on these fields.)
6. Kryloff, N., and N. Bogoliuboff: "Introduction to Nonlinear Mechanics," Princeton University Press, Princeton, N.J., 1943.   (A clear and careful exposition of the Kryloff-Bogoliuboff method, with interesting applications to electrical circuits and dynamics.)
7. Duffing, G.: "Erzwungene Schwingungen bei veranderlicher Eigenfrequenz und ihre technische Bedeutung," Vieweg-Verlag, Brunswick, Germany, 1918.   (A pioneer study of nonlinear vibration problems; discusses methods of approximation and questions of stability with technical applications.)
8. Lefshetz, S.: "Contributions to the Theory of Nonlinear Oscillations," vol. II, Princeton University Press, Princeton, N.J., 1952.   (Discussions of several topics of nonlinear differential equations by several mathematicians from the rigorous point of view.)
9. Pipes, L. A.: "Operational Methods in Nonlinear Mechanics," Department of Engineering, University of California, Los Angeles, 1951.   (Contains some useful methods for the solution of certain classes of nonlinear technical problems; operational methods are introduced to reduce the algebraic labor of the solution.)
10. Edson, W. A.: "Vacuum-tube Oscillators," John Wiley & Sons, Inc., New York, 1953, pp. 408–412.
11. McLachlan, N. W.: "Theory and Application of Mathieu Functions," Oxford University Press, London, 1947.   (Contains 226 references to the international literature.)

[1] See J. P. Den Hartog, Forced Vibrations with Combined Coulomb and Viscous Friction, *Transactions of the American Society of Mechanical Engineers*, vol. 53, pp. 107–115, 1931.

12. Kemble, E. C.: "The Fundamental Principles of Quantum Mechanics," McGraw-Hill Book Company, New York, 1937, pp. 90–112. (Contains a detailed exposition of the BWK procedure and an extensive list of references to papers on wave mechanics.)

13. McLachlan, N. W.: "Bessel Functions for Engineers," Oxford University Press, London, 1934, pp. 42–43.

14. Ford, L. R.: "Differential Equations," McGraw-Hill Book Company, New York, 1933, pp. 75–76.

The bibliographies in the above books and reports are quite extensive and cover the most important papers to 1950. The following technical papers are representative of recent publications in the field of nonlinear problems:

Scientific papers

15. Sim, A. C.: A Generalization of Reversion Formulae with Their Application to Nonlinear Differential Equations, *Philosophical Magazine*, ser. 7, vol. 42, pp. 237–240, March, 1951. (Discusses an ingenious method for the approximate solution of nonlinear differential equations similar to the reversion of a power series.)

16. Cartwright, M. L.: Forced Oscillations in Nonlinear Systems, *Journal of Research of the National Bureau of Standards*, vol. 45, no. 6, December, 1950. (A rigorous discussion of the van der Pol equation with a forcing term.)

17. Hayashi, C.: Forced Oscillations with Nonlinear Restoring Force, *Journal of Applied Physics*, vol. 24, no. 2, pp. 198–207, 1953. (An excellent discussion of forced oscillations of nonlinear systems; the equation is transformed so that topological methods can be used.)

18. Hayashi, C.: Stability Investigation of the Nonlinear Periodic Oscillations, *Journal of Applied Physics*, vol. 24, no. 3, pp. 238–246, 1953. (The stability of nonlinear periodic oscillations is discussed by solving the variational equation for small deviations from the periodic state.)

19. Hayashi, C.: Subharmonic Oscillations in Nonlinear Systems, *Journal of Applied Physics*, vol. 24, no. 5, pp. 521–529, 1953. (The possibility of subharmonic solutions and their stability is studied for systems having a nonlinear restoring force and driven by a harmonic forcing function.)

20. Pipes, L. A.: A Mathematical Analysis of a Series Circuit Containing a Nonlinear Capacitor, *Communication and Electronics*, July, 1953, pp. 238–244. (An extension of Duffing's method and its application to a technically important problem.)

21. Pipes, L. A.: Applications of Integral Equations to the Solution of Nonlinear Electric Circuit Problems, *Communication and Electronics*, September, 1953, pp. 445–450. (The formulation of some nonlinear differential equations in terms of integral equations and their solution by iterative procedures.)

22. Thomsen, J. S.: Graphical Analysis of Nonlinear Circuits Using Impedance Concepts, *Journal of Applied Physics*, vol. 24, no. 11, pp. 1379–1382, 1953. (An interesting extension of linear-impedance concepts to the calculation of the amplitudes of nonlinear oscillations.)

23. Jacobsen, L. S.: On a General Method of Solving Second Order Ordinary Differential Equations by Phase-plane Displacements, *Journal of Applied Mechanics*, vol. 19, no. 4, pp. 533–553, December, 1952. (This paper presents a powerful graphical procedure for the solution of complicated equations; the method is illustrated by several important examples.)

24. Klotter, K.: Nonlinear Vibration Problems Treated by the Averaging Method of W. Ritz, *Proceedings of the First National Congress of Applied Mechanics*, pp. 125–131, American Society of Mechanical Engineers, 1952.

25. Clauser, F. H.: The Behavior of Nonlinear Systems, *Journal of the Aeronautical Sciences*, vol. 23, no. 5, pp. 411–434, May, 1956. (An excellent expository paper. It presents a diversified group of nonlinear phenomena in terms of fundamental and well-understood concepts. Contains a comprehensive bibliography of recent papers.)

26. van der Pol, B.: Nonlinear Theory of Electric Oscillations, *Proceedings of the Institute of Radio Engineers*, vol. 22, pp. 1051–1086, 1934.

27. Cartwright, M. L.: Forced Oscillations in Nearly Sinusoidal Systems, *Proceedings of the Institution of Electrical Engineers*, vol. 95, pt. III, pp. 88–96, 1948.

28. Gillies, A. W.: The Application of Power Series to the Solution of Nonlinear Circuit Problems, *Proceedings of the Institution of Electrical Engineers*, vol. 96, pt. III, pp. 453–475, 1949.

29. Tucker, D. G.: Forced Oscillations in Oscillator Circuits and the Synchronization of Oscillators, *Journal of the Institution of Electrical Engineers*, vol. 92, pt. III, p. 226, 1946.

30. Pipes, L. A.: The Reversion Method of Solving Nonlinear Differential Equations, *Journal of Applied Physics*, vol. 23, p. 202, 1952.

31. Pipes, L. A.: A Mathematical Analysis of a Dielectric Amplifier, *Journal of Applied Physics*, vol. 23, no. 8, pp. 818–824, August, 1952.

32. Miles, James G.: Bibliography of Magnetic Amplifier Devices and the Saturable Reactor Art, *Transactions of the American Institute of Electrical Engineers*, vol. 70, pt. II, pp. 2104–2123, 1951.

33. Finzi, L. A., and Pitman, G. F., Jr.: Comparison of Methods of Analysis of Magnetic Amplifiers, *Proceedings of the National Electronics Conference*, vol. 8, p. 144, 1952.

34. Pipes, L. A.: Operational and Matrix Methods in Linear Variable Networks, *Philosophical Magazine*, vol. 25, pp. 585–600, 1938.

35. Barrow, W. L.: On the Oscillations of a Circuit Having a Periodically Varying Capacitance, *Proceedings of the Institute of Radio Engineers*, vol. 22, pp. 201–212, 1934.

36. Erdelyi, A.: Ueber die Freien Schwingungen in Kondensatorkreisen mit Periodisch Veralnderlicher Kapazitaet, *Physica*, vol. 5, no. 19, pp. 585–622, 1934.

37. Pipes, L. A.: Solution of Variable Circuits by Matrices, *Journal of the Franklin Institute*, vol. 224, pp. 767–777, 1937.

38. Bennett, W. R.: A General Review of Linear Varying Parameter and Nonlinear Circuit Analysis, *Proceedings of the Institute of Radio Engineers*, vol. 38, pp. 259–263, 1950.

39. Kingston, R. H.: Resonant Circuits with Time-Varying Parameters, *Proceedings of the Institute of Radio Engineers*, vol. 37, pp. 1478–1481, 1949.

40. Bura, P., and Tombs, D. M.: Resonant Circuit with Periodically-Varying Parameters, *Wireless Engineer*, pp. 95–100, April 1952, pp. 120–126, May, 1952.

41. Gadsen, C. P.: An Electrical Network with Varying Parameters, *Quarterly of Applied Mathematics*, vol. 8, no. 2, pp. 199–205, 1950.

42. Aseltine, J. A.: Transforms for Linear Time-Varying Systems, Report No. 52.1, University of California, Los Angeles, February, 1952.

43. Brillouin, L.: The B.W.K. Approximation and Hill's Equation, *Quarterly of Applied Mathematics*, vol. 7, no. 4, pp. 363–380, 1950.

44. Carson, J. R.: Notes on the Theory of Frequency Modulation, *Proceedings of the Institute of Radio Engineers*, vol. 10, no. 62, 1922.

16
Statistics and Probability

1 INTRODUCTION

In many branches of engineering and physics, processes are encountered which are not known with exactness. In the study of phenomena of this type, the applications of the methods of statistics and of the theory of probability are imperative. In this chapter, some of the *basic* elementary concepts and methods employed in the theory of statistics and probability will be given. The notion of probability is a very difficult concept whose fundamental aspects we shall not discuss. Simple cases and examples will be given to introduce its use in a reasonable way. For the reader who wishes to study the *vast subject* of statistics and probability further, a fairly complete and representative set of references to the more important modern treatises on the subject is given at the end of the chapter.

2 STATISTICAL DISTRIBUTIONS

When a series of measurements is made on a set of objects, the set of objects being measured is called a *population*. A population may be, for example, the animals living on a farm, the group of people living in a certain city, the automobiles produced by a factory in a year, the set of results of a physical or chemical experiment, etc.

If a *single quantity* is measured for each member of a population, this quantity is called a *variate*. Variates are usually denoted in the mathematical literature by symbols such as x or y. The possible observed values of a variate x will be a finite sequence or set of numbers $x_1, x_2, x_3, \ldots, x_n$. The numbers x_k, for example, may be the heights of a group of n individuals or the prices of a certain make of automobile in a dealer's showroom.

FREQUENCY

In a given population, the number of times a particular value x_r of a variate is observed is called the *frequency* f_r of x_r. For example, an automobile dealer may have nine cars in his showroom; three cars are priced at \$3,500, four at \$4,000, and two at \$2,000. In this case we have $x_1 = \$3,500$, $f_1 = 3$, $x_2 = \$4,000$, $f_2 = 4$, and $x_3 = \$2,000$, $f_3 = 2$. The total number of automobiles in the population is N, where

$$N = f_1 + f_2 + f_3 = 3 + 4 + 2 = 9 \tag{2.1}$$

The total value of the automobiles S is

$$S = f_1 x_1 + f_2 x_2 + f_3 x_3 = 3(3,500) + 4(4,000) + 2(2,000) = 30,500 \tag{2.2}$$

The average or mean price $\bar{x}$ is

$$\bar{x} = \frac{1}{N}(f_1 x_1 + f_2 x_2 + f_3 x_3) = \$3,388 = \frac{1}{N}\sum_{r=1}^{3} f_r x_r \qquad N = f_1 + f_2 + f_3 \tag{2.3}$$

THE MEAN OF A DISTRIBUTION

The most elementary statistical quantity defined by a distribution is the arithmetic mean $\bar{x}$ of the variate x, usually known as the *mean*. This is the sum of the values of x for each member of the population divided by N. Therefore,

$$\bar{x} = \frac{1}{N}\sum_{r=1}^{k} f_r x_r \qquad k = \text{number of frequencies} \tag{2.4}$$

Let us suppose that a is a typical value of the variate x, and let us define

$$d_r = x_r - a \qquad r = 1, 2, 3, \ldots, N \tag{2.5}$$

It is thus evident that d_r is the deviation of x_r from the typical value a. The variate d has a distribution f_r over the values d_r. The mean $\bar{d}$ of d for this distribution is

$$\bar{d} = \frac{1}{N} \sum_{r=1}^{k} f_r d_r \tag{2.6}$$

$\bar{d}$ is simply related to $\bar{x}$; this can be seen by writing (2.6) in the form:

$$\bar{d} = \frac{1}{N} \sum_{r=1}^{k} f_r(x_r - a) = \bar{x} - a \tag{2.7}$$

Therefore we have

$$\bar{x} = \bar{d} + a \tag{2.8}$$

The following relation is sometimes useful.

$$d_r - \bar{d} = (x_r - a) + (a - \bar{x}) = x_r - \bar{x} \qquad r = 1, 2, 3, \ldots, N \tag{2.9}$$

Equation (2.8) is useful in computing $\bar{x}$ quickly, provided that we choose a to be a number close to the mean.

3 SECOND MOMENTS AND STANDARD DEVIATION

In statistical investigations, it is of interest to know whether or not a distribution is closely grouped around its mean. One of the most useful quantities for estimating the spread of a distribution is the *standard deviation* σ. Before we define σ it is instructive to note that $\bar{d}$ as given by (2.7) is the *first moment* of a distribution about a given value a of the variate. We note from (2.8) that if $a = \bar{x}$, then the first moment $\bar{d} = 0$.

THE SECOND MOMENT

The *second moment* μ_2' of a distribution about a given value a of the variate is the average over the population of the *squares* of the differences $d_r = x_r - a$ between a and the values of the variate x_r. It is given by

$$\mu_2' = \frac{1}{N} \sum_{r=1}^{k} f_r d_r^2 = \frac{1}{N} \sum_{r=1}^{k} f_r(x_r - a)^2 \tag{3.1}$$

It is apparent that if most of the measured values of x for the population are approximately equal to a, μ_2' will be small, while if a differs considerably from the observed values of x, then μ_2' will be large since all the terms in (3.1) are positive. If we now take $a = \bar{x}$ in (3.1), that is, we take the second moment

about the *mean* $\bar{x}$ of the population, this second moment μ_2 is called the
variance. The variance is therefore given by

$$\mu_2 = \frac{1}{N} \sum_{r=1}^{n} f_r(x_r - \bar{x})^2 \tag{3.2}$$

The variance gives an estimate of the spread of values of x about their
mean. The variance is not directly comparable to $\bar{x}$ since it is of dimension
of x^2.

THE STANDARD DEVIATION

The positive square root of μ_2 is the *standard deviation* σ; since it is of the
same dimension as x, σ gives us an estimate of the *spread* of values of x which
can be compared directly with $\bar{x}$. By definition, we have therefore

$$\sigma^2 = \mu_2 = \frac{1}{N} \sum_{r=1}^{k} f_r(x_r - \bar{x})^2 = \frac{1}{N} \sum_{r=1}^{k} f_r(d_r - \bar{d})^2 \tag{3.3}$$

where we have made use of (2.9).

If (3.3) is expanded, we obtain the following result:

$$\sigma^2 = \frac{1}{N} \left(\sum_{r=1}^{k} f_r d_r^2 - 2d \sum_{r=1}^{k} f_r d_r + d^2 \sum_{r=1}^{k} f_r \right)$$

$$= \frac{1}{N} (N\mu - 2_2'Nd^2 + Nd^2) \tag{3.4}$$

It is thus evident that the variance μ_2 and the second moment μ_2' about any
value a are related by

$$\sigma^2 = \mu_2 = \mu_2' - d^2 \tag{3.5}$$

In order to calculate the standard deviation σ, it is usually easier to calculate the
second moment μ_2' about a simple value a and then use (3.5) instead of using
(3.3) directly.

Since σ^2 is the mean value of the squares of the deviations of the variate
from its average value $\bar{x}$, σ (the "root mean square") gives us an over-all
measure of the deviation of the set from the average $\bar{x}$ without reference to
sign. There are two other features of the population which are sometimes
found useful. Suppose that a frequency diagram has been constructed in
which the ordinates represent the number of readings lying in successive
intervals. The interval in which the ordinate attains its maximum clearly
corresponds to the most frequent or "most fashionable" value of x among the
set. This value is called the *mode*; in general, it is not identical with the

average or mean, but it will be if the frequency curve is symmetric about the mean value.   A frequency curve may have more than one mode, but we are here concerned only with cases in which a single mode exists.

We may also arrange our data in ascending order to magnitude and divide them into two sections halfway, so that as many measurements lie about this division as below it.   This position is called the *median*, and is such that it is useful in analyzing the data in many experimental observations.

4   DEFINITIONS OF PROBABILITY

A POSTERIORI PROBABILITY

Let us take an ordinary coin and flip it N times.   Let the number of times the "heads" side of the coin comes up be denoted by N_h; we may then write the following ratio:

$$R(N) = \frac{N_h}{N} \tag{4.1}$$

When this experiment is performed with a balanced coin, the result of this experiment has been found to be that

$$\lim_{N \to \text{large}} R(N) = \lim_{N \to \text{large}} \frac{N_h}{N} = \tfrac{1}{2} \tag{4.2}$$

Because of the result (4.2), we say that the *probability* of a coin coming up heads when it is flipped is $\tfrac{1}{2}$.

This simple case can be generalized to the following situation.   Consider an experiment that is performed N times.   Let the times that the experiment succeeds be N_s; then if the ratio

$$R(N) = \frac{N_s}{N} \tag{4.3}$$

approaches the limit

$$\lim_{N \to \infty} R(N) = P_s \tag{4.4}$$

we say that the probability of the success of the experiment is P_s.

Because in this case the probability P_s is *measured* by performing the experiment and is not *predicted*, P_s is called the *a posteriori* probability that the experiment will be successful.

A PRIORI PROBABILITY

Let us now perform another experiment that has n equally likely outcomes. Let us call the successes of the experiment n_s.   If we now *predict* that the ratio

$$r(n) = \frac{n_s}{n} \tag{4.5}$$

is such that

$$\lim_{n \to \infty} r(n) = p_s \tag{4.6}$$

Then p_s is called the *a priori* probability that the experiment will succeed.

Of course, the two types of probability should always give the same answer. If the a posteriori probability comes out to be different from the a priori probability, it will be concluded that an error has been made in the analysis. In the following discussion, no distinction will be made between the two types of probability.

5 FUNDAMENTAL LAWS OF PROBABILITY

In this section a simple discussion of the fundamental laws of probability will be given. Let $P(A)$ be the probability of an event called A occurring when a certain experiment is performed. By the definition of probability, $P(A)$ is a fraction that lies between 0 and 1. We also have by definition that if

$$P(A) = 1 \tag{5.1}$$

the event A is certain to happen. However if,

$$P(A) = 0 \tag{5.2}$$

the event A is certain *not* to happen.

COMPOUND EVENTS

To illustrate more complex possibilities, let us consider an experiment with N equally likely outcomes that involve two events A and B. Let us use the following notation:

$n_1 =$ the number of outcomes in which A only occurs

$n_2 =$ the number of outcomes in which B only occurs

$n_3 =$ the number of outcomes in which both A and B occur

$n_4 =$ the number of outcomes in which neither A nor B occurs.

$N = n_1 + n_2 + n_3 + n_4$

We see that by the use of the above notation we may define the following probabilities:

$$P(A) = \frac{n_1 + n_3}{N} = \text{probability of } A \text{ occurring} \tag{5.3}$$

$$P(B) = \frac{n_2 + n_3}{N} = \text{probability of } B \text{ occurring} \tag{5.4}$$

$$P(A + B) = \frac{n_1 + n_2 + n_3}{N} = \frac{\text{probability of } either \ A \text{ or } B \text{ (or both)}}{\text{occurring}} \tag{5.5}$$

$$P(AB) = \frac{n_3}{N} = \quad \text{the probability of the joint occurrence of } A \text{ and } B \tag{5.6}$$

$$P\left(\frac{A}{B}\right) = \frac{n_3}{n_2 + n_3} = \quad \text{the conditional probability of the event } A \text{ given that the event } B \text{ has occurred} \tag{5.7}$$

$$P\left(\frac{B}{A}\right) = \frac{n_3}{n_1 + n_3} = \quad \text{the conditional probability of the event } B \text{ given that the event } A \text{ has occurred} \tag{5.8}$$

From the above definitions, we see that

$$P(A + B) = \frac{n_1 + n_2 + n_3}{N} = \frac{n_1 + n_3}{N} + \frac{n_2 + n_3}{N} - \frac{n_3}{N} \tag{5.9}$$

or

$$P(A + B) = P(A) + P(B) - P(AB) \quad \text{(additive law of probability)} \tag{5.10}$$

We also have

$$P(AB) = \frac{n_3}{N} = \frac{n_2 + n_3}{N} \frac{n_3}{n_2 + n_3} = \frac{n_1 + n_3}{N} \frac{n_3}{n_1 + n_3} \tag{5.11}$$

or

$$P(AB) = P(B) P\left(\frac{A}{B}\right) = P(A) P\left(\frac{B}{A}\right) \quad \begin{array}{l}\text{(multiplicative law of} \\ \text{probability)}\end{array} \tag{5.12}$$

MUTUALLY EXCLUSIVE EVENTS

If A and B are mutually exclusive events so that the occurrence of one precludes the occurrence of the other, then the probability of their joint occurrence is zero and we have

$$P(AB) = 0 \tag{5.13}$$

For this case (5.10) reduces to the equation

$$P(A + B) = P(A) + P(B) = P(A \text{ or } B) \quad \begin{array}{l}\text{(additive law of} \\ \text{probability)}\end{array} \tag{5.14}$$

STATISTICALLY INDEPENDENT EVENTS

If the events A and B are *statistically independent*, then the probability of the occurrence of A does not depend on the occurrence of B and vice versa; we then have

$$P\left(\frac{B}{A}\right) = P(B) \quad \text{and} \quad P\left(\frac{A}{B}\right) = P(A) \tag{5.15}$$

In this case, Eq. (5.12) reduces to

$$P(AB) = P(A) P(B) \quad \text{(multiplicative law of probability)} \tag{5.16}$$

If the events A and B are statistically independent, then the additive law of probability (5.10) becomes

$$P(A + B) = P(A) + P(B) - P(A)P(B) \qquad (5.17)$$

As an example of the use of (5.17), let it be required to compute the probability that when one card is drawn from each of two decks, at least one will be an ace.   In this case the events A and B are the drawing of an ace out of a deck of 52 cards.   Therefore,

$$P(A) = \tfrac{4}{52} = \tfrac{1}{13} \qquad P(B) = \tfrac{1}{13} \qquad (5.18)$$

Since the events A and B are statistically independent, we have from (5.17) the following result:

$$P(A + B) = \tfrac{1}{13} + \tfrac{1}{13} - \tfrac{1}{13}\tfrac{1}{13} = \tfrac{25}{169} \qquad (5.19)$$

BAYES'S THEOREM

Some interesting relations among conditional probabilities may be deduced by means of Eq. (5.12).   If we solve (5.12) for $P(B/A)$, we obtain

$$P\left(\frac{B}{A}\right) = \frac{P(AB)}{P(A)} \qquad (5.20)$$

However, we also have from (5.12) the result

$$P(AB) = P(B)P\left(\frac{A}{B}\right) \qquad (5.21)$$

If we substitute this into (5.20), we obtain

$$P\left(\frac{B}{A}\right) = \frac{P(B)}{P(A)}P\left(\frac{A}{B}\right) \qquad (5.22)$$

Let us now write a relation similar to (5.22) but with B replaced by C. We thus obtain

$$P\left(\frac{C}{A}\right) = \frac{P(C)}{P(A)}P\left(\frac{A}{C}\right) \qquad (5.23)$$

We now divide Eq. (5.22) by (5.23) and obtain the result

$$\frac{P(B/A)}{P(C/A)} = \frac{P(B)P(A/B)}{P(C)P(A/C)} \qquad (5.24)$$

This result is a form of *Bayes's theorem* in the theory of probability.

6   DISCRETE PROBABILITY DISTRIBUTIONS

In Secs. 2 and 3 we discussed observed or empirical distributions.   The theory of probability is concerned with the *prediction* of distribution when some basic law is assumed to govern the behavior of the variate.   A simple example of a law of probability is that governing the behavior of a perfect die. The fact that the die is symmetric with respect to its six faces suggests very strongly that each of the faces is equally likely to end facing upward if we make an unbiased throw of the die. The probability p of any particular face appearing is thus equal to $\frac{1}{6}$. A similar situation is that involving the probability p that a particular card is on top of a thoroughly shuffled pack of 52 cards, $p = \frac{1}{52}$. The fact that if a fair die is thrown, the probability distribution is $p_k = \frac{1}{6}$ for $k = 1, 2, \ldots, 6$ is the *definition* of a *fair* die. The statistical problem then becomes one of deciding whether or not any particular die is fair.

TRIALS

When a card is drawn, a die is thrown, or the height of a man selected from a certain group of men is measured, we are said to be making a *trial*. Let us make a series of completely *independent* trials on a system, and let there be a finite number n of possible results of a trial. Let these n possible results be denoted by $x_1, x_2, x_3, \ldots, x_n$. If in the series of trials the result x_1 occurs m_1 times, the result x_2 occurs m_2 times, and in general the result x_r occurs m_r times $(r, 2, 3, \ldots, n)$, then

$$\sum_{r=1}^{n} m_r = N = \text{the total number of observed results in the total series of trials} \tag{6.1}$$

Now if as N becomes large, it is found that the proportion of any result x_r measured by x_r/N tends to a definite value p_r, so that

$$\lim_{N \to \infty} \frac{m_r}{N} = p_r \tag{6.2}$$

then p_r is the *probability* of the result x_r.

Since we cannot in practice make an infinite number of trials, we *cannot* determine p_r exactly by making a series of trials. It is therefore evident that a *law of probability* that specifies definite values of the probabilities p_r is never established with certainty by a series of trials, although it may be strongly suggested. The *assumption* of a law of probability corresponding to a series of trials is therefore a hypothesis which may or may not be confirmed by further trials.

If we divide (6.1) by N, we obtain

$$\sum_{r=1}^{n} \frac{m_r}{N} = 1 \tag{6.3}$$

We now let $N \to \infty$ in (6.3) and use the result (6.2) to obtain

$$\lim \sum_{r=1}^{n} \frac{m_r}{N} = \sum_{r=1}^{r} p_r = 1 \tag{6.4}$$

As an example of the above discussion, we note that the possible results of a throw of a die are $x_1 = 1$, $x_2 = 2$, $x_3 = 3$, $x_4 = 4$, $x_5 = 5$, $x_6 = 6$. For an *unbiased* die, we *assume the probability law* $p_r = \frac{1}{6}$ $(r = 1, 2, \ldots, 6)$. If in a large number of throws the values of m_r/N $(r = 1, 2, 3, \ldots, 6)$ do not tend to the value $\frac{1}{6}$, we conclude that *the die is biased*.

As an example of the use of an assumed law of probability let us consider the following question: "What is the probability that the face number 1 appears at least once in n throws of a die?"

Since we assume that the probability of the face 1 appearing in one throw is $\frac{1}{6}$, the probability of *not* obtaining a 1 in one throw is $q = 1 - \frac{1}{6} = \frac{5}{6}$. The probability of not obtaining a 1 in two throws is $q^2 = (\frac{5}{6})^2$ and of not obtaining a 1 in n throws is $q^n = (\frac{5}{6})^n$. The probability of throwing a 1 at least once in n throws p is therefore $p = 1 - q^n = 1 - (\frac{5}{6})^n$. In this calculation we assume that the result of each throw is independent of the result of any other throw and use (5.16).

7   ELEMENTS OF THE THEORY OF COMBINATIONS AND PERMUTATIONS

Since the mathematical theory of probability treats questions involving the relative frequency with which certain groups of objects may be conceived as arranged within a population, it is necessary to review the theory involving the number of ways in which various subgroups may be partitioned from the members of a larger group.

In dealing with objects in groups one is led to consider two kinds of arrangements, according to whether the order of the objects in the groups is or is not taken into account. Some elementary results in the theory of combinations and permutations will now be reviewed.

PERMUTATIONS AND COMBINATIONS

The number of arrangements, or *permutations*, of n objects is $n!$. This can be seen as follows since the first position can be occupied by any of the n objects, the second by any of the $n - 1$ remaining objects, etc. Therefore if we denote the number of permutations of n objects by $_nP_n$, we have

$$_nP_n = n(n - 1)(n - 2) \cdots (3)(2)(1) = n! \tag{7.1}$$

where $n! = (1)(2)(3) \cdots (n - 1)n$ is the *factorial* of n. By definition, $0! = 1$ and $1! = 1$.

THE NUMBER OF PERMUTATIONS OF n THINGS TAKEN r AT A TIME

The number of permutations of n different things taken r at a time is denoted by the symbol $_nP_r$ and is found easily as follows. The first place of any arrangement may have any one of the n things, so that there are n choices for the first place. The second place may have any one of the remaining $n - 1$ things, the third any of the remaining $(n - 2)$, and so on, to the rth place which has any of the remaining $(n - r + 1)$. It follows that

$$_nP_r = n(n - 1)(n - 2) \cdots (n - r + 1) = \frac{n!}{(n - r)!} = P(n,r) \qquad (7.2)$$

THE NUMBER OF COMBINATIONS OF n THINGS TAKING r AT A TIME

The combinations of n things taking r at a time is defined as the possible arrangements of r of the n things, no account being taken of the order of the arrangement. Thus the combinations of three things a, b, c, taking two at a time, are ab, ac, and bc. The combination ba is the same as ab, and so on.

The number of combinations of n things taking r at a time is denoted by the symbol $_nC_r$ and is found as follows. Any combination of the r things can be arranged in $r!$ ways since this is the number of arrangements of r things taking r at a time, or $_rP_r$. If every combination is so treated, we clearly obtain $_nP_r$ arrangements and it therefore follows that

$$_nC_r \, Xr! = _nP_r \qquad (7.3)$$

If we divide (7.3) by $r!$, we obtain

$$_nC_r = \frac{_nP_r}{r!} = \frac{n!}{r!(n - r)!} \qquad (7.4)$$

The quantity $_nC_r = n!/r!(n - r)!$ is called the *binomial coefficient*. It is shown in works on algebra that if n is a positive integer, we may write

$$(a + b)^n = a^n + _nC_1 a^{n-1} b + _nC_2 a^{n-2} b^2 + _nC_3 a^{n-3} b^3 + \cdots + _nC_n b^n \qquad (7.5)$$

Equation (6.5) for $n = 2, 3, 4, \ldots$, is known as the *binomial theorem*.

PROPERTIES OF THE BINOMIAL COEFFICIENT $_nC_r = C(n,r)$

The following interesting and useful properties of the binomial coefficient $_nC_r$ can be established:

$$_nC_r = \frac{n!}{r!(n - r)!} = _nC_{n-r} = C(n,r) \qquad (7.6)$$

$$2^n = 1 + _nC_1 + _nC_2 + \cdots + _nC_n \qquad (7.7)$$

$$0 = 1 - _nC_1 + _nC_2 - _nC_3 + \cdots \qquad (7.8)$$

$$_nC_r + _nC_{r-1} = _{n+1}C_r \qquad (7.9)$$

As a simple example of the use of Eq. (7.4), let it be required to solve the following problem. There are n points in a plane, no three being on a line. Find the number of lines passing through pairs of points.

We may solve this problem by realizing that since no three points lie on a line, any line can be considered as a combination of two of the n points. The number of lines is therefore $_nC_2 = n!/2!(n-2)! = n(n-1)/2$.

8 STIRLING'S APPROXIMATION FOR THE FACTORIAL

In many phases of the mathematical theory of probability, the factorial of large numbers occurs.   A classical result given by Stirling is of paramount importance in the computation of the factorial of large numbers.   The formula given by Stirling was obtained in the 1730s and may be stated in the following form:

$$\lim_{n\to\infty} n! = (2\pi)^{\frac{1}{2}} n^{n+\frac{1}{2}} e^{-n} \qquad (8.1)$$

The percent error of this formula for finite values of n is of the order of $100/12n$ percent error.

In addition to being a very useful method for the computation of factorial numbers, Stirling's formula is very useful in a theoretical sense.   Its use in the computation of the factorial or large numbers may be seen by considering the computation of 10!.   Instead of multiplying together a large number of integers, we have merely to calculate the factorial by the use of expression (8.1) by means of a table of logarithms, this involves far fewer operations.   For example, for $n = 10$ we obtain the value 3,598,696 by Stirling's expression (using seven-figure tables), while the exact value of 10! obtained by direct multiplication is 3,628,800.   The percentage error obtained by the use of Stirling's formula is of the order of $\frac{5}{6}$ percent too low.

A more exact formula for the computation of factorial numbers is the following one:

$$(2\pi)^{\frac{1}{2}} n^{n+\frac{1}{2}} e^{-n} < n! < (2\pi)^{\frac{1}{2}} n^{n+\frac{1}{2}} e^{-n}\, 1 + \frac{1}{12n} \qquad (8.2)$$

It is thus evident that the expressions $n!$ and $2\pi n^{n+\frac{1}{2}} e^{-n}$ differ only by a small percentage of error when the value of n is large.   The factor $1 + \frac{1}{4}n$ gives us an estimate of the degree of accuracy of the approximation.

In order to obtain this remarkable formula let us try to estimate the area under the curve $y = \ln x$.   This curve has the general form depicted in Fig. 8.1. If we integrate $\ln x$ by parts we obtain the result

$$A_n = \int_1^n \ln x\, dx = x \ln x - x \Big|_1^n = n \ln n - n + 1 \qquad (8.3)$$

A_n is the exact area under the curve shown in Fig. 8.1.

If we erect ordinates $y_1, y_2, \ldots, y_n$ at $x = x_1, x = x_2, \ldots, x = x_n$, the trapezoidal rule for the area under the curve T_n gives

$$T_n = \tfrac{1}{2}(y_1 + y_2) + \tfrac{1}{2}(y_2 + y_3) + \tfrac{1}{2}(y_3 + y_4) + \cdots + \tfrac{1}{2}y_n \qquad (8.4)$$

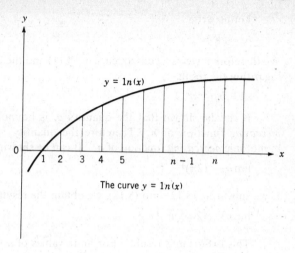

The curve $y = \ln(x)$

Fig. 8.1

Since we have

$$y_k = \ln(k) \qquad k = 1, 2, 3, \ldots, n \tag{8.5}$$

we obtain the following expression for T_n on substituting (8.5) into (8.4):

$$T_n = \tfrac{1}{2}\ln(1) + \ln(2) + \ln(3) + \cdots + \tfrac{1}{2}\ln(n)$$
$$= \ln(2) + \ln(3) + \ln(4) + \cdots + \ln(n-1) + \tfrac{1}{2}\ln(n) \tag{8.6}$$

By the definition of the factorial number $n!$, we have

$$n! = (1)(2)(3)(4) \cdots (n-1)(n) \tag{8.7}$$

Therefore if we take the natural logarithm of (8.7), we obtain

$$\ln(n!) = \ln(2) + \ln(3) + \ln(4) + \cdots + \ln(n-1) + \ln(n) \tag{8.8}$$

If we now compare (8.6) with (8.8) it can be seen that

$$T_n = \ln(n!) - \tfrac{1}{2}\ln(n) \tag{8.9}$$

The exact area A_n and the approximate area T_n are of the same order of magnitude. Since the trapezoids used in computing T_n lie *below* the curve $y = \ln(x)$, it is apparent that the true area A_n exceeds the trapezoidal area T_n. Let us write

$$A_n - T_n = a_n = [n\ln(n) - n + 1] - [\ln(n!) - \tfrac{1}{2}\ln(n)] \tag{8.10}$$

If we solve (8.10) for $\ln(n!)$, we obtain

$$\ln(n!) = (1 - a_n) + (n + \tfrac{1}{2})\ln(n) - n \tag{8.11}$$

Let us write

$$b_n = e^{1-a_n} \qquad (8.12)$$

We therefore have as a consequence of (8.11) and the definition of the natural logarithm the result

$$n! = b_n e^{-n} n^{n+\frac{1}{2}} \qquad (8.13)$$

It can be shown that the quantity a_n is bounded and is a monotonic *increasing* function of n. Therefore the quantity b_n is bounded and is a monotonic *decreasing* function of n. It can be shown that[†]

$$\lim_{n \to \infty} b_n = (2\pi)^{\frac{1}{2}} \qquad (8.14)$$

If we substitute (8.12) into (8.13), we obtain the result

$$\lim_{n \to \infty} n! = (2\pi)^{\frac{1}{2}} n^{n+\frac{1}{2}} e^{-n} \qquad (8.15)$$

This is Stirling's result. For finite values of n, (8.15) gives a result for $n!$ that is of the order of $100/12n$ percent too small. For more precise results, the inequality (8.2) may be used. The factor $(1 + \frac{1}{4}n)$ gives us an estimate of the degree of accuracy of the approximation.

9 CONTINUOUS DISTRIBUTIONS

In many problems encountered in the mathematical theory of probability, the possible values of a variate x are not discrete but lie in a continuous range. For example, the heights of adults normally take values in the continuous range from 4.5 to 6.5 ft. Even though observed heights are classified as taking one of a finite discrete set of values, any theoretical probability law should refer to the continuous range. If, for example, a law of probability predicted a distribution of heights at intervals of $\frac{1}{2}$ in., it would not be appropriate and adequate for comparison with a set of measurements made to the nearest centimeter.

A law of probability for a *continuous* variate is therefore expressed in terms of a *probability function* $\phi(x)$, such that $\phi(x)\,dx$ is the probability of a trial giving a result in the infinitesimal range $[x - (dx/2), x + (dx/2)]$ for all values of x. The probability $P(x_1, x_2)$ of the result lying between values x_1 and x_2 is then

$$P(x_1, x_2) = \int_{x_1}^{x_2} \phi(x)\,dx \qquad (9.1)$$

Since the total probability is unity, we have

$$\int \phi(x)\,dx = 1 \qquad (9.2)$$

[†] See R. Courant, "Differential and Integral Calculus," 2d ed., vol. I, pp. 361–364, Interscience Publishers, Inc., New York, 1938.

where the integral is taken over the complete range of the variate x. The equation (9.1) is analogous to Eq. (7.4) for discrete distributions; integration over the probability function replaces summation over the discrete probabilities.

10   EXPECTATION, MOMENTS, AND STANDARD DEVIATION, CHEBYSHEV'S INEQUALITY

The results of Secs. 2 and 3 apply equally well to probabilities.   The mean of a probability distribution for a variate x is called the *expectation* $E(x)$ and is defined by

$$E(x) = \frac{p_1 x_1 + p_2 x_2 + \cdots + p_n x_n}{p_1 + p_2 + \cdots + p_n} \tag{10.1}$$

Since the sum of the probabilities is unity, we have

$$p_1 + p_2 + p_3 + \cdots + p_n = 1 \tag{10.2}$$

Therefore the expectation $E(x)$ as given by (10.1) takes the form

$$E(x) = p_1 x_1 + p_2 x_2 + \cdots + p_n x_n = [x] \tag{10.3}$$

The expectation of a probability distribution of a *continuous variate* is defined by the equation

$$E(x) = \int x\phi(x)\,dx = [x] \tag{10.4}$$

where the integration takes place over the whole range of x.

The *second moment* about a of a discrete probability distribution μ_2' is defined by the equation

$$\mu_2' = \sum_{r=1}^{n} p_r(x_r - a)^2 \tag{10.5}$$

The *second moment* about a of a continuous probability distribution is defined by the equation

$$\mu_2' = \int \phi(x)(x - a)^2\,dx = [(x - a)^2] \tag{10.6}$$

The *variance* μ_2 is given by (10.5) or (10.6) with $a = [x]$. The *standard deviation* σ is given by

$$\sigma^2 = \mu_2 = [(x - [x])^2] \tag{10.7}$$

The standard deviation and the second moment $\mu_2'(a)$ are related in the same manner in (3.5) by the equation

$$\sigma^2 = \mu_2' - [d]^2 \tag{10.8}$$

where

$$d = x - a \qquad (10.9)$$

so that

$$[d] = [x] - a \qquad (10.10)$$

It is interesting to note that when $a = [x]$, then (10.10) becomes

$$[x - [x]] = [x] - [x] = 0 \qquad (10.11)$$

As an example of the above definitions consider the following illustration. Let us suppose that cards numbered 1, 2, 3, . . . , n are put in a box and thoroughly shuffled; then one card is drawn at random. If we assume that each card is equally likely to be drawn, so that the law of probability of drawing the cards is

$$p_r = \frac{1}{n} \qquad r = 1, 2, 3, \ldots, n \qquad (10.12)$$

then the expectation $E(x)$ of the number x on the card is

$$E(x) = [x] = \frac{1 + 2 + 3 + \cdots + n}{n} = \frac{n+1}{2} \qquad (10.13)$$

The second moment about $x = 0$ of the probability distribution of x is, by (10.5),

$$\mu_2' = [x^2] = n^{-1}(1^2 + 2^2 + \cdots + n^2) = \tfrac{1}{6}(n+1)(2n+1) \qquad (10.14)$$

The variance is given by (10.7) in the form

$$\sigma^2 = \mu_2 = [x^2] - [x]^2 = \tfrac{1}{12}(n^2 - 1) \qquad (10.15)$$

It can be seen that for large values of n we have

$$\left. \begin{array}{l} [x] = \dfrac{n}{2} \\[2mm] \sigma = \dfrac{n}{2(3)^{\frac{1}{2}}} \end{array} \right\} \quad (n \text{ large}) \qquad \begin{array}{l} (10.16) \\[4mm] (10.17) \end{array}$$

It is to be expected that the values of $[x]$ and σ will become proportional to n when n is large.

As an example of a continuous distribution, consider a variate x which is equally likely to take any value in the range $(0,X)$. In such a case the probability function $\phi(x)$ must equal a constant ϕ_0 in this range and zero elsewhere. In order to satisfy (9.2), we must have

$$\int_0^x \phi_0 \, dx = 1 \qquad \text{or} \qquad \phi(x) = \phi_0 = \frac{1}{X} \qquad (10.18)$$

The expectation $E(x)$ of x is, by (10.4),

$$E(x) = [x] = X^{-1} \int_0^x x \, dx = \frac{X}{2} \tag{10.19}$$

The variance is given by (10.6) with $a = [x] = X/2$, so that

$$\mu_2 = \int_0^x X^{-1} \left(x - \frac{X}{2} \right)^2 dx = \frac{X^2}{12} = \sigma^2 \tag{10.20}$$

Therefore the standard deviation is $\sigma = X/2(3)^{\frac{1}{2}}$.

CHEBYSHEV'S INEQUALITY

Let us consider a variate x with probability function $f(x)$. Let μ be the mean value and σ the standard deviation of x. We shall prove that if we select any number a, the probability that x differs from its mean value μ by more than a is less than σ^2/a^2. This means that x is unlikely to differ from μ by more than a few standard deviations, and that therefore the standard deviation σ gives an estimate of the *spread* of the distribution. For example, if $a = 2\sigma$, we find that the probability for x to differ from μ by more than 2σ is less than $\sigma^2/(2\sigma)^2 = \frac{1}{4}$.

Proof By the definition of the standard deviation σ, we have

$$\sigma^2 = \sum_{x=0}^{n} (x - \mu)^2 f(x) \tag{10.21}$$

If we now sum over values of x for which x exceeds $(\mu + a) = x_0$, we get *less* than σ^2, so that we have

$$\sigma^2 > \sum_{x=x_0}^{n} (x - \mu)^2 f(x) \tag{10.22}$$

If we replace $(x - \mu)$ by a in (10.22), the sum is further decreased and we have

$$\sigma^2 > \sum_{x_0}^{n} a^2 f(x) = a^2 \sum_{x_0}^{n} f(x) \tag{10.23}$$

We may write (10.23) in the form

$$\sum_{x=x_0}^{n} f(x) < \frac{\sigma^2}{a^2} \quad \text{(Chebyshev's inequality)} \tag{10.24}$$

However $f(x)$ is the sum of the probabilities of values of x which differ from μ by more than a, so that $x > (\mu + a)$ and (10.24) says that this probability is less than σ^2/a^2. This result is known in the literature as *Chebyshev's inequality*.

11 THE BINOMIAL DISTRIBUTION

Let us suppose that a trial can have only two possible results, one called a "success" with probability p, and the other called a "failure" with a probability of $q = (1 - p)$. If we make n independent trials of this type in succession, then the probability of a *particular sequence* of r successes and $(n - r)$ failures is $p^2 q^{n-r}$. However, the number of sequences with r successes and $(n - r)$ failures is

$$C(n,r) = \frac{n!}{[r!(n-r)!]} = \text{the binomial coefficient} \tag{11.1}$$

Therefore the probability of there being exactly r *successes* in n trials is

$$P(r) = C(n,r)p^r q^{n-r} = C(n,r)p^r(1-p)^{n-r} \tag{11.2}$$

It will be noticed that $P(r)$ is the term containing p^r in the binomial expansion of $(p + q)^n$. For this reason the probability distribution (11.2) is known as the *binomial distribution*.

Example 1 If the probability that a man aged 60 will live to 70 is $p = 0.65$, what is the probability that 8 out of 10 men now aged 60 will live to 70? This is exactly the problem proposed above, with $p = 0.65$, $r = 8$, and $n = 10$. The answer is, therefore,

$$P(8) = C(10,8)(0.65)^8(0.35)^2 = \left[\frac{10!}{8!2!}\right](0.65)^8(0.35)^2$$

$$= 45(0.65)^8(0.35)^2 = 0.176 \tag{11.3}$$

Example 2 What is the probability of throwing an ace exactly three times in four trials with a single die? In this case we have $n = 4$, $r = 3$, and since there is one chance in six of throwing an ace on a single trial, we have $p = \frac{1}{6}$. Therefore we have

$$P(3) = C(4,3)(\tfrac{1}{6})^3 \tfrac{5}{6} = \tfrac{5}{324} \tag{11.4}$$

Example 3 What is the probability of throwing a deuce exactly three times in three trials? In this case $n = 3$, $r = 3$, $p = \frac{1}{6}$, therefore we have the result

$$P(3) = C(3,3)(\tfrac{1}{6})^3(\tfrac{5}{6})^0 = (\tfrac{1}{6})^3 = \tfrac{1}{216}$$

THE MEAN OF THE BINOMIAL DISTRIBUTION

The expectation or the mean number of successes $E(x) = \bar{x}$ is given by equation (10.1) in the form

$$\bar{x} = \frac{\sum\limits_{r=0}^{n} rp_r}{\sum\limits_{r=0}^{n} p_r} \qquad \text{where } p_r = C(n,r)p^r q^{n-r} \tag{11.5}$$

We have

$$\sum_{r=0}^{n} p_r = \sum_{r=0}^{n} C(n,r) p^r q^{n-r} = (p+q)^n$$

$$= [p + (1-p)]^n = 1^n = 1 \qquad \text{since } q = 1 - p \qquad (11.6)$$

If we denote the partial derivative with respect to p by D_p, we note that

$$r p^r = p \, D_p \, p^r \qquad (11.7)$$

Since the denominator of (11.5) is unity, we have

$$\bar{x} = \sum_{r=0}^{n} C(n,r) r p^r q^{n-r} \qquad (11.8)$$

The series in (11.8) is summed by treating p and q as *independent* variables. If we now use the result (11.7), we may write (11.8) in the form

$$\bar{x} = p D_p \sum_{r=0}^{n} C(n,r) p^r q^{n-r}$$

$$= p D_p (p+q)^n = np(p+q)^{n-1} \qquad (11.9)$$

We now substitute $q = 1 - p$ in (11.9) and obtain the result

$$\bar{x} = np[p + (1-p)]^{n-1} = np \qquad (11.10)$$

Therefore the mean number of successes for the binomial distribution is $\bar{x} = np$.

THE STANDARD DEVIATION OF THE BINOMIAL DISTRIBUTION

We may compute the variance of the binomial distribution by the use of Eq. (10.8). The second moment μ_2' about $x = 0$ of the distribution of the number of successes can be found by a method similar to that used in computing the first moment above (we leave the derivation as an exercise for the reader and quote the result):

$$\mu_2' = np[1 + p(n-1)] \qquad (11.11)$$

As a consequence of (10.8) we have

$$\sigma^2 = \mu_2' - \bar{x}^2 = np[1 + p(n-1)] - (np)^2$$

$$= npq \qquad q = 1 - p \qquad (11.12)$$

Therefore the *standard deviation* of the *total* number of successes in n trials is

$$\sigma = (npq)^{\frac{1}{2}} \qquad (11.13)$$

12 THE POISSON DISTRIBUTION

The Poisson distribution is derived as the limit of the binomial distribution when the number of trials n is very large and the probability of success p is very small. For the binomial distribution we have seen that the mean number of successes is given by

$$\bar{x} = np = m \tag{12.1}$$

We now investigate the limiting form taken by the binomial distribution as the number of trials n tends to infinity and the probability p tends to zero in such a manner that

$$\lim_{\substack{n \to \infty \\ p \to 0}} (np) = m = \text{a constant} \qquad \text{or } p = \frac{m}{n} \tag{12.2}$$

If we substitute $p = m/n$ in the expression (11.2) for the binomial distribution, we obtain the equation

$$P(r) = n(n-1)(n-2) \cdots (n-r+1) \left(\frac{m}{n}\right)^r \frac{(1-m/n)^{n-r}}{r!} \tag{12.3}$$

This expression may be written in the form

$$P(r) = \frac{m^r}{r!} \left(1 - \frac{1}{n}\right)\left(1 - \frac{2}{n}\right) \cdots \left(1 - \frac{(r-1)}{n}\right)\left(1 - \frac{m}{n}\right)^n \left(1 - \frac{m}{n}\right)^{-r} \tag{12.4}$$

Now for any fixed r we have, as n tends to infinity,

$$\lim_{n \to \infty} \left(1 - \frac{1}{n}\right)\left(1 - \frac{2}{n}\right) \cdots \left(1 - \frac{(r-1)}{n}\right) = 1$$

$$\lim_{n \to \infty} \left(1 - \frac{m}{n}\right)^{-r} = 1$$

$$\lim_{n \to \infty} \left(1 - \frac{m}{n}\right)^n = e^{-m}$$

We therefore see that the limiting form of (12.4) as n tends to infinity is

$$P(r) = \frac{m^r e^{-m}}{r!} \tag{12.5}$$

This is known as the *Poisson distribution*.

We note that

$$\sum_{r=0}^{\infty} P(r) = e^{-m} \sum_{r=0}^{\infty} \frac{m^r}{r!} = e^{-m} e^m = 1 \tag{12.6}$$

as it should.

THE MEAN OF THE POISSON DISTRIBUTION

The mean $\bar{x}$ of the Poisson distribution may be computed by the equation

$$\bar{x} = \sum_{r=0}^{\infty} \frac{rm^r e^{-m}}{r!} = m e^{-m} \sum_{r=0}^{\infty} \frac{m^{r-1}}{(r-1)!} = m e^{-m} e^m = m \tag{12.7}$$

THE STANDARD DEVIATION OF THE POISSON DISTRIBUTION

The standard deviation of the Poisson distribution may be obtained by taking the limiting value of the standard deviation of the binomial distribution as we let p tend to zero. Equation (11.12) may be written in the form

$$\sigma^2 = np(1 - p) \tag{12.8}$$

We therefore have

$$\lim_{p \to 0} \sigma^2 = np = m \tag{12.9}$$

Therefore,

$$\sigma = m^{\frac{1}{2}} = (np)^{\frac{1}{2}} \tag{12.10}$$

We thus see that the standard deviation of the Poisson distribution is $m^{\frac{1}{2}}$.

13 THE NORMAL OR GAUSSIAN DISTRIBUTION

A most important probability distribution that arises as a limit of the binomial distribution is a continuous distribution known as the *normal* or *gaussian* distribution, with the probability function

$$\phi(x) = \frac{(2\pi)^{-\frac{1}{2}}}{\sigma} e^{-(x-\mu)^2/2\sigma^2} \qquad -\infty x < \infty \tag{13.1}$$

We shall show that μ is the *mean* of the distribution and σ is the *standard deviation* of the normal or gaussian distribution. If the function $\phi(x)$ is to be a probability density function, it is implied that

$$A = \int_{-\infty}^{+\infty} \phi(x)\, dx = 1 \tag{13.2}$$

In order to verify that (13.2) is true, let us compute

$$A = \int_{-\infty}^{+\infty} \frac{(2\pi)^{-\frac{1}{2}}}{\sigma} \int_{-\infty}^{+\infty} e^{-(x-\mu)^2/2\sigma^2}\, dx \tag{13.3}$$

In order to simplify (13.3), we make the substitution

$$y = \frac{x - \mu}{\sigma} \tag{13.4}$$

in the integral (13.3) and find that it is transformed into

$$A = (2\pi)^{-\frac{1}{2}} \int_{-\infty}^{+\infty} e^{-y^2}\, dy \tag{13.5}$$

From a table of definite integrals we obtain the result

$$\int_{-\infty}^{+\infty} e^{-y^2}\, dy = (2\pi)^{\frac{1}{2}} \tag{13.6}$$

If we now substitute (13.6) into (13.5), we find that $A = 1$.

THE MEAN VALUE OF THE NORMAL DISTRIBUTION

We may compute the mean $\bar{x}$ of the normal distribution by the use of Eq. (10.4), which in this case takes the form

$$\bar{x} = \int_{-\infty}^{+\infty} x\phi(x)\,dx = \frac{(2\pi)^{-\frac{1}{2}}}{\sigma} \int_{-\infty}^{+\infty} x\,e^{-(x-\mu)^2/2\sigma^2}\,dx \tag{13.7}$$

To simplify (13.7), let us introduce the following change in variable:

$$y = \frac{(x-\mu)}{\sigma} \qquad dx = \sigma\,dy \qquad x = (\sigma y + \mu) \tag{13.8}$$

With this change in variable, (13.7) becomes

$$\bar{x} = (2\pi)^{-\frac{1}{2}}\,\sigma \int_{-\infty}^{+\infty} y\,e^{-y^2/2}\,dy + \mu(2\pi)^{-\frac{1}{2}} \int_{-\infty}^{+\infty} e^{-y^2/2}\,dy \tag{13.9}$$

From a table of definite integrals, we have the results

$$\int_{-\infty}^{+\infty} y\,e^{-y^2/2}\,dy = 0 \qquad \int_{-\infty}^{+\infty} e^{-y^2/2}\,dy = (2\pi)^{\frac{1}{2}} \tag{13.10}$$

If we substitute the results (13.10) into (13.9), we obtain

$$\bar{x} = \mu \tag{13.11}$$

We have therefore verified that μ is the mean or expected value of the distribution $\phi(x)$ of (13.1).

THE STANDARD DEVIATION OF THE NORMAL DISTRIBUTION

The variance $\mu_2 = \sigma^2$ of the normal distribution may be obtained by the use of (10.6) by placing $a = \mu$ and thus obtaining

$$\mu_2 = \int_{-\infty}^{+\infty} (x-\mu)^2\,\phi(x)\,dx \tag{13.12}$$

If we now substitute $\phi(x)$ as given by (13.1) into (13.12) and make use of the change in variable

$$(x - \mu) = \sigma y \tag{13.13}$$

we obtain

$$\mu_2 = (2\pi)^{-\frac{1}{2}}\,\sigma^2 \int_{-\infty}^{+\infty} y^2\,e^{-y^2/2}\,dy = (2\pi)^{-\frac{1}{2}}\,I \tag{13.14}$$

The integral

$$I = \int_{-\infty}^{+\infty} y(y\,e^{-y^2/2})\,dy \tag{13.15}$$

may be integrated by parts to obtain

$$I = -y\,e^{-y^2/2}\Big|_{-\infty}^{+\infty} + \int_{-\infty}^{+\infty} e^{-y^2/2}\,dy = (2\pi)^{\frac{1}{2}} \tag{13.16}$$

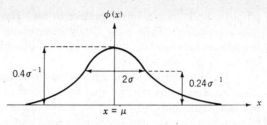

Fig. 13.1

If the above value for I is substituted into (13.14),

$$\mu_2 = (2\pi)^{-\frac{1}{2}}(2\pi)^{\frac{1}{2}} \sigma^2 = \sigma^2 \tag{13.17}$$

Therefore σ correctly represents the standard deviation of the normal or gaussian distribution (Fig. 13.1).

The *maximum* value of $\phi(x)$ is located at $x = \mu$; $\phi(\mu)$ is approximately $\frac{2}{5}\sigma$. When $x = \mu \pm \sigma$, the value of $\phi(x)$ is slightly less than $\frac{1}{4}\sigma$. It is easy to show that the points $x = \mu \pm \sigma$ are the points of *inflection* of the normal distribution curve $\phi(x)$. It is of great interest to know the probability that the deviation $y = x - \mu$ from the mean is within certain limits. The probability that y lies between the limits σt_1 and σt_2 is

$$\frac{(2\pi)^{-\frac{1}{2}}}{\sigma} \int_{\sigma t_1}^{\sigma t_2} e^{-y^2/2\sigma^2} \, dy = \int_{t_1}^{t_2} (2\pi)^{-\frac{1}{2}} e^{-t^2/2} \, dt \tag{13.18}$$

By placing $t_1 = -a$ and $t_2 = a$ in (13.18) and using a table of definite integrals, it can be shown that the probability that the deviation y exceeds 2σ is less than $\frac{1}{20}$ and the probability that y exceeds 3σ is of the order of 0.003. In statistics, an observation which is improbable is said to be *significant*. For a normally distributed variate, it is usual to regard a single observation with $|x| > \mu + 2\sigma$ as significant and one with $|x| > \mu + 3\sigma$ as *highly significant*.

The normal or gaussian distribution may be derived as a limiting form of the binomial distribution when n becomes very large and p remains constant.†

The normal or gaussian distribution (13.1) is sometimes also known as the *normal error law*, although some statisticians object that it is not in fact a normal occurrence in nature because it is only valid as a limiting case for exceptionally large numbers. Sufficiently large numbers are, however, available in problems of applied physics, such as the kinetic theory of gases and fluctuations in the magnitude of an electric current, and so the gaussian distribution seems more real to physicists and engineers than to social and biological scientists. It is interesting to note that the eminent mathematical

† See Harald Cramer, "The Elements of Probability Theory," pp. 97–101, John Wiley & Sons, Inc., New York, 1955.

physicist H. Poincaré, in the preface to his "*Thermodynamique*"† makes the laconic remark, "Everybody firmly believes in the law of errors because mathematicians imagine that it is a fact of observation, and observers that it is a theorem of mathematics."

The application of the gaussian distribution to the theory of errors is made reasonable by the fact that any process of random sampling tends to produce a gaussian distribution of sample values, even if the whole population from which the samples are drawn does not have a normal distribution. Hence it is usual to assume that the gaussian distribution will be a useful approximation to the distribution of errors of observation, provided that all causes of systematic error can be excluded.

THE AREA UNDER SOME PART OF THE GAUSSIAN DISTRIBUTION FUNCTION

The area under some part of the gaussian distribution function is frequently needed.   That is, we must evaluate the integral of $\phi(x)$ between finite values of x.   For example, let it be required to evaluate the integral I where

$$I = \int_{x_1}^{x_2} \phi(x)\,dx = \frac{(2\pi)^{-\frac{1}{2}}}{\sigma} \int_{x_1}^{x_2} e^{-(x-\mu)^2/2\sigma^2}\,dx \tag{13.19}$$

In order to simplify the above integral, let

$$v = \frac{x - \mu}{\sigma} \qquad dx = \sigma\,dv \tag{13.20}$$

With this change in variable, (13.19) becomes

$$I = (2\pi)^{-\frac{1}{2}} \int_{v_1}^{v_2} e^{-v^2/2}\,dv \tag{13.21}$$

where

$$v_1 = \frac{x_1 - \mu}{\sigma} \qquad v_2 = \frac{x_2 - \mu}{\sigma} \tag{13.22}$$

Now the function $F(z)$ given by

$$F(z) = (2\pi)^{-\frac{1}{2}} \int_{-\infty}^{z} e^{-y^2/2}\,dy \tag{13.23}$$

has been *extensively tabulated*.   It is easy to see that the integral is given by

$$I = F(v_2) - F(v_1) \tag{13.24}$$

in terms of the function $F(z)$.

PROBLEM ILLUSTRATING A TYPICAL STATISTICAL DISTRIBUTION

As an interesting typical problem involving a statistical distribution, consider that the heights of 500 men are measured to the nearest $\frac{1}{2}$ in.   The frequencies f_r of heights x_r (in inches) are given in the following table.

† Paris, 1892.

Heights of 500 men

x_r	f_r	x_r	f_r	x_r	f_r
58	1	64	18	70	16
58.5	0	64.5	23	70.5	15
59	0	65	31	71	10
59.5	1	65.5	34	71.5	9
60	2	66	35	72	7
60.5	0	66.5	36	72.5	2
61	1	67	41	73	3
61.5	4	67.5	41	73.5	1
62	5	68	39	74	0
62.5	8	68.5	35	74.5	1
63	11	69	29	75	0
63.5	17	69.5	23	75.5	1

Show that the distribution of the heights of the 500 men given in the above table is approximately normal, with mean 67 in. and standard deviation 2.5 in. Assuming this normal distribution of heights, which of the following measurements of heights, considered individually, are significant or highly significant: 63 in., 72 in., 69.5 in., 58.5 in., 73 in., 75 in.?

14 DISTRIBUTION OF A SUM OF NORMAL VARIATES

If two variates x and y are distributed normally, their sum $u = x + y$ is also distributed normally. To prove this, we shall refer each distribution to its mean and assume that the distribution functions of $v = x - \bar{x}$ and $w = y - \bar{y}$ are

$$\phi_1(v) = \frac{(2\pi)^{-\frac{1}{2}}}{\sigma_1} e^{-v^2/2\sigma_1^2} \tag{14.1}$$

$$\phi_2(w) = \frac{(2\pi)^{-\frac{1}{2}}}{\sigma_2} e^{-w^2/2\sigma_2^2} \tag{14.2}$$

The probability that $(v + w)$ takes a particular value z is found by integrating over the probabilities that v and w take particular values summing to z. This gives

$$\int \phi_1(v)\,\phi_2(z - v)\,dv = \frac{1}{2\pi\sigma_1\sigma_2} e^{-z^2/2(\sigma_1^2+\sigma_2^2)}$$

$$\times \int_{-\infty}^{+\infty} e^{-(v-z\sigma_1^2)^2/2\sigma_1^2+\sigma_2^2\,(\sigma_1^2+\sigma_2^2)}\,dv$$

$$= F(z) \tag{14.3}$$

If we now let

$$t = \frac{(\sigma_1^2 + \sigma_2^2)^{\frac{1}{2}}}{\sigma_1\sigma_2} \frac{(v - z\sigma_1^2)}{(\sigma_1^2 + \sigma_2^2)} \tag{14.4}$$

the integral in (14.4) becomes

$$\frac{e^{-t^2/2}\,\sigma_1\,\sigma_2\,dt}{(\sigma_1^2 + \sigma_2^2)^{\frac{1}{2}}} = \frac{(2\pi)^{\frac{1}{2}}\,\sigma_1\,\sigma_2}{(\sigma_1^2 + \sigma_2^2)^{\frac{1}{2}}} \tag{14.5}$$

Hence the distribution (14.4) is

$$F(z) = \frac{(2\pi)^{-\frac{1}{2}}}{(\sigma_1^2 + \sigma_2^2)^{\frac{1}{2}}}\,e^{-z^2/2(\sigma_1^2 + \sigma_2^2)} \tag{14.6}$$

We therefore see that $z = v + w$ is normally distributed about the mean value zero, with standard deviation given by

$$\sigma^2 = \sigma_1^2 + \sigma_2^2 \tag{14.7}$$

The distribution of $u = x + y$, therefore, has the expectation $\bar{u} = \bar{x} + \bar{y}$, with a standard deviation given by (14.8).

15   APPLICATIONS TO EXPERIMENTAL MEASUREMENTS

Let $x_1, x_2, \ldots, x_n$ be a set of n measurements.   The x_i, $i = 1, 2, \ldots, n$, quantities may be the n determinations of the velocity of light in a certain medium.   If we let z be a typical value of the measurements x_i, then d_i is the deviation of the measurement x_i from z where

$$d_i = z - x_i \qquad i = 1, 2, 3, \ldots, n \tag{15.1}$$

Let y represent the sum of the squares of the deviations of the measurements x_i from the typical value z, so that

$$y = \sum_{i=1}^n d_i^2 = \sum_{i=1}^n (z - x_i)^2 \tag{15.2}$$

We now wish to choose the typical value z so that the sum of the squares of the deviations y is a minimum.   We therefore minimize (15.2) in the usual way by computing the derivative

$$\frac{dy}{dz} = \sum_{i=1}^n 2(z - x_i) \tag{15.3}$$

On setting the derivative dy/dz equal to zero and solving for z, we find

$$z = \bar{x} = \sum_{i=1}^n \frac{x_i}{n} = \text{the arithmetic mean of } x_i \qquad i = 1, 2, 3, \ldots, n \tag{15.4}$$

That is, we find that the arithmetic mean of the measurements x_i is the typical value that minimizes y.   It will be noted that $\bar{x}$ gives the "best" typical value of the x_i quantities in the least-square sense.

The *minimum* value of y, y_{min} is obtained by substituting $z = \bar{x}$ in (15.2) and thus obtaining

$$y_{min} = \sum_{i=1}^{n} (\bar{x} - x_i)^2 \tag{15.5}$$

If we divide (15.5) by n, we obtain the *variance* σ^2 of the set of measurements x_i, so that

$$\sigma^2 = \frac{1}{n} \sum_{i=1}^{n} (\bar{x} - x_i)^2 = \text{the variance of } x_i \tag{15.6}$$

The square root of (15.6) is the *standard deviation* σ of the set of measurements x_i, so that

$$\sigma = \left[\sum_{i=1}^{n} \frac{(\bar{x} - x_i)^2}{n} \right]^{\frac{1}{2}} = \text{the standard deviation of } x_i \tag{15.7}$$

16 THE STANDARD DEVIATION OF THE MEAN

We now turn to a very important question that arises in the theory of statistics: "What is the relation between the standard deviation of the individual measurements and the standard deviation of the *mean* of a set of measurements?"

In order to answer this question, we will assume that we have a parent distribution of errors which follow the *Gauss distribution*. (The justification for assuming that the distribution is gaussian is that in practice errors do follow the gaussian law). We suppose first that we take N measurements or a sample of the parent distribution. We then take *another* sample or set of measurements. If we now compute the mean and standard deviation of both sets, these quantities will in general be different. That is, the mean and variance of a *sample* of N observations are *not* in general equal to the mean and variance of the *parent* distribution.

The process to be followed is to take M sets of N measurements, each with its own mean and standard deviation, and then compute the standard deviation of the means. The standard deviation of the means provides an indication of the reliability of any one of the means.

To facilitate the calculation let the following notation be introduced:

$\sigma = $ the standard deviation of the individual measurements

$\sigma_m = $ the standard deviation of the means (of the various samples)

$x_{si} = $ measurement i in the set s

$\bar{x}_s = $ the mean of the set s

$\bar{X} = $ the mean of all the measurements

$d_{si} = $ the deviation of x_{si}, $x_{si} - \bar{X} = d_{si}$

$D_s = \bar{x}_s - \bar{X} = $ the deviation of the mean $\bar{x}_s$

Since we are taking M sets of measurements with N measurements in each set, there will be $n = MN$ measurements in the total. The variance of the individual measurements is given by

$$\sigma^2 = \sum_{s=1}^{M} \sum_{i=1}^{N} \frac{d_{si}^2}{MN} \tag{16.1}$$

The variance of the means is given by

$$\sigma_m^2 = \sum_{s=1}^{m} \frac{D_s^2}{M} \tag{16.2}$$

The deviations D_s of the means can be expressed in terms of the deviations d_{si} of the individual observations as follows:

$$D_s = \bar{x}_s - \bar{X} = \frac{1}{N} \sum_{i=1}^{N} x_{si} - \bar{X} \tag{16.3}$$

We also have

$$d_{si} = x_{si} - \bar{X} \tag{16.4}$$

If we now substitute the square of (16.4) into (16.1), we obtain

$$\sigma^2 = \sum_{s=1}^{M} \sum_{i=1}^{N} \frac{(x_{si} - \bar{X})^2}{MN} \tag{16.5}$$

The substitution of the square of (16.3) into (16.2) gives the results

$$\sigma_m^2 = \frac{1}{M} \sum_{s=1}^{M} \left(\frac{1}{N} \sum_{i=1}^{N} x_{si} - \bar{X} \right)^2 \tag{16.6}$$

After some reductions, it can be seen that (16.6) may be written in the following form:

$$\sigma_m^2 = \sum_{s=1}^{M} \sum_{i=1}^{N} \frac{(x_{si} - \bar{X})^2}{MN^2} \tag{16.7}$$

If we now compare (16.5) with (16.7), we see that

$$\sigma_m^2 = \frac{\sigma^2}{N} \tag{16.8}$$

or therefore

$$\sigma_m = \frac{\sigma}{N^{\frac{1}{2}}} \tag{16.9}$$

Therefore the variance of the mean of a set of N measurements is simply the variance of the individual measurements divided by the number of measurements.

It may be mentioned that in order to reduce (16.6) to the form (16.7) it is necessary to assume that the cross-product terms are negligibly small. This is true for the gaussian distribution, and since experimental measurements so often obey the gaussian distribution, formula (16.8) is a very useful one.

PROBLEMS

1. What is the probability that number 1 appears at least in n throws of dice?

2. A hunter finds that on the average he kills once in three shots. He fires three times at a duck; on the assumption that his a priori probability of killing is $\frac{1}{3}$, what is the probability that he kills the duck?

3. Prove the following theorem: If in any continuously varying process a certain characteristic is present to the extent of one in T units, then the probability that the characteristic does not occur in a sample of t units is $e^{-t/T}$. (Consider the Poisson distribution.)

4. It is known that 100 liters of water have been polluted with 10^6 bacteria. If 1 cc of water is drawn off, what is the probability that the sample is not polluted?

5. An aircraft company carries on the average P passengers M miles for every passenger killed. What is the probability of a passenger completing a journey of m miles in safety?

6. Given the result of Prob. 5, estimate an apparently reasonable premium to pay in order that if a passenger is killed in such a flight, his heir would receive $10,000.

7. If during the weekend road traffic 100 cars per hr pass along a certain road, each taking 1 min to cover it, find the probability that at any given instant no car will be on this road. (NOTE: Evidently no car must have entered the road during the previous minute; however, on the average a car enters every $3,600/100 = 36$ sec. It is thus apparent that the required probability is $e^{-60/36}$.)

8. In a completed book of 1,000 pages, 500 typographical errors occur. What is the probability that four specimen pages selected for advertisement are free from errors?

9. Criticize the following statements:

(a) The sun rises once per day; hence the probability that it will not rise tomorrow is e^{-1}.

(b) The probability that it will rise at least once is $1 - e^{-1}$.

10. Let it be supposed that in the manufacture of firecrackers there is a certain defect that causes only three-fourths of them to explode when lighted. What is the probability that if we choose five of these firecrackers, they will all explode?

11. It has been observed in human reproduction that twins occur approximately once in 100 births. If the number of babies in a birth are assumed to follow a Poisson distribution, calculate the probability of the birth of quintuplets. What is the probability of the birth of octuplets?

12. A coin is tossed 10,000 times; the results are 5,176 heads and 4,824 tails. Is this a reasonable result for a symmetric coin, or is it fairly conclusive evidence that the coin is defective?

HINT: Calculate the total probability for *more* than 5,176 heads in 10,000 tosses. To do this a gaussian distribution may be assumed.

13. If a set of measurements is distributed in accordance with a gaussian distribution, find the probability that any single measurement will fall between $m - \sigma/2$ and $m + \sigma/2$.

14. The "probable error" of a distribution is defined as the error such that the probability of occurrence of an error whose absolute is less than this value is $\frac{1}{2}$. Determine the probable error for the gaussian distribution and express it as a multiple of σ.

15. An object undergoes a simple harmonic motion with amplitude A and angular frequency w according to the equation $x = A\sin wt$, where x represents the displacement of the object from equilibrium. Calculate the mean and standard deviation of the position and of the speed of the object.

16. Among a large number of eggs, 1 percent were found to be rotten. In a dozen eggs, what is the probability that none is rotten?

17. A certain quantity was measured N times, and the mean and its standard deviation were computed. If it is desired to increase the precision of the result (decrease σ) by a factor of 2, how many additional measurements should be made?

18. A group of *underfed* chickens were observed for 50 consecutive days and found to lay the following number of eggs:

Eggs laid	No. of days
0	10
1	13
2	13
3	8
4	4
5	2

Show that this is approximately a Poisson distribution. Calculate the mean and standard deviation directly from the data. Compare with the standard deviation predicted by the Poisson distribution.

19. On the average, how many times must a die be thrown until one gets a 6?
   HINT: Use the binomial distribution.

20. When 100 coins are tossed, what is the probability that exactly 50 are heads?
   HINT: Use the binomial distribution.

21. A bread salesman sells on the average 20 cakes on a round of his route. What is the probability that he sells an even number of cakes?
   HINT: Assume the Poisson distribution.

22. How thick should a coin be to have a $\frac{1}{3}$ chance of landing on an edge.
   HINT: In order to get an approximate solution to this problem, one may imagine the coin inscribed inside a sphere, where the center of the coin is the center of the sphere. The coin may be regarded as a right circular cylinder. A random point on the surface of the sphere is now chosen; if the radius from that point to the center strikes the edge, the coin is said to have fallen on an edge.

23. If a stick is broken into two at random, what is the average length of the smaller piece?

24. Shuffle an ordinary deck of 52 playing cards containing four aces and then turn up cards from the top until the first ace appears. On the average, how many cards are required to produce the first ace?

25. If n grains of wheat are scattered in a haphazard manner over a surface of S units of area, show that the probability that A units of area will contain R grains of wheat is

$$P = \left(\frac{An}{S}\right)^r \frac{e^{-An/s}}{R!}$$

Hint: ndS/S represents the infinitely small probability that the small space dS contains a grain of wheat. If the selected space be A units of area, we may suppose each dS to be a trial; the number of trials will therefore be A/dS. Hence we must substitute An/S for np in the Poisson distribution $P = (np)^r e^{-np}/r!$

26. A basketball player succeeds in making a basket three tries out of four. How many times must he try for a basket in order to have probability greater than 0.99 of making at least one basket?

27. An unpopular instructor who grades "on the curve" computes the mean and standard deviation of the grades in his class, and then he assumes a normal or gaussian distribution with this μ and σ. He then sets borderlines between the grades at C from $\mu - \sigma/2$ to $\mu + \sigma/2$, B from $\mu + \sigma/2$ to $\mu + 3\sigma/2$, A from $\mu + 3\sigma/2$ up. Find the percentages of the students receiving the various grades. Where should the borderlines be set to give the following percentages: A and F, 10 percent; B and D, 20 percent; C, 40 percent?

28. A popular lady receives an average of four telephone calls a day. What is the probability that on a given day she will receive no telephone calls? Just one call? Exactly four calls?

29. The Poisson distribution is used mainly for *small* values of $\mu = np$. For large values of μ, the Poisson distribution (as well as the binomial) is fairly well approximated by the normal or gaussian distribution. Show that

$$\lim_{\mu \to \text{large}} \frac{\mu^x e^{-u}}{x!} \approx \frac{e^{-(x-\mu)^2/2\mu}}{(2\pi\mu)^{\frac{1}{2}}} \qquad \mu \text{ large}$$

The peak of the graph of the Poisson distribution is at $x = \mu$; note that the approximating normal curve is shifted to have its center at $x = \mu$. The equation given above gives a good approximation in the central region (around $x = \mu$) where the probability is large.

REFERENCES

1939. von Mises, Richard: "Probability, Statistics and Truth," The Macmillan Company, New York.

1955. Cramer, Harald: "The Elements of Probability Theory," John Wiley & Sons, Inc., New York.

1957. Feller, William: "An Introduction to Probability Theory and Its Applications," 2d ed., vol. I, John Wiley & Sons, Inc., New York.

1960. Parzen, E.: "Modern Probability Theory and Its Applications," John Wiley & Sons, Inc., New York.

1963. Levy, H., and Roth, L.: "Elements of Probability," Oxford University Press, London.

1965. Fry, C. T.: "Probability and Its Engineering Uses," 2d ed., D. Van Nostrand Company, Inc., New York.

1966. Feller, William: "An Introduction to Probability Theory and Its Applications," vol. II, John Wiley & Sons, Inc., New York.

Table of Laplace Transforms

This Appendix contains 196 common Laplace transforms which are frequently encountered in applied mathematics. The familiar reader will note that the indexing method differs somewhat from the usual method. In this table we have introduced an index notation which hopefully will provide a more efficient means of using the tables to find a desired transform pair. The basic transforms have been cataloged by the standard numerical method where the numerical value denotes the order of the denominator of $g(s)$. For the more complex forms an alphabetic index has been used which is described in Table 1. The references at the end of Table 2 cite other works which contain extensive tables of Laplace transforms.

Table 1 Summary of index notation employed in cataloging transforms

Symbol	Description
T.X†	Basic theorems (nos. T.1 through T.15 correspond with those listed in Chap. 4)
0.X	Step and delta functions; powers of t
1.X	Denominator 1st order in s; numerator in polynomial form
2.X	Denominator 2d order in s; numerator in polynomial form
3.X	Denominator 3d order in s; numerator in polynomial form
4.X	Denominator 4th order in s; numerator in polynomial form
H.X	$g(s)$ contains hyperbolic functions
N.X	$g(s)$ and $f(t)$ forms for use in nonlinear problems
R.X	$g(s)$ contains functions of $s^{\frac{1}{2}}$
R-E.X	$g(s)$ contains functions of $s^{\frac{1}{2}}$ and $f(t)$; contains error functions
R-B.X	$g(s)$ contains functions of $s^{\frac{1}{2}}$ and $f(t)$; contains Bessel functions
SP.X	Special functions

† X denotes a numerical value.

Table 2 Laplace-transform pairs

No.	$g(s)$	$f(t)$	Notes
T.0	$g(s) = \displaystyle\int_0^\infty f(t)\,e^{-st}\,dt$	$f(t) = \dfrac{1}{2\pi j}\displaystyle\int_{c-j\infty}^{c+j\infty} g(s)\,e^{st}\,ds$	$j = (-1)^{\frac{1}{2}}$ $c = $ Real
T.1	$\dfrac{k}{s}$	k	$k = $ const
T.2	$kg(s)$	$kf(t)$	$k = $ const
T.3a	$sg(s) - f(0)$	$\dfrac{df}{dt}$	
T.3b	$s^2 g(s) - sf(0) - \dfrac{df}{dt}(0)$	$\dfrac{d^2 f}{dt^2}$	
T.4	$s^n g(s) - \displaystyle\sum_{i=1}^{n} s^{(i-1)}\dfrac{d^{(n-1)}f}{dt^{(n-1)}}\bigg|_{t=0}$	$\dfrac{d^n f}{dt^n}$	
T.5	$g_1(s)g_2(s)$	$\displaystyle\int_0^t f_1(t')f_2(t-t')\,dt'$ or $\displaystyle\int_0^t f_2(t')f_1(t-t')\,dt'$	$g_1(s) = \mathscr{L}\{f_1(t)\}$ $g_2(s) = \mathscr{L}\{f_2(t)\}$
T.6a	$g(s + a)$	$e^{-at}f(t)$	Real $a \geqslant 0$ $g(s) = \mathscr{L}\{f(t)\}$
T.6b	$g(s - a)$	$e^{at}f(t)$	See T.6a

Table 2   Laplace-transform pairs (*Continued*)

No.	$g(s)$	$f(t)$	Notes
T.7	$e^{-as}g(s)$	$f(t-a)\,u(t-a)$	$a = \text{Real} > 0$ $u(t) = \text{Heaviside}$ unit step
T.8	$e^{as}g(s)$	$f(t+a)\,u(t+a)$	See T.7
T.9	$\displaystyle\lim_{s\to\infty} sg(s)$	$\displaystyle\lim_{t\to 0} f(t)$	
T.10	$\displaystyle\lim_{s\to 0} sg(s)$	$\displaystyle\lim_{t\to\infty} f(t)$	
T.11	$\dfrac{1}{s}g(s) + \dfrac{1}{s}\displaystyle\int_{-\infty}^{0} f(t)\,dt$	$\displaystyle\int f(t)\,dt$	
T.12	$\dfrac{1}{s}g(s)$	$\displaystyle\int_{0}^{t} f(t')\,dt'$	
T.13	$(-1)^n \dfrac{d^n g(s)}{ds^n}$	$t^n f(t)$	$n = \text{integer}$
T.14	$\dfrac{\partial g(x,s)}{\partial x}$	$\dfrac{\partial f(x,t)}{\partial x}$	
T.15	$\displaystyle\int g(s,x)\,dx$	$\displaystyle\int f(t,x)\,dx$	
T.16	$\dfrac{1}{k}g\left(\dfrac{s}{k}\right)$	$f(kt)$	$k = \text{const}$
T.17	$\dfrac{N(s)}{D(s)}$	$\displaystyle\sum_{i=1}^{n} \dfrac{N(s_i)\,e^{s_i t}}{D'(s_i)}$ $D'(s_i) = \dfrac{dD(s)}{ds}\bigg|_{s=s_i}$	$N(s)$ and $D(s)$ are polynomials in s; the order of D must be at least one more than the order of N; D has n simple roots, s_i
T.18	$\displaystyle\int_{s}^{\infty} g(s')\,ds'$	$\dfrac{1}{t}f(t)$	
T.19	$\dfrac{1}{s}\displaystyle\int_{0}^{s} g(s')\,ds'$	$\displaystyle\int_{0}^{\infty} t^{-1}f(t)\,dt$	
0.1	$\dfrac{1}{s}$	$u(t) = \begin{cases} 0 & t < 0 \\ \frac{1}{2} & t = 0 \\ 1 & t > 0 \end{cases}$	$u(t) = \text{Heaviside}$ step function
0.2	1	$\delta(t) = \begin{cases} \infty & t = 0 \\ 0 & t \neq 0 \end{cases}$	$\delta(t) = \text{Dirac delta}$ function $\displaystyle\int_{-\infty}^{\infty} \delta(t)\,dt = 1$

Table 2   Laplace-transform pairs (*Continued*)

No.	$g(s)$	$f(t)$	Notes
0.3	$\dfrac{1}{s}e^{-as}$	$u(t-a) = \begin{cases} 0 & t < a \\ \frac{1}{2} & t = a \\ 1 & t > a \end{cases}$	a = Real
0.4	$\dfrac{1}{s}(1-e^{-as})$	$u(t) - u(t-a)$	
0.5	e^{-as}	$\delta(t-a)$	
0.6	s^n	$\dfrac{d^n}{dt^n}\delta(t)$	
0.7	$s^n e^{-as}$	$\dfrac{d^n}{dt^n}\delta(t-a)$	a = Real
0.8	$s^{(2n-1)/2}$	$(-1)^{n+1}\sqrt{\dfrac{2}{\pi}} \\ \times \dfrac{(1)(3)(5)\cdots(2n+1)}{(2t)^{(2n+1)/2}}$	$n = 0, 1, 2, 3, \ldots$ $t > 0$
0.9	$s^{-(2n+1)/2}$	$\sqrt{\dfrac{2}{\pi}}\dfrac{(2t)^{(2n-1)/2}}{(1)(3)(5)\cdots(2n-1)}$	$n = 1, 2, 3, \ldots$
0.10	$\dfrac{1}{s^{n+1}}$	$\dfrac{t^n}{n!}$	$n = 1, 2, 3, \ldots$
0.11	$\dfrac{1}{s^{n+1}}$	$\dfrac{t^n}{\Gamma(n+1)}$	n = all values except negative integers $\Gamma(x)$ = gamma function
1.1	$\dfrac{1}{s+a}$	e^{-at}	a = real or complex
1.2	$\dfrac{s}{s+a}$	$\delta(t) - ae^{-at}$	
2.1	$\dfrac{1}{s(s+a)}$	$\dfrac{1}{a}(1 - e^{-at})$	
2.2	$\dfrac{1}{(s+a)(s+b)}$	$\dfrac{e^{-at}}{b-a} + \dfrac{e^{-bt}}{a-b}$	
2.3	$\dfrac{s+a_1}{(s+a)(s+b)}$	$\dfrac{a_1-a}{b-a}e^{-at} + \dfrac{a_1-b}{a-b}e^{-bt}$	
2.4	$\dfrac{1}{s^2+a^2}$	$\dfrac{1}{a}\sin at$	

Table 2   Laplace-transform pairs (*Continued*)

No.	$g(s)$	$f(t)$	Notes
2.5	$\dfrac{s}{s^2 + a^2}$	$\cos at$	
2.6	$\dfrac{s + a_1}{s^2 + a^2}$	$\dfrac{1}{a}(a_1 + a^2)^{\frac{1}{2}} \sin(at + \psi)$	$\psi \equiv \tan^{-1}\dfrac{a}{a_1}$
2.7	$\dfrac{1}{s^2 - a^2}$	$\dfrac{1}{a}\sinh at$	
2.8	$\dfrac{s}{s^2 - a^2}$	$\cosh at$	
2.9	$\dfrac{s + a_1}{s^2 - a^2}$	$\dfrac{1}{a}(a_1 - a^2)^{\frac{1}{2}} \sinh(at + \psi)$	$\psi \equiv \tanh^{-1}\dfrac{a}{a_1}$
2.10	$\dfrac{1}{(s + a)^2}$	$t\,e^{-at}$	
2.11	$\dfrac{s + a_1}{(s + a)^2}$	$[1 + (a_1 - a)\,t]\,e^{-at}$	
2.12	$\dfrac{1}{(s + a)^2 + b^2}$	$\dfrac{1}{b}e^{-at}\sin bt$	Gardner and Barnes, 1956, no. 1.301
2.13	$\dfrac{s + a_1}{(s + a)^2 + b^2}$	$\dfrac{1}{b}[(a_1 - a)^2 + b^2]^{\frac{1}{2}}$ $\times e^{-at}\sin(bt + \psi)$	Same as no. 2.12, no. 1.303 $\psi \equiv \tan^{-1}\dfrac{b}{a_1 - a}$
2.14	$\dfrac{a}{(s + b)^2 + a^2}$	$e^{-bt}\sin at$	$a^2 > 0$
2.15	$\dfrac{s + b}{(s + b)^2 + a^2}$	$e^{-bt}\cos at$	$a^2 > 0$
2.16	$\dfrac{a}{(s + b)^2 - a^2}$	$e^{-bt}\sinh at$	
2.17	$\dfrac{s + b}{(s + b)^2 - a^2}$	$e^{-bt}\cosh at$	
2.18	$\dfrac{\omega\cos\phi \pm s\sin\phi}{s^2 + \omega^2}$	$\sin(\omega t \pm \phi)$	
2.19	$\dfrac{s\cos\phi \mp \omega\sin\phi}{s^2 + \omega^2}$	$\cos(\omega t \pm \phi)$	
2.20	$\dfrac{\omega\cos\phi \pm (s + b)\sin\phi}{(s + b)^2 + \omega^2}$	$e^{-bt}\sin(\omega t \pm \phi)$	
2.21	$\dfrac{(s + b)\cos\phi \mp \omega\sin\phi}{(s + b)^2 + \omega^2}$	$e^{-bt}\cos(\omega t \pm \phi)$	

Table 2   Laplace-transform pairs (*Continued*)

No.	$g(s)$	$f(t)$	Notes
22,2	$\dfrac{1}{s^2 + 2as + \omega_0^2}$	(a) $\underline{\omega_0^2 > a^2}$ $\dfrac{1}{\omega} e^{-at} \sin \omega t$ (b) $\underline{\omega_0^2 = a^2}$ $t\, e^{-at}$ (c) $\underline{\omega_0^2 < a^2}$ $\dfrac{1}{n - m}(e^{-mt} - e^{-nt})$	$\omega^2 \equiv \omega_0^2 - a^2$ $-m$ and $-n$ are roots of $s^2 + 2as +$ $\omega_0^2 = 0$
22,3	$\dfrac{s}{s^2 + 2as + \omega_0^2}$	(a) $\underline{\omega_0^2 > a^2}$ $-\dfrac{\omega_0}{\omega} e^{-at} \sin (\omega t - \phi)$ (b) $\underline{\omega_0^2 = a^2}$ $e^{-at}(1 - at)$ (c) $\underline{\omega_0^2 < a^2}$ $\dfrac{1}{n - m}(n e^{-mt} - m e^{-nt})$	$\omega^2 = \omega_0^2 - a^2$ $\phi = \tan^{-1}\dfrac{\omega}{a}$ m and n are defined in 2.22c
3.1	$\dfrac{1}{s(s + a)(s + b)}$	$\dfrac{1}{ab} + \dfrac{b e^{-at} - a e^{-bt}}{ab(a - b)}$	
3.2	$\dfrac{s + a_1}{s(s + a)(s + b)}$	$\dfrac{a_1}{ab} + \dfrac{a_1 - a}{a(a - b)} e^{-at} + \dfrac{a_1 - b}{b(b - a)} e^{-bt}$	
3.3	$\dfrac{s^2 + a_1 s + a_2}{s(s + a)(s + b)}$	$\dfrac{a_2}{ab} + \dfrac{a^2 - a_1 a + a_2}{a(a - b)} e^{-at}$ $+ \dfrac{b^2 - a_1 b + a_2}{b(b - a)} e^{-bt}$	
3.4	$\dfrac{1}{(s + a)(s + b)(s + c)}$	$\dfrac{e^{-at}}{(b - a)(c - a)} + \dfrac{e^{-bt}}{(a - b)(c - b)}$ $+ \dfrac{e^{-ct}}{(a - c)(b - c)}$	
3.5	$\dfrac{s + a_1}{(s + a)(s + b)(s + c)}$	$\dfrac{a_1 - a}{(b - a)(c - a)} e^{-at}$ $+ \dfrac{a_1 - b}{(a - b)(c - b)} e^{-bt}$ $+ \dfrac{a_1 - c}{(a - c)(b - c)} e^{-ct}$	Gardner and Barnes, 1956, no. 1.114
3.6	$\dfrac{s^2 + a_1 s + a_2}{(s + a)(s + b)(s + c)}$	$\dfrac{a^2 - a_1 a + a_2}{(b - a)(c - a)} e^{-at}$ $+ \dfrac{b^2 - a_1 b + a_2}{(a - b)(c - b)} e^{-bt}$ $+ \dfrac{c^2 - a_1 c + a_2}{(a - c)(b - c)} e^{-ct}$	Same as no. 3.5, no. 1.118

Table 2 Laplace-transform pairs (*Continued*)

No.	$g(s)$	$f(t)$	Notes
3.7	$\dfrac{1}{s^3 + a^3}$	$\dfrac{1}{3a^2}[e^{-at} - e^{\frac{1}{2}at}(\cos\frac{1}{2}\sqrt{3}\,at$ $- \sqrt{3}\sin\frac{1}{2}\sqrt{3}\,at)]$	
3.8	$\dfrac{s}{s^3 + a^3}$	$-\dfrac{1}{3a}[e^{-at} - e^{\frac{1}{2}at}(\cos\frac{1}{2}\sqrt{3}\,at$ $+ \sqrt{3}\sin\frac{1}{2}\sqrt{3}\,at)]$	
3.9	$\dfrac{s^2}{s^3 + a^3}$	$\frac{1}{3}[e^{-at} + 2e^{\frac{1}{2}at}\cos\frac{1}{2}\sqrt{3}\,at]$	
3.10	$\dfrac{1}{s(s^2 + a^2)}$	$\dfrac{1}{a^2}(1 - \cos at)$	
3.11	$\dfrac{s + a_1}{s(s^2 + a^2)}$	$\dfrac{a_1}{a^2} - \dfrac{(a_1^2 + a^2)^{\frac{1}{2}}}{a^2}\cos(at + \phi)$	$\phi \equiv \tan^{-1}\dfrac{a}{a_1}$
3.12	$\dfrac{s^2 + a_1 s + a_2}{s(s^2 + a^2)}$	$\dfrac{a_2}{a^2} - \dfrac{[(a_2 - a^2)^2 + a_1^2 a^2]^{\frac{1}{2}}}{a^2}$ $\times \cos(at + \phi)$	$\phi \equiv \tan^{-1}\dfrac{a_1 a}{a_2 - a^2}$
3.13	$\dfrac{1}{s[(s + a)^2 + b^2]}$	$\dfrac{1}{b_0^2} + \dfrac{1}{bb_0}e^{-at}\sin(bt + \phi)$	$b_0^2 \equiv a^2 + b^2$ $\phi \equiv \tan^{-1}\dfrac{b}{a}$ Same as no. 3.5, no. 1.305
3.14	$\dfrac{s + a_1}{s[(s + a)^2 + b^2]}$	$\dfrac{a_1}{b_0^2} + \dfrac{1}{bb_0}[(a_1 - a)^2 + b^2]^{\frac{1}{2}}e^{-at}$ $\times \sin(bt + \phi)$	$b_0^2 \equiv a^2 + b^2$ $\phi \equiv \tan^{-1}\dfrac{b}{a_1 - a}$ $+ \tan^{-1}\dfrac{b}{a}$ Same as no. 3.5, no. 1.307
3.15	$\dfrac{s^2 + a_1 s + a_2}{s[(s + a)^2 + b^2]}$	$\dfrac{a_2}{b_0^2} + \dfrac{1}{bb_0}[(a^2 - b^2 - a_1 a + a_2)^2$ $+ b^2(a_1 - 2a)^2]^{\frac{1}{2}}e^{-at}$ $\times \sin(bt + \phi)$	$b_0^2 \equiv a^2 + b^2$ $\phi \equiv \tan^{-1}$ $\times \dfrac{b(a_1 - 2a)}{a^2 - b^2 - a_1 a + a_2}$ $+ \tan^{-1}\dfrac{b}{a}$
3.16	$\dfrac{1}{(s + c)[(s + a)^2 + b^2]}$	$\dfrac{e^{-ct}}{(c - a)^2 + b^2}$ $+ \dfrac{e^{-at}\sin(bt + \phi)}{b[(c - a)^2 + b^2]^{\frac{1}{2}}}$	$\phi \equiv \tan^{-1}\dfrac{b}{a - c}$
3.17	$\dfrac{s^2 + 2\omega^2}{s(s^2 + 4\omega^2)}$	$\cos^2 \omega t$	
3.18	$\dfrac{1}{s(s + a)^2}$	$\dfrac{1}{a^2}[1 - (1 + at)e^{-at}]$	

Table 2   Laplace-transform pairs (*Continued*)

No.	$g(s)$	$f(t)$	Notes
3.19	$\dfrac{s+a_1}{s(s+a)^2}$	$\dfrac{a_1}{a^2}\left\{1-\left[1+\left(1-\dfrac{a}{a_1}\right)at\right]e^{-at}\right\}$	
3.20	$\dfrac{s^2+a_1 s+a_2}{s(s+a)^2}$	$\left[1+\dfrac{a_2}{a^2}(1-e^{-at})\right]$ $+\left(a_1-a-\dfrac{a_2}{a}\right)te^{-at}$	
3.21	$\dfrac{1}{(s+a)(s^2+\omega^2)}$	$\dfrac{1}{\omega(a^2+\omega^2)^{\frac{1}{2}}}$ $\times[e^{-at}\sin\beta t+\sin(\omega t-\beta)]$	$\beta\equiv\tan^{-1}\dfrac{\omega}{a}$
3.22	$\dfrac{(s-a)^2}{s(s+a)^2}$	$1-4at\,e^{-at}$	
3.23	$\dfrac{1}{s(s^2+2as+\omega_0^2)}$	(a) $\omega_0^2>a^2$ $\dfrac{1}{\omega_0^2}\left[1-\dfrac{\omega_0}{\omega}e^{-at}\right.$ $\left.\times\sin(\omega t+\phi)\right]$ (b) $\omega_0^2=a^2$ $\dfrac{1}{\omega_0^2}[1-e^{-at}(1+at)]$ (c) $\omega_0^2<a^2$ $\dfrac{1}{\omega_0^2}\left[1-\dfrac{\omega_0^2}{n-m}\right.$ $\left.\times\left(\dfrac{e^{-mt}}{m}-\dfrac{e^{-nt}}{n}\right)\right]$	$\tan\phi=\dfrac{\omega}{a}$ $\omega^2\equiv\omega_0^2-a^2$ m and n are roots of $s^2+2as+\omega_0^2=0$
3.24	$\dfrac{1}{s^2(s+a)}$	$\dfrac{t}{a}-\dfrac{1}{a^2}(1-e^{-at})$	
4.1	$\dfrac{1}{s^4+4a^4}$	$\dfrac{1}{4a^3}(\sin at\cosh at$ $-\cos at\sinh at)$	
4.2	$\dfrac{s}{s^4+4a^4}$	$\dfrac{1}{2a^2}\sin at\sinh at$	
4.3	$\dfrac{s^2}{s^4+4a^4}$	$\dfrac{1}{2a}(\sin at\cosh at$ $+\cos at\sinh at)$	
4.4	$\dfrac{s^3}{s^4+4a^4}$	$\cos at\cosh at$	
4.5	$\dfrac{1}{s^4-a^4}$	$\dfrac{1}{2a^3}(\sinh at-\sin at)$	

Table 2 Laplace-transform pairs (*Continued*)

No.	$g(s)$	$f(t)$	Notes
4.6	$\dfrac{s}{s^4 - a^4}$	$\dfrac{1}{2a^2}(\cosh at - \cos at)$	
4.7	$\dfrac{s^2}{s^4 - a^4}$	$\dfrac{1}{2a}(\sinh at + \sin at)$	
4.8	$\dfrac{s^3}{s^4 - a^4}$	$\tfrac{1}{2}(\cosh at + \cos at)$	
4.9	$\dfrac{a^2}{(s^2 + a^2)^2}$	$\dfrac{1}{2a}(\sin at - at\cos at)$	
4.10	$\dfrac{s}{(s^2 + a^2)^2}$	$\dfrac{t}{2a}\sin at$	
4.11	$\dfrac{s}{(s^2 + a^2)(s^2 + b^2)}$	$\dfrac{\cos bt - \cos at}{a^2 - b^2}$	$a \neq b$
4.12	$\dfrac{s}{[s^2 + (b + a)^2][s^2 + (b - a)^2]}$	$\dfrac{1}{2ab}\sin at \sin bt$	
4.13	$\dfrac{1}{s^2(s^2 + a^2)}$	$\dfrac{t}{a^2} - \dfrac{1}{a^3}\sin at$	
4.14	$\dfrac{1}{s^2(s^2 - a^2)}$	$\dfrac{1}{a^3}\sinh at - \dfrac{t}{a^2}$	
H.1	$\dfrac{\sinh bs^{\frac{1}{2}}}{s\cosh as^{\frac{1}{2}}}$	$\dfrac{b}{a} + \dfrac{2}{\pi}\displaystyle\sum_{n=1}^{\infty}\dfrac{(-1)^n}{n}$ $\times \sin k_n b\, e^{-(k_n^2 + c)t}$	$k_n = \dfrac{\pi n}{a}$
H.2	$\dfrac{\sinh b(s + c)^{\frac{1}{2}}}{s\cosh a(s + c)^{\frac{1}{2}}}$	$\dfrac{b}{a} + \dfrac{2}{\pi}\displaystyle\sum_{n=1}^{\infty}\dfrac{(-1)^n}{n}\sin k_n b$ $\times \left[\dfrac{c}{c + k_n^2} + \dfrac{k_n^2}{c + k_n^2}e^{-(k_n^2 + c)t}\right]$	$k_n = \dfrac{n\pi}{a}$
H.3	$\dfrac{\cosh x(RCs)^{\frac{1}{2}}}{s\cosh l(RCs)^{\frac{1}{2}}}$	$1 + \dfrac{4}{\pi}\displaystyle\sum_{n=1}^{\infty}\dfrac{(-1)^n}{2n - 1}$ $\times \cos m_n x\, e^{-m_n^2 t/RC}$	$m_n = \dfrac{2n - 1}{2}\dfrac{\pi}{l}$
H.4	$\dfrac{C^{\frac{1}{2}}\sinh x(RCs)^{\frac{1}{2}}}{R^{\frac{1}{2}}\cosh l(RCs)^{\frac{1}{2}}}$	$-\dfrac{2}{Rl}\displaystyle\sum_{n=1}^{\infty}(-1)^n$ $\times \sin m_n x\, e^{-m_n^2 t/RC}$	$m_n = \dfrac{2n - 1}{2}\dfrac{\pi}{l}$

Table 2   Laplace-transform pairs (*Continued*)

No.	$g(s)$	$f(t)$	Notes
H.5	$\dfrac{C^{\frac{1}{2}} \cosh x(RCs)^{\frac{1}{2}}}{sR^{\frac{1}{2}} \sinh l(RCs)^{\frac{1}{2}}}$	$\dfrac{1}{Rl} + \dfrac{2}{Rl} \displaystyle\sum_{n=1}^{\infty} (-1)^n$ $\times \cos k_n x \, e^{-k_n^2 t/RC}$	$k_n = \dfrac{n\pi}{l}$
H.6	$\dfrac{\alpha \cosh \alpha x}{sZ \cosh \alpha l}$	$\left(\dfrac{G}{R}\right)^{\frac{1}{2}} \dfrac{\cosh x(RG)^{\frac{1}{2}}}{\sinh l(RG)^{\frac{1}{2}}} - \dfrac{E}{Rl} e^{-Rt/L}$ $+ 2 \displaystyle\sum_{k=1}^{\infty} \dfrac{(-1)^k \cos m_k x}{l(m_k^2 + RG)} e^{-\rho t}$ $\times \left\{ \dfrac{1}{\beta}\left[\dfrac{m_k^2}{L} + \tfrac{1}{2}G\left(\dfrac{R}{L} - \dfrac{G}{l}\right)\right] \right.$ $\left. \times \sin \beta_k t - G \cos \beta_k t \right\}$	$\alpha = [(R + Ls) \times (G + Cs)]^{\frac{1}{2}}$ $Z = R + Ls$ $m_k = \dfrac{k\pi}{l}$ $\rho = \tfrac{1}{2}\left(\dfrac{R}{L} + \dfrac{G}{C}\right)$ $\beta_k = \left[\dfrac{m_k^2}{LC} - \tfrac{1}{4}\right.$ $\left. \times \left(\dfrac{R}{L} - \dfrac{G}{C}\right)^2\right]^{\frac{1}{2}}$
H.7	$\dfrac{\cosh \alpha(l - x)}{s \cosh \alpha l}$	$\dfrac{\cosh p(l - x)}{\cosh pl}$ $- \dfrac{\pi v^2}{l} e^{-\rho t} \displaystyle\sum_{n=\text{odd}}^{\infty} n \sin \dfrac{n\pi x}{2l}$ $\times \dfrac{\rho \sin \beta_n t + \beta_n \cos \beta_n t}{\beta_n(\rho^2 + \beta_n^2)}$	$\beta_n = \left(\dfrac{n^2 \pi^2 v^2}{4l^2} - \sigma^2\right)^{\frac{1}{2}}$ $\alpha = [(R + Ls) \times (G + SC)]^{\frac{1}{2}}$ $p = (RG)^{\frac{1}{2}}$ $\rho = \tfrac{1}{2}\left(\dfrac{R}{L} + \dfrac{G}{C}\right)$ $\sigma = \tfrac{1}{2}\left(\dfrac{R}{L} - \dfrac{G}{C}\right)$ $v = (LC)^{-\frac{1}{2}}$
H.8	$\dfrac{\sinh \alpha(l - x)}{sZ_0 \cosh \alpha l}$	$\dfrac{G}{p} \dfrac{\sinh p(l - x)}{\cosh pl}$ $+ \dfrac{2v^2}{l^2} e^{-\rho t} \displaystyle\sum_{n=\text{odd}}^{\infty} \cos \dfrac{n\pi x}{2l}$ $\times \left[\dfrac{G\rho \sin \beta_n t + G\beta_n \cos \beta_n t}{\beta_n(\rho^2 + \beta_n^2)}\right.$ $\left. - \dfrac{C \sin \beta_n t}{\beta_n}\right]$	Constants are defined in H.7 $Z_0 = \left(\dfrac{R + Ls}{G + Cs}\right)^{\frac{1}{2}}$
H.9	$\dfrac{\sinh \alpha(l - x)}{s \sinh \alpha l}$	$\dfrac{\sinh p(l - x)}{\sinh pl}$ $- \dfrac{2v^2 \pi}{l^2} e^{-\rho t} \displaystyle\sum_{n=\text{odd}}^{\infty} n \sin \dfrac{n\pi x}{l}$ $\times \dfrac{\rho \sin \beta_n t + \beta_n \cos \beta_n t}{\beta_n(\rho^2 + \beta_n^2)}$	Constants are defined in H.7

Table 2 Laplace-transform pairs (*Continued*)

No.	$g(s)$	$f(t)$	Notes

H.10 $\dfrac{1}{s}\tanh\dfrac{as}{2}$

Meander function

H.11 $\dfrac{m}{s^2}-\dfrac{ma}{2s}\left(\coth\dfrac{as}{2}-1\right)$

$\dfrac{m}{s^2}-\dfrac{ma}{s(e^{sa}-1)}$

Sawtooth function
Slope $= m$

Note alternate expression for $g(s)$

H.12 $\dfrac{1}{2s}\left(1+\coth\dfrac{as}{2}\right)$

Stepped function

H.13 $\dfrac{1}{s\sinh as}$

H.14 $\dfrac{\sinh bs^{\frac{1}{2}}}{s\sinh as^{\frac{1}{2}}}$ $\dfrac{b}{a}+\dfrac{2}{\pi}\displaystyle\sum_{n=1}^{\infty}\dfrac{(-1)^n}{n}$

$\times \sin\dfrac{n\pi b}{a}e^{-n^2\pi^2 t/a^2}$

N.1 $\omega T(\omega)$ $\sin\omega t$ $T(\omega)=\dfrac{1}{s^2+\omega^2}$

N.2 $\dfrac{1}{2s}[1-s^2\,T(2\omega)]$ $\sin^2\omega t$

N.3 $\dfrac{3\omega}{4}[T(\omega)-T(3\omega)]$ $\sin^3\omega t$

N.4 $sT(\omega)$ $\cos\omega t$

N.5 $\dfrac{1}{2s}[1+s^2\,T(2\omega)]$ $\cos^2\omega t$

N.6 $\tfrac{1}{4}[3sT(\omega)+sT(3\omega)]$ $\cos^3\omega t$

N.7 $\tfrac{1}{2}[(A+B)\,T(A+B)$
$+(A-B)\,T(A-B)]$ $\sin At\cos Bt$

N.8 $\dfrac{s}{2}[T(A+B)+T(A-B)]$ $\cos At\cos Bt$

Table 2   Laplace-transform pairs (*Continued*)

No.	$g(s)$	$f(t)$	Notes
N.9	$\dfrac{s}{2}[T(A-B)-T(A+B)]$	$\sin At \sin Bt$	
N.10	$\begin{aligned}&\tfrac{1}{4}[(A+B-C)\,T(A+B-C)\\&+(B+C-A)\,T(B+C-A)\\&+(C+A-B)\,T(C+A-B)\\&-(A+B+C)\,T(A+B+C)]\end{aligned}$	$\sin At \sin Bt \sin Ct$	
N.11	$\begin{aligned}&\tfrac{1}{4}[(A+B-C)\,T(A+B-C)\\&-(B+C-A)\,T(B+C-A)\\&+(C+A-B)\,T(C+A-B)\\&+(A+B+C)\,T(A+B+C)]\end{aligned}$	$\sin At \cos Bt \cos Ct$	
N.12	$\begin{aligned}\dfrac{s}{4}[&\dot{T}(B+C-A)\\&-T(A+B-C)\\&+T(C+A-B)\\&\quad -T(A+B+C)]\end{aligned}$	$\sin At \sin Bt \cos Ct$	
N.13	$\begin{aligned}\dfrac{s}{4}[&T(A+B-C)\\&+T(B+C-A)\\&+T(C+A-B)\\&\quad +T(A+B+C)]\end{aligned}$	$\cos At \cos Bt \cos Ct$	
N.14	$sT(a)\,T(b)$	$\dfrac{\cos at - \cos bt}{b^2 - a^2}$	$a \neq b$
N.15	$sT^2(\omega)$	$\dfrac{t\sin \omega t}{2\omega}$	
N.16	$T^2(\omega)$	$\dfrac{\sin \omega t}{2\omega^3} - \dfrac{t\cos \omega t}{2\omega^2}$	
N.17	$\dfrac{1}{s}T(\omega)$	$\dfrac{1-\cos \omega t}{\omega^2}$	
R.1	$e^{-a\sqrt{s}}$	$\dfrac{a}{2\sqrt{\pi t^3}}\,e^{-a^2/4t}$	$a > 0$
R.2	$\dfrac{e^{-a\sqrt{s}}}{\sqrt{s}}$	$\dfrac{1}{\sqrt{\pi t}}\,e^{-a^2/4t}$	$a > 0$
R.3	$\dfrac{1}{\sqrt{s+a}}$	$\dfrac{e^{-at}}{\sqrt{\pi t}}$	$a > 0$
R.4	$e^{-\alpha x}$	$\dfrac{y}{2t\sqrt{\pi t}}\,e^{-2\beta t - y^2/4t}$	$\begin{aligned}&\alpha = [R(Cs+G)]^{\frac{1}{2}}\\&\beta = \dfrac{G}{2C}\\&y = x(RC)^{\frac{1}{2}}\end{aligned}$

Table 2   **Laplace-transform pairs** (*Continued*)

No.	$g(s)$	$f(t)$	Notes
			For constants, see R.4
R.5	$\dfrac{e^{-ax}}{Z_0}$	$\dfrac{u(y^2 - 2t)}{4t^2\sqrt{\pi t}}\, e^{-2\beta t - y^2/4t}$	$Z_0 = \left(\dfrac{R}{Cs + G}\right)^{\pm}$ $u = \left(\dfrac{C}{R}\right)^{\pm}$
R.6	$\dfrac{g(s^{\pm})}{s^{\pm}}$	$\dfrac{t^{\pm}}{\pi^{\pm}}\displaystyle\int_0^{\infty} e^{-u^2/4t} f(u)\, du$	$t > 0$ $g(s) = Lf(t)$ Erdélyi, 1954, no. 33, p. 132
R-B.1	$\dfrac{1}{\sqrt{s^2 + a^2}}$	$J_0(at)$	
R-B.2	$\dfrac{(\sqrt{s^2 + a^2} - s)^n}{\sqrt{s^2 + a^2}}$	$\dfrac{1}{a^n} J_n(at)$	$n = 0, 1, 2, \ldots$
R-B.3	$\dfrac{(\sqrt{s^2 + a^2} - s)^n}{s\sqrt{s^2 + a^2}}$	$\dfrac{1}{a^n}\displaystyle\int_0^t J_n(at')\, dt'$	$n = 0, 1, 2, \ldots$
R-B.4	$\dfrac{1}{\sqrt{s^2 + a^2} + s}$	$\dfrac{1}{a}\dfrac{J_1(at)}{t}$	
R-B.5	$\dfrac{1}{(\sqrt{s^2 + a^2} + s)^n}$	$\dfrac{n}{a^n}\dfrac{J_n(at)}{t}$	$n = 1, 2, 3, \ldots$
R-B.6	$\dfrac{1}{s(\sqrt{s^2 + a^2} + s)^n}$	$\dfrac{n}{a^n}\displaystyle\int_0^t \dfrac{J_n(at)}{t}\, dt$	$n = 1, 2, 3, \ldots$
R-B.7	$\dfrac{1}{\sqrt{s^2 - a^2}}$	$I_0(at)$	
R-B.8	$\dfrac{(s - \sqrt{s^2 - a^2})^n}{\sqrt{s^2 - a^2}}$	$I_n(at)$	
R-B.9	$\dfrac{e^{-a\sqrt{s^2 + b^2}}}{\sqrt{s^2 + b^2}}$	$J_0(b\sqrt{t^2 - a^2})\, u(t - a)$	$a > 0$
R-B.10	$\dfrac{e^{-a\sqrt{s^2 - b^2}}}{\sqrt{s^2 - b^2}}$	$I_0(b\sqrt{t^2 - a^2})\, u(t - a)$	$a > 0$
R-B.11	$\dfrac{e^{-(x/v)\sqrt{(s+a)^2 - \sigma^2}}}{\sqrt{(s + a)^2 - \sigma^2}}$	$e^{-at} I_0\left(\sigma\sqrt{t^2 - \dfrac{x^2}{v^2}}\right)$ $\times\, u\left(t - \dfrac{x}{v}\right)$	$a > 0$

Table 2 Laplace-transform pairs (*Continued*)

No.	$g(s)$	$f(t)$	Notes
R-B.12	$\dfrac{1}{\sqrt{s(s+2a)}}$	$e^{-at} I_0(at)$	
R-B.13	$\dfrac{1}{\sqrt{(s+a)^2 - b^2}}$	$e^{-at} I_0(bt)$	
R-B.14	$\dfrac{q}{s(q-a)}$	$\dfrac{1}{1-a^2}\Big[e^{a^2 bt(1-a^2)} + a\, e^{-(bt/2)}$ $\times I_0\!\left(\dfrac{bt}{2}\right) + \dfrac{ab}{(1-a^2)^2}$ $\times e^{a^2 bt(1-a^2)}$ $\times \displaystyle\int_0^t e^{(a^2+1)bt/2(a^2-1)}$ $\times I_0\!\left(\dfrac{bt}{2}\right) dt$	$a^2 \neq 1$ $q = \left(\dfrac{s}{s+b}\right)^{\frac{1}{2}}$
R-B.15	$\dfrac{1}{s} e^{-\alpha x}$	$\Big[e^{-\rho x/v} + \dfrac{\sigma x}{v}\displaystyle\int_{x/v}^t e^{-\rho u}$ $\times \dfrac{I_1(\sigma\sqrt{u^2 - x^2/v^2}}{\sqrt{u^2 - x^2/v^2}}\, du\Big]$ $\times u\!\left(t - \dfrac{x}{v}\right)$	$\alpha = [(Ls+R)$ $\times (Cs+G)]^{\frac{1}{2}}$ $\rho = \frac{1}{2}\left(\dfrac{R}{L} + \dfrac{G}{C}\right)$ $\sigma = \frac{1}{2}\left(\dfrac{R}{L} - \dfrac{G}{C}\right)$ $v = \dfrac{1}{\sqrt{LC}}$
R-B.16	$e^{-\alpha x}$	$\Big[\delta\!\left(t - \dfrac{x}{v}\right)e^{-\rho x/v}$ $+ \dfrac{\sigma x}{vz}e^{-\rho t} I_1(\sigma z)\Big] u\!\left(t - \dfrac{x}{v}\right)$	For constants, see R-B.15 $z = \left(t^2 - \dfrac{x^2}{v^2}\right)^{\frac{1}{2}}$
R-B.17	$\dfrac{e^{-\alpha x}}{Z_0}$	$\Big\{\dfrac{1}{k}e^{-\rho x/v}\delta\!\left(t - \dfrac{x}{v}\right) + \dfrac{1}{k}e^{-\rho t}$ $\times \Big[\dfrac{\sigma t}{z}I_1(\sigma z) - \sigma I_0(\sigma z)\Big]\Big\}$ $\times u\!\left(t - \dfrac{x}{v}\right)$	For constants see R-B.15 $z = \left(t^2 - \dfrac{x^2}{v^2}\right)^{\frac{1}{2}}$ $Z_0 = \left(\dfrac{Ls+R}{Cs+G}\right)^{\frac{1}{2}}$ $k = \left(\dfrac{L}{C}\right)^{\frac{1}{2}}$
R-B.17	$\dfrac{e^{\alpha x}}{sZ_0}$	$\dfrac{1}{k}e^{-at} I_0(az)\, u\!\left(t - \dfrac{x}{v}\right)$	$\alpha = [(Ls+R)\,Cs]^{\frac{1}{2}}$ $Z_0 = \left(\dfrac{Ls+R}{Cs}\right)^{\frac{1}{2}}$ $k = \left(\dfrac{L}{C}\right)^{\frac{1}{2}}$ $a = \dfrac{R}{2L}$

Table 2   Laplace-transform pairs (*Continued*)

No.	$g(s)$	$f(t)$	Notes
R-B.18	$\dfrac{\sqrt{(s+2a)}}{\sqrt{s}+\sqrt{s+2a}}$	$\dfrac{\delta(t)}{2}+\dfrac{1}{2t}e^{-at}I_1(at)$	
R-B.19	$\dfrac{\sqrt{s+2a}}{s(\sqrt{s}+\sqrt{s+2a})}$	$1-\tfrac{1}{2}e^{-at}[I_0(at)+I_1(at)]$	
R-B.20	$\dfrac{\omega^{2n+1}[\sqrt{(s+a)^2+\omega^2}+(s+a)]^{-2n}}{k\sqrt{(s+a)^2+\omega^2}}$	$\dfrac{2}{L}e^{-at}J_{2n}(wt)$	$k=\left(\dfrac{L}{C}\right)^{\frac{1}{2}}$ $\omega=2\left(\dfrac{C}{L}\right)^{\frac{1}{2}}$ $a=\dfrac{R}{L}=\dfrac{G}{C}$ $n=$ positive integer
R-B.21	$\dfrac{s(2a)^n\sqrt{s/(s+2a)}/R}{(\sqrt{s+2a}+\sqrt{s})^{2n}}$	$\dfrac{a}{R}e^{-at}[I_{n-1}(at)-2I_n(at)$ $+I_{n+1}(at)]$	$a=\dfrac{2}{RC}$ $n=$ positive integer
R-B.22	$\dfrac{2(2a)^n\sqrt{s/(s+2a)}}{sR(\sqrt{s+2a}+\sqrt{s})^{2n}}$	$\dfrac{2}{R}e^{-at}I_n(at)$	$a=\dfrac{2}{RC}$ $n=$ positive integer
R-B.23	$\dfrac{(\sqrt{1+a}+\sqrt{a})^{-2n}}{s\sqrt{(1+a)}Z_1Z_2}$	$\dfrac{1}{k}\displaystyle\int_0^{\omega_c t}e^{-(b/\omega_c)u}J_{2n}(u)\,du$	$n=$ positive integer $a=\dfrac{LC}{4}(s+b)^2$ $Z_1=L(s+b)$ $\dfrac{1}{Z_2}=C(s+b)$ $k=\left(\dfrac{1}{C}\right)^{\frac{1}{2}}$ $\omega_c=\dfrac{2}{\sqrt{LC}}$
R-B.24	$\dfrac{\omega_c(\sqrt{s^2+\omega_c^2}-s/\omega_c)^{2n}}{sk\qquad\sqrt{s^2+\omega_c^2}}$	$\dfrac{1}{k}\displaystyle\int_0^{\omega_c t}J_{2n}(u)\,du$	For constants, see R-B.23
R-E.1	$\dfrac{1}{s}e^{-as^{\frac{1}{2}}}$	$1-\mathrm{erf}\left(\dfrac{a}{2t^{\frac{1}{2}}}\right)$	$a>0$ $\mathrm{erf}(y)$ $=\dfrac{2}{\pi^{\frac{1}{2}}}\displaystyle\int_0^u e^{-u^2}\,du$
R-E.2	$\dfrac{e^{-as^{\frac{1}{2}}}}{s+bs^{\frac{1}{2}}}$	$e^{b^2t+ab}\left[1-\mathrm{erf}\left(bt^{\frac{1}{2}}+\dfrac{a}{2t^{\frac{1}{2}}}\right)\right]$	$a>0$

Table 2 Laplace-transform pairs (*Continued*)

No.	$g(s)$	$f(t)$	Notes
R-E.3	$\dfrac{1}{s(1+s)^{\frac{1}{2}}}$	$\operatorname{erf}(t^{\frac{1}{2}})$	
R-E.4	$\dfrac{1}{s^{\frac{1}{2}}(s^{\frac{1}{2}}-a)}$	$e^{2at}[1+\operatorname{erf}(at^{\frac{1}{2}})]$	
R-E.5	$\dfrac{(s+b)^{\frac{1}{2}}}{s[(s+b)^{\frac{1}{2}}-a]}$	$\dfrac{b}{b-a^2}-\dfrac{a^2}{b-a^2}e^{(a^2-b)t}$ $+\dfrac{a\sqrt{b}}{b-a^2}\operatorname{erf}(\sqrt{bt})$ $-\dfrac{a^2}{b-a^2}e^{(a^2-b)t}\operatorname{erf}(a\sqrt{t})$	$b\neq a^2$
R-E.6	$\dfrac{s^{\frac{1}{2}}}{s+a}$	$\dfrac{1}{(\pi t)^{\frac{1}{2}}}+ja^{\frac{1}{2}}e^{-at}\operatorname{erf}(ja^{\frac{1}{2}}t^{\frac{1}{2}})$	$j=\sqrt{-1}$
R-E.7	$\dfrac{e^{-\alpha x}}{s}$	$\dfrac{1}{2}\left[e^{-y\sqrt{2\beta}}\operatorname{erf}\left(\dfrac{y}{2\sqrt{t}}-\sqrt{2\beta t}\right)\right.$ $\left.+e^{y\sqrt{2\beta}}\operatorname{erf}\left(\dfrac{y}{2\sqrt{t}}+\sqrt{2\beta t}\right)\right]$	$\alpha=\sqrt{R(Cs+G)}$ $y=x\sqrt{RC}$ $\beta=\dfrac{G}{2C}$ $\operatorname{erf}(y)=\dfrac{2}{\sqrt{\pi}}$ $\times\displaystyle\int_0^y e^{-u^2}\,du$
R-E.8	$\dfrac{e^{-\alpha x}}{sZ_0}$	$\dfrac{u}{\sqrt{\pi t}}e^{-y^2/4t-2\beta t}+\dfrac{u\sqrt{2\beta}}{2}$ $\times\left[e^{-y\sqrt{2\beta}}\operatorname{erf}\left(\dfrac{y}{2\sqrt{t}}-\sqrt{2\beta t}\right)\right.$ $\left.-e^{y\sqrt{2\beta}}\operatorname{erf}\left(\dfrac{y}{2\sqrt{t}}+\sqrt{2\beta t}\right)\right]$	For α,β,y see R-E.7 $u=\left(\dfrac{C}{R}\right)^{\frac{1}{2}}$ $Z_0=\left(\dfrac{R}{Cs+G}\right)^{\frac{1}{2}}$
R-E.9	$\dfrac{e^{-ys^{\frac{1}{2}}}}{1+(s/a)^{\frac{1}{2}}}$	$\sqrt{\dfrac{a}{\pi t}}e^{-y^2/4t}-ae^{y\sqrt{a}+at}$ $\times\operatorname{erf}\left(\dfrac{y}{2\sqrt{t}}+\sqrt{at}\right)$	
R-E.10	$\dfrac{e^{-ys^{\frac{1}{2}}}}{s[1+(s/a)^{\frac{1}{2}}]}$	$\operatorname{erf}\left(\dfrac{y}{2\sqrt{t}}\right)-e^{y\sqrt{a}+at}$ $\times\operatorname{erf}\left(\dfrac{y}{2\sqrt{t}}+\sqrt{at}\right)$	
R-E.11	$\dfrac{s^{\frac{1}{2}}e^{-ys^{\frac{1}{2}}}}{1+(s/a)^{\frac{1}{2}}}$	$\dfrac{(y-2t\sqrt{a})}{2t}\sqrt{\dfrac{a}{\pi t}}e^{-y^2/4t}$ $+a\sqrt{a}e^{y\sqrt{a}+at}\operatorname{erf}\left(\dfrac{y}{2\sqrt{t}}+\sqrt{at}\right)$	

Table 2 Laplace-transform pairs (*Continued*)

No.	$g(s)$	$f(t)$	Notes
R-E.12	$\dfrac{e^{-ys^{\frac{1}{2}}}}{s^{\frac{1}{2}}[1 + (s/a)^{\frac{1}{2}}]}$	$e^{y\sqrt{a}+at}\operatorname{erf}\left(\dfrac{y}{2\sqrt{t}} + \sqrt{at}\right)$	
R-E.13	$\dfrac{s^{\frac{1}{2}}}{1 + (s/a)^{\frac{1}{2}}}$	$\sqrt{a}\left[\delta(t) - \sqrt{\dfrac{a}{\pi t}} + ae^{at}\right.$ $\left.\times \operatorname{erf}(\sqrt{at})\right]$	
R-E.14	$\dfrac{1}{s^{\frac{1}{2}}[1 + (s/a)^{\frac{1}{2}}]}$	$\sqrt{a}\,e^{at}\operatorname{erf}(\sqrt{at})$	
R-E.15	$\dfrac{se^{-ys^{\frac{1}{2}}}}{1 + (s/a)^{\frac{1}{2}}}$	$\dfrac{y^2 - 2yt\sqrt{a} - 2t + 4at^2}{4t^2}$ $\times \sqrt{\dfrac{a}{\pi t}}\,e^{-y^2/4t}$ $- a^2 e^{y\sqrt{a}+at}\operatorname{erf}\left(\dfrac{y}{2\sqrt{t}} + \sqrt{at}\right)$	
R-E.16	$\dfrac{s}{1 + (s/a)^{\frac{1}{2}}}$	$-a\delta(t)\dfrac{2at-1}{2t}\sqrt{\dfrac{a}{\pi t}} - a^2 e^{at}$ $\times \operatorname{erf}(\sqrt{at})$	
R-E.17	$\dfrac{1}{1 + (s/a)^{\frac{1}{2}}}$	$\sqrt{\dfrac{a}{\pi t}} - ae^{at}\operatorname{erf}(\sqrt{at})$	
SP.1	$\dfrac{1}{(s + a)^n}$	$\dfrac{t^{n-1}}{(n-1)!}e^{-at}$	n = positive integer
SP.2	$\tan^{-1}\left(\dfrac{a}{s}\right)$	$\dfrac{\sin at}{t}$	
SP.3	$\ln\left(\dfrac{s+b}{s+a}\right)$	$\dfrac{e^{-at} - e^{-bt}}{t}$	
SP.4	$K_0(as^{\frac{1}{2}})$	$\dfrac{1}{2t}e^{-a^2/4t}$	$a > 0$ $K_0(y)$ is the modified Bessel function of the second kind of order zero
SP.5	$\dfrac{1}{s(s + a)^n}$	$\dfrac{1}{\Gamma(n)}\displaystyle\int_0^t e^{-u} u^{n-1}\,du$	$\Gamma(n)$ = gamma function n = positive integer
SP.6	$\dfrac{(2n)!}{s(s^2+2^2)(s^2+4^2)\cdots[s^2+(2n)^2]}$	$\sin^{2n} t$	

Table 2   Laplace-transform pairs (*Continued*)

No.	$g(s)$	$f(t)$	Notes
SP.7	$\dfrac{(2n+1)!}{(s^2+1^2)(s^2+3^2)\,\cdots\,[s^2+(2n+1)^2]}$	$\sin^{2n+1}t$	
SP.8	$\dfrac{\omega s^{\frac{1}{2}}}{s^2+\omega^2}$	$(2\omega)^{\frac{1}{2}}\left[\sin\omega t\,S\!\left(\dfrac{\sqrt{2\omega t}}{\pi}\right)\right.$ $\left.+\cos\omega t\,C\!\left(\dfrac{\sqrt{2\omega t}}{\pi}\right)\right]$	$S(y)$ and $C(y)$ are Fresnel integrals defined by $S(y)=\displaystyle\int_0^y\sin\dfrac{\pi u^2}{2}\,du$ $C(y)=\displaystyle\int_0^y\cos\dfrac{\pi u^2}{2}\,du$
SP.9	$\dfrac{1}{s}e^{-\alpha x}$	$0\qquad t<\dfrac{x}{v}$ $e^{-\rho x/v}\qquad t>\dfrac{x}{v}$	$\alpha=[(Ls+R)\\\times(G+Cs)]^{\frac{1}{2}}$ $\dfrac{R}{L}=\dfrac{G}{C}$ $\rho=\dfrac{R}{2L}+\dfrac{G}{2C}$ $v=\sqrt{\dfrac{L}{C}}$
SP.10	$e^{-\alpha x}$	$\delta\!\left(\dfrac{t-x}{v}\right)e^{-\rho x/v}$	Constants are defined in SP.9 $\delta(x)=$ Dirac delta function
SP.11	$\dfrac{s^{(1-n)/n}}{s^{1/n}-a}$	$e^{ant}\displaystyle\sum_{m=1}^{n-1}[1+\phi_m(t,a)]$	$\phi_m(t,a)=\dfrac{1}{\Gamma(m/n+1)}$ $\times\displaystyle\int_0^{a^m t^{m/n}}e^{-u^{m/n}}\,du$
SP.12	$\dfrac{(s-a)^n}{s^{n+1}}$	$L_n(at)=e^{-at}\dfrac{d^n}{dt^n}\!\left(\dfrac{t^n}{n!}e^{-at}\right)$ $=$ Laguerre polynomial of nth order	
SP.13	$\dfrac{1}{s}e^{-a/s}$	$J_0(2\sqrt{at})$	
SP.14	$\dfrac{1}{s}\cos\dfrac{1}{s}$	$\text{ber}\,(2\sqrt{t})$	ber is the Bessel real function of Lord Kelvin
SP.15	$\dfrac{1}{s}\sin\dfrac{1}{s}$	$\text{bei}\,(2\sqrt{t})$	bei is the Bessel imaginary function of Lord Kelvin
SP.16	$\left(\dfrac{1}{s}\right)^{n+1}e^{-1/s}$	$t^{n/2}J_n(2\sqrt{t})$	Real $n>-1$

Table 2   Laplace-transform pairs (*Continued*)

No.	$g(s)$	$f(t)$	Notes
SP.17	$\dfrac{e^{-as} - e^{-bs}}{s}$		
SP.18	$\dfrac{m}{s^2}(1 - e^{-as}) - \dfrac{ma}{s}e^{-as}$		
SP.19	$K_0(as)$	$\sqrt{t^2 - a^2}\, u(t - a)$	
SP.20	$\dfrac{1}{s}e^{j/s}$	$J_0(2\sqrt{-jt}) = \mathrm{ber}\,(2\sqrt{t})$ $+ j\,\mathrm{bei}\,(2\sqrt{t})$	$j = \sqrt{-1}$
SP.21	$\dfrac{1}{s}e^{-j/s}$	$J_0(2\sqrt{jt}) = \mathrm{ber}\,(2\sqrt{t})$ $- j\,\mathrm{bei}\,(2\sqrt{t})$	$j = \sqrt{-1}$
SP.22	$\dfrac{1}{s}\ln\left[\dfrac{1}{\sqrt{1 + s^2}}\right]$	$Ci(t) =$ integral cosine function	$Ci(t) = \displaystyle\int_{-\infty}^{t}\dfrac{\cos x}{x}\,dx$
SP.23	$\dfrac{1}{s}\cot^{-1}s$	$Si(t) =$ integral sine function	$Si(t) = \displaystyle\int_{0}^{t}\dfrac{\sin x}{x}\,dx$
SP.24	$\dfrac{1}{s}\ln(s - 1)$	$Ei(t) =$ exponential integral function	$Ei(t) = -\displaystyle\int_{\infty}^{-t}\dfrac{e^{-x}}{x}\,dx$
SP.25	$\dfrac{m}{s^2} - \dfrac{2mT}{s}\left(\dfrac{1}{e^{Ts} - 1} - \dfrac{1}{e^{2Ts} - 1}\right)$		$m =$ slope
SP.26	$\dfrac{1}{s}\dfrac{1}{e^{Ts} + 1}$		

REFERENCES

1954. A. Erdélyi (ed.): "Tables of Integral Transforms," vol. I, McGraw-Hill Book Company, New York.

1963. Ryshik, I. M., and I. S. Gradstein: "Tables of Series, Products, and Integrals," VEB Deutscher Verlag Der Wissenschaften, Berlin, p. 249 ff.

1965. Fodor, G.: "Laplace Transforms in Engineering," Akademiai Kiado, Budapest, p. 691 ff.

1965. McCollum, P. A., and B. F. Brown: "Laplace Transform Tables and Theorems," Holt, Rinehart and Winston, Inc., New York.

1966. Roberts, G. E., and H. Kaufman: "Table of Laplace Transforms," W. B. Saunders Company, Philadelphia.

Special Functions of Applied Mathematics

1 INTRODUCTION

In the solution of a great many types of problems in applied mathematics we are led to the solution of linear differential equations or sets of linear differential equations. Usually these equations are equations having constant coefficients, and in a majority of these cases we are led to solutions of the exponential type, which includes trigonometric and hyperbolic functions as special cases. This is the situation that arises when we study the oscillations of linear electrical or mechanical systems.

However, this is not always the case, and equations of other forms are encountered; because of the frequency of their appearance, their solutions have been classified and tabulated for future reference. In this appendix we will present some of the properties of these functions in view of their practical importance. The functions to be considered are Bessel, Legendre, gamma, beta, and error functions.

2 BESSEL'S DIFFERENTIAL EQUATION

As a starting point of the discussion, let us consider the linear differential equation

$$x^2 \frac{d^2 y}{dx^2} + x \frac{dy}{dx} + (x^2 - n^2) y = 0 \tag{2.1}$$

where n is a constant.

This equation is known in the literature as *Bessel's differential equation.* Since it is a linear differential equation of the second order, it must have two linearly independent solutions. The standard form of the general solution of (2.1) is

$$y = C_1 J_n(x) + C_2 Y_n(x) \tag{2.2}$$

where C_1 and C_2 are arbitrary constants and the function $J_n(x)$ is called the *Bessel function of order n of the first kind* and $Y_n(x)$ is the *Bessel function of order n of the second kind.* These functions have been tabulated and behave somewhat like trigonometric functions of damped amplitude. To see this qualitatively, let us transform the dependent variable by the substitution

$$y = \frac{u}{\sqrt{x}} \tag{2.3}$$

This transformation converts (2.1) into

$$\frac{d^2 u}{dx^2} + \left(1 - \frac{n^2 - \frac{1}{4}}{x^2}\right) u = 0 \tag{2.4}$$

In the special case in which

$$n = \pm \tfrac{1}{2} \tag{2.5}$$

this becomes

$$\frac{d^2 u}{dx^2} + u = 0 \tag{2.6}$$

Hence

$$u = C_1 \sin x + C_2 \cos x \tag{2.7}$$

and

$$y = C_1 \frac{\sin x}{\sqrt{x}} + C_2 \frac{\cos x}{\sqrt{x}} \tag{2.8}$$

where C_1 and C_2 are arbitrary constants. Also we see that as $x \to \infty$ in (2.4), and n is finite, we would expect the solution of (2.1) to behave qualitatively as (2.8) to a first approximation.

3   SERIES SOLUTION OF BESSEL'S DIFFERENTIAL EQUATION

If we introduce the operator

$$\theta = x\frac{d}{dx} \tag{3.1}$$

then Bessel's differential equation (2.1) may be written in the form

$$\theta^2 y + (x^2 - n^2)y = 0 \tag{3.2}$$

In order to solve this equation, let us assume an infinite-series solution in the form

$$y = x^r \sum_{s=0}^{s=\infty} C_s x^s = \sum_{s=0}^{s=\infty} C_s x^{r+s} \tag{3.3}$$

Now

$$\theta x^m = x\frac{d}{dx}x^m = xmx^{m-1} = mx^m \tag{3.4}$$

$$\theta^2 x^m = \theta(mx^m) = m\theta x^m = m(mx^m) = m^2 x^m \tag{3.5}$$

Hence, on substituting (3.3) into (3.2), we have

$$\theta^2 y + (x^2 - n^2)y = \sum_{s=0}^{s=\infty} [(r+s)^2 + x^2 - n^2]C_s x^{r+s} = 0 \tag{3.6}$$

If we now equate the coefficients of the various powers of x, x^r, x^{r+1}, x^{r+2}, etc., to zero in (3.6), we obtain the set of equations

$$C_s[(r+s)^2 - n^2] + C_{s-2} = 0 \tag{3.7}$$

This is valid for $s = 0, 1, 2, \ldots$ in view of the fact that

$$\begin{align} C_{-1} &= 0 \\ C_{-2} &= 0 \end{align} \tag{3.8}$$

since the leading coefficient in the expansion (3.3) is C_0.

Letting $s = 0$ in (3.7), we obtain

$$C_0(r^2 - n^2) = 0 \tag{3.9}$$

This equation is known as the *indicial equation*, and since

$$C_0 \neq 0 \tag{3.10}$$

it follows that

$$r = \pm n \tag{3.11}$$

and from the equation

$$C_1[(r+1)^2 - n^2] = 0 \tag{3.12}$$

it follows that

$$C_1 = 0 \qquad (3.13)$$

The relation between C_s and C_{s-2} now shows, taking $s = 3, 5, \ldots$ in succession, that all coefficients of odd rank vanish.

Taking first of all $r = n$, we may write (3.7) in the form

$$C_s = -\frac{C_{s-2}}{s(2n+s)} \qquad s = 2, 4, 6, \ldots \qquad (3.14)$$

From (3.14) we see that the coefficients C_2, C_4, C_6, etc., are all determined in terms of C_0. Inserting these values of the coefficient into the assumed form of solution (3.3), we obtain the solution

$$y = C_0\left[x^n - \frac{x^{n+2}}{2^2(n+1)} + \frac{x^{n+4}}{2^4(n+1)(n+2)2!} - \cdots \right.$$
$$\left. + \frac{(-1)^s x^{n+2s}}{2^{2s}(n+1)\cdots(n+s)s!} + \cdots\right] \qquad (3.15)$$

The coefficients are finite except when n is a negative integer. Excluding this case, we standardize the solution by taking

$$C_0 = \frac{1}{2^n \Gamma(n+1)} = \frac{1}{2^n \Pi(n)} \qquad (3.16)$$

in general and

$$C_0 = \frac{1}{2^n n!} \qquad (3.17)$$

when n is a positive integer. Inserting this value of C_0 into (3.15) and generalizing the factorial numbers when n is not an integer by writing

$$(n+s)! = \Pi(n+s) \qquad (3.18)$$

we obtain

$$J_n(x) = \sum_{s=0}^{s=\infty} \frac{(-1)^s}{\Pi(s)\,\Pi(n+s)}\left(\frac{x}{2}\right)^{n+2s} \qquad (3.19)$$

This series converges for any finite value of x and represents a function $J_n(x)$ of x that is known as the *Bessel function of the first kind of order n.* When n is not an integer, the second solution may be obtained by replacing n by $-n$ in accordance with (3.11). It is therefore

$$J_{-n}(x) = \sum_{s=0}^{s=\infty} \frac{(-1)^s}{\Pi(s)\,\Pi(s-n)}\left(\frac{x}{2}\right)^{2s-n} \qquad (3.20)$$

The leading terms of $J_n(x)$ and $J_{-n}(x)$ are, respectively, finite (nonzero) multiples of x^n and x^{-n}; the two functions are not mere multiples of each other, and hence the general solution of the Bessel differential equation may be expressed in the form

$$y = AJ_n(x) + BJ_{-n}(x) \qquad (3.21)$$

where A and B are arbitrary constants provided that n is not an integer.

However, when n is an integer, and since n appears in the differential equation only as n^2 there is no loss of generality in taking it to be a positive integer, $J_{-n}(x)$ is not distinct from $J_n(x)$. In this case the denominators of the first n terms of the series for $J_{-n}(x)$ contain the factors

$$\frac{1}{\Pi(s-n)} = 0 \qquad (3.22)$$

for $s = 0, 1, 2, \ldots, n-1$. Hence these terms vanish. Therefore

$$J_{-n}(x) = \sum_{s=n}^{s=\infty} \frac{(-1)^s}{\Pi(s)\,\Pi(s-n)} \left(\frac{x}{2}\right)^{2s-n} \qquad (3.23)$$

If we now let

$$r = s - n \qquad (3.24)$$

then

$$J_{-n}(x) = \sum_{r=0}^{r=\infty} \frac{(-1)^{r+n}}{\Pi(r+n)\,\Pi(r)} \left(\frac{x}{2}\right)^{2r+n} = (-1)^n J_n(x) \qquad (3.25)$$

for $n = 0, 1, 2, 3, \ldots$. In this case we no longer have two linearly independent solutions of the differential equation, and an independent second solution must be found.

4 THE BESSEL FUNCTION OF ORDER n OF THE SECOND KIND

In the preceding section we have seen that if n is not an integer, a general solution of the Bessel differential equation of order n is given by (3.21). If, however, n is an integer, then in view of (3.25) we have

$$y = AJ_n(x) + B(-1)^n J_n(x)$$
$$= [A + B(-1)^n]J_n(x)$$
$$= CJ_n(x) \qquad (4.1)$$

where C is an arbitrary constant. We therefore do not have the general solution of Bessel's differential equation, since such a solution must consist of two linearly independent functions multiplied by arbitrary constants. Consider the function

$$Y_n(x) = \frac{1}{\sin n\pi} [\cos n\pi J_n(x) - J_{-n}(x)] \tag{4.2}$$

Now if n is not an integer, the function $Y_n(x)$ is dependent on $J_n(x)$, and since it is a linear combination of $J_n(x)$ and $J_{-n}(x)$, it is a solution of Bessel's differential equation of order n. If now n is an integer, because of the relation (3.25), we have

$$Y_n(x) = \frac{0}{0} \tag{4.3}$$

Hence, when n is an integer, we define $Y_n(x)$ to be

$$Y_n(x) = \lim_{r \to n} \frac{J_r(x) \cos r\pi - J_{-r}(x)}{\sin r\pi} \tag{4.4}$$

With this definition of $Y_n(x)$ we have, on carrying out the limiting process,

$$\frac{\pi}{2} Y_0(x) = J_0(x) \left(\log \frac{x}{2} + \gamma \right) + \left(\frac{x}{2} \right)^2 - \frac{(1 + \frac{1}{2})(x/2)^4}{(2!)^2}$$

$$+ \left(1 + \frac{1}{2} + \frac{1}{3} \right) \frac{(x/2)^6}{(3!)^2} - \cdots \tag{4.5}$$

where γ is Euler's constant defined by

$$\gamma = \lim_{n \to \infty} \left(1 + \frac{1}{2} + \frac{1}{3} + \cdots + \frac{1}{n} - \log n \right) = 0.5772157 \tag{4.6}$$

Also, when n is any positive integer, we have

$$\pi Y_n(x) = 2J_n(x) \left(\log \frac{x}{2} + \gamma \right) - \sum_{r=0}^{\infty} (-1)^r \frac{(x/2)^{n+2r}}{r!(n+r)!}$$

$$\times \left(\sum_{m=1}^{n+r} m^{-1} + \sum_{m=1}^{r} m^{-1} \right) - \sum_{r=0}^{n-1} \left(\frac{x}{2} \right)^{-n+2r} \frac{(n-r-1)!}{r!} \tag{4.7}$$

where, for $r = 0$, instead of

$$\sum_{m=1}^{n+r} m^{-1} + \sum_{m=1}^{r} m^{-1}$$

we write

$$\sum_{m=1}^{n} m^{-1}$$

The presence of the logarithmic term in the function $Y_n(x)$ shows that these functions are infinite at $x = 0$. The general solution of Bessel's differential equation may now be written in the form

$$y = C_1 J_n(x) + C_2 Y_n(x) \tag{4.8}$$

where C_1 and C_2 are arbitrary constants.

5   VALUES OF $J_n(x)$ AND $Y_n(x)$ FOR LARGE AND SMALL VALUES OF x

In Sec. 2 we saw that the transformation

$$y = \frac{u}{\sqrt{x}} \tag{5.1}$$

transformed Bessel's differential equation into the form

$$\frac{d^2 u}{dx^2} + \left(1 - \frac{n^2 - \frac{1}{4}}{x^2}\right) u = 0 \tag{5.2}$$

We would suspect qualitatively that for large values of x the Bessel functions would behave as the solutions of the equation obtained from (5.2) by neglecting the $1/x^2$ term, that is, as solutions of the equation

$$\frac{d^2 u}{dx^2} + u = 0 \tag{5.3}$$

and hence as

$$y = C_1 \frac{\sin x}{\sqrt{x}} + C_2 \frac{\cos x}{\sqrt{x}} \tag{5.4}$$

More precise analysis shows that

$$\lim_{x \to \infty} J_n(x) \cong \frac{\cos(x - \pi/4 - n\pi/2)}{\sqrt{\pi x/2}} \tag{5.5}$$

$$\lim_{x \to \infty} Y_n(x) \cong \frac{\sin(x - \pi/4 - n\pi/2)}{\sqrt{\pi x/2}} \tag{5.6}$$

That is, for large values of the argument x, the Bessel functions behave like trigonometric functions of decreasing amplitude.

From the series expansions of the functions $J_n(x)$ and $Y_n(x)$ we also have the following behavior for small values of x:

$$\lim_{x \to 0} J_n(x) \cong \frac{x^n}{2^n \, \Pi(n)} \tag{5.7}$$

The value of $Y_n(x)$ is always infinite at $x = 0$. For small values of x this function is of the order $1/x^n$ if $n \neq 0$ and of the order $\log x$ if $n = 0$.

6   RECURRENCE FORMULAS FOR $J_n(x)$

Some important recurrence relations involving the function $J_n(x)$ may be
obtained directly from the series expansion for the function.   From (3.19)
we have

$$J_n(x) = \sum_{s=0}^{s=\infty} \frac{(-1)^s}{\Pi(s)\,\Pi(n+s)} \left(\frac{x}{2}\right)^{n+2s} \tag{6.1}$$

If we write

$$J_n' = \frac{d}{dx} J_n(x) \tag{6.2}$$

we have

$$xJ_n' = \sum_{s=0}^{s=\infty} \frac{(-1)^s(n+2s)}{\Pi(s)\,\Pi(n+s)} \left(\frac{x}{2}\right)^{n+2s}$$

$$= nJ_n + x \sum_{s=1}^{\infty} \frac{(-1)^s}{\Pi(s-1)\,\Pi(n+s)} \left(\frac{x}{2}\right)^{n+2s-1} \tag{6.3}$$

If in the last summation we place

$$s = r + 1 \tag{6.4}$$

we obtain

$$xJ_n' = nJ_n - x \sum_{r=0}^{\infty} \frac{(-1)^r}{\Pi(r)\,\Pi(n+1+r)} \left(\frac{x}{2}\right)^{n+1+2r}$$

$$= nJ_n - xJ_{n+1} \tag{6.5}$$

In the same manner, we can prove that

$$xJ_n' + nJ_n = xJ_{n-1} \tag{6.6}$$

If we add (6.5) to (6.6), we have

$$2J_n' = J_{n-1} - J_{n+1} \tag{6.7}$$

If we place $n = 0$ and use Eq. (3.25), we have

$$J_0' = -J_1 \tag{6.8}$$

If we multiply (6.5) by x^{-n-1}, we obtain

$$x^{-n}J_n' = x^{-n-1}nJ_n - x^{-n}J_{n+1} \tag{6.9}$$

Hence

$$\frac{d}{dx}(x^{-n}J_n) = -x^{-n}J_{n+1} \tag{6.10}$$

Similarly, we may prove that

$$\frac{d}{dx}(x^n J_n) = 8^n J_{n-1} \tag{6.11}$$

If we subtract (6.6) from (6.5), we obtain

$$\frac{2n}{x} J_n = J_{n-1} + J_{n+1} \tag{6.12}$$

Many other recurrence formulas may be obtained.

7 EXPRESSIONS FOR $J_n(x)$ WHEN n IS HALF AN ODD INTEGER

The case when n is half an odd integer is of importance because these particular Bessel functions can be expressed in finite form by elementary functions.

If we place $n = \frac{1}{2}$ in the general series for $J_n(x)$ given by (3.19), we obtain

$$J_{\frac{1}{2}}(x) = \sum_{s=0}^{s=\infty} \frac{(-1)^s}{\Pi(s)\Pi(s+\frac{1}{2})} \left(\frac{x}{2}\right)^{2s+\frac{1}{2}} \tag{7.1}$$

Now since

$$\Pi(r) = r\Pi(r-1) \tag{7.2}$$

and

$$\Pi(s) = s! \qquad \text{if } s = 1, 2, 3, \ldots \tag{7.3}$$

we have

$$J_{\frac{1}{2}}(x) = \frac{x^{\frac{1}{2}}}{2^{\frac{1}{2}}\Pi(\frac{1}{2})}\left(1 - \frac{x^2}{3!} + \frac{x^4}{5!} - \cdots\right) \tag{7.4}$$

However, we have

$$\Pi\left(\frac{1}{2}\right) = \frac{\sqrt{\pi}}{2} \tag{7.5}$$

and

$$\frac{\sin x}{x} = \left(1 - \frac{x^2}{3!} + \frac{x^4}{5!} - \cdots\right) \tag{7.6}$$

Hence from (7.4) we have

$$J_{\frac{1}{2}}(x) = \sqrt{\frac{2}{\pi x}}\sin x \tag{7.7}$$

If we place $n = -\frac{1}{2}$ in the general series for $J_n(x)$, we may also show that

$$J_{-\frac{1}{2}}(x) = \sqrt{\frac{2}{\pi x}}\cos x \tag{7.8}$$

If in the recurrence formula (6.12) we place $n = \frac{1}{2}$, we obtain

$$\frac{J_{\frac{1}{2}}}{x} = J_{-\frac{1}{2}}(x) + J_{\frac{3}{2}}(x) \tag{7.9}$$

Hence

$$J_{\frac{3}{2}}(x) = \frac{1}{x} J_{\frac{1}{2}}(x) - J_{-\frac{1}{2}}(x)$$

$$= \sqrt{\frac{2}{\pi x}} \left(\frac{\sin x}{x} - \cos x \right) \tag{7.10}$$

If in (6.12) we let $n = \frac{3}{2}$, we obtain

$$\frac{3}{x} J_{\frac{3}{2}} = J_{\frac{1}{2}} + J_{\frac{5}{2}} \tag{7.11}$$

or

$$J_{\frac{5}{2}} = \frac{3}{x} J_{\frac{3}{2}} - J_{\frac{1}{2}} = \sqrt{\frac{2}{\pi x}} \left(\frac{3 - x^2}{x^2} \sin x - \frac{3}{x} \cos x \right) \tag{7.12}$$

In the same way we may show that

$$J_{-\frac{3}{2}} = \sqrt{\frac{2}{\pi x}} \left(- \sin x - \frac{\cos x}{x} \right) \tag{7.13}$$

$$J_{-\frac{5}{2}} = \sqrt{\frac{2}{\pi x}} \left(\frac{3}{x} \sin x + \frac{3 - x^2}{x^2} \cos x \right), \text{ etc.} \tag{7.14}$$

8   THE BESSEL FUNCTIONS OF ORDER n OF THE THIRD KIND, OR HANKEL FUNCTIONS OF ORDER n

In some physical investigations we encounter complex combinations of Bessel functions of the first and second kinds so frequently that it has been found convenient to tabulate these combinations and thus define new functions.

These new functions are defined by the equations

$$\left. \begin{array}{l} H_n^{(1)}(x) = J_n(x) + j Y_n(x) \\ H_n^{(2)}(x) = J_n(x) - j Y_n(x) \end{array} \right\} \quad j = \sqrt{-1} \tag{8.1} \tag{8.2}$$

and are called *Bessel functions of order n of the third kind*, or *Hankel functions of order n*.   These functions are complex quantities.

9   DIFFERENTIAL EQUATIONS WHOSE SOLUTIONS ARE EXPRESSIBLE IN TERMS OF BESSEL FUNCTIONS

In applied mathematics differential equations whose solutions are expressible in terms of Bessel functions are frequently encountered.   Lists of these equations have been compiled.† Several of these differential equations will be listed here for reference.

In order to shorten the writing, the notation

$$Z_n(x) = C_1 J_n(x) + C_2 Y_n(x) \qquad (9.1)$$

where C_1 and C_2 are arbitrary constants, will be introduced. In terms of this notation the following differential equations have the indicated solutions:

$$\frac{d^2 y}{dx^2} + \left(b^2 - \frac{4n^2 - 1}{4x^2}\right) y = 0 \qquad y = x^{\frac{1}{2}} Z_n(bx) \qquad (9.2)$$

$$\frac{d^2 y}{dx^2} + \frac{1 - 2a}{x}\frac{dy}{dx}$$
$$+ \left[(bcx^{c-1})^2 + \frac{a^2 - n^2 c^2}{x^2}\right] y = 0 \qquad y = x^a Z_n(bx^c) \quad (9.3)$$

$$\frac{d^2 y}{dx^2} + bxy = 0 \qquad y = x^{\frac{1}{2}} Z_{\frac{1}{3}}[\tfrac{2}{3}(bx^3)^{\frac{1}{2}}] \qquad (9.4)$$

$$\frac{d^2 y}{dx^2} + bx^2 y = 0 \qquad y = x^{\frac{1}{2}} Z_{\frac{1}{4}}\left(\frac{x^2}{2} b^{\frac{1}{2}}\right) \qquad (9.5)$$

$$\frac{d^2 y}{dx^2} + (k^2 e^{2x} - n^2) y = 0 \qquad y = Z_n(k\, e^x) \qquad (9.6)$$

Equation (9.4) is known in the mathematical literature as *Stokes's equation* and is of considerable importance in mathematical physics because of its occurrence in the theory of diffraction and refraction of waves as well as in the theory of transmission across potential barriers.‡

10 MODIFIED BESSEL FUNCTIONS

Let us consider the differential equation

$$\frac{d^2 y}{dx^2} + \frac{1}{x}\frac{dy}{dx} + \left(-1 - \frac{n^2}{x^2}\right) y = 0 \qquad (10.1)$$

This equation is of the form (9.3) with

$$a = 0 \qquad b = j \qquad c = 1 \qquad (10.2)$$

Hence $J_n(jx)$ is a solution of this equation. The function

$$I_n(x) = j^{-n} J_n(jx) \qquad (10.3)$$

---

† See E. Jahnke and F. Emde, "Tables of Functions," pp. 146–147, Dover Publications, Inc., New York, 1943; W. G. Bickley, "Bessel Functions and Formulae," Cambridge University Press, New York, 1953.
‡ See "Tables of Modified Functions of Order One-third and Their Derivatives," Harvard University Press, Cambridge, Mass., 1945.

is taken as the standard form for one of the fundamental solutions of (10.1). The function $I_n(x)$ defined in this manner is a real function and is known as the *modified Bessel function of the first kind of order n.*   Another fundamental solution of Eq. (10.1) is known as the *modified Bessel function of the second kind* and is defined by

$$K_n(x) = \frac{\pi/2}{\sin n\pi} [I_{-n}(x) - I_n(x)] \qquad n \text{ nonintegral} \qquad (10.4)$$

The general solution of Eq. (10.1) may be written in the form

$$y = AI_n(x) + BK_n(x) \qquad n \text{ nonintegral} \qquad (10.5)$$

where A and B are arbitrary constants.

In contrast to the Bessel function $J_n(x)$ and $Y_n(x)$ the functions $I_n(x)$ and $K_n(x)$ are not of the oscillating type, but their behavior is similar to that of the exponential functions.   For large values of x we have

$$I_0(x) \cong \frac{e^x}{\sqrt{2\pi x}} \qquad (10.6)$$

$$K_0(x) \cong \sqrt{\frac{\pi}{2x}} e^{-x} \qquad (10.7)$$

For small values of x we have

$$I_0(x) \cong 1 \qquad (10.8)$$

$$K_0(x) \cong -\log\frac{x}{2} \qquad (10.9)$$

11   THE ber AND bei FUNCTIONS

In determining the distribution of alternating currents in wires of circular cross section, the following differential equation is encountered:

$$\frac{d^2 y}{dx^2} + \frac{1}{x}\frac{dy}{dx} - jy = 0 \qquad j = \sqrt{-1} \qquad (11.1)$$

This equation is a special case of Eq. (9.3) with

$$a = 0 \qquad n = 0 \qquad c = 1 \qquad (11.2)$$

and

$$b^2 = -j \qquad (11.3)$$

Hence

$$b = \sqrt{-j} = j\sqrt{j} = j^{\frac{3}{2}} \qquad (11.4)$$

The general solution of Eq. (11.1) may therefore be written in the form

$$y = AJ_0(j^{\frac{3}{2}}x) + K_0(j^{\frac{3}{2}}x) \qquad (11.5)$$

where A and B are arbitrary constants. The functions $J_0(j^{\frac{3}{2}}x)$ and $Y_0(j^{\frac{3}{2}}x)$ are complex functions. Decomposing them into their real and imaginary parts, we obtain

$$J_0(j^{\frac{3}{2}}x) = \text{ber}(x) + j\,\text{bei}(x) \tag{11.6}$$

and

$$K_0(j^{\pm}x) = \text{ker}(x) + j\,\text{kei}(x) \tag{11.7}$$

These equations define the functions $\text{ber}(x)$, $\text{bei}(x)$ and $\text{ker}(x)$, $\text{kei}(x)$.

12 EXPANSION IN SERIES OF BESSEL FUNCTIONS

In Appendix C it is pointed out that the expansion of an arbitrary function into a Fourier series is only a special case of the expansion of an arbitrary function in a series of orthogonal functions under certain restrictions. It will now be shown that it is possible to expand an arbitrary function in a series of Bessel functions. If in Eq. (2.4) we place ax instead of x, we obtain

$$\frac{d^2 u}{dx^2} + \left(a^2 - \frac{n^2 - \frac{1}{4}}{x^2}\right)u = 0 \tag{12.1}$$

This equation has one solution,

$$u = \sqrt{x}\,J_n(ax) \tag{12.2}$$

In the same manner

$$v = \sqrt{x}\,J_n(bx) \tag{12.3}$$

satisfies the equation

$$\frac{d^2 v}{dx^2} + \left(b^2 - \frac{n^2 - \frac{1}{4}}{x^2}\right)v = 0 \tag{12.4}$$

If we multiply (12.1) by v and (12.4) by u and subtract the second product from the first, we obtain

$$(b^2 - a^2)uv = u''v - v''u \tag{12.5}$$

Let us now integrate both members of (12.5) with respect to x from 0 to x. We thus obtain

$$(b^2 - a^2)\int_0^x uv\,dx = \int_0^x (u''v - v''u)\,dx \tag{12.6}$$

However, we have

$$\frac{d}{dx}(vu' - uv') = u''v - v''u \tag{12.7}$$

Hence

$$(b^2 - a^2) \int_0^x uv\,dx = \int_0^x \frac{d}{dx}(vu' - uv')\,dx$$

$$= (vu' - uv')\Big|_0^x \qquad (12.8)$$

That is,

$$(b^2 - a^2) \int_0^x xJ_n(ax)J_n(bx)\,dx = x[aJ_n(bx)J_n'(ax) - bJ_n(ax)J_n'(bx)] \quad (12.9)$$

If we now differentiate the last equation with respect to b and then set

$$b = a \qquad (12.10)$$

we obtain

$$2a \int_0^x xJ_n^2(ax)\,dx = x[axJ_n'^2(ax) - J_n(ax)J_n'(ax)$$
$$- axJ_n(ax)J_n''(ax)] \qquad (12.11)$$

From (12.9) we have

$$(b^2 - a^2) \int_0^1 xJ_n(ax)J_n(bx)\,dx = aJ_n(b)J_n'(a) - bJ_n(a)J_n'(b) \qquad (12.12)$$

Now the second member vanishes if a and b are roots of the equation

$$J_n(\alpha) = 0 \qquad (12.13)$$

That is, if a and b are distinct positive zeros of $J_n(\alpha)$, we have

$$(b^2 - a^2) \int_0^1 xJ_n(ax)J_n(bx)\,dx = 0 \qquad (12.14)$$

and since

$$a \neq b \qquad (12.15)$$

we have

$$\int_0^1 xJ_n(ax)J_n(bx)\,dx = 0 \qquad (12.16)$$

We are now in a position to expand an arbitrary function $F(x)$ in the interval from $x = 0$ to $x = 1$ in a series of the form

$$F(x) = \sum_{s=1}^{s=\infty} C_s J_n(\alpha_s x) \qquad (12.17)$$

where the α_s are the successive positive roots of (12.13). To obtain the general coefficient C_k of this expansion, we multiply both members of (12.17) by $xJ_n(\alpha_k x)\,dx$ and integrate from $x = 0$ to $x = 1$. We have by virtue of (12.16)

$$\int_0^1 xJ_n(\alpha_k x)F(x)\,dx = C_k \int_0^1 xJ_n^2(\alpha_k x)\,dx \qquad (12.18)$$

The last integral, which is independent of x, may be evaluated by means of (12.11), (12.3), and (6.9). Its value is

$$\int_0^1 xJ_n^2(\alpha_k x)\,dx = \tfrac{1}{2}J_{n+1}^2(\alpha_k) \tag{12.19}$$

Hence the typical coefficient of the series expansion (12.17) is given by

$$C_s = \frac{2}{J_{n+1}^2(\alpha_s)} \int_0^1 xJ_n(\alpha_s x)\,F(x)\,dx \tag{12.20}$$

This expansion is analogous to the expansion of an arbitrary function in a Fourier series.

13 THE BESSEL COEFFICIENTS

The functions $J_n(x)$ of *integral* order are sometimes called the *Bessel coefficients*. These functions occur as coefficients in the expansion of the following function in powers of t:

$$F(t) = e^{x(t-t^{-1})/2} = e^{xt/2}\,e^{-xt^{-1}/2} \tag{13.1}$$

This function of t can be expressed as a series of the form

$$F(t) = \sum_{n=-\infty}^{+\infty} A_n t^n = e^{xt/2}\,e^{-xt^{-1}/2} \tag{13.2}$$

This is an infinite series of ascending and descending powers of t and is known in the mathematical literature as a *Laurent series* (see Chap. 1, Sec. 8). In order to obtain the coefficients A_n, we take the product of the following two expansions:

$$e^{xt/2} = 1 + \frac{x}{2}\frac{t}{1!} + \left(\frac{x}{2}\right)^2\frac{t^2}{2!} + \left(\frac{x}{2}\right)^3\frac{t^3}{3!} + \cdots$$

$$e^{-xt^{-1}/2} = 1 - \frac{x}{2}\frac{t^{-1}}{1!} + \left(\frac{x}{2}\right)^2\frac{t^{-2}}{2!} - \left(\frac{x}{2}\right)^3\frac{t^{-3}}{3!} + \cdots \tag{13.3}$$

The constant term A_0 of the expansion (13.2) is

$$A_0 = 1 - \frac{(x/2)^2}{(1!)^2} + \frac{(x/2)^4}{(2!)^2} - \cdots = J_0(x) \tag{13.4}$$

The term A_n is

$$A_n = \frac{(x/2)^n}{n!}\left[1 - \frac{(x/2)^2}{1\times(n+1)} + \frac{(x/2)^4}{1\times 2(n+1)(n+2)} - \cdots\right] = J_n(x) \tag{13.5}$$

Similarly the coefficient A_{-n} is

$$A_{-n} = \frac{(-x/2)^n}{n!}\left[1 - \frac{(x/2)^2}{1 \times (n+1)} + \frac{(x/2)^4}{1 \times 2(n+1)(n+2)} - \cdots\right] = J_{-n}(x) \quad (13.6)$$

Accordingly we have

$$e^{x(t-t^{-1})/2} = \sum_{n=-\infty}^{+\infty} J_n(x)\, t^n$$

$$= J_0(x) + J_1(x)(t - t^{-1}) + J_2(x)(t^2 + t^{-2}) + \cdots \quad (13.7)$$

If we place $t = e^{j\theta}, j = (-1)^{\frac{1}{2}}$ in the result (13.7), we obtain the following series:

$$e^{jx\sin\theta} = J_0(x) + 2J_2(x)\cos 2\theta + 2J_4(x)\cos 4\theta + \cdots$$
$$+ 2j[J_1(x)\sin\theta + J_3(x)\sin 3\theta + \cdots] \quad (13.8)$$

Since $e^{jx\sin\theta} = \cos(x\sin\theta) + j\sin(x\sin\theta)$, if the real and imaginary parts of (13.8) are separated, the results are

$$\cos(x\sin\theta) = J_0(x) = 2J_2(x)\cos 2\theta + 2J_4(x)\cos 4\theta + \cdots \quad (13.9)$$

and

$$\sin(x\sin\theta) = 2[J_1(x)\sin\theta + J_3(x)\sin 3\theta + \cdots] \quad (13.10)$$

If in (13.9) and (13.10) θ is replaced by its complement, $\pi/2 - \theta$, the following equations are obtained:

$$\cos(x\cos\theta) = J_0(x) - 2J_2(x)\cos 2\theta + 2J_4(x)\cos 4\theta - \cdots \quad (13.11)$$

$$\sin(x\cos\theta) = 2[J_1(x)\cos\theta - J_3(x)\cos 3\theta + \cdots] \quad (13.12)$$

These series are usually called *Jacobi series* in the mathematical literature. Each of these series may be regarded as a Fourier series. If we multiply all terms of (13.9) by $\cos n\theta\, d\theta$ and all terms of (13.10) by $\sin n\theta\, d\theta$ and integrate between 0 and π, the following results are obtained as a consequence of the orthogonality relations of the sines and cosines:

$$\frac{1}{\pi}\int_0^\pi \cos(x\sin\theta)\cos n\theta\, d\theta = \begin{cases} J_n(x) & n \text{ even or zero} \\ 0 & n \text{ odd} \end{cases} \quad (13.13)$$

and

$$\frac{1}{\pi}\int_0^\pi \sin(x\sin\theta)\sin n\theta\, d\theta = \begin{cases} J_n(x) & n \text{ odd} \\ 0 & n \text{ even} \end{cases} \quad (13.14)$$

From this it can be deduced that

$$J_n(x) = \frac{1}{\pi}\int_0^\pi [\cos(x\sin\theta)\cos n\theta + \sin(x\sin\theta)\sin n\theta]\, d\theta \quad (13.15)$$

whether n is an odd or even integer. Therefore

$$J_n(x) = \frac{1}{\pi} \int_0^\pi \cos(n\theta - x\sin\theta)\, d\theta \qquad n \text{ any integer} \qquad (13.16)$$

These integrals are originally due to Bessel. The integral (13.16) occurred in an astronomy problem involving what is known as the *eccentric anomaly*.

14 LEGENDRE'S DIFFERENTIAL EQUATION

In the last sections we discussed the solutions of Bessel's differential equation or Bessel functions. Another differential equation that arises very frequently in various branches of applied mathematics is Legendre's differential equation. This equation occurs in the process of obtaining solutions of Laplace's equation in spherical coordinates and hence is of great importance in mathematical applications to physics and engineering. The following sections are devoted to the study of the solutions of Legendre's differential equation and to a discussion of their most important properties.

The differential equation

$$(1 - x^2)\frac{d^2 y}{dx^2} - 2x\frac{dy}{dx} + n(n + 1)y = 0 \qquad (14.1)$$

is known in the literature as *Legendre's differential equation of degree n*. We shall consider here only the important special case in which the parameter n is zero or a positive integer. As in the case of Bessel's differential equation let us assume an infinite-series solution of this differential equation in the form

$$y = x^m \sum_{r=0}^{r=\infty} a_r x^r = \sum_{r=0}^{r=\infty} a_r x^{m+r} \qquad (14.2)$$

For (14.2) to be a solution of (14.1), it is necessary that when (14.2) is substituted into (14.1) the coefficient of every power of x vanish. Equating the coefficient of the power x^{m+r-2} to zero, we obtain

$$(m + r)(m + r - 1)a_r + (n - m - r + 2)(n + m + r - 1)a_{r-2} = 0 \qquad (14.3)$$

Since the leading coefficient in the series (14.2) is a_0, we have

$$a_{-1} = 0 \qquad a_{-2} = 0 \qquad (14.4)$$

in (14.3).

With this stipulation, placing $r = 0$ in (14.3), we have

$$m(m - 1)a_0 = 0 \qquad (14.5)$$

Placing $r = 1$ in (14.3), we obtain

$$(m + 1)ma_1 = 0 \qquad (14.6)$$

Equation (14.5) gives $m = 0$ or $m = 1$, with a_0 arbitrary in any case. Let us take $m = 0$; then a_1 is arbitrary. Placing this value of m in (14.3), we have

$$a_r = -\frac{(n - r + 2)(n + r - 1)}{r(r - 1)} a_{r-2} \tag{14.7}$$

This enables us to determine any coefficient from the one which precedes it by two terms. We therefore have

$$y = a_0\left[1 - \frac{n(n + 1)}{2!}x^2 + \frac{n(n - 2)(n + 1)(n + 3)}{4!}x^4 - \cdots\right]$$

$$+ a_1\left[x - \frac{(n - 1)(n + 2)}{3!}x^3\right.$$

$$\left. + \frac{(n - 1)(n - 3)(n + 2)(n + 4)}{5!}x^5 - \cdots\right] \tag{14.8}$$

It may be shown by the ratio test that each of these series converges in the interval $(-1, +1)$. Had we taken the possibility $m = -1$ in (14.6), we would not have obtained anything new but only the second series in (14.8).

Since a_0 and a_1 are *arbitrary*, this is the general solution of Legendre's equation. We notice that the first series reduces to a polynomial when n is an even integer and the second series reduces to a polynomial when n is an odd integer. Now if we give the arbitrary coefficients a_0 or a_1, as the case may be, such a numerical value that the polynomial becomes equal to unity when x is unity, we obtain the following system of polynomials:

$$\begin{array}{ll} P_0(x) = 1 & P_4(x) = \tfrac{1}{8}(35x^4 - 30x^2 + 3) \\ P_1(x) = x & P_5(x) = \tfrac{1}{8}(63x^5 - 70x^3 + 15x) \\ P_2(x) = \tfrac{1}{2}(3x^2 - 1) & P_6(x) = \tfrac{1}{16}(231x^6 - 315x^4 + 105x^2 - 5) \\ P_3(x) = \tfrac{1}{2}(5x^3 - 3x) & \end{array} \tag{14.9}$$

These are called *Legendre polynomials*. Each satisfies a Legendre differential equation in which n has the value indicated by the subscript.

The general polynomial $P_n(x)$ is given by the series

$$P_n(x) = \sum_{r=0}^{N} (-1)^r \frac{(2n - 2r)!}{2^n r!(n - r)!(n - 2r)!} x^{n-2r} \tag{14.10}$$

where $N = n/2$ for n even and $N = (n - 1)/2$ for n odd.

It is thus seen that the Legendre polynomial $P_n(x)$ is even or odd according as its degree n is even or odd. Since

$$P_n(1) = 1 \tag{14.11}$$

we therefore conclude that

$$P_n(-1) = (-1)^n \tag{14.12}$$

15 RODRIGUES' FORMULA FOR THE LEGENDRE POLYNOMIALS

An important formula for $P_n(x)$ may be deduced directly from Legendre's differential equation.   Let

$$v = (x^2 - 1)^n \tag{15.1}$$

Then

$$\frac{dv}{dx} = 2nx(x^2 - 1)^{n-1} \tag{15.2}$$

Hence

$$(1 - x^2)\frac{dv}{dx} + 2nxv = 0 \tag{15.3}$$

If we differentiate (15.3) with respect to x, we obtain

$$(1 - x^2)\frac{d^2 v}{dx^2} + 2(n - 1)x\frac{dv}{dx} + 2nv = 0 \tag{15.4}$$

If we now differentiate this equation r times in succession, we have

$$(1 - x^2)\frac{d^2 v_r}{dx^2} + 2(n - r - 1)x\frac{dv_r}{dx} + (r + 1)(2n - r)v_r = 0 \tag{15.5}$$

where

$$v_r = \frac{d^r v}{dx^r} \tag{15.6}$$

In particular, if $r = n$, (15.5) reduces to

$$(1 - x^2)\frac{d^2 v_n}{dx^2} - 2x\frac{dv_n}{dx} + (n + 1)nv_n = 0 \tag{15.7}$$

This is Legendre's equation (14.1).   Hence v_n satisfies Legendre's equation. But since v_n is

$$v_n = \frac{d^n v}{dx^n} = \frac{d^n}{dx^n}(x^2 - 1)^n \tag{15.8}$$

v_n is a polynomial of degree n, and since Legendre's equation has one and only one distinct solution of that form, $P_n(x)$, it follows that $P_n(x)$ is a constant multiple of v_n.   Hence we have

$$P_n(x) = C\frac{d^n}{dx^n}(x^2 - 1)^n \tag{15.9}$$

To determine the constant C we merely consider the highest power of x on each side of the equation, that is,

$$\frac{(2n)!}{2^n(n!)^2}x^n = C\frac{d^n}{dx^n}x^{2n}$$

$$= C\frac{(2n)!}{n!}x^n \tag{15.10}$$

Hence

$$C = \frac{1}{2^n n!} \tag{15.11}$$

Substituting this value of C into (15.9), we obtain

$$P_n(x) = \frac{1}{2^n n!} \frac{d^n}{dx^n} (x^2 - 1)^n \tag{15.12}$$

This is *Rodrigues' formula* for the Legendre polynomials.

16   LEGENDRE'S FUNCTION OF THE SECOND KIND

The *general* solution of Legendre's equation may be written in the form

$$y = AP_n(x) + BQ_n(x) \tag{16.1}$$

where A and B are arbitrary constants and $Q_n(x)$ is called *Legendre's function of the second kind*.   This function is obtained by methods that are beyond the scope of this discussion.   It is defined by the following series when $|x| < 1$:

$$Q_n(x) = b_1 \left[x - \frac{(n-1)(n+2)}{3!} x^3 \right.$$
$$\left. + \frac{(n-1)(n-3)(n+2)(n+4)}{5!} x^5 - \cdots \right] \tag{16.2}$$

if n is even and

$$Q_n(x) = b_0 \left[1 - \frac{n(n+1)}{2!} x^2 + \frac{n(n-2)(n+1)(n+3)}{4!} x^4 - \cdots \right] \tag{16.3}$$

if n is odd, where

$$b_1 = (-1)^{n/2} \frac{2^n [(n/2)!]^2}{n!}$$
$$b_0 = (-1)^{(n+1)/2} \frac{2^{n-1} \{[(n-1)/2]!\}^2}{n!} \tag{16.4}$$

If, however, $|x| > 1$, the above series do not converge.   In this case the following series in descending powers of x is taken as the definition of $Q_n(x)$:

$$Q_n(x) = \sum_{r=0}^{r=\infty} \frac{2^n (n+r)!(n+2r)!}{r!(2n+2r+1)!} x^{-n-2r-1} \tag{16.5}$$

Both $P_n(x)$ and $Q_n(x)$ are special cases of a function known as the *hypergeometric function*.   The function $P_n(x)$ is the more important and occurs more frequently in the literature of applied mathematics.

17 THE GENERATING FUNCTION FOR $P_n(x)$

The Legendre polynomial $P_n(x)$ is the coefficient of Z^n in the expansion of

$$
\begin{aligned}
\phi &= (1 - 2xZ + Z^2)^{-\frac{1}{2}} \\
&= (1 + Z^2 - 2xZ)^{-\frac{1}{2}}
\end{aligned}
\tag{17.1}
$$

in ascending powers of Z. This may be verified for the lower powers of n by expanding (17.1) by the binomial theorem. To prove it for the general term, we write

$$
\phi = \sum_{n=0}^{n=\infty} A_n Z^n
\tag{17.2}
$$

Now it is obvious from the nature of the binomial expansion that A_n is a polynomial in x of degree n. Also, if we place $x = 1$ in (17.1), we obtain

$$
\phi = (1 - 2Z + Z^2)^{-\frac{1}{2}} = \frac{1}{1 - Z}
$$
$$
= 1 + Z + Z^2 + Z^3 + \cdots + Z^n + \cdots
\tag{17.3}
$$

Hence A_n is equal to 1, when $x = 1$. Now, if we can show that A_n satisfies Legendre's equation, it will be identical with $P_n(x)$ as the A_n's are the only polynomials of degree n that satisfy the equation and have the value 1 when $x = 1$. From (17.1) we obtain by differentiation

$$
(1 - 2Zx + Z^2)\frac{\partial \phi}{\partial Z} = (x - Z)\phi
\tag{17.4}
$$

and

$$
Z\frac{\partial \phi}{\partial Z} = (x - Z)\frac{\partial \phi}{\partial x}
\tag{17.5}
$$

If we now substitute from (17.2) into (17.4) and equate the coefficients of Z^{n-1} on both sides of the equation, we obtain

$$
nA_n - (2n - 1)xA_{n-1} + (n - 1)A_{n-2} = 0
\tag{17.6}
$$

Substituting into (17.5) from (17.2) and equating the coefficients of the power Z^{n-1} on both sides, we obtain

$$
x\frac{dA_{n-1}}{dx} - \frac{dA_{n-2}}{dx} = (n - 1)A_{n-1}
\tag{17.7}
$$

If in (17.7) we replace n by $n + 1$, we obtain

$$
x\frac{dA_n}{dx} - \frac{dA_{n-1}}{dx} = nA_n
\tag{17.8}
$$

Now if we differentiate (17.6) with respect to x and eliminate dA_{n-2}/dx by (17.7), we have

$$\frac{dA_n}{dx} - x\frac{dA_{n-1}}{dx} = nA_{n-1} \tag{17.9}$$

We now multiply (17.8) by $-x$ and add it to (17.9) and obtain

$$(1 - x^2)\frac{dA_n}{dx} = n(A_{n-1} - xA_n) \tag{17.10}$$

Differentiating (17.10) with respect to x and simplifying the result by means of (17.8), we finally obtain

$$(1 - x^2)\frac{d^2 A_n}{dx^2} - 2x\frac{dA_n}{dx} + n(n + 1)A_n = 0 \tag{17.11}$$

This shows that A_n is a solution of Legendre's equation. Hence, for the reasons stated above, it is the same as $P_n(x)$. We therefore have

$$A_n = P_n(x) \tag{17.12}$$

The above formulas for the A_n's are therefore valid for $P_n(x)$ and give important relations connecting Legendre polynomials of different orders. From (17.1) and (17.2) we have the important relation

$$\phi = \frac{1}{\sqrt{1 - 2xZ + Z^2}} = \sum_{n=0}^{n=\infty} P_n(x)Z^n \tag{17.13}$$

This equation is valid in the ranges

$$-1 \leqslant x \leqslant 1 \quad \text{and} \quad |Z| < 1 \tag{17.14}$$

because of the region of convergence of the binomial expansion (17.2). The function ϕ is called the *generating function* for $P_n(x)$. This result is of great importance in potential theory.

18   THE LEGENDRE COEFFICIENTS

If we let

$$x = \cos\theta = \frac{e^{j\theta} + e^{-j\theta}}{2} \quad j = \sqrt{-1} \tag{18.1}$$

and substitute this into (17.13), we have

$$[1 - Z(e^{j\theta} + e^{-j\theta}) + Z^2]^{-\frac{1}{2}} = \sum_{n=0}^{n=\infty} P_n(\cos\theta)Z^n \tag{18.2}$$

Now we have

$$[1 - Z(e^{j\theta} + e^{-j\theta}) + Z^2]^{-\frac{1}{2}} = (1 - Ze^{j\theta})^{-\frac{1}{2}}(1 - Ze^{-j\theta})^{-\frac{1}{2}} \tag{18.3}$$

By the binomial theorem we obtain

$$(1 - Z e^{j\theta})^{-\frac{1}{2}} = 1 + \frac{Z e^{j\theta}}{2} + \frac{1 \times 3}{2 \times 4} Z^2 e^{2j\theta} + \cdots$$

$$+ \frac{1 \times 3 \times \cdots \times (2n - 1)}{2 \times 4 \times \cdots \times 2n} Z^n e^{jn\theta} + \cdots \quad (18.4)$$

and

$$(1 - Z e^{-j\theta})^{-\frac{1}{2}} = 1 + \frac{Z e^{-j\theta}}{2} + \frac{1 \times 3}{2 \times 4} Z^2 e^{-2j\theta} + \cdots$$

$$+ \frac{1 \times 3 \times \cdots \times (2n - 1) Z^n e^{-jn\theta}}{2 \times 4 \times \cdots \times 2n} + \cdots \quad (18.5)$$

Multiplying (18.4) and (18.5) and picking out the coefficient of Z^n, we have

$$P_n(\cos \theta) = \frac{1 \times 3 \times \cdots \times (2n - 1)}{2 \times 4 \times \cdots \times 2n} 2 \cos n\theta$$

$$+ \frac{1 \times 1 \times 3 \times \cdots \times (2n - 3)}{2 \times 2 \times 4 \times \cdots \times (2n - 2)} 2 \cos (n - 2) \theta + \cdots \quad (18.6)$$

Every coefficient is positive so that P_n is numerically greatest when each cosine is equal to unity, that is, when $\theta = 0$. But since

$$P_n (\cos 0) = P_n(1) = 1 \quad (18.7)$$

it follows that

$$|P_n (\cos \theta)| \leqslant 1 \qquad n = 0, 1, 2, \ldots \quad (18.8)$$

The first few functions $P_n (\cos \theta)$ are

$$P_0(\cos \theta) = 1$$
$$P_1(\cos \theta) = \cos \theta$$
$$P_2(\cos \theta) = \tfrac{1}{4}(3 \cos 2\theta + 1) \quad (18.9)$$
$$P_3(\cos \theta) = \tfrac{1}{8}(5 \cos 3\theta + 3 \cos \theta)$$
$$P_4(\cos \theta) = \tfrac{1}{64}(35 \cos 4\theta + 20 \cos 2\theta + 9)$$

19 THE ORTHOGONALITY OF $P_n(x)$

Like the trigonometric functions $\cos mx$ and $\sin mx$, the Legendre polynomials $P_n(x)$ are orthogonal functions. Because of this property it is possible to expand an arbitrary function in a series of Legendre polynomials.

We shall now establish the orthogonality property

$$\int_{-1}^{+1} P_m(x) P_n(x) \, dx = 0 \qquad \text{if } m \neq n \quad (19.1)$$

To do this, recall that $P_n(x)$ satisfies the Legendre differential equation (14.1). This equation may be written in the form

$$\frac{d}{dx}[(1-x^2)P_n'(x)] + n(n+1)P_n(x) = 0 \tag{19.2}$$

If we now multiply this by $P_m(x)$ and integrate between the limits -1 and $+1$, we obtain

$$\int_{-1}^{+1} P_m(x)\frac{d}{dx}[(1-x^2)P_n'(x)]\,dx$$
$$+ n(n+1)\int_{-1}^{+1} P_m(x)P_n(x)\,dx = 0 \tag{19.3}$$

Now we may integrate the first term by parts in the form

$$\int_{-1}^{+1} P_m(x)\frac{d}{dx}[(1-x^2)P_n'(x)]\,dx = \{P_m(x)[(1-x^2)P_n'(x)]\}_{-1}^{+1}$$
$$- \int_{-1}^{+1}(1-x^2)P_n'(x)P_m'(x)\,dx \tag{19.4}$$

The first term of (19.4) vanishes at both limits because of the factor $1-x^2$; hence (19.3) reduces to

$$-\int_{-1}^{+1}(1-x^2)P_n'(x)P_m'(x)\,dx + n(n+1)\int_{-1}^{+1}P_m(x)P_n(x)\,dx = 0 \tag{19.5}$$

If in (19.5) we interchange n and m, we obtain

$$-\int_{-1}^{+1}(1-x^2)P_m'(x)P_n'(x)\,dx + m(m+1)\int_{-1}^{+1}P_n(x)P_m(x)\,dx = 0 \tag{19.6}$$

Subtracting (19.6) from (19.5), we get

$$(n-m)(n+m+1)\int_{-1}^{+1}P_m(x)P_n(x)\,dx = 0 \tag{19.7}$$

This establishes (19.1).

If $n = m$, Eq. (19.1) fails to hold. We shall now show that

$$\int_{-1}^{+1}[P_n(x)]^2\,dx = \frac{2}{2n+1} \qquad n = 0, 1, 2, \ldots \tag{19.8}$$

To do this, we square both sides of (17.13) and obtain

$$(1 - 2xZ + Z^2)^{-1} = \left[\sum_{n=0}^{\infty} P_n(x)Z^n\right]^2 \tag{19.9}$$

We now integrate both sides of this equation with respect to x over the interval $(-1,1)$ and observe that the product terms on the right vanish in view of the orthogonality property (19.1). We thus obtain

$$\int_{-1}^{+1}\frac{dx}{1-2xZ+Z^2} = \sum_{n=0}^{n=\infty} Z^{2n}\int_{-1}^{+1}[P_n(x)]^2\,dx \tag{19.10}$$

if $|Z| < 1$.   But the integral on the left has the value

$$\int_{-1}^{+1} \frac{dx}{1 - 2xZ + Z^2} = \frac{1}{Z} \ln \frac{1 + Z}{1 - Z}$$

$$= 2\left(1 + \frac{Z^2}{3} + \frac{Z^4}{5} + \cdots + \frac{Z^{2n}}{2n + 1} + \cdots\right)$$

$$= \sum_{n=0}^{\infty} Z^{2n} \int_{-1}^{+1} [P_n(x)]^2 \, dx \qquad |Z| < 1 \tag{19.11}$$

Equating the coefficient of the power Z^{2n} on both sides of (19.11), we have

$$\int_{-1}^{+1} [P_n(x)]^2 \, dx = \frac{2}{2n + 1} \qquad n = 0, 1, 2, \ldots \tag{19.12}$$

20   EXPANSION OF AN ARBITRARY FUNCTION IN A SERIES OF LEGENDRE POLYNOMIALS

If $F(x)$ is sectionally continuous in the interval $(-1,1)$ and if its derivative $F(x)$ is sectionally continuous in every interval interior to $(-1,1)$, it may be shown that $F(x)$ may be expanded in a series of the form

$$F(x) = \sum_{n=0}^{n=\infty} a_n P_n(x) \tag{20.1}$$

To obtain the general coefficient a_m, we multiply both sides of (20.1) by $P_m(x)$ and integrate over the interval $(-1,1)$.   We then obtain

$$\int_{-1}^{1} F(x) P_m(x) \, dx = a_m \int_{-1}^{1} [P_m(x)]^2 \, dx$$

$$= \frac{2a_m}{2m + 1} \tag{20.2}$$

in view of (19.1) and (19.8).   The general coefficient of the expansion (20.1) is given by

$$a_n = \frac{2n + 1}{2} \int_{-1}^{+1} F(x) P_n(x) \, dx \tag{20.3}$$

The expansion (20.1) is similar to an expansion of an arbitrary function into a Fourier series.

21   ASSOCIATED LEGENDRE POLYNOMIALS

In the solution of some potential problems it is convenient to use certain polynomials closely related to the Legendre polynomials.   We shall discuss them briefly in this section.

If we differentiate Legendre's equation

$$(1 - x^2)\frac{d^2 y}{dx^2} - 2x\frac{dy}{dx} + n(n + 1)y = 0 \tag{21.1}$$

m times with respect to x and write

$$v = \frac{d^m y}{dx^m} \tag{21.2}$$

we obtain

$$(1 - x^2)\frac{d^2 v}{dx^2} - 2x(m + 1)\frac{dv}{dx} + (n - m)(n + m + 1)v = 0 \tag{21.3}$$

Since P_n is a solution of Legendre's equation (21.1), this equation is satisfied by

$$v = \frac{d^m}{dx^m}P_n(x) \tag{21.4}$$

If now in (21.3) we let

$$w = v(1 - x^2)^{m/2} \tag{21.5}$$

we obtain

$$(1 - x^2)\frac{d^2 w}{dx^2} - 2x\frac{dw}{dx} + \left[n(n + 1) - \frac{m^2}{1 - x^2}\right]w = 0 \tag{21.6}$$

This equation differs from Legendre's equation in an added term involving m. It is called the *associated Legendre equation*. By Eq. (21.5) we see that it is satisfied by

$$w = (1 - x^2)^{m/2}\frac{d^m}{dx^m}P_n(x) \tag{21.7}$$

This value of w is the *associated Legendre polynomial*, and it is denoted by $P_n^m(x)$. We therefore have

$$P_n^m(x) = (1 - x^2)^{m/2}\frac{d^m}{dx^m}P_n(x) \tag{21.8}$$

We notice that, if $m > n$, we have

$$P_n^m(x) = 0 \tag{21.9}$$

22 THE GAMMA FUNCTION

The gamma function $\Gamma(n)$ has been defined by Euler to be the definite integral

$$\Gamma(n) = \int_0^\infty x^{n-1} e^{-x}\, dx \qquad n > 0 \tag{22.1}$$

This definite integral converges when n is positive and therefore defines a function of n for positive values of n. By direct integration it is evident that

$$\Gamma(1) = \int_0^\infty e^{-x} dx = 1 \tag{22.2}$$

By an integration by parts the following identity may be established:

$$\Gamma(n+1) = \int_0^\infty x^n e^{-x} dx = n \int_0^\infty x^{n-1} e^{-x} dx + (-x^n e^{-x})\Big|_0^\infty$$
$$= n \int_0^\infty x^{n-1} e^{-x} dx \tag{22.3}$$

Comparing the result with (22.1), we have

$$\Gamma(n+1) = n\Gamma(n) \tag{22.4}$$

This is the fundamental recursion relation satisfied by the gamma function. From this relation it is evident that if the value of $\Gamma(n)$ is known for n between any two successive positive integers, the value of $\Gamma(n)$ for any positive value of n may be found by successive applications of (22.4). Equation (22.4) may be used to define $\Gamma(n)$ for values of n for which the definition (22.1) fails. We may write (22.4) in the form

$$\Gamma(n) = \frac{\Gamma(n+1)}{n} \tag{22.5}$$

Then if

$$-1 < n < 0 \tag{22.6}$$

formula (22.5) gives us $\Gamma(n)$ since $n+1$ is positive. We may then find $\Gamma(n)$ where $-2 < n < -1$ since now $n+1$ on the right-hand side of (22.5) is known, and so on indefinitely. We then have in (22.1) and (22.5) the complete definition of $\Gamma(n)$ for all values of n except $n = 0, -1, -2, \ldots$.

23   THE FACTORIAL; GAUSS'S PI FUNCTION

From Eq. (22.2) we have

$$\Gamma(1) = 1 \tag{23.1}$$

Now, by the use of (22.4), we obtain

$$\Gamma(2) = 1 \times \Gamma(1) = 1$$
$$\Gamma(3) = 2 \times \Gamma(2) = 2 \times 1$$
$$\Gamma(4) = 3 \times \Gamma(3) = 3 \times 2 \times 1 \tag{23.2}$$
$$\cdots \cdots \cdots \cdots \cdots$$
$$\Gamma(n+1) = n!$$

provided n is a positive integer. From this it is convenient to define 0! in the form

$$0! = \Gamma(1) = 1 \tag{23.3}$$

Gauss's pi function is defined in terms of the gamma function by the equation

$$\Pi(n) = \Gamma(n+1) \tag{23.4}$$

We thus see that if n is a positive integer

$$\Pi(n) = n! \tag{23.5}$$

If we place $n = 0$ in Eq. (22.5), we have

$$\Gamma(0) = \frac{\Gamma(1)}{0} = \frac{1}{0} = \infty \tag{23.6}$$

By repeated application of (22.5) it is seen that the gamma function becomes infinite when n is zero or a negative integer.

24   THE VALUE OF $\Gamma(\frac{1}{2})$; GRAPH OF THE GAMMA FUNCTION

If in the fundamental integral (22.1) we make the substitution

$$x = y^2 \tag{24.1}$$

we obtain

$$\Gamma(n) = 2 \int_0^\infty y^{2n-1} e^{-y^2} \, dy \tag{24.2}$$

If now $n = \frac{1}{2}$, we have

$$\Gamma(\tfrac{1}{2}) = 2 \int_0^\infty e^{-y^2} \, dy \tag{24.3}$$

By making use of Appendix F, Eq. (11.22), we obtain

$$\Gamma(\tfrac{1}{2}) = 2 \frac{\sqrt{\pi}}{2} = \sqrt{\pi} \tag{24.4}$$

From this result and (22.5) we obtain

$$\Gamma(-\tfrac{1}{2}) = \frac{\Gamma(\tfrac{1}{2})}{-\tfrac{1}{2}} = -2\sqrt{\pi} \tag{24.5}$$

$$\Gamma(-\tfrac{3}{2}) = \frac{\Gamma(-\tfrac{1}{2})}{-\tfrac{3}{2}} = -\tfrac{2}{3}(-2\sqrt{\pi}) = \frac{4\sqrt{\pi}}{3} \tag{24.6}$$

etc.   Figure 24.1 represents the graph of $\Gamma(n)$.

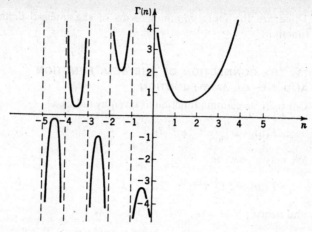

Fig. 24.1

25 THE BETA FUNCTION

The beta function $\beta(m,n)$ is defined by the definite integral

$$\beta(m,n) = \int_0^1 x^{m-1}(1-x)^{n-1}\,dx \qquad \begin{cases} m > 0 \\ n > 0 \end{cases} \tag{25.1}$$

This integral converges and thus defines a function of m and n provided that m and n are positive.

If we let

$$x = 1 - y \tag{25.2}$$

in (25.1), we obtain

$$\beta(m,n) = \int_0^1 (1-y)^{m-1}\,y^{n-1}\,dy = \beta(n,m) \tag{25.3}$$

OTHER FORMS OF THE BETA FUNCTION

If in (25.1) we let $x = \sin^2\phi$, we obtain

$$\beta(m,n) = 2\int_0^{\pi/2} (\sin\phi)^{2m-1}(\cos\phi)^{2n-1}\,d\phi \tag{25.4}$$

The substitution $x = y/a$ in (25.1) gives

$$\beta(m,n) = \frac{1}{a^{m+n-1}} \int_0^a y^{m-1}(1-y)^{n-1}\,dy \tag{25.5}$$

If $x = y/(1+y)$ in (25.1), we obtain

$$\beta(m,n) = \int_0^\infty \frac{y^{n-1}\,dy}{(1+y)^{m+n}} \tag{25.6}$$

These are the more common forms of the integral definition of the beta function.

26 THE CONNECTION OF THE BETA FUNCTION AND THE GAMMA FUNCTION

Consider the gamma function as given by (24.2),

$$\Gamma(n) = 2 \int_0^\infty y^{2n-1} e^{-y^2} \, dy \tag{26.1}$$

We may also write

$$\Gamma(m) = 2 \int_0^\infty x^{2m-1} e^{-x^2} \, dx \tag{26.2}$$

and hence

$$\Gamma(m)\,\Gamma(n) = 4 \left(\int_0^\infty x^{2m-1} e^{-x^2} \, dx \right) \left(\int_0^\infty y^{2n-1} e^{-y^2} \, dy \right)$$
$$= 4 \int_0^\infty \int_0^\infty x^{2m-1} y^{2n-1} e^{-(x^2+y^2)} \, dx \, dy \tag{26.3}$$

If we now consider this integral as a surface integral in the first quadrant of the xy plane and introduce the polar coordinates

$$x = r \cos \theta$$
$$y = r \sin \theta \tag{26.4}$$

and introduce the surface element ds in the form

$$ds = r \, dr \, d\theta \tag{26.5}$$

then (26.3) becomes

$$\Gamma(m)\,\Gamma(n) = 4 \int_0^{\pi/2} \int_0^\infty r^{(2m+n-1)} (\cos \theta)^{2m-1} (\sin \theta)^{2n-1} e^{-r^2} r \, d\theta \, dr$$
$$= 4 \int_0^{\pi/2} (\cos \theta)^{2m-1} (\sin \theta)^{2n-1} \, d\theta \int_0^\infty r^{2(m+n)-1} e^{-r^2} \, dr \tag{26.6}$$

Now from (25.4) we have

$$\beta(n,m) = 2 \int_0^{\pi/2} (\cos \theta)^{2m-1} (\sin \theta)^{2n-1} \, d\theta$$
$$= \beta(m,n) \tag{26.7}$$

and from (26.2) we have

$$\Gamma(m+n) = 2 \int_0^\infty r^{2(m+n)-1} e^{-r^2} \, dr \tag{26.8}$$

Hence (26.6) may be written in the form

$$\Gamma(m)\,\Gamma(n) = \beta(m,n)\,\Gamma(m+n) \tag{26.9}$$

or

$$\beta(m,n) = \frac{\Gamma(m)\,\Gamma(n)}{\Gamma(m+n)} \tag{26.10}$$

This formula is very useful for the evaluation of certain classes of definite integrals. For example, from (26.7) and (26.10) we obtain

$$\int_0^{\pi/2} (\cos\theta)^{2m-1}(\sin\theta)^{2n-1}\,d\theta = \frac{\Gamma(m)\,\Gamma(n)}{2\Gamma(m+n)} \qquad \begin{cases} m>0 \\ n>0 \end{cases} \tag{26.11}$$

If in (26.11) we let

$$2m - 1 = r \quad \text{or} \quad m = \frac{r+1}{2}$$
$$2n - 1 = 0 \quad \text{or} \quad n = \tfrac{1}{2} \tag{26.12}$$

we obtain

$$\int_0^{\pi/2} (\cos\theta)^r\,d\theta = \frac{\Gamma(r+1/2)}{\Gamma(r/2+1)}\frac{\sqrt{\pi}}{2} \qquad r > -1 \tag{26.13}$$

In a similar manner we obtain

$$\int_0^{\pi/2} (\sin\theta)^r\,d\theta = \frac{\Gamma(r+1/2)}{\Gamma(r/2+1)}\frac{\sqrt{\pi}}{2} \qquad r > -1 \tag{26.14}$$

In a similar manner many other integrals may be evaluated in terms of the gamma functions. If a table of gamma functions is available, then the computation of these integrals is considerably simplified.

27 AN IMPORTANT RELATION INVOLVING GAMMA FUNCTIONS

Substituting (25.6) into the relation (26.10), we obtain

$$\int_0^\infty \frac{y^{n-1}\,dy}{(1+y)^{m+n}} = \frac{\Gamma(m)\Gamma(n)}{\Gamma(m+n)} \qquad \begin{cases} m>0 \\ n>0 \end{cases} \tag{27.1}$$

If we now let

$$m = 1 - n \qquad 0 < n < 1 \tag{27.2}$$

in (27.1), we obtain

$$\int_0^\infty \frac{y^{n-1}\,dy}{1+y} = \frac{\Gamma(1-n)\,\Gamma(n)}{\Gamma(1)} \tag{27.3}$$

Now in Chap. 1 it is shown that

$$\int_0^\infty \frac{y^{n-1}\,dy}{1+y} = \frac{\pi}{\sin n\pi} \qquad 0 < n < 1 \tag{27.4}$$

Hence, since

$$\Gamma(1) = 1 \tag{27.5}$$

we have from (27.3) the important relation

$$\Gamma(n)\,\Gamma(1-n) = \frac{\pi}{\sin n\pi} \qquad 0 < n < 1 \tag{27.6}$$

28 THE ERROR FUNCTION OR PROBABILITY INTEGRAL

Another very important function that occurs frequently in various branches of applied mathematics is the *error function*, erf(x), or the probability integral defined by

$$\text{erf}(x) = \frac{2}{\sqrt{\pi}} \int_0^x e^{-n^2}\, dn \tag{28.1}$$

This integral occupies a central position in the theory of probability and arises in the solution of certain partial differential equations of physical interest.

From the definition of erf(x) we have

$$\text{erf}(-x) = -\text{erf}(x) \tag{28.2}$$

$$\text{erf}(0) = 0 \tag{28.3}$$

$$\text{erf}(\infty) = \frac{1}{\sqrt{\pi}} \int_0^\infty e^{-n^2}\, dn = \frac{2}{\sqrt{\pi}} \frac{\sqrt{\pi}}{2} = 1 \tag{28.4}$$

$$\text{erf}(jy) = \frac{2j}{\sqrt{\pi}} \int_0^y e^{n^2}\, dn \qquad j = \sqrt{-1} \tag{28.5}$$

PROBLEMS

1. Prove that

$$\frac{d}{dx}[x^n J_n(ax)] = ax^n J_{n-1}(ax)$$

2. Show that the differential equation

$$\frac{x\,d^2 y}{dx^2} + (1+n)\frac{dy}{dx} + y = 0$$

is satisfied by $y = x^{-n/2} J_n(2x^{\frac{1}{2}})$.

3. Show that

$$J_{\frac{5}{2}}(x) = \left(\frac{2}{\pi x}\right)^{\frac{1}{2}} \left[(3 - x^2)\frac{\sin x}{x^2} - \frac{3\cos x}{x}\right]$$

4. Show that

$$J_n(x+y) = \sum_{k=-\infty}^{k=+\infty} J_k(x) J_{n-k}(y)$$

for integral values of n.

5. By multiplying the expansions for $e^{x(t-t^{-1})/2}$ and $e^{-x(t-t^{-1})/2}$ show that

$$[J_0(x)]^2 + 2[J_1(x)]^2 + 2[J_2(x)]^2 + \cdots = 1$$

6. Show that

$$\int_0^\infty e^{-pt} J_0(at)\, dt = (p^2 + a^2)^{-\frac{1}{2}} \qquad p > 0$$

7. Obtain the recurrence formulas of Sec. 6 by differentiating the relation

$$e^{x(t-t^{-1})/2} = \sum_{n=-\infty}^{n=-\infty} J_n(x)\, t^n$$

with respect to x, or to t, and comparing the coefficients of corresponding powers of t.

8. By modifying the variables in the above relation prove that

$$e^{x(t+t^{-1})/2} = \sum_{n=-\infty}^{+\infty} I_n(x)\, t^n$$

Replace t by t^{-1} and deduce that $I_n(x) = I_{-n}(x)$.

9. Prove that

   (a) $\cos(x\cosh\theta) = J_0(x) - 2J_2(x)\cosh 2\theta + 2J_4(x)\cosh 4\theta - \cdots$.

   (b) $\sin(x\cosh\theta) = 2[J_1(x)\cos\theta - J_3(x)\cosh 3\theta + \cdots]$.

10. Show that

   (a) $\sinh(x\cosh\theta) = 2[I_1(x)\cosh\theta + I_3(x)\cosh 3\theta + \cdots]$.

   (b) $\cosh(x\cosh\theta) = I_0(x) + 2[I_2(x)\cosh 2\theta + I_4(x)\cosh 4\theta + \cdots]$.

11. Show that $e^{x\cos\theta} = I_0(x) + 2I_1(x)\cos\theta + 2I_2(x)\cos\theta + \cdots$.

12. From the Jacobi series deduce that

$$1 = J_0(x) + 2J_2(x) + 2J_4(x) + \cdots$$

$$\cos x = J_0(x) - 2J_2(x) + 2J_4(x) - \cdots$$

13. A simple pendulum is performing small oscillations about the vertical while the length of the pendulum is increasing at a constant rate. Obtain the equation of motion of this pendulum, and express the solution in terms of Bessel functions.

14. A variable mass $m(t)$ is attached to a spring of constant spring constant k. One end of the spring is fixed, and the mass is performing oscillations on a smooth horizontal plane. Discuss the oscillations of the mass if its magnitude is $m = (a + bt)^{-1}$, where a and b are positive constants and t is the time.

15. A nonlinear electrical circuit has a potential $E\cos wt$ impressed upon it. The current produced by this potential can be expressed in the form $i = A e^{bE\cos wt}$, where A and b are positive constants. Prove that the mean value of the current is $AI_0(bE)$ and that the root-mean-square current is $A[I_0(2bE)]^{\frac{1}{2}}$. I_0 is the modified Bessel function of zeroth order.

16. Prove that

$$\int_0^\infty J_0(bx)\, dx = \frac{1}{b}$$

17. Find the complete solution of the differential equation $d^2y/dx^2 + x^{-\frac{1}{2}}y = 0$.

18. Show that $x/2 = J_1(x) + 3J_3(x) + 5J_5(x) + \cdots$.

19. Show that $x \sin x/2 = 2^2 J_2(x) - 4^2 J_4(x) + 6^2 J_6(x) - \cdots$.

20. Show that $x \cos x/2 = 1^2 J_1(x) - 3^2 J_3(x) + 5^2 J_5(x) - \cdots$.

21. Find the general solution of the equation $d^2 y/dx^2 + (1/x)dy/dx - k^2 y = a$, where k and a are constants.

22. Show that

$$\int_{-1}^{1} P_n(x)\, dx = 0 \qquad n = 1, 2, 3, \ldots$$

23. Establish the orthogonality property of the Legendre polynomials (19.1) by using Rodrigues' formula for $P_n(x)$ and successive integration by parts.

24. Show that

$$x^2 = \tfrac{2}{3} P_2(x) + \tfrac{1}{3} P_0(x)$$
$$x^3 = \tfrac{2}{5} P_3(x) + \tfrac{3}{5} P_1(x)$$

25. Show that

$$\int_{-1}^{1} x P_n(x) P_{n-1}(x)\, dx = \frac{2n}{4n^2 - 1}$$

26. Prove that

$$\frac{dP_{n+1}}{dx} - \frac{dP_{n-1}}{dx} = (2n + 1) P_n$$

27. Using Rodrigues' formula, integrate by parts to show that

$$\int_{-1}^{1} x^m P_n(x)\, dx = 0 \qquad \text{if } m < n$$

28. Show that, if $R_m(x)$ is a polynomial of degree m less than n, we have

$$\int_{-1}^{+1} P_n(x) R_m(x)\, dx = 0$$

29. Show that

$$\Gamma\left(\frac{2k+1}{2}\right) = \frac{1 \times 3 \times 5 \times \cdots \times (2k-1)}{2^k} \sqrt{\pi}$$

where k is a positive integer.

30. Show that

$$\int_{0}^{1} \frac{x^n\, dx}{\sqrt{1 - x^2}} = \frac{1 \times 3 \times 5 \times \cdots \times (n-1)}{2 \times 4 \times 6 \times \cdots \times n} \frac{\pi}{2}$$

if n is an even positive integer.

31. Show that

$$\int_{0}^{1} \frac{x^n\, dx}{\sqrt{1 - x^2}} = \frac{2 \times 4 \times 6 \times \cdots \times (n-1)}{1 \times 3 \times 5 \times \cdots \times n}$$

if n is an odd positive integer.

32. Show that

$$\int_{0}^{1} \frac{dx}{\sqrt{1 - x^n}} = \frac{\sqrt{\pi}}{n} \frac{\Gamma(1/n)}{\Gamma(1/n + \tfrac{1}{2})}$$

33. Evaluate the following definite integrals:

(a) $\int_0^\infty e^{-x^4}\,dx.$

(b) $\int_0^\infty 4x^4\,e^{-x^4}\,dx.$

(c) $\int_0^{\pi/2} \dfrac{\sqrt{\sin^8 x}}{\sqrt{\cos x}}\,dx.$

34. Show that

$$\int_0^{\pi/2} \sqrt{\tan\theta}\,d\theta = \frac{\Gamma(\tfrac14)\,\Gamma(\tfrac34)}{2}$$

35. Show that

$$\frac{d^n\,\Gamma(y)}{dy^n} = \int_0^\infty x^{y-1}\,e^{-x}\,(\log y)^n\,dx$$

36. Show that by a suitable change in variable we have

$$\Gamma(n) = \int_0^1 \left(\log\frac{1}{y}\right)^{n-1}\,dy$$

37. Evaluate the integral

$$\int_0^x e^{-n^2}\,dn$$

by expanding the integral in series, and show that

$$\int_0^x e^{-n^2}\,dn = x - \frac{x^3}{3\times1!} + \frac{x^5}{5\times2!} - \frac{x^7}{7\times3!} + \frac{x^9}{9\times4!} - R$$

where $R < x^{11}/1{,}320.$

38. Show by integrating by parts that

$$\int_0^x e^{-n^2}\,dn = \frac{\sqrt{\pi}}{2} - \frac{e^{-x^2}}{2x}\left(1 - \frac{1}{2x^2} + \frac{1\times3}{2^2x^4}\right) + \frac{1\times3\times5}{2^3}\int_x^\infty \frac{e^{-n^2}\,dn}{n^6}$$

Show how this expression may be used to compute the value of erf(x) for large values of x.

REFERENCES

1893. Byerly, W. E.: "Fourier's Series and Spherical Harmonics," Ginn and Company, Boston.

1911. Wilson, E. B.: "Advanced Calculus," chap. 14, Ginn and Company, Boston.

1922. Watson, G. N.: "Theory of Bessel Functions," Cambridge University Press, New York.

1926. Woods, F. S.: "Advanced Calculus," chap. 7, Ginn and Company, Boston.

1927. Jeans, J. H.: "The Mathematical Theory of Electricity and Magnetism," Cambridge University Press, New York.

1927. Whittaker, E. T., and G. N. Watson: "A Course in Modern Analysis," 4th ed., Cambridge University Press, New York.

1931. Gray, A., G. B. Mathews, and T. M. Macrobert: "A Treatise on Bessel Functions," The Macmillan Company, New York.
1934. McLachlan, N. W.: "Bessel Functions for Engineers," Oxford University Press, New York.
1939. Smythe, W. R.: "Static and Dynamic Electricity," McGraw-Hill Book Company, New York.
1941. Churchill, R. V.: "Fourier Series and Boundary Value Problems," McGraw-Hill Book Company, New York.
1946. Relton, F. E.: "Applied Bessel Functions," Blackie & Son, Ltd., Glasgow.
1953. Bickley, W. G.: "Bessel Functions and Formulae," Cambridge University Press, New York.

appendix **C**

Infinite Series, Fourier Series, and Fourier Integrals

1 INFINITE SERIES

The following sections will be devoted to the exposition of some of the properties of infinite series and particular attention will be given to power series. The subject of infinite series is of extreme importance in applied mathematics because it makes possible the numerical solution of many important physical problems. The solutions of certain differential equations that occur frequently in the mathematical solution of many physical problems are expressed in terms of infinite series, and a study of the properties of these solutions requires a knowledge of the manner in which infinite series may be manipulated. Hence it is essential that students of applied science acquire an intelligent understanding of the subject.

In this appendix some of the fundamental notions and concepts of infinite series will be discussed. The algebra and calculus of series will be developed, and some of the practical uses of series will be used as illustrations of the general principles.

2 DEFINITIONS

In this section we shall consider some fundamental definitions of the subject of infinite series.

SEQUENCE

A sequence is a succession of terms formed according to some fixed rule or law. For example,

$$1, 4, 9, 16, 25 \tag{2.1}$$

and

$$x, x^2, \frac{x^3}{1 \times 2}, \frac{x^4}{1 \times 2 \times 3} \tag{2.2}$$

are sequences.

SERIES

A series is the indicated sum of the terms of a sequence. That is, from the foregoing sequences we obtain the series

$$1 + 4 + 9 + 16 + 25 \tag{2.3}$$

and

$$x + x^2 + \frac{x^3}{1 \times 2} + \frac{x^4}{1 \times 2 \times 3} \tag{2.4}$$

If the number of terms is limited, the sequence or series is said to be finite. If the number of terms is unlimited, the sequence or series is said to be an infinite sequence or series.

The general term, or nth term, is the expression that indicates the law of formation of the terms of the series. For example, in the preceding illustrations suitable general terms are

$$n^2 \quad \text{and} \quad \frac{nx^n}{n!}$$

where $n!$ is the factorial number given by

$$n! = 1 \times 2 \times 3 \times \cdots \times (n-1) \times n \tag{2.5}$$

3 THE GEOMETRIC SERIES

Consider the series of n terms

$$S_n = a + ar + ar^2 + \cdots + ar^{n-1} \tag{3.1}$$

This series is called the geometric series. A simple expression for S_n may be obtained for the sum S_n of the geometric series in the following manner: Multiply (3.1) by r. We thus obtain

$$rS_n = ar + ar^2 + ar^3 + \cdots + ar^n \tag{3.2}$$

Let us now subtract (3.1) from (3.2). This gives

$$rS_n - S_n = ar^n - a \tag{3.3}$$

Hence

$$S_n = \frac{a(1 - r^n)}{1 - r} = \frac{a}{1 - r} - \frac{ar^n}{1 - r} \tag{3.4}$$

Now if $|r| < 1$, then r^n decreases in absolute value as n increases so that we have

$$\lim_{n \to \infty} r^n = 0 \tag{3.5}$$

From (3.4) we then have

$$\lim_{n \to \infty} S_n = \frac{a}{1 - r} = S \tag{3.6}$$

Hence, if $|r| < 1$, the sum S_n of a geometric series approaches a limit as the number of terms is increased indefinitely. In this case the infinite series $a + ar + ar^2 + \cdots$ is said to be *convergent*.

If $|r| > 1$, then r^n will become infinite as n increases indefinitely. Hence, from (3.4), $|S_n|$ will increase without limit. In this case the series is said to be *divergent*.

If $r = -1$, we encounter an interesting situation. In this case the geometric series becomes

$$a - a + a - a + \cdots \tag{3.7}$$

In this case, if n is even, S_n is zero. If n is odd, S_n is a. As n increases indefinitely, the absolute value of S_n does not increase indefinitely but still S_n does not approach a limit. A series of this sort is called an *oscillating series*. It also is a divergent series.

If we place $a = 1$ and $r = \frac{1}{2}$ in the general geometric series, we obtain

$$S_n = 1 + \frac{1}{2} + \frac{1}{4} + \cdots + \frac{1}{2^{n-1}} \tag{3.8}$$

and we have

$$\lim_{n \to \infty} S_n = \frac{1}{1 - \frac{1}{2}} = 2 \tag{3.9}$$

4 CONVERGENT AND DIVERGENT SERIES

Let us consider the series

$$S_n = u_1 + u_2 + u_3 + \cdots + u_n \tag{4.1}$$

The variable S_n denoting the sum of the series is a function of n. If we now allow the number of terms, n, to increase without limit, one of two things may happen.

Case I: S_n approaches a limit S indicated by

$$\lim_{n \to \infty} S_n = S \tag{4.2}$$

In this case the infinite series is said to be *convergent* and to converge to the value of S or to have the sum S.

Case II: In this case S_n *approaches no limit.* The infinite series is then said to be *divergent.* For example, the series

$$1 + 2 + 3 + 4 + 5 + \cdots$$
$$2 - 2 + 2 - 2 + \cdots$$

are said to be divergent.

In applied mathematics, convergent series are of the utmost importance; it is thus necessary to have a means of testing a series for convergence or divergence.

5 GENERAL THEOREMS

The following theorems, whose proofs are omitted, are of importance in the study of the convergence of series:

Theorem I If S_n is a variable which always increases as n increases but which never exceeds some definite fixed number A, then, as n increases without limit, S_n will approach a limit S which is not greater than A.

This statement may be illustrated by Fig. 5.1. The points determined by the values S_1, S_2, S_3, etc., approach the point S where

$$\lim_{n \to \infty} S_n = S \tag{5.1}$$

and S is less than or equal to A.

This theorem enables us to establish the convergence of certain series. For example, let us consider the series

$$1 + 1 + \frac{1}{1 \times 2} + \frac{1}{1 \times 2 \times 3} + \cdots + \frac{1}{n!} + \cdots \tag{5.2}$$

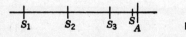

Fig. 5.1

If we temporarily neglect the first term, we may write

$$S_n = 1 + \frac{1}{1 \times 2} + \frac{1}{1 \times 2 \times 3} + \cdots + \frac{1}{n!} \tag{5.3}$$

Now let us consider the series defined by

$$U_n = 1 + \frac{1}{2} + \frac{1}{2 \times 2} + \cdots + \frac{1}{2^{n-1}} \tag{5.4}$$

Since the corresponding terms of the series S_n are less than the corresponding terms of the series U_n with the exception of the first two terms, it is obvious that

$$S_n < U_n \tag{5.5}$$

Now the series U_n is a geometric series with $a = 1$ and $r = \frac{1}{2}$. Hence, $U_n < 2$ no matter how large n may be.

It follows therefore that S_n is a variable that always increases as n increases but remains less than 2. Hence S_n approaches a limit as n becomes infinite, and this limit is less than 2. It is thus apparent that the series (5.2) is convergent and that its value is less than 3. It will be shown later that the sum of the infinite series (5.2) is the constant $e = 2.71828 \cdots$, the base of the natural-logarithm system.

Another fundamental theorem of great importance in testing for convergence will now be stated.

Theorem II If S_n is a variable which always decreases as n increases but which is never less than a certain number B, then, as n increases without limit, S_n will approach a limit which is not less than B.

Let us consider the convergent series

$$S = u_1 + u_2 + u_3 + \cdots + u_n + \cdots \tag{5.6}$$

for which

$$\lim_{n \to \infty} S_n = S \tag{5.7}$$

Now consider that the points determined by the values S_1, S_2, S_3, etc., are plotted on a directed line. Then these points as n increases will approach the point determined by S. It is thus evident that

$$\lim_{n \to \infty} u_n = 0 \tag{5.8}$$

That is, in a convergent series, the terms must approach zero as a limit. If, however, the nth term of a series does *not* approach zero as n becomes infinite, we know at once that the series is divergent.

Although (5.8) is a necessary condition for convergence, it is not *sufficient*. That is, even if the nth term does approach zero, we cannot state that the series is convergent.   Consider the series for which

$$S_n = 1 + \frac{1}{2} + \frac{1}{3} + \cdots + \frac{1}{n} \tag{5.9}$$

Here we have

$$\lim_{n\to\infty} u_n = \lim_{n\to\infty} \frac{1}{n} = 0 \tag{5.10}$$

Therefore (5.8) is fulfilled.   However, we shall prove later that this series is divergent.

Although the use of the preceding theorems in determining the convergence or divergence of series is fundamental, we shall now turn to the development of special tests that are, as a rule, easier to apply than these theorems.

6   THE COMPARISON TEST

In many cases the question of the convergence or divergence of a given series may be answered by comparing the given series with one whose character is known.

Let it be required to test the series

$$U = u_1 + u_2 + u_3 + \cdots + u_n + \cdots \tag{6.1}$$

where all the terms of (6.1) are positive.   Now if a convergent series of positive terms

$$V = v_1 + v_2 + v_3 + \cdots + v_n + \cdots \tag{6.2}$$

can be found whose terms are never less than the corresponding terms in the series (6.1), then (6.1) is a convergent series and its sum does not exceed that of (6.2).

To prove this statement, let

$$U_n = u_1 + u_2 + u_3 + \cdots + u_n \tag{6.3}$$

and

$$V_n = v_1 + v_2 + v_3 + \cdots + v_n \tag{6.4}$$

Since by hypothesis (6.2) is convergent, we have

$$\lim_{n\to\infty} V_n = V \tag{6.5}$$

Now since

$$V_n < V \quad \text{and} \quad U_n < V_n \tag{6.6}$$

it follows that

$$U_n < V \tag{6.7}$$

Therefore by Theorem I of Sec. 5, U_n approaches a limit, and the series (6.1) is convergent.

As an example of this test, consider the series

$$U = 1 + \frac{1}{2^2} + \frac{1}{3^3} + \frac{1}{4^4} + \cdots \tag{6.8}$$

This series can be compared with the series

$$V = 1 + \frac{1}{2} + \frac{1}{2^2} + \frac{1}{2^3} + \frac{1}{2^4} + \cdots \tag{6.9}$$

The latter series is a geometric series that is known to be convergent. Now the terms of (6.9) are never less than the corresponding terms of (6.8). Hence it follows that the series (6.8) is convergent.

TEST FOR DIVERGENCE

By the use of the comparison principle, it is also possible to test a series for divergence. Let

$$U = u_1 + u_2 + u_3 + \cdots \tag{6.10}$$

be a series of positive terms to be tested which are never less than the corresponding terms of a series of positive terms

$$W = w_1 + w_2 + w_3 + \cdots \tag{6.11}$$

which is known to be divergent. Then (6.10) is a divergent series.

By the use of this principle, we may prove that the harmonic series

$$U = 1 + \frac{1}{2} + \frac{1}{3} + \frac{1}{4} + \cdots + \frac{1}{n} + \cdots \tag{6.12}$$

is divergent. This may be done by rewriting (6.12) in the form

$$U = 1 + \tfrac{1}{2} + (\tfrac{1}{3} + \tfrac{1}{4}) + (\tfrac{1}{5} + \tfrac{1}{6} + \tfrac{1}{7} + \tfrac{1}{8})$$
$$+ (\tfrac{1}{9} + \cdots + \tfrac{1}{16}) + \cdots \tag{6.13}$$

Let us now compare this series with the series

$$W = \tfrac{1}{2} + \tfrac{1}{2} + (\tfrac{1}{4} + \tfrac{1}{4}) + (\tfrac{1}{8} + \tfrac{1}{8} + \tfrac{1}{8} + \tfrac{1}{8})$$
$$+ (\tfrac{1}{16} + \cdots + \tfrac{1}{16}) + \cdots \tag{6.14}$$

Now the terms of (6.13) are never less than the terms of (6.14). But the series (6.14) is divergent since the sum of the terms in each pair of parentheses is $\tfrac{1}{2}$; hence the sum of these terms increases without limit as the number of terms becomes infinite. Hence the series (6.13) is divergent.

7 CAUCHY'S INTEGRAL TEST

The comparison test requires that at least a few types of convergent series be known.   For the establishment of such types and for the test of many series of positive terms, Cauchy's integral test is useful.   This test may be stated in the following manner.   Let

$$u_1 + u_2 + u_3 + \cdots = \sum_{n=1}^{\infty} u_n \tag{7.1}$$

be a series of positive terms such that

$$u_{n+1} < u_n \tag{7.2}$$

Now if there exists a positive function $f(n)$, decreasing for $n > 1$ and such that $f(n) = u_n$, then the given series converges if the integral

$$I = \int_1^{\infty} f(n)\, dn \tag{7.3}$$

exists; the series diverges if the integral does not exist.

The proof of this test is deduced simply from Fig. 7.1.   We may think of each term u_n of the series as representing the area of a rectangle of base unity and height $f(n)$.   The sum of the areas of the first n inscribed rectangles is less than the integral

$$\int_1^{n+1} f(n)\, dn \tag{7.4}$$

Hence

$$u_2 + u_3 + u_4 + \cdots + u_{n+1} < \int_1^{n+1} f(n)\, dn \tag{7.5}$$

Now, since $f(x)$ is positive, we have

$$\int_1^{n+1} f(n) < \int_1^{\infty} f(n)\, dn = I \tag{7.6}$$

Hence we have

$$\sum_{n=2}^{n=\infty} u_n < \int_1^{\infty} f(n)\, dn = I \tag{7.7}$$

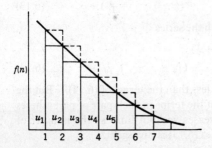

Fig. 7.1

and if the integral I exists, the series converges. We also have, from the figure, the relation

$$\sum_1^\infty u_n > \int_1^\infty f(n)\,dn \tag{7.8}$$

and hence, if the integral does not exist, the series diverges.

As an example of the application of this test, consider the series

$$\frac{1}{1^p} + \frac{1}{2^p} + \frac{1}{3^p} + \cdots + \frac{1}{n^p} \cdots = \sum_{n=1}^{n=\infty} \frac{1}{n^p} \tag{7.9}$$

To apply Cauchy's integral test, we let

$$f(n) = \frac{1}{n^p} \tag{7.10}$$

Now we have

$$I = \int_0^\infty \frac{dn}{n^p} = \begin{cases} \dfrac{1}{1-p} n^{1-p} \bigg|_1^\infty & \text{if } p \neq 1 \\[2mm] \log n \bigg|_1^\infty & \text{if } p = 1 \end{cases} \tag{7.11}$$

It is thus seen that I exists if $p > 1$ and does not exist if $p \leqslant 1$. Hence the series converges if $p > 1$ and diverges if $p \leqslant 1$. This series is a very useful one to use in comparison with others.

8 CAUCHY'S RATIO TEST

In the infinite geometric series

$$a + ar + ar^2 + \cdots + ar^n + \cdots \tag{8.1}$$

the ratio of the consecutive general terms ar^n and ar^{n+1} is the common ratio r. We have seen that this series is convergent when $|r| < 1$ and divergent for other values. We shall now consider a ratio test that may be applied to any series. Let

$$u_1 + u_2 + u_3 + \cdots + u_n + u_{n+1} + \cdots \tag{8.2}$$

be an infinite series of positive terms.

Consider consecutive general terms u_n and u_{n+1}, and form the test ratio.

$$\left| \frac{u_{n+1}}{u_n} \right| = \text{test ratio} \tag{8.3}$$

Now find the limit of this test ratio when n becomes infinite. Let this be

$$\rho = \lim_{n \to \infty} \left| \frac{u_{n+1}}{u_n} \right| \tag{8.4}$$

provided the limit exists.   We then have:

1. When $\rho < 1$, the series is convergent.
2. When $\rho > 1$, the series is divergent.
3. When $\rho = 1$, the test gives no information.

Proof   1. When $\rho < 1$, by definition of a limit we can choose n so large, say, $n \geqslant m$, that the ratio u_{n+1}/u_n will differ from ρ by as little as we please and will therefore be less than a proper fraction $r < 1$.   Hence

$$
\begin{aligned}
u_{m+1} &< u_m r \\
u_{m+2} &< u_m r^2 \\
u_{m+3} &< u_m r^3 \\
&\cdots \cdots
\end{aligned}
\tag{8.5}
$$

It therefore follows that after the term u_m each term of the series (8.2) is less than the corresponding term of the geometric series

$$
u_m(1 + r + r^2 + \cdots)
\tag{8.6}
$$

But since $r < 1$, the series (8.6), and therefore also the series (8.2), is convergent.
    2. If $\rho > 1$, the same line of reasoning as in (1) shows that the series is divergent.
    3. If $\rho = 1$, the test fails.   For example, consider the p series given by (7.9) above.   In this case, we have

$$
\text{Test ratio} = \left| \frac{u_{n+1}}{u_n} \right| = \left(\frac{u}{n+1} \right)^p = \left(1 - \frac{1}{n+1} \right)^p
\tag{8.7}
$$

and we have

$$
\rho = \lim_{n \to \infty} \left(1 - \frac{1}{n+1} \right)^p = (1)^p = 1
\tag{8.8}
$$

Hence we have $\rho = 1$ no matter what value p may have.   But in Sec. 7 it was demonstrated that when $p > 1$ the series converges and when $p \leqslant 1$ the series diverges.   It is thus evident that ρ can equal unity for both convergent and divergent series.
    We note that for convergence it is not enough that the test ratio is less than unity for all values of n.   The test requires that the limit of the test ratio shall be less than unity.   For example, in the series

$$
1 + \frac{1}{2} + \frac{1}{3} + \frac{1}{4} + \cdots = \sum_{n=1}^{\infty} \frac{1}{n}
\tag{8.9}
$$

the test ratio is always less than unity.   The *limit*, however, equals unity.

9 ALTERNATING SERIES

A series whose terms are alternately positive and negative is called an *alternating series*. Consider the alternating series

$$S = u_1 - u_2 + u_3 - u_4 + \cdots \tag{9.1}$$

If each term of the series is numerically less than the one preceding it and if

$$\lim_{n \to \infty} u_n = 0 \tag{9.2}$$

then the series is convergent.

Proof When n is even, S_n may be written in the form

$$S_n = (u_1 - u_2) + (u_3 - u_4) + \cdots + (u_{n-1} - u_n) \tag{9.3}$$

$$S_n = u_1 - (u_2 - u_3) - \cdots - (u_{n-2} - u_{n-1}) - u_n \tag{9.4}$$

Each expression in parentheses is positive. Therefore, when n increases through even values, (9.3) shows that S_n increases and (9.4) shows that S_n is always less than u_1. Therefore S_n approaches a limit L. But S_{n+1} also approaches this limit L, since

$$S_{n+1} = S_n + u_{n+1} \tag{9.5}$$

and

$$\lim_{n \to \infty} u_{n+1} = 0 \tag{9.6}$$

Hence, when n increases through all integral values, $S_n \to L$ and the series is convergent. An important consequence of this proof is given in the following statement:

The error made by terminating a convergent alternating series at any term does not exceed numerically the value of the first of the terms discarded.

For example, it will be shown in a later section that the sum of the series

$$S = 1 - \tfrac{1}{2} + \tfrac{1}{3} - \tfrac{1}{4} + \tfrac{1}{5} - \cdots = \ln 2 = 0.693 \tag{9.7}$$

to three decimal places. Now the sum of the first 10 terms of the series (9.7) is 0.646, and the value of the series differs from this by less than one-eleventh.

10 ABSOLUTE CONVERGENCE

A series is said to be *absolutely*, or unconditionally, convergent when the series formed from it by making all its terms positive is convergent. Other convergent series are said to be conditionally convergent. For example, the series

$$1 - \frac{1}{2^2} + \frac{1}{3^3} - \frac{1}{4^4} + \frac{1}{5^5} - \cdots + (-1)^{n-1} \frac{1}{n^n} + \cdots \tag{10.1}$$

is absolutely convergent since the series

$$1 + \frac{1}{2^2} + \frac{1}{3^3} + \frac{1}{4^4} + \frac{1}{5^5} + \cdots + \frac{1}{n^n} + \cdots \qquad (10.2)$$

is convergent. The alternating series

$$1 - \tfrac{1}{2} + \tfrac{1}{3} - \tfrac{1}{4} + \tfrac{1}{5} - \cdots = \ln 2 \qquad (10.3)$$

is conditionally convergent since the harmonic series is divergent.

In a conditionally convergent series, it is *not* always allowable to change the order of the terms or to group the terms together in parentheses in an arbitrary manner. These operations may alter the sum of such a series, or may change a convergent series into a divergent series, or vice versa. As an example, let us again consider the convergent series

$$1 - \frac{1}{2} + \frac{1}{3} - \frac{1}{4} + \cdots + \frac{1}{2n+1} - \frac{1}{2n+2} + \cdots \qquad (10.4)$$

The sum of this series is equal to the limit of the expression

$$S = \sum_{n=0}^{n=\infty} \left(\frac{1}{2n+1} - \frac{1}{2n+2} \right) \qquad (10.5)$$

Let us write the terms of this series in another way, putting two negative terms after each positive term in the following manner:

$$S_1 = 1 - \frac{1}{2} - \frac{1}{4} + \frac{1}{3} - \frac{1}{6} - \frac{1}{8} + \cdots + \frac{1}{2n+1}$$
$$- \frac{1}{4n+2} - \frac{1}{4n+4} + \cdots \qquad (10.6)$$

This series converges, and its sum is given by

$$S = \sum_{n=0}^{n=\infty} \left(\frac{1}{2n+1} - \frac{1}{4n+2} - \frac{1}{4n+4} \right) \qquad (10.7)$$

Now from the identity

$$\frac{1}{2n+1} - \frac{1}{4n+2} - \frac{1}{4n+4} = \frac{1}{2} \left(\frac{1}{2n+1} - \frac{1}{2n+2} \right) \qquad (10.8)$$

it is evident that the sum of the series (10.6) is half the sum of the series (10.4).

In general, given a series that is convergent but not absolutely convergent, it is possible to arrange the terms in such a way that the new series converges toward any preassigned number A whatsoever.

A series with some positive and some negative terms is convergent if the series deduced from it by making all the signs positive is convergent. The proof of this statement is omitted.

11 POWER SERIES

A series of the form

$$S = a_0 + a_1 x + a_2 x^2 + \cdots + a_n x^n + \cdots \tag{11.1}$$

where the coefficients $a_0, a_1, a_2, \ldots$ are independent of x, is called a power series in x. A power series in x may converge for all values of x or for no value of x except $x = 0$; or it may converge for some values of x different from 0 and be divergent for other values.

INTERVAL OF CONVERGENCE

Let us take the ratio of the $(n + 1)$st to the nth term of the power series (11.1). We thus obtain

$$\frac{a_{n+1} x^{n+1}}{a_n x^n} = \frac{a_{n+1}}{a_n} x \tag{11.2}$$

Let us consider the case where the coefficients of the series are such that

$$\lim_{n \to \infty} \left| \frac{a_{n+1}}{a_n} \right| = L \tag{11.3}$$

where L is a definite number. We thus see that the test ratio of Sec. 8 is given by

$$\rho = xL \tag{11.4}$$

By Cauchy's ratio test we have two cases:

1. If $L = 0$, the series (11.1) converges for all values of x since in this case $\rho = 0$.
2. If L is not zero, the series will converge when xL is numerically less than 1, that is, when x lies in the interval

$$-\frac{1}{|L|} < x < \frac{1}{|L|} \tag{11.5}$$

and will diverge for values of x outside this interval. This interval is called the interval of convergence. The end points of this interval must be examined separately.

As an example, consider the power series

$$x - \frac{x^2}{2^2} + \frac{x^3}{3^2} - \frac{x^4}{4^2} + \cdots + (-1)^{n-1} \frac{x^n}{n^2} + \cdots \tag{11.6}$$

Here we have

$$L = \lim_{n \to \infty} \left| \frac{a_{n+1}}{a_n} \right| = \lim_{n \to \infty} \frac{n^2}{(n+1)^2} = +1 \tag{11.7}$$

Hence this series converges when x is numerically less than 1 and diverges when x is numerically greater than 1. We must now examine the end points. If we place $x = 1$ in (11.6), we get

$$1 - \frac{1}{2^2} + \frac{1}{3^2} - \frac{1}{4^2} + \cdots \tag{11.8}$$

This is an alternating series that converges. If we now place $x = -1$ in (11.6), we obtain

$$-1 - \frac{1}{2^2} - \frac{1}{3^2} - \frac{1}{4^2} - \cdots \tag{11.9}$$

This is a convergent series, as can be seen by comparison with the p series of (7.9) when $p > 1$.

We thus see that the series (11.6) has the interval of convergence

$$-1 \leqslant x \leqslant 1 \tag{11.10}$$

12   THEOREMS REGARDING POWER SERIES

In this section we shall state some fundamental theorems that are of importance in the use of power series in applied mathematics. The proofs of these theorems belong in a book on analysis rather than one devoted to applied mathematics. Accordingly the proofs will be omitted, and the interested reader is advised to refer to the works listed at the end of this chapter.

Theorem I   Sum of two convergent power series   Let

$$S_1(x) = a_0 + a_1 x + a_2 x^2 + \cdots + a_n x^n + \cdots \tag{12.1}$$

be a power series convergent for $|x| < r_1$ and

$$S_2(x) = b_0 + b_1 x + b_2 x^2 + \cdots + b_n x^n + \cdots \tag{12.2}$$

be a second power series convergent for $|x| < r_2$. Now if we let r be the smaller of the two numbers r_1 and r_2 so that both series converge for $|x| < r$, then the sum of $S_1(x)$ and $S_2(x)$ is represented by the power series

$$S_1(x) + S_2(x) = (a_0 + b_0) + (a_1 + b_1) x + \cdots$$
$$+ (a_n + b_n) x^n + \cdots \tag{12.3}$$

The series (12.3) is convergent at least for $|x| < r$. That is, two power series may be added term by term and the resulting series converges to $S_1(x) + S_2(x)$ for those values of x for which both of the given series converge. For example, let

$$S_1(x) = e^x = 1 + x + \frac{x^2}{2!} + \cdots + \frac{x^n}{n!} + \cdots \tag{12.4}$$

and

$$S_2(x) = e^{-x} = 1 - x + \frac{x^2}{2!} - \cdots + (-1)^n \frac{x^n}{n!} + \cdots \tag{12.5}$$

Then
$$S_1(x) + S_2(x) = e^x + e^{-x}$$
$$= 2\left[1 + \frac{x^2}{2!} + \cdots + \frac{1 + (-1)^n}{2}\frac{x^n}{n!} + \cdots\right]$$
$$= 2\cosh x \qquad (12.6)$$

In this case both series $S_1(x)$ and $S_2(x)$ converge for all values of x, and hence their sum converges for all values of x.

Theorem II The product of two convergent power series   If

$$S_1(x) = \sum_{n=0}^{n=\infty} a_n x^n \qquad (12.7)$$

and

$$S_2(x) = \sum_{n=0}^{n=\infty} b_n x^n \qquad (12.8)$$

both converge for $|x| < r$ and the product $S_1(x)S_2(x)$ is given by the power series

$$S_1(x)S_2(x) = a_0 b_0 + (a_0 b_1 + a_1 b_0)x$$
$$+ (a_0 b_2 + a_1 b_1 + a_2 b_0)x^2 + \cdots \qquad (12.9)$$

which also converges at least for $|x| < r$.

That is, one power series may be multiplied by another to obtain the expansion of the product of two functions.   The coefficients are formed just as in the multiplication of two polynomials.   The product series converges whenever the two given series are convergent.

As an illustration of this theorem, let us consider the two series

$$S_1(x) = \sinh x = x + \frac{x^3}{3!} + \frac{x^5}{5!} + \cdots + \frac{x^{2n-1}}{(2n-1)!} + \cdots \qquad (12.10)$$

$$S_2(x) = \cosh x = 1 + \frac{x^2}{2!} + \frac{x^4}{4!} + \cdots + \frac{x^{2n-2}}{(2n-2)!} \qquad (12.11)$$

We then have

$$S_1(x)S_2(x) = \sinh x \cosh x$$
$$= x + \left(\frac{1}{2!} + \frac{1}{3!}\right)x^3 + \left(\frac{1}{4!} + \frac{1}{3!2!} + \frac{1}{5!}\right)x^5 + \cdots$$
$$= x + \tfrac{2}{3}x^3 + \tfrac{2}{15}x^5 + \cdots$$
$$= \frac{1}{2}\left[2x + \frac{(2x)^3}{3!} + \frac{(2x)^5}{5!} + \cdots\right]$$
$$= \tfrac{1}{2}\sinh 2x \qquad (12.12)$$

Since the original series converge for all values of x, so does the product series.

Theorem III   The quotient of two convergent series   If

$$S_1(x) = \sum_{n=0}^{n=\infty} a_n x^n \tag{12.13}$$

and

$$S_2(x) = \sum_{n=0}^{n=\infty} b_n x^n \tag{12.14}$$

both converge for $|x| < r$, and if $b_0 \neq 0$, then the quotient is represented by the series

$$\frac{S_1(x)}{S_2(x)} = \frac{a_0}{b_0} + \frac{a_1 b_0 - a_0 b_1}{b_0^2} x$$

$$+ \frac{a_2 b_0^2 - a_1 b_0 b_1 + a_0 b_1^2 - a_0 b_0 b_2}{b_0^3} x^2 + \cdots \tag{12.15}$$

obtained by dividing the series for $S_1(x)$ by that for $S_2(x)$. In this case no conclusion can be drawn concerning the region of convergence of the quotient series from a knowledge of the regions of convergence of the series $S_1(x)$ and $S_2(x)$.

This may be illustrated by considering the two series

$$S_1(x) = \sin x = x - \frac{x^3}{3!} + \frac{x^5}{5!} - \cdots + \frac{(-1)^{n-1} x^{2n-1}}{(2n-1)!} + \cdots \tag{12.16}$$

$$S_2(x) = \cos x = 1 - \frac{x^2}{2!} + \frac{x^4}{4!} - \cdots + \frac{(-1)^{n-1} x^{2n-2}}{(2n-2)!} + \cdots \tag{12.17}$$

If we now divide $S_1(x)$ by $S_2(x)$, we have

$$\frac{S_1(x)}{S_2(x)} = \frac{\sin x}{\cos x} = x + \frac{x^3}{3} + \frac{2x^5}{15} + \cdots = \tan x \tag{12.18}$$

Now although the series for $\sin x$ and $\cos x$ converge for all values of x, the series for $\tan x$ is convergent only for $|x| < \pi/2$.

Theorem IV   Substitution of one series into another   Let the series

$$z = a_0 + a_1 y + a_2 y^2 + \cdots + a_n y^n + \cdots \tag{12.19}$$

converge for $|y| < r_1$, and let the series

$$y = b_0 + b_1 x + b_2 x^2 + \cdots + b_n x^n + \cdots \tag{12.20}$$

converge for $|x| < r_2$.

Now if $|b_0| < r_1$, then we may substitute for y in the first series its value in terms of x from the second series and thus obtain z as a power series in x. This will converge if x is sufficiently small.

In the special case that the given series for z converges for *all* values of y, the series for z in terms of x may then always be found, and this series will converge for all values of $|x| < r_2$.

As an example, let us consider the expansion of $e^{\cos x}$ as a power series in x. In this case we have

$$z = e^y = 1 + \frac{y}{1!} + \frac{y^2}{2!} + \cdots + \frac{y^n}{n!} + \cdots \tag{12.21}$$

and

$$y = \cos x = 1 - \frac{x^2}{2!} + \frac{x^4}{4!} - \frac{x^6}{6!} + \cdots \tag{12.22}$$

In this case the series (12.21) and (12.22) converge for all values of x. We now form the various powers of y and substitute into (12.21)

$$y^2 = 1 - x^2 + \tfrac{1}{3}x^4 - \cdots$$
$$y^3 = 1 - \tfrac{3}{2}x^2 + \tfrac{7}{8}x^4 - \cdots \tag{12.23}$$
$$y^4 = 1 - 2x^2 + \tfrac{5}{3}x^4 - \cdots$$

Hence

$$e^3 = 1 + \left(1 - \frac{x^2}{2} + \frac{x^4}{24} - \cdots\right)$$
$$+ \frac{1}{2}\left(1 - x^2 + \frac{x^4}{3} - \cdots\right)$$
$$+ \frac{1}{6}\left(1 - \frac{3x^2}{2} + \frac{7}{7}x^4 - \cdots\right)$$
$$+ \tfrac{1}{24}(1 - 2x^2 + \tfrac{5}{3}x^4 - \cdots)$$
$$= (1 + 1 + \tfrac{1}{2} + \tfrac{1}{6} + \tfrac{1}{24} + \cdots)$$
$$- (\tfrac{1}{2} + \tfrac{1}{2} + \tfrac{1}{4} + \tfrac{1}{12} + \cdots)x^2$$
$$+ (\tfrac{1}{24} + \tfrac{1}{6} + \tfrac{7}{48} + \tfrac{5}{72} + \cdots)x^4 + \cdots \tag{12.24}$$

Hence

$$e^y = e^{\cos x} = 2\tfrac{17}{24} - 1\tfrac{1}{3}x^2 + \tfrac{61}{144}x^4 - \cdots \tag{12.25}$$

It should be noted that the coefficients in this series are really infinite series and that the final values here given are only the approximate values found by taking the first few terms of each series. This situation arises whenever $b_0 \neq 0$ in (12.20). However, it is sometimes possible to make a preliminary change that simplifies the final result. In the above case we could write

$$e^{\cos x} = e^{\cos x - 1 + 1} = e^{\cos x - 1}e$$

$$= e^u e = e\left(1 + u + \frac{u^2}{2!} + \frac{u^3}{3!} + \cdots\right) \tag{12.26}$$

where

$$u = \cos x - 1 = -\frac{x^2}{2!} + \frac{x^4}{4!} - \frac{x^6}{6!} + \cdots \tag{12.27}$$

Raising u to the various required powers and substituting the result into (12.26), we have

$$e^{\cos} = e\left(1 - \frac{x^2}{2} + \frac{x^4}{6} - \frac{31}{720}x^6 + \cdots\right) \tag{12.28}$$

The coefficients are now exact, and the computation of the successive terms is much simpler than by the previous method.

Theorem V   Differentiation of a power series   If

$$S_1(x) = a_0 + a_1 x + a_2 x^2 + \cdots + a_n x^n + \cdots \tag{12.29}$$

converges for $|x| < r$, the derivatives of $S_1(x)$ may be obtained by term-by-term differentiation of the series (12.29) in the form

$$\frac{d}{dx} S_1(x) = a_1 + 2a_2 x + \cdots + na_n x^{n-1} + \cdots \tag{12.30}$$

and the series (12.30) is also convergent for $|x| < r$.

As an example, consider

$$\sin x = x - \frac{x^3}{3!} + \frac{x^5}{5!} - \cdots + (-1)^{n-1}\frac{x^{2n-1}}{(2n-1)!} + \cdots \tag{12.31}$$

We have

$$\frac{d}{dx}\sin x = 1 - \frac{x^2}{2!} + \frac{x^4}{4!} - \cdots + (-1)^{n-1}\frac{x^{2n-2}}{(2n-2)!} + \cdots$$

$$= \cos x \tag{12.32}$$

In this case both series converge for all values of x.

Theorem VI   Integration of a power series   If

$$S_1(x) = \sum_{n=0}^{n=\infty} a_n x^n \tag{12.33}$$

converges for $|x| < r$, the integral of $S_1(x)$ may be found by integrating the series (12.33) term by term and we have

$$\int_0^x S_1(x)\,dx = \sum_{n=0}^{n=\infty} \frac{a_n}{n+1}x^{n+1} \tag{12.34}$$

The new series converges for $|x| < r$.

For example, we have

$$\frac{1}{1+x^2} = 1 - x^2 + x^4 - \cdots + (-1)^{n-1} x^{2n-2} + \cdots \qquad (12.35)$$

and therefore,

$$\int_0^x \frac{dx}{1+x^2} = x - \frac{x^3}{3} + \frac{x^5}{5} - \cdots + (-1)^{n-1} \frac{x^{2n-1}}{2n-1} + \cdots$$

$$= \tan^{-1} x \qquad (12.36)$$

Since the series (12.35) is convergent for $|x| < 1$, the series for $\tan^{-1} x$ is also convergent in this interval.

Theorem VII   Equality of power series   If we have

$$\sum_{n=0}^{n=\infty} a_n x^n = \sum_{n=0}^{n=\infty} b_n x^n \qquad (12.37)$$

for $|x| < r$, then the coefficients of like powers of the two series must be equal. That is, we must have

$$a_s = b_s \qquad s = 0, 1, 2, 3, \ldots \qquad (12.38)$$

It then follows that if a function is expanded in a certain interval by different methods, the series obtained must be identical.

Theorem VIII   Reversion of a power series   In certain applications of mathematical analysis to physical problems it is necessary to *reverse* a power series.   Let the given series be

$$y = ax + bx^2 + cx^3 + dx^4 + ex^5 + fx^6 + gx^7 + \cdots \qquad (12.39)$$

where it is stipulated that the coefficient a is not equal to zero.
     It is required to find the coefficients of the series

$$x = Ay + By^2 + Cy^3 + Dy^4 + Ey^5 + Fy^6 + Gy^7 + \cdots \qquad (12.40)$$

In order to determine the coefficients A, B, C, etc., the series (12.40) is substituted into the series (12.39), and the coefficients of like powers of y are equated to each other.   Following this procedure, the following equations are obtained:†

$$A = \frac{1}{a} \qquad B = -\frac{b}{a^3} \qquad C = \frac{2b^2 - ac}{a^5}$$

$$D = \frac{5abc - a^2 d - 5b^3}{a^7}$$

$$E = \frac{6a^2 bd + 3a^2 c^2 + 14b^4 - a^3 e - 21ab^2 c}{a^9} \qquad (12.41)$$

$$F = \frac{7a^3 be + 7a^3 cd + 84ab^3 c - a^4 f - 28a^2 b^2 d - 28a^2 bc^2 - 42b^5}{a^{11}}$$

$$G = \frac{8a^4 bf + 8a^4 ce + 4a^4 d^2 + 120a^2 b^3 d + 180a^2 b^2 c^2 + 132b^6 - a^5 g - 36a^3 b^2 e - 72a^3 bcd - 12a^3 c^3 - 330ab^4 c}{a^{13}}$$

13 SERIES OF FUNCTIONS AND UNIFORM CONVERGENCE

Consider a series of the form

$$S(x) = u_0(x) + u_1(x) + \cdots + u_n(x) + \cdots \qquad (13.1)$$

whose terms are continuous functions of a variable x in an interval (a,b) and which converges for every value of x inside that interval. This series does not necessarily represent a continuous function, as one might be tempted to believe. For example, let us consider the series

$$S(x) = x^2 + \frac{x^2}{1 + x^2} + \cdots + \frac{x^2}{(1 + x^2)^n} + \cdots \qquad (13.2)$$

Now if $x \neq 0$, this series is a geometric progression whose ratio is $1/(1 + x^2)$, and hence the sum of the series is

$$S(x) = \frac{x^2}{1 - [1/(1 + x^2)]} = \frac{x^2(1 + x^2)}{x^2} = 1 + x^2 \qquad x \neq 0 \qquad (13.3)$$

If we call $S_n(x)$ the sum of the first n terms of the series, then we have

$$S(0) = \lim_{n \to \infty} S_n(0) = 0 \qquad (13.4)$$

since every term of the series is zero when $x = 0$. However, we also have

$$\lim_{x \to 0} S(x) = \lim_{x \to 0} (1 + x^2) = 1 \qquad (13.5)$$

In this example the function $S(x)$ approaches a definite limit as x approaches zero, but that limit is different from the value of the function for $x = 0$. Thus $S(x)$ is not continuous at $x = 0$.

---

† Additional coefficients are given in "Smithsonian Mathematical Formulae and Tables of Elliptic Functions," p. 116, Smithsonian Institution, Washington, D.C., 1922.

Since a large number of the functions that occur in mathematics are defined by series, it has been found necessary to study the properties of the functions given in the form of a series.   The first question which arises is that of determining whether or not the sum of a given series is a continuous function of the variable.   This has led to the development of the very important notion of *uniform convergence*.

A series of the type (13.1) each of whose terms is a function of x which is defined in an interval (a,b) is said to be *uniformly convergent* in that interval if it converges for every value of x between a and b, and if, corresponding to any arbitrary preassigned positive number δ, a positive integer N, independent of x, can be found such that the absolute value of the remainder R_n of the given series

$$R_n = u_{n+1}(x) + u_{n+2}(x) + \cdots \tag{13.6}$$

is less than δ for every value of $n \geqslant N$ and for every value of x that lies in the interval (a,b).

THE WEIERSTRASS M TEST FOR UNIFORM CONVERGENCE

It would seem at first thought very difficult to determine whether or not a given series is uniformly convergent in a given interval.   The following theorem, due to the German mathematician Weierstrass, enables us to show in many cases that a given series converges uniformly.   Let

$$u_0(x) + u_1(x) + \cdots + u_n(x) + \cdots \tag{13.7}$$

be a series each of whose terms is a continuous function of x in an interval (a,b), and let

$$M_0 + M_1 + M_2 + \cdots + M_n + \cdots \tag{13.8}$$

be a convergent series whose terms are positive constants.   Then if

$$|u_n| \leqslant M_n \tag{13.9}$$

for all values of x in the interval (a,b) and for all values of n, the series (13.7) converges uniformly in the interval (a,b).

Proof   It is evident from (13.9) that

$$|u_{n+1} + u_{n+2} + \cdots| \leqslant M_{n+1} + M_{n+2} + \cdots \tag{13.10}$$

for all values of x between a and b.   If n is chosen so large that the remainder R_n of the series (13.8) is less than δ for all values of n greater than N, we shall also have

$$|u_{n+1} + u_{n+2} + \cdots| < \delta \tag{13.11}$$

whenever n is greater than N for all values of x in the interval (a,b).

As an example, let it be required to examine the series

$$\frac{\sin x}{1} + \frac{\sin 2x}{2^2} + \cdots + \frac{\sin nx}{n^2} + \cdots \tag{13.12}$$

for uniform convergence.

In this case since

$$|\sin nx| \leqslant 1 \qquad \text{for all values of } x \tag{13.13}$$

we may take

$$1 + \frac{1}{2^2} + \frac{1}{3^2} + \cdots + \frac{1}{n^2} + \cdots \tag{13.14}$$

for the M series. Since the series (13.14) converges, it follows that the series (13.12) converges uniformly in any interval.

14 INTEGRATION AND DIFFERENTIATION OF SERIES

Two theorems concerning the integration and differentiation of uniformly convergent series will now be stated without proof.

I Any series of continuous functions that converges uniformly in an interval (a,b) may be integrated term by term, provided the limits of integration are finite and lie in the interval (a,b).

II Any convergent series may be differentiated term by term if the resulting series converges uniformly.

For example, the series

$$S(x) = \frac{\sin x}{1} + \frac{\sin 2x}{2^2} + \frac{\sin 3x}{3^2} + \cdots + \frac{\sin nx}{n^2} + \cdots \tag{14.1}$$

has been shown to be uniformly convergent in any interval and hence defines a continuous function of x, $S(x)$, in that interval. The term-by-term derivative of (14.1) gives

$$\cos x + \frac{\cos 2x}{2} + \frac{\cos 3x}{3} + \cdots + \frac{\cos nx}{n} + \cdots \tag{14.2}$$

In this case we cannot find the proper M series to test (14.2) for uniform convergence since the series

$$1 + \frac{1}{2} + \frac{1}{3} + \cdots + \frac{1}{n} + \cdots$$

is divergent. The series (14.2) converges in the interval $(0,\pi)$. However, we have no assurance that it converges to the derivative of $S(x)$.

15 TAYLOR'S SERIES

We now consider a method by which we may expand a given function $f(x)$ into a power series. This section will be devoted to a derivation of Taylor's formula and a discussion of Taylor's series.

Let us consider

$$\int_{x_0}^{x_0+h} f'(x)\,dx = f(x_0 + h) - f(x_0) \tag{15.1}$$

Let us now change the variable of integration from x to t by means of the equation

$$x = (x_0 + h) - t \tag{15.2}$$

The relation between h and t is made clear by Fig. 15.1. Introducing this new variable of integration into (15.1), we have

$$\int_{x_0}^{x_0+h} f'(x)\,dx = -\int_h^0 f'(x_0 + h - t)\,dt = \int_0^h f'(x_0 + h - t)\,dt \tag{15.3}$$

We now apply the formula of integration by parts

$$\int u\,dv = uv - \int v\,du \tag{15.4}$$

to (15.3). We have here

$$\begin{aligned} u &= f'(x_0 + h - t) & dv &= dt \\ du &= -f''(x_0 + h - t)\,dt & v &= t \end{aligned} \tag{15.5}$$

Hence

$$\int_0^h f'(x_0 + h - t)\,dt = tf'(x_0 + h - t)\Big|_0^h \\ + \int_0^h tf''(x_0 + h - t)\,dt = hf'(x_0) + \int_0^h tf''(x_0 + h - t)\,dt \tag{15.6}$$

Integrating by parts again, we obtain

$$\int_0^h tf''(x_0 + h - t)\,dt = \frac{h^2}{2!}f''(x_0) + \int_0^h \frac{t^2}{2!}f'''(x_0 + h - t)\,dt \tag{15.7}$$

After n integrations by parts, we have

$$\int_{x_0}^{x_0+h} f'(x)\,dx = hf'(x_0) + \frac{h^2}{2!}f''(x_0) + \frac{h^3}{3!}f'''(x_0) + \cdots$$
$$+ \frac{h^n}{n!}f^{(n)}(x_0) + \int_0^h \frac{t^n}{n!}f^{(n+1)}(x_0 + h - t)\,dt$$
$$= f(x_0 + h) - f(x_0) \tag{15.8}$$

Fig. 15.1 x_0 x (x_0+h)

$\longleftarrow (h-t) \longrightarrow \!\!\mid\!\!\longleftarrow \quad t \quad \longrightarrow$

We may write the last integral of (15.8) in the form

$$\int_0^h \frac{t^n}{n!} \phi(t)\, dt = \frac{1}{n!} \int_0^h t^n \phi(t)\, dt = I \tag{15.9}$$

Now the integral I may be regarded as representing the area under the curve $y = t^n \phi(t)$ from the point $t = 0$ to $t = h$. If $\phi(t)$ is a continuous function of t, there will be some point t_0 such that $0 < t_0 < h$ for which we shall have

$$I = \frac{1}{n!} \int_0^h t^n \phi(t)\, dt = \frac{\phi(t_0)}{n!} \int_0^h t^n\, dt$$

$$= \frac{h^{n+1}}{(n+1)!} \phi(t_0)$$

$$= \frac{h^{n+1}}{(n+1)!} f^{(n+1)}(x_0 + \theta h) \qquad 0 < \theta < 1 \tag{15.10}$$

where $\theta h = h - t_0$. Hence we may write (15.8) in the form

$$f(x_0 + h) = f(x_0) + \frac{h}{1!} f'(x_0) + \frac{h^2}{2!} f''(x_0) + \cdots$$

$$+ \frac{h^n}{n!} f^{(n)}(x_0) + \frac{h^{n+1}}{(n+1)!} f^{(n+1)}(x_0 + \theta h) \qquad 0 < \theta < 1 \tag{15.11}$$

This is known as Taylor's formula with the Lagrangian form of the remainder. In this derivation of Taylor's formula, it was assumed that $f(x)$ possesses a continuous nth derivative. The term

$$R_{n+1} = \frac{h^{n+1}}{(n+1)!} f^{(n+1)}(x_0 + \theta h) \tag{15.12}$$

is called the remainder after $n + 1$ terms. It may happen that $f(x)$ possesses derivatives of all orders and that

$$\lim_{n \to \infty} R_{n+1} = 0 \tag{15.13}$$

In that case, we have the *convergent* infinite series

$$f(x_0 + h) = f(x_0) + \frac{h f'(x_0)}{1!} + \cdots + \frac{h^n f^{(n)}(x_0)}{n!} + \cdots \tag{15.14}$$

If we place $x_0 = 0$ and $h = x$ in (15.14), we obtain

$$f(x) = f(0) + \frac{x f'(0)}{1!} + \frac{x^2 f''(0)}{2!} + \cdots + \frac{x^n f^{(n)}(0)}{n!} + \cdots \tag{15.15}$$

This series is called Maclaurin's series.

As an example, let us obtain the Maclaurin-series expansion of the function $f(x) = e^x$. In this case, we have

$$f(0) = 1 \qquad f'(0) = 1 \qquad \cdots \qquad f^{(n)}(0) = 1 \tag{15.16}$$

Hence

$$e^x = 1 + \frac{x}{1!} + \frac{x^2}{2!} + \frac{x^3}{3!} + \cdots + \frac{x^n}{n!} + \cdots \tag{15.17}$$

This series is seen to converge for all values of x.

As another example, let it be required to expand the function $f(x) = \cos x$ into a Maclaurin series. In this case we have

$$f(x) = \cos x \qquad f'(x) = -\sin x \qquad f''(x) = -\cos x$$
$$f^{(n)}(x) = \cos\left(x + \frac{n\pi}{2}\right) \tag{15.18}$$

Substituting this into (15.15), we have

$$\cos x = 1 - \frac{x^2}{2!} + \frac{x^4}{4!} - \frac{x^6}{6!} + \cdots + (-1)^{n-1}\frac{x^{2n-2}}{(2n-2)!} + \cdots \tag{15.19}$$

This series is seen to converge for all values of x.

THE BINOMIAL SERIES

If we consider the function

$$f(x) = (1 + x)^n \tag{15.20}$$

and expand it in a Maclaurin series in powers of x, we have

$$f'(x) = n(1 + x)^{n-1} \qquad f''(x) = n(n - 1)(1 + x)^{n-2}$$
$$f'''(x) = n(n - 1)(n - 2)(1 + x)^{n-3}$$
$$\cdots \cdots \cdots \cdots \cdots \cdots \cdots \cdots \cdots \cdots \cdots$$
$$f^{(r)}(x) = n(n - 1)(n - 2) \cdots (n - r + 1)(1 + x)^{n-r} \tag{15.21}$$
$$\cdots \cdots \cdots \cdots \cdots \cdots \cdots \cdots \cdots \cdots \cdots$$

On substituting this into (15.15), we have

$$(1 + x)^n = 1 + nx + \frac{n(n - 1)x^2}{2!} + \frac{n(n - 1)(n - 2)x^3}{3!}$$
$$+ \cdots + \frac{n(n - 1)(n - 2) \cdots (n - r + 1)x^r}{r!} + \cdots \tag{15.22}$$

This series is convergent if $|x| < 1$ and divergent when $|x| > 1$, as may be seen by applying the ratio test. Equation (15.22) expresses the binomial theorem. If n is a positive integer, the series is finite. We may also write

$$(a + b)^n = a^n(1 + x)^n \qquad \text{if } x = \frac{b}{a}$$

$$= a^n + na^{n-1}b + \frac{n(n - 1)}{2!}a^{n-2}b^2 + \cdots$$

$$+ \frac{n!\,a^{n-r}b^r}{(n - r)!r!} + \cdots \qquad \text{valid for } |b| < |a| \tag{15.23}$$

16 SYMBOLIC FORM OF TAYLOR'S SERIES

TAYLOR'S-SERIES EXPANSION OF FUNCTIONS OF TWO OR MORE VARIABLES

A very useful and convenient form of Taylor's expansion may be obtained by the use of the symbolic operator D defined by

$$D_x = \frac{d}{dx} \qquad D_x^2 = \frac{d^2}{dx^2} \qquad \cdots \qquad D_x^n = \frac{d^n}{dx^n} \tag{16.1}$$

By the use of the derivative operator D_x, defined by (16.1), we may write Taylor's expansion in the form

$$f(x_0 + h) = f(x_0) + \frac{h}{1!} D_x f(x_0) + \frac{h^2}{2!} D_x^2 f(x_0) + \cdots$$

$$+ \frac{h^n}{n!} D_x^n f(x_0) + \cdots \tag{16.2}$$

where

$$D_x^r f(x_0) = \frac{d^r}{dx^r} f(x) \qquad \text{at } x = x_0 \tag{16.3}$$

We may write (16.2) in the form

$$f(x_0 + h) = \left(1 + \frac{hD_x}{1!} + \frac{h^2 D_x^2}{2!} + \cdots + \frac{h^n D_x^n}{n!} + \cdots\right) f(x_0) \tag{16.4}$$

However, if we place $x = hD_x$ in (15.17), we obtain

$$e^{hD_x} = 1 + \frac{hD_x}{1!} + \frac{h^2 D_x^2}{2!} + \cdots + \frac{h^n D_x^n}{n!} + \cdots \tag{16.5}$$

Hence we may write (16.4) in the symbolic form

$$f(x_0 + h) = e^{hD_x} f(x_0) \tag{16.6}$$

This form of Taylor's expansion is compact and easy to remember. By the use of (16.6) we may deduce the form of Taylor's expansion for a function of two or more variables.

Let $F(x,y)$ be a function of the two independent variables x and y, and let it have continuous partial derivatives of all orders. Now if we hold y constant, we have by (16.6)

$$F(x + h, y) = e^{hD_x} F(x,y) \tag{16.7}$$

where in this case D_x^r has the significance

$$D_x^r = \frac{\partial^r}{\partial x^r} \tag{16.8}$$

since we are holding y constant.

In the same way, if we hold x constant, we have

$$F(x, y + k) = e^{kD_y} F(x,y) \tag{16.9}$$

where

$$D_y^r = \frac{\partial^r}{\partial y^r} \tag{16.10}$$

If we operate on (16.7) with e^{kD_y}, we obtain

$$
\begin{aligned}
e^{kD_y} F(x + h, y) &= F(x + h, y + k) \\
&= e^{kD_y} e^{hD_x} F(x,y) \\
&= e^{hD_x + kD_y} F(x,y)
\end{aligned} \tag{16.11}
$$

We thus have the important result that

$$
\begin{aligned}
F(x + h, y + k) &= e^{hD_x + kD_y} F(x,y) \\
&= \left[1 + \frac{hD_x + kD_y}{1!} + \frac{(hD_x + kD_y)^2}{2!} + \cdots \right. \\
&\qquad \left. + \frac{(hD_x + kD_y)^n}{n!} + \cdots \right] F(x,y) \\
&= F(x,y) + h\frac{\partial F}{\partial x} + k\frac{\partial F}{\partial y} + \frac{1}{2!} \\
&\qquad \times \left(h^2 \frac{\partial^2 F}{\partial x^2} + 2hk \frac{\partial^2 F}{\partial x\,\partial y} + k^2 \frac{\partial^2 F}{\partial y^2} \right) + \cdots
\end{aligned} \tag{16.12}
$$

Equation (16.12) is Taylor's expansion of a function of two variables. The symbolic derivation given above is based on the fact that the operators D_x and D_y commute with constants and satisfy many of the basic laws of algebra and hence may be treated in many ways as if they were algebraic quantities. This matter is discussed in greater detail in Chaps. 3 and 7.

By the same reasoning, if we have a function $F(x_1, x_2, x_3, \ldots, x_n)$ of the n variables $(x_1, x_2, \ldots, x_n)$ that has continuous partial derivatives of all orders, we have

$$
\begin{aligned}
F(x_1 + h_1, x_2 + h_2, \ldots, x_n + h_n) \\
= e^{h_1 D_1 + h_2 D_2 + \cdots + h_n D_n} F(x_1, x_2, \ldots, x_n)
\end{aligned} \tag{16.13}
$$

where

$$D_m^r = \frac{\partial^r}{\partial x_m^r} \qquad D_m^r D_n^s = \frac{\partial^{r+s}}{\partial x_m^r \, \partial x_n^s} \qquad \text{etc.} \tag{16.14}$$

This is Taylor's expansion of a function of n variables.

17 EVALUATION OF INTEGRALS BY MEANS OF POWER

In applied mathematics we frequently encounter definite integrals in which the indefinite integral cannot be found in closed form. Such integrals as, for example,

$$\int_0^1 \sin x^2 \, dx \qquad \int_0^1 e^{-x^2} \, dx$$

and many others are frequently encountered in the investigation of physical problems. By the use of the power-series expansions of the integrand, we may find the value of these integrals to any desired accuracy.

For example, let us consider the integral

$$\int_0^1 \sin x^2 \, dx = I \tag{17.1}$$

If we let $x^2 = u$, we have the Maclaurin expansion for $\sin u$:

$$\sin u = u - \frac{u^3}{3!} + \frac{u^5}{5!} - \cdots \tag{17.2}$$

Hence we have

$$\sin x^2 = x^2 - \frac{x^6}{3!} + \frac{x^{10}}{5!} - \cdots \tag{17.3}$$

Hence we have

$$I = \int_0^1 \sin x^2 \, dx = \int_0^1 \left(x^2 - \frac{x^6}{3!} + \frac{x^{10}}{5!} \right) dx \, \text{approx.}$$

$$= \left(\frac{x^3}{3} - \frac{x^7}{42} + \frac{x^{11}}{1{,}320} \right) \Big|_0^1 = 0.3333 - 0.0238 + 0.0008$$

$$= 0.3103 \tag{17.4}$$

Using the fact that the error is less than the first term neglected, we may show that this result is correct to four decimal places.

In certain investigations, we encounter the integral

$$F(k,\phi) = \int_0^\phi \frac{du}{\sqrt{1 - k^2 \sin^2 u}} \qquad k^2 < 1 \tag{17.5}$$

This integral is called an *elliptic integral of the first kind*. The integrand of this integral may be expanded by the binomial theorem in the form

$$(1 - k^2 \sin^2 u)^{-\frac{1}{2}} = 1 + \frac{k^2}{2} \sin^2 u + \frac{3k^4}{8} \sin^4 u + \cdots \tag{17.6}$$

This series is convergent for $|k| < 1$ and for any value of u. If we place $u = \pi/2$, we get the convergent series

$$1 + \frac{k^2}{2} + \frac{3k^4}{8} + \frac{5k^6}{16} + \cdots \tag{17.7}$$

Since $|\sin u| \leqslant 1$, the terms of the series (17.6) are less than the corresponding terms of the series (17.7). It follows, therefore, by Weierstrass's M test that the series (17.6) is uniformly convergent in any interval of u. We may therefore integrate term by term. The integration is facilitated by the recursion formula

$$\int \sin^n u \, du = -\frac{\sin^{n-1} u \cos u}{n} + \frac{n-1}{n} \int \sin^{n-2} u \, du \qquad (17.8)$$

given by No. 263 of Peirce's tables of integrals. If, in particular, $\phi = \pi/2$, we have

$$K(k) = \int_0^{\pi/2} \frac{du}{\sqrt{1 - k^2 \sin^2 u}} \qquad k^2 < 1 \qquad (17.9)$$

This is the complete elliptic integral of the first kind. The integration may then be facilitated by Wallis's formula

$$\int_0^{\pi/2} \sin^n u \, du = \frac{1 \times 3 \times 5 \times \cdots \times (n-1)}{2 \times 4 \times 6 \times \cdots \times n} \frac{\pi}{2} \qquad (17.10)$$

if n is an even integer (Peirce No. 483). We then have

$$K = \frac{\pi}{2}\left[1 + \left(\frac{1}{2}\right)^2 k^2 + \left(\frac{1 \times 3}{2 \times 4}\right)^2 k^4 + \left(\frac{1 \times 3 \times 5}{2 \times 4 \times 6}\right)^2 k^6 + \cdots \right] \qquad (17.11)$$

This series may be used to compute K for various values of k. If, for example, $k = \sin 10°$, we have

$$K = \frac{\pi}{2}(1 + 0.00754 + 0.00012 + \cdots) = 1.5828 \qquad (17.12)$$

We may sometimes obtain the power-series expansion of a function most readily by the evaluation of an integral. For example, we have

$$\frac{d}{dx} \sin^{-1} x = \frac{1}{\sqrt{1 - x^2}} \qquad (17.13)$$

Hence

$$\sin^{-1} x = \int_0^x \frac{du}{\sqrt{1 - u^2}} \qquad (17.14)$$

Now, by the binomial theorem, we have

$$(1 - u^2)^{-\frac{1}{2}} = 1 + \frac{u^2}{2} + \frac{1 \times 3}{2 \times 4} u^4 + \frac{1 \times 3 \times 5}{2 \times 4 \times 6} u^6 + \cdots \qquad (17.15)$$

This series converges when $|u| < 1$. If we now substitute this into (17.14) and integrate term by term, we have

$$\sin^{-1} x = x + \frac{1}{2}\frac{x^3}{3} + \frac{1 \times 3}{2 \times 4}\frac{x^5}{5} + \frac{1 \times 3 \times 5}{2 \times 4 \times 6}\frac{x^7}{7} + \cdots \qquad (17.16)$$

This series converges when $|x| < 1$.

If we let $x = \frac{1}{2}$, we have

$$\sin^{-1}\frac{1}{2} = \frac{\pi}{6} = \frac{1}{2} + \frac{1 \times 1}{2 \times 3}\left(\frac{1}{2}\right)^3 + \frac{1 \times 3 \times 1}{2 \times 4 \times 5}\left(\frac{1}{2}\right)^5 + \cdots \qquad (17.17)$$

or

$$\pi = 3.1415 \cdots \qquad (17.18)$$

18   APPROXIMATE FORMULAS DERIVED FROM MACLAURIN'S SERIES

It frequently happens in applied mathematics that by using a few terms of the power series by which a function is represented we obtain an approximate formula that has some degree of accuracy. Such approximate formulas are of great utility.

For example, by the use of the binomial series we may write down at once the following approximate formulas valid for small values of x:

$$(1 + x)^n \doteq 1 + nx \doteq 1 + nx + \tfrac{1}{2}n(n - 1)x^2 \qquad (18.1)$$
$$\text{First approx.} \qquad \text{Second approx.}$$

where the symbol $\doteq$ means "approximately equal to." Also, we have

$$(1 + x)^{-n} \doteq 1 - nx \doteq 1 - nx + \tfrac{1}{2}n(n + 1)x^2 \qquad (18.2)$$

From the Maclaurin series for the sine we have

$$\sin x \doteq x - \frac{x^3}{6} \qquad (18.3)$$

We might inquire for what value of x we may use the approximation

$$\sin x \doteq x \qquad (18.4)$$

such that the result is accurate to three decimal places. Since the Maclaurin series for the sine is an alternating series, we know that if only the first term is retained the value of the remaining series is numerically less than the term $x^3/6$. Hence we *must* have

$$\left|\frac{x^3}{6}\right| < 0.0005 \qquad (18.5)$$

so that the result of the approximation shall be valid to three decimal places. Hence

$$|x| < \sqrt[3]{0.003}$$
$$\text{or} \quad |x| < 0.1442 \text{ radian} \qquad (18.6)$$

This corresponds to

$$|x| < 8.2° \qquad (18.7)$$

19 USE OF SERIES FOR THE COMPUTATION OF FUNCTIONS

In many cases the series expansion of a function gives a direct means for the computation of the numerical value of a function for a certain given value of the argument. In this way, tables of functions may be computed.

For example, let it be required to compute the value of sin 10°. This may be done by the use of the Maclaurin-series expansion of the sine.

$$\sin x = x - \frac{x^3}{3!} + \frac{x^5}{5!} - \frac{x^7}{7!} + \cdots \tag{19.1}$$

In this case $x = 10° = \pi/18$ radians. Hence we have

$$\sin\left(\frac{\pi}{18}\right) = \frac{\pi}{18} - \left(\frac{\pi}{18}\right)^3 \frac{1}{3!} + \left(\frac{\pi}{18}\right)^5 \frac{1}{5!} - \left(\frac{\pi}{18}\right)^7 \frac{1}{7!} + \cdots$$
$$= 0.1736 \tag{19.2}$$

Since this is an alternating series, we know that the error introduced by stopping at this term is less than

$$\left(\frac{\pi}{18}\right)^9 \frac{1}{9!}$$

As another example, let us consider a series useful in the computation of logarithms. The Maclaurin-series expansion of the function $\ln(1 + x)$ is easily shown to be

$$\ln(1 + x) = x - \frac{x^2}{2} + \frac{x^3}{3} - \frac{x^4}{4} + \cdots \tag{19.3}$$

If we let $x = -x$ in (19.3), we have

$$\ln(1 - x) = -x - \frac{x^2}{2} - \frac{x^3}{3} - \frac{x^4}{4} - \cdots \tag{19.4}$$

Now we have

$$\ln\frac{1+x}{1-x} = \ln(1 + x) - \ln(1 - x)$$
$$= 2\left(x + \frac{1}{3}x^3 + \frac{1}{5}x^5 + \frac{x^7}{7} + \cdots\right) \tag{19.5}$$

This series converges when $|x| < 1$.

Now if we let

$$x = \frac{1}{2n+1} \qquad n > 0 \tag{19.6}$$

we have

$$\frac{1+x}{1-x} = \frac{n+1}{n} \tag{19.7}$$

Then $|x| < 1$ for all values of $n > 0$. If we substitute this into (19.5), we have

$$\ln(n+1) = \ln n + 2\left[\frac{1}{2n+1} + \frac{1}{3}\frac{1}{(2n+1)^3} + \frac{1}{5}\frac{1}{(2n+1)^5} + \cdots\right] \quad (19.8)$$

This series converges for all positive values of n and is well adapted to computation. For example, if we let $n = 1$, we have

$$\ln 2 = 2\left(\frac{1}{3} + \frac{1 \times 1}{3 \times 3^3} + \frac{1 \times 1}{5 \times 3^5} + \cdots\right)$$

$$= 0.69315 \quad (19.9)$$

If we let $n = 2$, we have

$$\ln 3 = \ln 2 + 2\left(\frac{1}{5} + \frac{1 \times 1}{3 \times 5^3} + \frac{1 \times 1}{5 \times 5^5} + \cdots\right)$$

$$= 1.09861 \quad (19.10)$$

In this way we may compute the natural logarithms of any number. If we wish the Briggs or common logarithms of numbers, written $\log n$, we have

$$\log n = \frac{\ln n}{\ln 10} = \frac{\ln n}{2.30259} \quad (19.11)$$

For example,

$$\log 2 = \frac{\ln 2}{\ln 10} = \frac{0.69315}{2.30259} = 0.3010 \cdots \quad (19.12)$$

20 EVALUATION OF A FUNCTION TAKING ON AN INDETERMINATE FORM

a THE FORM 0/0

It sometimes happens that we encounter functions of the form $(\sin x)/x$, $(1 - \cos x)/x^2$, etc., and we wish to investigate the limiting value of these functions as the variable approaches a critical value. That is, we are given a function of the form $f(x)/\phi(x)$ such that $f(a) = 0$ and $\phi(a) = 0$. The function is indeterminate when $x = a$. It is then required to find

$$\lim_{x \to a} \frac{f(x)}{\phi(x)} \quad (20.1)$$

Now we have

$$\frac{f(a+b)}{\phi(a+b)} = \frac{f(a) + f'(a)b + f''(a)(b^2/2!) + \cdots}{\phi(a) + \phi'(a)b + \phi''(a)(b^2/2!) + \cdots} \quad (20.2)$$

by Taylor's-series expansion for $f(a + b)$ and $\phi(a + b)$.   Now

$$\lim_{x \to a} \frac{f(x)}{\phi(x)} = \lim_{b \to 0} \frac{f(a + b)}{\phi(a + b)} \tag{20.3}$$

But since by hypothesis $f(a) = 0$ and $\phi(a) = 0$, we have from (20.2) and (20.3)

$$\lim_{x + a} \frac{f(x)}{\phi(x)} = \frac{f'(a)}{\phi'(a)} \tag{20.4}$$

This is known as *L'Hospital's rule*. If $f'(a)/\phi'(a)$ is again indeterminate, we apply the rule again.   For example,

$$\lim_{x \to 0} \frac{1 - \cos x}{x^2} = \lim_{x \to 0} \frac{\sin x}{2x} = \lim_{x \to 0} \frac{\cos x}{2} = \frac{1}{2} \tag{20.5}$$

This form could have also been evaluated by the use of the first few terms of the Maclaurin expansion for the cosine; that is, if x is small, we have

$$\cos x \doteq 1 - \frac{x^2}{2} \tag{20.6}$$

hence

$$1 - \cos x \doteq \frac{x^2}{2} \tag{20.7}$$

and we therefore have

$$\lim_{x \to 0} \frac{1 - \cos x}{x^2} = \lim_{x \to 0} \frac{x^2/2}{x^2} = \frac{1}{2} \tag{20.8}$$

b **THE FORM** ∞/∞

The indeterminate form ∞/∞ may be brought under the form 0/0 by the device of writing the quotient in the form

$$\lim_{x \to a} \frac{f(x)}{\phi(x)} = \lim_{x \to a} \frac{1/\phi(x)}{1/f(x)} \tag{20.9}$$

Now since by hypothesis $\phi(a) = \infty$ and $f(a) = \infty$, we again have the form 0/0 and we may then apply L'Hospital's rule.

   For example, consider

$$\lim_{x \to \pi/2} \frac{\sec 3x}{\sec 5x} = \lim_{x \to \pi/2} \frac{1/\sec 5x}{1/\sec 3x}$$

$$= \lim_{x \to \pi/2} \frac{\cos 5x}{\cos 3x} = \lim_{x \to \pi/2} \frac{-5 \sin 5x}{-3 \sin 3x}$$

$$= -\tfrac{5}{3} \tag{20.10}$$

It may be shown that indeterminate forms of the form ∞/∞ may be brought under the same rule as L'Hospital's rule for the form $0/0$. The proof of this statement is rather lengthy, but it may be made plausible by the following heuristic process: Consider

$$\lim_{x \to a} \frac{f(x)}{\phi(x)} \tag{20.11}$$

where $f(a) = \infty$ and $\phi(a) = \infty$. Then we may write the quotient in the form

$$\lim_{x \to a} \frac{f(x)}{\phi(x)} = \lim_{x \to a} \frac{1/\phi(x)}{1/f(x)} \tag{20.12}$$

where this is now of the form $0/0$, and hence we may apply L'Hospital's rule to it. On carrying out the differentiation, we have

$$\lim_{x \to a} \frac{f(x)}{\phi(x)} = \lim_{x \to a} \frac{-\phi'(x)/\phi^2(x)}{-f'(x)/f^2(x)}$$
$$= \lim_{x \to a} \left\{ \left[\frac{f(x)}{\phi(x)} \right]^2 \frac{\phi'(x)}{f'(x)} \right\} \tag{20.13}$$

Now since the limit of a product is equal to the product of the limits, we have

$$\lim_{x \to a} \frac{f(x)}{\phi(x)} = \lim_{x \to a} \frac{f^2(x)}{\phi^2(x)} \lim_{x \to a} \frac{\phi'(x)}{f'(x)} \tag{20.14}$$

and hence

$$\lim_{x \to a} \frac{f(x)}{\phi(x)} = \lim_{x \to a} \frac{f'(x)}{\phi'(x)} \tag{20.15}$$

This is the same rule as that for the evaluation of indeterminate forms of the form $0/0$. As an example of this rule, consider

$$\lim_{x \to 0} \frac{\ln x}{\csc x} = \lim_{x \to 0} \frac{1/x}{-\csc x \cot x}$$
$$= \lim_{x \to 0} \frac{-\sin^2 x}{x \cos x} = \lim_{x \to 0} \frac{-x^2}{x} = 0 \tag{20.16}$$

c FORM $0 \times \infty$

If a function $f(x) \times \phi(x)$ takes on the indeterminate form $0 \times \infty$ for $x = a$, we may write the given function in the form

$$f(x) \times \phi(x) = \frac{f(x)}{1/\phi(x)} = \frac{\phi(x)}{1/f(x)} \tag{20.17}$$

This causes it to take on one of the forms $0/0$ or ∞/∞ which we may evaluate by the methods of part a or b above.   For example, consider

$$\lim_{x\to 0} x \ln x = \lim_{x\to 0} \frac{\ln x}{1/x} = \lim_{x\to 0} \frac{1/x}{-1/x^2} \tag{20.18}$$
$$= 0$$

d FORM $\infty - \infty$

It is possible in general to transform the expression into a fraction that will assume the form $0/0$ or ∞/∞ for the critical value of the argument.

$$\lim_{x\to \pi/2} (\sec x - \tan x) = \lim_{x\to \pi/2} \frac{1 - \sin x}{\cos x}$$

$$= \lim_{x\to \pi/2} \frac{-\cos x}{-\sin x} = 0 \tag{20.19}$$

e FORMS 1^∞, 0^0, ∞^0

In general it is possible to transform these forms into the cases discussed in parts *a* and *b*.

Example 1

$$\lim_{x\to 0} (\cos x)^{1/x^2} = 1^\infty \tag{20.20}$$

Let

$$u = (\cos x)^{1/x^2} \tag{20.21}$$

$$\lim_{x\to 0} \ln u = \lim_{x\to 0} \frac{\ln (\cos x)}{x^2} = \lim_{x\to 0} \frac{-\sin x/\cos x}{2x}$$

$$= \lim_{x\to 0} \frac{-\sec^2 x}{2} = -\frac{1}{2} \tag{20.22}$$

Hence

$$\lim_{x\to 0} u = e^{-\frac{1}{2}} \tag{20.23}$$

Example 2

$$\lim_{x\to 0} x^{\sin x} = 0^0 \tag{20.24}$$

$$\lim_{x\to 0} x^{\sin x} = \lim_{x\to 0} x^x \tag{20.25}$$

Let

$$u = x^x \tag{20.26}$$

$$\lim_{x\to 0} \ln u = \lim_{x\to 0} x \ln x = \lim_{x\to 0} \frac{\ln x}{1/x}$$

$$= \lim_{x\to 0} \frac{1/x}{-1/x^2} = 0 \tag{20.27}$$

Hence

$$\lim_{x \to 0} u = e^0 = 1 \tag{20.28}$$

Example 3

$$\lim_{x \to 0} \left(\frac{1}{x} \right)^{\sin x} = \infty^0 \tag{20.29}$$

Let

$$u = \left(\frac{1}{x} \right)^{\sin x} \tag{20.30}$$

Hence

$$\lim_{x \to 0} \ln u = \lim_{x \to 0} \sin x \ln \frac{1}{x} = \lim_{x \to 0} \frac{\cos x}{-1/x} = 0 \tag{20.31}$$

Hence

$$\lim_{x \to 0} u = e^0 = 1 \tag{20.32}$$

Example 4

$$\lim_{n \to \infty} \left(1 + \frac{1}{n} \right)^n = 1^\infty \tag{20.33}$$

Let

$$u = \left(1 + \frac{1}{n} \right)^n \qquad \frac{1}{n} = x \tag{20.34}$$

$$\lim_{n \to \infty} \ln u = \lim_{x \to 0} \frac{\ln (1 + x)}{x}$$

$$= \lim_{x \to 0} \frac{1/(1 + x)}{1} = 1 \tag{20.35}$$

Hence

$$\lim_{n \to \infty} u = \lim_{x \to 0} u = e^1 = e \tag{20.36}$$

The above examples are typical of indeterminate forms that may be brought under parts a and b by the use of logarithmic transformations.

21   FOURIER SERIES AND INTEGRALS

In the following sections the mathematical representation of periodic phenomena will be considered. The use of complex numbers for this purpose will be developed. This leads naturally to a consideration of the complex form of the Fourier representation of periodic functions. This form of Fourier series is extensively used by physicists but is not generally used by engineers. From the standpoint of utility the complex form of Fourier series has marked advantages over the more cumbersome form involving sines and cosines. It is widely used in the modern literature in studying the response of electrical circuits and mechanical vibrating systems to the action of periodic potentials and forces. A brief discussion of the Fourier integral is given. The treatment of the Fourier series and the Fourier integral is heuristic; an account of the rigorous investigations of Dirichlet on the subject belongs in a book on analysis.

SIMPLE HARMONIC VIBRATIONS

The simplest periodic process that occurs in nature is described mathematically by the sine or the cosine function. Such processes as the oscillations of a pendulum through small amplitudes, the vibrations of a tuning fork, and other similar physical phenomena are of this type. If the process repeats itself f times a second, the function representing the simple oscillation may be either

$$u = A \sin 2\pi ft \qquad \text{or} \qquad u = A \cos 2\pi ft \tag{21.1}$$

where t is time measured in seconds from a suitable starting point, A is called the amplitude, and f the frequency of the vibration. Besides the frequency f, it is customary to speak of the angular frequency of the vibration ω defined by

$$\omega = 2\pi f \tag{21.2}$$

A phenomenon described by a simple sine or cosine function is called a *simple harmonic vibration.*

We may simplify the computations involving these functions considerably by using imaginary exponentials instead of trigonometric functions. We have from the Euler formula

$$e^{j\omega t} = \cos \omega t + j \sin \omega t \qquad j = \sqrt{-1} \tag{21.3}$$

If we let

$$Z = A e^{j\omega t} \tag{21.4}$$

then Z represents a complex number whose representative point in the complex plane describes a circle of radius A with angular velocity ω. The projections on the real and imaginary axes are, respectively,

$$\begin{aligned} x &= \operatorname{Re} Z = A \cos \omega t \\ y &= \operatorname{Im} Z = A \sin \omega t \end{aligned} \tag{21.5}$$

where Re means "the real part of" and Im means "the imaginary part of." Since physical calculations deal with real quantities, the final results of a computation using complex numbers must be translated into real magnitudes. This may be done simply, since an equation involving complex numbers means that the real as well as the imaginary parts satisfy the equation.

It sometimes happens that the square of the amplitude of the oscillation is of importance. This may be readily obtained from its complex representation by multiplying it by its conjugate. That is, since

$$\bar{Z} = A e^{-j\omega t} \tag{21.6}$$

then

$$Z\bar{Z} = A e^{j\omega t} A e^{-j\omega t} = A^2 \tag{21.7}$$

Consider two vibrations of the same frequency but of different phase δ of the form

$$
\begin{aligned}
u_1 &= A_1 \cos \omega t = \operatorname{Re} A_1 e^{j\omega t} \\
u_2 &= A_2 \cos (\omega t + \delta) = \operatorname{Re} A_2 e^{j(\omega t + \delta)}
\end{aligned}
\tag{21.8}
$$

Hence

$$u_1 + u_2 = \operatorname{Re} (A_1 e^{j\omega t} + A_2 e^{j(\omega t + \delta)}) \tag{21.9}$$

We also have

$$A_1 e^{j\omega t} + A_2 e^{j(\omega t + \delta)} = e^{j\omega t}(A_1 + A_2 e^{j\delta}) \tag{21.10}$$

and

$$
\begin{aligned}
A_1 + A_2 e^{j\delta} &= A_1 + A_2 \cos \delta + jA_2 \sin \delta \\
&= M e^{j\phi}
\end{aligned}
\tag{21.11}
$$

where

$$M = \sqrt{(A_1 + A_2 \cos \delta)^2 + A_2^2 \sin^2 \delta} \tag{21.12}$$

and

$$\phi = \tan^{-1} \frac{A_2 \sin \delta}{A_1 + A_2 \cos \delta} \tag{21.13}$$

Hence we have from (21.9)

$$
\begin{aligned}
u_1 + u_2 &= \operatorname{Re} M e^{j(\omega t + \phi)} \\
&= M \cos (\omega t + \phi)
\end{aligned}
\tag{21.14}
$$

The amplitude M of the resulting vibration is given as a third side of the triangle having the amplitudes A_1 and A_2 as adjacent sides and δ as exterior angle, as shown in Fig. 21.1. The difference in phase ϕ between the resultant vibration and u_1 is the angle between the sides M and A_1 of this triangle. It is noted that the construction in Fig. 21.1 corresponds exactly with the addition of two complex numbers in the complex plane. From this consideration we obtain the following rule:

To obtain the resultant of two vibrations having equal frequencies but differing in phase, add the corresponding complex numbers. The amplitude

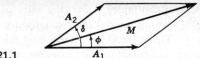

Fig. 21.1

and phase of the resultant are given by the length and direction, respectively, of the complex number representing this sum.

This method is most frequently used in electrical engineering and is sometimes referred to as the "vector diagram" because of the manner in which the quantities are combined. This diagram may be drawn for any instant of time, but since the phase difference is constant, the entire triangle rotates as a rigid figure with the angular speed ω as time advances. It is therefore possible to choose any position such as the real axis, for example, for the line of the first vibration.

If more than two vibrations of the same frequency but of different phases are to be compounded, we obtain the magnitude and phase resultant by plotting the individual complex numbers representing the several vibrations and adding them.

22   REPRESENTATION OF MORE COMPLICATED PERIODIC PHENOMENA; FOURIER SERIES

Let us consider an arbitrary process that is repeated every T sec. Let this process be represented by

$$u = F(t) \tag{22.1}$$

Since, by hypothesis, the process repeats itself every T sec, we have

$$F(t + T) = F(t) \tag{22.2}$$

Let us make the proviso that $F(t)$ is single-valued and finite and has a finite number of discontinuities and a finite number of maxima and minima in the interval of one oscillation, T. Under these conditions, which in the mathematical literature are known as Dirichlet conditions, the function $F(t)$ may be represented over a complete period and hence from $t = -\infty$ to $t = +\infty$, except at the discontinuities, by a series of simple harmonic functions, the frequencies of which are integral multiples of the fundamental frequency. Such series are called *Fourier series* after their discoverer. For a proof of the possibility of developing $F(t)$ in a Fourier series under these very general conditions, the reader may consult the literature cited in the references at the end of this appendix. Dirichlet's treatment of the subject is long and difficult and has no place in a book on applied mathematics. In his treatment the sum of n terms of the series is taken, and it is shown that when n becomes infinitely great the sum approaches $F(t)$ provided the above conditions are satisfied.

At a discontinuity in $F(t)$ the value of the series is the mean of the values of $F(t)$ on both sides of the discontinuity.

The method of determining the coefficients will now be given. It is most convenient to start from the complex representation and write

$$F(t) = a_0 + a_1 r^{j\omega t} + a_2 e^{2j\omega t} + \cdots + a_n e^{jn\omega t} + \cdots$$
$$+ a_{-1} e^{-j\omega t} + a_{-2} e^{-2j\omega t} + \cdots + a_{-n} e^{-jn\omega t} + \cdots$$
$$= \sum_{n=-\infty}^{n=+\infty} a_n e^{jn\omega t} \tag{22.3}$$

where

$$\omega = \frac{2\pi}{T} \tag{22.4}$$

Since the left member of (22.3) is real, the coefficients of the series on the right must be such that no imaginary terms occur. To determine a_0, we integrate both sides over one complete period, that is, from 0 to $T = 2\pi/\omega$. We thus obtain

$$\int_0^{2\pi/\omega} F(t)\,dt = \int_0^{2\pi/\omega} \left(\sum_{n=-\infty}^{n=+\infty} a_n e^{jn\omega t} \right) dt$$
$$= \sum_{n=-\infty}^{n=+\infty} a_n \int_0^{2\pi/\omega} e^{jn\omega t}\,dt \tag{22.5}$$

where we have assumed term-by-term integration permissible.

The integral of the general term is

$$\int_0^{2\pi/\omega} e^{jn\omega t}\,dt = \frac{1}{jn\omega} e^{jn\omega t} \Big|_0^{2\pi/\omega} = \frac{1}{jn\omega}(e^{j2n\pi} - 1) = 0 \tag{22.6}$$

if $n \neq 0$. If $n = 0$, we have

$$\int_0^{2\pi/\omega} dt = \frac{2\pi}{\omega} = T \tag{22.7}$$

Hence (22.5) reduces to

$$\int_0^T F(t)\,dt = a_0 T \tag{22.8}$$

or

$$a_0 = \frac{1}{T} \int_0^T F(t)\,dt = \overline{F(t)} \tag{22.9}$$

where $\overline{F(t)}$ denotes the mean value of $F(t)$.

To determine the other coefficients, we multiply both sides of (22.3) by $e^{-jn\omega t}$ and integrate as before from $t = 0$ to $t = T = 2\pi/\omega$. Again all terms on the right are equal to zero because of the periodicity of the imaginary exponentials, except the a_n term, which contains no exponential factor. This gives the value T on integration.   We then have

$$\int_0^T F(t)e^{-jn\omega t}\,dt = a_n T \tag{22.10}$$

or

$$a_n = \frac{1}{T}\int_0^T F(t)e^{-jn\omega t}\,dt \tag{22.11}$$

Equation (22.11) gives the coefficient of the general term in the expression (22.3).   The coefficient a_0 is a special case of (22.11).   We have also from (22.11) the relation

$$a_{-n} = \frac{1}{T}\int_0^T F(t)e^{jn\omega t}\,dt \tag{22.12}$$

We thus see that a_n and a_{-n} are conjugate imaginaries, and we have

$$a_{-n} = \bar{a}_n \tag{22.13}$$

The usual real form of the Fourier series may be obtained in the following manner.   Equation (22.3) may be written in the following form:

$$
\begin{aligned}
F(t) &= \sum_{n=-\infty}^{n=-1} a_n e^{jn\omega t} + a_0 + \sum_{n=1}^{n=\infty} a_n e^{jn\omega t} \\
&= \sum_{n=\infty}^{n=1} a_{-n} e^{-jn\omega t} + a_0 + \sum_{n=1}^{n=\infty} a_n e^{jn\omega t} \\
&= a_0 + \sum_{n=1}^{n=\infty} (a_n e^{jn\omega t} + a_{-n} e^{-jn\omega t})
\end{aligned}
\tag{22.14}
$$

By using Euler's relation, this may be written in the form

$$F(t) = a_0 + \sum_{n=1}^{n=\infty} (a_n + a_{-n})\cos n\omega t + \sum_{n=1}^{n=\infty} j(a_n - a_{-n})\sin n\omega t \tag{22.15}$$

If we now let

$$A_n = a_n + a_{-n} \qquad B_n = j(a_n - a_{-n}) \qquad \frac{A_0}{2} = a_0 \tag{22.16}$$

we obtain

$$F(t) = \frac{A_0}{2} + \sum_{n=1}^{n=\infty} A_n \cos n\omega t + \sum_{n=1}^{n=\infty} B_n \sin n\omega t \tag{22.17}$$

This is the usual real form of the Fourier series.   By using (22.16) and (22.11), we obtain the coefficients A_n and B_n directly in terms of $F(t)$.   We thus obtain

$$
\begin{aligned}
A_n = a_n + a_{-n} &= \frac{1}{T}\int_0^T F(t)(e^{-jn\omega t} + e^{jn\omega t})\,dt \\
&= \frac{2}{T}\int_0^T F(t)\cos n\omega t\,dt
\end{aligned}
\tag{22.18}
$$

We also have

$$B_n = j(a_n - a_{-n}) = \frac{1}{T} \int_0^T F(t) j(e^{-jn\omega t} - e^{jn\omega t}) \, dt$$

$$= \frac{2}{T} \int_0^T F(t) \sin n\omega t \, dt \tag{22.19}$$

The $A_0/2$ term is introduced in the series (22.17) so that Eq. (22.18) giving the general term A_n will be applicable for A_0 as well. In either the complex or the real form of Fourier series the constant term is always equal to the mean value of the function.

A third form of the Fourier series involving phase angles may be obtained from (22.17) by letting

$$A_n \cos n\omega t + B_n \sin n\omega t = C_n \cos (n\omega t - \phi_n)$$

$$= C_n \cos n\omega t \cos \phi_n + C_n \sin n\omega t \sin \phi_n \tag{22.20}$$

Equating the coefficients of like cosine and sine terms, we have

$$A_n = C_n \cos \phi_n \qquad B_n = C_n \sin \phi_n \tag{22.21}$$

and hence

$$C_n = \sqrt{A_n^2 + B_n^2} \qquad \phi_n = \tan^{-1} \frac{B_n}{A_n} \tag{22.22}$$

In this case the series takes the form

$$F(t) = \frac{A_0}{2} + \sum_{n=1}^{n=\infty} C_n \cos (n\omega t - \phi_n) \tag{22.23}$$

or

$$F(t) = \frac{A_0}{2} + \sum_{n=1}^{n=\infty} C_n \sin \left(n\omega t + \frac{\pi}{2} - \phi_n \right) \tag{22.24}$$

The complex form of the Fourier series has many advantages over the real form involving sines and cosines. It is much simpler to perform processes of differentiation and integration with the form (22.3) than with the real forms, and also no harmonic phase angles appear explicitly in the complex form but are contained in the complex character of the coefficients.

23 EXAMPLES OF FOURIER EXPANSIONS OF FUNCTIONS

Let it be required to obtain the Fourier expansion of the function of Fig. 23.1.

This function is assumed to continue in the same fashion in both directions. The origin of the time is arbitrarily chosen as indicated.

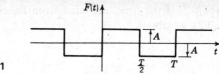

Fig. 23.1

The coefficients of the complex Fourier series are given by Eq. (22.11). We thus have

$$a_n = \frac{1}{T}\int_0^T F(t)e^{-jn\omega t}\,dt = \frac{A}{T}\int_0^{T/2} e^{-jn\omega t}\,dt - \frac{A}{T}\int_{T/2}^T e^{-jn\omega t}\,dt$$

$$= \frac{A}{T}\left(\int_0^{\pi/\omega} e^{-jn\omega t}\,dt - \int_{\pi/\omega}^{2\pi/\omega} e^{-jn\omega t}\,dt\right)$$

$$= \frac{A}{Tjn\omega}\left(e^{-jn\omega t}\Big|_{\pi/\omega}^{0} + e^{-jn\omega t}\Big|_{\pi/\omega}^{2\pi/\omega}\right)$$

$$= \begin{cases} 0 & \text{if } n = 0, \text{ or } n \text{ is even} \\ \dfrac{2A}{jn\pi} & \text{if } n \text{ is odd} \end{cases} \tag{23.1}$$

The complex Fourier-series expansion of this function is therefore

$$F(t) = \frac{2A}{j\pi}\sum_{n=-\infty}^{n=+\infty}\frac{e^{jn\omega t}}{n} \qquad n \text{ odd} \tag{23.2}$$

The coefficients of the real Fourier series are given by (22.18). We thus have

$$A_n = a_n + a_{-n} = \frac{2A}{\pi j}\left(\frac{1}{n} - \frac{1}{n}\right) = 0$$

$$B_n = j(a_n - a_{-n}) = \frac{2A}{\pi}\left(\frac{1}{n} + \frac{1}{n}\right) = \frac{4A}{n\pi} \tag{23.3}$$

Hence the real Fourier-series expansion of this function is

$$F(t) = \frac{4A}{\pi}\sum_{n=1}^{n=\infty}\frac{\sin n\omega t}{n} \qquad n \text{ odd} \tag{23.4}$$

It is interesting to determine the character of the Fourier-series expansion of the function of Fig. 23.1 if we had taken the origin of time at a point t_0 to the right of the origin in Fig. 23.1, where

$$t_0 = \frac{\theta}{\omega} \tag{23.5}$$

We may obtain the required result from (23.2) if we introduce the change of variable

$$t = t' + \frac{\theta}{\omega} \tag{23.6}$$

where t' is the time measured from the new origin. Substituting this into (23.2), we obtain

$$F\left(t' + \frac{\theta}{\omega}\right) = \frac{2A}{\pi j} \sum_{n=-\infty}^{n=+\infty} \frac{e^{jn\theta} e^{jn\omega t'}}{n} \qquad n \text{ odd} \tag{23.7}$$

The coefficient of the general term is now

$$a_n = \frac{2A}{\pi jn} e^{jn\theta} \qquad n \text{ odd} \tag{23.8}$$

If $\theta = \pi/2$, the origin of time is chosen at the center of the positive half cycle. In this case we have

$$a_n = \frac{2A}{\pi jn} e^{jn(\pi/2)} \qquad n \text{ odd} \tag{23.9}$$

Hence in this case we have

$$A_n = a_n + a_{-n} = \frac{2A}{\pi jn} [e^{jn(\pi/2)} - e^{-jn(\pi/2)}]$$

$$= \frac{2A}{\pi jn}\left(2j\sin\frac{n\pi}{2}\right) = \frac{4A}{n\pi}\sin\frac{n\pi}{2} \tag{23.10}$$

$$B_n = j(a_n - a_{-n}) = \frac{2A}{\pi n}[e^{jn(\pi/2)} + e^{-jn(\pi/2)}] = \frac{4A}{n\pi}\cos\frac{n\pi}{2} = 0 \tag{23.11}$$

Hence in this we have

$$F\left(t' + \frac{T}{4}\right) = \frac{4A}{\pi} \sum_{n=1}^{n=\infty} \frac{\sin(n\pi/2)}{n}\cos n\omega t' \qquad n \text{ odd} \tag{23.12}$$

for the real form of the Fourier expansion of the function of Fig. 23.1.

As a more complicated example of the Fourier-series expansion of a function, let us consider the function defined by Fig. 23.2. This function consists essentially of a series of positive and negative pulses that have a fundamental period of τ sec. The time origin has been chosen so that each pulse comes in the center of a half period. This choice of origin makes the Fourier series simpler. The duration of each pulse is denoted by a.

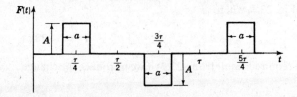

Fig. 23.2

To obtain the general coefficient of the complex Fourier series of this function, we use the general equation (22.11). We then have

$$a_n = \frac{A}{\tau}\left(\int_{\tau/4-a/2}^{\tau/4+a/2} e^{-jn\omega t}\, dt - \int_{3\tau/4-a/2}^{3\tau/4+a/2} e^{-jn\omega t}\, dt\right) \tag{23.13}$$

where

$$\tau = \frac{2\pi}{\omega} \tag{23.14}$$

We notice that the second integral is identical with the first except for the limits of integration, which are advanced by $\tau/2 = \pi/\omega$. We may therefore combine the above integral into a single integral and obtain

$$a_n = \frac{\omega A}{2\pi}(1 - e^{-jn\pi})\int_{\pi/2\omega-a/2}^{\pi/2\omega+a/2} e^{-jn\omega t}\, dt \tag{23.15}$$

Evaluating the integral, we obtain

$$a_n = \frac{A(1 - e^{-jn\pi})e^{-jn(\pi/2)}}{2\pi jn}(e^{jn(\omega a/2)} - e^{-jn(\omega a/2)}) \tag{23.16}$$

The factor $1 - e^{-jn\pi}$ vanishes for $n = 0$ and for even values of n. For odd integral values of n it is equal to 2. For these values we also have

$$e^{-jn(\pi/2)} = j(-1)^{(n+1)/2} \qquad n \text{ odd} \tag{23.17}$$

If we let

$$\delta = \frac{2a}{\tau} = \frac{\omega a}{\pi} \tag{23.18}$$

the parameter δ is the ratio of the duration of the pulse a to the fundamental half period. δ is a real number that varies from zero to unity depending on the relative width of the pulse as compared with the fundamental half period. If $\delta = 1$, we have the function of Fig. 23.1.

With this notation, (23.16) becomes

$$a_n = j(-1)^{(n+1)/2}\frac{2A\sin(n\delta\pi/2)}{n\pi} \qquad n \text{ odd} \tag{23.19}$$

It is easy to show that if $\delta = 1$ this reduces to

$$a_n = \frac{2A}{jn\pi} \tag{23.20}$$

as it should.

24 SOME REMARKS ABOUT CONVERGENCE OF FOURIER SERIES

It remains to say something about the convergence of Fourier series. To do this, it is convenient to examine the coefficients of the real form of the Fourier series. These are given by (22.18) and (22.19) in the form

$$A_n = \frac{2}{T} \int_0^T F(t) \cos n\omega t\, dt \tag{24.1}$$

and

$$B_n = \frac{2}{T} \int_0^T F(t) \sin n\omega t\, dt \tag{24.2}$$

From these equations it is plausible that the coefficients A_n and B_n must diminish indefinitely as n increases because of the more and more rapid fluctuation in sign of $\cos n\omega t$ and $\sin n\omega t$ and the consequent more complete canceling of the various elements of the definite integrals (24.1) and (24.2). This result is known as the *Riemann-Lebesgue theorem*.†

Stokes has formulated more definite results. The following statement must be understood to refer to a function that satisfies the Dirichlet conditions, but care is necessary in particular cases to determine whether discontinuities of $F(t)$ or its derivatives are introduced at the terminal points of the various segments.

1. If $F(t)$ has a finite number of isolated discontinuities in a period, the coefficients converge ultimately toward zero like the members of the sequence

$$1, \tfrac{1}{2}, \tfrac{1}{3}, \tfrac{1}{4}, \ \cdots \tag{24.3}$$

2. If $F(t)$ is everywhere continuous while its first derivative $F'(t)$ has a finite number of isolated discontinuities, the convergence is ultimately like that of the sequence

$$1, \frac{1}{2^2}, \frac{1}{3^2}, \frac{1}{4^2}, \ \cdots \tag{24.4}$$

3. If $F(t)$ and $F'(t)$ are continuous while $F''(t)$ is discontinuous at isolated points, the sequence of comparison is

$$1, \frac{1}{2^3}, \frac{1}{3^3}, \frac{1}{4^3}, \ \cdots \tag{24.5}$$

4. In general, if $F(t)$ and its derivatives up to the order $n-1$ are continuous while the nth derivative (in a period) has a finite number of isolated discontinuities, the convergence is ultimately as

$$1, \frac{1}{2^{n+1}}, \frac{1}{3^{n+1}}, \frac{1}{4^{n+1}}, \ \cdots \tag{24.6}$$

† Cf. Titchmarsh, 1937 (see References).

The nature of the proof of these statements, which is simple, may be briefly indicated for the sine series.   If we integrate by parts, we have

$$B_n = \frac{2}{T} \int_0^T F(t) \sin n\omega t \, dt$$

$$= -\frac{1}{n} \left[\frac{F(t) \cos n\omega t}{\pi} \right]_0^T + \frac{1}{\pi n} \int_0^T F'(t) \cos n\omega t \, dt \qquad (24.7)$$

The integrated term is to be calculated separately for each of the segments lying between the points of discontinuity of $F(t)$, if there are any that lie in the range extending from $t = 0$ to $t = T$.   If there is no discontinuity of $F(t)$ even at the points $t = 0$ and $t = T$, the first term vanishes.   If there is a discontinuity, then there is for all values of n an upper limit to the coefficient of $1/n$ in the first part of (24.7).   Let us denote this limit by M.   The definite integral in the second term tends ultimately to zero as n increases because of the fluctuations in the sign of $\cos n\omega t$.   Hence B_n is comparable with M/n.

If there is no discontinuity in $F(t)$, we have

$$B_n = \frac{1}{\pi n} \int_0^T F'(t) \cos n\omega t \, dt \qquad (24.8)$$

If we again integrate by parts, we obtain

$$B_n = \frac{1}{n^2} \left[\frac{F'(t) \sin n\omega t}{\pi \omega} \right]_0^T - \frac{1}{\pi \omega n^2} \int_0^T F''(t) \sin n\omega t \, dt \qquad (24.9)$$

In the integrated term of (24.9) we must take into account the discontinuities of $F'(t)$ in the interval, if there are any.   If $F'(t)$ has discontinuities, denote by M the upper limit of the coefficient of $1/n^2$; we then see that B_n is ultimately comparable with M/n^2 since the second term of (24.9) vanishes because of the fluctuation of $\sin n\omega t$.   This outlines the method of proof, and the course of the argument is apparent.

The preceding statements concerning the convergence of Fourier series are very useful.   We thus know beforehand how well or how poorly the series will converge.   We also have a partial check on the numerical work in some problems.

Differentiating a Fourier series makes the convergence poorer, while integrating the series increases its rate of convergence.   When a Fourier series has been differentiated until it converges as $1/n$, it cannot be further differentiated.

25   EFFECTIVE VALUES AND THE AVERAGE OF A PRODUCT

The determination of the root-mean-square, or effective, value of a periodic function is a common problem in electrical-circuit theory and in the theory of mechanical vibrations.   The manner in which this may be done by the use of the complex Fourier-series expansion of the function will be demonstrated.

Suppose that we have a periodic function $F(t)$ whose Fourier-series expansion is given by

$$F(t) = \sum_{n=-\infty}^{n=+\infty} a_n e^{jn\omega t} \tag{25.1}$$

By definition the root-mean-square, or effective, value of the function F_E over a period T is given by

$$F_E^2 = \frac{1}{T} \int_0^T F^2(t)\, dt \qquad T = \frac{2\pi}{\omega} \tag{25.2}$$

To obtain $F^2(t)$, we use the series expansion (25.1) and obtain

$$F^2(t) = \left(\sum_{n=-\infty}^{n=+\infty} a_n e^{jn\omega t} \right) \left(\sum_{r=-\infty}^{r=+\infty} a_r e^{jr\omega t} \right)$$

$$= \sum_{n=-\infty}^{n=+\infty} \sum_{r=-\infty}^{r=+\infty} a_n a_r e^{j(n+r)\omega t} \tag{25.3}$$

It is necessary to use two different indices in the multiplication to avoid confusion. Substituting (25.3) into (25.2), we obtain

$$F_E^2 = \frac{1}{T} \int_0^T \sum_{n=-\infty}^{n=+\infty} \sum_{r=-\infty}^{r=+\infty} a_n a_r e^{j(n+r)\omega t}\, dt \tag{25.4}$$

Carrying out term-by-term integration, we obtain

$$F_E^2 = \frac{1}{T} \sum_{n=-\infty}^{n=+\infty} \sum_{r=-\infty}^{r=+\infty} a_n a_r \int_0^{2\pi/\omega} e^{j(n+r)\omega t}\, dt \tag{25.5}$$

However, we have

$$\int_0^{2\pi/\omega} e^{jm\omega t}\, dt = \frac{e^{jm\omega t}}{jm\omega} \Big|_0^{2\pi/\omega} = 0 \qquad \text{if } m \text{ is any integer} \neq 0$$

$$= \frac{2\pi}{\omega} = T \qquad \text{if } m = 0 \tag{25.6}$$

It follows, therefore, that all the integrals in (25.5) vanish except those for which

$$r = -n \tag{25.7}$$

and we have

$$F_E^2 = \frac{1}{T} \sum_{n=-\infty}^{n=+\infty} a_n a_{-n} T = \sum_{n=-\infty}^{n=+\infty} a_n a_{-n} \tag{25.8}$$

This result may be put into a different form by recognizing that a_{-n} is the conjugate of a_n; hence the quantity in the summation sign is the square of the magnitude of a_n. Since summation over negative values of n gives the same result as summation over positive values of n, we have

$$F_E^2 = 2 \sum_{n=1}^{\infty} |a_n|^2 + a_0^2 \tag{25.9}$$

Another problem that occurs frequently in electrical-circuit theory and the theory of mechanical vibrations is the problem of determining the average value over a period of the product of two periodic functions having the same period. Suppose we have the two periodic functions, having the period T,

$$F_1(t) = \sum_{n=-\infty}^{n=+\infty} a_n e^{jn\omega t}$$
$$T = \frac{2\pi}{\omega} \tag{25.10}$$
$$F_2(t) = \sum_{r=-\infty}^{r=+\infty} b_r e^{jr\omega t}$$

We wish to compute

$$P = \frac{1}{T} \int_0^T F_1(t) F_2(t) \, dt \tag{25.11}$$

Substituting (25.10) into (25.11) and interchanging the order of summation and integration, we have

$$P = \frac{1}{T} \sum_{n=-\infty}^{n=+\infty} \sum_{r=-\infty}^{r=+\infty} a_n b_r \int_0^T e^{j(n+r)\omega t} \, dt \tag{25.12}$$

This is the same integral that we evaluated in (25.5). The result is

$$P = \sum_{n=-\infty}^{n=+\infty} a_n b_{-n} \tag{25.13}$$

This is a very concise form for the average of the product.

26 MODULATED VIBRATIONS AND BEATS

A very interesting type of oscillation occurs in radiotelephony. There we encounter *modulated vibrations*. These are oscillations in which the maximum amplitude itself is a periodic function of the time. The amplitude changes slowly compared with the frequency of the actual vibration. The latter vibration is called the *carrier wave* and has a frequency of the order of 10^6 cycles per sec, while the frequency of modulation is the frequency of the radiated tone and is of the order of 1,000 cycles per sec. This type of oscillation may be represented by the equation

$$u = A(1 + K \sin \omega_1 t) \sin \omega_2 t \qquad \omega_2 \gg \omega_1 \tag{26.1}$$

By a familiar trigonometric formula this may be written in the form

$$u = A \sin \omega_2 t - \frac{AK}{2} \cos (\omega_2 + \omega_1) t + \frac{AK}{2} \cos (\omega_2 - \omega_1) t \qquad (26.2)$$

This signifies the combination of two vibrations of equal amplitude and having angular frequencies $\omega_2 + \omega_1$ and $\omega_2 - \omega_1$ which are very close together. In practice this type of vibration may be produced in either of two ways. One method is to modulate a carrier frequency, and another is to combine two oscillations that differ very little in frequency. This latter case is known in the theory of sound as *beats*. The angular frequency of the beats is $\omega_2 - \omega_1$.

Equation (26.1) represents the simplest case of a class of functions called *almost-periodic*. If, for example, ω_1 and ω_2 are not commensurable, then the function u has no definite period; that is, no fixed interval T exists such that the value of u is repeated at a time $T + t$.

27 THE PROPAGATION OF PERIODIC DISTURBANCES IN THE FORM OF WAVES

Let us consider a process that varies as either the real or imaginary part of the function

$$u(x,t) = A e^{j\omega(t-x/v)} \qquad (27.1)$$

In this case, as t (the time) increases, the argument of the function changes. If, however, the coordinate x increases in such a way that the argument of the exponential function remains constant; that is, if

$$t - \frac{x}{v} = \text{const} \qquad (27.2)$$

then the phase of the function $u(x,t)$ is unaltered. We thus see that (27.1) represents a disturbance that travels along the x axis with a phase velocity of

$$\frac{dx}{dt} = v \qquad (27.3)$$

Now let us consider a given instant of time t_0. For this value of t we have

$$u(x,t_0) = A e^{j\omega(t_0-x/v)} \qquad (27.4)$$

The value of the function at a given point x_1 is given at this instant by

$$u(x_1,t_0) = A e^{j\omega(t_0-x_1/v)} \qquad (27.5)$$

If we now move along the x axis to the first new point x_2 such that the function at x_2 resumes its value at x_1, we have

$$e^{j\omega(t_0-x_1/v)} = e^{j\omega(t_0-x_2/v)} \qquad (27.6)$$

or

$$e^{-j(\omega x_1/v)} = e^{-j(\omega x_2/v)} \tag{27.7}$$

and hence

$$e^{j(\omega/v)(x_2-x_1)} = 1 \tag{27.8}$$

or

$$\frac{\omega}{v}(x_2 - x_1) = 2\pi \tag{27.9}$$

and hence

$$x_2 - x_1 = \lambda = \frac{2\pi v}{\omega} = vT \tag{27.10}$$

The distance λ that gives the separation of the successive points of equal phase is called the *wavelength*. If f is the frequency of the oscillation, we have from (27.10)

$$\frac{\lambda}{T} = \lambda f = v \tag{27.11}$$

that is,

Wavelength × frequency = velocity of propagation of phase   (27.12)

A process that varies in the form (27.1) is called a *plane wave* since u is constant in any plane perpendicular to the direction of propagation x. The simple plane wave (27.1) is a particular integral of a partial differential equation that is easily deduced. If u is differentiated twice with respect to t and twice with respect to x, we obtain

$$\frac{\partial^2 u}{\partial t^2} = -\omega^2 A e^{j\omega(t-x/v)}$$

$$\frac{\partial^2 u}{\partial x^2} = -\frac{\omega^2}{v^2} A e^{j\omega(t-x/v)} \tag{27.13}$$

and hence

$$\frac{\partial^2 u}{\partial x^2} = \frac{1}{v^2}\frac{\partial^2 u}{\partial t^2}. \tag{27.14}$$

This is called the *wave equation in one dimension*. It is fundamental in the study of many important physical phenomena. A detailed discussion of the wave equation is given in Chap. 12.

28 THE FOURIER INTEGRAL

In this section we shall consider the limiting form of the Fourier series as the fundamental period T is made infinite. We shall see that when this is done the series passes into an integral. A rigorous derivation of the Fourier integral is beyond the scope of this discussion and will be found in the works quoted in the references at the end of this chapter. However, the heuristic derivation given here shows the general trend of the argument.

We start with the complex Fourier-series expansion of the periodic function $F(t)$:

$$F(t) = \sum_{n=-\infty}^{n=+\infty} a_n e^{jn\omega t} \qquad T = \frac{2\pi}{\omega} \tag{28.1}$$

where the coefficients a_n are given by

$$a_n = \frac{1}{T} \int_0^T F(u) e^{-jn\omega u} du \tag{28.2}$$

Now, because of the like periodicity of $F(t)$ and $e^{-jn\omega u}$, we can take the range of integration in (28.2) form $-T/2$ to $+T/2$ instead of from 0 to T. We thus have

$$a_n = \frac{1}{T} \int_{-T/2}^{T/2} F(u) e^{-jn\omega u} du \qquad \omega = \frac{2\pi}{T} \tag{28.3}$$

Substituting this into (28.1), we obtain

$$F(t) = \sum_{n=-\infty}^{n=+\infty} \frac{1}{T} \int_{-T/2}^{+T/2} F(u) e^{(2\pi nj/T)(t-u)} du \tag{28.4}$$

Let us now place

$$\frac{1}{T} = \Delta s \tag{28.5}$$

This gives

$$F(t) = \sum_{n=+\infty}^{n=-\infty} \Delta s \int_{-T/2}^{+T/2} F(u) e^{2\pi nj(t-u)\,\Delta s} du \tag{28.6}$$

Now the definite integral

$$\int_0^\infty \phi(s)\, ds$$

may be defined as the limit, as Δs approaches zero, of the sum

$$\sum_{n=0}^{n=\infty} \phi(n\,\Delta s)\, \Delta s \tag{28.7}$$

Also, we have

$$\int_{-\infty}^{+\infty} \phi(s)\, ds = \int_{-\infty}^{0} \phi(s)\, ds + \int_0^\infty \phi(s)\, ds = \lim_{\Delta s \to 0} \sum_{n=-\infty}^{n=+\infty} \phi(n\,\Delta s)\, \Delta s \tag{28.8}$$

From this it follows that as T grows beyond all bounds the expression (28.6) passes over into the Fourier integral

$$F(t) = \int_{-\infty}^{+\infty} ds \int_{-\infty}^{+\infty} F(u) e^{2\pi js(t-u)} du$$

$$= \int_{-\infty}^{+\infty} e^{2\pi jst} ds \int_{-\infty}^{+\infty} F(u) e^{-2\pi jsu} du \qquad (28.9)$$

The second form of the identity (28.9) shows that the function $F(t)$ may be expressed by a continuous series of harmonics, one for each value of s. The possibility of such a representation is of great importance in the analytical treatment of functions that are otherwise not expressible by a unified mathematical expression.

The limitations on $F(t)$ that allow the above formal procedure to be valid will now be given:

1. $F(t)$ must be a single-valued function of the real variable t throughout the range $-\infty < t < \infty$. It may, however, have a finite number of finite discontinuities.
2. At a point of discontinuity t_0 the function will be given the mean value

$$F(t_0) = \tfrac{1}{2}[F(t_0 + 0) + F(t_0 - 0)]$$

3. The integral

$$\int_{-\infty}^{+\infty} |F(t)| \, dt$$

must exist.

We noted that when we expand a function into a Fourier series in a certain range then the function is defined by the series outside this range in a periodic manner. However, by the Fourier integral, we obtain analytical expressions for discontinuous functions that represent the function throughout the infinite range $-\infty < t < +\infty$. The following example will make this clear. Let us suppose that $F(t)$ is the single pulse given by Fig. 28.1.

The pulse has a height equal to A and begins at $t = -a$ and ends at $t = a$. Hence we have

$$F(t) = \begin{cases} 0 & t < -a \\ 0 & t > a \\ A & -a < t < a \end{cases} \qquad (28.10)$$

Substituting this value of $F(t)$ into (28.9), we obtain

$$F(t) = \int_{-\infty}^{+\infty} e^{2\pi jst} ds \int_{-a}^{+a} A e^{-2\pi jsu} du \qquad (28.11)$$

If we let

$$2\pi s = v \qquad (28.12)$$

we have

$$F(t) = \frac{A}{2\pi j} \int_{-\infty}^{+\infty} \frac{(e^{jv(t+a)} - e^{jv(t-a)}) \, dv}{v} \tag{28.13}$$

This is the Fourier-integral representation of the function (28.10).

Another form of the Fourier integral may be obtained from (28.9) by using Euler's relation on the complex exponentials and the fact that the cosine is an even function and the sine is an odd function of its argument. We thus obtain the real form of the Fourier integral

$$F(t) = 2 \int_0^\infty ds \int_{-\infty}^{+\infty} F(u) \cos 2\pi s(t - u) \, du \tag{28.14}$$

The Fourier integral is of great importance in the field of electrical communication and forms the basis of a powerful method for the solution of partial differential equations due to Cauchy.

FOURIER TRANSFORMS

In the modern literature of the mathematical theory of communication engineering, wave propagation, and other branches of applied mathematics, the expression (28.9) is written in a slightly different form.

Let us introduce the new variable

$$\omega = 2\pi s \tag{28.15}$$

In terms of the variable ω (28.9) is transformed to

$$F(t) = \frac{1}{2\pi} \int_{-\infty}^{+\infty} e^{j\omega t} \, d\omega \int_{-\infty}^{+\infty} F(u) e^{-j\omega u} \, du \tag{28.16}$$

It is convenient to write

$$g(\omega) = \frac{1}{2\pi} \int_{-\infty}^{+\infty} F(u) e^{-j\omega u} \, du \tag{28.17}$$

Then, as a consequence of (28.16), we have

$$F(t) = \int_{-\infty}^{+\infty} g(\omega) e^{j\omega t} \, d\omega \tag{28.18}$$

Since the letter chosen as a variable of integration in a definite integral is of no importance, the letter t may be used instead of the letter u and we write (28.17) in the form

$$g(\omega) = \frac{1}{2\pi} \int_{-\infty}^{+\infty} F(t) e^{-j\omega t} \, dt \tag{28.19}$$

The relations (28.18) and (28.19) are known in the literature as *Fourier transforms*. The relations are symmetrical. Many writers enhance the symmetry by writing the factor $1/\sqrt{2\pi}$ before each integral rather than having

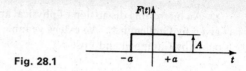

Fig. 28.1

no factor for the integral (28.18) and a factor of $1/2\pi$ for the integral (28.19). This notation introduces a change in the scale factor and is of small importance. In the literature the expression (28.19) is usually called the *Fourier transform of the function $F(t)$*.

The Fourier transform of the pulse function of Fig. 28.1 is given by

$$g(\omega) = \frac{A}{2\pi} \int_{-a}^{+a} e^{-j\omega t}\, dt = \frac{A \sin \omega a}{\pi \omega} \tag{28.20}$$

The Fourier-integral representation of the pulse function as given by (28.18) is therefore

$$
\begin{aligned}
F(t) &= \frac{A}{\pi} \int_{-\infty}^{+\infty} \frac{\sin \omega a}{\omega} e^{j\omega t}\, d\omega \\
&= \frac{A}{\pi} \int_{-\infty}^{+\infty} \frac{\sin \omega a \cos \omega t}{\omega}\, d\omega + \frac{jA}{\pi} \int_{-\infty}^{+\infty} \frac{\sin \omega a \sin \omega t}{\omega}\, d\omega
\end{aligned} \tag{28.21}
$$

Since the integrand of the second integral of (28.21) is an odd function of ω, we have

$$\int_{-\infty}^{+\infty} \frac{\sin \omega a \sin \omega t}{\omega}\, d\omega = 0 \tag{28.22}$$

Hence the pulse function is represented analytically by

$$F(t) = \frac{A}{\pi} \int_{-\infty}^{+\infty} \frac{\sin \omega a \cos \omega t}{\omega}\, d\omega \tag{28.23}$$

The fact that (28.23) actually represents the pulse function of Fig. 28.1 may be seen by using the trigonometric identity

$$\sin \omega a \cos \omega t = \tfrac{1}{2}[\sin \omega(a+t) + \sin \omega(a-t)] \tag{28.24}$$

and the following result (see Appendix F, Sec. 11):

$$\int_0^\infty \frac{\sin k\omega}{\omega}\, d\omega = \begin{cases} -\dfrac{\pi}{2} & k < 0 \\[2mm] 0 & k = 0 \\[2mm] \dfrac{\pi}{2} & k > 0 \end{cases} \tag{28.25}$$

An interesting discussion of physical applications of the Fourier integral is given by Guillemin.† An extensive table of Fourier transforms is given in a paper by Campbell and Foster.‡

PROBLEMS

Test the following series for convergence or divergence:

1. $\dfrac{1}{1} + \dfrac{1}{\sqrt{2^3}} + \dfrac{1}{\sqrt{3^3}} + \cdots + \dfrac{1}{\sqrt{n^3}} + \cdots$

2. $\dfrac{1}{1} + \dfrac{1}{\sqrt[3]{2}} + \dfrac{1}{\sqrt[3]{3}} + \cdots + \dfrac{1}{\sqrt[3]{n}} + \cdots$

3. $\dfrac{1}{3} + \dfrac{1}{\sqrt{3}} + \dfrac{1}{\sqrt[3]{3}} + \cdots + \dfrac{1}{\sqrt[n]{3}} + \cdots$

4. $\dfrac{1}{2\ln 2} + \dfrac{1}{3\ln 3} + \dfrac{1}{4\ln 4} + \cdots + \dfrac{1}{(n+1)\ln(n+1)} + \cdots$

5. $\dfrac{1}{2} + \dfrac{1}{5} + \dfrac{1}{10} + \cdots + \dfrac{1}{1+n^2} + \cdots$

For what values of the variable x are the following series convergent?

6. $1 + x + x^2 + x^3 + \cdots$

7. $x - \dfrac{x^2}{2} + \dfrac{x^3}{3} - \dfrac{x^4}{4} + \cdots$

8. $x + \dfrac{x^2}{\sqrt{2}} + \dfrac{x^3}{\sqrt{3}} + \cdots$

9. $1 + x + \dfrac{x^2}{2!} + \dfrac{x^3}{3!} + \cdots$

10. $1 - x + \dfrac{x^2}{2^2} - \dfrac{x^3}{3^2} + \cdots$

11. $\dfrac{ax}{2} + \dfrac{a^2 x^2}{5} + \dfrac{a^3 x^3}{10} + \cdots + \dfrac{a^n x^n}{n^2 + 1} + \cdots \quad a > 0$

Using the binomial series, find approximately the values of the following numbers:

12. $\sqrt{98}$ **13.** $\sqrt[5]{35}$ **14.** $\dfrac{1}{\sqrt[6]{30}}$ **15.** $\sqrt[4]{630}$

Verify the following expansions of functions by Maclaurin's series, and determine for what values of the variable they are convergent:

16. $e^x = 1 + x + \dfrac{x^2}{2!} + \cdots + \dfrac{x^{n-1}}{(n-1)!} + \cdots$

---

† E. A. Guillemin, "Communication Networks," vol. II, chap. XI, John Wiley & Sons, Inc., New York, 1935.

‡ G. A. Campbell and R. M. Foster, Fourier Integrals for Practical Applications, *Bell Telephone System Technical Publications, Monograph* B-584, September, 1931.

17. $\ln(1+x) = x - \dfrac{x^2}{2} + \dfrac{x^3}{3} - \dfrac{x^4}{4} + \cdots$

18. $\sin^{-1} x = x + \dfrac{1 \times x^3}{2 \times 3} + \dfrac{1 \times 3 \times x^5}{2 \times 4 \times 5} + \cdots$

Verify the following expansions:

19. $\tan x = x + \dfrac{x^3}{3} + \dfrac{2x^5}{15} + \dfrac{17x^7}{315} + \cdots$

20. $\cosh x = \dfrac{1}{2}(e^x + e^{-x}) = 1 + \dfrac{x^2}{2!} + \dfrac{x^4}{4!} + \dfrac{x^6}{6!} + \cdots$

21. $\ln \cos x = -\dfrac{x^2}{2} - \dfrac{x^4}{12} - \dfrac{x^6}{45} \cdots$

Compute the values to four decimal places of the following functions by substituting directly in their power-series expansion:

22. e

23. $\tan^{-1} \frac{1}{5}$

24. $\cos 10°$

25. $\sqrt{e}$

Obtain the following expansions:

26. $e^{-x} \cos x = 1 - x + \frac{1}{3}x^3 - \frac{1}{6}x^4 + \cdots$

27. $\dfrac{\cos x}{\sqrt{1+x}} = 1 - \frac{1}{2} - \frac{1}{8}x^2 - \dfrac{x^3}{16} + \dfrac{49x^4}{384} + \cdots$

28. $\dfrac{\ln(1+x)}{1 + \sin x} = x - \frac{3}{2}x^2 + \frac{11}{6}x^3 - \frac{23}{12}x^4 + \cdots$

Using series, find approximately the values of the following integrals to four decimals:

29. $\displaystyle \int_0^{\frac{1}{2}} \dfrac{\cos x\, dx}{1 + x}$

30. $\displaystyle \int_0^{\frac{1}{2}} e^x \ln(1+x)\, dx$

31. $\displaystyle \int_0^{\frac{1}{2}} \dfrac{e^{-x^2} dx}{\sqrt{1 - x^2}}$

32. $\displaystyle \int_0^1 e^{-x^2}\, dx$

33. $\displaystyle \int_0^1 e^{-x} \cos \sqrt{x}\, dx$

34. How many terms of the Maclaurin's series for $\sin x$ must be taken to give $\sin 45°$ correct to five decimal places?

35. How many terms of the series $\ln(1+x)$ must be taken to give $\ln 1.2$ correct to five decimals?

36. Verify the approximate formula $\ln(10 + x) \doteq 2.303 + x/10$.

Evaluate each of the following indeterminate forms:

37. $\displaystyle \lim_{x \to \infty} \dfrac{\ln x}{x^n}$

38. $\displaystyle \lim_{\theta \to \pi/2} \dfrac{\tan 3\theta}{\tan \theta}$

39. $\displaystyle \lim_{x \to 0} \left(\dfrac{1}{\sin^2 x} - \dfrac{1}{x^2} \right)$

40. $\lim\limits_{x \to \infty} \dfrac{x + \ln x}{x \ln x}$

41. $\lim\limits_{x \to 1} \left(\dfrac{2}{x^2 - 1} - \dfrac{1}{x - 1} \right)$

42. $\lim\limits_{x \to 2} \dfrac{x^2 - 4}{x^2} \tan \dfrac{\pi x}{4}$

43. $\lim\limits_{x \to 0} \left(\dfrac{1}{\sin^3 x} - \dfrac{1}{x^3} \right)$

Test the following series for uniform convergence:

44. $\dfrac{\cos x}{1^2} + \dfrac{\cos 3x}{3^2} + \dfrac{\cos 5x}{5^2} + \cdots + \dfrac{\cos (2n - 1) x}{(2n - 1)^2} + \cdots$

45. $\dfrac{\sin x}{1} + \dfrac{\sin 3x}{3} + \dfrac{\sin 5x}{5} + \cdots + \dfrac{\sin (2n - 1) x}{2n - 1} + \cdots$

46. Show that if $F(t)$ is an even function, that is, $F(t) = F(-t)$, then its real Fourier series expansion contains no sine terms.

47. Show that if $F(t)$ is an odd function so that $F(-t) = -F(t)$, then its real Fourier series expansion contains no cosine terms and no constant term.

48. Expand the function $F(t) = kt$ in the interval $-T/2 < t < T/2$ in a complex Fourier series. Plot the function defined by the series outside this range.

49. Expand t^2 in the interval $0 < t < T$ in a complex and a real Fourier series.

50. Expand the function e^{at} in a complex and a real Fourier series in the interval $0 < t < T$.

51. Show that, if A is a constant, then in the range $0 < t < \pi$

$$A = \frac{4A}{\pi} \left(\sin t + \frac{\sin 3t}{3} + \frac{\sin 5t}{5} + \cdots \right)$$

52. Show that if $F(t) = \phi(-t)$ for $-T/2 < t < 0$ and $F(t) = \phi(t)$ for $0 < t < T/2$, then the real Fourier series for $F(t)$ contains no sine terms.

53. Show that if $F(t) = -\phi(-t)$ for $-T/2 < t < 0$ and $F(t) = \phi(t)$ for $0 < t < T/2$, then the real Fourier series for $F(t)$ contains no cosine terms.

54. Show by using the results of Probs. 52 and 53 that a function $F(t)$ may be expanded in the range $0 < t < T/2$ either in terms of sines alone or in terms of cosines alone.

55. Expand the function of period 12 defined by the following equations in the interval $-6 < t < 6$:

$$F(t) = 0 \qquad \text{for } -6 \leqslant t \leqslant -3$$
$$F(t) = t + 3 \qquad \text{for } -3 < t \leqslant 0$$
$$F(t) = 3 - t \qquad \text{for } 0 < t \leqslant 3$$
$$F(t) = 0 \qquad \text{for } 3 < t \leqslant 6$$

Plot the function.

56. Prove that the numerical value of $\sin t$, $|\sin t|$, is an even function of period π, and find the Fourier series that represents it.

57. Find the Fourier series that represents $|\cos t|$.

58. Show that if a is not an integer

$$\sin ax = \frac{2 \sin a\pi}{\pi} \left(\frac{\sin x}{1^2 - a^2} - \frac{2 \sin 2x}{2^2 - a^2} + \frac{3 \sin 3x}{3^2 - a^2} - \cdots \right)$$

59. Show that

$$\cosh ax = \frac{2a \sinh a\pi}{\pi} \left(\frac{1}{2a^2} - \frac{\cos x}{1^2 + a^2} + \frac{\cos 2x}{2^2 + a^2} - \frac{\cos 3x}{3^2 + a^2} + \cdots \right)$$

60. The quantity $g(\omega)$ given by

$$g(\omega) = \int_{-\infty}^{+\infty} F(t) e^{-j\omega t} dt$$

is the complex Fourier transform of $F(t)$.   Use the results of Sec. 28 to show that the inverse transform of $g(\omega)$ is given by

$$F(t) = \frac{1}{2\pi} \int_{-\infty}^{+\infty} g(\omega) e^{j\omega t} d\omega$$

61. Compute the complex Fourier transform of the function $F(t) = e^{-|t|}$.

62. Use the results of Sec. 28 to deduce that, if

$$g(p) = \int_0^\infty f(x) \sin px \, dx$$

then

$$f(x) = \frac{2}{\pi} \int_0^\infty g(p) \sin xp \, dp$$

under suitable restrictions.   The quantity $g(p)$ is said to be the Fourier sine transform of $f(x)$, and $f(x)$ is said to be the inverse Fourier sine transform of $g(p)$.

63. Deduce that, under suitable restrictions, if

$$g(p) = \int_0^\infty f(x) \cos px \, dx$$

then

$$f(x) = \frac{2}{\pi} \int_0^\infty g(p) \cos xp \, dp$$

$g(p)$ is the Fourier cosine transform of $f(x)$, and $f(x)$ is the inverse Fourier cosine transform of $g(p)$.

64. Show that if

$$F(\theta) = \frac{a_0}{2} + \sum_{n=1}^\infty (a_n \cos n\theta + b_n \sin n\theta)$$

then

$$\frac{1}{\pi} \int_{-\pi}^{+\pi} [F(\theta)]^2 \, d\theta = \frac{a_0^2}{2} + \sum_{n=1}^\infty (a_n^2 + b_n^2)$$

65. By the use of the Fourier cosine integral, show that, for $x > 0$,

$$e^{-bx} = \frac{2b}{\pi} \int_0^\infty \frac{\cos ux \, du}{u^2 + b^2}$$

REFERENCES

1893. Byerly, W. E.: "Fourier's Series and Spherical Harmonics," Ginn and Company, Boston.

1904. Goursat, E.: "A Course in Mathematical Analysis," Ginn and Company, Boston.

1908. Bromwich, T. J.: "Theory of Infinite Series," The Macmillan Company, New York.

1911. Wilson, E. B.: "Advanced Calculus," Ginn and Company, Boston.

1927. Whittaker, E. T., and G. N. Watson: "A Course in Modern Analysis," chap. 9, Cambridge University Press, New York.

1937. Titchmarsh, E. C.: "Introduction to the Theory of Fourier Integrals," Oxford University Press, New York.

1941. Churchill, R. V.: "Fourier Series and Boundary Value Problems," McGraw-Hill Book Company, New York.

1951. Sneddon, I. N.: "Fourier Transforms," McGraw-Hill Book Company, New York.

1951. Tranter, C. J.: "Integral Transforms in Mathematical Physics," John Wiley & Sons, Inc., New York.

1954. Erdélyi, A. (ed.): "Tables of Integral Transforms," McGraw-Hill Book Company, New York.

appendix D
The Solution of Transcendental and Polynomial Equations

1   INTRODUCTION

In applied mathematics the need frequently arises for solving numerically transcendental or higher-degree algebraic equations; for example, in determining the natural frequencies of oscillation of a uniform prismatic bar built in at one end and with the other end supported, it is necessary to solve the transcendental equation

$$\tan \theta = \tanh \theta \tag{1.1}$$

Transcendental equations occur very frequently in determining the natural frequencies and modes of oscillation of electrical and mechanical systems.   Higher-degree polynomial equations also arise frequently in practice.   Algebraic formulas exist for the solution of the general quadratic, cubic, and quartic equations with literal coefficients.   However, no formulas exist for the solution of a general algebraic equation with literal coefficients if it is of higher degree than the fourth.   The formulas for the cubic and quartic equations are sometimes laborious to apply to given cases.   The importance of being able to solve numerically equations of this type is very great, and this appendix will be devoted to a brief discussion of possible methods of solution.

2 GRAPHICAL SOLUTION OF TRANSCENDENTAL EQUATIONS

As an example, let us solve the transcendental equation

$$\cos x \cosh x + 1 = 0 \qquad (2.1)$$

This equation occurs in determining the natural frequencies of oscillation of a clamped cantilever beam.

Equation (2.1) is satisfied by an infinite number of values of x.   Let us write Eq. (2.1) in the form

$$\cos x = -\frac{1}{\cosh x} \qquad (2.2)$$

If we plot the curves

$$y_1 = -\frac{1}{\cosh x} \qquad (2.3)$$

and

$$y_2 = \cos x \qquad (2.4)$$

the roots of (2.1) are given by the abscissas of the points of intersection of these two curves as shown in Fig. 2.1.

From the figure, we find the first three roots to be approximately

$$x_1 = 1.87 \qquad x_2 = \frac{3\pi}{2} \qquad x_3 = \frac{5\pi}{2} \qquad (2.5)$$

Since

$$\lim_{x \to \infty} \frac{1}{\cosh x} = 0 \qquad (2.6)$$

the higher roots are given with satisfactory accuracy by the equation

$$x_r = (r - \tfrac{1}{2})\pi \qquad r = 2, 3, 4, \ldots \qquad (2.7)$$

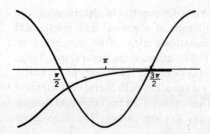

Fig. 2.1

This example illustrates the general principle involved in the graphical solution of transcendental equations.   That is, if we wish to solve the equation

$$F(x) = 0 \tag{2.8}$$

we write it in the form

$$F_1(x) = F_2(x) \tag{2.9}$$

This may usually be done in many ways.   We then draw the curves

$$y_1 = F_1(x)$$
$$y_2 = F_2(x) \tag{2.10}$$

The real roots of $F(x) = 0$ are evidently the abscissas of the points of intersection of these curves.   The larger the scale of the graph and the more carefully the drawing is performed, the greater the accuracy of the roots. Having once found the approximate location of the roots, the accuracy may be improved by an iterative process called the *Newton-Raphson method*.

3   THE NEWTON-RAPHSON METHOD

Let us consider the function

$$y = F(x) \tag{3.1}$$

Let us draw this curve as in Fig. 3.1.   The point $x = A$ is a root of the equation

$$F(x) = 0 \tag{3.2}$$

Let us draw the tangent to the curve at the point P.   This tangent will intersect the x axis at a point x_1.
From the figure we have

$$\tan \phi = \frac{F(x_0)}{x_0 - x_1} = F'(x_0) \tag{3.3}$$

Hence

$$x_1 = x_0 - \frac{F(x_0)}{F'(x_0)} \tag{3.4}$$

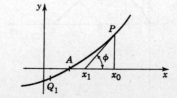

Fig. 3.1

If we now set up the sequence

$$x_{r+1} = x_r - \frac{F(x_r)}{F'(x_r)} \tag{3.5}$$

it is apparent from the figure that this sequence tends to the root A.   If we start on the other side of A where the arc of the curve is convex to the x axis, the first step carries us to the opposite side of A where the arc is concave to the x axis; after this the sequence tends to the root as before.

This discussion is based on the following two assumptions:

1. That the slope of the curve does not become zero along the arc Q_1, P.
2. That the curve has no inflection point along Q_1, P.

More precisely, we can say that, if $F(x)$ has only one root between two points Q_1 and P while $F'(x)$ and $F''(x)$ are never zero between these two points, then the Newton-Raphson process will succeed if we begin it at one of the points for which $F(x)$ and $F''(x)$ have the same sign.

It is sometimes more convenient to use the formula

$$x_{r+1} = x_r = \frac{F(x_r)}{F'(x_0)} \tag{3.6}$$

instead of (3.5).

This means that in the successive steps of the process we replace the tangents calculated at x_1, x_2, etc., by lines parallel to the tangent at P.   This saves the trouble of calculating $F'(x_r)$ at each stage.

As an example of the method, let it be required to determine the solution of the equation

$$x = \sin x + \frac{\pi}{2} \tag{3.7}$$

To obtain a rough estimate of the root, we draw the curves

$$y_1 = x - \frac{\pi}{2} \quad \text{and} \quad y_2 = \sin x \tag{3.8}$$

as in Fig. 3.2.

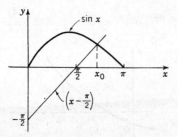

Fig. 3.2

From the graph, we obtain

$$x_0 = 2.3 \text{ radians} \tag{3.9}$$

as a rough estimate of the root.   With this value for x_0, we begin the iterative process.   Here we have

$$F(x) = x - \sin x - \frac{\pi}{2} \tag{3.10}$$

$$F'(x) = 1 - \cos x \tag{3.11}$$

The first approximation is

$$x_1 = x_0 - \frac{F(x_0)}{F'(x_0)} = x_0 - \frac{x_0 - \sin x_0 - \pi/2}{1 - \cos x_0} \tag{3.12}$$

Now

$$x_0 = 2.3 \text{ radians} = 132° \tag{3.13}$$

$$\sin x_0 = 0.7431 \qquad \cos x_0 = -0.669 \tag{3.14}$$

Hence

$$x_1 = 2.3 - \frac{2.3 - 0.743 - 1.57}{1 + 0.669} = 2.308 \tag{3.15}$$

This is a very good approximation to the root.   If more significant figures are desired, the iterative process of (3.6) may be repeated.

4   SOLUTION OF CUBIC EQUATIONS

In the study of the natural frequencies of undamped electrical and mechanical systems with three degrees of freedom, we have to determine the solution of the equation

$$Z^3 + A_2 Z^2 + A_1 Z + A_0 = 0 \tag{4.1}$$

We may eliminate the Z^2 term by the substitution

$$Z = x - \frac{A_2}{3} \tag{4.2}$$

We then obtain

$$x^3 + \left(A_1 - \frac{A_2^2}{3} \right) x + \left(A_0 - \frac{A_1 A_2}{3} + \frac{2}{27} A_2^3 \right) = 0 \tag{4.3}$$

This may be written in the form

$$x^3 - qx - r = 0 \tag{4.4}$$

There are two principal cases to consider.

a **THE CASE WHERE** $27r^2 > 4q^3$

In this case the cubic has one real root and two complex roots.   Then, if q and r are *both* positive, we find ϕ such that

$$\cos\phi = \left(\frac{3}{q}\right)^{\frac{3}{2}} \frac{r}{2} \tag{4.5}$$

Then the real root is given by the equation

$$x_0 = \frac{2}{\sqrt{3}} q^{\frac{1}{2}} \cosh\frac{\phi}{3} \tag{4.6}$$

Dividing (4.4) by $x - x_0$, we reduce the equation to a quadratic and obtain the pair of complex roots by solving the resulting quadratic.

If q is *negative* and r is *positive*, we find ϕ such that

$$\sinh\phi = \left(\frac{3}{-q}\right)^{\frac{3}{2}} \frac{r}{2} \tag{4.7}$$

Then the real root is given by the equation

$$x = \frac{2}{\sqrt{3}} (-q)^{\frac{1}{2}} \sinh\frac{\phi}{3} \tag{4.8}$$

It may be noted that we may always suppose that r is positive since if we change the sign of r we merely change the sign of the roots.

b **THE CASE WHERE** $27r^2 < 4q^3$

In this case the cubic equation has three real roots.   We now find the smallest positive angle ϕ such that

$$\cos\phi = \left(\frac{3}{q}\right)^{\frac{3}{2}} \frac{r}{2} \tag{4.9}$$

Then the real roots are given by

$$x_1 = \frac{2}{\sqrt{3}} q^{\frac{1}{2}} \cos\frac{\phi}{3}$$
$$x_2 = -\frac{2}{\sqrt{3}} q^{\frac{1}{2}} \cos\frac{\pi - \phi}{3} \tag{4.10}$$
$$x_3 = -\frac{2}{\sqrt{3}} q^{\frac{1}{2}} \cos\frac{\pi + \phi}{3}$$

For example, let it be required to solve

$$Z^3 + 9Z^2 + 23Z + 14 = 0 \tag{4.11}$$

In this case we remove the Z^2 term by writing

$$Z = x - 3 \tag{4.12}$$

and the equation becomes

$$x^3 - 4x - 1 = 0 \tag{4.13}$$

We now have

$$27 \times 1^2 < 4 \times 4^3 \tag{4.14}$$

Hence this equation comes under case b, and the roots are all real. We now compute

$$\frac{\phi}{3} = 23° 41' \tag{4.15}$$

and, substituting into (4.10), we obtain

$$\begin{aligned} x_1 &= 2.11 \\ x_2 &= -1.86 \\ x_3 &= -0.254 \end{aligned} \tag{4.16}$$

Hence the roots of (4.11) are

$$\begin{aligned} Z_1 &= -0.89 \\ Z_2 &= -4.86 \\ Z_3 &= -3.254 \end{aligned} \tag{4.17}$$

5 GRAEFFE'S ROOT-SQUARING METHOD

In the last section we considered some algebraic formulas for the solution of the cubic equation. There also exists a formula solution for the quartic equation.† These formulas are, in general, laborious to use in numerical computations. No formulas exist for the solution of a general algebraic equation with literal coefficients if it is of higher degree than the fourth.

In this section we shall discuss a method for the numerical solution of an algebraic equation of any degree. Before considering this method, it is well to recall the following properties concerning the nature of algebraic equations:

1. The equation

$$x^n + a_1 x^{n-1} + a_2 x^{n-2} + \cdots + a_n = 0 \tag{5.1}$$

   where the coefficients a_r are real numbers, has exactly n roots. Some of the roots may be repeated roots.
2. If n is a positive odd integer, the equation always has one real root.

† See, for example, E. R. Dickson, "Elementary Theory of Equations," p. 31, John Wiley & Sons, Inc., New York, 1922.

3. The number of positive roots either is equal to the number of variations of signs of the a's or is less than this number of variations by an even integer (Descartes' rule of signs).
4. The complex roots occur in conjugate complex pairs.

The method we shall consider was suggested by Dandelin in 1826 and independently by Graeffe in 1837. It is of great use, especially in the case of equations possessing complex roots. The fundamental principle of the method is to form a new equation whose roots are some high power of the roots of the given equation. That is, if the roots of the original equation are $x_1, x_2, \ldots, x_n$, the roots of the new equation are $x_1^s, x_2^s, \ldots, x_n^s$. If s is large, then the high powers of the roots will be *widely separated*. If the roots are very widely separated, they may be obtained by a simple process.

Let the roots of Eq. (5.1) be $-r_1, -r_2, \ldots, -r_n$. These values are the roots of the equation with the signs reversed and are called the *Encke roots*. (We shall assume at present that they are real and unequal.) Since the r quantities are the roots of the equation with the signs reversed, it may be factored in the form

$$(x + r_1)(x + r_2)(x + r_3) \cdots (x + r_n) = 0 \tag{5.2}$$

If we use the convenient notation

$$[r_i] = r_1 + r_2 + \cdots + r_n \qquad = \text{sum of Encke roots}$$

$$[r_i r_j] = r_1 r_2 + r_1 r_3 + \cdots \qquad = \text{sum of products of Encke roots taken two at a time}$$

$$[r_i r_j r_k] = r_1 r_2 r_3 + r_1 r_2 r_4 + \cdots = \text{sum of products of} \tag{5.3}$$
$$\text{Encke roots taken three at a time}$$

$$[r_1 r_2 \cdots r_n] = \text{product of all the roots}$$

then on multiplying the various factors of (5.2), we obtain

$$x^n + [r_i] x^{n-1} + [r_i r_j] x^{n-2} + \cdots + [r_1 r_2 \cdots r_n] = 0 \tag{5.4}$$

SQUARING THE ROOTS

A simple device by which one may obtain a new equation whose Encke roots are the squares of the Encke roots of the original equation will now be explained. Let us write

$$F(x) = (x + r_1)(x + r_2) \cdots (x + r_n) = 0 \tag{5.5}$$

Then

$$F(-x) = (-x + r_1)(-x + r_2) \cdots (-x + r_n) = 0 \tag{5.6}$$

and

$$F(x) F(-x) = (r_1^2 - x^2)(r_2^2 - x^2)(r_3^2 - x^2) \cdots (r_n^2 - x^2) = 0 \tag{5.7}$$

If in (5.7) we let

$$y = -x^2 \qquad (5.8)$$

we obtain

$$F(x)\,F(-x) = (y + r_1^2)(y + r_2^2)\,\cdots\,(y + r_n^2) = 0 \qquad (5.9)$$

Now the roots of (5.9) are $-r_1^2,\ -r_2^2,\ \ldots,\ -r_n^2$, and hence the Encke roots are $r_1^2, r_2^2, \ldots, r_n^2$, and hence they are the squares of the Encke roots of (5.5). If we now write $F(x)$ in the form

$$F(x) = x^n + a_1 x^{n-1} + a_2 x^{n-2} + \cdots + a_n = 0 \qquad (5.6a)$$

then

$$F(x)\,F(-x) = (x^n + a_1 x^{n-1} + \cdots + a_n)$$
$$[(-x)^n + a_1(-x)^{n-1} + \cdots + a_n] = 0 \qquad (5.7a)$$

Carrying out the multiplication and writing

$$-x^2 = y \qquad (5.8a)$$

we obtain

$$y^n + (a_1^2 - 2a_2)\,y^{n-1} + (a_2^2 - 2a_1 a_3 + 2a_4)\,y^{n-2} + \cdots = 0 \qquad (5.9a)$$

This may be written in the form

$$y^n + \left\{\begin{array}{c} a_1^2 \\ -2a_2 \end{array}\right\} y^{n-1} + \left\{\begin{array}{c} a_2^2 \\ -2a_1 a_3 \\ +2a_4 \end{array}\right\} y^{n-2} + \left\{\begin{array}{c} a_3^2 \\ -2a_2 a_4 \\ +2a_1 a_5 \\ -2a_6 \end{array}\right\} y^{n-3} + \cdots = 0 \qquad (5.10)$$

We notice that the coefficients of (5.9a) are found from the coefficients of the original equation (5.6a) by the following simple rule:

The coefficient of any power of y is formed by adding to the square of the corresponding coefficient in the original equation the doubled product of every pair of coefficients which stand equally far from it on either side. These products are taken with signs alternately negative and positive.

If a power of x is absent, then it is taken with a coefficient equal to zero. To facilitate the process, a table is constructed in the following manner:

$F(x) = x^n + a_1 x^{n-1} + a_2 x^{n-2} + \cdots + a_n = 0$					
1	a_1	a_2	a_3	a_4	a_5
	a_1^2	a_2^2	a_3^2	a_4^2	
	$-2a_2$	$-2a_1 a_3$	$-2a_2 a_4$	$-2a_3 a_5$	etc.
		$+2a_4$	$+2a_1 a_5$	$+2a_2 a_6$	
			$-2a_6$	$-2a_1 a_7$	
				$+2a_8$	
1	b_1	b_2	b_3	b_4	b_5

The b's are the coefficients of the equation whose Encke roots are the squares of the Encke roots of the original equation.

If we now repeat this process several times, we finally arrive at an equation whose Encke roots are the mth powers of the Encke roots of the original equation. This equation has the form

$$x^n + [r_i^m] x^{n-1} + [r_i^m r_j^m] x^{n-2} + \cdots = 0 \tag{5.11}$$

If now

$$r_1 > r_2 > r_3 > \cdots > r_n \tag{5.12}$$

then

$$r_1^m \gg r_2^m \gg r_3^m \cdots \gg r_n^m \tag{5.13}$$

That is, if the roots differ in magnitude in the manner (5.12), then the mth powers of the roots, where m is a large number, are widely separated. Hence

$$[r_i^m] = r_1^m + r_2^m + \cdots + r_n^m \doteq r_1^m \tag{5.14}$$

for a sufficiently large m. We also have

$$[r_i^m r_j^m] = r_1^m r_2^m + r_1^m r_3^m + \cdots \doteq r_1^m r_2^m \tag{5.15}$$

Hence from (5.14) we obtain

$$\log |r_1| \doteq \frac{1}{m} \log [r_i^m] \tag{5.16}$$

and from (5.15) we have

$$\log |r_2| \doteq \frac{1}{m} \log [r_i^m r_j^m] - \frac{1}{m} \log [r_i^m] \tag{5.17}$$

Equation (5.16) determines the absolute value of the largest root r_1, and Eq. (5.17) determines the absolute value of the second largest root r_2, and so on.

In the solution of equations by this method it is very necessary to know when to stop the root-squaring process. The time to stop is when another doubling of m produces new coefficients $[r_i^{2m}]$, $[r_i^{2m} r_j^{2m}]$ that are practically the squares of the corresponding coefficients $[r_i^m]$, $[r_i^m r_j^m]$ in the equation already obtained. To illustrate the general theory, let us consider the solution of the equation

$$F(x) = x^3 + 9x^2 + 23x + 14 = 0 \tag{5.18}$$

which has three real roots.

We construct the following table:

	x^3	x^2	x	c
p	1	9	23	14
p^2	1	35	277	196
p^4	1	671	63,009	38,416
p^8	1	324,223	3.9185×10^9	1.4757×10^9
p^{16}	1	9.728×10^{10}	1.535×10^{12}	2.177×10^{18}
p^{32}	1	9.433×10^{21}	2.357×10^{38}	$(2.177 \times 10^{18})^2$
p^{64}	1	8.898×10^{43}	$(2.357 \times 10^{38})^2$	$(2.177 \times 10^{18})^4$

In this table p^2 denotes the equation whose Encke roots are the squares of the Encke roots of p, etc.

If we stop at this stage, we have

$$\log r_1^{64} = \log (8.898 \times 10^{43}) = 43.9493 \qquad (5.19)$$

$$\log |r_1| = 0.68670 \qquad (5.20)$$

and hence

$$r_1 = \pm 4,860 \qquad (5.21)$$

This is the magnitude of the numerically greatest root.   We also have

$$\log (r_1 r_2)^{64} = 2 \log (2.357 \times 10^{38})$$
$$= 76.74492 \qquad (5.22)$$

Hence

$$\log |r_2| = \tfrac{1}{64}(76.7449 - 43.9943)$$
$$= 0.51243 \qquad (5.23)$$

Therefore

$$r_2 = \pm 3.254 \qquad (5.24)$$

Finally, we have

$$\log (r_1 r_2 r_3)^{64} = 4 \log (2.177 \times 10^{18})$$
$$= 73.352 \qquad (5.25)$$

and

$$\log |r_3| = \tfrac{1}{64}(73.352 - 76.7449)$$
$$= 0.94698 - 1 \qquad (5.26)$$

and therefore

$$r_3 = \pm 0.885 \qquad (5.27)$$

It is not possible to determine the signs of the roots by this process. Descartes' rule of signs shows that all the roots are negative, and they are therefore

$$x_1 = -4.860$$
$$x_2 = -3.254 \qquad\qquad (5.28)$$
$$x_3 = -0.885$$

COMPLEX ROOTS

We shall now discuss briefly the modifications which must be introduced in the above procedure in case the equation under consideration has complex roots. To illustrate the general procedure, let the equation under consideration be of the fifth degree, and let it have the following Encke roots:

$$r_1, Z_1, Z_1, r_2, r_3$$

where

$$|r_1| > |Z| > |r_2| > |r_3| \qquad\qquad (5.29)$$

We carry out the root-squaring process as before and obtain an equation whose Encke roots are the mth powers of the Encke roots of the original equation. For a sufficiently large m we have

$$[r_1^m] \doteq r_1^m \qquad\qquad (5.30)$$

as before, since by hypothesis r_1 is the dominant root numerically. Hence this root may be determined as before. We now have

$$[r_i^m r_j^m] \doteq r_1^m Z^m + r_1^m \bar{Z}^m$$
$$= r_1^m (Z^m + \bar{Z}^m) \qquad\qquad (5.31)$$

If we let

$$Z = R e^{j\phi} \qquad\qquad (5.32)$$

we then have

$$[r_i^m r_j^m] = 2 r_1^m R^m \cos m\phi \qquad\qquad (5.33)$$

This shows that the coefficient of the x^3 term will fluctuate in sign as the number m takes in succession a set of increasing values because of the cosine term. We also have

$$[r_i^m r_j^m r_k^m] \doteq r_1^m Z^m \bar{Z}^m = r_1^m R^{2m} \qquad\qquad (5.34)$$

Proceeding in this manner, we see that when m is large enough so that only the dominant part of the coefficient of each power of x is retained the equation whose Encke roots are the mth powers of the roots of the original equation is

$$x^5 + r_1^m x^4 + (2r_1^m R^m \cos m\phi) x^3$$

$$+ r_1^m R^{2m} x^2 + r_1^m R^{2m} r_2^m x + r_1^m R^{2m} r_2^m r_3^m = 0 \qquad (5.35)$$

From this we may find r_1, r_2, r_3, and R. To obtain the angle of the complex root, we write

$$Z = R e^{j\phi} = u + jv = R(\cos\phi + j\sin\phi) \qquad (5.36)$$

But we know that the sum of the Encke roots of the original equation is given by

$$a_1 = r_1 + Z + \bar{Z} + r_2 + r_3$$

$$= r_1 + 2u + r_2 + r_3 \qquad (5.37)$$

Hence

$$u = \frac{a_1 - r_1 - r_2 - r_3}{2} = R\cos\phi \qquad (5.38)$$

or

$$\phi = \cos^{-1}\left\{\frac{u}{R}\right\} \qquad (5.39)$$

In the above example we saw that the fluctuation of the coefficient of the x^3 term indicated the presence of a pair of complex roots. If two coefficients fluctuate in sign, the presence of two pairs of complex roots may be inferred and the analysis modified accordingly. Rather than consider any fixed rules, it is well to consider the general nature of the root-squaring process in solving equations by this method.

THE CASE OF REPEATED ROOTS

The nature of the process in case the equation has repeated roots may be illustrated by the consideration of a special case. As before, let the Encke roots of the equation be denoted by r_1, r_2, r_3, . . . , r_m. Let the root r_2 be equal to the root r_3,

$$r_2 = r_3 \qquad (5.40)$$

That is, the equation has the repeated root r_2. In this case again the equation whose Encke roots are the mth powers of those of the given equation is

$$x^n + [r_i^m] x^{n-1} + [r_i^m r_j^m] x^{n-2} + [r_i^m r_j^m r_k^m] x^{n-3} + \cdots = 0 \qquad (5.41)$$

If m is sufficiently large so that we may retain only the dominant term in each coefficient, we have

$$x^n + r_1^m x^{n-1} + 2r_1^m r_2^m x^{n-2} + r_1^m r_2^{2m} x^{n-3} + \cdots = 0 \qquad (5.42)$$

We notice that the coefficient of the term x^{n-2} does not follow the usual law that when m is doubled the coefficient is approximately squared. In this case, when m is doubled, the new coefficient is approximately *half* the square of the old one. This is an indication of a repeated root. To compute r_2, let

$$b_1 = r_1^m$$
$$b_2 = 2r_1^m r_2^m \qquad\qquad (5.43)$$
$$b_3 = r_1^m r_2^{2m}$$

If we divide b_3 by b_1, we obtain

$$\frac{b_3}{b_1} = r_2^{2m} \qquad\qquad (5.44)$$

and hence

$$|r_2| = \sqrt[2m]{\frac{b_3}{b_1}} \qquad\qquad (5.45)$$

The rest of the roots are computed as before. The foregoing discussion is the general theory of Graeffe's root-squaring method. In general, it is better to keep the basic concept of the method of root squaring in mind rather than to formulate elaborate rules for special cases.

PROBLEMS

1. Determine the roots of the equation $\tanh x = \tan x$.
2. Determine the roots of the equation $\tan x = x$.
3. Find the roots of $e^x = 5x$.
4. Solve the equation $x^3 - 2x - 1 = 0$.
5. Solve the equation $x^3 - 97x - 202 = 0$.
6. By using Graeffe's method, solve the equation $x^2 - 2x + 2 = 0$.
7. Solve the equation $x^3 + x^2 + x + 1 = 0$.
8. Find the roots of the equation $x^3 - 5x^2 + 6x - 1 = 0$:

    (a) By the formula for the solution of a cubic equation.

    (b) By Graeffe's method.

9. Solve $\tan x = 10/x$.
10. The equation $x^3 - 2x - 5 = 0$ has a real root between 2 and 3. Determine the magnitude of this root by the use of the Newton-Raphson method.
11. Compute to four decimal places the real root of the equation $x^2 + 4\sin x = 0$ by the use of the Newton-Raphson method.
12. Determine the roots of the equation $x \tan x = 10$ to three decimal places.
13. Find the smallest positive root of the equation $x^x + 2x = 6$ to three decimal places.

REFERENCES

1918. Burnside, R., and F. Panton: "Theory of Equations," 8th ed., vol. 1, Hodges & Figgis & Co., Dublin.
1924. Whittaker, E. T., and G. Robinson: "The Calculus of Observations," Blackie & Son, Ltd., Glasgow.
1934. Cowley, W. L.: "Advanced Practical Mathematics," Pitman Publishing Corporation, New York.
1936. Doherty, R. E., and G. Keller: "Mathematics of Modern Engineering," John Wiley & Sons, Inc., New York.
1949. Milne, W. E.: "Numerical Calculus," Princeton University Press, Princeton, N.J.
1950. Scarborough, J. B.: "Numerical Mathematical Analysis," Oxford University Press, New York.
1952. Hartree, D. R.: "Numerical Analysis," Oxford University Press, New York.
1954. Soroka, W. W.: "Analog Methods in Computation and Simulation," McGraw-Hill Book Company, New York.

appendix **E**

Vector and Tensor Analysis

1 INTRODUCTION

The equations of applied mathematics express relations between quantities that are capable of measurement in terms of certain defined units. The simplest types of physical quantities are completely defined by a certain simple number. Examples of such quantities are mass, temperature, length. Such quantities are called *scalars*.

There are, however, other physical quantities that are not scalars. Such quantities as the displacement of a point, the velocity of a particle, a mechanical force require three numbers to specify them completely. The three numbers that are required to specify the quantities are scalars and could be, for example, the components of the displacement of the particle with respect to an arbitrary cartesian coordinate reference frame.

We could carry out the various mathematical operations with these scalar quantities, but we would be neglecting the fact that from a physical point of view a displacement, for example, is one entity, and also we are introducing a foreign element into the question, that is, the coordinate system.

Accordingly it has been found convenient to introduce a mathematical discipline that enables us to study quantities of this type without recourse to a definite coordinate system.

It is only when we come to evaluate formulas numerically that it will be necessary to introduce a definite coordinate system. The mathematical technique that enables us to do this is *vector analysis*.

2 THE CONCEPT OF A VECTOR

A physical quantity possessing both magnitude (length) and direction is called a *vector*. Typical examples are force, velocity, acceleration, momentum. It is customary to represent vectors by letters in boldface type and scalars in lightface italics.

A vector may be indicated graphically by an arrow drawn between two points. It is thus evident that rectilinear displacements of a point and all physical quantities that can be represented by such displacements in the same manner that a scalar can be represented by the points of a straight line are vectors.

3 ADDITION AND SUBTRACTION OF VECTORS; MULTIPLICATION OF A VECTOR BY A SCALAR

A vector having been defined as an entity that behaves in the same manner as the rectilinear displacement of a point, vector addition is reduced to a composition of linear displacements.

Consider two vectors **A** and **B**, as shown in Fig. 3.1. The vector **C**, which is obtained by moving a point along **A** and then along **B**, is called the resultant, or sum, of the vectors **A** and **B**, and we write

$$\mathbf{C} = \mathbf{A} + \mathbf{B} \tag{3.1}$$

From the nature of the definition of vector addition it is apparent that

$$\mathbf{B} + \mathbf{A} = \mathbf{A} + \mathbf{B} \tag{3.2}$$

and that therefore vector addition is commutative. If the vectors **A** and **B** are situated as shown in Fig. 3.2, then the resultant vector **C** is obtained by completing the parallelogram formed by the two vectors.

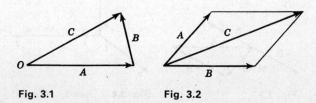

Fig. 3.1   Fig. 3.2

In general, the sum of the vectors **A** and **B** is obtained by placing the origin of **B** at the terminus of **A**. Then the vector **C** extending from the origin of **A** to the terminus of **B** is defined as the sum, or resultant, of **A** and **B**.

To add several vectors **A**, **B**, and **C**, first find the sum of **A** and **B** and the sum of **A** + **B** and **C**. It is seen that the result is the same as finding the sum of **B** and **C** first and then the sum of **A** and **B** + **C**. Thus vector addition is associative.

$$\mathbf{A} + (\mathbf{B} + \mathbf{C}) = (\mathbf{A} + \mathbf{B}) + \mathbf{C}$$

To subtract **B** from **A**, add −**B** to **A**, as shown in Fig. 3.3.

EQUALITY OF VECTORS

Two vectors are said to be equal if they have the same magnitude and the same direction.

MULTIPLICATION OF A VECTOR BY A SCALAR

By the product of a vector **a** by a scalar n we understand a vector whose magnitude is equal to the magnitude of the product of the magnitudes of **a** and n and has the same direction as **a** or the opposite direction, depending on whether the scalar n is positive or negative. We thus write

$$\mathbf{A} = n\mathbf{a} \tag{3.3}$$

to denote this new vector.

UNIT VECTORS

A vector having unit magnitude is called a *unit vector*. The most common unit vectors are those which have the directions of a right-handed cartesian coordinate system as shown in Fig. 3.4. The vector **i** is a unit vector having the x direction of the coordinate system; the unit vector **j** has the y direction; and the unit vector **k** has the z direction.

THE COMPONENTS OF A VECTOR

The components of a vector **A** are any vectors whose sum is **A**. The components most frequently used are those parallel to the axes x, y, and z. These

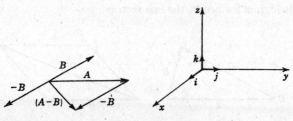

Fig. 3.3   **Fig. 3.4**

are called the *rectangular components* of the vector. If A_x, A_y, and A_z are the projections of **A** on the axes x, y, z, respectively, we may write

$$\mathbf{A} = \mathbf{i}A_x + \mathbf{j}A_y + \mathbf{k}A_z \tag{3.4}$$

If a vector **A** is given in magnitude and direction, then its components along the three axes of a cartesian reference frame are given by

$$A_x = |\mathbf{A}| \cos(\mathbf{A},\mathbf{i})$$
$$A_y = |\mathbf{A}| \cos(\mathbf{A},\mathbf{j}) \tag{3.5}$$
$$A_z = |\mathbf{A}| \cos(\mathbf{A},\mathbf{k})$$

where $|\mathbf{A}|$ denotes the magnitude of the vector. If, conversely, the three components of the vector are assigned, the vector **A** is uniquely specified as the diagonal of the rectangular parallelepiped whose edges are the vectors $\mathbf{i}A_x$, $\mathbf{j}A_y$, and $\mathbf{k}A_z$. Its magnitude is

$$|\mathbf{A}| = \sqrt{A_x^2 + A_y^2 + A_z^2} \tag{3.6}$$

and its direction is given by the three cosines which can be found from (3.5) and (3.6).

4 THE SCALAR PRODUCT OF TWO VECTORS

The scalar, or dot, product of two vectors is defined as a scalar quantity equal in magnitude to the product of the magnitudes of the two given vectors and the cosine of the angle between them. The scalar product of the two vectors **A** and **B** is thus given by the equation

$$\mathbf{A} \cdot \mathbf{B} = |\mathbf{A}||\mathbf{B}| \cos(\mathbf{A},\mathbf{B}) \tag{4.1}$$

The cosine of the angle between the directions of the two vectors becomes $+1$ when the directions are the same, -1 when they are opposite, and 0 when they are perpendicular.

It is clear from the definition of the dot product that the commutative law of multiplication holds, that is,

$$\mathbf{A} \cdot \mathbf{B} = \mathbf{B} \cdot \mathbf{A} \tag{4.2}$$

The fact that the distributive law of multiplication holds can be seen with the aid of Fig. 4.1. We have from the figure

$$\mathbf{P} \cdot \mathbf{R} + \mathbf{Q} \cdot \mathbf{R} = (OA)R + (AB)R$$
$$= (OB)R = (\mathbf{P} + \mathbf{Q}) \cdot \mathbf{R} \tag{4.3}$$

From the fundamental unit vectors we can form the following scalar products:

$$\mathbf{i} \cdot \mathbf{i} = \mathbf{j} \cdot \mathbf{j} = \mathbf{k} \cdot \mathbf{k} = 1$$
$$\mathbf{i} \cdot \mathbf{j} = \mathbf{j} \cdot \mathbf{k} = \mathbf{k} \cdot \mathbf{i} = 0 \tag{4.4}$$

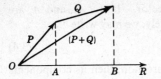

Fig. 4.1

As a consequence of these results, we obtain

$$\mathbf{A} \cdot \mathbf{B} = (\mathbf{i}A_x + \mathbf{j}A_y + \mathbf{k}A_z) \cdot (\mathbf{i}B_x + \mathbf{j}B_y + \mathbf{k}B_z)$$
$$= A_x B_x + A_y B_y + A_z B_z \tag{4.5}$$

We may also write

$$A^2 = \mathbf{A} \cdot \mathbf{A} = A_x^2 + A_y^2 + A_z^2 = |\mathbf{A}|^2 \tag{4.6}$$

5   THE VECTOR PRODUCT OF TWO VECTORS

The vector, or cross, product of two vectors is defined to be a vector perpendicular to the plane of the two given vectors in the sense of advance of a right-handed screw rotated from the first to the second of the given vectors through the smaller angle between their positive directions.

The meaning of this definition is made clear in Fig. 5.1.   The magnitude of this vector is equal to the product of the magnitudes of the two given vectors times the sine of the angle between them.

The vector, or cross, product is denoted by $\mathbf{A} \times \mathbf{B}$.   As is clear from the definition and the figure, the commutative law does not hold for this type of multiplication; instead we have

$$\mathbf{A} \times \mathbf{B} = - \mathbf{B} \times \mathbf{A} \tag{5.1}$$

Also we have

$$\mathbf{A} \times \mathbf{A} = 0 \tag{5.2}$$

It follows from the definition that if the vector product of two vectors vanishes, the vectors are parallel provided neither vector is zero.

$\mathbf{A} \times \mathbf{B}$

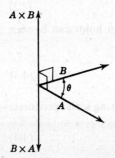

$\mathbf{B} \times \mathbf{A}$

Fig. 5.1

VECTOR REPRESENTATION OF SURFACES

Let us consider a plane surface such as shown in Fig. 5.2.   Since this surface has a magnitude represented by its area and a direction specified by its normal, it is a vector quantity.   A certain ambiguity exists as to the positive sense of the normal.   In order to remove this ambiguity, the following conventions are adopted:

If the surface is part of a closed surface, the outward-drawn normal is taken as positive.

If the surface is not part of a closed surface, the positive sense in describing the periphery is connected with the positive direction of the normal by the rule that a right-handed screw rotated in the plane of the surface in the positive sense of describing the periphery advances along the positive normal.   For example, in Fig. 5.2, if the periphery of the surface is described in the sense *ABC*, the positive sense of the normal, and therefore the direction of the vector **S** representing the surface, is upward.

If a surface is not plane, it may be divided into a number of elementary surfaces each of which is plane to any desired degree of approximation.   In this case the vector representative of the entire surface is the sum of the vectors representing its elements.   Two surfaces, considered as vectors, are equal if the representative vectors are equal.   Therefore two plane surfaces are equal if they have equal areas and are normal in the same direction even if they have different shapes.

A curved surface may be replaced by a plane surface perpendicular to its representative vector having an area equal to the magnitude of this vector. The vector representing a closed surface is zero because the projection of the entire surface on any plane is zero; since as much of the projected area is negative as positive, it therefore follows that the vector representing the entire surface has zero components along the three axes x, y, z, and consequently it equals zero.

THE DISTRIBUTIVE LAW OF VECTOR MULTIPLICATION

To prove that the distributive law of multiplication holds for the vector product, consider the prism of Fig. 5.3.   The edges of this prism are the vectors **P**, **Q**,

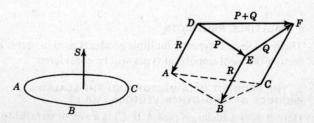

Fig. 5.2   Fig. 5.3

$\mathbf{P} + \mathbf{Q}$, and $\mathbf{R}$.   The vectors representing the faces of the closed prism are

$$\overline{ABED} = \mathbf{R} \times \mathbf{P}$$

$$\overline{BCFE} = \mathbf{R} \times \mathbf{Q}$$

$$\overline{ACFD} = (\mathbf{P} + \mathbf{Q}) \times \mathbf{R}$$

$$\overline{ABC} = \tfrac{1}{2}(\mathbf{Q} \times \mathbf{P})$$

$$\overline{DEF} = \tfrac{1}{2}(\mathbf{P} \times \mathbf{Q})$$

Therefore the representative vector for the entire polyhedral surface is

$$\mathbf{R} \times \mathbf{P} + \mathbf{R} \times \mathbf{Q} + (\mathbf{P} + \mathbf{Q}) \times \mathbf{R} + \tfrac{1}{2}(\mathbf{Q} \times \mathbf{P}) + \tfrac{1}{2}(\mathbf{P} \times \mathbf{Q}) = 0 \qquad (5.3)$$

or

$$(\mathbf{P} + \mathbf{Q}) \times \mathbf{R} = \mathbf{P} \times \mathbf{R} + \mathbf{Q} \times \mathbf{R} \qquad (5.4)$$

By the definition of the vector product we have the following relations for the unit vectors:

$$\mathbf{i} \times \mathbf{j} = \mathbf{k} \qquad \mathbf{j} \times \mathbf{k} = \mathbf{i} \qquad \mathbf{k} \times \mathbf{i} = \mathbf{j}$$

$$\mathbf{i} \times \mathbf{i} = \mathbf{j} \times \mathbf{j} = \mathbf{k} \times \mathbf{k} = 0 \qquad (5.5)$$

If we write the vectors $\mathbf{P}$ and $\mathbf{Q}$ in terms of their rectangular components,

$$\mathbf{P} = \mathbf{i}P_x + \mathbf{j}P_y + \mathbf{k}P_z$$

$$\mathbf{Q} = \mathbf{i}Q_x + \mathbf{j}Q_y + \mathbf{k}Q_z \qquad (5.6)$$

and realize that the distributive law holds for a vector product, we obtain

$$\mathbf{P} \times \mathbf{Q} = (\mathbf{i}P_x + \mathbf{j}P_y + \mathbf{k}P_z) \times (\mathbf{i}Q_x + \mathbf{j}Q_y + \mathbf{k}Q_z)$$

$$= \mathbf{i}(P_y Q_z - P_z Q_y) + \mathbf{j}(P_z Q_x - P_x Q_z) + \mathbf{k}(P_x Q_y - P_y Q_x) \qquad (5.7)$$

in view of the properties of the unit vectors expressed by (5.5).   This expression can be represented in a compact fashion by the determinant

$$\mathbf{P} \times \mathbf{Q} = \begin{vmatrix} \mathbf{i} & \mathbf{j} & \mathbf{k} \\ P_x & P_y & P_z \\ Q_x & Q_y & Q_z \end{vmatrix} \qquad (5.8)$$

6 MULTIPLE PRODUCTS

There are several types of multiple products used in vector analysis, and in this section the most important types will be considered.

a THE PRODUCT OF A VECTOR AND THE SCALAR PRODUCT OF TWO OTHER VECTORS, A(B·C)

Here $\mathbf{B} \cdot \mathbf{C}$ is a scalar so that $\mathbf{A}(\mathbf{B} \cdot \mathbf{C})$ is a vector parallel to $\mathbf{A}$ and is of course an entirely different vector from $(\mathbf{A} \cdot \mathbf{B})\mathbf{C}$.

b SCALAR PRODUCT OF A VECTOR AND THE VECTOR PRODUCT OF TWO OTHER VECTORS†

Consider $\mathbf{A} \cdot \mathbf{B} \times \mathbf{C}$.   In this case we have the important relation

$$\mathbf{A} \cdot (\mathbf{B} \times \mathbf{C}) = \mathbf{B} \cdot (\mathbf{C} \times \mathbf{A}) = \mathbf{C} \cdot (\mathbf{A} \times \mathbf{B}) \tag{6.1}$$

The proof of this relation follows from the fact that each of these expressions represents the volume of the parallelepiped whose edges are $\mathbf{A}, \mathbf{B}, \mathbf{C}$.   Furthermore, all three expressions give this volume with the positive sign provided the vectors $\mathbf{A}, \mathbf{B}, \mathbf{C}$, in that order, form a right-handed system.

c VECTOR PRODUCT OF A VECTOR AND THE VECTOR PRODUCT OF TWO OTHER VECTORS

Consider the product

$$\mathbf{q} = \mathbf{a} \times (\mathbf{b} \times \mathbf{c}) \tag{6.2}$$

By the definition of a cross product $\mathbf{q}$ is perpendicular to the vector $\mathbf{a}$ and also to the vector $\mathbf{b} \times \mathbf{c}$.   Accordingly the vector $\mathbf{q}$ lies in the plane of $\mathbf{b}$ and $\mathbf{c}$ and may be expressed in the form

$$\mathbf{q} = u\mathbf{b} + v\mathbf{c} \tag{6.3}$$

where u and v are scalar multipliers.

Now we have

$$\mathbf{q} \cdot \mathbf{a} = u(\mathbf{b} \cdot \mathbf{a}) + v(\mathbf{c} \cdot \mathbf{a}) = 0 \tag{6.4}$$

or

$$v = -\frac{u(\mathbf{b} \cdot \mathbf{a})}{\mathbf{c} \cdot \mathbf{a}} \tag{6.5}$$

Hence

$$\begin{aligned} \mathbf{q} &= u\mathbf{b} - \frac{u(\mathbf{b} \cdot \mathbf{a})\,\mathbf{c}}{\mathbf{c} \cdot \mathbf{a}} \\ &= \frac{u}{\mathbf{c} \cdot \mathbf{a}} [\mathbf{b}(\mathbf{a} \cdot \mathbf{c}) - \mathbf{c}(\mathbf{a} \cdot \mathbf{b})] \end{aligned} \tag{6.6}$$

Now let

$$n = \frac{u}{\mathbf{c} \cdot \mathbf{a}} \tag{6.7}$$

where n is some scalar.   To find the magnitude of n, consider a set of cartesian coordinate axes oriented so that the vectors $\mathbf{a}$, $\mathbf{b}$, and $\mathbf{c}$ have the following components in terms of these axes:

$$\begin{aligned} \mathbf{a} &= \mathbf{i}a_x + \mathbf{j}a_y + \mathbf{k}a_z \\ \mathbf{b} &= \mathbf{i}b_x \\ \mathbf{c} &= \mathbf{i}c_x + \mathbf{j}c_y \end{aligned} \tag{6.8}$$

† This product is commonly called the *scalar triple product* and is denoted by $(\mathbf{ABC})$.

In this case we have

$$b \times c = k b_x c_y \tag{6.9}$$

$$\begin{aligned}
a \times (b \times c) &= (i a_x + j a_y + k a_z) \times (k b_x c_y) \\
&= -j(a_x b_x c_y) + i(a_y b_x c_y) \\
&= i b_x(a_y c_y) + i b_x a_x c_x - i b_x a_x c_x - j(a_x b_x c_y) \\
&= i b_x(a_x c_x + a_y c_y) - (i c_x + j c_y)(a_x b_x) \\
&= b(a \cdot c) - c(a \cdot b) \tag{6.10}
\end{aligned}$$

Hence, comparing (6.9) and (6.10), it is seen that the constant n is equal to 1.

7 DIFFERENTIATION OF A VECTOR WITH RESPECT TO TIME

The differential coefficient or derivative of a vector $\mathbf{A}$ with respect to a scalar variable t, say, the time, is defined as a limit by the equation

$$\frac{d\mathbf{A}}{dt} = \lim_{\Delta t \to 0} \frac{\mathbf{A}(t + \Delta t) - \mathbf{A}(t)}{\Delta t} \tag{7.1}$$

Since division by a scalar does not alter vectorial properties, the derivative of a vector with respect to a scalar variable is itself a vector.

As an example of differentiation of a vector with respect to a scalar, consider a particle p moving along a curve c, as shown in Fig. 7.1. Let the position of the particle with respect to the origin of a cartesian reference frame be denoted by $\mathbf{r}$.   In that case the velocity of the particle is given by

$$\mathbf{v} = \frac{d\mathbf{r}}{dt} = \frac{d\mathbf{r}}{ds}\frac{ds}{dt} \tag{7.2}$$

where ds is a differential of arc measured along the curve, as shown in Fig. 7.2. Now

$$\frac{d\mathbf{r}}{ds} = \lim_{\Delta s \to 0} \frac{\Delta \mathbf{r}}{\Delta s} = \mathbf{T} \tag{7.3}$$

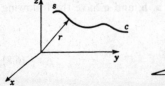

Fig. 7.1

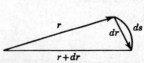

Fig. 7.2

where **T** is a unit vector tangent to the curve defining the path of the particle. Hence we have

$$\mathbf{v} = \frac{d\mathbf{r}}{dt} = \mathbf{T}\frac{ds}{dt} \tag{7.4}$$

The quantity ds/dt is the speed of the particle. The acceleration of the particle is defined as the time derivative of the velocity. We thus have

$$\mathbf{a} = \frac{d\mathbf{v}}{dt} = \frac{d}{dt}\mathbf{T}\frac{ds}{dt} = \frac{d\mathbf{T}}{dt}\frac{ds}{dt} + \mathbf{T}\frac{d^2 s}{dt^2} \tag{7.5}$$

Now

$$\frac{d\mathbf{T}}{dt} = \frac{d\mathbf{T}}{ds}\frac{ds}{dt} \tag{7.6}$$

Consider Fig. 7.3.
We may write

$$\frac{d\mathbf{T}}{ds} = \frac{d\mathbf{T}}{d\phi}\frac{d\phi}{ds} \tag{7.7}$$

$$\frac{d\mathbf{T}}{d\phi} = \mathbf{n} = \text{unit inward normal to the curve} \tag{7.8}$$

$$\frac{d\phi}{ds} = \frac{1}{\rho} = \frac{1}{\text{radius of curvature of path}} \tag{7.9}$$

Substituting these expressions into (7.6), we obtain

$$\frac{d\mathbf{T}}{dt} = \frac{\mathbf{n}}{\rho}\frac{ds}{dt} \tag{7.10}$$

Substituting this into (7.5), we finally obtain

$$\mathbf{a} = \mathbf{T}\frac{d^2 s}{dt^2} + \frac{\mathbf{n}}{\rho}\left(\frac{ds}{dt}\right)^2 \tag{7.11}$$

We thus see that the acceleration of the particle consists of two terms. The first term depends upon the rate of change of speed of the particle and is directed along the tangent to the particle; the second term is the centripetal

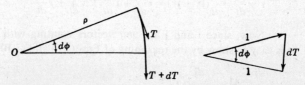

Fig. 7.3

acceleration of the particle and depends on the radius of curvature ρ of the particle and the square of the speed.   This acceleration is directed in a direction normal to the curve and toward the center of curvature.

Since the derivatives of vectors with respect to a scalar variable are deduced by a limiting process from subtraction of vectors and division by scalars, which are operations subject to the rules of ordinary algebra, it follows that the rules of the differential calculus can be extended at once to the differentiation of a sum of vectors,

$$\frac{d}{dt}(\mathbf{A} + \mathbf{B}) = \frac{d\mathbf{A}}{dt} + \frac{d\mathbf{B}}{dt} \tag{7.12}$$

or the product of a scalar and a vector,

$$\frac{d}{dt}(u\mathbf{a}) = \mathbf{a}\frac{du}{dt} + u\frac{d\mathbf{a}}{dt} \tag{7.13}$$

And, similarly,

$$\frac{d}{dt}(\mathbf{A} \cdot \mathbf{B}) = \mathbf{B} \cdot \frac{d\mathbf{A}}{dt} + \mathbf{A} \cdot \frac{d\mathbf{B}}{dt} \tag{7.14}$$

$$\frac{d}{dt}(\mathbf{A} \times \mathbf{B}) = \frac{d\mathbf{A}}{dt} \times \mathbf{B} + \mathbf{A} \times \frac{d\mathbf{B}}{dt} \tag{7.15}$$

APPLICATION TO MECHANICS—CORIOLIS' ACCELERATION

An interesting application of vector differentiation of importance to mechanics will now be discussed.   Consider a small insect on a rotating plane sheet of cardboard.   The cardboard will be assumed to be rotating about a fixed point O with a uniform angular evlocity w.   It will be supposed that the insect may move with respect to the cardboard.   In order to describe the velocity and acceleration of the moving insect, let a cartesian coordinate system x, y be drawn on the rotating cardboard with the center of rotation as the origin.   Let the position vector of the insect be

$$\mathbf{r} = \mathbf{i}x + \mathbf{j}y \tag{7.16}$$

where $\mathbf{i}$ and $\mathbf{j}$ are unit vectors in the x and y directions that rotate with the cardboard.   The velocity of the insect $\mathbf{v}$ relative to fixed space is

$$\mathbf{v} = \frac{d\mathbf{r}}{dt} = \frac{d}{dt}(\mathbf{i}x + \mathbf{j}y) = \mathbf{i}\frac{dx}{dt} + \mathbf{j}\frac{dy}{dt} + x\frac{d\mathbf{i}}{dt} + y\frac{d\mathbf{i}}{dt} \tag{7.17}$$

Now, since $\mathbf{i}$ and $\mathbf{j}$ are *unit* vectors rotating with angular velocity w, it is easy to show by the reasoning of Eqs. (7.6) and (7.10) that

$$\frac{d\mathbf{j}}{dt} = w\mathbf{j} \qquad \frac{d\mathbf{j}}{dt} = -w\mathbf{i} \tag{7.18}$$

Hence (7.17) may be written in the form

$$\mathbf{v} = (\dot{x} - wy)\mathbf{i} + (\dot{y} + wx)\mathbf{j} \tag{7.19}$$

where the dots indicate differentiation with respect to the time in the usual manner. The acceleration **a** of the insect is given by

$$\mathbf{a} = \frac{d\mathbf{v}}{dt} = (\ddot{x} - w\dot{y})\mathbf{i} + (\ddot{y} + w\dot{x})\mathbf{j} + (\dot{x} - wy)\frac{d\mathbf{i}}{dt} + (\dot{y} + wx)\frac{d\mathbf{j}}{dt}$$

$$= (\ddot{x} - 2w\dot{y} - w^2 x)\mathbf{i} + (\ddot{y} + 2w\dot{x} - w^2 y)\mathbf{j} \tag{7.20}$$

Equation (7.20) indicates that the acceleration of the insect consists of three separate terms of the form

$$\mathbf{a} = \mathbf{a}_1 + \mathbf{a}_2 + \mathbf{a}_3 \tag{7.21}$$

where

$$\begin{aligned} \mathbf{a}_1 &= \ddot{x}\mathbf{i} + \ddot{y}\mathbf{j} \\ \mathbf{a}_2 &= 2w(-\dot{y}\mathbf{i} + \dot{x}\mathbf{j}) \\ \mathbf{a}_3 &= -w^2(x\mathbf{i} + y\mathbf{j}) = -w^2\,\mathbf{r} \end{aligned} \tag{7.22}$$

$\mathbf{a}_1$ is the acceleration of the insect with respect to the rotating coordinate system. If the angular velocity w were zero and the card were stationary, this would be the only acceleration. $\mathbf{a}_2$ is the so-called *Coriolis' acceleration*. It depends on the angular velocity w of the rotating card and on the *velocity* of the particle *relative to* the rotating coordinate axis. This relative velocity $\mathbf{v}_r$ is given by

$$\mathbf{v}_r = \dot{x}\mathbf{i} + \dot{y}\mathbf{j} \tag{7.23}$$

If the scalar product of the relative velocity $\mathbf{v}_r$ and the Coriolis' acceleration $\mathbf{a}_2$ is taken, the following result is obtained:

$$\mathbf{a}_2 \cdot \mathbf{v}_r = 0 \tag{7.24}$$

This indicates that the Coriolis' acceleration is perpendicular to the relative velocity.

The acceleration $\mathbf{a}_3$ is the centripetal acceleration. Its magnitude is proportional to the square of the angular velocity w of the frame of reference and to the distance of the particle from the center of rotation. It is directed radially toward the center of rotation. Equation (7.20) may be generalized, and the acceleration of a particle in a rotating reference plane such as the earth can be computed. The Coriolis' acceleration causes deviation of falling bodies and projectiles and explains the motion of Foucault's pendulum.†

---

† See J. L. Synge and B. A. Griffith, "Principles of Mechanics," 1st ed., chap. 13, McGraw-Hill Book Company, New York, 1942.

8　THE GRADIENT

Let $\phi(x,y,z)$ be a scalar function of position in space, that is, of the coordinates x, y, z. If the coordinates x, y, z are increased by dx, dy, dz, respectively, we have

$$d\phi = \frac{\partial \phi}{\partial x} dx + \frac{\partial \phi}{\partial y} dy + \frac{\partial \phi}{\partial z} dz \tag{8.1}$$

If we denote by $d\mathbf{r}$ the vector representing the displacement specified by dx, dy, dz, then

$$d\mathbf{r} = \mathbf{i}\, dx + \mathbf{j}\, dy + \mathbf{k}\, dz \tag{8.2}$$

In vector analysis a certain vector differential operator ∇ (read "del") defined by

$$\nabla = \mathbf{i}\frac{\partial}{\partial x} + \mathbf{j}\frac{\partial}{\partial y} + \mathbf{k}\frac{\partial}{\partial z} \tag{8.3}$$

plays a very prominent role. The gradient of a scalar function $\phi(x,y,z)$ is defined by

$$\operatorname{grad} \phi = \mathbf{i}\frac{\partial \phi}{\partial x} + \mathbf{j}\frac{\partial \phi}{\partial y} + \mathbf{k}\frac{\partial \phi}{\partial z} \tag{8.4}$$

Operating with ∇ on the scalar function $\phi(x,y,z)$, we get

$$\nabla \phi = \mathbf{i}\frac{\partial \phi}{\partial x} + \mathbf{j}\frac{\partial \phi}{\partial y} + \mathbf{k}\frac{\partial \phi}{\partial z} = \text{a vector} \tag{8.5}$$

This is just the expression (8.4) defined as the gradient of ϕ. Now, from (8.1) and (8.2), we have

$$d\phi = \left(\mathbf{i}\frac{\partial \phi}{\partial x} + \mathbf{j}\frac{\partial \phi}{\partial y} + \mathbf{k}\frac{\partial \phi}{\partial z}\right) \cdot (\mathbf{i}\, dx + \mathbf{j}\, dy + \mathbf{k}\, dz)$$

$$= (\nabla \phi) \cdot d\mathbf{r} \tag{8.6}$$

The equation

$$\phi(x,y,z) = \text{const} \tag{8.7}$$

represents a certain surface, and as we change the value of the constant, we obtain a family of surfaces.

Consider the surfaces of Fig. 8.1. If dn denotes the distance along the normal from the point P to the surface S_2, we may write

$$dn = \mathbf{n} \cdot d\mathbf{r} \tag{8.8}$$

where $\mathbf{n}$ is the unit normal to the surface S_1 at P. We have

$$d\phi = \frac{\partial \phi}{\partial n} dn = \frac{\partial \phi}{\partial n} \mathbf{n} \cdot d\mathbf{r} = (\nabla \phi) \cdot d\mathbf{r} \tag{8.9}$$

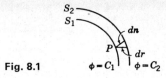

Fig. 8.1

and, in particular, if $d\mathbf{r}$ lies in the surface S_1, we have

$$d\phi = (\nabla\phi)\cdot d\mathbf{r} = 0 \tag{8.10}$$

showing that the vector $\nabla\phi$ is normal to the surface $\phi = \text{const.}$ Since the vector $d\mathbf{r}$ is arbitrary, we have from (8.9)

$$\nabla\phi = \left(\frac{\partial\phi}{\partial n}\right)\mathbf{n} \tag{8.11}$$

Hence $\nabla\phi$ is a vector whose magnitude is equal to the maximum rate of change of ϕ with respect to the space variables and has the direction of that change.

9   THE DIVERGENCE AND GAUSS'S THEOREM

The scalar product of the vector operator ∇ and a vector $\mathbf{A}$ gives a scalar that is called the divergence of $\mathbf{A}$; that is,

$$\nabla\cdot\mathbf{A} = \frac{\partial A_x}{\partial x} + \frac{\partial A_y}{\partial y} + \frac{\partial A_z}{\partial z} = \text{div }\mathbf{A} \tag{9.1}$$

This quantity has an important application in hydrodynamics.

Consider a fluid of density $\rho(x,y,z,t)$ and velocity $\mathbf{v} = \mathbf{v}(x,y,z,t)$, and let

$$\mathbf{V} = \mathbf{v}\rho \tag{9.2}$$

If $\mathbf{S}$ is the representative vector of the area of a plane surface, then $\mathbf{V}\cdot\mathbf{S}$ is the mass of fluid flowing through the surface S in a unit time.

Consider a small fixed rectangular parallelepiped of dimensions dx, dy, dz, as shown in Fig. 9.1.   The mass of fluid flowing in through face F_1 per unit time is

$$V_y\,dx\,dz = (\rho v)_y\,dx\,dz \tag{9.3}$$

and that flowing out through face F_2 is

$$V_{y+dy}\,dx\,dz = \left(V_y + \frac{\partial V_y}{\partial y}\,dy\right)dx\,dz \tag{9.4}$$

Hence the net increase of mass of fluid inside the parallelepiped per unit time is

$$V_y\,dx\,dz - \left(V_y + \frac{\partial V_y}{\partial y}\,dy\right)dx\,dz = -\frac{\partial V_y}{\partial y}\,dx\,dz\,dy \tag{9.5}$$

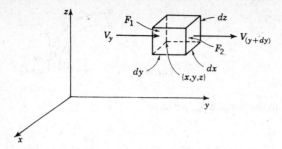

Fig. 9.1

Considering the net increase of mass of fluid per unit time entering through the other two pairs of faces, we obtain

$$-\left(\frac{\partial V_x}{\partial x} + \frac{\partial V_y}{\partial y} + \frac{\partial V_z}{\partial z}\right) dx\,dy\,dz = -(\nabla \cdot \mathbf{V})\,dx\,dy\,dz \qquad (9.6)$$

as the total increase in mass in the parallelepiped per unit time.

But by the principle of conservation of matter, this must be equal to the time rate of increase of density multiplied by the volume of the parallelepiped. Hence

$$-(\nabla \cdot \mathbf{V})\,dx\,dy\,dz = \left(\frac{\partial \rho}{\partial t}\right) dx\,dy\,dz \qquad (9.7)$$

or

$$\nabla \cdot \mathbf{V} = -\frac{\partial \rho}{\partial t} \qquad (9.8)$$

This is known in hydrodynamics as the *equation of continuity*. If the fluid is incompressible, then

$$\nabla \cdot \mathbf{V} = -\frac{\partial \rho}{\partial t} = 0 \qquad (9.9)$$

The name *divergence* originated in this interpretation of $\nabla \cdot \mathbf{V}$. For since $-\nabla \cdot \mathbf{V}$ represents the excess of the inward over the outward flow, or the convergence of the fluid, so $\nabla \cdot \mathbf{V}$ represents the excess of the outward over the inward flow, or the divergence of the fluid.

SURFACE INTEGRAL

Consider a surface s as shown in Fig. 9.2. Divide the surface into the representative vectors $d\mathbf{s}_1, d\mathbf{s}_2, \ldots$, etc. Let $\mathbf{V}_i$ be the value of the vector function of position $\mathbf{V}_i(x, y, z)$ at $d\mathbf{s}_i$. Then

$$\lim_{\substack{\Delta s_i \to 0 \\ n \to \infty}} \sum_{i=1}^{n} \mathbf{V}_i \cdot d\mathbf{s}_i = \iint_s \mathbf{V} \cdot d\mathbf{s} \qquad (9.10)$$

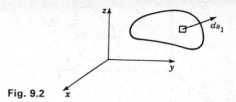

Fig. 9.2

is known as the *surface integral* of **V** over the surface *s*. The sign of the integral depends on which face of the surface is taken as positive. If the surface is closed, the outward normal is taken as positive.

Since

$$d\mathbf{s} = \mathbf{i}\, ds_x + \mathbf{j}\, ds_y + \mathbf{k}\, ds_z \tag{9.11}$$

we have

$$\iint_s \mathbf{V} \cdot d\mathbf{s} = \iint (V_x\, ds_x + V_y\, ds_y + V_z\, ds_z) \tag{9.12}$$

The surface integral of a vector **V** is called the flux of **V** throughout the surface.

GAUSS'S THEOREM

This is one of the most important theorems of vector analysis. It states that the volume integral of the divergence of a vector field **A** taken over any volume *V* is equal to the surface integral of **A** taken over the closed surface surrounding the volume *V*; that is,

$$\iiint_v (\nabla \cdot \mathbf{A})\, dv = \iint_s \mathbf{A} \cdot d\mathbf{s} \tag{9.13}$$

To prove Gauss's theorem, let us expand the left-hand side of Eq. (9.13). We then have

$$\iiint_v (\nabla \cdot \mathbf{A})\, dv = \iiint_v \left(\frac{\partial A_x}{\partial x} + \frac{\partial A_y}{\partial y} + \frac{\partial A_z}{\partial z}\right) dx\, dy\, dz$$

$$= \iiint_v \frac{\partial A_x}{\partial x}\, dx\, dy\, dz + \iiint_v \frac{\partial A_y}{\partial y}\, dx\, dy\, dz$$

$$+ \iiint_v \frac{\partial A_z}{\partial z}\, dx\, dy\, dz \tag{9.14}$$

Let us consider the first integral on the right. Integrate with respect to x, that is, along a strip of cross section $dy\, dz$ extending from P_1 to P_2 of Fig. 9.3. We thus obtain

$$\iiint_v \frac{\partial A_x}{\partial x}\, dx\, dy\, dz = \iint [A_x(x_2, y, z) - A_x(x_1, y, z)]\, dy\, dz \tag{9.15}$$

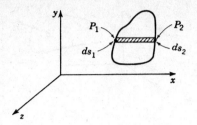

Fig. 9.3

Here (x_1, y, z) are the coordinates of P_1, and $(x_2 y, z)$ are the coordinates of P_2. Now at P_1 we have

$$dy\,dz = -ds_x \tag{9.16}$$

At P_2 we have

$$dy\,dz = ds_x \tag{9.17}$$

Therefore

$$\int\int\int_v \frac{\partial A_x}{\partial x}\,dx\,dy\,dz = \int\int_s A_x\,ds_x \tag{9.18}$$

where the surface integral on the right is evaluated over the entire surface. In the same manner we obtain

$$\int\int\int_v \frac{\partial A_y}{\partial y}\,dx\,dy\,dz = \int\int_s A_y\,ds_y \tag{9.19}$$

$$\int\int\int_v \frac{\partial A_z}{\partial z}\,dx\,dy\,dz = \int\int_s A_z\,ds_z \tag{9.20}$$

If we now add (9.18), (9.19), and (9.20), we obtain Gauss's theorem:

$$\int\int\int_v (\nabla \cdot \mathbf{A})\,dv = \int\int_s (A_x\,ds_x + A_y\,ds_y + A_z\,ds_z)$$

$$= \int\int_s \mathbf{A} \cdot d\mathbf{s} \tag{9.21}$$

GREEN'S THEOREM

By the use of Gauss's theorem we are able to make some important transformations.   Consider

$$\mathbf{A} = u\,\nabla w \tag{9.22}$$

that is, let the vector field **A** be the product of a scalar function u and the gradient of another scalar function w.   Consider

$$\nabla \cdot \mathbf{A} = \frac{\partial A_x}{\partial x} + \frac{\partial A_y}{\partial y} + \frac{\partial A_z}{\partial z} = \frac{\partial}{\partial x}\left(u\frac{\partial w}{\partial x}\right) + \frac{\partial}{\partial y}\left(u\frac{\partial w}{\partial y}\right) + \frac{\partial}{\partial z}\left(u\frac{\partial w}{\partial z}\right)$$

$$= u\left(\frac{\partial^2 w}{\partial x^2} + \frac{\partial^2 w}{\partial y^2} + \frac{\partial^2 w}{\partial z^2}\right) + \frac{\partial u}{\partial x}\frac{\partial w}{\partial x} + \frac{\partial u}{\partial y}\frac{\partial w}{\partial y} + \frac{\partial u}{\partial z}\frac{\partial w}{\partial z}$$

$$= u\,\nabla^2 w + \nabla u \cdot \nabla w \tag{9.23}$$

If we place this value of $\nabla \cdot \mathbf{A}$ in the left-hand side of Gauss's theorem (9.21), we obtain

$$\iiint_v (u\,\nabla^2 w + \nabla u \cdot \nabla w)\,dv = \iint_s (u\,\nabla w)\cdot d\mathbf{s} \tag{9.24}$$

This transformation is referred to as the *first form of Green's theorem.*

In (9.24) if we interchange the functions u and w, we obtain

$$\iiint_v (w\,\nabla^2 u + \nabla u \cdot \nabla w)\,dv = \iint_s (w\,\nabla u)\cdot d\mathbf{s} \tag{9.25}$$

If we now subtract Eq. (9.25) from Eq. (9.24), we obtain

$$\iiint_v (u\,\nabla^2 w - w\,\nabla^2 u)\,dv = \iint_s (u\,\nabla w - w\,\nabla u)\cdot d\mathbf{s} \tag{9.26}$$

This transformation is referred to as the *second form of Green's theorem.* The transformations (9.24) and (9.26) are of extreme importance in the fields of electrodynamics and hydrodynamics.

10   THE CURL OF A VECTOR FIELD AND STOKES'S THEOREM

THE LINE INTEGRAL

Let **A** be a vector field in space and AB (Fig. 10.1) a curve described in the sense A to B.   Let the continuous curve AB be subdivided into infinitesimal vector elements $d\mathbf{l}_1, d\mathbf{l}_2, \ldots, d\mathbf{l}_n$.   Take the scalar products $\mathbf{A}_1 \cdot d\mathbf{l}_1$,

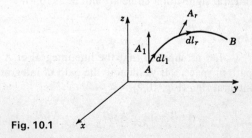

Fig. 10.1

$\mathbf{A}_2 \cdot d\mathbf{l}_2, \ldots, \mathbf{A}_n \cdot d\mathbf{l}_n$, where $\mathbf{A}_1, \mathbf{A}_2, \ldots, \mathbf{A}_n$ are the values that the vector field $\mathbf{A}$ takes at the junction points of the vectors $d\mathbf{l}_1, d\mathbf{l}_2, \ldots, d\mathbf{l}_n$. The sum of these scalar products, that is,

$$\sum_A^B \mathbf{A}_r \cdot d\mathbf{l}_r = \int_A^B \mathbf{A} \cdot d\mathbf{l} \tag{10.1}$$

summed up along the entire length of the curve, is known as the *line integral* of $\mathbf{A}$ along the curve AB. It is obvious that the line integral from B to A is the negative of that from A to B.

In terms of cartesian components we can write

$$\int_A^B \mathbf{A} \cdot d\mathbf{l} = \int_A^B (A_x \, dx + A_y \, dy + A_z \, dz) \tag{10.2}$$

If $\mathbf{F}$ represents the force on a moving particle, then the line integral of $\mathbf{F}$ over the path described by the particle is the work done by the force.

Let $\mathbf{A}$ be the gradient $\nabla\phi$ of a scalar function of position; that is,

$$\mathbf{A} = \nabla\phi \tag{10.3}$$

Then

$$\int_A^B \mathbf{A} \cdot d\mathbf{l} = \int_A^B (\nabla\phi) \cdot d\mathbf{l} = \int_A^B \left(\frac{\partial\phi}{\partial x} dx + \frac{\partial\phi}{\partial y} dy + \frac{\partial\phi}{\partial z} dz \right) \tag{10.4}$$

But we have

$$\frac{\partial\phi}{\partial x} dx + \frac{\partial\phi}{\partial y} dy + \frac{\partial\phi}{\partial z} dz = d\phi \tag{10.5}$$

the total differential of ϕ. We thus have

$$\int_A^B \mathbf{A} \cdot d\mathbf{l} = \int_A^B d\phi = \phi_B - \phi_A \tag{10.6}$$

where ϕ_B and ϕ_A are the values of ϕ at the points B and A, respectively. It follows from this that the line integral of the gradient of any scalar function of position ϕ around a closed curve vanishes, because if the curve is closed, the points A and B are coincident and the line integral is equal to $\phi_A - \phi_A$, which is zero. The line integral around a closed curve is denoted by an integral sign with a circle around it, as follows:

$$\oint \mathbf{A} \cdot d\mathbf{l}$$

Let us suppose that the line integral of $\mathbf{A}$ vanishes about *every* closed path in space. If we denote the path of integration by a subscript under the integral sign (Fig. 10.2),

$$\int_{ACB} \mathbf{A} \cdot d\mathbf{l} - \int_{ADB} \mathbf{A} \cdot d\mathbf{l} = \int_{ACB} \mathbf{A} \cdot d\mathbf{l} + \int_{BDA} \mathbf{A} \cdot d\mathbf{l} = 0 \tag{10.7}$$

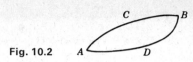

Fig. 10.2

and therefore

$$\int_{ACB} \mathbf{A} \cdot d\mathbf{l} = \int_{ADB} \mathbf{A} \cdot d\mathbf{l} \tag{10.8}$$

This shows that the line integral of $\mathbf{A}$ from A to B is independent of the path followed. It is apparent, therefore, that it can depend only upon the initial point A and the final point B of the path, that is,

$$\int_A^B \mathbf{A} \cdot d\mathbf{l} = \phi_B - \phi_A \tag{10.9}$$

Now if we take the two points A and B very close together, we have

$$\mathbf{A} \cdot d\mathbf{l} = d\phi = (\nabla \phi) \cdot d\mathbf{l} \tag{10.10}$$

or

$$(\mathbf{A} - \nabla \phi) \cdot d\mathbf{l} = 0 \tag{10.11}$$

As this is true for all directions, the vector $\mathbf{A} - \nabla \phi$ can have no component in any direction, and hence it must vanish. Therefore

$$\mathbf{A} = \nabla \phi \tag{10.12}$$

That is, if the line integral of $\mathbf{A}$ vanishes about every closed path, $\mathbf{A}$ must be the gradient of some scalar function ϕ.

THE CURL OF A VECTOR FIELD

If $\mathbf{A}$ is a vector field, the curl, or rotation, of $\mathbf{A}$ is defined as the vector function of space obtained by taking the vector product of the operator ∇ and $\mathbf{A}$. That is,

$$\begin{aligned}
\operatorname{curl} \mathbf{A} &= \nabla \times \mathbf{A} \\
&= \mathbf{i}\left(\frac{\partial A_z}{\partial y} - \frac{\partial A_y}{\partial z}\right) + \mathbf{j}\left(\frac{\partial A_x}{\partial z} - \frac{\partial A_z}{\partial x}\right) + \mathbf{k}\left(\frac{\partial A_y}{\partial x} - \frac{\partial A_x}{\partial y}\right)
\end{aligned} \tag{10.13}$$

This may be written conveniently in the following determinantal form:

$$\nabla \times \mathbf{A} = \begin{vmatrix} \mathbf{i} & \mathbf{j} & \mathbf{k} \\ \dfrac{\partial}{\partial x} & \dfrac{\partial}{\partial y} & \dfrac{\partial}{\partial z} \\ A_x & A_y & A_z \end{vmatrix} \tag{10.14}$$

If $\mathbf{A} = \nabla \phi$,

$$\nabla \times \mathbf{A} = \nabla \times (\nabla \phi) = \mathbf{i}\left(\frac{\partial^2 \phi}{\partial y \, \partial z} - \frac{\partial^2 \phi}{\partial z \, \partial y}\right)$$

$$+ \mathbf{j}\left(\frac{\partial^2 \phi}{\partial z \, \partial x} - \frac{\partial^2 \phi}{\partial x \, \partial z}\right) + \mathbf{k}\left(\frac{\partial^2 \phi}{\partial x \, \partial y} - \frac{\partial^2 \phi}{\partial y \, \partial x}\right) = 0 \quad (10.15)$$

It thus follows that if $\mathbf{A}$ is the gradient of a scalar, the curl of $\mathbf{A}$ vanishes.

LINE INTEGRAL IN A PLANE

To show the connection between the line integral and the curl of a vector field, let us compute the line integral of a vector field $\mathbf{A}$ around an infinitesimal rectangle of side Δx and Δy lying in the xy plane, as shown in Fig. 10.3. That is, we shall compute $\oint \mathbf{A} \cdot d\mathbf{l}$ around this rectangle. We may write down the various contributions to this integral as follows:

Along AB: $A_x \, \Delta x$

Along BC: $\left(A_y + \dfrac{\partial A_y}{\partial x} \Delta x\right) \Delta y$

Along CD: $-\left(A_x + \dfrac{\partial A_x}{\partial y} \Delta y\right) \Delta x$

Along DA: $-A_y \, \Delta y$

where we have made use of the fact that Δx and Δy are infinitesimals. Adding the various contributions, we obtain

$$\oint_{ABCD} \mathbf{A} \cdot d\mathbf{l} = \left(\frac{\partial A_y}{\partial x} - \frac{\partial A_x}{\partial y}\right) \Delta x \, \Delta y \qquad (10.16)$$

In view of (10.13), this may be written in the form

$$\oint_{ABCD} \mathbf{A} \cdot d\mathbf{l} = (\nabla \times \mathbf{A})_z \, ds_{xy} \qquad (10.17)$$

where $\nabla \times \mathbf{A}_z$ is the z component of the curl of $\mathbf{A}$ and ds_{xy} is the area of the rectangle $ABCD$.

Consider now a closed curve in the xy plane, as shown in Fig. 10.4. Divide the space inside C by a network of lines joining a network of infinitesimal rectangles.

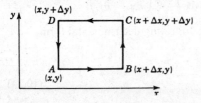

Fig. 10.3

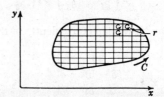

Fig. 10.4

If we take the sum of the line integrals around the various meshes, we obtain

$$\sum_{r=1}^{\infty} \oint_r \mathbf{A} \cdot d\mathbf{l} = \sum_{r=1}^{\infty} (\nabla \times \mathbf{A})_z \, ds_{xy} \tag{10.18}$$

Now it is easily seen that the contributions to the line integrals of adjoining meshes neutralize each other because they are traversed in opposite directions; the only contributions which are not neutralized are those on the periphery of the surface. Hence

$$\sum_{r=1}^{\infty} \oint_r \mathbf{A} \cdot d\mathbf{l} = \oint_C \mathbf{A} \cdot d\mathbf{l} \tag{10.19}$$

where the line integral on the right is taken along the boundary curve C in the positive sense.

Now the sum on the right of (10.18) reduces to the following integral:

$$\sum (\nabla \times \mathbf{A})_z \, ds_{xy} = \iint_S (\nabla \times \mathbf{A})_z \, ds_{xy} \tag{10.20}$$

Hence, substituting this into (10.18), we obtain

$$\oint_C \mathbf{A} \cdot d\mathbf{l} = \iint_S (\nabla \times \mathbf{A})_z \, ds_{xy} \tag{10.21}$$

That is, the line integral of a vector field $\mathbf{A}$ about the contour C of a plane surface S is equal to the surface integral of the normal component of the curl of $\mathbf{A}$ to the surface throughout the surface s.

Consider now the triangular surface of Fig. 10.5. It is easy to see that

$$\oint_{ABC} \mathbf{A} \cdot d\mathbf{l} = \oint_{OBC} \mathbf{A} \cdot d\mathbf{l} + \oint_{OCA} \mathbf{A} \cdot d\mathbf{l} + \oint_{OAB} \mathbf{A} \cdot d\mathbf{l} \tag{10.22}$$

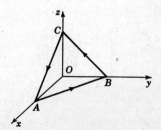

Fig. 10.5

since the contribution along the lines OA, OB, and OC cancel each other. However, as a consequence of (10.21), we have

$$\int_{OBC} \mathbf{A} \cdot d\mathbf{l} = \iint_{OBC} (\nabla \times \mathbf{A})_x \, dy \, dz$$

$$\int_{OCA} \mathbf{A} \cdot d\mathbf{l} = \iint_{OCA} (\nabla \times \mathbf{A})_y \, dz \, dx \qquad (10.23)$$

$$\int_{OAB} \mathbf{A} \cdot d\mathbf{l} = \iint_{OAB} (\nabla \times \mathbf{A})_z \, dx \, dy$$

If we add these equations and make use of (10.22), we obtain

$$\oint_{ABC} \mathbf{A} \cdot d\mathbf{l} = \iint_{OBC} (\nabla \times \mathbf{A})_x \, dy \, dz$$

$$+ \iint_{OCA} (\nabla \times \mathbf{A})_y \, dz \, dx + \iint_{OAB} (\nabla \times \mathbf{A})_z \, dx \, dy \quad (10.24)$$

Now we may write

$$d\mathbf{s} = \mathbf{i} \, ds_x + \mathbf{j} \, ds_y + \mathbf{k} \, ds_z$$
$$= \mathbf{i} \, dy \, dz + \mathbf{j} \, dx \, dz + \mathbf{k} \, dx \, dy \qquad (10.25)$$

for the projections of the representative surface vector $\mathbf{s}$ of the plane ABC to the yz, xz, and xy planes. Using this notation, (10.24) becomes

$$\oint_{ABC} \mathbf{A} \cdot d\mathbf{l} = \iint_{ABC} (\nabla \times \mathbf{A}) \cdot d\mathbf{s} \qquad (10.26)$$

Consider now the open surface of Fig. 10.6. We can regard the surface s of this open surface as being made up of an infinite number of elementary triangular surfaces. If we label r the typical triangle, we have from (10.26)

$$\sum_r \oint_r \mathbf{A} \cdot d\mathbf{l} = \sum_r \iint_r (\nabla \times \mathbf{A}) \cdot d\mathbf{s} \qquad (10.27)$$

Now the sum of the line integrals of the elementary triangles reduces to a line integration about the periphery C of the closed surface S since the

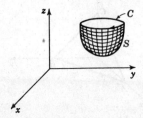

Fig. 10.6

line integrations along adjacent triangles are described in opposite senses and hence cancel. We therefore have

$$\sum \oint_r \mathbf{A} \cdot d\mathbf{l} = \iint_s \mathbf{A} \cdot d\mathbf{l} \tag{10.28}$$

In the limit the summation on the right of (10.27) reduces to an integral, and we have

$$\sum \oint_r (\nabla \times \mathbf{A}) \cdot d\mathbf{s} = \iint_s (\nabla \times \mathbf{A}) \cdot d\mathbf{s} \tag{10.29}$$

As a consequence of (10.19), we then have

$$\oint_c \mathbf{A} \cdot d\mathbf{l} = \iint_s (\nabla \times \mathbf{A}) \cdot d\mathbf{s} \tag{10.30}$$

This relation is known as *Stokes's theorem*. It states that the surface integral of the curl of a vector field **A** taken over any surface S is equal to the line integral of **A** around the periphery of the surface.

If the surface to which Stokes's theorem is applied is a closed surface, the length of the periphery is zero and then

$$\oiint_s (\nabla \times \mathbf{A}) \cdot d\mathbf{s} = 0 \tag{10.31}$$

By the use of Stokes s theorem we see that, if $\nabla \times \mathbf{A} = 0$ everywhere, **A** is the gradient of a scalar function, because if $\nabla \times \mathbf{A} = 0$, then the line integral of **A** around any closed curve vanishes. This is just the condition that **A** should be the gradient of a scalar function.

11 SUCCESSIVE APPLICATIONS OF THE OPERATOR ∇

It frequently happens in various applications of vector analysis that we must operate successively with the operator ∇. For example, since the curl of a vector field **A**, $\nabla \times \mathbf{A}$, is a vector field **B**, it is possible to take the curl of **B**, that is,

$$\nabla \times \mathbf{B} = \nabla \times (\nabla \times \mathbf{A}) \tag{11.1}$$

If we expand this equation in terms of the cartesian coordinates of **A**. we obtain

$$\nabla \times (\nabla \times \mathbf{A}) = \mathbf{i} \left[\frac{\partial}{\partial y} \left(\frac{\partial A_y}{\partial x} - \frac{\partial A_x}{\partial y} \right) - \frac{\partial}{\partial z} \left(\frac{\partial A_x}{\partial z} - \frac{\partial A_z}{\partial x} \right) \right] + \text{, etc.}$$

$$= \mathbf{i} \left(\frac{\partial^2 A_y}{\partial y\, \partial x} + \frac{\partial^2 A_z}{\partial z\, \partial x} - \frac{\partial^2 A_x}{\partial y^2} - \frac{\partial^2 A_x}{\partial z^2} \right) + \text{, etc.}$$

$$= \mathbf{i} \left[\frac{\partial}{\partial x} \left(\frac{\partial A_x}{\partial x} + \frac{\partial A_y}{\partial y} + \frac{\partial A_z}{\partial z} \right) \right.$$

$$\left. - \left(\frac{\partial^2}{\partial x^2} + \frac{\partial^2}{\partial y^2} + \frac{\partial^2}{\partial z^2} \right) A_x \right] + \text{, etc.}$$

$$= \nabla(\nabla \cdot \mathbf{A}) - \nabla^2 \mathbf{A} \tag{11.2}$$

In the same manner the following vector identities may be established by expanding ∇ and the other vectors concerned in terms of their components:

$$\nabla \cdot (u\mathbf{A}) = u(\nabla \cdot \mathbf{A}) + \mathbf{A} \cdot (\nabla u) \qquad (11.3)$$

$$\nabla \times (u\mathbf{A}) = u(\nabla \times \mathbf{A}) + (\nabla u) \times \mathbf{A} \qquad (11.4)$$

$$\nabla \cdot (\mathbf{A} \times \mathbf{B}) = \mathbf{B} \cdot (\nabla \times \mathbf{A}) - \mathbf{A} \cdot (\nabla \times \mathbf{B}) \qquad (11.5)$$

$$\nabla(\mathbf{A} \cdot \mathbf{B}) = \mathbf{A} \times (\nabla \times \mathbf{B}) + (\mathbf{A} \cdot \nabla)\mathbf{B} + \mathbf{B} \times (\nabla \times \mathbf{A}) + (\mathbf{B} \cdot \nabla)\mathbf{A} \qquad (11.6)$$

$$\nabla \times (\mathbf{A} \times \mathbf{B}) = (\mathbf{B} \cdot \nabla)\mathbf{A} - (\mathbf{A} \cdot \nabla)\mathbf{B} - \mathbf{B}(\nabla \cdot \mathbf{A}) + \mathbf{A}(\nabla \cdot \mathbf{B}) \qquad (11.7)$$

$$\nabla \times (\nabla u) = 0 \qquad (11.8)$$

$$\nabla \cdot (\nabla \times \mathbf{A}) = 0 \qquad (11.9)$$

In the above set of equations the symbol $(\mathbf{A} \cdot \nabla)\mathbf{B}$ stands for the vector

$$(\mathbf{A} \cdot \nabla)\mathbf{B} = \mathbf{i}\left(A_x \frac{\partial B_x}{\partial x} + A_y \frac{\partial B_x}{\partial y} + A_z \frac{\partial B_x}{\partial z}\right)$$
$$+ \mathbf{j}\left(A_x \frac{\partial B_y}{\partial x} + A_y \frac{\partial B_y}{\partial y} + A_z \frac{\partial B_y}{\partial z}\right)$$
$$+ \mathbf{k}\left(A_x \frac{\partial B_z}{\partial x} + A_y \frac{\partial B_z}{\partial y} + A_z \frac{\partial B_z}{\partial z}\right) \quad (11.10)$$

The above formulas are very useful in applications of vector analysis to various branches of engineering and physics.

12 ORTHOGONAL CURVILINEAR COORDINATES

Many calculations in applied mathematics can be simplified by choosing instead of a cartesian coordinate system another kind of system that takes advantage of the relations of symmetry involved in the particular problem under consideration.

Let these new coordinates be denoted by u_1, u_2, u_3. These are defined by specifying the cartesian coordinates x, y, z as functions of u_1, u_2, u_3, as follows:

$$x = x(u_1, u_2, u_3)$$
$$y = y(u_1, u_2, u_3) \qquad (12.1)$$
$$z = z(u_1, u_2, u_3)$$

We shall confine ourselves to the case when the three families of surfaces $u_1 = \text{const}$, $u_2 = \text{const}$, $u_3 = \text{const}$ are orthogonal to one another. In this case the line element ds is given by

$$ds^2 = h_1^2\, du_1^2 + h_2^2\, du_2^2 + h_3^2\, du_3^2 \qquad (12.2)$$

where h_1, h_2, h_3 may be functions of u_1, u_2, u_3. We shall also adopt the convention that the new coordinate system shall be right-handed like the old.

Consider now the infinitesimal parallelepiped whose diagonal is the line element ds and whose faces coincide with the planes u_1 or u_2 or $u_3 = $ const (see Fig. 12.1).

The lengths of its edges are $h_1 du_1$, $h_2 du_2$, $h_3 du_3$, and its volume is $h_1 h_2 h_3\, du_1\, du_2\, du_3$. Furthermore let $\phi(u_1,u_2,u_3)$ be a scalar function and $\mathbf{A}$ be a vector field with components A_1, A_2, A_3 in the three directions in which the coordinates u_1, u_2, u_3 increase. The u_1 component of the gradient of ϕ we can compute at once since by definition

$$(\text{grad } \phi)_1 = \lim_{du_1 \to 0} \frac{\phi(A) - \phi(O)}{h_1\, du_1}$$

$$= \frac{1}{h_1} \frac{\partial \phi}{\partial u_1} \tag{12.3}$$

We also have similar relations for the directions 2 and 3.

In order to calculate the divergence of a vector field $\mathbf{A}$, we use Gauss's theorem:

$$\iiint (\nabla \cdot \mathbf{A})\, dV = \iint_s \mathbf{A} \cdot d\mathbf{s} \tag{12.4}$$

The contribution to the integral $\iint \mathbf{A} \cdot d\mathbf{s}$ through the area $OBHC$, taken in the direction of the outward normal, is $-A_1 h_2 h_3\, du_2\, du_3$, while that through the area $AFGJ$ is

$$A_1 h_2 h_3\, du_2\, du_3 + \frac{\partial}{\partial u_1}(A_1 h_2 h_3)\, du_1\, du_2\, du_3$$

From these and the corresponding expressions for the other two pairs of surfaces, we have by (12.4)

$$\lim_{V \to 0} \iiint_V (\nabla \cdot \mathbf{A})\, dv = \lim_{V \to 0} \iint \mathbf{A} \cdot d\mathbf{s}$$

$$= \iint \mathbf{A} \cdot d\mathbf{s} \tag{12.5}$$

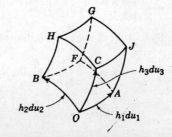

Fig. 12.1

We thus obtain

$$\nabla \cdot \mathbf{A} = \frac{1}{h_1 h_2 h_3} \left[\frac{\partial}{\partial u_1} (h_2 h_3 A_1) + \frac{\partial}{\partial u_2} (h_3 h_1 A_2) + \frac{\partial}{\partial u_3} (h_1 h_2 A_3) \right] \qquad (12.6)$$

If $\mathbf{A} = \nabla \phi$,

$$\nabla^2 \phi = \frac{1}{h_1 h_2 h_3} \left[\frac{\partial}{\partial u_1} \left(\frac{h_2 h_3}{h_1} \frac{\partial \phi}{\partial u_1} \right) + \frac{\partial}{\partial u_2} \left(\frac{h_3 h_1}{h_2} \frac{\partial \phi}{\partial u_2} \right) + \frac{\partial}{\partial u_3} \left(\frac{h_1 h_2}{h_3} \frac{\partial \phi}{\partial u_3} \right) \right] \quad (12.7)$$

The components of the curl of $\mathbf{A}$ may be found by Stokes's theorem,

$$\iint_S (\nabla \times \mathbf{A}) \cdot d\mathbf{s} = \oint \mathbf{A} \cdot d\mathbf{l} \qquad (12.8)$$

For example, the component 1 of the curl of $\mathbf{A}$ is obtained by applying Stokes's theorem to the surface $OBHC$.   We calculate

$$\oint_{OBHC} \mathbf{A} \cdot d\mathbf{l} = \int_O^B \mathbf{A} \cdot d\mathbf{l} + \int_B^H \mathbf{A} \cdot d\mathbf{l} + \int_H^C \mathbf{A} \cdot d\mathbf{l} + \int_C^O \mathbf{A} \cdot d\mathbf{l}$$

$$= A_2 h_2 \, du_2 + A_3 h_3 \, du_3 + \frac{\partial}{\partial u_2} (A_3 h_3) \, du_3 \, du_2$$

$$- \left[A_2 h_2 \, du_2 + \frac{\partial}{\partial u_3} (A_2 h_2) \, du_2 \, du_3 \right] - A_3 h_3 \, du_3$$

$$= \left[\frac{\partial}{\partial u_2} (A_3 h_3) - \frac{\partial}{\partial u_3} (A_2 h_2) \right] du_2 \, du_3 \qquad (12.9)$$

By Stokes's theorem this equals the 1 component of the curl of $\mathbf{A}$, $(\nabla \times \mathbf{A})_1$, multiplied by the area of the face $OBHC$.   That is,

$$(\nabla \times \mathbf{A})_1 \, h_2 h_3 \, du_2 \, du_3 = \left[\frac{\partial}{\partial u_2} (A_3 h_3) - \frac{\partial}{\partial u_3} (A_2 h_2) \right] du_2 \, du_3 \qquad (12.10)$$

Hence

$$(\nabla \times \mathbf{A})_1 = \frac{1}{h_2 h_3} \left[\frac{\partial}{\partial u_2} (A_3 h_3) - \frac{\partial}{\partial u_3} (A_2 h_2) \right] \qquad (12.11)$$

By a cyclic change of the indices we obtain

$$(\nabla \times \mathbf{A})_2 = \frac{1}{h_3 h_1} \left[\frac{\partial}{\partial u_3} (h_1 A_1) - \frac{\partial}{\partial u_1} (h_3 A_3) \right] \qquad (12.12)$$

$$(\nabla \times \mathbf{A})_3 = \frac{1}{h_1 h_2} \left[\frac{\partial}{\partial u_1} (h_2 A_2) - \frac{\partial}{\partial u_2} (h_1 A_1) \right] \qquad (12.13)$$

If we introduce unit vectors i_1, i_2, i_3 along the directions 1, 2, and 3, we may write symbolically

$$\nabla \times \mathbf{A} = \frac{1}{h_1 h_2 h_3} \begin{vmatrix} h_1 i_1 & h_2 i_2 & h_3 i_3 \\ \dfrac{\partial}{\partial u_1} & \dfrac{\partial}{\partial u_2} & \dfrac{\partial}{\partial u_3} \\ h_1 A_1 & h_2 A_2 & h_3 A_3 \end{vmatrix} \tag{12.14}$$

In the case of cartesian coordinates we have $u_1 = x$, $u_2 = y$, $u_3 = z$, $h_1 = h_2 = h_3 = 1$, and $i_1 = \mathbf{i}$, $i_2 = \mathbf{j}$, $j_3 = \mathbf{k}$. In this case (12.14) reduces to (10.14). We shall now apply these general formulas to two special cases that are particularly important in applications.

a CYLINDRICAL COORDINATES

The position of a point in space may be determined by the cylindrical coordinate system of Fig. 12.2. In this case we have

$$x = r \cos \theta$$
$$y = r \sin \theta \tag{12.15}$$
$$z = z$$

$$ds^2 = dr^2 + r^2 d\theta^2 + dz^2 \tag{12.16}$$

We have, therefore, in this case,

$$\begin{aligned} u_1 &= r & h_1 &= 1 \\ u_2 &= \theta \quad \text{and} & h_2 &= r \\ u_3 &= z & h_3 &= 1 \end{aligned} \tag{12.17}$$

By (12.3) we obtain

$$\operatorname{grad}_r \phi = \frac{\partial \phi}{\partial r} \qquad \operatorname{grad}_\theta \phi = \frac{1}{r}\frac{\partial \phi}{\partial \theta} \qquad \operatorname{grad}_z \phi = \frac{\partial \phi}{\partial z} \tag{12.18}$$

By (12.6) we have

$$\nabla \cdot \mathbf{A} = \frac{1}{r}\frac{\partial}{\partial r}(r A_r) + \frac{1}{r}\frac{\partial A_\theta}{\partial \theta} + \frac{\partial A_z}{\partial z} \tag{12.19}$$

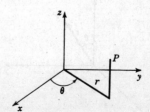

Fig. 12.2

The Laplacian operator as given by (12.7) gives

$$\nabla^2 \phi = \frac{1}{r}\frac{\partial}{\partial r}\left(r\frac{\partial \phi}{\partial r}\right) + \frac{1}{r^2}\frac{\partial^2 \phi}{\partial \theta^2} + \frac{\partial^2 \phi}{\partial z^2} \tag{12.20}$$

We obtain the components of the curl of **A** by (12.11), (12.12), and (12.13):

$$(\nabla \times \mathbf{A})_r = \frac{1}{r}\frac{\partial A_z}{\partial \theta} - \frac{\partial A_\theta}{\partial z} \tag{12.21}$$

$$(\nabla \times \mathbf{A})_\theta = \frac{\partial A_r}{\partial z} - \frac{\partial A_z}{\partial r} \tag{12.22}$$

$$(\nabla \times \mathbf{A})_z = \frac{1}{r}\left[\frac{\partial}{\partial r}(rA_\theta) - \frac{\partial A_r}{\partial \theta}\right] \tag{12.23}$$

b SPHERICAL POLAR COORDINATES

Another very important coordinate system is that of spherical polar coordinates, as shown in Fig. 12.3. In this case we have

$$
\begin{aligned}
x &= r\sin\theta\cos\phi \\
y &= r\sin\theta\sin\phi \\
z &= r\cos\theta \\
ds^2 &= dr^2 + r^2\sin^2\theta\,d\phi^2 + r^2\,d\theta^2
\end{aligned}
\tag{12.24}
$$

We have therefore

$$
\begin{aligned}
u_1 &= r & h_1 &= 1 \\
u_2 &= \theta \quad \text{and} & h_2 &= r \\
u_3 &= \phi & h_3 &= r\sin\theta
\end{aligned}
\tag{12.25}
$$

$$(\text{grad } V)_r = \frac{\partial V}{\partial r} \qquad (\text{grad } V)_\theta = \frac{1}{r}\frac{\partial V}{\partial \theta} \qquad (\text{grad } V)_\phi = \frac{1}{r\sin\theta}\frac{\partial V}{\partial \phi} \tag{12.26}$$

$$\nabla \cdot \mathbf{A} = \frac{1}{r^2}\frac{\partial}{\partial r}(r^2 A_r) + \frac{1}{r\sin\theta}\frac{\partial}{\partial \theta}(\sin\theta A_\theta) + \frac{1}{r\sin\theta}\frac{\partial A_\phi}{\partial \phi} \tag{12.27}$$

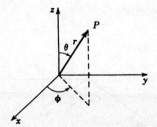

Fig. 12.3

$$\nabla^2 V = \frac{1}{r^2}\frac{\partial}{\partial r}\left(r^2 \frac{\partial V}{\partial r}\right) + \frac{1}{r^2 \sin\theta}\frac{\partial}{\partial \theta}\left(\sin\theta \frac{\partial V}{\partial \theta}\right) + \frac{1}{r^2 \sin^2\theta}\frac{\partial^2 V}{\partial \phi^2} \tag{12.28}$$

$$(\nabla \times \mathbf{A})_r = \frac{1}{r\sin\theta}\left[\frac{\partial}{\partial \theta}(\sin\theta A_\phi) - \frac{\partial A_\theta}{\partial \phi}\right]$$

$$(\nabla \times \mathbf{A})_\theta = \frac{1}{r}\left[\frac{1}{\sin\theta}\frac{\partial A_r}{\partial \phi} - \frac{\partial (rA_\phi)}{\partial r}\right] \tag{12.29}$$

$$(\nabla \times \mathbf{A})_\phi = \frac{1}{r}\left[\frac{\partial}{\partial r}(rA_\theta) - \frac{\partial A_r}{\partial \theta}\right]$$

13   APPLICATION TO HYDRODYNAMICS

Consider a region of space containing a fluid of density $\rho(x,y,z,t)$. Let V be the volume inside an arbitrary closed surface s located in this region. Let $Q(t)$ be the mass of fluid inside the volume V at any instant.   Then

$$Q(t) = \iiint_V \rho \, dV \tag{13.1}$$

If v denotes the velocity of a typical particle of the fluid, the rate at which the mass of fluid inside V is *increasing* is

$$\frac{dQ}{dt} = -\iint_s (\rho\mathbf{v})\cdot d\mathbf{s} \tag{13.2}$$

Now, if we differentiate (13.1) with respect to time and equate the result to (13.2), we have

$$\frac{dQ}{dt} = \iiint_V \left(\frac{\partial \rho}{\partial t}\right) dV = -\iint_s (\rho\mathbf{v})\cdot d\mathbf{s} \tag{13.3}$$

But, by Gauss's theorem, we have

$$\iint_s (\rho\mathbf{v})\cdot d\mathbf{s} = \iiint_V \nabla\cdot(\rho\mathbf{v}) \, dV \tag{13.4}$$

Substituting this into (13.3) and transposing, we obtain

$$\iiint_V \left[\frac{\partial \rho}{\partial t} + \nabla\cdot(\rho\mathbf{v})\right] dV = 0 \tag{13.5}$$

Now since the integrand is continuous and the volume V is arbitrary, we conclude that

$$\frac{\partial \rho}{\partial t} + \nabla\cdot(\rho\mathbf{v}) = 0 \tag{13.6}$$

This is the basic equation of hydrodynamics and is known as the *equation of continuity*.

If the fluid is incompressible, then ρ is a constant and we have

$$\frac{\partial \rho}{\partial t} = 0 = \rho(\nabla \cdot \mathbf{v}) \tag{13.7}$$

If the flow is irrotational, then we have

$$\nabla \times \mathbf{v} = 0 \tag{13.8}$$

and we know from Sec. 10 that there exists a scalar function ϕ such that

$$\mathbf{v} = \nabla \phi \tag{13.9}$$

If the fluid is incompressible, then from (13.7) we have

$$\nabla \cdot \mathbf{v} = 0 \tag{13.10}$$

and hence ϕ, called the *velocity potential*, satisfies the equation

$$\nabla \cdot (\nabla \phi) = \nabla^2 \phi = 0 \tag{13.11}$$

On a fixed boundary of the fluid, the velocity has no normal component, and if $\partial/\partial n$ denotes differentiation with respect to the normal direction of the boundary, we must have

$$\frac{\partial \phi}{\partial n} = 0 \tag{13.12}$$

as a consequence of (13.9) .

Let us denote any general property of a particle of the fluid, such as its pressure, density, etc., by the function $H(x,y,z,t)$. Then by $\partial H/\partial t$ is meant the variation of H at a *particular point* in *space* as a function of the time t. If we take the total differential of $H(x,y,z,t)$, we obtain

$$dH = \frac{\partial H}{\partial x}dx + \frac{\partial H}{\partial y}dy + \frac{\partial H}{\partial z}dz + \frac{\partial H}{\partial t}dt \tag{13.13}$$

$$\frac{dH}{dt} = \frac{\partial H}{\partial x}\frac{dx}{dt} + \frac{\partial H}{\partial y}\frac{dy}{dt} + \frac{\partial H}{\partial z}\frac{dz}{dt} + \frac{\partial H}{\partial t} \tag{13.14}$$

The quantity dH/dt is the variation of H when we fix our attention *on the same particle of fluid*.

The velocity of the particle is given by

$$\mathbf{v} = \mathbf{i}\frac{dx}{dt} + \mathbf{j}\frac{dy}{dt} + \mathbf{k}\frac{dz}{dt} \tag{13.15}$$

By definition we have

$$\nabla H = \mathbf{i}\frac{\partial H}{\partial x} + \mathbf{j}\frac{\partial H}{\partial y} + \mathbf{k}\frac{\partial H}{\partial z} \tag{13.16}$$

Hence

$$\mathbf{v} \cdot \nabla H = \frac{\partial H}{\partial x}\frac{dx}{dt} + \frac{\partial H}{\partial y}\frac{dy}{dt} + \frac{\partial H}{\partial z}\frac{dz}{dt} \tag{13.17}$$

$$\frac{dH}{dt} = \frac{\partial H}{\partial t} + \mathbf{v} \cdot \nabla H \tag{13.18}$$

EULER'S EQUATION OF MOTION

To obtain the equation of motion of a frictionless fluid, we consider the forces acting on an element of fluid whose volume is $dx\,dy\,dz$ and whose mass is $\rho\,dx\,dy\,dz$. As a consequence of the pressure of the fluid, p, there will be a force in the x direction on the element of fluid under consideration, of magnitude

$$p\,dy\,dz - \left(p + \frac{\partial p}{\partial x}dx\right)dy\,dz = -\frac{\partial p}{\partial x}dx\,dy\,dz \tag{13.19}$$

Let us also consider the action of an external force $\mathbf{F}$ acting on unit *mass* of fluid. The external force acting on the element under consideration in the x direction is $F_x \rho\,dx\,dy\,dz$. The acceleration of the element in the x direction is dv_x/dt. Hence, by Newton's second law of motion, we have

$$F_x \rho\,dx\,dy\,dz - \frac{\partial p}{\partial x}dx\,dy\,dz = \frac{dv_x}{dt}\rho\,dx\,dy\,dz \tag{13.20}$$

By this reasoning, and considering the y and z directions, we obtain Euler's equations of motion:

$$\frac{dv_x}{dt} = F_x - \frac{1}{\rho}\frac{\partial p}{\partial x}$$

$$\frac{dv_y}{dt} = F_y - \frac{1}{\rho}\frac{\partial p}{\partial y} \tag{13.21}$$

$$\frac{dv_z}{dt} = F_z - \frac{1}{\rho}\frac{\partial p}{\partial z}$$

These three scalar equations may be combined into the single vector equation

$$\frac{d\mathbf{v}}{dt} = \mathbf{F} - \frac{1}{\rho}\nabla p \tag{13.22}$$

As a consequence of Eq. (13.18), we have

$$\frac{d\mathbf{v}}{dt} = \frac{d\mathbf{v}}{\partial t} + (\mathbf{v} \cdot \nabla)\mathbf{v} \tag{13.23}$$

Now by the vector identity (11.6), we have

$$(\mathbf{v} \cdot \nabla)\mathbf{v} = \tfrac{1}{2}\nabla v^2 + (\nabla \times \mathbf{v}) \times \mathbf{v} \tag{13.24}$$

Hence, by the use of the two equations above, we may write Euler's equation of motion in the following form:

$$\frac{\partial \mathbf{v}}{dt} + \tfrac{1}{2}\nabla v^2 + (\nabla \times \mathbf{v}) \times \mathbf{v} = \mathbf{F} - \frac{1}{\rho}\nabla p \tag{13.25}$$

If the external force is *conservative*, it has a potential V and we have

$$F = -\nabla V \tag{13.26}$$

If the motion of the fluid is *irrotational*, $\nabla \times \mathbf{v} = 0$ and $\mathbf{v} = \nabla \phi$. In this case Eq. (13.25) becomes

$$\nabla\left(\frac{\partial \phi}{\partial t}\right) + \tfrac{1}{2}\nabla v^2 + \nabla V + \frac{1}{\rho}\nabla p = 0 \tag{13.27}$$

or

$$\nabla w = 0 \tag{13.28}$$

where

$$w = \frac{\partial \phi}{\partial t} + \frac{v^2}{2} + V + \frac{p}{\rho} \tag{13.29}$$

Now if $d\mathbf{r}$ denotes an *arbitrary* path in the fluid at any instant, we have

$$d\mathbf{r} = \mathbf{i}\,dx + \mathbf{j}\,dy + \mathbf{k}\,dz \tag{13.30}$$

Let us form the scalar product of ∇w and $d\mathbf{r}$. We then obtain

$$\nabla w \cdot d\mathbf{r} = \left(\frac{\partial w}{\partial x} + \mathbf{j}\frac{\partial w}{\partial y} + \mathbf{k}\frac{\partial w}{\partial z}\right)(\mathbf{i}\,dx + \mathbf{j}\,dy + \mathbf{k}\,dz)$$

$$= \frac{\partial w}{\partial x}dx + \frac{\partial w}{\partial y}dy + \frac{\partial w}{\partial z}dz = dw = 0 \tag{13.30a}$$

as a consequence of Eq. (13.28). Hence

$$w = \beta(t) \tag{13.31}$$

where $\beta(t)$ is an arbitrary function of *time*, since we have integrated along an arbitrary path in the fluid at any instant. That is, we have

$$w = \frac{\partial \phi}{\partial t} + \frac{v^2}{2} + V + \int \frac{dp}{\rho} = \beta(t) \tag{13.32}$$

Equation (13.32) is known as *Bernoulli's equation*.

In the special case where ρ is constant and the motion is steady so that $\partial \phi / \partial t = 0$, this equation takes the form

$$\frac{v^2}{2} + V + \frac{p}{\rho} = \beta \tag{13.33}$$

where β is now a constant.   This equation states that, *per unit mass* of fluid, the sum of the kinetic energy, the potential energy V, and the pressure energy p/ρ has a constant value β for all points of the fluid.

Usually in aerodynamics the variations in V are so small that they can be neglected so that we can write

$$\frac{\rho v^2}{2} + p = p_0 \tag{13.34}$$

where p_0 is the pressure when the fluid is at rest.   It can be seen from this that in this case the pressure diminishes as the velocity increases in the form

$$p = p_0 - \frac{\rho v^2}{2} \tag{13.35}$$

Airplanes are equipped with instruments for measuring p and p_0 so that the velocity of the machine relative to the air can be determined by the equation

$$v = \sqrt{\frac{2(p_0 - p)}{\rho}} \tag{13.36}$$

14   THE EQUATION OF HEAT FLOW IN SOLIDS

Consider a region inside a solid body such as a large block of metal.   Let a closed surface s be situated inside this region.   Let the volume inside the surface s be denoted by V.

In suitable units the amount of heat H inside the volume V of this body is given by

$$H = \iiint\limits_{V} (uc\rho)\, dV \tag{14.1}$$

where $u = u(x,y,z,t) =$ temperature of body

$\quad c =$ specific heat of body

$\quad \rho =$ density of body

It is an empirical fact that the rate of flow of heat *into* the volume V may be expressed in the following form:

$$\frac{dH}{dt} = \iint\limits_{s} \left(k\frac{\partial u}{\partial n} \right) ds \tag{14.2}$$

where k is a constant known as the *thermal conductivity* and $\partial u/\partial n$ is the derivative of the temperature with respect to the outward drawn normal to the surface s.

If we differentiate (14.1) with respect to t and equate the result to (14.2), we obtain

$$\frac{dH}{dt} = \iiint_V \left(c\rho \frac{\partial u}{\partial t} \right) dV = \iint_s \left(k \frac{\partial u}{\partial n} \right) ds \tag{14.3}$$

If $d\mathbf{n}$ is a vector drawn in the direction of the outward drawn normal to the surface s, we have

$$du = \nabla u \cdot d\mathbf{n} \tag{14.4}$$

where du is the total derivative of the temperature and represents the change in temperature as one moves through the distance $d\mathbf{n}$. If we divide both sides of (14.4) by the differential dn, we obtain

$$\frac{\partial u}{\partial n} = \nabla u \cdot \mathbf{n} \tag{14.5}$$

where $\mathbf{n}$ represents a unit vector in the normal direction. We then have

$$\iint_s \left(k \frac{\partial u}{\partial n} \right) ds = \iint_s k(\nabla u) \cdot \mathbf{n}\, ds = \iint_s (k\, \nabla u) \cdot d\mathbf{s} \tag{14.6}$$

If we let

$$\mathbf{q} = k\, \nabla u \tag{14.7}$$

we have, in view of (14.3) and (14.6),

$$\iiint_V \left(c\rho \frac{\partial u}{\partial t} \right) dV = \iint_s \mathbf{q} \cdot d\mathbf{s} = \iiint_V (\nabla \cdot \mathbf{q})\, dV \tag{14.8}$$

where we have used Gauss's theorem to transform the last integral. Transposing, we may write

$$\iiint_V \left(c\rho \frac{\partial u}{\partial t} - \Delta \cdot \mathbf{q} \right) dV = 0 \tag{14.9}$$

Since the integral is continuous and the volume V is arbitrary, the integrand must vanish and we obtain

$$\nabla \cdot \mathbf{q} = c\rho \frac{\partial u}{\partial t} \tag{14.10}$$

or

$$\nabla \cdot (k\, \nabla u) = c\rho \frac{\partial u}{\partial t} \tag{14.11}$$

If k is a constant, we have

$$\nabla \cdot (k\, \nabla u) = k\, \nabla^2 u = c\rho \frac{\partial u}{\partial t} \tag{14.12}$$

or

$$\frac{\partial u}{\partial t} = \frac{k}{c\rho} \nabla^2 u = h^2 \nabla^2 u \tag{14.13}$$

where

$$h^2 = \frac{k}{c\rho} \tag{14.14}$$

Equation (14.13) was derived by Fourier in 1822 and is sometimes called the *heat-flow*, or *diffusion, equation*.

15 THE GRAVITATIONAL POTENTIAL

Consider two particles at Q and P of masses m_1 and m_2, respectively. Then, according to the Newtonian law of gravitation, there is a force of attraction between them given in magnitude by the equation

$$F = k \frac{m_1 m_2}{r^2} \tag{15.1}$$

where k is a constant depending upon the units and r is the distance between the particles as shown in Fig. 15.1.

If we choose the unit of mass as that of a particle which, placed at a unit distance from one of equal mass, attracts it with unit force, Eq. (15.1) becomes

$$F = \frac{m_1 m_2}{r^2} \tag{15.2}$$

If we denote the vector QP by $\mathbf{r}$, we may express the force per unit mass at P due to the attracting particle at Q by

$$\mathbf{F} = -\frac{m\mathbf{r}}{r^2} = \nabla \frac{m}{r} \tag{15.3}$$

$\mathbf{F}$ is called the *intensity* of force or the *gravitational field* of force at the point P, and we note that it may be expressed as the gradient of the scalar m/r. It follows therefore that $\mathbf{F}$ satisfies the equation

$$\nabla \times \mathbf{F} = \nabla \times \nabla \frac{m}{r} = 0 \tag{15.4}$$

and is therefore a conservative field of force.

Let us suppose that the particle m is stationary at Q and that another of unit mass moves under the attraction of the former from infinity up to P

Fig. 15.1

along any path.   The work done by the force of attraction during an infinitesimal displacement $d\mathbf{r}$ of the unit mass is $\mathbf{F} \cdot d\mathbf{r}$.   The total work done by the force while the unit particle moves from infinity up to P is

$$\int_{\infty}^{P} \mathbf{F} \cdot d\mathbf{r} = \int_{\infty}^{P} \nabla \frac{m}{r} \cdot d\mathbf{r} = \frac{m}{r} \Big|_{\infty}^{P} = \frac{m}{P} \tag{15.5}$$

This is independent of the path by which the particle comes to P and is called the *potential* at P due to the particle of mass at Q.   Let us denote it by V.   We then have

$$V = \frac{m}{P} \tag{15.6}$$

while the intensity of force at P due to it is

$$\mathbf{F} = \nabla \frac{m}{P} = \nabla V \tag{15.7}$$

That is, the intensity at any point is equal to the gradient of the potential.

If we now suppose that there are n particles of masses $m_1, m_2, \ldots, m_n$, relative to which P has position vectors $\mathbf{r}_1, \ldots, \mathbf{r}_n$, respectively, then the force of attraction per unit mass at P due to the system is the vector sum of the intensities due to each, that is,

$$\mathbf{F} = \nabla \frac{m_1}{r_1} + \nabla \frac{m_2}{r_2} + \cdots + \nabla \frac{m_n}{r_n} = \nabla \sum_{s=1}^{n} \frac{m_s}{r_s} \tag{15.8}$$

Now, by the same argument as before, the work done by the attracting forces on a particle of unit mass while it moves from infinity up to P is

$$\int_{\infty}^{P} \mathbf{F} \cdot d\mathbf{r} = \sum_{s=1}^{n} \frac{m_s}{r_s} = V \tag{15.9}$$

Therefore the potential at P due to a system of particles is the sum of the potentials due to each.   The potential V is a scalar function of position and, except at the points Q_s where the masses are situated, it satisfies

$$\nabla^2 V = \nabla^2 \sum_{s=1}^{n} \frac{m_s}{r_s} = \sum_{s=1}^{n} \nabla^2 \frac{m_s}{r_s} = 0 \tag{15.10}$$

which is Laplace's equation.   Since $\mathbf{F} = \nabla V$, this means that at points excluding matter we must have

$$\nabla \cdot \mathbf{F} = 0 \tag{15.11}$$

CONTINUOUS DISTRIBUTION OF MATTER

Let us now suppose that the attracting matter forms a continuous body filling the space bounded by the closed surface s.   We divide the body into an infinite number of elements of mass $\rho\,dV$, where ρ is the density of the body and dV is the element of volume of the body, as shown in Fig. 15.2.

If P is *outside* the body, the sum in Eq. (15.9) passes into the integral

$$V_P = \iiint \frac{\rho\,dV}{r} \tag{15.12}$$

where r is the distance from the element of volume dV to the point P. This gives the potential at P due to the entire body.

If the point P is *inside* the body, the integrand (15.12) becomes infinite. In this case we define the potential in the following manner: Surround the point P by a closed spherical surface s_0, and consider the potential due to the matter in the space between s_0 and s. The integrand is finite everywhere, since P is outside the region. Now let the surface s_0 decrease indefinitely, converging to the point P as a limit.

We then consider

$$V_P = \lim_{s_0 \to 0} \iiint_{s_0}^{s} \frac{\rho\,dv}{r} \tag{15.13}$$

Since the volume of s_0 is of the same order as r^3, where r is the radius of the sphere s_0, while the integrand becomes infinite like $1/r$ if ρ is finite, the value of the above integral tends to a definite limit, which is called the *potential at P due to the whole body.*

POISSON'S EQUATION

We have seen that the potential function satisfies $\nabla^2 V = 0$, or Laplace's equation, in the region outside matter. Let us now consider the equation satisfied by the potential in a region inside matter. Consider a point P inside a body of density ρ. Surround the point P by a sphere s_0 of radius a, as shown in Fig. 15.3.

The gravitational field at the point P is the vector sum of the gravitational field $\mathbf{F}_o$ produced by the matter outside s_0 and $\mathbf{F}_i$ produced by the matter inside s_0. That is, we write

$$\mathbf{F}_P = \mathbf{F}_o + \mathbf{F}_i \tag{15.14}$$

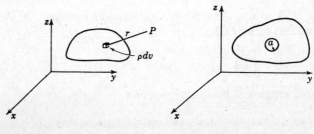

Fig. 15.2   **Fig. 15.3**

Now let us calculate $\nabla \cdot \mathbf{F}_P$, that is,

$$\nabla \cdot \mathbf{F}_P = \nabla \cdot \mathbf{F}_o + \nabla \cdot \mathbf{F}_i \tag{15.15}$$

But since $\mathbf{F}_o$ is produced by matter outside s_0, we have from (15.11)

$$\nabla \cdot \mathbf{F}_o = 0 \tag{15.16}$$

Hence we have

$$\nabla \cdot \mathbf{F}_P = \nabla \cdot \mathbf{F}_i \tag{15.17}$$

Consider the sphere of radius a; its mass is therefore equal to

$$m = \tfrac{4}{3}\pi\rho a^3 \tag{15.18}$$

Now as a tends to zero, the intensity at the surface of this sphere is equal to

$$F_s = \frac{m}{a^2} = \frac{4}{3}\pi\rho a \tag{15.19}$$

in magnitude and in the direction of the inwardly drawn normal to the surface. By Gauss's theorem we then have

$$\lim_{a \to 0} \iiint (\nabla \cdot \mathbf{F}_i)\, dV = \lim_{a \to 0} (\nabla \cdot \mathbf{F}) \tfrac{4}{3}\pi a^3$$

$$= \lim_{a \to 0} \iint \mathbf{F}_s \cdot d\mathbf{s} = \lim_{a \to 0} (-\tfrac{4}{3}\pi\rho a \times 4\pi a^2) \tag{15.20}$$

Hence

$$\nabla \cdot \mathbf{F}_P = \nabla \cdot \mathbf{F}_i = -4\pi\rho \tag{15.21}$$

But

$$\mathbf{F}_P = \nabla V_P$$

Hence by (15.21) the equation satisfied by the potential in a region containing matter is

$$\nabla^2 V = -4\pi\rho = \nabla \cdot \mathbf{F} \tag{15.22}$$

This relation is known as *Poisson's equation*. In view of (15.13) we see that we may take

$$V = \iiint \frac{\rho\, dv}{r} \tag{15.23}$$

as a solution of Poisson's equation.†

† For a more rigorous derivation of Poisson's integral, refer to O. D. Kellogg, "Foundations of Potential Theory," Springer-Verlag OHG, Berlin, 1929.

GAUSS'S LAW OF GRAVITATION

As a consequence of Poisson's equation we may prove that the surface integral of the gravitational force $\mathbf{F}$ over a closed surface s drawn in the field is equal to -4π times the total mass enclosed by the surface.

To establish this, apply Gauss's theorem to Eq. (15.22); we then have

$$\iint_s \mathbf{F} \cdot d\mathbf{s} = \iiint_v (\nabla \cdot \mathbf{F})\, dv = -4\pi \iiint_v \rho\, dv \tag{15.24}$$

But $\iiint \rho\, dv$ is the total mass enclosed by the surface s. This theorem has many applications in potential theory and in the theory of electrostatics.

16 MAXWELL'S EQUATIONS

The study of electrodynamics affords one of the most important applications of vector analysis. According to the modern point of view, by an *electromagnetic field* is understood the domain of the five vectors $\mathbf{E}$, $\mathbf{B}$, $\mathbf{D}$, $\mathbf{H}$, and $\mathbf{J}$. These vectors satisfy the differential equations

$$\nabla \times \mathbf{E} = -\frac{\partial \mathbf{B}}{\partial t} \tag{16.1}$$

$$\nabla \times \mathbf{H} = \mathbf{J} + \frac{\partial \mathbf{D}}{\partial t} \tag{16.2}$$

$$\nabla \cdot \mathbf{B} = 0 \tag{16.3}$$

$$\nabla \cdot \mathbf{D} = \rho \tag{16.4}$$

and in a homogeneous isotropic medium we have the additional relations

$$\mathbf{D} = K\mathbf{E} \tag{16.5}$$

$$\mathbf{B} = \mu\mathbf{H} \tag{16.6}$$

$$\mathbf{J} = \sigma\mathbf{E} \tag{16.7}$$

The above set of differential equations are the fundamental equations of Maxwell written in a rationalized mks system of units. In this system of units we have

$\mathbf{E} =$ electric intensity, volts per m

$\mathbf{B} =$ magnetic induction, webers per m

$\mathbf{D} =$ electric displacement, coulombs per m^2

$\mathbf{H} =$ magnetic intensity, amp per m

$\mathbf{J} =$ current density, amp per m^2

$\sigma =$ electric conductivity, 1/ohm-m

$K = K_r K_0 =$ electric inductive capacity of medium

$\mu = \mu_r \mu_0 =$ magnetic inductive capacity of medium

$K_r =$ dielectric constant

$\mu_r =$ permeability

$K_0 = 8.854 \times 10^{-12}$ farad per m

$\mu_0 = 4\pi \times 10^{-7} = 1.257 \times 10^{-6}$ henry per m

$\rho =$ charge density, coulombs per m^3

$c = 1/\sqrt{K_0 \mu_0} = 2.998 \times 10^8$ m per sec

The solution of electrodynamic problems depends on the solution of these equations in special cases. For example, we may discuss briefly the following special cases.

ELECTROSTATICS AND MAGNETOSTATICS

In the case that the field vectors are independent of the time and in a region where $\mathbf{J} = 0$, we see that all the terms involving partial derivatives with respect to time vanish and the electric vectors and magnetic vectors become independent of each other. We then have

$$\nabla \times \mathbf{E} = 0 \tag{16.8}$$

$$\nabla \cdot \mathbf{D} = \rho \tag{16.9}$$

$$\mathbf{D} = K\mathbf{E} \tag{16.10}$$

and

$$\nabla \times \mathbf{H} = 0 \tag{16.11}$$

$$\nabla \cdot \mathbf{B} = 0 \tag{16.12}$$

Since the curls of $\mathbf{E}$ and $\mathbf{H}$ vanish, we see that both these fields may be derived from potential functions. It is conventional to write

$$\mathbf{E} = -\nabla V_E \tag{16.13}$$

$$\mathbf{H} = -\nabla V_M \tag{16.14}$$

where V_E and V_M are the electric and magnetic scalar potentials, respectively.

In view of (16.9) and (16.12) we see that these potentials satisfy the equations

$$\nabla^2 V_E = -\frac{\rho}{K} \tag{16.15}$$

and

$$\nabla^2 V_M = 0 \tag{16.16}$$

That is, the electric potential satisfies Poisson's equation, and the magnetic potential satisfies Laplace's equation. If the charge density ρ vanishes, then both V_E and V_M satisfy Laplace's equation and the solution of electrostatic and magnetostatic problems reduces to the solution of this equation subject to the proper boundary conditions.

17   THE WAVE EQUATION

Let us eliminate the magnetic-intensity vector **H** from the Maxwell field equations.   To do this, let us take the curl of both sides of Eq. (16.1); we then have

$$\nabla \times (\nabla \times \mathbf{E}) = -\mu \frac{\partial}{\partial t} (\nabla \times \mathbf{H}) \tag{17.1}$$

in view of the fact that $\mathbf{B} = \mu\mathbf{H}$ and that the operators $\nabla \times$ and $\partial/\partial t$ commute.
Equation (16.2) may be written in the form

$$\nabla \times \mathbf{H} = \sigma\mathbf{E} + K\frac{\partial \mathbf{E}}{\partial t} \tag{17.2}$$

Substituting this into (17.1) and making use of the identity

$$\nabla \times (\nabla \times \mathbf{E}) = \nabla(\nabla \cdot \mathbf{E}) - \nabla^2 \mathbf{E} \tag{17.3}$$

we obtain

$$\nabla(\nabla \cdot \mathbf{E}) - \nabla^2 \mathbf{E} = -\mu K \frac{\partial^2 \mathbf{E}}{\partial t^2} - \mu\sigma \frac{\partial \mathbf{E}}{\partial t} \tag{17.4}$$

Now, in view of (16.4), we have

$$\nabla \cdot \mathbf{E} = \frac{\rho}{K} \tag{17.5}$$

and it may be shown that in free space or in a conducting medium ρ is independent of the field distribution and may be taken to be equal to zero. Accordingly, Eq. (17.4) becomes

$$\nabla^2 \mathbf{E} = \mu K \frac{\partial^2 \mathbf{E}}{\partial t^2} + \mu\sigma \frac{\partial \mathbf{E}}{\partial t} \tag{17.6}$$

As a consequence of (16.7) we also have

$$\nabla^2 \mathbf{J} = \mu K \frac{\partial^2 \mathbf{J}}{\partial t^2} + \mu\sigma \frac{\partial \mathbf{J}}{\partial t} \tag{17.7}$$

Eliminating **E** from the Maxwell equations, we also obtain

$$\nabla^2 \mathbf{H} = \mu K \frac{\partial^2 \mathbf{H}}{\partial t^2} + \mu\sigma \frac{\partial \mathbf{H}}{\partial t} \tag{17.8}$$

It is thus seen that the three vectors **H**, **E**, and **J** satisfy equations of the same form.

In free space we have the conductivity $\sigma = 0$ and $\mu_r = K_r = 1$; we then obtain

$$\nabla^2 \mathbf{E} = \mu_0 K_0 \frac{\partial^2 \mathbf{E}}{\partial t^2} \tag{17.9}$$

and

$$\nabla^2 \mathbf{H} = \mu_0 K_0 \frac{\partial^2 \mathbf{H}}{\partial t^2} \tag{17.10}$$

If we let

$$c = \frac{1}{\sqrt{\mu_0 K_0}} \tag{17.11}$$

We may write (17.9) in the form

$$\nabla^2 \mathbf{E} = \frac{1}{c^2} \frac{\partial^2 \mathbf{E}}{\partial t^2} \tag{17.12}$$

This equation is the general wave equation in vector form and is discussed in Chap. 12. It is there shown that this equation governs the propagation of various entities, such as electric waves, displacements of tightly stretched strings, deflections of membranes, etc. It will be shown that such disturbances are propagated with a velocity equal to c.

Equations (17.9) and (17.10) are taken as the starting point in the theory of electromagnetic waves. In this case, c is the velocity of light in free space and is approximately 3×10^8 m per sec.

18 THE SKIN-EFFECT, OR DIFFUSION, EQUATION

In metals σ is of the order of $10^7 (1/\text{ohm-m})$ and $\epsilon_r \approx 1$, and it may be seen that the term involving the second derivative in (17.7) may be neglected for frequencies of the order of 10^{10} cycles per sec and lower. In such a case this equation reduces to

$$\nabla^2 \mathbf{J} = \mu \sigma \frac{\partial \mathbf{J}}{\partial t} \tag{18.1}$$

We thus see that the current density in metals satisfies an equation of the same form as is satisfied by the equation of heat flow in solids as given in Sec. 14. In the electrical-engineering literature this equation is called the *skin-effect equation*. The solution of this equation in various special cases is considered in Chap. 11.

19 TENSORS (QUALITATIVE INTRODUCTION)

In many physical problems vector notation is not sufficiently general to express the relations between the quantities involved. For example, in electrostatics, if a dielectric medium is isotropic, the relation between the electric-displacement vector $\mathbf{D}$ and the electric-intensity vector $\mathbf{E}$ is given by the vector equation

$$\mathbf{D} = K\mathbf{E} \tag{19.1}$$

where the scalar K is the electric inductive capacity of the medium (see Sec. 16). If, however, the dielectric medium is *not* isotropic, the electric-displacement vector $\mathbf{D}$ does not have the same direction as the vector $\mathbf{E}$, so that the scalar K must be replaced by a more general quantity to effect the required change in direction and magnitude. Such a quantity is called a *tensor*.

A similar situation arises in the formulation of elastic problems. If $\mathbf{F}$ is the stress vector in an isotropic medium and $\mathbf{X}$ is the strain vector, then the relation

$$\mathbf{F} = k\mathbf{X} \tag{19.2}$$

where k is a scalar, exists. If the medium is *not* isotropic, $\mathbf{F}$ and $\mathbf{X}$ are no longer in the same direction and Eq. (19.2) takes on a more general form in which k is a tensor.

After this qualitative introduction, the simpler analytical properties of tensors will be discussed in the following sections.

20   COORDINATE TRANSFORMATIONS

The fundamental definitions involved in the calculus of tensors or tensor analysis are intimately connected with the subject of coordinate transformations. In this section some of the fundamental properties of these transformations will be discussed.

LINEAR TRANSFORMATIONS AND THE SUMMATION CONVENTION

Consider a set of variables (u^1, u^2, u^3) related to another set of variables (w^1, w^2, w^3). (The superscripts denote *different* variables and *not* the powers of the variables involved.) Let the relation between the two sets of variables be of the form

$$
\begin{aligned}
w^1 &= c_1^1 u^1 + c_2^1 u^2 + c_3^1 u^3 \\
w^2 &= c_1^2 u^1 + c_2^2 u^2 + c_3^2 u^3 \\
w^3 &= c_1^3 u^1 + c_2^3 u^2 + c_3^3 u^3
\end{aligned}
\tag{20.1}
$$

where the coefficients c_j^i are constants. In this case the variables (u^1, u^2, u^3) are related to the variables (w^1, w^2, w^3) by a *linear* transformation. This transformation may be expressed by the equation

$$w^r = \sum_{n=1}^{n=3} c_n^r u^n \qquad r = 1, 2, 3 \tag{20.2}$$

For convenience, in tensor analysis, the following *summation convention* has been adopted:

The repetition of an index in a term once as a subscript and once as a superscript will denote a summation with respect to that index over its range. In (20.2) the range of n is from 1 to 3, so that, by the above summation convention, (20.2) may be written in the form

$$w^r = c^r_n u^n \qquad r = 1, 2, 3 \tag{20.3}$$

The determinant $C = |c^r_s|$ is the determinant of the linear transformation (20.1). If $C \neq 0$, the transformation will have an inverse. Let b^s_r be the *cofactor* of c^r_s, in the determinant C, divided by C. Then, by Chap. 2, Eq. (4.8), we have

$$c^r_m b^m_s = b^r_m c^m_s = \delta^r_s = \begin{cases} 0 & \text{if } r \neq s \\ 1 & \text{if } r = s \end{cases} \tag{20.4}$$

The symbol δ^r_s is called a *Kronecker delta* and has the values indicated in (20.4). In order to solve the system of Eqs. (20.1), multiply both sides by b^i_r and obtain

$$b_r w^r = b^i_r c^r_n u^n = \delta^i_n u^n = u^i \tag{20.5}$$

Therefore the solution of the system (20.3) is

$$u^i = b^i_r w^r \tag{20.6}$$

and therefore the transformation (20.1) is *reversible*.

SUCCESSIVE LINEAR TRANSFORMATIONS

Consider the two linear transformations

$$y^r = b^r_s x^s \tag{20.7}$$

and

$$z^r = a^r_k y^k \tag{20.8}$$

where all the indices range from 1 to n. If (20.7) is substituted into (20.8), the result is

$$z^r = a^r_k b^k_s x^s = c^r_s x^s \tag{20.9}$$

where

$$c^r_s = a^r_k b^k_s \tag{20.10}$$

is the transformation matrix of the linear transformation (20.9). These linear transformations may also be expressed by means of the matrix notation of Chap. 3 in the following form:

$$(y) = [b](x) \qquad (z) = [a](y) \tag{20.11}$$

Therefore

$$(z) = [a][b](x) \qquad [c] = [a][b] \tag{20.12}$$

where the matrices (x), (y), (z) are column matrices of order n and the matrices $[a]$, $[b]$, and $[c]$ are square matrices of order n.

FUNCTIONAL TRANSFORMATIONS, CURVILINEAR COORDINATES

Consider a three-dimensional Euclidean space. The position of a point P in this space can be specified by its coordinates referred to an orthogonal cartesian system of axes. Let (y^1, y^2, y^3) denote such a system of coordinates. Let

$$x^r = F^r(y^1, y^2, y^3) \qquad r = 1, 2, 3 \tag{20.13}$$

be a transformation of coordinates from the rectangular cartesian coordinates (y^1, y^2, y^3) to some general coordinates (x^1, x^2, x^3), not necessarily rectangular cartesian coordinates. For example, (x^1, x^2, x^3) may be spherical coordinates. The inverse of the transformation of coordinates (20.13) is the transformation that takes one from the coordinates (x^1, x^2, x^3) to the rectangular cartesian coordinates (y^1, y^2, y^3). Let

$$y^r = G^r(x^1, x^2, x^3) \tag{20.14}$$

be the inverse of the coordinate transformation (20.13). When this transformation exists, it is seen that to each set of values y^r there is a unique set of values of x^r, and vice versa. Hence the variables x^r determine a point in space uniquely and they are called *curvilinear coordinates*.

SPHERICAL POLAR COORDINATES

The spherical polar coordinates of Sec. 12 may be taken as a special example of the general transformations (20.13) and (20.14). In this notation, consider the spherical polar coordinates of Fig. 20.1. In this figure y^1, y^2, y^3 are the cartesian coordinates of a point P, and x^1, x^2, x^3 are the spherical polar coordinates of P. x^1 is the radius vector from the origin to the point P, and

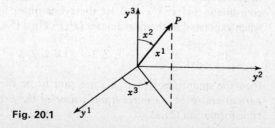

Fig. 20.1

x^2 and x^3 are the angles shown in Fig. 20.1.   In this special case the transformation (20.13) takes the form

$$x^1 = [(y^1)^2 + (y^2)^2 + (y^3)^2]^{\frac{1}{2}}$$
$$x^2 = \cos^{-1}\frac{y^3}{[(y^1)^2 + (y^2)^2 + (y^3)^2]^{\frac{1}{2}}} \qquad (20.15)$$
$$x^3 = \tan^{-1}\left(\frac{y^2}{y^1}\right)$$

The inverse transformation (20.14) in this case takes the following form:

$$y^1 = x^1 \sin x^2 \cos x^3$$
$$y^2 = x^1 \sin x^1 \sin x^3 \qquad (20.16)$$
$$y^3 = x^1 \cos x^2$$

21 SCALARS, CONTRAVARIANT VECTORS, AND COVARIANT VECTORS

The fundamental types of tensors will now be defined with respect to functional transformations from one set of curvilinear coordinates to another.   The discussion of tensors with respect to general functional transformations follows the same fundamental principles and may be carried out by considering a space of n dimensions.

a SCALARS

Let a Euclidean space be specified by the curvilinear coordinates (x^1, x^2, x^3). Let new curvilinear coordinates $(\bar{x}^1, \bar{x}^2, \bar{x}^3)$ be introduced by the transformation

$$\bar{x}^r = F^r(x^1, x^2, x^3) = \bar{x}^r(x^1, x^2, x^3) \qquad r = 1, 2, 3 \qquad (21.1)$$

Let a quantity have the value N in the *old* variables x^r and the value $\bar{N}$ in the *new* variables $\bar{x}^r$.   Then, if the quantity has the *same* value in both sets of variables so that

$$\bar{N} = N \qquad (21.2)$$

the quantity N is called a scalar, or an *invariant*, or a tensor of order zero.

b CONTRAVARIANT VECTORS

Let three quantities have the values (A^1, A^2, A^3) when expressed in terms of the coordinates (x^1, x^2, x^3), and let these quantities have the values $(\bar{A}^1, \bar{A}^2, \bar{A}^3)$ when expressed in the coordinates $(\bar{x}^1, \bar{x}^2, \bar{x}^3)$.   If

$$\bar{A}^m = \frac{\partial \bar{x}^m}{\partial x^i} A^i \qquad m = 1, 2, 3 \qquad i = 1, 2, 3 \qquad (21.3)$$

then the quantities (A^1, A^2, A^3) are said to be the components of a *contravariant vector* or a *contravariant tensor of the first rank* with respect to the transformation (21.1).

Taking differentials on both sides of (21.1),

$$d\bar{x}^r = \frac{\partial \bar{x}^r}{\partial x^s} dx^s \tag{21.4}$$

Since the components of an ordinary vector in three-dimensional space transform as the differentials of the coordinates, it is seen that the components of an ordinary vector are actually the components of a *contravariant* tensor of rank 1.

If three quantities have the values (A_1, A_2, A_3) in the system of coordinates (x_1, x_2, x_3) and the values $(\bar{A}_1, \bar{A}_2, \bar{A}_3)$ in the system of coordinates $(\bar{x}_1, \bar{x}_2, \bar{x}_3)$, and if

$$\bar{A}_m = \frac{\partial x^i}{\partial \bar{x}^m} A_i \tag{21.5}$$

then (A_1, A_2, A_3) are the components of a *covariant vector* or a covariant tensor of rank 1.

TENSORS OF HIGHER RANKS

In the above three classes of quantities, scalars or invariants, contravariant and covariant vectors, there exist components in any two coordinate systems, and the components in any two coordinate systems are related by the defining laws of transformation. There are other quantities whose components in any two coordinate systems are related by characteristic laws of transformation. These laws of transformation are involved in the definitions of tensors of higher ranks. There are three varieties of second-rank tensors:

$$\bar{A}^{mn} = \frac{\partial \bar{x}^m}{\partial x^i} \frac{\partial \bar{x}^n}{\partial x^j} A^{ij} \qquad \text{contravariant tensor of second rank} \tag{21.6}$$

$$\bar{A}_{mn} = \frac{\partial x^i}{\partial \bar{x}^m} \frac{\partial x^j}{\partial \bar{x}^n} A_{ij} \qquad \text{covariant tensor of second rank} \tag{21.7}$$

$$\bar{A}^m_n = \frac{\partial \bar{x}^m}{\partial x^j} \frac{\partial x^j}{\partial \bar{x}^n} A^i_j \qquad \text{mixed tensor of second rank} \tag{21.8}$$

The A's with the bars denote components in the coordinates $(\bar{x}_1, \bar{x}_2, \bar{x}_3)$, and those without the bars denote components in the coordinates (x_1, x_2, x_3).

Tensors of higher rank are defined by similar laws; for example, a mixed tensor of rank 4 is

$$\bar{A}^m_{npq} = \frac{\partial \bar{x}^m}{\partial x^i} \frac{\partial x^j}{\partial \bar{x}^n} \frac{\partial x^k}{\partial \bar{x}^p} \frac{\partial x^h}{\partial \bar{x}^q} A^i_{jhk} \tag{21.9}$$

A very useful mixed tensor of the second rank is the Kronecker delta δ^m_n. This is defined by the equation

$$\delta^m_n = \begin{cases} 1 & \text{if } m = n \\ 0 & \text{if } m \neq n \end{cases} \tag{21.10}$$

To determine whether δ_n^m is a mixed tensor of the second rank, we investigate whether it satisfies the definition (21.8).

$$\bar{\delta}_n^m = \frac{\partial \bar{x}^m}{\partial x^i} \frac{\partial x^j}{\partial \bar{x}^n} \delta_j^i = \frac{\partial \bar{x}^m}{\partial x^i} \frac{\partial x^i}{\partial \bar{x}^n} = \frac{\partial \bar{x}^m}{\partial \bar{x}^n} \tag{21.11}$$

But

$$\frac{\partial \bar{x}^m}{\partial \bar{x}^n} = \begin{cases} 1 & \text{if } m = n \\ 0 & \text{if } m \neq n \end{cases} \tag{21.12}$$

since $\bar{x}^m$ and $\bar{x}^n$ are *independent* variables.   Hence

$$\bar{\delta}_n^m = \delta_n^m \qquad \text{an invariant} \tag{21.13}$$

We thus see that δ_n^m has the same components in all coordinate systems.   The relation

$$\frac{\partial \bar{x}^m}{\partial x^i} \frac{\partial x^i}{\partial \bar{x}^n} = \delta_n^m \tag{21.14}$$

is a very useful one.

22   ADDITION, MULTIPLICATION, AND CONTRACTION OF TENSORS

a   ADDITION

It can be easily shown that the sum or difference of two or more tensors of the same rank and type is a tensor of the same rank and type.   For example, if

$$A^{mn} + B^{mn} = C^{mn} \tag{22.1}$$

it follows from (22.1) that C^{mn} is a contravariant tensor of the second rank.

b   THE OUTER PRODUCT

If the contravariant tensor A^m is multiplied by the covariant tensor B_n, the product is the following mixed tensor of the second rank:

$$C_n^m = A^m B_n \tag{22.2}$$

It is easy to see that C_n^m transforms like (21.8).   The product (22.2) is called the *outer product*.   It may be obtained with tensors of any rank or type. For example, the tensor A_n^m may be multiplied by the tensor B_{pq} to obtain the following outer product:

$$C_{npq}^m = A_n^m B_{pq} \tag{22.3}$$

c   CONTRACTION

Consider the mixed tensor A_n^m, and let the index m be set equal to n so that $m = n$.   The law of transformation (21.8) now gives the result

$$\bar{A}_m^m = \frac{\partial \bar{x}^m}{\partial x^i} \frac{\partial x^j}{\partial \bar{x}^m} A_j^i = A_i^i = A_m^m \tag{22.4}$$

and hence A_n^m is a scalar. This process of summing over a pair of contravariant and covariant indices is called *contraction*. As another example, consider the mixed tensor A_{npq}^m, and let $m = q$. Let $B_{np} = A_{npm}^m$. The law of transformation now gives

$$\bar{B}_{np} = \bar{A}_{npm}^m = \frac{\partial \bar{x}^m}{\partial x^i} \frac{\partial x^j}{\partial \bar{x}^n} \frac{\partial x^k}{\partial \bar{x}^p} \frac{\partial x^h}{\partial \bar{x}^m} A_{jkh}^i$$

$$= \frac{\partial x^i}{\partial \bar{x}^n} \frac{\partial x^k}{\partial \bar{x}^p} \delta_i^h = \frac{\partial x^j}{\partial \bar{x}^n} \frac{\partial x^k}{\partial \bar{x}^p} A_{jki}^i \tag{22.5}$$

Therefore A_{npm}^m is a covariant tensor of rank 2. The process of contraction always reduces the rank of a mixed tensor by 2.

d THE INNER PRODUCT

If two tensors are multiplied together and then contracted, the result is called the *inner product*. For example, the inner product is

$$A^m B_m = C \qquad \text{a scalar} \tag{22.6}$$

The inner product of A^{mn} and B_{rpq} is

$$A^{mn} B_{npq} = C_{pq}^m \tag{22.6a}$$

The product $A^m B_m$ is equivalent to the scalar product of two vectors in rectangular coordinates. In tensor analysis the length L of the tensor A^m or A_m is defined by the equation

$$L^2 = A^m A_m = |A|^2 \tag{22.7}$$

The scalar product of the two vectors $\mathbf{A}_m$ and $\mathbf{B}_m$ is

$$A_m B^m = |A||B| \cos \theta = (A_m A^m)^{\frac{1}{2}} (B_m B^m)^{\frac{1}{2}} \cos \theta \tag{22.8}$$

where θ is the angle between the vectors $\mathbf{A}_m$ and $\mathbf{B}_m$. Hence

$$\cos \theta = \frac{A_m B^m}{[(A_m A^m)(B_m B^m)]^{\frac{1}{2}}} \tag{22.9}$$

If $\mathbf{A}_m$ and $\mathbf{B}_m$ are orthogonal, then

$$A_m B^m = 0 \tag{22.10}$$

23 ASSOCIATED TENSORS

a THE QUOTIENT LAW OF TENSORS

It can be shown by applying the law of transformation of tensors that, if the following relation exists,

$$A(r,s,t) B^{st} = C^r \tag{23.1}$$

where C^r is known to be a tensor and B^{st} is a known arbitrary tensor, then $A(r,s,t)$ is a tensor of the rank

$$A(r,s,t) = A^r_{st} \tag{23.2}$$

This law is of importance in recognizing tensors.

b THE METRIC TENSOR

Let the cartesian coordinates of the point P be y^r. The element of length dS of such a coordinate system is given by

$$\overline{dS}^2 = (dy^1)^2 + (dy^2)^2 + (dy^3)^2 = dy^i \, dy^i \tag{23.3}$$

Since the cartesian coordinates are related to the general coordinates x^i by the functional relations

$$y^i = y^i(x^1, x^2, x^3) \tag{23.4}$$

we have

$$dy^i = \frac{\partial y^i}{\partial x^m} dx^m \tag{23.5}$$

It therefore can be seen that

$$\overline{dS}^2 = g_{mn} \, dx^m \, dx^n \tag{23.6}$$

where

$$g_{mn} = \sum_{i=1}^{3} \frac{\partial y^i}{\partial x^m} \frac{\partial y^i}{\partial x^n} = \frac{\partial y^i}{\partial x^m} \frac{\partial y^i}{\partial x^n} \tag{23.7}$$

This equation shows that g_{mn} is *symmetric*, and since dS is an *invariant* for arbitrary values of the contravariant vector dx^m, it follows from the *quotient law* that g_{mn} is a double covariant tensor. The tensor g_{mn} is called the *fundamental*, or *metric*, *tensor*. Let g denote the *determinant* $|g_{mn}|$ and G^{mn} denote the *cofactor* of g_{mn} in g. We now define the quantity g^{mn} by the equation

$$g^{mn} = \frac{G^{mn}}{g} \tag{23.8}$$

Now, from the theory of determinants [see Chap. 2] we have

$$g_{mn} g^{mr} = \delta^r_n \tag{23.9}$$

Hence by the *quotient law* it can be seen that g^{mn} is a double contravariant tensor of rank 2.

Two vectors related by the following equations,

$$A^m = g^{mn} A_n \tag{23.10}$$

and

$$A_m = g_{mn} A^n \tag{23.11}$$

are called *associated vectors*. In the literature of tensor analysis it is sometimes said that A^m and A_m are the same vector, A^m being the *contravariant*
components and A_m the *covariant* ones. Besides the fundamental, or metric,
tensor g_{mn} and the second fundamental tensor g^{mn} there is the third associated
tensor, g_s^r, defined by the equation

$$g_s^r = g_{sm} g^{rm} = \delta_s^r \tag{23.12}$$

This relation follows from (23.9).

Associated tensors of any rank may be obtained in the same manner as
(23.10) and (23.11). For example,

$$A^{mn} = g^{mp} g^{nq} A_{pq} \tag{23.13}$$

If we are given a contravariant vector A^r, we can form an invariant A
by means of the equation

$$A - (g_{mn} A^m A^n)^{\frac{1}{2}} \tag{23.14}$$

A is called the *magnitude* of the vector **A**. In the same manner the magnitude
B of the covariant vector B_m is defined by the equation

$$B = (g^{mn} B_m B_n)^{\frac{1}{2}} \tag{23.15}$$

A unit vector is one whose magnitude is unity. If (23.6) is divided by
$\overline{dS^2}$, we have

$$g_{mn} \frac{dx^m}{dS} \frac{dx^n}{dS} = 1 \tag{23.16}$$

From this it can be concluded that dx^m/dS is a *unit vector*. By the use of
(22.9) and (23.11) it can be shown that the angle θ between the vectors A^m and
B^m may be expressed in the form

$$\cos \theta = \frac{g_{mn} \mathbf{A}^m \mathbf{B}^n}{A B} \tag{23.17}$$

24 DIFFERENTIATION OF AN INVARIANT

Let ϕ be an *invariant* function of a parameter t; then, in the new variables $\bar{x}^r$,
we have $\bar{\phi} = \phi$, and it can be shown that

$$\frac{d\bar{\phi}}{dt} = \frac{d\phi}{dt} \tag{24.1}$$

so that $d\phi/dt$ is also an *invariant*. If ϕ is an invariant function of x^r, we have

$$\frac{\partial \bar{\phi}}{\partial \bar{x}^r} = \frac{\partial \phi}{\partial x^m} \frac{\partial x^m}{\partial \bar{x}^r} \tag{24.2}$$

This shows that $\partial \phi/\partial x^r$ is a *covariant vector* since it transforms like one.

25 DIFFERENTIATION OF TENSORS: THE CHRISTOFFEL SYMBOLS

In Sec. 24 it has been shown that the derivative of a scalar point function is a *covariant vector*. However, the derivative of a covariant vector is *not* a tensor, for if

$$\bar{A}_m = \frac{\partial x^h}{\partial \bar{x}^m} A_h \tag{25.1}$$

the derivative of $\bar{A}_m$ with respect to $\bar{x}^n$ is

$$\frac{\partial \bar{A}_m}{\partial \bar{x}^n} = \frac{\partial^2 x^h}{\partial \bar{x}^n \partial \bar{x}^m} A_h + \frac{\partial x^h}{\partial \bar{x}^m} \frac{\partial A_h}{\partial \bar{x}^n} \tag{25.2}$$

The presence of the second derivative shows that $\partial A_m/\partial x^n$ does *not* transform like a tensor.

In order to define "derivatives" of tensors that preserve the proper tensor character, it is necessary to introduce the following notation:[†]

$$[mn,p] = \frac{1}{2}\left(\frac{\partial g_{np}}{\partial x^m} + \frac{\partial g_{pm}}{\partial x^n} - \frac{\partial g_{mn}}{\partial x^p}\right) \tag{25.3}$$

and

$$\begin{Bmatrix} r \\ mn \end{Bmatrix} = g^{rp}[mn,p] \tag{25.4}$$

The quantity $[mn,p]$ is called the *Christoffel symbol of the first kind*, and $\begin{Bmatrix} r \\ mn \end{Bmatrix}$ is called the *Christoffel symbol of the second kind*.

a THE INTRINSIC DERIVATIVE OF A VECTOR

It can be shown that the quantity

$$\frac{\delta A^r}{\delta t} = \frac{dA^r}{dt} + \begin{Bmatrix} r \\ mn \end{Bmatrix} A^m \frac{dx^n}{dt} \tag{25.5}$$

is a *contravariant vector*. This vector is defined to be the *intrinsic derivative* of the vector A^r with respect to the parameter t.

The quantity

$$\frac{\delta A_r}{\delta t} = \frac{dA_r}{dt} - \begin{Bmatrix} m \\ rn \end{Bmatrix} A_m \frac{dx^n}{dt} \tag{25.6}$$

can be shown to be a covariant vector and is defined to be the *intrinsic derivative* of A_r with respect to the parameter t.

† See A. J. McConnel, "Applications of the Absolute Differential Calculus," pp. 143–150, Blackie & Son, Ltd., Glasgow, 1931.

b THE COVARIANT DERIVATIVE OF A VECTOR

The *covariant derivative* of the contravariant vector A^r with respect to the coordinate x^s is defined by the equation

$$A^r_s = \frac{\partial A^r}{\partial x^s} + \begin{Bmatrix} r \\ ms \end{Bmatrix} A^m \tag{25.7}$$

This quantity can be shown to be a mixed tensor contravariant in r and covariant in s. The *covariant derivative* of the covariant vector A_r with respect to the coordinate x^s is defined by the equation

$$A_{r,s} = \frac{\partial A_r}{\partial x^s} - \begin{Bmatrix} m \\ rs \end{Bmatrix} A_m \tag{25.8}$$

This quantity can be shown to be a tensor covariant in both r and s.

26 INTRINSIC AND COVARIANT DERIVATIVES OF TENSORS OF HIGHER ORDER

The intrinsic derivative of the tensor A^r_{sp} with respect to the parameter t is defined by the equation

$$\frac{\delta A^r}{\delta t^{sp}} = \frac{dA^r}{dt^{sp}} + \begin{Bmatrix} r \\ mn \end{Bmatrix} A^m_{sp} \frac{dx^n}{dt} - \begin{Bmatrix} m \\ sn \end{Bmatrix} A^r_{mp} \frac{dx^n}{dt} - \begin{Bmatrix} m \\ np \end{Bmatrix} A^r_{sm} \frac{dx^n}{dt} \tag{26.1}$$

It can be shown that this quantity is a tensor of the type A^r_{sp}.

The *covariant derivative* of the tensor A^r_{sp} with respect to the variable x^n is defined by the equation

$$A^r_{st,n} = \frac{\partial A^r_{st}}{\partial x^n} + \begin{Bmatrix} r \\ mn \end{Bmatrix} A^m_{st} - \begin{Bmatrix} m \\ sn \end{Bmatrix} A^r_{mt} - \begin{Bmatrix} m \\ tn \end{Bmatrix} A^r_{sm} \tag{26.2}$$

This is a tensor with one covariant index more than A^r_{st}. The definitions of the intrinsic and covariant derivatives are quite general. Terms of the forms (25.5) and (25.7) appear in the derivatives of a mixed tensor for each contravariant index, and terms of the forms (25.6) and (25.8) appear in the derivatives of mixed tensors for each covariant index.

If the coordinate system is a rectangular cartesian one, the g_{mn} quantities are all constants and the Christoffel systems are identically zero. In such case the *intrinsic* derivatives are identical to the ordinary derivatives, and the *covariant* derivatives are ordinary partial derivatives.

The fundamental, or metric, tensor g_{rs} has components that are constants in cartesian systems, and therefore

$$g_{rs} = 0 \tag{26.3}$$

in such a system. Since g_{rs} is a tensor, it follows that, in *every* coordinate system, the following equation must be satisfied:

$$g_{rs,m} = 0 \tag{26.4}$$

Therefore the covariant derivative of the metric tensor g_{rs} is zero. It can also be shown that the covariant derivative of the associated tensor g^{rs} is zero.

27 APPLICATION OF TENSOR ANALYSIS
TO THE DYNAMICS OF A PARTICLE

In the brief introduction to the basic ideas and definitions of tensor analysis given in this appendix, space does not permit the discussion of numerous applications to applied mathematics. These applications will be found in the treatises on tensor analysis listed in the references at the end of this chapter. For purposes of illustration, however, a brief discussion of the application of tensor analysis to the dynamics of a particle will be given in this section.

Consider a particle of mass M that is moving in space. Let the position of this particle be given by a system of curvilinear coordinates x^r. As the time t varies, the particle will describe a certain curve in space, called the *trajectory* of the particle. The equations of this curve may be written in the form

$$x^r = x^r(t) \qquad r = 1, 2, 3 \tag{27.1}$$

If we transform to another curvilinear coordinate system $\bar{x}^r$, in accordance with the transformation

$$\bar{x}^r = \bar{x}^r(x^1, x^2, x^3) \tag{27.2}$$

then the trajectory of the particle may be obtained in terms of t by substituting (27.1) into (27.2).

Consider the quantity

$$v^r = \frac{dx^r}{dt} \tag{27.3}$$

In the new coordinates $\bar{x}^r$ the corresponding quantities are

$$\bar{v}^r = \frac{d\bar{x}^r}{dt} = \frac{\partial \bar{x}^r}{\partial x^s}\frac{dx^s}{dt} = \frac{\partial \bar{x}^r}{\partial x^s}v^s \tag{27.4}$$

This equation expresses the fact that the v^r are the components of a *contravariant vector*. In dynamics this vector is called the *velocity* of the particle. The velocity vector v^r is a function of the time t. If the *intrinsic derivative* with respect to t is taken, the following result is obtained:

$$a^r = \frac{\delta v^r}{\delta t} = \frac{d^2 x^r}{dt^2} + \begin{Bmatrix} r \\ mn \end{Bmatrix}\frac{dx^m}{dt}\frac{dx^n}{dt} \tag{27.5}$$

where $\begin{Bmatrix} r \\ mn \end{Bmatrix}$ are the Christoffel symbols of the coordinate system x^r.

If the coordinates are rectangular cartesian, the Christoffel symbols are identical to zero and the vector $\acute{a}^r$ becomes

$$a^r = \frac{d^2 x^r}{dt^2} \tag{27.6}$$

These are the components of acceleration along the three coordinate axes; (27.5) is called the *acceleration vector* of the particle. If the mass of the particle M remains constant, it is obviously a quantity independent of the coordinate system and the time and is, therefore, an *invariant*.

Newton's second law of motion may be expressed in the form

$$F^r = Ma^r \tag{27.7}$$

where the contravariant vector F^r is called the *force vector*. This vector completely specifies the magnitude and direction of the force acting upon the particle. In cartesian coordinates (27.7) takes the well-known form

$$F^r = M \frac{d^2 x^r}{dt^2} \tag{27.8}$$

and F^1, F^2, F^3 are the components of the force along the three axes of coordinates. In curvilinear coordinates this equation takes the form

$$F^r = M \frac{d^2 x^r}{dt^2} + M \left\{ {r \atop mn} \right\} \frac{dx^m}{dt} \frac{dx^n}{dt} \tag{27.9}$$

WORK AND ENERGY: LAGRANGE'S EQUATIONS

If the axes $\bar{x}^r$ are rectangular cartesian, a force whose components in these coordinates are $\bar{F}^r$ and whose point of application is moved through a small displacement $\delta\bar{x}^r$ does an amount of work given by

$$\delta W = \bar{F}^{\cdot 1} \delta\bar{x}^1 + \bar{F}^2 \delta\bar{x}^2 + \bar{F}^3 \delta\bar{x}^3 \tag{27.10}$$

In orthogonal cartesian coordinates the associated vector $\bar{F}_r$ has exactly the same components as $\bar{F}^r$, and the expression for δW becomes

$$\delta W = \bar{F}_r \delta\bar{x}^r \tag{27.11}$$

From the relation connecting $\bar{x}^r$ and x^r it can be seen that δW can also be written in the form

$$\partial W = F_r \delta x^r \tag{27.12}$$

Since δW is an *invariant*, it can be concluded from the above equation that the F_r are the components of a covariant vector. Since the components of F_r in an orthogonal cartesian coordinate system are those of the force vector, it follows that F_r is the *covariant force vector* in the coordinates x^r.

The covariant and contravariant components of this vector are related to each other by the following formulas:

$$F_r = g_{rs} F^s \qquad F^r = g^{rs} F_s \tag{27.13}$$

If the expression $F_r dx^r$ is a *perfect differential*, then the force is said to be *conservative* and a function V may be defined by the equation

$$V = - \int F_r dx^r \tag{27.14}$$

The function V is called the *force potential*. From (27.14) it follows that

$$F_r = -\frac{\partial V}{\partial x^r} \tag{27.15}$$

The *kinetic energy* T of a particle is defined as $\frac{1}{2}Mv^2$, where v is the magnitude of the velocity. Therefore

$$T = \tfrac{1}{2}Mv^2 = \tfrac{1}{2}Mg_{mn} v^m v^n = \tfrac{1}{2}Mg_{mn} \dot{x}^m \dot{x}^n \tag{27.16}$$

where g_{mn} is the metric tensor of the curvilinear coordinates x^m.

LAGRANGE'S EQUATIONS

If T is differentiated partially with respect to $\dot{x}^r$, the result is

$$\frac{\partial T}{\partial \dot{x}^r} = Mg_{rm} \dot{x}^m \tag{27.17}$$

Therefore

$$\frac{d}{dt}\left(\frac{\partial T}{\partial \dot{x}^n}\right) = M\left(g_{rm} x^m + \frac{\partial g_{rm}}{\partial x^n}\dot{x}^m \dot{x}^n\right) \tag{27.18}$$

Also, if T is differentiated partially with respect to x^r, the result is

$$\frac{\partial T}{\partial x^r} = \tfrac{1}{2}M\frac{\partial g_{mn}}{\partial x^r}\dot{x}^m \dot{x}^n \tag{27.19}$$

From (27.18) and (27.19) it can be deduced that

$$\frac{d}{dt}\left(\frac{\partial T}{\partial \dot{x}^r}\right) - \frac{\partial T}{\partial x^r} = M\left[g_{rm}x^m + \frac{1}{2}\left(\frac{\partial g_{rm}}{\partial x^n} + \frac{\partial g_{rn}}{\partial x^m} - \frac{\partial g_{mn}}{\partial x^r}\right)\dot{x}^m \dot{x}^n\right] \tag{27.20}$$

$$= M(g_{rm}x^m + [mn,r]\,\dot{x}^m \dot{x}^n)$$

where $[mn,r]$ is the first Christoffel symbol. Therefore

$$\frac{d}{dt}\left(\frac{\partial T}{\partial \dot{x}^r}\right) - \frac{\partial T}{\partial x^r} = Mg_{rx}\left(\ddot{x}^s + \begin{Bmatrix} s \\ mn \end{Bmatrix}\dot{x}^m \dot{x}^n\right) \tag{27.21}$$

As a consequence of (27.5) it can be seen that the right-hand member of (27.21) is equal to $Mg_{rs}a^s = Ma_r$. Therefore (27.21) may be written in the form

$$\frac{d}{dt}\left(\frac{\partial T}{\partial \dot{x}^r}\right) - \frac{\partial T}{\partial x^r} = Ma_r = F_r \tag{27.22}$$

These are Lagrange's equations. If the force system is conservative, these equations may be written in the form

$$\frac{d}{dt}\left(\frac{\partial T}{\partial \dot{x}^r}\right) - \frac{\partial T}{\partial x^r} + \frac{\partial V}{\partial x^r} = 0 \tag{27.23}$$

PROBLEMS

1. The point of application of a force $\mathbf{F} = 5, 10, 15$ lb is displaced from the point $(1,0,3)$ to the point $(3,-1,-6)$. Find the work done by the force.

2. Find the scalar product of two diagonals of a unit cube. What is the angle between them?

3. A force $\mathbf{F}$ acts at a distance $\mathbf{r}$ from the origin. Show that the torque L about any axis through the origin is $L = (\mathbf{r} \times \mathbf{F}) \cdot \mathbf{a}$ when $\mathbf{a}$ is a unit vector in the direction of the axis.

4. Show that the lines joining the midpoints of the opposite sides of a quadrilateral bisect each other.

5. Show that the bisectors of the angles of a triangle meet at a point.

6. What is the cosine of the angle between the vectors $\mathbf{A} = 4i + 6j + 2k$ and $\mathbf{B} = i - 2j + 3k$?

7. Let $\mathbf{r}$ be the radius vector from the origin to any point and a constant vector. Find the gradient of the scalar product of $\mathbf{a}$ and $\mathbf{r}$.

8. A plane central field $\mathbf{A}$ is defined by $\mathbf{A} = \mathbf{r}F(r)$. Determine $F(r)$ so that the field may be irrotational and solenoidal.

9. A central field $\mathbf{A}$ in space is defined by $\mathbf{A} = \mathbf{r}F(r)$. Determine $F(r)$ so that the field may be irrotational and solenoidal.

10. If $\mathbf{r}$ is a unit vector of variable direction, the position vector of a moving point may be written $\mathbf{r} = r\mathbf{r}$. Find by vector methods the components of the acceleration $\mathbf{F}$ parallel and perpendicular to the radius vector of a particle moving in the xy plane.

11. Prove that the vector $\nabla\phi$ is perpendicular to the surface $\phi(x,y,z) = $ const.

12. Find ∇u if $u = \ln(x^2 + y^2 + z^2)$.

13. Show that if $\mathbf{r}$ is the position vector of any point of a closed surface s, then

$$\iint_s (\mathbf{r} \cdot d\mathbf{s}) = 3V$$

where V is the volume bounded by s.

14. Show that a vector field $\mathbf{A}$ is uniquely determined within a region V bounded by a surface S when its divergence and curl are given throughout V and the normal component of the curl is given on S.

15. With every point of a curve in space there is associated a unit vector t the direction of which is that of the velocity of a point describing the curve. This vector therefore has the direction of the tangent. Prove that $t \cdot dt/ds = 0$, where s is the length of the arc of the curve measured from a fixed point on it. What is the geometrical meaning of dt/ds?

16. The expression for $(ds)^2$ is $(ds)^2 = (u^2 + v^2)[(du)^2 + (dv)^2] + u^2v^2(d\phi)^2$ in parabolic coordinates. What is the form assumed by Laplace's equation in these coordinates?

17. Write the heat-flow equation in cylindrical and spherical coordinates.

18. What form does the general wave equation take in cylindrical coordinates?

19. Write Maxwell's equations in cylindrical coordinates.

20. Show that the potential due to a solid sphere of mass M and uniform density at an *external* point at a distance r from the center is M/r and, hence, that the gravitational intensity is M/r^2 toward the center.

21. Show that the gravitational field inside a homogeneous spherical shell is zero.

22. Find the potential inside a solid uniform sphere and the gravitational force inside the sphere.

23. Find the covariant components of the acceleration vector in spherical coordinates.

24. Show that the velocity vector is given by $(1/M)g^{rs}(\partial T/\partial \dot{x}s)$.

25. If ϕ is an invariant function of x^r and $\dot{x}^r$, show that the partial derivative of ϕ with respect to $\dot{x}^r$ is a covariant vector.

26. If a particle is moving with uniform velocity in a straight line prove that $\delta v^r/\delta t = 0$.

27. Find the contravariant components of the acceleration vector in cylindrical coordinates.

REFERENCES

1901. Gibbs, J. Willard, and Edwin Bidwell Wilson: "Vector Analysis," Yale University Press, New Haven, Conn.

1911. Coffin, Joseph George: "Vector Analysis," John Wiley & Sons, Inc., New York.

1928. Weatherburn, C. E.: "Advanced Vector Analysis with Application to Mathematical Physics," George Bell & Sons, Ltd., London.

1928. Weatherburn, C. E.: "Elementary Vector Analysis with Applications to Geometry and Physics," George Bell & Sons, Ltd., London.

1931. McConnel, A. J.: "Applications of the Absolute Differential Calculus," Blackie & Son, Ltd., Glasgow.

1931. Wills, A. P.: "Vector Analysis with an Introduction to Tensor Analysis," Prentice-Hall, Inc., Englewood Cliffs, N.J.

1932. Gans, Richard: "Vector Analysis with Applications to Physics," Blackie & Son, Ltd., Glasgow.

1946. Brillouin, L.: "Les Tenseurs en mécanique et en élasticité," Dover Publications, Inc., New York.

1951. Sokolnikoff, I. S.: "Tensor Analysis," John Wiley & Sons, Inc., New York.

appendix F
Partial Differentiation and the Calculus of Variations

1 INTRODUCTION

A great many of the fundamental laws of the various branches of science are expressed most simply in terms of differential equations. For example, the motion of a particle of mass M, when acted upon by a force whose components along the three axes of a cartesian reference frame are F_x, F_y, and F_z, is given by the three differential equations

$$M\frac{d^2 x}{dt^2} = F_x \qquad M\frac{d^2 y}{dt^2} = F_y \qquad M\frac{d^2 z}{dt^2} = F_z \qquad (1.1)$$

where x, y, z is the position of the particle at any time t. In this case the motion is given by a system of ordinary differential equations.

The physical laws governing the distribution of temperature in solids, the propagation of electricity in cables, and the distribution of velocities in moving fluids are expressed in terms of partial differential equations. It is therefore necessary that the student of applied mathematics should have a clear idea of the fundamental definitions and operations involving partial differentiation.

2   PARTIAL DERIVATIVES

A quantity $F(x,y,z)$ is said to be a function of the three variables x, y, z if the value of F is determined by the values of x, y, and z. If, for example, x, y, and z are the cartesian coordinates of a certain point in space, then $F(x,y,z)$ may be the temperature at that point, and as x, y, and z take on other values, $F(x,y,z)$ will give the temperature in the region under consideration.

CONTINUITY

The function $F(x,y,z)$ is continuous at a point (a,b,c) for which it is defined if

$$\lim_{\substack{x \to a \\ y \to b \\ z \to c}} F(x,y,z) = F(a,b,c) \tag{2.1}$$

independently of the manner in which x approaches a, y approaches b, and z approaches c.

Now, given $F(x,y,z)$, it is possible to hold y and z constant and allow x to vary; this reduces F to a function of x only which may have a derivative defined and computed in the usual way.   This derivative is called the *partial derivative of F* with respect to x.   Therefore, by definition,

$$\frac{\partial F}{\partial x} = \lim_{h \to 0} \frac{F(x+h,\,y,\,z) - F(x,y,z)}{h} \tag{2.2}$$

The symbol $\partial F/\partial x$ denotes the partial derivative.   Sometimes the alternative notations are used:

$$\frac{\partial F}{\partial x} = F_x = \left(\frac{dF}{dx}\right)_{y,\,z} \tag{2.3}$$

Again, if we hold x and z constant, we make F a function of y alone whose derivative is the partial derivative of F with respect to y; this is written

$$\frac{\partial F}{\partial y} = F_y = \left(\frac{dF}{dy}\right)_{x,\,z} = \lim_{k \to 0} \frac{F(x,\,y+k,\,z) - F(x,y,z)}{k} \tag{2.4}$$

In the same manner we define the partial derivative with respect to z:

$$\frac{\partial F}{\partial z} = F_z = \left(\frac{dF}{dz}\right)_{x,\,y} = \lim_{q \to 0} \frac{F(x,\,y,\,z+q) - F(x,y,z)}{q} \tag{2.5}$$

If $F(x,y,z)$ has partial derivatives at each point of a domain, then those derivatives are themselves functions of x, y, and z and may have partial derivatives which are called the *second partial derivatives of the function F*. For example,

$$\frac{\partial}{\partial x}\left(\frac{\partial F}{\partial x}\right) = \frac{\partial^2 F}{\partial x^2} = F_{xx}$$

$$\frac{\partial}{\partial y}\left(\frac{\partial F}{\partial x}\right) = \frac{\partial^2 F}{\partial y\,\partial x} = F_{yx}$$

$$\frac{\partial}{\partial y}\left(\frac{\partial F}{\partial z}\right) = \frac{\partial^2 F}{\partial y\,\partial z} = F_{yz}$$

(2.6)

etc.

ORDER OF DIFFERENTIATION

$\partial^2 F/(\partial x\,\partial y)$ denotes the derivative of $\partial F/\partial y$ with respect to x, while $\partial^2 F/(\partial y\,\partial x)$ denotes the derivative of $\partial F/\partial x$ with respect to y. It may be shown that if $F(x,y)$ and its derivatives $\partial F/\partial x$ and $\partial F/\partial y$ are *continuous* then the order of differentiation is immaterial, and we have

$$\frac{\partial^2 F}{\partial x\,\partial y} = \frac{\partial^2 F}{\partial y\,\partial x}, \text{ etc.}$$

(2.7)

In general $\partial^{p+q} F/(\partial x^p\,\partial y^q)$ signifies the result of differentiating $F(x,y,z)$ p times with respect to x and q times with respect to y, the order of differentiation being immaterial. The extension to any number of variables is obvious.

3 THE SYMBOLIC FORM OF TAYLOR'S EXPANSION

In Appendix C, Sec. 16, we wrote Taylor's expansion of a function of one variable in the form

$$e^{hD_x} f(x) = f(x + h)$$

(3.1)

where

$$D_x = \frac{d}{dx}$$

(3.2)

and e is the base of the natural logarithms. The symbolic expansion of a function $F(x,y)$ of two variables written in the form

$$e^{(hD_x + kD_y)} F(x,y) = F(x + h, y + k)$$

(3.3)

where

$$D_x = \frac{\partial}{\partial x} \qquad D_y = \frac{\partial}{\partial y}$$

(3.4)

is to be interpreted by substituting

$$u = hD_x + kD_y$$

(3.5)

in the Maclaurin expansion

$$e^u = 1 + \frac{u}{1!} + \frac{u^2}{2!} + \cdots + \frac{u^n}{n!} + \cdots \tag{3.6}$$

and operating with the result on $F(x,y)$. Terms of the type D_x^r and $D_x^r D_y^s$, etc., are interpreted by

$$D_x^r = \frac{\partial^r}{\partial x^r} \qquad D_x^r D_y^s = \frac{\partial^{r+s}}{\partial x^r \, \partial y^s} \tag{3.7}$$

The justification of (3.1) depends on the fact that the operators D_x and D_y satisfy certain laws of algebra and commute with constants, as discussed in Chap. 7. This form of Taylor's expansion is of great usefulness in applied mathematics.

4 DIFFERENTIATION OF COMPOSITE FUNCTIONS

As a simple case of composite functions, let us consider

$$F = F(x,y) \tag{4.1}$$

where x and y are both functions of the independent variable t, that is,

$$x = x(t) \qquad y = y(t) \tag{4.2}$$

Now if t is given an increment Δt, then x and y receive increments Δx, Δy, and F receives an increment ΔF given by

$$\Delta F = F(x + \Delta x, y + \Delta y) - F(x,y) \tag{4.3}$$

Now, by Taylor's expansion, we have

$$F(x + \Delta x, y + \Delta y) = e^{\Delta x \, D_x + \Delta y \, D_y} F(x,y)$$

$$= F(x,y) + \frac{\partial F}{\partial x}\Delta x + \frac{\partial F}{\partial y}\Delta y + \delta_1 \Delta x + \delta_2 \Delta y \tag{4.4}$$

where

$$\lim_{\substack{\Delta x \to 0 \\ \Delta y \to 0}} \delta_1 = 0 \qquad \lim_{\substack{\Delta x \to 0 \\ \Delta y \to 0}} \delta_2 = 0 \tag{4.5}$$

Hence

$$\Delta F = \frac{\partial F}{\partial x}\Delta x + \frac{\partial F}{\partial y}\Delta y + \delta_1 \Delta x + \delta_2 \Delta y \tag{4.6}$$

Dividing this by Δt and taking the limit as $\Delta t \to 0$, we have

$$\lim_{\Delta t \to 0} \frac{\Delta F}{\Delta t} = \lim_{\Delta t \to 0}\left(\frac{\partial F}{\partial x}\frac{\Delta x}{\Delta t} + \frac{\partial F}{\partial y}\frac{\Delta y}{\Delta t} + \delta_1 \frac{\Delta x}{\Delta t} + \delta_2 \frac{\Delta y}{\Delta t}\right) \tag{4.7}$$

Now as $\Delta t \to 0$, $\Delta x \to 0$, and $\Delta y \to 0$, and if t is the only independent variable, we have

$$\lim_{\Delta t \to 0} \frac{\Delta F}{\Delta t} = \frac{dF}{dt} \qquad \lim_{\Delta t \to 0} \frac{\Delta x}{\Delta t} = \frac{dx}{dt}, \text{ etc.} \qquad (4.8)$$

Therefore, provided the functions $x(t)$ and $y(t)$ are differentiable, (4.7) becomes

$$\frac{dF}{dt} = \frac{\partial F}{\partial x}\frac{dx}{dt} + \frac{\partial F}{\partial y}\frac{dy}{dt} \qquad (4.9)$$

If there are other independent variables besides t, then we must use the notation

$$\lim_{\Delta t \to 0} \frac{\Delta F}{\Delta t} = \frac{\partial F}{\partial t} \qquad \lim_{\Delta t \to 0} \frac{\Delta x}{\Delta t} = \frac{\partial x}{\partial t}, \text{ etc.} \qquad (4.10)$$

and we have

$$\frac{\partial F}{\partial t} = \frac{\partial F}{\partial x}\frac{\partial x}{\partial t} + \frac{\partial F}{\partial y}\frac{\partial y}{\partial t} \qquad (4.11)$$

This formula may be extended to the case where F is a function of any number of variables $x, y, z, \ldots$ and $x, y, z, \ldots$, etc., are functions of the variables $t, r, s, p, \ldots$, etc.

The results may be stated in the following form: If F is a function of the n variables $x_1, x_2, \ldots, x_n$ so that

$$F = F(x_1, x_2, x_3, \ldots, x_n) \qquad (4.12)$$

and each variable x is a function of the single variable t so that

$$x_r = x_r(t) \qquad r = 1, 2, 3, \ldots, n \qquad (4.13)$$

then

$$\frac{dF}{dt} = \frac{\partial F}{\partial x_1}\frac{dx_1}{dt} + \frac{\partial F}{\partial x_2}\frac{dx_2}{dt} + \cdots + \frac{\partial F}{\partial x_n}\frac{dx_n}{dt} \qquad (4.14)$$

If, however, each variable x is a function of the p variables $t_1, t_2, \ldots, t_p$ so that

$$x_r = x_r(t_1, t_2, \ldots, t_p) \qquad r = 1, 2, 3, \ldots, n \qquad (4.15)$$

then

$$\frac{\partial F}{\partial t_1} = \frac{\partial F}{\partial x_1}\frac{\partial x_1}{\partial t_1} + \frac{\partial F}{\partial x_2}\frac{\partial x_2}{\partial t_1} + \cdots + \frac{\partial F}{\partial x_n}\frac{\partial x_n}{\partial t_1}$$
$$\frac{\partial F}{\partial t_s} = \frac{\partial F}{\partial x_1}\frac{\partial x_1}{\partial t_s} + \frac{\partial F}{\partial x_2}\frac{\partial x_2}{\partial t_s} + \cdots + \frac{\partial F}{\partial x_n}\frac{\partial x_n}{\partial t_s}, \text{ etc.} \qquad (4.16)$$

SECOND AND HIGHER DERIVATIVES

As an illustration of the manner in which higher derivatives may be computed from these fundamental formulas, let us differentiate (4.9) on the assumption that x and y are functions of the single variable t. We therefore have

$$\frac{d^2 F}{dt^2} = \frac{d}{dt}\left(\frac{\partial F}{\partial x}\right)\frac{dx}{dt} + \frac{\partial F}{\partial x}\frac{d^2 x}{dt^2} + \frac{d}{dt}\left(\frac{\partial F}{\partial y}\right)\frac{\partial y}{dt} + \frac{\partial F}{\partial y}\frac{d^2 y}{dt^2} \tag{4.17}$$

Now since $\partial F/\partial x$ and $\partial F/\partial y$ are functions of x and y, we apply (4.9) to $\partial F/\partial x$ and $\partial F/\partial y$ instead of to F and obtain

$$\frac{d}{dt}\left(\frac{\partial F}{\partial x}\right) = \frac{\partial^2 F}{\partial x^2}\frac{dx}{dt} + \frac{\partial^2 F}{\partial x\,\partial y}\frac{dy}{dt} \tag{4.18}$$

and

$$\frac{d}{dt}\left(\frac{\partial F}{\partial y}\right) = \frac{\partial^2 F}{\partial x\,\partial y}\frac{dx}{dt} + \frac{\partial^2 F}{\partial y^2}\frac{dy}{dt} \tag{4.19}$$

Substituting these in (4.17), we have

$$\frac{d^2 F}{dt^2} = \frac{\partial^2 F}{\partial x^2}\left(\frac{dx}{dt}\right)^2 + 2\frac{\partial^2 F}{\partial x\,\partial y}\frac{dx}{dt}\frac{dy}{dt} + \frac{\partial^2 F}{\partial y^2}\left(\frac{dy}{dt}\right)^2 + \frac{\partial F}{\partial x}\frac{d^2 x}{dt^2} + \frac{\partial F}{\partial y}\frac{d^2 y}{dt^2} \tag{4.20}$$

Expressions for the third and higher derivatives may be found in a similar manner.

5 CHANGE OF VARIABLES

An important application of Eq. (4.11) is its use in changing variables; for example, let

$$F = F(x,y) \tag{5.1}$$

and it is desired to replace x and y by the polar coordinates r and θ given by

$$x = r\cos\theta \qquad y = r\sin\theta \tag{5.2}$$

Then F becomes a function of r and θ, and we have by (4.11)

$$\frac{\partial F}{\partial r} = \frac{\partial F}{\partial x}\frac{\partial x}{\partial r} + \frac{\partial F}{\partial y}\frac{\partial y}{\partial r} = \frac{\partial F}{\partial x}\cos\theta + \frac{\partial F}{\partial y}\sin\theta$$

$$\frac{\partial F}{\partial \theta} = \frac{\partial F}{\partial x}\frac{\partial x}{\partial \theta} + \frac{\partial F}{\partial y}\frac{\partial y}{\partial \theta} = \frac{\partial F}{\partial x}(-r\sin\theta) + \frac{\partial F}{\partial y}r\cos\theta \tag{5.3}$$

Solving these equations for $\partial F/\partial x$ and $\partial F/\partial y$, we have

$$\frac{\partial F}{\partial x} = \frac{\partial F}{\partial r}\cos\theta - \frac{\partial F}{\partial \theta}\frac{\sin\theta}{r}$$

$$\frac{\partial F}{\partial y} = \frac{\partial F}{\partial r}\sin\theta + \frac{\partial F}{\partial \theta}\frac{\cos\theta}{r} \tag{5.4}$$

The second derivatives may be computed by Eq. (4.20).

6 THE FIRST DIFFERENTIAL

For simplicity, let us consider

$$F = F(x, y) \tag{6.1}$$

a function of two variables x and y. Now let us give x an increment Δx and y an increment Δy. Then, as was stated in (4.6), F takes an increment ΔF, where

$$\Delta F = \frac{\partial F}{\partial x} \Delta x + \frac{\partial F}{\partial y} \Delta y + \delta_1 \Delta x + \delta_2 \Delta y \tag{6.2}$$

provided $\partial F/\partial x$ and $\partial F/\partial y$ are continuous.

In general the third term is an infinitesimal of higher order than the first term, and the fourth term is in general a higher-order infinitesimal than the second. We take the first two terms of (6.2) and call them the *differential* of F and write

$$dF = \frac{\partial F}{\partial x} \Delta x + \frac{\partial F}{\partial y} \Delta y \tag{6.3}$$

The definition is completed by saying that if x and y are *independent* variables

$$dx = \Delta x \qquad dy = \Delta y \tag{6.4}$$

then (6.3) takes the form

$$dF = \frac{\partial F}{\partial x} dx + \frac{\partial F}{\partial y} dy \tag{6.5}$$

This expression is called the *total differential* of $F(x, y)$. This definition may be extended to the case where F is a function of the n *independent variables* $x_1, x_2, \ldots, x_n$ to obtain

$$dF = \frac{\partial F}{\partial x_1} dx_1 + \frac{\partial F}{\partial x_2} dx_2 + \cdots + \frac{\partial F}{\partial x_n} dx_n \tag{6.6}$$

This definition (6.5) has been based on the assumption that x and y are independent variables. Let us now examine the case where this is not true. Let us suppose that x and y are functions of the three independent variables u, v, w, so that

$$\begin{aligned} x &= x(u,v,w) \\ y &= y(u,v,w) \end{aligned} \tag{6.7}$$

Now since $u, v,$ and w are independent, we have

$$\begin{aligned} dx &= \frac{\partial x}{\partial u} du + \frac{\partial x}{\partial v} dv + \frac{\partial x}{\partial w} dw \\ dy &= \frac{\partial y}{\partial u} du + \frac{\partial y}{\partial v} dv + \frac{\partial y}{\partial w} dw \end{aligned} \tag{6.8}$$

And since F is a function of u, v, and w, we have

$$dF = \frac{\partial F}{\partial u}\,du + \frac{\partial F}{\partial v}\,dv + \frac{\partial F}{\partial w}\,dw \tag{6.9}$$

But, by Eqs. (4.16), we have

$$\frac{\partial F}{\partial u} = \frac{\partial F}{\partial x}\frac{\partial x}{\partial u} + \frac{\partial F}{\partial y}\frac{\partial y}{\partial u}$$

$$\frac{\partial F}{\partial v} = \frac{\partial F}{\partial x}\frac{\partial x}{\partial v} + \frac{\partial F}{\partial y}\frac{\partial y}{\partial v} \tag{6.10}$$

$$\frac{\partial F}{\partial w} = \frac{\partial F}{\partial x}\frac{\partial x}{\partial w} + \frac{\partial F}{\partial y}\frac{\partial y}{\partial w}$$

Substituting these equations into (6.9), we have

$$dF = \frac{\partial F}{\partial x}\left(\frac{\partial x}{\partial u}\,du + \frac{\partial x}{\partial v}\,dv + \frac{\partial x}{\partial w}\,dw\right) + \frac{\partial F}{\partial y}\left(\frac{\partial y}{\partial u}\,du + \frac{\partial y}{\partial v}\,dv + \frac{\partial y}{\partial w}\,dw\right)$$

$$= \frac{\partial F}{\partial x}\,dx + \frac{\partial F}{\partial y}\,dy \tag{6.11}$$

This is the same as (6.5). It thus follows that the differential of a function F of the variables $x_1, x_2, \ldots, x_n$ has the form (6.6) *whether the variables $x_1, x_2, \ldots, x_n$ are independent or not.*

Let us now consider the case where

$$F(x_1, x_2, \ldots, x_n) = C \tag{6.12}$$

where C is a constant. This relation cannot exist when $x_1, x_2, \ldots, x_n$ are independent variables unless $F = C$. Let us suppose that $x_1, x_2, \ldots, x_n$ are functions of *independent* variables $u_1, u_2, \ldots, u_m$. Hence F may be regarded as a function of the variables $u_1, u_2, \ldots, u_m$, and we write

$$F(u_1, u_2, \ldots, u_m) = C \tag{6.13}$$

Hence

$$dF = \frac{\partial F}{\partial u_1}\,du_1 + \frac{\partial F}{\partial u_2}\,du_2 + \cdots + \frac{\partial F}{\partial u_m}\,du_m \tag{6.14}$$

Now, since $u_1, u_2, \ldots, u_m$ are *independent* variables, u_1 may be changed without changing the value of the other variables or the value of F since $F = C$. Therefore

$$F(u_1 + \Delta u_1, u_2, u_3, \ldots, u_m) = C \tag{6.15}$$

Hence

$$\frac{\partial F}{\partial u_1} = \lim_{\Delta u_1 \to 0} \frac{F(u_1 + \Delta u_1, u_2, \ldots, u_m) - F(u_1, u_2, \ldots, u_m)}{\Delta u_1}$$

$$= 0 \tag{6.16}$$

In the same manner we may prove that

$$\frac{\partial F}{\partial u_r} = 0 \qquad r = 2, 3, \ldots, m \tag{6.17}$$

Hence, as a consequence of (6.14), we have

$$dF = 0 \tag{6.18}$$

and by (6.6) we have

$$dF = \frac{\partial F}{\partial x_1} dx_1 + \frac{\partial F}{\partial x_2} dx_2 + \cdots + \frac{\partial F}{\partial x_n} dx_n = 0 \tag{6.19}$$

7   DIFFERENTIAL OF IMPLICIT FUNCTIONS

If we have the relation

$$F(x,y) = 0 \tag{7.1}$$

we are accustomed to say that this equation defines y as an *implicit* function of x and is equivalent to the equation

$$y = \phi(x) \tag{7.2}$$

If the functional relation (7.1) is simple, then we can actually solve (7.1) to obtain y in the form (7.2).   For example, consider

$$x^2 + y^2 - a^2 = 0 \tag{7.3}$$

This equation may be solved for y to give

$$y = \pm\sqrt{a^2 - x^2} \tag{7.4}$$

However, if Eq. (7.1) is complicated, it is in general not possible to solve it for y.   It may be shown that y in (7.1) satisfies the definition of a function of x in the sense that when x is given (7.1) determines a value of y. It is convenient to be able to differentiate (7.1) with respect to either x or y without solving the equation explicitly for x or y.

To differentiate (7.1), let us take its first differential.   As a special case of (6.19) we have

$$dF = \frac{\partial F}{\partial x} dx + \frac{\partial F}{\partial y} dy = 0 \tag{7.5}$$

and hence

$$\frac{dy}{dx} = -\frac{\partial F/\partial x}{\partial F/\partial y} = y' \tag{7.6}$$

This may be written in the form

$$\frac{\partial F}{\partial x} + \frac{\partial F}{\partial y} y' = 0 \tag{7.7}$$

To obtain the second derivative, let

$$\phi = \left(\frac{\partial F}{\partial x} + \frac{\partial F}{\partial y} y' \right) \tag{7.8}$$

Applying (7.7) to ϕ, we have

$$\frac{\partial \phi}{\partial x} + \frac{\partial \phi}{\partial y} y' = 0 \tag{7.9}$$

Now

$$\frac{\partial \phi}{\partial x} = \frac{\partial^2 F}{\partial x^2} + \frac{\partial^2 F}{\partial x \, \partial y} y' + \frac{\partial F}{\partial y} \frac{\partial y'}{\partial x} \tag{7.10}$$

and

$$\frac{\partial \phi}{\partial y} y' = \frac{\partial^2 F}{\partial x \, \partial y} y' + \frac{\partial^2 F}{\partial y^2} (y')^2 + y' \frac{\partial F}{\partial y} \frac{\partial y'}{\partial y} \tag{7.11}$$

Substituting in (7.9), we obtain

$$\frac{\partial^2 F}{\partial x^2} + 2 \frac{\partial^2 F}{\partial x \, \partial y} y' + \frac{\partial^2 F}{\partial y^2} (y')^2 + \frac{\partial F}{\partial y} y'' = 0 \tag{7.12}$$

since

$$\frac{\partial y'}{\partial x} + y' \frac{\partial y'}{\partial y} = \frac{dy'}{dx} = y''$$

Repeating this process, we may find the derivatives y''', y'''', etc., provided the partial derivatives of $F(x,y)$ exist and provided that $\partial F / \partial y \neq 0$.

ONE EQUATION, MORE THAN TWO VARIABLES
The equation

$$F(x,y,z) = 0 \tag{7.13}$$

defines any one of the variables, for example, x, in terms of the other two. If we take the differential of (7.13), we have

$$dF = \frac{\partial F}{\partial x} dx + \frac{\partial F}{\partial y} dy + \frac{\partial F}{\partial z} dz = 0 \tag{7.14}$$

If we place $y = $ const, then $dy = 0$, and we have

$$\left(\frac{dz}{dx} \right)_y = - \frac{\partial F / \partial x}{\partial F / \partial z} \tag{7.15}$$

where the subscript denotes that y is held constant.   This is less ambiguous than the notation $\partial z/\partial x$.   If $x =$ const, we have

$$\left(\frac{dy}{dz}\right)_x = -\frac{\partial F/\partial z}{\partial F/\partial y} \tag{7.16}$$

If $z =$ const, we obtain

$$\left(\frac{dx}{dy}\right)_z = -\frac{\partial F/\partial y}{\partial F/\partial x} \tag{7.17}$$

Multiplying Eqs. (7.15), (7.16), and (7.17) together, we have

$$\left(\frac{dx}{dy}\right)_z \left(\frac{dy}{dz}\right)_x \left(\frac{dz}{dx}\right)_y = -1 \tag{7.18}$$

This is sometimes written in the form

$$\left(\frac{\partial x}{\partial y}\right) \left(\frac{\partial y}{\partial z}\right) \left(\frac{\partial z}{\partial x}\right) = -1 \tag{7.19}$$

The absurdity of using ∂x, ∂y, ∂z as symbols for *differentials* which may be canceled is apparent from Eq. (7.19).

8   MAXIMA AND MINIMA

Quite frequently in the application of mathematics to science it is necessary to determine the maximum or minimum values of a function of one or more variables.

Let us consider a function F of the single variable x, so that

$$F = F(x) \tag{8.1}$$

A *maximum* of $F(x)$ is a value of $F(x)$ which is *greater* than those immediately preceding or immediately following, while a *minimum* of $F(x)$ is a value of $F(x)$ which is *less* than those immediately preceding or following. In defining and discussing maxima and minima of $F(x)$, it is assumed that $F(x)$ and all necessary derivatives are continuous and single-valued functions of x.

In order to determine whether the function $F(x)$ has a maximum or a minimum at a point $x = a$, we may use Taylor's expansion of a function of one variable, in the form

$$F(x + h) = F(x) + \frac{h}{1!} F'(x) + \frac{h^2}{2!} F''(x) + \frac{h^3}{3!} F'''(x) + \cdots \tag{8.2}$$

Let $x = a$ be the critical point under consideration, and write

$$\Delta(h) = F(a + h) - F(a) = \frac{h}{1!} F'(a) + \frac{h^2}{2!} F''(a) + \frac{h^3}{3!} F'''(a) + \cdots \tag{8.3}$$

$\Delta(h)$ is thus the change in the value of the function when the argument of the function is changed by h. This is illustrated graphically, in the case that $F(a)$ is a maximum, by Fig. 8.1.

Now, if $x = a$ is a point at which $F(x)$ has either a maximum or a minimum, we shall have for h sufficiently small

$$\text{Sign of } \Delta(h) = \text{sign of } \Delta(-h) \tag{8.4}$$

since for a maximum, if we move to the left of the critical point or to the right of the critical point, the function will decrease and for a minimum point the function will increase. Now if

$$F'(x) \neq 0 \tag{8.5}$$

then for h sufficiently small

$$\Delta(h) \cong hF'(a) \tag{8.6}$$

since the higher-order terms in (8.3) may be neglected. In the same way

$$\Delta(-h) \cong -hF'(a) \tag{8.7}$$

Hence, in order for (8.4) to be satisfied, we must have

$$F'(a) = 0 \tag{8.8}$$

at *either* a *maximum* or a *minimum*. Now at a maximum $\Delta(h)$ must be negative and at a minimum $\Delta(h)$ must be positive for either positive or negative values of h.

Hence, if (8.8) is satisfied, for h sufficiently small,

$$\Delta(h) \cong \frac{h^2}{2!} F''(a) \tag{8.9}$$

Since h^2 is always positive, then it is evident that at a *maximum* we must have

$$F''(a) < 0 \tag{8.10}$$

and at a *minimum*

$$F''(a) > 0 \tag{8.11}$$

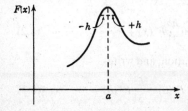

Fig. 8.1

Let us suppose that

$$F''(a) = 0 \quad \text{and} \quad F'''(a) \neq 0 \tag{8.12}$$

Then for h sufficiently small

$$\Delta(h) \cong \frac{h^3}{3!} F'''(a) \tag{8.13}$$

Since, if $F'''(a) \neq 0$, the expression (8.13) changes sign with h, we cannot have a maximum or a minimum. If, however,

$$F'(a) = 0 \qquad F''(a) = 0 \qquad F'''(a) = 0 \tag{8.14}$$

then for h sufficiently small

$$\Delta(h) = \frac{h^4}{4!} F''''(a) \tag{8.15}$$

Hence $F(a)$ will be a *maximum* if

$$F''''(a) < 0 \tag{8.16}$$

and a *minimum* if

$$F''''(a) > 0, \text{ etc.} \tag{8.17}$$

MAXIMA AND MINIMA OF FUNCTIONS OF TWO VARIABLES

We define the maximum and minimum values of a function $F(x,y)$ of two variables x, y in the following manner:

$F(a,b)$ is a *maximum* of $F(x,y)$ when, for all sufficiently small positive or negative values of h and k,

$$\Delta(h,k) = F(a + h, b + k) - F(a,b) < 0 \tag{8.18}$$

$F(a,b)$ is a *minimum* of $F(x,y)$ when, for all small positive or negative values of h and k,

$$\Delta(h,k) = F(a + h, b + k) - F(a,b) > 0 \tag{8.19}$$

By Taylor's expansion of a function of two variables, we have, if the necessary derivatives exist and are continuous,

$$F(x + h, y + k) = F(x,y) + h\frac{\partial F}{\partial x} + k\frac{\partial F}{\partial y}$$
$$+ \frac{1}{2!}\left(h^2\frac{\partial^2 F}{\partial x^2} + 2hk\frac{\partial^2 F}{\partial x\, \partial y} + k^2\frac{\partial^2 F}{\partial y^2}\right) + \cdots \tag{8.20}$$

Letting $F_x = \partial F/\partial x$, $F_y = \partial F/\partial y$,

$$A = \frac{\partial^2 F}{\partial x^2} \qquad B = \frac{\partial^2 F}{\partial x\, \partial y} \qquad C = \frac{\partial^2 F}{\partial y^2} \tag{8.21}$$

evaluated at the point $x = a$, $y = b$. Then

$$\Delta(h,k) = hF_x(a,b) + kF_y(a,b) + \frac{1}{2!}(h^2 A + 2hkB + k^2 C) + \cdots \qquad (8.22)$$

It is thus evident that for small values of h and k, in order for $\Delta(h,k)$ to have the same sign independently of the signs of h and k, it is necessary for the coefficients of h and k in (8.22) to vanish. This gives

$$\frac{\partial F}{\partial x} = 0 \qquad \frac{\partial F}{\partial y} = 0 \qquad (8.23)$$

evaluated at $x = a$, $y = b$ as a required condition for *either* a maximum or a minimum. If the conditions (8.23) are satisfied, then $\Delta(h,k)$ reduces to

$$\Delta(h,k) = \frac{1}{2!}(h^2 A + 2hkB + k^2 C) + \cdots \qquad (8.24)$$

To facilitate the discussion, we make use of the identity

$$Ah^2 + 2Bhk + Ck^2 = \frac{(Ah + Bk)^2 + (AC - B^2)k^2}{A} \qquad (8.25)$$

We may then write (8.24) in the form

$$\Delta(h,k) = \frac{(Ah + Bk)^2 + (AC - B^2)k^2}{2!A} \qquad (8.26)$$

The sign of $\Delta(h,k)$ given by (8.26) is *independent* of the signs of h and k provided that

$$AC - B^2 > 0 \qquad (8.27)$$

or

$$AC - B^2 = 0 \qquad (8.28)$$

This may be seen since $(Ah + Bk)^2$ is always positive or zero; therefore, if $AC - B^2$ is negative, the numerator of (8.26) will be positive when $k = 0$ and negative when $Ah + Bk = 0$.

Therefore a *second* condition for a maximum or a minimum of $F(x,y)$ is that at the point in question

$$AC > B^2 \qquad \text{or} \qquad \frac{\partial^2 F}{\partial x^2}\frac{\partial^2 F}{\partial y^2} > \left(\frac{\partial^2 F}{\partial x\,\partial y}\right)^2 \qquad (8.29)$$

An investigation of the exceptional cases when (8.28) is satisfied or when

$$A = B = C = 0 \qquad (8.30)$$

is beyond the scope of this discussion. The reader is referred to Goursat and Hedrick.† When condition (8.29) is satisfied at $x = a$, $y = b$, we see that $F(a,b)$ will have a *maximum* when

$$\frac{\partial^2 F}{\partial x^2} < 0 \qquad \frac{\partial^2 F}{\partial y^2} < 0 \tag{8.31}$$

evaluated at $x = a$, $y = b$. $F(a,b)$ will have a *minimum* when

$$\frac{\partial^2 F}{\partial x^2} > 0 \qquad \frac{\partial^2 F}{\partial y^2} > 0 \tag{8.32}$$

evaluated at $x = a$, $y = b$.
If

$$AC - B^2 < 0 \tag{8.33}$$

then $F(x,y)$ has *neither* a maximum nor a minimum. By a similar course of reasoning we obtain the conditions for maxima and minima of functions of three or more variables.

As an example of the above theory, let it be required to examine

$$F(x,y) = x^2 y + xy^2 - axy \tag{8.34}$$

for maximum and minimum values. Here

$$\frac{\partial F}{\partial x} = (2x + y - a)y \qquad \frac{\partial F}{\partial y} = (2y + x - a)x \tag{8.35}$$

$$\frac{\partial^2 F}{\partial x^2} = 2y \qquad \frac{\partial^2 F}{\partial y^2} = 2x \qquad \frac{\partial^2 F}{\partial x \, \partial y} = 2x + 2y - a \tag{8.36}$$

The conditions (8.23) are

$$(2x + y - a)y = 0 \qquad (2y + x - a)x = 0 \tag{8.37}$$

Condition (8.29) is

$$4xy > (2x + 2y - a)^2 \tag{8.38}$$

The system of Eqs. (8.37) has the four solutions

$$x = 0 \qquad x = a \qquad x = 0 \qquad x = \frac{a}{3}$$
$$\tag{8.39}$$
$$y = 0 \qquad y = 0 \qquad y = a \qquad y = \frac{a}{3}$$

† Goursat and Hedrick, 1904 (see References).

Only the last values satisfy (8.38), and a maximum or a minimum of $F(x,y)$ is located at

$$x = \frac{a}{3} \qquad y = \frac{a}{3} \tag{8.40}$$

If a is positive, $\partial^2 F/\partial x^2$ is positive when $y = a/3$; therefore

$$F\left(\frac{a}{3}, \frac{a}{3}\right) = -\frac{a^3}{27} \tag{8.41}$$

is a minimum. If a is negative, $\partial^2 F/\partial x^2$ is negative when $y = a/3$; hence $-a^3/27$ is a maximum.

LAGRANGE'S METHOD OF UNDETERMINED MULTIPLIERS

In a great many practical problems it is desired to find the maximum or minimum value of a function of certain variables when the variables are not independent but are connected by some given relation.

For example, let it be required to find the maximum or minimum value of the function

$$u = u(x,y,z) \tag{8.42}$$

and let the variables x, y, and z be connected by the relation

$$\phi(x,y,z) = 0 \tag{8.43}$$

In principle it should be possible to solve (8.43) for one variable in terms of the other two. This variable may then be eliminated by substitution into (8.42). Then the maximum or minimum values of u may be investigated by the methods discussed above since, by the elimination, u has been reduced to a function of two variables. In many cases the solution of (8.43) for any of the variables may be extremely difficult or impossible. To deal with such cases, Lagrange used an ingenious device which is known in the literature as the *method of undetermined multipliers*. This method will now be discussed.

If u is to have a maximum or minimum, it is necessary that

$$\frac{\partial u}{\partial x} = 0 \qquad \frac{\partial u}{\partial y} = 0 \qquad \frac{\partial u}{\partial z} = 0 \tag{8.44}$$

Hence

$$\frac{\partial u}{\partial x} dx + \frac{\partial u}{\partial y} dy + \frac{\partial u}{\partial z} dz = 0 \tag{8.45}$$

Differentiating the functional relation (8.43), we obtain

$$\frac{\partial \phi}{\partial x} dx + \frac{\partial \phi}{\partial y} dy + \frac{\partial \phi}{\partial z} dz = 0 \tag{8.46}$$

Let us now multiply (8.46) by the parameter λ and add the resulting equation to (8.45).   We thus obtain

$$\left(\frac{\partial u}{\partial x} + \lambda\frac{\partial \phi}{\partial x}\right) dx + \left(\frac{\partial u}{\partial y} + \lambda\frac{\partial \phi}{\partial y}\right) dy + \left(\frac{\partial u}{\partial z} + \lambda\frac{\partial \phi}{\partial z}\right) dz = 0 \qquad (8.47)$$

This equation will be satisfied if

$$\frac{\partial u}{\partial x} + \lambda\frac{\partial \phi}{\partial x} = 0 \qquad \frac{\partial u}{\partial y} + \lambda\frac{\partial \phi}{\partial y} = 0 \qquad \frac{\partial u}{\partial z} + \lambda\frac{\partial \phi}{\partial z} = 0 \qquad (8.48)$$

The three Eqs. (8.48) and Eq. (8.43) furnish four equations to determine the proper values of the variables x, y, z and λ to assure the maximum or minimum value of the function u.

As an example of the use of this method, let it be required to find the volume of the largest parallelepiped that can be inscribed in the ellipsoid

$$\frac{x^2}{a^2} + \frac{y^2}{b^2} + \frac{z^2}{c^2} = 1 \qquad (8.49)$$

In this case the function to be maximized is

$$u = 8xyz \qquad \bullet \qquad (8.50)$$

and Eqs. (8.48) are

$$8yz + \frac{2\lambda}{a^2}x = 0$$

$$8xz + \frac{2\lambda}{b^2}y = 0 \qquad (8.51)$$

$$8xy + \frac{2\lambda}{c^2}z = 0$$

Multiplying the first equation by x, the second one by y, and the last one by z, and adding them, we obtain

$$3u + 2\lambda\left(\frac{x^2}{a^2} + \frac{y^2}{b^2} + \frac{z^2}{c^2}\right) = 0 \qquad (8.52)$$

In view of (8.49) this becomes

$$3u + 2\lambda = 0 \qquad (8.53)$$

Hence

$$2\lambda = -3u \qquad (8.54)$$

From the first equation (8.51) multiplied by x, we have

$$u\left(1 - \frac{3}{a^2}x^2\right) = 0 \qquad (8.55)$$

or

$$x = \frac{a}{\sqrt{3}} \tag{8.56}$$

Similarly

$$y = \frac{b}{\sqrt{3}} \qquad z = \frac{c}{\sqrt{3}} \tag{8.57}$$

Hence the required maximum volume is

$$u = \frac{8abc}{3\sqrt{3}} \tag{8.58}$$

9   DIFFERENTIATION OF A DEFINITE INTEGRAL

It is frequently required to differentiate a definite integral with respect to its limits or with respect to some parameter.   Let $F(x,u)$ be a continuous function of x, u, and consider

$$\phi(u) = \int_{a(u)}^{b(u)} F(x,u)\, dx \tag{9.1}$$

where u is a parameter appearing in the integrand and we assume that the limits a and b of the definite integral are continuous functions of the parameter u, so that

$$a = a(u) \qquad b = b(u) \tag{9.2}$$

The integral therefore defines a function $\phi(u)$ of the parameter u.   We shall now show that differentiation of the function $\phi(u)$ yields the important equation

$$\frac{d\phi}{du} = \int_a^b \frac{\partial F}{\partial u}\, dx + F(b,u)\frac{db}{du} - F(a,u)\frac{da}{du} \tag{9.3}$$

To establish this equation, let u be given an increment Δu in (9.1). Hence

$$\phi(u + \Delta u) = \int_{a+\Delta a}^{b+\Delta b} F(x,\, u+\Delta u)\, dx \tag{9.4}$$

where Δa and Δb are the increments that a and b take when u is increased by Δu.   Then

$$\Delta\phi = \phi(u + \Delta u) - \phi(u)$$

$$= \int_{a+\Delta a}^{a} F(x,\, u'+\Delta u)\, dx + \int_a^b [F(x,\, u+\Delta u) - F(x,u)]\, dx$$
$$+ \int_b^{b+\Delta b} F(x,\, u+\Delta u)\, dx \tag{9.5}$$

Now by the concept of the definite integral of a continuous function, between the limits x_1 and x_2, we have

$$\int_{x_1}^{x_2} F(x)\,dx = F(x_0)(x_2 - x_1) \tag{9.6}$$

where x_0 is some intermediate point between x_2 and x_1 given by

$$x_1 < x_0 < x_2 \tag{9.7}$$

This may be seen intuitively by the concept of the integral (9.6) giving the area under the curve $F(x)$ between the points x_1 and x_2. In this case $F(x_0)$ is a mean ordinate such that when it is multiplied by the length $x_2 - x_1$ it gives the same area as that given by the integral. We may apply Eq. (9.6) to the first integral of (9.5) and obtain

$$\int_{a+\Delta a}^{a} F(x, u + \Delta u)\,dx = F(t_1, u + \Delta u)[a - (a + \Delta a)]$$
$$= -F(t_1, u + \Delta u)\,\Delta a \tag{9.8}$$

where

$$a + \Delta a > t_1 > a \tag{9.9}$$

In the same way the last integral of (9.5) may be expressed in the form

$$\int_{b}^{b+\Delta b} F(x, u + \Delta u)\,dx = \Delta b = \Delta b\,F(t_2, u + \Delta u) \tag{9.10}$$

where

$$b < t_2 < b + \Delta b \tag{9.11}$$

Now

$$\lim_{\Delta u \to 0} \int_{a}^{b} \frac{[F(x, u + \Delta u) - F(x,u)]\,dx}{\Delta u} = \int_{a}^{b} \frac{\partial F}{\partial u}\,dx \tag{9.12}$$

Hence, if we divide (9.5) by Δu, using the results (9.8), (9.10), and (9.12) and realizing that

$$\lim_{\Delta u \to 0} t_1 = a \qquad \lim_{\Delta u \to 0} t_2 = b \tag{9.13}$$

we have

$$\frac{d\phi}{du} = \int_{a}^{b} \frac{\partial F}{\partial u}\,dx + F(b,u)\frac{db}{du} - F(a,u)\frac{da}{du} \tag{9.14}$$

This is the required result.

In the special case that a and b are fixed, we have

$$\frac{d\phi}{du} = \int_{a}^{b} \frac{\partial F}{\partial u}\,dx \tag{9.15}$$

If $F(x)$ does not contain the parameter u, and

$$b = u \qquad a = \text{const} \tag{9.16}$$

we have

$$\frac{\partial}{\partial b} \int_a^b F(x)\, dx = F(b) \tag{9.17}$$

In the same way, differentiating with respect to the lower limit gives

$$\frac{\partial}{\partial a} \int_a^b F(x)\, dx = -F(a) \tag{9.18}$$

These equations are useful in evaluating certain definite integrals.

10   INTEGRATION UNDER THE INTEGRAL SIGN

The possibility of differentiating under the integral sign leads to the converse possibility of integration.   Let

$$\phi(u) = \int_a^b F(x,u)\, dx \tag{10.1}$$

where a and b are constants.   Multiply by du and integrate with respect to u between u_0 and u.   We then have

$$\int_{u_0}^u \phi(u')\, du' = \int_{u_0}^u du' \int_a^b F(x,u')\, dx = Q(u) \tag{10.2}$$

The integrations are to be carried out first with respect to x and then with respect to u.   Now let us consider

$$P(u) = \int_a^b dx \int_{u_0}^u F(x,u')\, du' \tag{10.3}$$

We wish to show that

$$P(u) = Q(u) \tag{10.4}$$

Let us differentiate (10.3) with respect to u.   By the results of the last section, when we carry out the differentiation on the right under the integral sign, we obtain

$$\frac{\partial}{\partial u} \int_{u_0}^u F(x,u')\, du' = F(x,u) \tag{10.5}$$

where the differentiation has been carried out with respect to the upper limit. Hence

$$\frac{dP}{du} = \int_a^b F(x,u)\, dx = \phi(u) \tag{10.6}$$

Therefore

$$\int_{u_0}^{u} \frac{dP}{du'} du' = \int_{u_0}^{u} \phi(u') \, du' = P(u) - P(u_0) \tag{10.7}$$

But, from (10.3), we have

$$P(u_0) = 0 \tag{10.8}$$

Hence

$$P(u) \overset{.}{=} \int_{u_0}^{u} \phi(u') \, du' = Q(u) \tag{10.9}$$

or

$$\int_{a}^{b} dx \int_{u_0}^{u} F(x,u') \, du' = \int_{u_0}^{u} du' \int_{a}^{b} F(x,u') \, dx \tag{10.10}$$

as was to be proved. This shows the possibility of interchanging the order of multiple integrations.

As an example of the use of the concept of integrating under the integral sign, consider

$$F(x,u) = x^u \qquad \text{where } u > -1 \tag{10.11}$$

Now

$$\int_{0}^{1} F(x,u) \, dx = \int_{0}^{1} x^u \, dx = \frac{1}{u+1} \tag{10.12}$$

Now multiply by du, and integrate between a and b.   Then

$$\int_{a}^{b} du \int_{0}^{1} x^u \, dx = \int_{a}^{b} \frac{du}{u+1} = \ln \frac{b+1}{a+1} \tag{10.13}$$

But

$$\int_{a}^{b} du \int_{0}^{1} x^u \, dx = \int_{0}^{1} dx \int_{a}^{b} x^u \, du = \ln \frac{b+1}{a+1} \tag{10.14}$$

Now

$$\int_{a}^{b} x^u \, du = \frac{x^b - x^a}{\ln x} \tag{10.15}$$

Hence

$$\int_{0}^{1} \frac{x^b - x^a}{\ln x} \, dx = \ln \frac{b+1}{a+1} \tag{10.16}$$

11   EVALUATION OF CERTAIN DEFINITE INTEGRALS

The evaluation of various definite integrals will illustrate some of the general principles discussed in Sec. 10.   Let us consider the integral

$$I = \int_{0}^{\infty} \frac{\sin bx}{x} \, dx \tag{11.1}$$

If we change the sign of b, then the sign of the integral is changed. Placing $b = 0$ causes the integral to vanish. However, if we let

$$bx = y \qquad b > 0 \tag{11.2}$$

then we obtain

$$I = \int_0^\infty \frac{\sin y}{y} \, dy \tag{11.3}$$

This shows that the integral does not depend on b but is a constant. Considered as a function of b, it has a discontinuity at $b = 0$.

Let us consider the integral

$$\int_0^\infty e^{-kx} \, dx = -\frac{e^{-kx}}{k} \Big|_0^\infty = \frac{1}{k} \qquad \text{if } k > 0 \tag{11.4}$$

If we let k be the complex number

$$k = a + jb \tag{11.5}$$

then we have

$$\int_0^\infty e^{-(a+jb)x} \, dx = \frac{1}{a + jb} = \frac{a - jb}{a^2 + b^2} \qquad a > 0 \tag{11.6}$$

Separating the real and imaginary parts, we obtain the two integrals

$$\int_0^\infty e^{-ax} \cos bx \, dx = \frac{a}{a^2 + b^2} \tag{11.7}$$

$$\int_0^\infty e^{-ax} \sin bx \, dx = \frac{b}{a^2 + b^2} \tag{11.8}$$

Let us integrate (11.7) with respect to b and obtain

$$\int_0^b db \int_0^\infty e^{-ax} \cos bx \, dx = \int_0^\infty e^{-ax} \, dx \int_0^b \cos bx \, db$$

$$= \int_0^\infty e^{-ax} \frac{\sin bx}{x} \, dx = a \int_0^b \frac{db}{a^2 + b^2} = \tan^{-1} \frac{b}{a} \tag{11.9}$$

Placing $a = 0$ in (11.9), we obtain the result

$$\int_0^\infty \frac{\sin bx}{x} \, dx = \frac{\pi}{2} \qquad b > 0 \tag{11.10}$$

The integral

$$I_1 = \int_0^\infty e^{-x^2} \, dx \tag{11.11}$$

occurs very frequently in many branches of applied mathematics, particularly in the theory of probability.   This integral represents the area of the so-called "probability curve"

$$u = e^{-x^2} \tag{11.12}$$

Since the indefinite integral cannot be found except by a development in series, we are led to employ a certain device to evaluate the definite integral. Since the variable of integration in a definite integral is of no importance, we have

$$I_1 = \int_0^\infty e^{-x^2}\, dx \qquad I_1 = \int_0^\infty e^{-y^2}\, dy \tag{11.13}$$

Multiplying these integrals together, we obtain

$$I_1^2 = \int_0^\infty e^{-x^2}\, dx \int_0^\infty e^{-y^2}\, dy$$
$$= \int_0^\infty \int_0^\infty e^{-x^2-y^2}\, dx\, dy \tag{11.14}$$

It is permissible to introduce e^{-x^2} under the sign of integration since x and y are to be considered as independent variables.   If we consider x and y as the coordinates of a cartesian reference frame and let z be a vertical coordinate, then the double integral (11.4) will represent the volume of a solid of revolution bounded by the surface

$$z = e^{-x^2-y^2} \tag{11.15}$$

We may find this volume by introducing polar coordinates.   Then the element of area in the xy plane is

$$ds = r\, dr\, d\theta \qquad \text{where } x^2 + y^2 = r^2 \tag{11.16}$$

There may be some question concerning the validity of this process since the double integral (11.14) is the limit

$$I_1^2 = \lim_{R\to\infty} \int_0^R \int_0^R e^{-x^2-y^2}\, dx\, dy \tag{11.17}$$

which represents the volume over a square in the xy plane of sides equal to R (see Fig. 11.1).

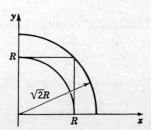

Fig. 11.1

It is easy to see that this volume is greater than that of the figure of revolution over the circle of radius R and less than that over the circle of radius $\sqrt{2}\,R$. Hence if the integral

$$\int_0^{\pi/2} \int_0^R e^{-r^2} r\,dr\,d\theta \tag{11.18}$$

approaches a limit for $R = \infty$, we have

$$I_1^2 = \int_0^{\pi/2} \int_0^\infty e^{-r^2} r\,dr\,d\theta \tag{11.19}$$

The integration with respect to θ merely multiplies by $\pi/2$, while in the integral

$$\int_0^R e^{-r^2} r\,dr = -\tfrac{1}{2} e^{-r^2} \Big|_0^R \tag{11.20}$$

the fact that we have an exact differential makes integration possible. Passing to the limit, we have

$$I_1^2 = \lim_{R\to\infty} \frac{\pi}{2} \int_0^R e^{-r^2} r\,dr = \frac{\pi}{4} \lim_{R\to\infty} (1 - e^{-R^2}) = \frac{\pi}{4} \tag{11.21}$$

We therefore have the desired result

$$I_1 = \int_0^\infty e^{-x^2}\,dx = \frac{\sqrt{\pi}}{2} \tag{11.22}$$

If in this integral we make the change in variable

$$x = \sqrt{au} \qquad a > 0 \tag{11.23}$$

then we obtain

$$I_1 = \int_0^\infty e^{-au^2} \sqrt{a}\,du = \frac{\sqrt{\pi}}{2} \tag{11.24}$$

or

$$\int_0^\infty e^{-au^2}\,du = \frac{1}{2}\sqrt{\frac{\pi}{a}} \qquad a > 0 \tag{11.25}$$

To illustrate a slightly different device, let us consider the integral

$$I = \int_0^\infty e^{-x^2 + a^2/x^2}\,dx \tag{11.26}$$

This integral may be differentiated with respect to a to obtain

$$\frac{dI}{da} = +2 \int_0^\infty \frac{a}{x^2} e^{-x^2 + a^2/x^2}\,dx \tag{11.27}$$

If we now change the variable of integration by putting

$$x = \frac{a}{y} \qquad dx = -\frac{a\,dy}{y^2} = -\frac{x^2}{a}\,dy \tag{11.28}$$

then (11.27) becomes

$$\frac{dI}{da} = -2 \int_0^\infty e^{-y^2 + a^2/y^2} \, dy = -2I \tag{11.29}$$

This is a linear differential equation with constant coefficients for I. Its general solution is

$$I = C e^{-2a} \tag{11.30}$$

where C is an arbitrary constant to be determined.   Placing $a = 0$, I reduces to

$$\int_0^\infty e^{-x^2} \, dx = \frac{\sqrt{\pi}}{2} \tag{11.31}$$

Therefore

$$C = \frac{\sqrt{\pi}}{2} \tag{11.32}$$

Hence we have finally

$$\int_0^\infty e^{-x^2 - a^2/x^2} \, dx = \frac{\sqrt{\pi}}{2} e^{-2a} \qquad a > 0 \tag{11.33}$$

The above examples illustrate typical procedures by which certain definite integrals may be evaluated.

12   THE ELEMENTS OF THE CALCULUS OF VARIATIONS

We have seen that a necessary condition for a function $F(x)$ to have a maximum or a minimum at a certain point is that the first derivative of the function shall vanish at that point; also a necessary condition for a maximum or a minimum of a function of several variables is that all its partial derivatives of the first order should vanish.

We now consider the following question: Given a definite integral whose integrand is a function of x, y and of the first derivatives $y' = dy/dx$,

$$I = \int_{x_0}^{x_1} F(x, y, y') \, dx \tag{12.1}$$

for what function $y(x)$ is the value of this integral a maximum or a minimum? In contrast to the simple maximum or minimum problem of the differential calculus the function $y(x)$ is not known here but is to be determined in such a way that the integral is a maximum or a minimum.   In applied mathematics we meet problems of this type very frequently.   A very simple example is given by the question, "What is the shortest curve that can be drawn between two given points?"   In a plane, the answer is obviously a straight line. However, if the two points and their connecting curve are to lie on a given arbitrary surface, then the analytic equation of this curve, which is called a *geodesic*, may be found only by the solution of the above problem, which is called the *fundamental problem of the calculus of variations*.

It will now be shown that the maximum or minimum problem of the calculus of variations may be reduced to the determination of the extreme value of a known function. To show this, consider functions $\bar{y}$ of x that are "neighboring" functions to the required function $y(x)$.

The function $\bar{y}$ is obtained as follows: Let ϵ be a small quantity, and let $n(x)$ be an arbitrary function of x which is continuous and whose first two derivatives are continuous in the range of integration. We then introduce into the integral (12.1) in place of y and y' the neighboring functions

$$\bar{y} = y + \epsilon n \qquad\qquad (12.2)$$

$$\bar{y}' = y' + \epsilon n' \qquad\qquad (12.3)$$

We stipulate, however, that these functions $\bar{y}$ coincide with the function $y(x)$ at the end points of the range of integration, as shown in Fig. 12.1. It must therefore be required that the arbitrary function $n(x)$ vanish at the end points of the interval.

If we substitute the function $\bar{y}$ into the integral, then the integral becomes a function of ϵ. We then require that $y(x)$ should make the integral a maximum or a minimum; that is, the function $I(\epsilon)$ must have a maximum or minimum value for $\epsilon = 0$. That is,

$$I(\epsilon) = \int_{x_0}^{x_1} F(x, y + \epsilon n, y' + \epsilon n')\, dx \qquad\qquad (12.4)$$

should be a maximum or minimum for $\epsilon = 0$.

This gives us a simple method of determining the extreme value of a given integral. The condition is

$$\left(\frac{dI}{d\epsilon}\right)_{\epsilon=0} = 0 \qquad\qquad (12.5)$$

We expand the integrand function F in a Taylor's series in the form

$$F(x, y + \epsilon n, y' + \epsilon n') = F(x, y, y') + \epsilon n \frac{\partial F}{\partial y} + \epsilon n' \frac{\partial F}{\partial y'}$$

$$+ \text{ terms in } \epsilon^2, \epsilon^3, \ldots \qquad (12.6)$$

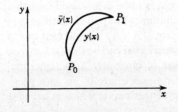

Fig. 12.1

Therefore

$$I(\epsilon) = \int_{x_0}^{x_1} \left[F(x,y,y') + \epsilon n \frac{\partial F}{\partial y} + \epsilon n' \frac{\partial F}{\partial y'} + \text{terms in } \epsilon^2, \epsilon^3, \ldots \right] dx \quad (12.7)$$

If we differentiate (12.7) inside the integral sign with respect to ϵ, we obtain

$$\frac{dI}{d\epsilon} = \int_{x_0}^{x_1} \left(n \frac{\partial F}{\partial y} + n' \frac{\partial F}{\partial y'} + \text{terms in } \epsilon, \epsilon^2, \ldots \right) dx \quad (12.8)$$

This expression must vanish for $\epsilon = 0$. Since the terms in ϵ, ϵ^2, . . . vanish for $\epsilon = 0$, we have the condition

$$\int_{x_0}^{x_1} \left(n \frac{\partial F}{\partial y} + n' \frac{\partial F}{\partial y'} \right) dx = 0 \quad (12.9)$$

The second term of (12.9) may be transformed by integration by parts into the form

$$\int_{x_0}^{x_1} n' \frac{\partial F}{\partial y'} dx = n \frac{\partial F}{\partial y'} \Big|_{x_0}^{x_1} - \int_{x_0}^{x_1} n \frac{d}{dx} \left(\frac{\partial F}{\partial y'} \right) dx \quad (12.10)$$

The first term vanishes since $n(x)$ must be zero at the limits. Hence, substituting back into (12.9), we obtain

$$\int_{x_0}^{x_1} n \left(\frac{\partial F}{\partial y} - \frac{d}{dx} \frac{\partial F}{\partial y'} \right) dx = 0 \quad (12.11)$$

Now, since $n(x)$ is *arbitrary*, the only way that the integral (12.11) can vanish is for the term in parentheses to vanish; hence we have

$$\frac{\partial F(x,y,y')}{\partial y} - \frac{d}{dx} \frac{\partial F(x,y,y')}{\partial y'} = 0 \quad (12.12)$$

This equation must be satisfied by y if y is to make the integral (12.1) either a maximum or a minimum. It is known in the literature as the *Euler-Lagrange differential equation*.

The investigation whether this equation leads to a maximum or a minimum is difficult but is seldom necessary in applied mathematics.

As an example of the application of Eq. (12.12), consider the problem of determining the curve between two given points A and B (cf. Fig. 12.2) which

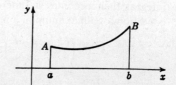

Fig. 12.2

by revolution about the x axis generates the surface of least *area*. The area of the surface s is given by the equation

$$s = 2\pi \int_a^b y\, ds = 2\pi \int_a^b y\sqrt{1+y'^2}\, dx \tag{12.13}$$

Here we have

$$F = y\sqrt{1+y'^2} \tag{12.14}$$

Therefore Eq. (12.12) becomes

$$\sqrt{1+y'^2} - \frac{d}{dx}\left[\frac{yy'}{(1+y'^2)^{\frac{1}{2}}}\right] = 0 \tag{12.15}$$

This reduces to

$$1 + y'^2 - yy'' = 0 \tag{12.16}$$

To integrate this equation, let

$$y' = p \qquad y'' = p\frac{dp}{dy} \tag{12.17}$$

The equation then becomes

$$\frac{p\,dp}{1+p^2} = \frac{dy}{y} \tag{12.18}$$

and finally we have

$$y = c_1 \cosh\frac{x - c_2}{c_1} \tag{12.19}$$

This is the equation of the catenary curve. The constants c_1 and c_2 must now be determined so that the curve (12.19) will pass through the points A and B as shown in Fig. 12.2.

This is the shape that a soap film assumes when stretched between two concentric parallel circular frames. It is obvious that, in this case, the surface s is a minimum.

The case where the function F is a function of several dependent variables y_k and their derivatives is of great importance. In this case we proceed as in the case of one variable and introduce as neighboring functions

$$\bar{y}_1 = y_1 + \epsilon_1 n_1 \qquad \bar{y}_2 = y_2 + \epsilon_2 n_2 \qquad \cdots \qquad \bar{y}_k = y_k + \epsilon_k n_k \tag{12.20}$$

where the functions $n_r(x)$ again vanish at the limits of the integral. The integral then becomes a function of the variables $\epsilon_1, \epsilon_2, \ldots, \epsilon_k$. The condition for a maximum or a minimum is

$$\frac{\partial I}{\partial \epsilon_r} = \int_{x_0}^{x_1} n_r \left(\frac{\partial F}{\partial y_r} - \frac{d}{dx}\frac{\partial F}{\partial y_r'}\right) dx = 0 \qquad r = 1, 2, \ldots, k \tag{12.21}$$

where $\epsilon_1 = \epsilon_2 = \cdots = \epsilon_r = \cdots = \epsilon_k = 0$.

It follows, therefore, that, as before, the coefficient of each of the functions n within the integral sign must vanish so that we have

$$\frac{d}{dx}\frac{\partial F}{\partial y_r'} - \frac{\partial F}{\partial y_r} = 0 \qquad r = 1, 2, \ldots, k \qquad (12.22)$$

We thus see that the Euler-Lagrange equations hold for each of the independent variables.

In literature the notation

$$\delta I = I(\epsilon) - I(0) = \epsilon \frac{dI}{d\epsilon} \qquad (12.23)$$

for ϵ small is frequently used. δI is termed the *variation* of the integral. The condition that the integral have a maximum or a minimum is then expressed in the form

$$\delta \int_{x_0}^{x_1} F(x,y,y')\,dx = \int_{x_0}^{x_1} \delta F(x,y,y')\,dx = 0 \qquad (12.24)$$

δF is called the *variation* of F.

THE BRACHISTOCHRONE BETWEEN TWO POINTS

Perhaps the earliest problem in the calculus of variations was proposed in 1696 by the Swiss mathematician John Bernoulli. He proposed the following problem of the *brachistochrone*:

It is required to determine the equation of the plane curve down which a particle acted upon by gravity alone would descend from one fixed point to another in the shortest possible time.

Let A be the upper point and B the lower one. Assume the x axis of a cartesian reference frame to be measured vertically downward, and let A be the origin of coordinates as shown in Fig. 12.3.

Let s be the length of the required curve at any point measured from A. Let v be the velocity of the particle at the same point and t its time of descent from A to that point. We wish to determine the curve that will make T a minimum, where T is the total time of descent from A to B. Now, from mechanics, we have

$$dt = \frac{ds}{v} \qquad (12.25)$$

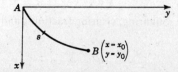

Fig. 12.3

where

$$ds = \sqrt{dx^2 + dy^2} = \sqrt{1 + y'^2}\, dx \qquad (12.26)$$

We know that the particle loses no energy in passing from one point to another of a smooth curve, since the loss of gravitational potential energy is transformed into kinetic energy; hence

$$v = \sqrt{2gx} \qquad (12.27)$$

where it is assumed that the particle starts from rest and g is the acceleration due to gravity.

Therefore (12.25) becomes

$$dt = \frac{\sqrt{1 + y'^2}}{\sqrt{2gx}}\, dx \qquad (12.28)$$

and the total time T is

$$T = \frac{1}{\sqrt{2g}} \int_{x=0}^{x=x_0} \frac{\sqrt{1 + y'^2}}{\sqrt{x}}\, dx \qquad (12.29)$$

We must therefore minimize the integral

$$I = \int_{x=0}^{x=x_0} \frac{\sqrt{1 + y'^2}}{\sqrt{x}}\, dx \qquad (12.30)$$

Hence we have

$$F = \frac{\sqrt{1 + y'^2}}{\sqrt{x}} \qquad (12.31)$$

In this case the Euler-Lagrange equation becomes

$$\frac{d}{dx}\frac{\partial F}{\partial y'} = \frac{d}{dx}\frac{y'}{\sqrt{x(1 + y'^2)}} = 0 \qquad (12.32)$$

Hence

$$\frac{y'}{\sqrt{x(1 + y'^2)}} = c_1 \qquad (12.33)$$

where c_1 is an arbitrary constant.

For simplicity, let

$$c_1 = \frac{1}{\sqrt{a}} \qquad (12.34)$$

Squaring, clearing fractions, and transposing, we have

$$y'^2 - \frac{xy'^2}{a} = \frac{x}{a} \qquad (12.35)$$

Solving for y', we obtain

$$y' = \frac{\sqrt{x}}{\sqrt{a-x}} = \frac{dy}{dx} \tag{12.36}$$

Hence

$$dy = \frac{\sqrt{x}\,dx}{\sqrt{a-x}} \tag{12.37}$$

Therefore

$$y = \int \frac{\sqrt{x}\,dx}{\sqrt{a-x}} + c_2 \tag{12.38}$$

Integrating, we obtain

$$y = a\sin^{-1}\sqrt{\frac{x}{a}} - \sqrt{ax - x^2} + c_2 \tag{12.39}$$

The arbitrary constant c_2 is zero since we have $x = 0$ at $y = 0$. Hence the equation of the curve is

$$y = a\sin^{-1}\sqrt{\frac{x}{a}} - \sqrt{ax - x^2} \tag{12.40}$$

This is the equation of a cycloid where a is the diameter of the generating circle. The constant a is determined by the condition that the cycloid must pass through the point x_0, y_0.

A more extended discussion of the calculus of variations is beyond the scope of this book. In recent years the application of the calculus of variations to problems of engineering and physics has proved of great value. Those interested will find the works listed in the references of this appendix.

13 SUMMARY OF FUNDAMENTAL FORMULAS OF THE CALCULUS OF VARIATIONS

By the extension of the principles of Sec. 12 the Euler equations that cause various integrals to attain stationary values can be obtained. The following summary of results that are useful in various applications is given here for reference.

1. One *dependent* variable y; one *independent* variable x. Integrand contains first-order derivative.

$$\delta I = \delta \int_{x_1}^{x_2} F(x, y, y_x)\,dx = 0 \qquad y_x = \frac{dy}{dx} \tag{13.1}$$

$$\frac{\partial F}{\partial y} - \frac{d}{dx}\left(\frac{\partial F}{\partial y_x}\right) = 0 \tag{13.2}$$

2. One *dependent* variable y, one *independent* variable x. Integrand contains derivatives $y_1, y_2, y_3, \ldots, y_n$, where $y_k = d^k y/dx^k$.

$$\delta I = \delta \int_{x_1}^{x_2} F(x, y, y_1, y_2, y_3, \ldots, y_n)\, dx = 0 \tag{13.3}$$

$$\frac{\partial F}{\partial y} - \frac{d}{dx}\left(\frac{\partial F}{\partial y_1}\right) + \frac{d^2}{dx^2}\left(\frac{\partial F}{\partial y_2}\right) - \cdots + (-1)^n \frac{d^n}{dx^n}\left(\frac{\partial F}{\partial y_n}\right) = 0 \tag{13.4}$$

3. Several *dependent* variables $x_1, x_2, x_3, \ldots, x_n$, one *independent* variable t, first-order derivatives $\dot{x}_k = dx_k/dt$.

$$\delta I = \delta \int_{t_1}^{t_2} F(t, x_1, x_2, x_3, \ldots, x_n, \dot{x}_1, \dot{x}_2, \dot{x}_3, \ldots, \dot{x}_n)$$
$$dt = 0 \tag{13.5}$$

$$\frac{\partial F}{\partial x_k} - \frac{d}{dt}\left(\frac{\partial F}{\partial \dot{x}_k}\right) = 0 \qquad k = 1, 2, 3, \ldots, n \tag{13.6}$$

4. Higher derivatives, n *dependent* variables, one independent variable. In this case there are n Euler equations of a type similar to (13.4), one for each dependent variable.

5. One *dependent* variable u, two *independent* variables x and y.

$$\delta I = \delta \int_{x_1}^{x_2} \int_{y_1}^{y_2} F(x, y, u, u_x, u_y, u_{xx}, u_{xy}, u_{yy})\, dx\, dy = 0 \tag{13.7}$$

where u_x, u_y are partial derivatives with respect to x and y and u_{xx}, u_{yy} are second partial derivatives with respect to x and y; u_{xy} is the mixed second partial derivative with respect to x and y.

$$F_u - \frac{\partial}{\partial x} F_{u_x} - \frac{\partial}{\partial y} F_{u_y} + \frac{\partial^2}{\partial x^2} F_{u_{xx}} + 2 \frac{\partial^2}{\partial x\, \partial y} F_{u_{xy}} + \frac{\partial^2}{\partial y^2} F_{u_{yy}} = 0 \tag{13.8}$$

where $F_u = \partial F/\partial u$, $F_{u_{xx}} = \partial F/\partial u_{xx}$, $F_{u_{xy}} = \partial F/\partial u_{xy}$, etc.

14   HAMILTON'S PRINCIPLE; LAGRANGE'S EQUATIONS

There are several possible formulations of the laws of mechanics. The most elementary formulation involves the concept of force and leads to the Newtonian equations which are discussed in elementary textbooks on mechanics. Another alternative and very important formulation is *Hamilton's principle*. This formulation is based on the energy concept and is a very useful one, particularly in cases where Newton's laws are difficult to apply. Hamilton's principle assumes that the mechanical system under consideration is characterized by a kinetic-energy function T and a potential-energy function V. In the case of a mechanical system having n degrees of freedom the configuration of the system may be described in terms of the n general coordinates $q_1, q_2, q_3, \ldots, q_n$. The kinetic energy will be a function of the coordinates q_k ($k = 1, 2, 3, \ldots, n$) as well as the generalized velocities $\dot{q}_k$ (except when the q's are cartesian coordinates).

Only conservative systems will be considered in this discussion. The potential energy V will be a function of the q's but will not depend on the velocities $\dot{q}_k$. We therefore have

$$T = T(q_1, q_2, q_3, \ldots, q_n, \dot{q}_1, \dot{q}_2, \dot{q}_3, \ldots, \dot{q}_n) \tag{14.1}$$

$$V = V(q_1, q_2, q_3, \ldots, q_n) \tag{14.2}$$

Hamilton's principle *postulates* that the integral

$$I = \int_{t_1}^{t_2} (T - V)\, dt \tag{14.3}$$

where t_1 and t_2 are two instants in time, shall have a stationary value. The integrand $T - V$ is called the *Lagrangian function*. Therefore Hamilton's principle states that the motion of the dynamical system under consideration takes place in such a manner that

$$\delta I = \delta \int_{t_1}^{t_2} (T - V)\, dt = \delta \int_{t_1}^{t_2} L\, dt = 0 \tag{14.4}$$

The integrand of (14.4) contains the n *dependent* variables q_k and their time derivatives $\dot{q}_k$ and the single *independent* variable t. The requirement that this integral be stationary leads to a system of n Euler equations of the form (13.6). These equations are

$$\frac{\partial L}{\partial q_k} - \frac{d}{dt}\left(\frac{\partial L}{\partial \dot{q}_k}\right) = 0 \qquad k = 1, 2, 3, \ldots, n \tag{14.5}$$

or

$$\frac{\partial T}{\partial q_k} - \frac{\partial V}{\partial q_k} - \frac{d}{dt}\left(\frac{\partial T}{\partial \dot{q}_k}\right) = 0 \qquad k = 1, 2, 3, \ldots, n \tag{14.6}$$

These are the Lagrangian equations of motion of the system which were deduced in Chap. 6, Sec. 5, directly from Newton's laws without the use of the calculus of variations.

THE PRINCIPLE OF LEAST ACTION

In a conservative system the sum of the potential and kinetic energy of the system is a constant c. Therefore

$$T + V = c \tag{14.7}$$

Hence

$$T - V = T - (c - T) = 2T - c \tag{14.8}$$

Hamilton's principle states that

$$\delta \int_{t_1}^{t_2} (T - V)\, dt = \delta \int_{t_1}^{t_2} (2T - c)\, dt = \delta \int_{t_1}^{t_2} 2T\, dt = 0 \tag{14.9}$$

since the variation of a constant is zero. The integral

$$A = 2 \int_{t_1}^{t_2} T \, dt \tag{14.10}$$

is called the *action*. As a consequence of (14.9), we have

$$\delta A = 0 \tag{14.11}$$

In classical mechanics this is the celebrated *principle of least action*. It appears to have been first announced by De Maupertuis in 1744.†

Hamilton's principle is very useful in obtaining the equations of motion of continuous systems. For example, consider a uniform string of mass m per unit length stretched between two supports from $x = 0$ to $x = s$ under the action of a constant tension k. If the string is performing small oscillations in one plane of amplitude $u(x,t)$, it can be shown that, if the deflection u is small, the kinetic energy T and the potential energy V of the string may be expressed in the following form:

$$T = \frac{m}{2} \int_0^s u_t^2 \, dx \qquad V = \frac{k}{2} \int_0^s u_x^2 \, dx \tag{14.12}$$

where u_t and u_x denote partial derivatives with respect to t and x, respectively. Hence Hamilton's principle states that

$$\delta \int_{t_1}^{t_2} (T - V) \, dt = \delta \int_{t_1}^{t_2} \int_0^s \left(\frac{m}{2} u_t^2 - \frac{k}{2} u_x^2 \right) dx \, dt \tag{14.13}$$

This is an integral of the type (13.7) with

$$F = \frac{m}{2} u_t^2 - \frac{k}{2} u_x^2 \tag{14.14}$$

Equation (13.8) in this case reduces to the equation

$$\frac{\partial F}{\partial u} - \frac{\partial}{\partial t} \left(\frac{\partial F}{\partial u_t} \right) - \frac{\partial}{\partial x} \left(\frac{\partial F}{\partial u_x} \right) = -m u_{tt} + k u_{xx} = 0 \tag{14.15}$$

or

$$u_{tt} = a^2 u_{xx} \qquad a = \left(\frac{k}{m} \right)^{\frac{1}{2}} \tag{14.16}$$

This is the equation of motion of the string. It is the equation for the propagation of waves in one dimension and is discussed in Chap. 12.

† See Cornelius Lanczos, "The Variational Principles of Mechanics," University of Toronto Press, Toronto, Canada, 1949.

15   VARIATIONAL PROBLEMS WITH ACCESSORY CONDITIONS: ISOPERIMETRIC PROBLEMS

Problems in the calculus of variations sometimes arise that involve making one integral stationary while one or more other integrals involving the same variable and the same limits are to be kept constant. For example, the problem may be to bend a wire of fixed length so that it will enclose the maximum plane area. There are many problems of this sort. As a class they are called *isoperimetric* problems.

For example, let it be required to find the stationary value of the integral

$$I = \int_{x_1}^{x_2} F(x, y, y_x)\, dx \tag{15.1}$$

provided that the integral

$$I_1 = \int_{x_1}^{x_2} F_1(x, y, y_x)\, dx = C_1 \qquad \text{a const} \tag{15.2}$$

To solve this problem, we consider the integral

$$I_0 = \int_{x_1}^{x_2} F_0(x, y, y_x)\, dx \tag{15.3}$$

where

$$F_0 = F + \mu F_1 \tag{15.4}$$

and μ is a *parameter* to be determined.

It is clear that, since the integral I_1 is to remain constant, the integral I_0 will be stationary if I is stationary. The Euler equation that makes I_0 stationary is

$$\frac{\partial F_0}{\partial y} - \frac{d}{dx}\left(\frac{\partial F_0}{\partial y_x}\right) = 0 \tag{15.5}$$

The parameter μ can now be eliminated by the use of Eqs. (15.2) and (15.5). As an example, let it be required to find the shape of the plane curve that must be assumed by a wire of fixed length S_0 so that it encloses the maximum area. To solve this problem, it is convenient to express the shape of the curve in plane polar coordinates. We therefore seek the curve $r = r(\theta)$ that maximizes the integral

$$I = \frac{1}{2}\int_0^{2\pi} r^2\, d\theta \tag{15.6}$$

and has a fixed length

$$I_1 = \int_0^{2\pi} (r^2 + r_\theta^2)^{\frac{1}{2}}\, d\theta = S_0 \tag{15.7}$$

In this case we have

$$F_0 = \frac{r^2}{2} + \mu(r^2 + r_\theta^2)^{\frac{1}{2}} \tag{15.8}$$

and

$$\frac{\partial F_0}{\partial r} - \frac{d}{d\theta}\left(\frac{\partial F_0}{\partial r_\theta}\right) = r + \mu r(r^2 + r_\theta^2)^{-\frac{1}{2}} - \frac{d}{d\theta}[\mu r_\theta(r^2 + r_\theta^2)^{-\frac{1}{2}}] = 0 \qquad (15.9)$$

This equation may be written in the form

$$\frac{rr_{\theta\theta} - 2r_\theta^2 - r^2}{(r^2 + r_\theta^2)^{\frac{3}{2}}} = \frac{1}{\mu} \qquad (15.10)$$

The absolute value of the left-hand member of (15.10) will be recognized as the *curvature* of $r(\theta)$ expressed in polar coordinates. Hence $\rho = |\mu|$, a constant, and the curve is a circle. The value of the parameter μ may now be determined in terms of the given fixed length S_0 by the use of (15.7).

PROBLEMS

1. Find the first and second partial derivatives of the function $\tan^{-1}(x/y)$.

2. Show that, if $u = \ln(x^2 + y^2) + \tan^{-1}(y/x)$, then $\partial^2 u/\partial x^2 + \partial^2 u/\partial y^2 = 0$.

3. Show that, if $u = \tan(y + ax) + \sqrt{y - ax}$, then $\partial^2 u/\partial x^2 = a^2(\partial^2 u/\partial y^2)$.

4. If $u = F_1(x + jy) + F_2(x - jy)$, where $j = \sqrt{-1}$, show that

$$\frac{\partial^2 u}{\partial x^2} + \frac{\partial^2 u}{\partial y^2} = 0$$

5. If F is a function of x and y and $x + y = 2e^\theta \cos\phi$, $x - y = 2je^\theta \sin\phi$, where $j = \sqrt{-1}$, prove that $\partial^2 F/\partial\theta^2 + \partial^2 F/\partial\phi^2 = 4xy(\partial^2 F/\partial x \partial y)$.

6. Change the independent variable from x to z in

$$x^2\left(\frac{d^2 y}{dx^2}\right) + ax\left(\frac{dy}{dx}\right) + by = 0$$

where $x = e^z$, and show that the equation is transformed to

$$\frac{d^2 y}{dz^2} + (a - 1)\left(\frac{dy}{dz}\right) + by = 0$$

7. Find the maximum value of $V = xyz$ subject to the condition that

$$\frac{x^2}{a^2} + \frac{y^2}{b^2} + \frac{z^2}{c^2} = 1$$

What is the geometric interpretation of this problem?

8. Divide 24 into three parts such that the continued product of the first, the square of the second, and the cube of the third may be a maximum.

9. Find the points on the surface $xyz = a^3$ which are nearest the origin.

10. Show that the necessary conditions for the maximum and minimum values of $\phi(x,y)$, where x and y are connected by an equation $F(x,y) = 0$, are that x and y should satisfy the two equations $F(x,y) = 0$ and

$$\frac{\partial\phi}{\partial x}\frac{\partial F}{\partial y} - \frac{\partial\phi}{\partial y}\frac{\partial F}{\partial x} = 0$$

11.   Determine the equation of the shortest curve between two points in the xy plane.

12.   Find the equation of the shortest line on the surface of a sphere, and prove that it is a great circle.

13.   Show that the shortest lines on a right-circular cylinder are helices.

14.   Find the equation of the shortest line on a cone of revolution.

15.   Find the curve of given length between two fixed points which generates the minimum surface of revolution.

16.   A cubic centimeter of brass is to be worked into a solid of revolution.   It is desired to make its moment of inertia about the axis as small as possible.   What shape should be given?

17.   Find the curve of length 3 that joins the points $(-1,1)$ and $(1,1)$ which, when rotated about the x axis, will give minimum area.

18.   Find the surface of revolution which encloses the maximum volume for a given surface area.

19.   Use Hamilton's principle to derive the equation of motion of a particle moving in a straight line with potential energy $V = 1/x$, where x is the distance of the particle from some arbitrarily chosen origin.

20.   Assuming that a string with fixed ends will hang in such a manner that its center of gravity has the lowest possible position, find the curve in which a string of given length will hang.

21.   Prove by Hamilton's principle that the string in Prob. 20 will hang as stated.

22.   Use Hamilton's principle to obtain the polar equations for the motion of a particle in space.

23.   Find the surface of revolution which encloses the maximum volume in a given area.

24.   Given a volume of homogeneous attracting gravitating matter, find the shape of the solid of revolution constituting the volume so that the attraction at a point on the axis of revolution shall be a maximum.

25.   Determine the shape of a curve of given length, so chosen that the area enclosed between the curve and its chord will be a maximum.

26.   Consider a vibrating membrane, like a drumhead whose position of equilibrium is in the (x,y) plane.   Let $u(x,y,t)$ be the small deflection of the membrane while it is performing oscillations.   Find the potential energy V and the kinetic energy T under the assumption that the membrane is under constant tension P and has a mass m per unit area.   Determine the equation of motion by Hamilton's principle.

27.   Show that, if the integrand of (12.1) does not contain x explicitly so that it takes the form

$$I = \int_{x_0}^{x_1} F(y, y_x)\, dx$$

then the Euler-Lagrange equation (12.12) takes the following form: $F - y(\partial F/\partial y_x) = \text{const.}$ This is a very useful result that can be used when the integrand F does not depend explicitly on x.

REFERENCES

1904. Goursat, E., and E. Hedrick: "A Course in Mathematical Analysis," vol. I, Ginn and Company, Boston.

1911. Wilson, E. B.: "Advanced Calculus," Ginn and Company, Boston.

1925. Bliss, G. A.: "Calculus of Variations," The University of Chicago Press, Chicago.

1927. Forsythe, A. R.: "Calculus of Variations," Cambridge University Press, New York.

1931. Bolza, O.: "Lectures on the Calculus of Variations," The University of Chicago Press, Chicago; reprinted, Stechert-Hafner, Inc., New York.

1933. Osgood, W. F.: "Advanced Calculus," chap. 17, The Macmillan Company, New York.

1950. Fox, C.: "An Introduction to the Calculus of Variations," Oxford University Press, New York.

1951. Polya, G., and G. Szego: "Isoperimetric Inequalities in Mathematical Physics," Princeton University Press, Princeton, N.J.

1952. Hildebrand, F. B.: "Methods of Applied Mathematics," Prentice-Hall, Inc., Englewood Cliffs, N.J.

1952. Weinstock, R. P.: "Calculus of Variations," McGraw-Hill Book Company, Inc., New York.

1953. Courant, R., and D. Hilbert: "Methods of Mathematical Physics," Interscience Publishers, Inc., New York.

1962. Akhiezer, N. I.: "The Calculus of Variations," Blaisdell Publishing Company, New York; translated by A. H. Frink.

appendix G
Answers to Selected Problems

CHAPTER 1

3. $\dfrac{e^{\pm Ja}}{\pm 2ja}$

4. $z_n = n\pi \qquad n = 0, \pm 1, \pm 2, \ldots$

6. $1 + \dfrac{4j}{\pi} \displaystyle\sum_{n=1}^{\infty} (-1)^n \dfrac{\sin(2n-1)\,\pi/2m}{2n-1}$

7. HINT: Integrate $e^{az}/\sinh \pi z$ around the rectangle of sides of $y = 0, y = 1, x = \pm R$, indented at the origin and at $+j$.

13. HINT: Integrate $e^{az}/\cosh \pi z$ in the same manner as described in the Hint to Prob. 7 above.

22. $\epsilon_1 = u_{xx}\,\Delta x + u_{xy}\,\Delta y \qquad \epsilon_2 = u_{yy}\,\Delta y + u_{xy}\,\Delta x$
$\epsilon_3 = v_{xx}\,\Delta x + v_{xy}\,\Delta y \qquad \epsilon_4 = v_{yy}\,\Delta y + v_{xy}\,\Delta x$

24. *(a)* $u = x^2 - y^2 \qquad v = 2xy \qquad$ *(c)* $u = \dfrac{x}{x^2 + y^2} \qquad v = \dfrac{-y}{x^2 + y^2}$

(e) $u = \sinh x \cos y \qquad v = \cosh x \sin y$

26. $v = \tan^{-1}\dfrac{y}{x}$ $\qquad \omega = \ln z$

31. See Chap. 4, Sec. 5.

33. $\dfrac{1}{(\pi t)^{\frac{1}{2}}}$

34. (a) 0 $\qquad (b)$ $\dfrac{-a^3 + 3ab - 2b^3}{(a-b)^2}$

$\quad (c)$ 0 $\qquad (d)$ $\dfrac{e^{bt}(bt+1)}{a+b} - \dfrac{ae^{-at} + be^{bt}}{(a+b)^2}$

38. $-\sum\limits_{n=0}^{\infty} z^{-n-2}$

40. $\dfrac{a^4}{6}$

43. $\omega = e^z$

45. $\operatorname{erf}(t^{\frac{1}{2}})$

CHAPTER 2

1. $ce^{-x} + (x - 1)$

4. $ce^{-x} + (x - 2)^2$

7. $\cos at \cosh at$ $\qquad a = \dfrac{k^{\frac{1}{4}}}{2^{\frac{1}{2}}}$

9. $e^{-t} - e^{-2t}$

12. $\dfrac{2}{3^{\frac{1}{2}}} \sin\left(3t + \dfrac{\pi}{4}\right) + \cos 2t - \sin 3t$

14. $c_1 \epsilon^{-t} + c_2 + c_3(t - 1) + 16 - 16t + 8t^2 - 2t^3 + \frac{1}{2}t^4$

17. $\dfrac{kt}{2b} \sin bt$

18. (a) e^x $\qquad (b)$ $\frac{2.2}{7} x^{-5} + \frac{6}{7}x^2$

19. (a) $3e^{5x}$ $\qquad (d)$ $3e^{3x} - 2e^x - e^{-x}$

20. (a) $(x + 1)e^{-x}$ $\qquad (c)$ e^x

26. (a) $\frac{1}{2}e^x - e^{2x} + \frac{1}{2}e^{3x}$ $\qquad (c)$ $e^x - e^{-2x} + \sin 3x$

$\quad (d)$ $e^{-2x} + e^{2x} + e^{3x} + e^{-x}$

$\quad (e)$ $e^x \sin x$ $\qquad (g)$ $(3 - 2x)e^{-4x}$

$\quad (i)$ $(x - 2)^2 e^{-2x}$

27. (a) $\dfrac{\sin x}{\pi^{\frac{1}{2}}}$ $\qquad \dfrac{\sin 2x}{\pi^{\frac{1}{2}}}, \ \ldots$

$\quad (d)$ $\dfrac{1}{(2\pi)^{\frac{1}{2}}}$ $\qquad \dfrac{\cos x}{\pi^{\frac{1}{2}}}$ $\qquad \dfrac{\sin x}{\pi^{\frac{1}{2}}}, \ \ldots$

28. (b) $\sin \dfrac{(2n+1)}{2l}\pi x$ $\qquad n = 0, 1, 2, \ldots$

CHAPTER 3

1. $x = 3 \qquad y = 2 \qquad z = 1$

2. 446

3. $x = 4/3, \qquad y = -2, \qquad z = 4, \qquad \omega = 3$

4. HINT: Use, for example, (4.8).

7. $\lambda_1 = 5.049 \quad \begin{Bmatrix} 0.3280 \\ 0.5910 \\ 0.7370 \end{Bmatrix}_1 \quad \lambda_2 = 0.6431 \quad \begin{Bmatrix} 0.7370 \\ 0.3280 \\ -0.5910 \end{Bmatrix}_2$

$\lambda_3 = 0.3080 \quad \begin{Bmatrix} -0.5910 \\ 0.7370 \\ -0.3280 \end{Bmatrix}_3$

11. $1, -2, 3$

12. $1, 5, 10$

16. $c = -2 \qquad x_2 = -\dfrac{x_1}{2} \qquad x_3 = -\dfrac{3x_1}{2}$

23. (a) $\{x\} = e^t \begin{Bmatrix} 1 \\ -1 \end{Bmatrix} + e^{5t} \begin{Bmatrix} 1 \\ 3 \end{Bmatrix}$

(b) $\{x\} = e^t \begin{Bmatrix} 1 \\ -4 \end{Bmatrix} + 4e^{6t} \begin{Bmatrix} 1 \\ 1 \end{Bmatrix}$

25. (a) $\lambda_1 = 3 \qquad \lambda_2 = 2 \qquad \{x\}_1 = \begin{Bmatrix} 2 \\ 1 \end{Bmatrix} \qquad \{x\}_2 = \begin{Bmatrix} 1 \\ 1 \end{Bmatrix}$

(c) $\lambda_1 = 1 \qquad \lambda_2 = -1 \qquad \{x\}_1 = \begin{Bmatrix} 1 \\ 1 \end{Bmatrix} \qquad \{x\}_2 = \begin{Bmatrix} 0 \\ 1 \end{Bmatrix}$

(d) $\lambda_1 = 8 \qquad \lambda_2 = 4 \qquad \lambda_3 = -2 \qquad \{x\}_1 = \begin{Bmatrix} 1 \\ 0 \\ -1 \end{Bmatrix} \qquad \{x\}_2 = \begin{Bmatrix} 1 \\ -2 \\ -3 \end{Bmatrix} \qquad \{x\}_3 = \begin{Bmatrix} 1 \\ 0 \\ -3 \end{Bmatrix}$

CHAPTER 4

5. (a) $e^{-t} - e^{-2t}$

(d) $\dfrac{2}{3^{\frac{1}{2}}} \sin\left(3t + \dfrac{\pi}{4}\right) + \cos 2t - \sin 3t$

(f) $c_1 e^{-t} + c_2 + c_3(t - 1) + 16 - 16t + 8t^2 - 2t^3 + \frac{1}{2}t^4$

(h) $\dfrac{kt}{2b} \sin bt$

11. (a) $\dfrac{1}{s^2}$ (c) $\dfrac{1}{s + a}$ (d) $\dfrac{a}{s^2 + \omega^2}$

12. (a) $3e^{5t}$ (d) $3e^{3t} - 2e^t - e^{-t}$

13. (a) $\frac{1}{2}e^t - e^{2t} + \frac{1}{2}e^{3t}$ (c) $e^t - e^{-2t} + \sin 3t$

(e) $e^t \sin t$ (g) $(3 - 2t)e^{-4t}$

(i) $(t - 2)^2 e^{-2t}$

CHAPTER 5

2. $\dfrac{CE_0}{I_0}(1 - \cos\omega_0 t) - \dfrac{CE_0}{T_0}\left\{1 - \left(\dfrac{1 + T_0^2}{LC}\right)^{\frac{1}{2}} \cos\left[\omega_0(t - T_0) + \phi\right]\right\} u(t - T_0)$

$$\omega_0^2 = \frac{1}{LC} \qquad \phi = \tan^{-1}\omega_0 T_0$$

6. $i_1 = \dfrac{L_2 E}{b}\left[\dfrac{R_2}{L_2} + \left(\dfrac{b}{a} - \dfrac{R_2}{L_2}\right)e^{-bt/a}\right]$

$i_2 = -\left(\dfrac{ME}{a}\right)e^{-bt/a}$

11. $i_{ss} = \dfrac{\omega_1/L}{\omega_1^2 - \omega_0^2}E_1\cos\omega_1 t + \dfrac{\omega_2/L}{\omega_2^2 - \omega_0^2}E_2\cos\omega_2 t$

14. $0 \leqslant \omega \leqslant \left(\dfrac{2}{LC}\right)^{\frac{1}{2}}$

21. $i_1 = \frac{50}{31}\sin t \qquad i_2 = \frac{20}{31}\sin t$

23. $i_1 = \frac{1}{40}\sin 2t \qquad i_2 = \frac{1}{4}\sin 2t - 5\cos 2t$

25. $i_1 = \frac{170}{65}\sin t \qquad i_2 = \frac{200}{65}\sin t \qquad i_3 = \frac{400}{65}\sin t$

27. $i = \dfrac{E}{R}e^{-t/RC}$

30. $i = \dfrac{E}{\omega_0 L}e^{-Rt/2L}\sin\omega_0 t \qquad \omega_0^2 = \dfrac{1}{LC} - \dfrac{R^2}{4L^2} > 0$

32. $i_1 = \dfrac{E_0}{\omega L}(1 - \cos\omega t) + \dfrac{E_0}{R}\sin\omega t$

$i_2 = \dfrac{E_0}{\omega L}(1 - \cos\omega t)$

CHAPTER 6

1. $[k]^{-1} = \dfrac{1}{Mgl(\alpha + 1)(\alpha + 3)}\begin{bmatrix} \alpha^2 + 3\alpha + 1 & \alpha + 1 & 1 \\ \alpha + 1 & (\alpha + 1)^2 & \alpha + 1 \\ 1 & \alpha + 1 & \alpha^2 + 3\alpha + 1 \end{bmatrix}$

$\alpha = \dfrac{Mgl}{kh^2}$

3. $R > 2(Mk)^{\frac{1}{2}}$

6. $M\ddot{x} + ax = 0 \qquad M\ddot{y} + ay = 0$

8. $M\ddot{y}_1 + \dfrac{3T}{2d}y_1 - \dfrac{T}{2d}y_2 = 0 \qquad M\ddot{y}_2 + \dfrac{3T}{2d}y_2 - \dfrac{T}{2d}y_1 = 0$

$$\omega_1^2 = \frac{T}{Md} \qquad \{y\}_1 = \begin{Bmatrix} 1 \\ 1 \end{Bmatrix} \qquad \omega_2^2 = \frac{2T}{Md} \qquad \{y\}_2 = \begin{Bmatrix} 1 \\ -1 \end{Bmatrix}$$

$$\eta = y_1 + y_2 \qquad \zeta = y_1 - y_2$$

15. $\omega_1^2 = \dfrac{2kh^2}{Ml^2} \qquad \omega_2^2 = \dfrac{4kh^2}{Ml^2}$

CHAPTER 7

1.

x	$F(x)$	$\Delta F(x)$	$\Delta^2 F(x)$	$\Delta^3 F(x)$
1	1.0000	−0.5000	+0.3333	−0.2499
2	0.5000	−0.1667	+0.0834	−0.0501
3	0.3333	−0.0833	+0.0333	−0.0166
4	0.2500	−0.0500	+0.0167	
5	0.2000	−0.0333		
6	0.1667			

3. $A = 6.109$

7. $c_1 + c_2 4^x + \dfrac{m^x}{(m-4)(m-1)}$

8. $c_1 4^x + \dfrac{x}{5} - \dfrac{26}{25}$

9. $\dfrac{\sin 1 \cos (x-1)}{2(\cos 1 - 1)} - \dfrac{\sin (x-1)}{2}$

13.

$$\omega = 2\left(\frac{\tau}{mh}\right)^{\frac{1}{2}} \sin \frac{n\pi}{2(N+1)} \qquad n = 0, 1, 2, \ldots$$

15. $\omega^2 = \dfrac{4c}{J} \sin^2 \dfrac{r\pi}{2(n+1)}$

$$\theta_n = \frac{a \sin \omega t \sin n\pi/(n+1)}{\sin \pi/(n+1)}$$

CHAPTER 8

1. (a) $\dfrac{1}{1 + RCs}$ (c) $\dfrac{1}{1 + RC_2}$

(e) $\dfrac{1}{1 + 3LCs^2 + L^2 C^2 s^4}$

5. (a) $\dfrac{1}{RC} e^{-t/RC}$ (c) $\dfrac{1}{RC_2} e^{-t/RC_2}$

(e) $\dfrac{1}{5^{\frac{1}{2}} LC\omega_1 \omega_2} (\omega_2 \sin \omega_1 t - \omega_1 \sin \omega_2 t)$ $\omega_1^2 = \dfrac{3 - 5^{\frac{1}{2}}}{2LC}$ $\omega_2^2 = \dfrac{3 + 5^{\frac{1}{2}}}{2LC}$

10. Unstable.

11. Unstable (oscillatory).

12. Unstable.

13. (a) $K_p = \infty$ $K_v = \dfrac{K}{12}$ $K_a = 0$

(c) $K_p = 2K \times 10^{-4}$ $K_v = K_a = 0$

14. (a) 84   (b) 14.225

(d) Stable all K.   (f) 0

(h) 35.555

CHAPTER 9

4. $v(x,y) = 2v_0 \left[1 - \dfrac{1}{\pi} \tan^{-1} \left(\dfrac{2ry}{x^2 + y^2 - r^2} \right) \right]$

8. $v(r) = \sum\limits_{n=0}^{\infty} A_n r^n P_n(\cos \theta)$ $r < 1$

$$A_n = \begin{cases} 0 & n = 0, 2, 4, \ldots \\ (-1)^n \dfrac{2n + 1}{2} \dfrac{(-1)(-3) \, \cdots \, (-n + 2)}{(n + 1)(n - 1) \, \cdots \, (2)} & n = 1, 3, 5, \ldots \end{cases}$$

10. $v(r) = \sum\limits_{n=0}^{\infty} A_n r^n P_n(\cos \theta)$ $r < 1$

$$A_n = \begin{cases} 0 & n \neq 1 \\ u_0 & n = 1 \end{cases}$$

CHAPTER 10

2.

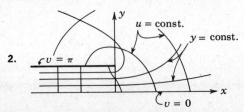

5. $\omega = -\dfrac{q}{2\pi K} \ln\left(\dfrac{e^{\pi z/a} - e^{j\pi b/a}}{e^{\pi z/a} + e^{j\pi b/a}}\right)$

8. $\sigma = \dfrac{qa}{\pi(x^2 + a^2)}$

13. $p(r) = p_0 - \dfrac{\rho k^2}{2r^2}$

16. $\sin k$ at $z = a$ plus source at $z = -a$ if $k > 0$

19. $A^2 - \left(\dfrac{Ak}{y}\right)\sin\dfrac{2Ay}{k} + k^2\sin^2\dfrac{Ay}{k} = 0.2p_0$

21. $(b)\ -1, 1 \qquad (f)\ 0$

22. $ad - bc \neq 0$

23. $(b)\ \left|\omega + \dfrac{2j}{3}\right| = \tfrac{1}{3} \qquad (d)\ \left|\omega + \dfrac{1+j}{2}\right| = \dfrac{1}{2^{\frac{1}{4}}}$

26. $\omega = -\dfrac{(z - 1 - 2j)^4 - j}{-j(z - 1 - 2j)^4 + 1}$

28. $T(x,y) = \dfrac{T_0}{\theta_0}\tan^{-1}\dfrac{y}{x}$

29. $H(x,y) = \dfrac{2}{\pi}\tan^{-1}\dfrac{\tanh y}{\tan x}$

CHAPTER 11

5. $u = u_0 e^{-kx/h}$

8. $u(x,t) = \displaystyle\sum_{n=1}^{\infty} A_n \sin\dfrac{n\pi x}{s} e^{-(hn\pi/s)^2 t}$

$A_n = \dfrac{4c}{(n\pi)^3}[1 + (-1)^{n+1}]$

10. $v(x,y,t) = A\sin\dfrac{\pi x}{a}\sin\dfrac{\pi y}{b} e^{-\theta t}$

$\theta = h^2\left[\left(\dfrac{\pi}{a}\right)^2 + \left(\dfrac{\pi}{b}\right)^2\right]$

13. $v(x,t) = \displaystyle\sum_{n=1}^{\infty} A_n \sin\dfrac{n\pi x}{s} e^{-(hn\pi/s)^2 t}$

$A_n = \dfrac{2v_0}{n\pi}[1 + (-1)^{n+1}]$

15. $v(x,t) = v_0 + (v_s - v_0)\dfrac{x}{s} + \displaystyle\sum_{n=1}^{\infty} A_n \sin\dfrac{n\pi x}{s} e^{-(hn\pi/s)^2 t}$

$A_n = \dfrac{2}{s}\displaystyle\int_0^s \left[F(x) - v_0 + (v_0 - v_s)\dfrac{x}{s}\right]\sin\dfrac{n\pi x}{s}\,dx$

17. $v(x,y,t) = \sum\limits_{m=1}^{\infty} \sum\limits_{n=1}^{\infty} B_{mn} \sin\dfrac{m\pi x}{a} \sin\dfrac{n\pi x}{b} e^{-\theta_{mn}t}$

$$\theta_{mn} = h^2\left[\left(\dfrac{m\pi}{a}\right)^2 + \left(\dfrac{n\pi}{b}\right)^2\right]$$

$$B_{mn} = \dfrac{4(-1)^{m+n}}{mn\pi^2}$$

20. $B_{mnr} = \dfrac{8v_0}{m\kappa r\pi^3}[1 + (-1)^{m+1}][1 + (-1)^{n+1}][1 + (-1)^{r+1}]$

CHAPTER 12

2. $y(x,t) = \sum\limits_{n=1}^{\infty} A_n \sin\dfrac{n\pi ct}{s} \sin\dfrac{n\pi x}{s}$

$$A_n = \dfrac{4ks^3}{(n\pi)^4 c}[1 + (-1)^{n+1}]$$

4. $a = \left(\dfrac{2T}{m}\right)^{\ddagger}\dfrac{\pi}{F_0}$ $\omega_{12} = \omega_{21} = \left(\dfrac{5F_0}{2}\right)^{\ddagger}$

$\omega_{13} = \omega_{31} = (5F_0)^{\ddagger}$

6. $c^2\left(\dfrac{\partial^2 u}{\partial r^2} + \dfrac{1}{r}\dfrac{\partial u}{\partial r} + \dfrac{1}{r^2}\dfrac{\partial^2 u}{\partial \theta^2}\right) = \dfrac{\partial^2 u}{\partial t^2}$

10. Cf. Chap. 13, Sec. 12, Prob. *a*.

11. Cf. Chap. 13, Sec. 12, Prob. *c*.

14. (*a*) $u(x,t) = A(\cos t \sin x - \cos 2t \sin 2x)$

(*b*) $u(x,t) = \dfrac{8}{\pi}\sum\limits_{n=1}^{\infty}\dfrac{(-1)^{n+1}}{[2(2n-1)]^2}\cos[2(2n-1)t]\sin[2(2n-1)x]$

(*c*) $u(x,t) = \dfrac{0.8}{\pi}\sum\limits_{n=1}^{\infty}\dfrac{1}{(2n-1)^3}\cos[(2n-1)t]\sin[(2n-1)x]$

15. (*c*) $u(x,y) = ke^{x^2+y^2+c(x-y)}$

16. (*a*) $u(x,y) = f_1(x) + f_2(x+y)$

(*b*) $u(x,y) = f_1(x+y) + f_2(2x-y)$

17. (*a*) $u(x,y,t) = \dfrac{0.64}{\pi^6}\sum\limits_{\substack{m=1 \\ (m,n=\text{odd})}}^{\infty}\sum\limits_{n=1}^{\infty}\dfrac{1}{m^3 n^3}\cos\pi ct(m^2+n^2)^{\ddagger}[\sin m\pi x \sin n\pi y]$

CHAPTER 13

1. $y_{\max} = \dfrac{\omega_0 l^2}{8H}$ at $x = \dfrac{l}{2}$

2. $y(x) = \dfrac{x}{Hl}[P_a(l-a) + P_b(l-b)] - \dfrac{1}{H}[P_a(x-a)\,u(x-a) + P_b(x-b)\,u(x-b)]$

   $u(x - x_0) = $ Heaviside unit step at $x = x_0$

5. $y(x) = \dfrac{P_\beta}{2k}e^{-\beta x}(\cos\beta x + \sin\beta x)$

   $\beta^4 = \dfrac{k}{4EI}$

7. $\omega_0 = \left(\dfrac{4.73}{l}\right)^2 \sqrt{\dfrac{EI}{m}}$

11. $\dfrac{ab}{\sqrt{b^2 - bc}}\sin\sqrt{b^2 - bc}\,t$

14. $E(x,s) = E_0[e^{-kx} + e^{-k(x+2l)} + \cdots] - E_0[e^{-k(2l-x)} + e^{-k(4l-x)} + \cdots]$

   $k = \dfrac{s + R/L}{v} \qquad v = \sqrt{LC}$

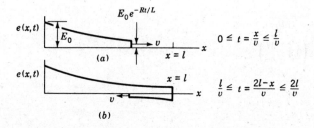

$$0 \le t = \frac{x}{v} \le \frac{l}{v}$$

$$\frac{l}{v} \le t = \frac{2l-x}{v} \le \frac{2l}{v}$$

15. $u(x,t) = \dfrac{F_0 c}{AE}\displaystyle\sum_{n=0}^{\infty}\left\{\left[t - \dfrac{2nl + x}{c}\right]u\left(t - \dfrac{2nl + x}{c}\right) + \left[t - \dfrac{2(n+2)\,l - x}{c}\right]\right.$

   $\left. \times\, u\left[t - \dfrac{(2n+2)\,l - x}{c}\right]\right\}$

19. $v(x,t) = \dfrac{v_0\,x}{l} + \dfrac{2v_0}{\pi}\displaystyle\sum_{n=1}^{\infty}\dfrac{(-1)^n}{n}\sin\dfrac{n\pi(l-x)}{l}e^{-n^2\pi^2 h^2 t/l^2}$

34. (a) $y(x,t) = \left(t - \dfrac{x^2}{2}\right)u\left(t - \dfrac{x^2}{2}\right)$

   (b) $u(x,t) = xf(x + t - 1)\,u(x + t - 1)$

   (c) $y(x,t) = (1 + x^2)u(t - x^2) + (1 + t)u(x^2 - t)$

35. $u(x,t) = -v_0\,tu\left(\dfrac{x}{a} - t\right) - v_0\left(\dfrac{x}{a}\right)u\left(t - \dfrac{x}{a}\right)$

   $a^2 = \dfrac{E}{\rho}$

36. $T(x,t) = \dfrac{v_0}{1 + \sigma}\, \text{erfc}\left[\dfrac{x}{2(k_2 t)^{\frac{1}{2}}}\right]$

$\sigma = \dfrac{k_2\, \kappa_2^{\frac{1}{2}}}{k_1\, \kappa_1^{\frac{1}{2}}}$

41. $\phi = -j\omega\, \dfrac{Aa}{k\lambda}\, J_1(ka)\, e^{j\omega t - \lambda x}$

$\lambda = \left[k^2 - \left(\dfrac{\omega}{c}\right)^2\right]^{\frac{1}{2}}$

$F = -2\pi\rho\omega^2\, Aa^2\, e^{j\omega t} \displaystyle\int_0^\infty (k\lambda)^{-1} J_1^2(ka)\, dk$

CHAPTER 14

3. $y(2) = 0.316$

4. $y(2) = 0.348$

6. $\omega = \dfrac{c}{s\sqrt{3}\,\sqrt{\frac{1}{30} + (M/\rho s)}}$

7. $\omega = \dfrac{2}{s^2}\left(\dfrac{EI}{m}\right)^{\frac{1}{2}}\left(0.2 + \dfrac{M}{ms}\right)^{-\frac{1}{2}}$

9. $\omega_1 = \dfrac{3.2c}{l}$

CHAPTER 16

1. $1 - \left(\frac{5}{6}\right)^n$

2. $\frac{19}{27}$

4. $e^{-10} = 4.5 \times 10^{-5}$

5. $\sim e^{-m/MP}$

10. $\left(\frac{3}{4}\right)^5 = 0.237$

19. 6

20. 0.08

21. 0.50

22. 35%

23. $\dfrac{l}{4}$

24. 9.6 cards

Index

Index